AF606429

The Hydrogen Bond

The Hydrogen Bond

Recent developments in theory and experiments

III. Dynamics, thermodynamics and special systems

P. Schuster

Institut für Theoretische Chemie und Strahlenchemie der Universität Wien

G. Zundel

Institut für Physikalische Chemie der Universität München

C. Sandorfy

Département de chimie, Université de Montréal

1976

NORTH-HOLLAND PUBLISHING COMPANY
AMSTERDAM · NEW YORK · OXFORD

North-Holland ISBN: 0 7204 0315 4
American Elsevier ISBN: 0 444 10805 X

Published by:

North-Holland Publishing Company - Amsterdam, New York, Oxford

Sole distributors for the U.S.A. and Canada:

American Elsevier Publishing Company, Inc.
52 Vanderbilt Avenue
New York, N.Y. 10017

PRINTED IN THE NETHERLANDS

EDITORS' PREFACE

The methods and techniques applied to studies of hydrogen bonding have reached a high degree of refinement and specialization and have branched into a variety of different fields. Consequently it has become almost impossible for an individual or a small group of scientists to give a comprehensive and detailed review on this subject. Fifteen years ago, Pimentel and McClellan covered all the material that was then available. In the series of volumes we present, almost forty scientists contribute their most recent results and ideas in their own fields, surveying the present stage of knowledge on the nature and properties of the hydrogen bond.

One of the editors' aims was to combine findings and views of experimentalists and theoreticians. Whereas only modest advance could be made towards a unified theory covering all of the interesting aspects and the various results in this widespread field, we hope that the treatise presented will initiate and encourage discussions and cooperation between scientists working in the various fields.

The Reader familiar with the subject, will probably notice the absence of two of the many important and promising approaches. Both of these fields are developing extremely fast and we therefore felt that at this time it would be inappropriate to prepare an extensive survey on these. We are referring to hydrogen bonding in excited states and to the application of relaxation kinetics to hydrogen bond systems which have been covered only briefly.

Our present volumes do not include investigations with a strong tie-in to biology. On the one hand, interpretation of experimental facts is often very difficult in molecular biophysics and, further, at the present time views are changing much faster than in other fields. On the other hand, an impressive amount of important material has been already accumulated on hydrogen bonds in biology, in sufficient quantity as to be presented in a separate volume.

Wien/München/Montréal,
September 1975

P. Schuster
G. Zundel
C. Sandorfy

CONTENTS

Preface v

Volume I. Theory

Ch. 1. Introduction, E. Lippert 1

PART A. THEORY OF THE HYDROGEN BOND 23

Ch. 2. Energy surfaces for hydrogen bonded systems, P. Schuster 25
Ch. 3. Calculated vibrational spectra of hydrogen bonded systems, P. Janoschek 165
Ch. 4. Proton motions in hydrogen bonds, J. Brickmann . . 217
Ch. 5. Model studies on proton correlations in hydrogen bonds, E. Weidemann 245
Ch. 6. Dynamical properties of hydrogen bonded systems, G.L. Hofacker, Y. Maréchal and M. Ratner 295
Ch. 7. Hydrogen bond statistics, J.W. Perram 359

Volume II. Structure and spectroscopy

PART B. INVESTIGATIONS ON STRUCTURE AND STEREOCHEMISTRY . . 391

Ch. 8. X-Ray and neutron diffraction studies of hydrogen bonded systems, I. Olovsson and P.-G. Jönsson . . . 393
Ch. 9. Neutron diffraction studies of hydrogen bonding in α-amino acids, Th.F. Koetzle and M.S. Lehmann . . 457
Ch. 10. The hydrated proton in solids, J.-O. Lundgren and I. Olovsson 471
Ch. 11. The angular dependence of hydrogen bond interactions, W.A.P. Luck 527

PART C. VIBRATIONAL AND RESONANCE SPECTROSCOPIC STUDIES . . 563

Ch. 12. Vibrational spectroscopy of the hydrogen bond, D. Hadži and S. Bratos 565

Ch. 13. Anharmonicity and hydrogen bonding, C. Sandorfy . 613
Ch. 14. Spectroscopic studies of hydrated proton species, $H^+(H_2O)_n$, in crystalline compounds, J.M. Williams . 655
Ch. 15. Easily polarizable hydrogen bonds – Their interactions with the environment – IR continuum and anomalous large proton conductivity, G. Zundel . . 683
Ch. 16. Long wave length vibrational spectroscopy of hydrogen bonding, W.G. Rothschild 767
Ch. 17. High resolution nuclear magnetic resonance studies of hydrogen bonding, E.E. Tucker and E. Lippert . . . 791
Ch. 18. Magnetic resonance studies of hydrogen bonding in solids, R. Blinc 831

Volume III. Dynamics, thermodynamics and special systems

PART D. DYNAMICS OF HYDROGEN BONDED SYSTEMS 889

Ch. 19. Incoherent neutron scattering experiments on hydrogen bonded systems, J.A. Janik 891
Ch. 20. Dielectric properties of hydrogen bonded systems, L. Sobczyk, H. Engelhardt and K. Bunzl 937
Ch. 21. Nuclear magnetic relaxation in hydrogen bonded liquids, H.G. Hertz and M.D. Zeidler 1027

PART E. MATRIX ISOLATION TECHNIQUE 1063

Ch. 22. Matrix-isolation studies of hydrogen bonding, H.E. Hallam 1065

PART F. INVESTIGATIONS ON FERROELECTRICS WITH HYDROGEN BONDS 1107

Ch. 23. Ferroelectric hydrogen bonded systems, V.H. Schmidt. 1109
Ch. 24. Coherent neutron scattering for observations on collective proton motions in systems of hydrogen bonds, H. Stiller 1169

PART G. THERMODYNAMICS OF HYDROGEN BONDED SYSTEMS . . . 1197

Ch. 25. Spectroscopic and calorimetric studies of hydrogen bonding, A.D. Sherry 1199

Ch. 26. Vapor pressure studies on hydrogen and deuterium bonding, H. Wolff 1225

PART H. HYDROGEN BONDING AND ADSORPTION ON SURFACES . . . 1261

Ch. 27. Hydrogen bonds in systems of adsorbed molecules, H. Knözinger 1263

PART I. WATER AND ICE 1365

Ch. 28. Hydrogen bonds in liquid water. W.A.P. Luck . . . 1367
Ch. 29. The hydrogen bond in ice, E. Whalley 1425

Author index 1471

Subject index *(also included, without pagination, at the end of volumes I and II)* 1529

PART D

DYNAMICS OF HYDROGEN BONDED SYSTEMS

CHAPTER 19

INCOHERENT NEUTRON SCATTERING EXPERIMENTS ON HYDROGEN BONDED SYSTEMS

J. A. JANIK

Institute of Nuclear Physics, 31–342 Kraków, Poland

Contents

19.1. Introduction 893
19.2. Theoretical and experimental aspects of the incoherent neutron scattering method 894
19.3. Incoherent inelastic neutron scattering results with hydrogen bonded systems 897
19.3.1. General remarks 897
19.3.2. Ice and water 898
19.3.3. Water of crystallization 902
19.3.4. Hydronium perchlorate ($H_3O \cdot ClO_4$) 903
19.3.5. Ammonium salts 905
19.3.6. KH_2PO_4 908
19.3.7. $NaHF_2$ 909
19.3.8. HCl 909
19.4. Quasielastic and elastic neutron incoherent scattering results with hydrogen bonded systems 910
19.4.1. General remarks 910
19.4.2. Water 912
19.4.3. Hydronium perchlorate 915
19.4.4. $(NH_4)_2SO_4$ 918
19.4.5. KH_2PO_4 920
19.5. Total neutron scattering experiments with hydrogen bonded systems 925
19.6. Final remarks 927
Appendix 929
References 933

The hydrogen bond – recent developments in theory and experiments
Eds. P. Schuster et al. © *North-Holland Publ. Co., Amsterdam, 1976*

19.1. Introduction

Neutron incoherent inelastic scattering has proved to be a powerful tool, which gives information concerning the frequency spectrum of vibrations in condensed systems. Its superiority over optical spectroscopy methods, connected with a neutron wave length much smaller than that of photons used in IR and Raman spectroscopy, lies in the fact that vibrational frequencies from the whole Brillouin zone are accessible for neutrons. Optical spectroscopy, on the other hand, limits our information to the centre of the Brillouin zone, except when the investigated system is disordered or when combination bands, exciton+phonons, are at our disposal.

Most information obtained so far by the incoherent inelastic neutron scattering method (IINS) concerns vibrational motions in which H-nuclei take part. This is a result of the fact that protons are approximately one order of magnitude more powerful incoherent scatterers of neutrons than other nuclei. The application of IINS for studies of H-bonded systems is therefore natural, and indeed among frequency spectra of vibrations in hydrogenous substances there are many connected with protons taking part in H-bonds.

The incoherent scattering of neutrons may also be applied to the study of stochastic motions of nuclei (again with preference of proton motions). These are, for instance, all kinds of motions connected with an activation process, such as translatory or rotational jumps of molecules over certain activation barriers. Such motions give rise to a broadening of an originally monoenergetic neutron "line". This broadening, when studied with sufficiently good resolution and in a properly chosen neutron momentum transfer range, may give information concerning both relaxation times for a given type of stochastic motion and jump lengths. The corresponding method is called the quasielastic neutron scattering method (QNS) and may be treated as yielding analogous and also complementary information to IR and Raman line profile studies in optical spectroscopy.

There exist several articles concerning neutron scattering by H-bonding systems. In the book "Thermal Neutron Scattering" edited by Egelstaff [1965], the reader may find information concerning the theory and experimental technique and also, especially in ch. 8 by Larsson and ch. 10 by Janik and Kowalska, some information connected with H-bonding systems. In Hamilton and Ibers' [1968] book "Hydrogen Bonding in Solids", ch. 4 concerns IINS as applied to studies of H-bonds. The Proceedings of the 2nd Summer School on Solid State Physics held in Hercegnovi in 1968, edited by Jović in the form of a book "Dynamic and Magnetic Properties of Solids and Liquids", contain a chapter on "Neutron Scattering from Hydrogen Bonds" written by Plesser and Stiller [1970]. Finally, *in this book*, the "Coherent Neutron Scattering for Observations on Collective Proton Motions in Systems of Hydrogen Bonds" is discussed in ch. 24 by Stiller.

The present chapter will contain the following items: section 19.2 will provide information concerning the theoretical and experimental aspects of both the IINS and QNS methods. In section 19.3 results obtained by the IINS method will be discussed. Here the reader will find information concerning torsional and translatory frequencies of molecules which take part in H-bonds. In section 19.4 incoherent elastic and quasielastic scattering data will be given, together with such problems as the shape of the H-bond potential curve and the geometry of proton jumps. Section 19.5 will give information concerning H-bonds obtained by total neutron scattering experiments. In the Appendix the reader will find a list of substances with H-bonds investigated so far by neutron incoherent scattering methods.

19.2. Theoretical and experimental aspects of the incoherent neutron scattering method

Theoretical formulas of this section are written following the book of Egelstaff [1965], and taking into account the theoretical introduction of ch. 24 by Stiller.

As Stiller pointed out, the cross-section for neutron scattering by a certain type of atom into solid angle dΩ and energy interval $\mathrm{d}E = \hbar\,\mathrm{d}\omega$ may be written

$$\mathrm{d}^2\sigma/\mathrm{d}\Omega\,\mathrm{d}\omega = (\boldsymbol{k}/\boldsymbol{k}_0)S(\boldsymbol{\kappa}, \omega), \tag{19.1}$$

where $\boldsymbol{k}_0$ and $\boldsymbol{k}$ stand for the incident and scattered neutron wave vector respectively, and $\boldsymbol{\kappa} = \boldsymbol{k}_0 - \boldsymbol{k}$.

The scattering function $S(\boldsymbol{\kappa}, \omega)$ may in turn be written

$$S(\boldsymbol{\kappa}, \omega) = a_{\text{inc}}^2 S_{\text{inc}}(\boldsymbol{\kappa}, \omega) + a_{\text{coh}}^2 S_{\text{coh}}(\boldsymbol{\kappa}, \omega), \tag{19.2}$$

where a_{inc} and a_{coh} are the incoherent and the coherent scattering lengths respectively, and the space and time Fourier transforms of S_{inc} and S_{coh} are the so-called self and distinct correlation functions $G_s(\boldsymbol{r} - \boldsymbol{r}_0, t - t_0)$ and $G_d(\boldsymbol{r} - \boldsymbol{r}_0, t - t_0)$ respectively.

In ch. 24 Stiller is concerned with the coherent neutron scattering only, i.e., with the scattering pattern which, after subtraction of the incoherent background, may be represented by the formula

$$\mathrm{d}^2\sigma/\mathrm{d}\Omega\,\mathrm{d}\omega = (k/k_0) a_{\text{coh}}^2 S_{\text{coh}}(\boldsymbol{\kappa}, \omega). \tag{19.3}$$

In the present chapter, on the other hand, we are concerned with the scattering by protons for which $a_{\text{inc}} = 2.52 \times 10^{-12}$ cm, whereas $a_{\text{coh}} = -0.38 \times 10^{-12}$ cm, so that we may assume $a_{\text{inc}}^2 \gg a_{\text{coh}}^2$ and hence

$$\mathrm{d}^2\sigma/\mathrm{d}\Omega\,\mathrm{d}\omega = (k/k_0) a_{\text{inc}}^2 S_{\text{inc}}(\boldsymbol{\kappa}, \omega). \tag{19.4}$$

Thus, as the neutron scattering by proton systems is predominantly incoherent, it will give us information about the average space time behaviour of individual atoms and not about the correlations among them (as is clearly pointed out in ch. 24).

Very often it is convenient to discuss, besides the S and G functions, the intermediate scattering function $I(\boldsymbol{\kappa}, t - t_0)$, which is the space Fourier transform of G

$$I(\boldsymbol{\kappa}, t - t_0) = \frac{1}{2\pi\hbar} \int \mathrm{d}t \exp[\mathrm{i}\boldsymbol{\kappa} \cdot (\boldsymbol{r} - \boldsymbol{r}_0)] G(\boldsymbol{r} - \boldsymbol{r}_0, t - t_0), \tag{19.5}$$

and also another function $\tilde{I}(\boldsymbol{\kappa}, t - t_0)$, introduced by Schofield [1960a, b] by the formula

$$\tilde{I}(\boldsymbol{\kappa}, t - t_0) = I(\boldsymbol{\kappa}, t - t_0 + \mathrm{i}\hbar/2k_B T) \tag{19.6}$$

(k_B is the Boltzmann constant and T the temperature). We shall denote the time Fourier transform of this new function $\tilde{I}$ by $\tilde{S}(\boldsymbol{\kappa}, \omega)$. It may be proved that

$$S(\boldsymbol{\kappa}, \omega) = \exp(-\hbar\omega/2k_B T) \tilde{S}(\boldsymbol{\kappa}, \omega), \tag{19.7}$$

so that (in our incoherent case)

$$\mathrm{d}^2\sigma/\mathrm{d}\Omega\,\mathrm{d}\omega = (k/k_0) a_{\text{inc}}^2 \exp(-\hbar\omega/2k_B T) \tilde{S}_{\text{inc}}(\boldsymbol{\kappa}, \omega). \tag{19.8}$$

In section 19.4 we prefer to use the formula (19.8) instead of (19.4) because the function $\tilde{S}_{\mathrm{inc}}(\boldsymbol{\kappa}, \omega)$, being symmetric, is more convenient for description of the quasielastic neutron scattering process than $S_{\mathrm{inc}}(\boldsymbol{\kappa}, \omega)$.

At present we do not wish to stress differences between formulas (19.4) and (19.8), but rather to conclude that measurements of $\mathrm{d}^2\sigma/\mathrm{d}\Omega\,\mathrm{d}\omega$ for an incoherent scatterer give us $S_{\mathrm{inc}}(\boldsymbol{\kappa}, \omega)$ or $\tilde{S}_{\mathrm{inc}}(\boldsymbol{\kappa}, \omega)$ functions, which contain information concerning the dynamical behaviour of an individual atom–in our case, an individual proton in the scatterer. One should once more remember here that

$$S_{\mathrm{inc}}(\boldsymbol{\kappa}, \omega) = \frac{1}{2\pi\hbar} \int \mathrm{d}\boldsymbol{r}\, \mathrm{d}t\, \exp\{\mathrm{i}[\boldsymbol{\kappa} \cdot (\boldsymbol{r} - \boldsymbol{r}_0) - \omega(t - t_0)]\} G_{\mathrm{s}}(\boldsymbol{r} - \boldsymbol{r}_0, t - t_0), \tag{19.9}$$

where the self-correlation function $G_{\mathrm{s}}(\boldsymbol{r} - \boldsymbol{r}_0, t - t_0)$ is the probability that given an atom in $\boldsymbol{r}_0$, t_0, the same atom will be in $\boldsymbol{r}$, t. More specified formulas will be used in sections 19.3 and 19.4.

Now, let us discuss briefly the experimental apparatus. Figure 24.1 shows schematically the principle of such apparatus. Neutrons of a given $\boldsymbol{k}_0$ (selected by the monochromator) are scattered by the sample at an angle θ, and their energy distribution after the scattering (which means also the $\boldsymbol{k}$ distribution) is measured by the analyser. The results are proportional to the cross-section $\mathrm{d}^2\sigma/\mathrm{d}\Omega\,\mathrm{d}E$.

Practical adaptations of this principle may differ because different types of monochromators and/or analysers may be applied. One may use, for instance, monocrystals of known materials (e.g., graphite) for both the analyser and the monochromator, and apply the Bragg reflection for selecting neutrons of a desired energy. Or one may apply a polycrystalline filter (e.g., beryllium) either as the "monochromator" or as the "analyser". In this case no real neutron "line" is obtained but a "cut-off" of the Maxwell spectrum at energies of several meV, which in many examples–especially in section 19.3–will be treated as equivalent to a "line". Again, one may apply (usually for the analyser) the time-of-flight technique, i.e., use the interrupted neutron beam and a delayed detection of neutrons after scattering. In all these apparatus the energy resolution $(\Delta\omega)/\omega$ lies in the range from ca. 2% to ca. 15%. This is sufficient for many of the IINS spectra but very often insufficient for QNS results. Therefore, the existing QNS data must be discussed very carefully, stressing the fact that in future the new high flux reactors (Grenoble, Dubna) should enable the full power of the QNS method to be exploited. At present there exists only one apparatus (Alefeld et al. [1969]) which possessing a resolution of ca. three orders of magnitude better is especially suitable for QNS experiments.

19.3. Incoherent inelastic neutron scattering results with hydrogen bonded systems

19.3.1. General remarks

In this section we will be interested in vibrations of nuclei (mostly protons) in the scattering system. If a vibration may be treated as harmonic, one can easily write down a specified formula for the G_s function and then, applying formulas (19.9) and (19.4), obtain formula (19.10) for $d^2\sigma/d\Omega\, d\omega$ per one proton per one vibration, in the case of an incoherent scatterer, in one phonon approximation, and for a cubic symmetry.

$$\frac{d^2\sigma}{d\Omega\, d\omega} = a_{inc}^2 \frac{k}{k_0} \kappa^2 e^{-2W} \frac{g(\omega)}{\omega} \frac{(n+\frac{1}{2}) \pm \frac{1}{2}}{2M} |u(\omega)|^2, \tag{19.10}$$

where e^{-2W} is the Debye–Waller factor, n is the number of phonons in the quantum state under consideration (energy $\hbar\omega$ and impulse $\hbar \boldsymbol{q}$, $\boldsymbol{q}$ being connected with $\boldsymbol{k}_0$ and $\boldsymbol{k}$ by the formula (24.7c)), M is the mass of the scattering nucleus (proton), $\boldsymbol{u}$ is the polarization factor of the given vibration, and $g(\omega)$ is the density of vibrational states corresponding to the frequency ω. Signs + and − correspond to neutron energy gain and neutron energy loss cases respectively (anti-Stokes and Stokes scattering). For harmonic crystals we replace n by $\langle n \rangle$ and use the Planck formula

$$\langle n \rangle = [\exp(\hbar\omega/k_B T) - 1]^{-1}. \tag{19.11}$$

In practical applications of formula (19.10) for $g(\omega)$ determination one must be careful, because its form will be different for non-cubic symmetry and also because formula (19.11) is questionable (e.g., for liquids). Nevertheless, formula (19.10) is widely used for the determination of what may be called the effective frequency distribution $g_{eff}(\omega)$, which in the case of cubic symmetry and harmonic crystals is

$$g_{eff}(\omega) = a_{inc} |\boldsymbol{u}(\omega)|^2 \frac{e^{-2W}}{2M} g(\omega), \tag{19.12}$$

i.e.,

$$g_{eff}(\omega) = \frac{d^2\sigma/d\Omega\, d\omega}{\kappa^2} \frac{\boldsymbol{k}_0}{\boldsymbol{k}} \omega [\exp(\hbar\omega/k_B T) - 1]. \tag{19.13}$$

Formula (19.13) is also used for non-cubic systems, since the $g_{\text{eff}}(\omega)$ function obviously reproduces the positions of peaks of the $g(\omega)$ function without many distortions.

In the following sections we will discuss IINS results concerning some selected substances. The quality of the results obtained so far is different for different substances. In many cases we have at our disposal just frequencies of vibrations, which are in a way raw data for possible future applications to H-bond problems. In a few cases more sophisticated speculations concerning the H-bond are made.

19.3.2. Ice and water

As seen from the list in the Appendix, a large number of authors studied ice or water using the neutron method.

In this section the recent results of Blanckenhagen [1972] will be described. They are representative of most of the investigations performed so far for water by IINS.

The experimental conditions were as follows: incident neutron energy -5 meV; energy resolution of scattered neutrons -4% in the elastic scattering region and 15% in the energy transfer region of ca. 150 meV; data were obtained for several scattering angles from $\Theta = 20°$ to $\Theta = 104°$, and for several temperatures, -5, $+2$, $+23$ and $+95$ °C. In the same experiment QNS spectra were measured; they will be discussed in section 19.4.

Some exemplarily chosen time-of-flight spectra of scattered neutrons are presented in fig. 19.1. From these spectra the $g_{\text{eff}}(\omega)$ were calculated by applying formula (19.13) and with corrections made for the multiple scattering and multi-phonon scattering in the sample. From $g_{\text{eff}}(\omega)$ the frequency distribution functions $g(\omega)$ were calculated by using formula (19.12) and the following approximations: (a) $|\boldsymbol{u}(\omega)|^2 = 1$, (b) the Debye–Waller factor e^{-2W} is taken from integrated QNS peaks versus κ^2 dependence by using a formula $e^{-2W} = \exp(-u_\kappa^2\kappa^2)$, where u_κ is the displacement of the scattering proton along $\boldsymbol{\kappa}$, and (c) the mass M is equal to 18 for translational motions and 2.32 for torsional motions of H_2O molecules, the latter value being the result of considerations based upon the Sachs and Teller mass tensor concept (see, e.g., Egelstaff [1965]). The $g(\omega)$ functions so obtained are presented in fig. 19.2.

The main features of these functions are as follows: for ice there appear peaks at 54 cm^{-1}, at ca. 256 cm^{-1} and at ca. 696 cm^{-1} (split into two components); for water the peaks are broader but they appear at nearly the same energies as for ice.

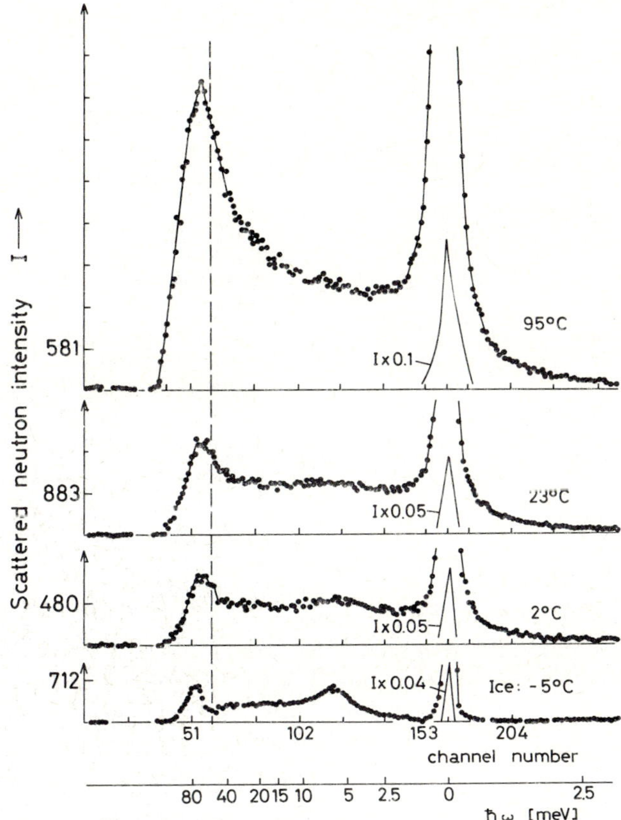

Fig. 19.1. IINS time-of-flight spectra for H_2O at various temperatures (Blanckenhagen [1972]).

Even without any theoretical model the observed similarity of the ice and water spectra indicates that a local solid-like order of a certain range must exist in water for times sufficiently long to be observed by neutrons (ca. 10^{-12} s). The H-bonding is no doubt responsible for this local order. It is natural to expect, on the other hand, that some H-bonds are broken when ice melts. This breaking of H-bonds must contribute to the specific heat jumps at melting, and we may write, following Eucken [1948]

$$C_v = C_v^{dyn} + C_v^{assoc}, \tag{19.14}$$

where C_v^{dyn} is specific heat connected with filling vibrational energy levels and C_v^{assoc} is that connected with the energy used for breaking of H-bonds. The specific heat measured calorimetrically contains both contributions.

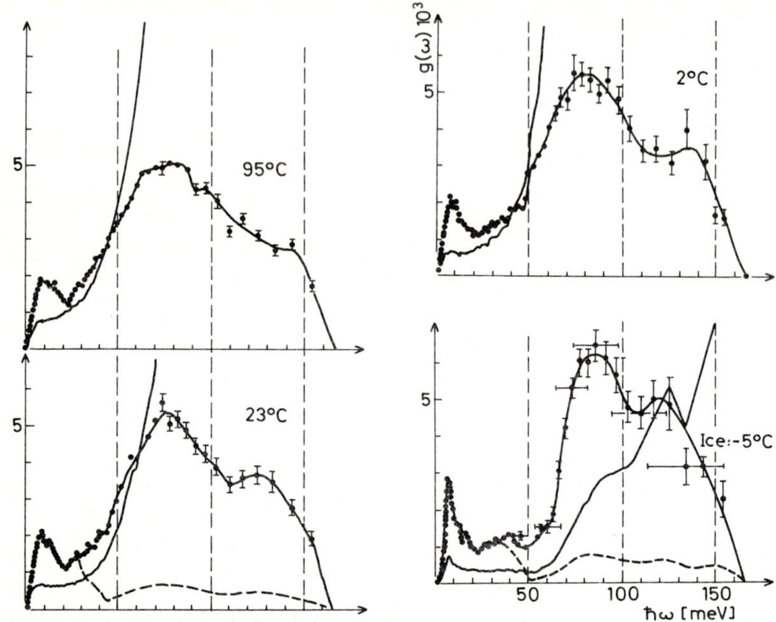

Fig. 19.2. $g(\omega)$ functions for H_2O at various temperatures (Blanckenhagen [1972]). Solid line with experimental points–after corrections for multiphonon and multiple scattering. Solid line–uncorrected distribution. Interrupted line–after all corrections as described in text (1 meV = 8.06 cm^{-1}).

But with neutrons, as pointed out by Szkatuła and Fuliński [1967], we measure the C_v^{dyn} part separately, through the $g(\omega)$ measurements, as

$$C_v^{dyn} = 3N \int_0^\infty d\omega\, g(\omega) \left(\frac{\partial \langle n(\omega, T) \rangle}{\partial T} \right)_v . \tag{19.15}$$

Figure 19.3 presents the C_v and C_v^{dyn} comparison. C_v are tabulated calorimetric data and the C_v^{dyn} data are from the IINS measurement under discussion. One may see that there is indeed a large C_v^{assoc} contribution for water caused predominantly by the demolition of the short range order (i.e., demolition of H-bonds). In principle, it is possible to evaluate from such a comparison the H-bond energy and the number of H-bonds broken at a given temperature. These two kinds of information were not drawn by the author of the paper under discussion. An estimation based on earlier, less accurate data (Szkatuła and Fuliński [1967]) for the energy necessary to break a H-bond gives a value of ca. 2 kcal/mole, and for the percentage of H-bonds broken at melting gives a value of ca. 30%.

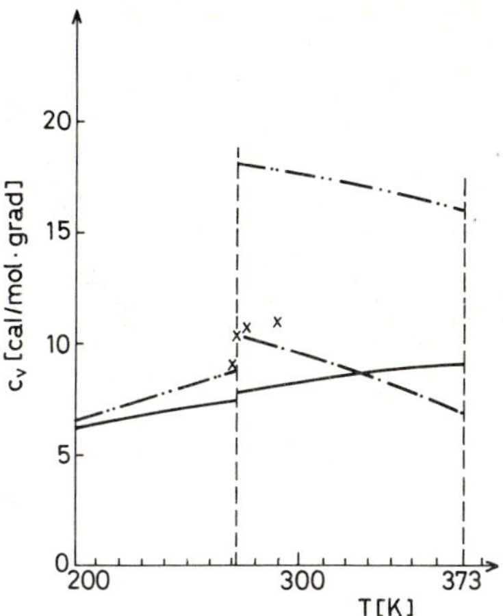

Fig. 19.3. Specific heat for H_2O (Blanckenhagen [1972]). – – · –. C_v as measured calorimetrically. ———C_v^{dyn} as measured by IINS (Blanckenhagen [1972]). × × × C_v^{dyn} as measured by IINS (Szkatuła and Fuliński [1967]). –·–·–$C_v - C_v^{dyn}$ for liquid H_2O.

The only model calculation made so far with an attempt to obtain the space-time behaviour of water molecules is the molecular dynamics study by Rahman and Stillinger [1971]. The authors perform calculations for 216 molecules interacting by an effective pair potential, which consists of an electrostatic part (four electric charges in a tetrahedral configuration) and a Lennard-Jones part. It is perhaps a little surprising that, in rough agreement with the experiment of Blanckenhagen [1972] under discussion, the frequency spectrum of their velocity auto-correlation function, which is a natural generalization of the phonon density function when passing from solids into liquids, shows maxima at ca. 53, 200 and 400 cm^{-1}, the last one being rather broad and extending to ca. 1280 cm^{-1}.

The interpretation of the maxima observed for ice is as follows: the peak at 54 cm^{-1} is connected with one acoustical and one optical translatory mode (H-bond bending); the broad peak at 256 cm^{-1} contains contributions from higher optical translatory modes (H-bond stretching); the broad and split peak at 696 cm^{-1} is caused by torsional vibrations of H_2O molecules. In fact, there should appear three torsional peaks corresponding to rotatory motions around three molecular axes, but due to the rather poor instrumental resolution in the 100 meV region, we observe a splitting into two components only.

It should be noted here that some authors obtained more peaks for ice than those discussed above. Burgman et al. [1968], for instance, have peaks at 52, 69, 86, 210, 275 and 632 cm^{-1}.

The molecular dynamics calculation quoted above is not the only theoretical approach to the dynamics of the H_2O system. One should mention here, for instance, the force constants approach by Forslind [1954], which reproduces fairly well the ice frequency spectrum, and also when applied to D_2O it reproduces phonon dispersion relations for heavy ice as obtained in neutron coherent inelastic scattering experiments by Renker [1969].

19.3.3. Water of crystallization

IINS spectra of hydrates show a general character resembling that of ice. Practically in all hydrates studied so far there is a group of peaks interpreted as caused by optic translatory motions of H_2O molecules and appearing in the 130–350 cm^{-1} region, and another group interpreted as caused by H_2O torsions and appearing in the 400–800 cm^{-1} region. These frequencies are rather high and give evidence of strong coupling of a H_2O molecule to its neighbours by the H-bond. It should be pointed out that in a molecular crystal with no H-bond these types of vibrations occur at much lower frequencies; for a methyl iodide crystal, for instance, they are in the 30–120 cm^{-1} region (Janik et al. [1968]).

Even more similarities to the ice spectrum are shown by that obtained for the polycrystalline $Li_2SO_4 \cdot H_2O$ (Prask et al. [1966], Bajorek et al. [1968] and Thaper et al. [1969]), in which a low frequency group, in the 70–100 cm^{-1} region appears. As Rannev et al. [1968] pointed out, H_2O molecules form in $Li_2SO_4 \cdot H_2O$ zig-zag chains of H-bonds similar to those in ice, thus probably leading to similar shapes of acoustic phonon branches. In $Li_2SO_4 \cdot H_2O$ the shape of this low frequency part of the $g_{eff}(\omega)$ spectrum is very similar to the corresponding part of the far IR spectrum (Mikuli et al. [1972]) indicating that acoustic vibrations for this substance are observed in the IR. It is possible only if there exists some disorder in the H_2O zig-zag chain, i.e., if there occur proton jumps which cause some defects. Two mechanisms of such jumps are discussed, one connected with displacements of protons along a H-bond and the other connected with rotatory displacement of the whole H_2O molecule.

An interesting attempt to obtain more information concerning the type of torsional motion of H_2O molecules in hydrates was made by Thaper et al. [1970]. The authors applied the IINS method to monocrystals for which, instead of formula (19.10), a similar one, but with the scalar product squared $|\boldsymbol{\kappa} \cdot \boldsymbol{u}(\omega)|^2$ should be written. (For cubic symmetry, the mean value of $|\boldsymbol{\kappa} \cdot \boldsymbol{u}(\omega)|^2$ leads, after taking into account some numerical factors, to formula (19.10). It is clear that if we set $\boldsymbol{\kappa}$ perpendicular to the plane of the H_2O molecule, only twisting and waving torsional motions will be observed, whereas if $\boldsymbol{\kappa}$ is parallel to the H–H vector only the rocking torsional motion will be observed. Figure 19.4 presents the scattered neutron intensity obtained in two such settings for $Ba(ClO_3)_2{\cdot}H_2O$ and $K_2C_2O_4{\cdot}H_2O$ monocrystals. The authors claim that of the 395 and 457 cm^{-1} peaks obtained for $Ba(ClO_3)_2{\cdot}H_2O$ one corresponds to twisting and the other to waving (it is not possible, however, to say more, i.e., which is which?). The 477 cm^{-1} peak obtained for the same substance may only be connected with rocking. For $K_2C_2O_4{\cdot}H_2O$ the peak at 738 cm^{-1} is connected either with waving or twisting, whereas the 644 cm^{-1} peak must be connected with rocking.

19.3.4. Hydronium perchlorate ($H_3O{\cdot}ClO_4$)

The IINS spectra of hydronium salts resemble those discussed in the two preceding sections, this being a result of the fact that a H-bonded H_3O^+ unit undergoes similar translatory and torsional motions as a H_2O unit in hydrates. The reason why the hydronium perchlorate is presented sepa-

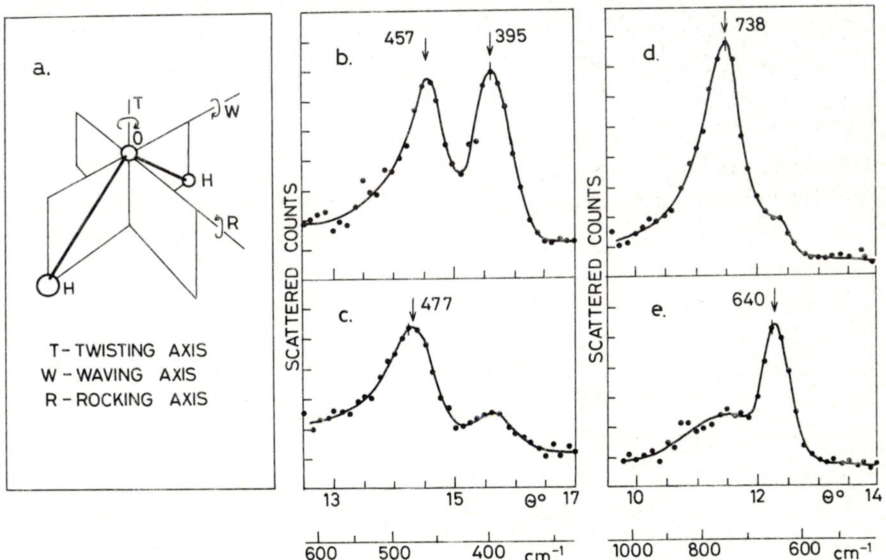

Fig. 19.4. IINS spectra for single crystals of $Ba(ClO_3)_2{\cdot}H_2O$ ((b) $\boldsymbol{\kappa} \perp \overrightarrow{HH}$, (c) $\boldsymbol{\kappa} \| \overrightarrow{HH}$) and $K_2C_2O_4{\cdot}H_2O$ ((d) $\boldsymbol{\kappa} \perp \overrightarrow{HH}$, (e) $\boldsymbol{\kappa} \| \overrightarrow{HH}$) (Thaper et al. [1970]).

rately is connected with the fact that the QNS spectra of this substance will be discussed in section 19.4.

Hydronium perchlorate (melting point at +50 °C) has two solid phases separated by a phase transition point (−25°C). Their structures are known from X-ray studies (Lee and Carpenter [1959] and Nordman [1962]), being orthorhombic (Pnma) for phase I and monoclinic ($P2_1/n$) for phase II. In phase II H-atoms have well-defined positions lying nearly on lines connecting oxygen of the H_3O^+ group with oxygens of three neighbouring ClO_4^- groups. In phase I an orientational disorder was suggested for protons of the H_3O^+ groups.

We shall briefly describe here the results of an IINS experiment performed for this substance by Janik et al. [1973a]. The experimental conditions were as follows: the neutron incident energy was variable and selected by the time-of-flight technique; neutron energy after scattering was approximately fixed by a cut-off of a Be filter–the mean value of the transmitted energy equalled 3.7 meV. The neutron energy loss (Stokes) spectrum was thus measured. The energy resolution in the 200–600 cm^{-1} region was ca. 10%. Measurements were performed for five temperatures, −150 and −50 °C for phase II, −15 and +21 °C for phase I and +60 °C for the melted substance.

The results are presented in fig. 19.5. When transformed to the $g_{eff}(\omega)$ form by formula (19.13) they show the following features. For phase II there are translatory peaks at ca. 60 cm^{-1} (probably acoustic), ca. 140 cm^{-1} and ca. 200 cm^{-1} (translatory optic); a H_3O^+ torsional peak appears at 560 cm^{-1}. For phase I all peaks are significantly broadened; the torsional peak shifts towards lower energies (520 cm^{-1}); for the melted substance the peak remains broad but shifts back to higher energies (ca. 600 cm^{-1}).

The sharp torsional peak at 560 cm^{-1} in phase II confirms the statement of a well-defined situation of the H_3O^+ group, corresponding, probably, to a uniaxial torsional oscillation of this group. In phase II the situation of the H_3O^+ group becomes much less defined, leading to a broadening of the peak and also, due to weaker acting forces, the average frequency is reduced to 520 cm^{-1}.

If one assumes the high torsional barrier approximation and consequently the formula

$$\omega_{tors} = 2n\sqrt{U_{act}B} \tag{19.16}$$

(ω_{tors} is the torsional frequency, n is the symmetry number, U_{act} is the barrier height and B is the rotational constant), one obtains for the torsional barrier U_{act} in phase II the value of 4.6 kcal/mole, and in phase I

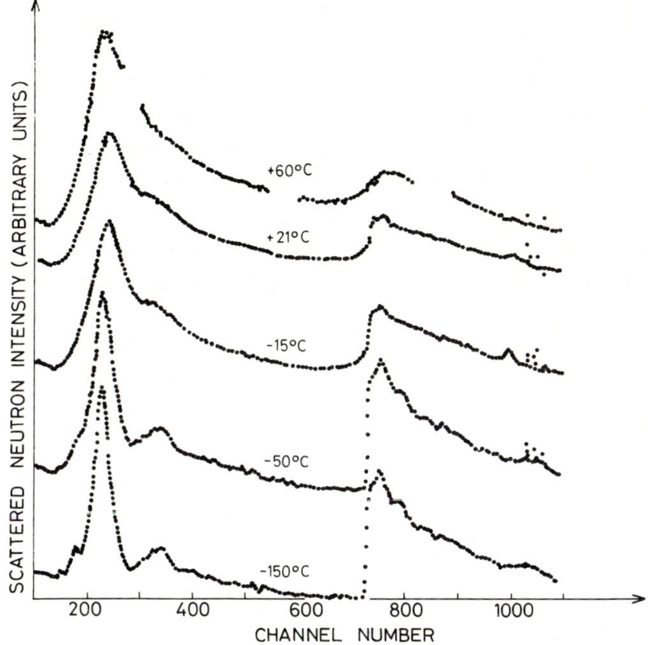

Fig. 19.5. IINS spectra for $H_3O{\cdot}ClO_4$ at various temperatures. Temperatures −150 and −50°C correspond to phase II; temperatures −15 and +21°C correspond to phase I; temperature +60°C corresponds to the liquid (Janik et al. [1973a]).

3.9 kcal/mole. These values may be compared with those obtained by the NMR (O'Reilly et al. [1971]) which are 4.8 and 4.2 kcal/mole respectively.

For the melted substance the position of the torsional peak practically coincides with that of water, this being connected with the fact that H_3O^+ groups no longer exist, but the substance is rather a solution of $HClO_4$ molecules in water.

19.3.5. Ammonium salts

In most of the ammonium salts there exist torsional and translatory IINS peaks similar to those described in preceding sections. Their frequency range indicates the existence of H-bonds. It should, however, be pointed out that some of the ammonium compounds listed in the Appendix have IINS spectra indicating a high degree of rotational freedom of NH_4^+ groups, certainly being by no means understood in the framework of H-bonding. This situation occurs, for instance, in NH_4ClO_4 and NH_4PF_6 (see, e.g., Janik et al. [1964, 1971]).

In this section we will be concerned with two representatives of ammonium compounds, NH_4Cl and $(NH_4)_2SO_4$.

Ammonium chloride has three solid phases: phase III of CsCl structure, with ordered configurations of NH_4^+ groups; phase II of CsCl structure, with disordered configurations of NH_4^+ groups; and phase I of NaCl structure, with relatively high rotational freedom of NH_4^+ groups. Transition temperatures are: phase III/phase II 243 K, phase II/phase I 457 K.

IINS spectra at different temperatures reported by various authors (see, e.g., Bajorek et al. [1964]) show no significant change when passing from phase III to phase II. For NH_4I where the corresponding transition temperatures are lower, it was proved that at the phase II/phase I transition the peak connected with the NH_4^+ torsion disappears, indicating a great rotational freedom of NH_4^+ groups in phase I.

There exist two theoretical approaches to understand the dynamics of molecules in the ordered phase III of NH_4Cl. They are both based upon a model with a pair interacting potential which is a composition of electrostatic and Lennard-Jones parts. The first one may be called a rigid ion model (Parliński [1968, 1969]); the NH_4^+ ion is approximated in it by a rigid tetrahedron in which, in proton positions, there are positive electric charges $+\frac{1}{4}e$; negative electric charges $-e$ are in Cl^- positions. The model yields 8 of a total of 18 force constants; the remaining 10 must be obtained by fitting to experimental data. The NH_4Cl frequency spectrum obtained in this theory reproduces with an accuracy of ca. 10% the IINS measured $g_{\text{eff}}(\omega)$ peaks, i.e., a very narrow NH_4^+ torsional one at 395 cm^{-1} and a broad one at 176 cm^{-1} composed of Cl^- and NH_4^+ translations, the latter being predominant in neutron measured $g_{\text{eff}}(\omega)$. The narrowness of the torsional peak is a result of a very flat dispersion curve for NH_4^+ torsions in all important directions of the reciprocal space. The translatory peak, on the other hand, is so broad because corresponding phonon branches (both acoustic and optic) show significant dispersion and slightly different behaviour for different directions in the reciprocal space. It should be pointed out that dispersion relations have been measured for NH_4Cl monocrystals with the coherent inelastic scattering method by Teh and Brockhouse [1971] and the results also agree with an accuracy of ca. 10%.

The second theoretical calculation for NH_4Cl was made by Cowley [1971], who introduces the shell model approximation. The shell model yields 10 of a total of 23 force constants; the remaining 13 must be fitted to experimental data. With this model one obtains a much better agreement

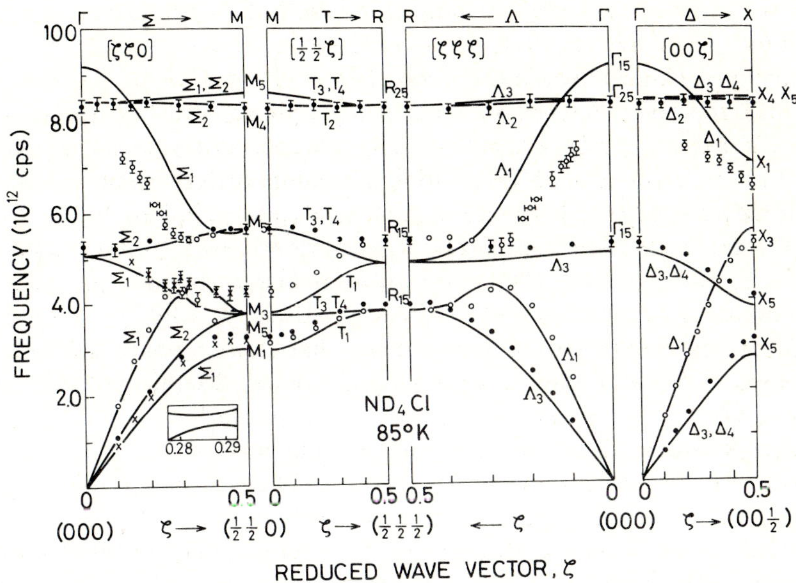

Fig. 19.6a. Phonon dispersion curves for NH_4Cl (Cowley [1971] and Teh and Brockhouse [1971]). Comparison with rigid NH_4^+ ion model of Parliński (solid lines).

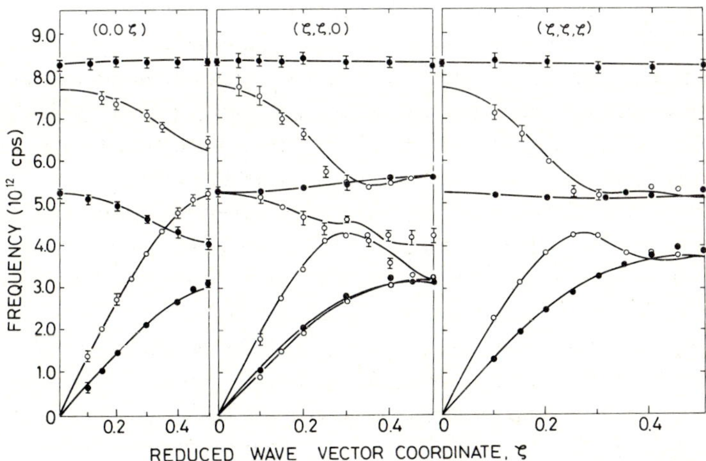

Fig. 19.6b. Phonon dispersion curves for NH_4Cl (Cowley [1971] and Teh and Brockhouse [1971]). Comparison with shell model of Cowley (solid lines).

with neutron data, though with more adjustable parameters than Parliński. The comparison of the neutron coherent scattering experiment and the two theories is shown in fig. 19.6.

It is interesting to note that models which are essentially based upon an electrostatic interaction describe the situation reasonably well both for NH_4Cl and (as seen in section 19.3.2) for water. Thus, it seems that an electrostatic interaction approach may work in some H-bonded systems.

Another ammonium compound, whose IINS spectra will briefly be described here, is $(NH_4)_2SO_4$. This substance will also be discussed in section 19.4 in connection with a QNS experiment. We shall describe here the IINS experiment performed by Rush and Taylor [1964], the experimental conditions of which were as follows: the Be-filtered beam was used as incident neutrons; their mean energy was ca. 3.4 meV; scattered neutrons were analysed by the time-of-flight technique; IINS spectra were measured at two temperatures, 296 and 172 K, one corresponding to the paraelectric phase and the other to the ferroelectric phase.

The high energy gain part of the spectra at both temperatures is clearly split into two peaks. It is, however, difficult to conclude with certainty whether this splitting is due to the existence of two different crystallographic situations of NH_4^+ molecules in the orthorhombic lattice and thus corresponds to NH_4^+ torsions of these two types of molecules, or whether it may also be that the lower of the two peaks corresponds to translatory NH_4^+ motions and the higher one to unresolved torsions. The two peak positions for the temperature 172 K are 335 and 200 cm^{-1}. The peak at 335 cm^{-1} shifts to 305 cm^{-1} when the temperature increases to 296 K. This peak can be assigned with certainty to torsional vibrations. It shifts very little when passing from the paraelectric to ferroelectric phase, thus the barrier to NH_4^+ rotation is similar in the two phases and amounts to ca. 3 kcal/mole. This result disagrees with NMR data (Miller et al. [1962]), which suggested an increase in the barrier from ca. 2 to ca. 6 kcal/mole when passing from the paraelectric to ferroelectric phase.

19.3.6. KH_2PO_4

IINS experiments were made for this substance in order to obtain an experimental proof for proton fluctuations between two equilibrium positions in the paraelectric phase (see, e.g., Pelah et al. [1964]). Unambiguous proof cannot be obtained, however, until the elastic incoherent neutron scattering has been applied. These elastic neutron scattering experiments will be discussed in section 19.4.

19.3.7. $NaHF_2$

IINS appeared to be a good tool for detection of the H-bond bending mode F–H↕–F (Ghosh et al. [1972]). In this mode mainly protons are moving, thus causing a great intensity of the corresponding inelastic peak. The peak appears at the frequency 675 cm^{-1} and shifts to 480 cm^{-1} after deuteration (the deuteration reduces its intensity, of course, owing to the reduction of the neutron scattering cross-section).

19.3.8. HCl

This crystal consists of parallel planar zig-zag chains of HCl units connected by H-bonds. Its vibrations were studied by the IR (Anderson et al. [1964] and Arnold and Heastie [1967]) and the IINS technique (Boutin and Safford [1964]). Trevino et al. [1968] made a determination of a set of seven force constants, giving the best fit to IR data, i.e., to data corresponding to the centre of the Brillouin zone. Using this set of force constants the authors were able to calculate the expected IINS spectrum. The thus calculated spectrum is then compared with IINS experimental data. The comparison is shown in fig. 19.7.

It is interesting to quote here the selected force constant values. The dominant one is that corresponding to the H-bond stretch (F_1 = 0.0641 mdyn/Å); force constants corresponding to H···Cl–H and Cl···H–Cl angle bends are 0.0428, 0.0151 and 0.0109 mdyn/Å; the inter-

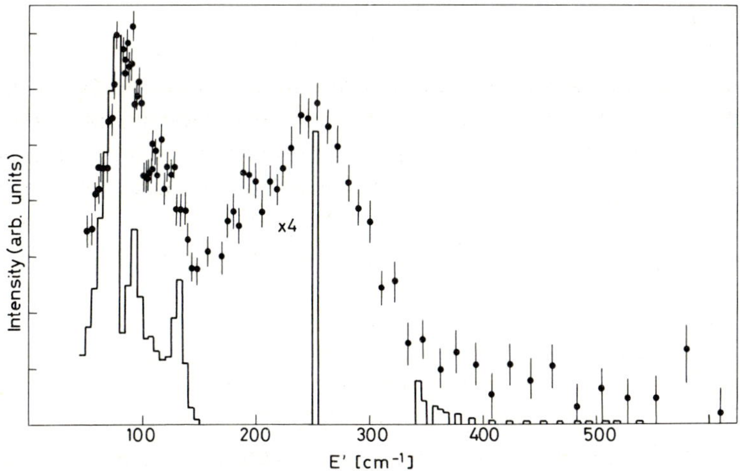

Fig. 19.7. IINS spectrum for polycrystalline HCl as compared with calculations based upon force constants fitting (Trevino et al. [1968]).

chain force constants are nearly zero, except that corresponding to the nearest chlorine stretching, which is 0.0301 mdyn/Å. Hence, these studies favour the hydrogen being bound to chlorine with one chain only, and the interchain forces being due only to chlorine–chlorine interaction.

19.4. Quasielastic and elastic neutron incoherent scattering results with hydrogen bonded systems

19.4.1. General remarks

The scattering function $\tilde{S}_{\text{inc}}(\boldsymbol{\kappa}, \omega)$ in formula (19.8) not only represents the vibrational spectrum but also reflects properties of random motions of protons in the scattering system. If there are no such motions, $\tilde{S}_{\text{inc}}$ at a given $\boldsymbol{\kappa}$ has in the neighbourhood of $\omega = 0$ the form of a δ-function. The scattered neutron intensity, however, must be broadened by the instrumental resolution, which in the majority of cases has a Gaussian form. If now, molecules in our scatterer perform a translatory diffusion, $\tilde{S}_{\text{inc}}(\boldsymbol{\kappa}, \omega)$ has near $\omega = 0$ the form of a Lorentz function. The actually measured scattered neutron intensity, being additionally broadened by the Gaussian instrumental function, thus has the form of a Lorentzian–Gaussian convolution. The width of $\tilde{S}_{\text{inc}}$ (i.e., the full width at half maximum) depends on $\boldsymbol{\kappa}$ differently for different models assumed for diffusion. If the diffusion is a continuous process this width is given by the formula (Vineyard [1958])

$$\Delta E = \Delta\omega = 2\hbar D\kappa^2 \tag{19.17}$$

(D is the diffusion coefficient). If the diffusion is a combination of alternative intermolecular vibrations occurring during time τ_0' and continuous processes occurring during time τ_1', and if we assume additionally that $\tau_0' \gg \tau_1'$, we have the formula (Singwi and Sjölander [1960])

$$\Delta\omega = \frac{2\hbar}{\tau_0'}\left(1 - \frac{e^{-2W}}{1 + \kappa^2 D\tau_0'}\right). \tag{19.18}$$

This formula has in the $\kappa \to 0$ limit the form (19.17), and in the $\kappa \to \infty$ limit the form

$$\Delta\omega = \frac{2\hbar}{\tau_0'}. \tag{19.19}$$

Let us now consider the case when no translatory diffusion takes place, but the molecule may rotate, i.e., the scattering nucleus moves on a sphere around the centre-of-mass of the molecule. Such a situation occurs in plastic crystals, for example. The $\tilde{S}_{\text{inc}}$ function is now (near $\omega = 0$) a composition of a δ-component (in practice broadened by the Gaussian instrumental function) and another component which has a different form for different rotational models. If we assume a large jump rotation, i.e., that the molecule performs torsional vibrations during time τ_0 and then continuously rotates over a large angle during time τ_1, we obtain in the $\tau_0 \gg \tau_1$ approximation for the second component a Lorentzian function whose width is (Sköld [1968], Larsson [1971, 1973] and Rościszewski [1973])

$$\Delta\omega = \frac{2\hbar}{\tau_0} \tag{19.20}$$

in the whole κ-region.

In molecular liquids the situation is usually quite complicated because both processes, translatory diffusion and rotations, occur. The diffusion causes a broadening of the δ-component in addition to an instrumental one; this component is now a Lorentzian–Gaussian convolution. The two components contribute to the $\tilde{S}_{\text{inc}}$ function differently in different κ-regions. In the $\kappa \to 0$ limit the rotational component vanishes, so that we observe the translatory diffusion broadening only, which in this limit is given by the formula (19.17). In the $\kappa \to \infty$ limit, on the other hand, there is for large jumps only one component, whose width is (Larsson [1971, 1973] and Rościszewski [1973])

$$\Delta\omega = \frac{2\hbar}{\tau_0 + \tau_0'}. \tag{19.21}$$

It should be stressed here that, as mentioned in section 19.2 one should have a sufficiently good instrumental resolution in order to detect experimentally the real shape of the scattering function. With a poor resolution one may, for instance, be inclined to interpret the data in frames of one Lorentzian component only, thus leading to entirely wrong numerical results. Moreover, even studied with a good resolution, the real situation may differ from the simplified one, described as above. For instance, if τ_0 is of the same order as τ_1, and/or τ_0' is of the same order as τ_1', the shapes of the translatory and the rotational components will not be simple Lorentzians but rather sums of Lorentzians, thus yielding more complications in interpretation.

These remarks explain why there exist relatively few really reliable data, or rather relatively few properly interpreted quasielastic and elastic neutron scattering results. In discussion of H-bonded systems we will limit ourselves in subsequent sections to four examples: water, hydronium perchlorate ($H_3O{\cdot}ClO_4$), ammonium sulphate ($(NH_4)_2SO_4$) and KH_2PO_4.

19.4.2. Water

QNS measurements were made for water by many authors (Cribier and Jacrot [1960], Egelstaff et al. [1960], Hughes et al. [1960], Larsson et al. [1960], Stiller and Danner [1960], Kottwitz et al. [1962], Larsson and Dahlborg [1962a, 1964], Golikov et al. [1964], Pope and Nations [1964], Sköld et al. [1964], Pelah and Imry [1966], Safford et al. [1969] and Blanckenhagen [1972]). We will describe here those of Blanckenhagen [1972] as being representative of the subject under discussion. The experimental conditions of the measurements were already described in section 19.3.2. The quasielastic lines measured at various angles, i.e., at various κ-values, are presented for two different temperatures 95 and 23°C in fig. 19.8. We may see that the instrumental line (as measured by the scattering from a vanadium sample, which scatters almost perfectly incoherently) is well represented by a Gaussian curve. The scattering from a water sample gives a broadened line; the broadening increases with the temperature for a given κ-value and also increases with κ for a given temperature. The shapes of QNS lines are well represented by a convolution of a single Lorentzian by the instrumental Gaussian, for any κ-value studied. It is not surprising that we have single Lorentzians at low κ-values. As explained above, translatory random motions are only seen at these values. We are also entitled to apply the formula (19.17) for low κ-region and to obtain diffusion coefficients for water in various temperatures. These diffusion coefficients are presented in fig. 19.9 together with those obtained by the NMR (spin echo) method (Simpson and Carr [1958] and Mills [1971]). There is a good agreement at temperatures higher than ca. 40°C, whereas at lower temperatures the QNS data are up to 70% greater than those from NMR. The reason for this discrepancy is not clear, as the rotational contribution in QNS data should not be present, firstly because we have taken into consideration only the low κ-values ($\kappa = 0.53\ \text{Å}^{-1}$), and secondly, even at larger κ, QNS shapes show no indication of a deviation from single Lorentzians. It should also be pointed out that the H_2O torsional frequency as measured, for instance, by IINS (see section 19.3.2) is quite high, thus giving rather rare reorientational jumps.

QNS results for all κ-values are presented in fig. 19.10 in the form of the width ΔE of the corresponding Lorentzian (after deconvolution) versus κ^2. One may see that although all dependences deviate from straight lines,

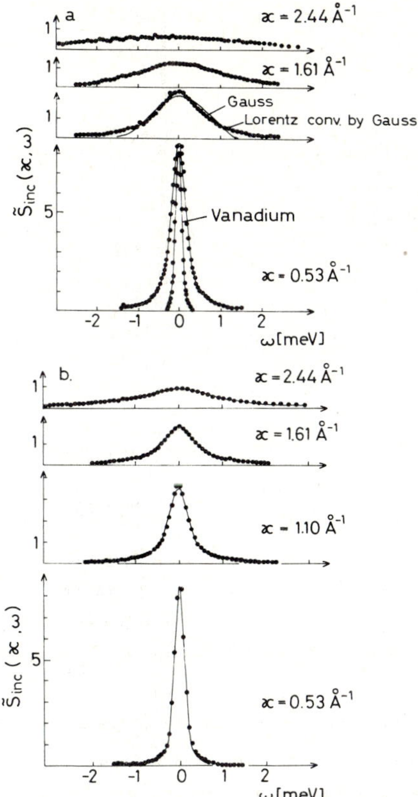

Fig. 19.8. QNS spectra for water at two temperatures: (a) 95°C, (b) 23°C. Curves correspond to the best fits of Lorentz functions convoluted by a Gaussian, except for the vanadium run to which a Gaussian curve is fitted (Blanckenhagen [1972]).

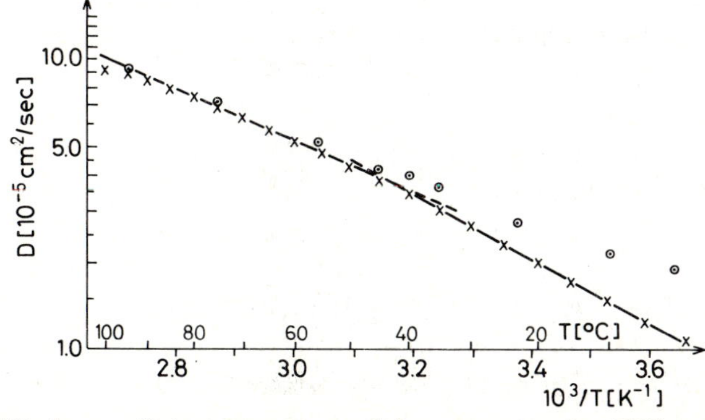

Fig. 19.9. Diffusion coefficients as measured for water by the QNS method (⊙ ⊙ ⊙) compared to those measured by NMR spin echo method (+ + +) (Blanckenhagen [1972]).

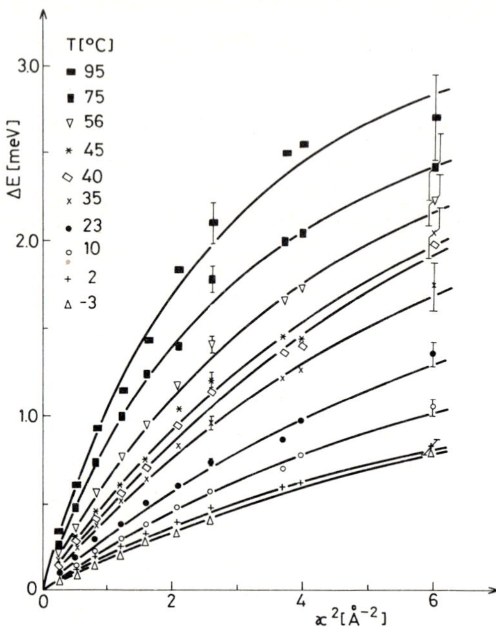

Fig. 19.10. QNS line width versus κ^2 for water at various temperatures. Solid lines correspond to the jump diffusion model (Balckenhagen [1972]).

certainly violating formula (19.17), there is no saturation at large κ as suggested by formulas (19.18) and (19.19). Therefore, an attempt was made to fit the curves of fig. 19.10 by a composition of two processes, the first corresponding to formula (19.18) and the second, a continuous process, corresponding to formula (19.17) but with a different parameter D'. It might be that this parameter D' corresponds now to a continuous rotational diffusion, which, although less important, still contributes a little to neutron data. So we have now a three-parameter model with parameters D, D' and τ_0'. The description by this model of ΔE versus κ^2 dependences presented in fig. 19.10 is quite good except for the highest temperatures. The τ_0' values obtained are in the range 10^{-12} s and roughly agree with relaxation times obtained from ultrasonic absorption and correlation times obtained from NMR. They are, by a factor of ca. 5, smaller than the dielectric relaxation time (Hall [1948] and Hertz [1967]). It is perhaps a little surprising that we should expect any coincidence between our τ_0', which is obviously of translatory origin, and those other relaxation times which are mostly rotational. On the other hand, it is possible that during a translatory jump the molecule reorients, thus giving rise to some rotational QNS broadening and making the above-mentioned coincidence of relaxation time not just fortuitous.

This picture still contains some inconsistencies, however. For instance, if reorientations occur only during translatory jumps, then rotational diffusion cannot be a continuous process, which was an explanation for introducing the D' parameter into our three-parameter model. We should also say that the jump diffusion mechanism is questionable in the light of the molecular dynamics study based upon the electrostatic plus Lennard-Jones interaction, and mentioned in section 19.3.2 (Rahman and Stillinger [1971]). This molecular dynamics study, which successfully describes the features of IINS spectrum for water, does not favour the jump diffusion mechanism. The deviation of ΔE versus κ^2 dependences from straight lines is, according to this study, not an indication for a jump mechanism but only an indication that the so-called Gaussian approximation, i.e., an approximation of the self-correlation functions by a Gaussian, is not valid.

We should also note that the QNS linewidth data of different authors for water show differences up to a factor of 2.

19.4.3. Hydronium perchlorate

We will discuss here an analysis of the quasielastic part of the neutron data, whose inelastic part was described in section 19.3.4 (Janik et al. [1973a, b]). One may see in fig. 19.5 that differences between phases I and II are reflected not only in inelastic parts of the spectra but also in the neighbourhood of the elastic "line", which in our case has the form of the Be cut-off (see the description of experimental conditions given in section 19.3.4). In phase II this part of the spectrum has the form of a sharp edge, whereas in phase I there is a distinct broadening with a clearly seen tail, suggesting a Lorentz component connected with the quasielastic scattering.

It is possible to transform the Be-filtered spectrum into a form proportional to the scattering function by using the following method (Turchin [1963] and Dahlborg et al. [1972]): if neutrons, after being scattered by the sample, are transmitted through the Be filter, the observed number of registered neutrons (N_{obs}) is a convolution of the filter-edge function (N) and a function ϕ which contains both the instrumental and the sample scattering law contributions

$$N_{obs}(\tau) = \int_0^\infty N(t')\,\phi(t'-\tau)\,\mathrm{d}t'; \qquad (19.22)$$

here τ stands for the neutron time of flight. We replace now the function N by a step function

$$N(t') = \begin{cases} N_0 & \text{for } t' \geqslant \tau_{Be} \\ 0 & \text{for } t' < \tau_{Be}, \end{cases} \tag{19.23}$$

where τ_{Be} is the neutron time of flight corresponding to the Be edge. Hence

$$N_{obs}(\tau) = N_0 \int_{\tau_{Be}}^{\infty} \phi(t' - \tau)\,dt' = N_0 \int_{\tau_{Be}-\tau}^{\infty} \phi(t)\,dt$$

and

$$\frac{d\,N_{obs}}{d\tau} = N_0 \phi(\tau_{Be} - \tau). \tag{19.24}$$

In this way we have shown that by differentiation of the vicinity of the elastic region of data obtained by the Be-filter method we obtain directly the scattering function broadened by the instrumental resolution. In order to have this function in its symmetric form we must introduce several corrections; some of these corrections are connected with a simple transformation of variable τ into the variable E, while others are connected with passing from the scattered neutron intensity to the $\tilde{S}_{inc}$ form. After all these corrections we expect that the curve obtained by differentiation will be a convolution of the scattering function and the Gaussian resolution.

Figure 19.11 presents the results: in the low temperature phase the peak represents an elastic scattering process; its form is nearly Gaussian, representing the instrumental resolution. One may see that the differential method leads to a very good resolution ($\Delta\omega/\omega$ = ca. 1.4%). In phase I the quasielastic scattering appears and the peak has a complicated form. According to what was explained in section 19.4.1 the peak should consist of two components, the Lorentz one, broadened by the resolution, and the δ one, also resolution broadened. Solid lines in fig. 19.10 present the best theoretical fits. From the parameters of the Lorentz components it is possible to obtain the time τ_0, i.e., the time between proton jumps, applying formula (19.20). The times τ_0 for the two temperatures of phase I are $\tau_0(-15\,°C) = 4.0 \times 10^{-12}$ s and $\tau_0(+21\,°C) = 3.0 \times 10^{-12}$ s.

At first one is inclined to interpret these times as times between orientational jumps of the H_3O^+ group. We must note, however, that according to what was explained in section 19.3.4 the barriers to such jumps are relatively high (4–5 kcal/mole) and, moreover, not very different in the two phases. Hence, the only conclusion is that an

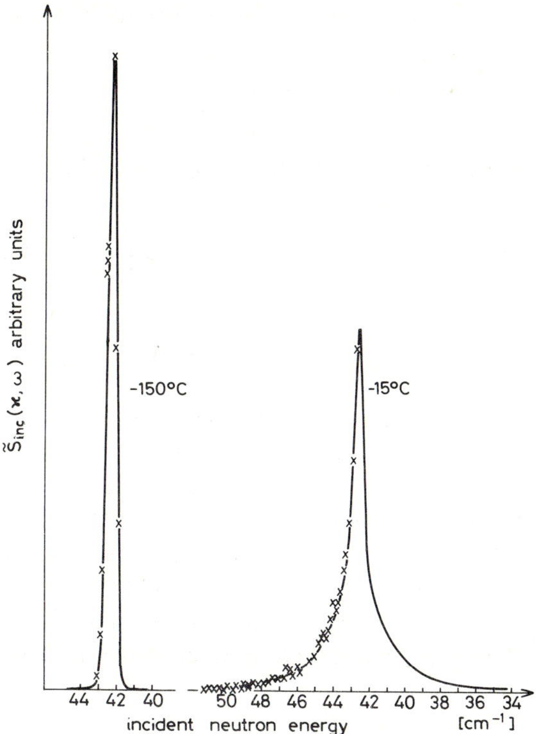

Fig. 19.11. Elastic and quasielastic neutron scattering for $H_3O{\cdot}ClO_4$ at two temperatures. Temperature −150°C corresponds to phase II; the experimental results are fitted to Gaussian curves. Temperature −15°C corresponds to phase I; the experimental results are fitted to a sum–the Gaussian curve plus a Lorentzian convoluted by Gaussian (Janik et al. [1973b]).

additional motion is switched on when passing from phase II to phase I. It is possible that this motion is connected with proton jumps along each H-bond in a double-well potential. If we assume that these jumps occur by an activation process (and not by tunnelling) we may estimate the barrier separating the two minima on the basis of the Eyring formula (Glasstone et al. [1941])

$$\tau_0 = \frac{h}{k_B T} \exp\left(\frac{\Delta H}{RT}\right) \exp\left(-\frac{\Delta S}{R}\right), \tag{19.25}$$

where ΔH stands for activation enthalpy and is equalled to U_{act}, and ΔS stands for activation entropy which, for the sake of simplicity, is equalled to zero in this rough estimation. Using for τ_0 the value 4.0×10^{-12} s (i.e., that for $-15\,°C$) we obtain for $\Delta H = U_{act}$ a value of 1.6 kcal/mole.

An additional support for the proton jumping in a double-well mechanism of quasielastic scattering is obtained from the thermodynamic data. The measured latent heat of phase transition (II → I) leads to an entropy change of 5.25 cal/mole K, which is roughly consistent with the number of H_3O^+ configurations, being a result of the proton jumping model.

19.4.4. $(NH_4)_2SO_4$

We will present here results of Kim et al. [1970]. The experimental conditions were as follows: the incident neutron energy was 4.9 meV (wave length 4.1 Å); the scattered neutron energy was analysed by the time-of-flight technique; the energy resolution of elastically scattered neutrons was ca. 4%.

Typical examples of the scattering patterns are presented in fig. 19.12. As pointed out in section 19.4.1 (and analogically to the $H_3O{\cdot}ClO_4$ case discussed in section 19.4.3) the spectrum in the neighbourhood of the incident energy should consist of an elastic (δ) part and a Lorentzian component, if NH_4 groups are performing rotational jumps. Both components should, of course, be broadened by the instrumental resolution. It is shown in fig. 19.12b how the total spectrum was decomposed into those two parts. By trial and error procedure the best Lorentzian convoluted by Gaussian line was established, such that, when subtracted from the total spectrum, the resulting curve gave the nearest possible approach to the resolution function (as measured separately with a zirconium hydride scatterer).

From the width of the so determined Lorentzian the τ_0, i.e., the time between rotational jumps, was evaluated as being $\tau_0 = 2.0 \times 10^{-11}$ s at room temperature. By repeating measurements at various temperatures (ranging from ca. 100 to ca. 400 K), and by assuming the Arrhenius activation formula

$$\tau_0 = \text{const} \times \exp\,(U_{act}/RT), \tag{19.26}$$

the barrier to rotation value for NH_4 groups was determined as being equal to 2.3 kcal/mole. This value is in rough agreement with that discussed for the same substance in section 19.3.5 in connection with the IINS experiments.

The authors of the paper under discussion make an attempt to obtain some information concerning the geometry of NH_4 motion. For this it is

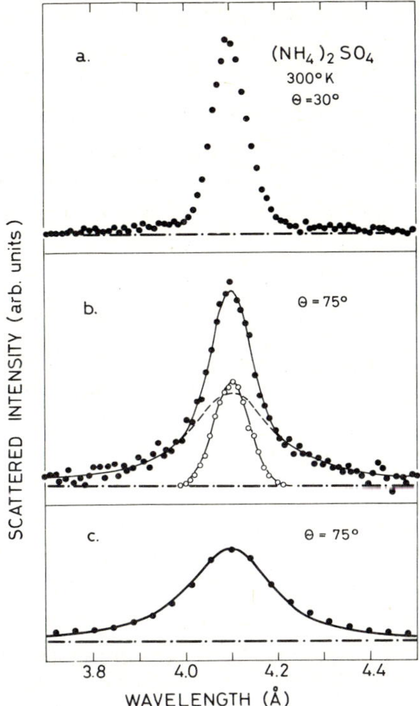

Fig. 19.12. Elastic and quasielastic neutron scattering for $(NH_4)_2SO_4$ at two scattering angles, which corresponds to two κ-values ((a) and (b)–black circles). The open circles correspond to resolution measurement with ZrH_2. (c) presents the separated quasielastic peak, with a theoretical curve computed as described in the text (Kim et al. [1970]).

necessary to apply a specified formula for the scattering function. Generally speaking, such a formula, as given for instance in the paper of Larsson [1973], is quite complicated. However, it is natural to apply here the formula given by Sköld [1968] for rotational jumps of methane molecule around the threefold symmetry axis

$$\tilde{S}_{\text{inc}}(\boldsymbol{\kappa}, \omega) \approx \pi\left(1 + \frac{3 \sin \kappa R}{\kappa R}\right)\delta(\omega) + 3\left(1 - \frac{\sin \kappa R}{\kappa R}\right)\frac{2\pi/\tau_0}{\omega^2 + (2\pi/\tau_0)^2}. \tag{19.27}$$

In this formula R stands for the jump length, being equal to the inter-proton distance. The first term corresponds to the elastic part and the second term to the Lorentz component (see discussion above). It is clear now that measurements of intensities of elastic and/or quasielastic components are able to yield information concerning the geometry of motion. As a matter of fact we may use formula (19.27) with either of two assumptions: (a) the NH_4 group perform reorientations about only one of the threefold axes, or (b) the NH_4 group may reorient about all four of the threefold axes. Figure 19.13 presents the intensities of form factors of the first and second terms in formula (19.27) versus κ, as compared with model (a) and model (b) calculations. It may be concluded that the uniaxial rotation of the NH_4 group in $(NH_4)_2SO_4$ is not supported by this experiment; the NH_4 groups seem rather to reorient about all four axes.

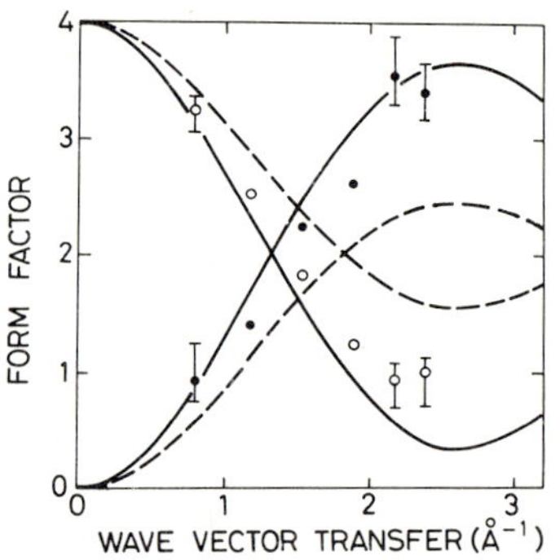

Fig. 19.13. Intensities of the elastic (○) and the quasielastic (●) form factors versus κ for $(NH_4)_2SO_4$. Solid lines are theoretical curves computed for many-axes reorientation; interrupted lines are theoretical curves computed for one-axis reorientation (Kim et al. [1970]).

19.4.5. KH_2PO_4

As was pointed out in section 19.3.6 it is still an open question whether protons do or do not tunnel in the H-bonds. The KH_2PO_4 crystal is known to be ferroelectric below and paraelectric above 122 K. In the ferroelectric phase the protons are arranged antisymmetrically in each H-bond. It is supposed that in the paraelectric phase the protons may either appear in one or the other of the equilibrium positions of the double-well potential, or are situated in potential minima in the centres between oxygens (Bacon and Pease [1953]).

Thermodynamic data, or more specifically the entropy change at the transition being equal to 0.422 R, favour the two sites hypothesis with additional conditions that, firstly, two protons cannot occupy the same H-bond and, secondly, of the four protons around a PO_4 group only two are on sites close to it.

We will describe here the neutron incoherent elastic scattering measurements of Grimm et al. [1970], Plesser and Stiller [1970] and Arsić-Eskinja et al. [1972], whose aim was to elucidate the problem of the time-averaged density distribution of an individual proton in KH_2PO_4. In the paper of Arsić-Eskinja et al. [1972] the coherent neutron scattering measurements with the same substance are discussed; the aim of these measurements was to seek a soft mode in KH_2PO_4. The results of this coherent scattering are described in detail in ch. 24 by Stiller.

The experimental conditions of the incoherent scattering measurements were as follows: the incident neutron energy was 30.2 meV; the scattered neutron energy was analysed by Bragg reflection from a monocrystal; the energy resolution for elastic scattering was 5%; KH_2PO_4 samples were in the form of single crystals mounted so that the $\boldsymbol{\kappa}$-vector was in the a–c plane and could change its angle Θ with the c-direction, as shown in fig. 19.14; measurements were performed in different temperatures.

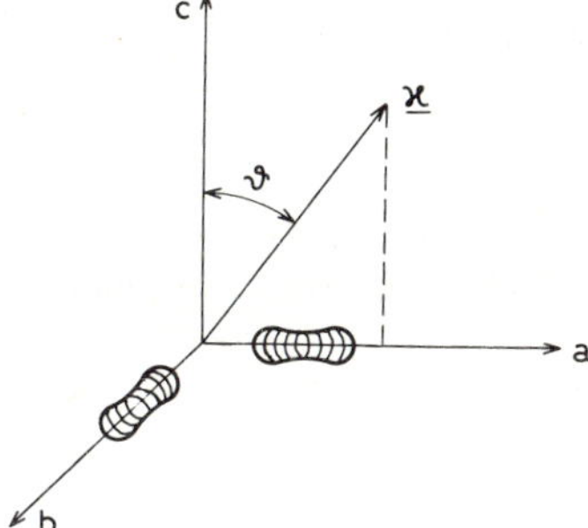

Fig. 19.14. The geometry of elastic incoherent neutron scattering measurements by Arsić-Eskinja et al. [1972] for monocrystals of KH_2PO_4.

The authors are considering three possibilities, as shown in fig. 19.15: (a) the proton is in a parabolic potential, whose minimum is in the centre of the bond and which is elongated in the bond direction; (b) the proton is in a parabolic, spherically symmetric potential with its minimum closer to one oxygen; and (c) the proton is in a double-minimum potential.

In order to calculate the $\tilde{S}_{\text{inc}}$ functions corresponding to the three situations as above, we need the respective proton wave functions, which for case (a) are

$$\psi_{\mathrm{a}} = \left(\frac{1}{2\sqrt{2}\pi^{3/2}u_{\perp}^{2}u_{\parallel}}\right)^{1/2} \exp\left\{-\frac{1}{4}\left[\frac{(x-x_0)^2}{u_{\parallel}^2} + \frac{(y-y_0)^2+(z-z_0)^2}{u_{\perp}^2}\right]\right\}, \tag{19.28}$$

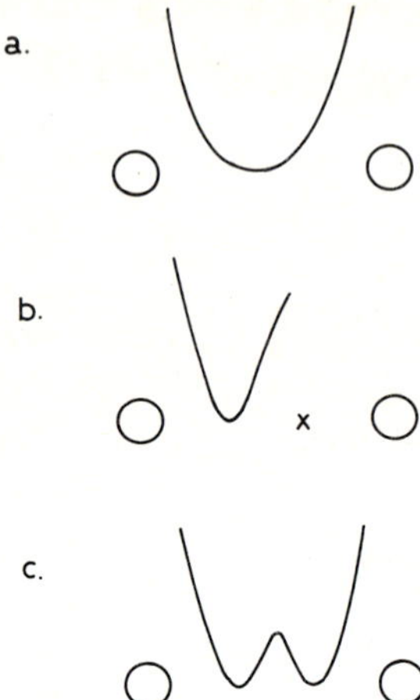

Fig. 19.15. Three possibilities of the shape of hydrogen potential curve in the paraelectric phase of KH_2PO_4. Circles denote oxygen. × indicates that the potential may be on the right side (Arsić-Eskinja et al. [1972]).

for case (b)

$$\psi_b^{\pm} = \left(\frac{1}{(2\pi)^{1/2}u}\right)^{3/2} \exp\left[-\frac{(x - x_0 \pm \frac{1}{2}l)^2 + (y - y_0)^2 + (z - z_0)^2}{4u^2}\right] \tag{19.29}$$

and for case (c)

$$\psi_{c1} = a_1(\varphi_l - \varphi_r), \qquad \psi_{c2} = a_2(\varphi_l + \varphi_r), \qquad \varphi_{l(r)} = \psi_b^{+(-)}. \tag{19.30}$$

In these formulas $u_\parallel^2$ and $u_\perp^2$ are the mean square amplitudes of protons parallel and perpendicular to the H-bond direction respectively; $u^2 = u_\parallel^2 + u_\perp^2$; and the connection between a_1 and a_2 is $2a_1^2(1 - L) = 2a_2^2(1 + L)$, L being equal to (φ_l, φ_r).

Now we are able to calculate $\tilde{S}_{inc}$ functions, but as our interest will be limited to the elastic component only, we will deal with S_{inc} rather than $\tilde{S}_{inc}$. The corresponding S_{inc} functions for case (a) are

$$\begin{aligned} S_{a,inc}(\boldsymbol{\kappa}, \omega) = \{&\exp[-(\kappa_x^2 + \kappa_z^2)u_\perp^2 - \kappa_y^2 u_\parallel^2] \\ &+ \exp[-\kappa_x^2 u_\parallel^2 - (\kappa_y^2 + \kappa_z^2)u_\perp^2]\}\delta(\omega), \end{aligned} \tag{19.31}$$

for case (b)

$$S_{b,inc}(\boldsymbol{\kappa}, \omega) = \exp[-\kappa^2 u^2]\delta(\omega) \tag{19.32}$$

and for case (c)

$$S_{c,inc}(\boldsymbol{\kappa}, \omega) = \exp[-\kappa^2 u^2]\{[C_1 + C_2 \cos(\boldsymbol{\kappa}\cdot\boldsymbol{l}) + C_3 \cos(\tfrac{1}{2}\boldsymbol{\kappa}\cdot\boldsymbol{l})]\delta(\omega) + [C_1' + C_2' \cos(\boldsymbol{\kappa}\cdot\boldsymbol{l})][p_1\delta(\omega+\omega_{12}) + p_2\delta(\omega-\omega_{12})]\}. \quad (19.33)$$

Here

$$C_1 = C_2(3 + 2L^2) + C_3$$
$$C_1' = 2(a_1^4 + a_2^4 - a_1^2 a_2^2)$$
$$C_2 = 2(p_1 a_1^4 + p_2 a_2^4)$$
$$C_2' = -2a_1^2 a_2^2$$
$$C_3 = 8L(-p_1 a_1^4 + p_2 a_2^2).$$

p_1, p_2 are the thermal occupations of ψ_{c1} and ψ_{c2} respectively, and l is the distance between the two proton sites. The second term in (19.33) represents the quasielastic component connected with random proton jumps in the double-minimum potential. In the experiments under discussion only elastic scattering is measured, so we shall drop the quasielastic component in what follows.

We shall now rewrite formulas (19.31), (19.32) and (19.33) in the form corresponding to the experimental geometry of fig. 19.13,

$$S_{a,inc}^{elastic}(\kappa, \Theta) = \exp(-\kappa^2 u_\perp^2)[1 + \exp(-\kappa^2 \Delta u^2 \sin^2\Theta)], \quad (19.34)$$

$$S_{b,inc}^{elastic}(\kappa, \Theta) = \exp(-\kappa^2 u^2) \quad (19.35)$$

and

$$S_{c,inc}^{elastic}(\kappa, \Theta) = \exp(-\kappa^2 u^2)[C_1'' + C_2 \cos(\kappa l \sin\Theta) + C_3 \cos(\tfrac{1}{2}\kappa l \sin\Theta)]. \quad (19.36)$$

Here C_1'' stands for $2C_1 + C_2 + C_3$.

Figure 19.16 presents the results of measurements of $S_{inc}^{elastic}$ variation with Θ. It is obvious that case (b) is not in agreement with the experimental data. It is, however, difficult to conclude definitely whether case (a) or (c) occurs. Certainly case (c), with parameters as taken from the neutron diffraction work of Bacon and Pease [1953], does not agree with this experiment. However, if we choose parameters giving the best fit to formula (19.36) then we obtain agreement with the experiment, being as good as for case (a). It should be stressed that if we take all κ-values into account and compare the (a)-case and (c)-case fits to the experiment, we obtain least squares deviation 5.1×10^{-3} for the (a)-fit and 2.0×10^{-3} for the (c)-fit, thus favouring the double-minimum hypothesis. The best fit

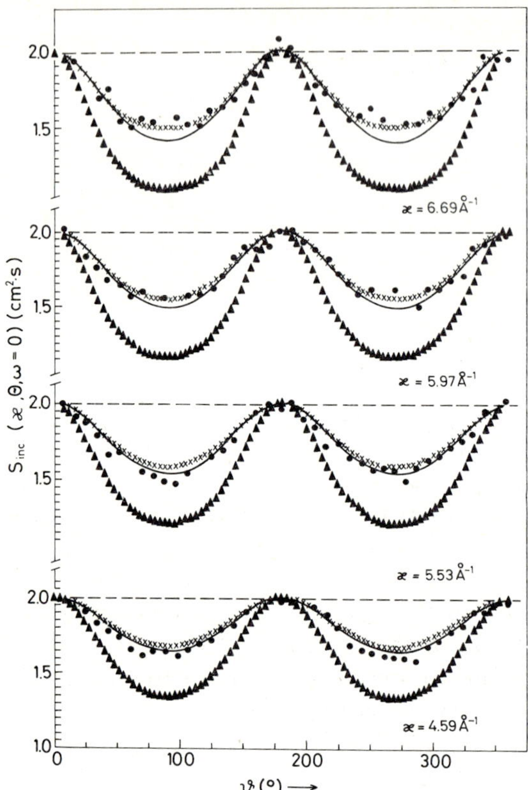

Fig. 19.16. $S_{\text{inc}}^{\text{elastic}}$ versus Θ dependence for paraelectric phase of KH_2PO_4. ●–experimental points. ▲–model (c) calculations with parameters as taken from Bacon and Pease [1953]. +–model (c) calculations with parameters from the best fit to the experimental points. Solid line–the best fit to the model (a). Interrupted line–the model (b) calculation (Arsić-Eskinja [1972]).

parameters for case (a) are $u_{\perp}^2 = 0.028$ Å^2, $u_{\parallel}^2 = 0.048$ Å^2 and for case (c) $u^2 = 0.028$ Å^2, $l = 0.28$ Å and $L = 0.71$.

Figure 19.17 shows the $\Delta u^2 = u_{\perp}^2 - u_{\parallel}^2$, as measured with parameter fit corresponding to case (a), versus temperature dependence. (Application of the (a)-case fit to this purpose does not mean a real acceptance of its reality; however, the Δu^2 parameter is certainly a good one for description of an extension of the proton cloud.) We may see that when the temperature decreases, the proton cloud which is in the paraelectric phase elongated in the direction of H-bond ($\Delta u^2 < 0$), passes the $\Delta u^2 = 0$ in the vicinity of the transition point (122 K), and in the ferroelectric phase

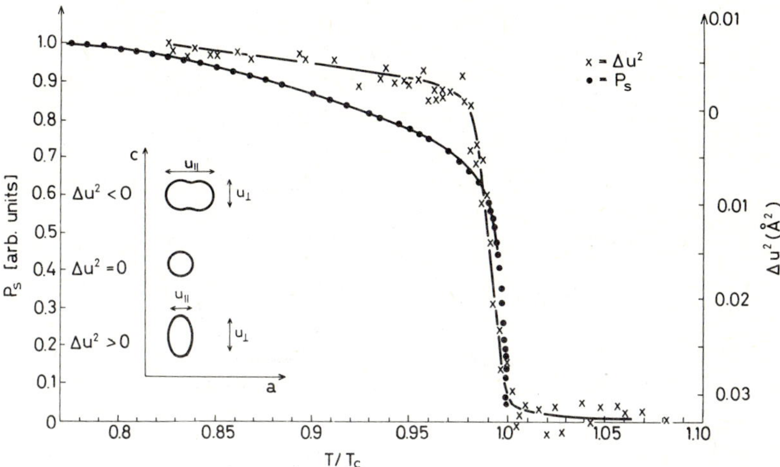

Fig. 19.17. Elongation of proton cloud $\Delta u^2 = u_\perp^2 - u_\parallel^2$ in KH_2PO_4 at different temperatures as obtained from neutron incoherent elastic scattering and compared to spontaneous polarization data (Arsić-Eskinja [1972]).

reaches the elongation in the direction perpendicular to the H-bond (in the asymmetric position of the minimum of course).

In the conclusion, we may say that the experiments described favour the double-minimum hypothesis, give the geometrical parameters of the potential, and show the behaviour of the shape of the proton cloud with the temperature. A connection between these results and the collective behaviour of atoms in KH_2PO_4 is discussed in ch. 24 by Stiller.

It is natural to expect that an improvement of neutron methods, and especially an improvement of their resolution both in energy and $\boldsymbol{\kappa}$, will lead in future to similar or analogical studies of H-bond as those discussed in this section.

19.5. Total neutron scattering experiments with hydrogen bonded systems

Rush et al. [1962] pointed out that the total neutron scattering cross-section σ (as measured by transmission measurements by using the formula

$$I = I_0 \, e^{-\sigma n x}, \tag{19.37}$$

I_0 and I being the intensities of incident and transmitted beam respectively, n is the number of scattering molecules per unit volume, and x is the sample thickness) is a linear function of the neutron wave length λ, for λ sufficiently large. When calculated per one proton in the molecule (after a proper subtraction of contributions from other nuclei), the slope of σ versus λ dependence is a good measure of the barrier-to-rotation characteristic for the proton motion. This may be explained in the following way: if the barrier is high, the torsional levels are separated by larger distances than for the low barrier; hence, the corresponding density of states is smaller, thus leading to smaller values of the scattering cross-section. It is impossible, however, to give an analytic formula connecting the barrier with the slope. Therefore, a calibration is necessary, which of course must be based on reliable barrier data, not easily obtainable as yet. Rush et al. [1962] presented such a calibration for ammonium salts, and obtained barrier data for some of those salts – for instance, for $(NH_4)_2SO_4$, which was discussed in sections 19.3.5 and 19.4.4 in connection with IINS and QNS experiments.

In this section we will briefly describe experiments of Fischer [1970] as representative of total neutron scattering application to H-bonded systems. The experiments were performed with liquid methyl and ethyl alcohols. Neutrons of wave length from 5 to 15 Å were used; they were selected from the "white" beam by the time-of-flight method. In order to distinguish between methyl, methylene and hydroxyl groups, partly deuterated samples were studied, as for instance, CD_3OH, CH_3OD, CH_3CH_2OD, CD_3CH_2OH and CH_3CD_2OH. In doing so, the author made use of the fact that the neutron–proton scattering is stronger than the neutron–deuteron scattering by more than one order of magnitude.

Typical experimental results are shown in fig. 19.18. When calculated per one proton (after a proper subtraction of contributions from other nuclei), one obtained the following slopes S (in barns/Å):

(a) for methyl alcohol

$S_1 = 3.4 \pm 0.8$ barns/Å for an OH proton

$S_2 = 11.6 \pm 0.5$ barns/Å for a CH_3 proton,

(b) for ethyl alcohol

$S_1 = 1.3 \pm 0.8$ barns/Å for an OH proton

$S_2 = 9.1 \pm 0.5$ barns/Å for a CH_2 proton

$S_3 = 9.8 \pm 0.4$ barns/Å for a CH_3 proton.

One may conclude from these results that in methyl alcohol there is a wide freedom of rotation of CH_3 groups (low barrier) and a great

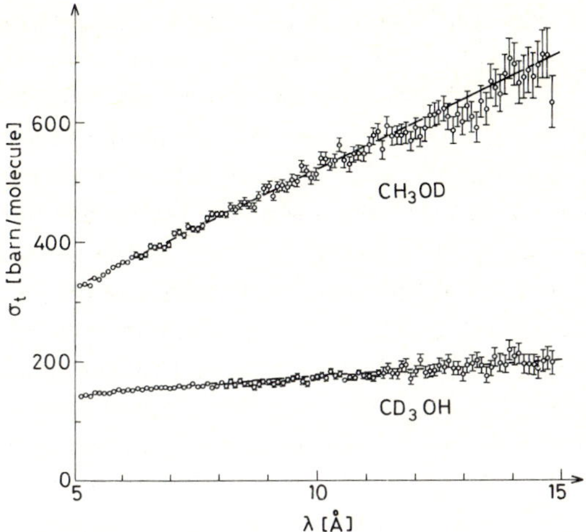

Fig. 19.18. Total neutron scattering cross-section versus neutron wave length for CH_3OD and CD_3OH (Fischer [1970]).

hindrance of rotation of OH groups (high barrier), and in ethyl alcohol CH_3 and CH_2 groups rotate with much freedom, whereas OH groups are again hindered. These facts seem to agree well with a picture of H-bond chains in which protons of OH groups participate only. It may also be concluded that OH groups in ethyl alcohol are more hindered than OH groups in methyl alcohol, this fact being supported by NMR results of Grüner [1969]. It should also be pointed out that the conclusion of a relative freedom of CH_3 groups is in agreement with IINS measurements performed with CD_3OH and CH_3OD by Aldred et al. [1967].

In concluding, one may say that the total neutron scattering slope method may in future be a more widely used tool, auxiliary to NMR, IINS and QNS measurements, especially when a proper deuteration has been applied.

19.6. Final remarks

The author feels that he owes the reader some final explanations. First of all, the selection of examples chosen for a more detailed discussion in sections 19.3 and 19.4 could not be very seriously defended. The author felt that more recent papers should be favoured and also those in which some new methodical ideas, not yet exploited in the past, but with prospects of being exploited in the future, be applied. Also, examples of theoretical generalizations, based either upon phenomenological force constants ideas or models assuming certain types of interaction, could have been chosen in a different way. These selections were especially difficult in the case of experiments with water; here, however, it is the author's opinion that a wider presentation of experimental facts and theories should rather be given in a monograph concerning water.

The author feels that incoherent inelastic scattering experiments as applied to vibration frequencies determinations should in future be performed with monocrystals, where properly chosen directions of $\boldsymbol{\kappa}$ in respect to vibration amplitude $\boldsymbol{u}$ may be used, and/or with partially deuterated samples with the aim of observing motions of particular groups, such as CH_3, OH, etc. This latter future aspect is especially important for liquids. There is no doubt that most valuable information concerning phonons, including those which are involved in H-bond dynamics, may be obtained with coherent inelastic scattering of neutrons.

As this information concerns the incoherent elastic and quasielastic neutron scattering, the author's opinion is that it will be more and more applied to H-bond problems in the future, because it gives not only time parameters of random jumping motions but also their geometrical characteristics. It is especially important here to exploit apparatus with very good resolutions, and in the case of solids, if possible, monocrystals. Such a special, but also crucial problem as the shape of the potential curve (double minimum or single minimum) for the proton, which takes part in a H-bond, will probably be studied very widely in the future with neutrons. So far, the author had at his disposal for presentation in this chapter only one really good quality paper of this kind; this was the paper of Arsić-Eskinja et al. [1972] discussed in detail in section 19.4.5.

It should be made clear to the reader that the quality of neutron data is constantly improving, especially owing to improvements of energy and momentum resolutions. Older, bad resolution data, however, especially those concerning the shape of the quasielastic peak, were often interpreted incorrectly.

Finally, it is the author's feeling that the experimental facts discussed in this chapter, although connected with substances with H-bonds, are not, except for some few cases, interpreted as selectively elucidating the H-bond concept and problematics. When read together with other chapters of this book, however, and especially those dealing with general theoretical and ideological aspects of the H-bond, these facts should yield experimental material for a better understanding of the nature of the H-bond in the future.

Appendix

List of substances with hydrogen bonds investigated so far by neutron incoherent scattering methods

Compounds	Authors	References
Ice and water	Hughes et al.	1960, Phys. Rev. **119**, 872.
	Larsson et al.	1960, IAEA Symp. Vienna, 329.
	Egelstaff et al.	1960, IAEA Symp. Vienna, 309.
	Woods et al.	1960, IAEA Symp. Vienna, 487.
	Stiller and Danner	1960, IAEA Symp. Vienna, 363.
	Cribier and Jacrot	1960, IAEA Symp. Vienna, 347.
	Mikke	1960, IAEA Symp. Vienna, 351.
	Singwi et al.	1962, IAEA Symp. Chalk River **1**, 215.
	Goldman and Federighi	1962, IAEA Symp. Chalk River **1**, 389.
	Kottwitz and Leonard	1962, IAEA Symp. Chalk River **1**, 359.
	Kottwitz et al.	1962, IAEA Symp. Chalk River **1**, 373.
	Sakamoto et al.	1962, J. Phys. Soc. Japan, Suppl. BII **17**, 371.
	Larsson and Dahlborg	1962a, IAEA Symp. Chalk River **1**, 317.
	Larsson and Dahlborg	1962b, J. Nucl. Energy **16AB**, 81.
	Egelstaff et al.	1962, IAEA Symp. Chalk River **1**, 343.
	Yip and Osborn	1963, Phys. Rev. **131**, 2547.
	Pope and Nations	1964, IAEA Symp. Bombay **2**, 141.
	Sköld et al.	1964, Atomic Energy Rept. No. AE133 (Stockholm).
	Larsson and Dahlborg	1964, Physica **30**, 1561.
	Golikov et al.	1964, IAEA Symp. Bombay **2**, 201.
	Pelah and Imry	1966, Phys. Letters **21**, 248.
	Rush et al.	1966, J. Chem. Phys. **45**, 1312.
	Szkatuła and Fuliński	1967, Physica **36**, 35.
	Nakahara	1968, J. Nucl. Sci. Technol. **5**, 635.
	Brugger	1968, Nucl. Sci. Eng. **33**, 187.
	Bajorek et al.	1968, IAEA Symp. Copenhagen **2**, 143.
	Prask et al.	1968, J. Chem. Phys. **48**, 3367.
	Harling	1968, IAEA Symp. Copenhagen **1**, 507.
	Burgman et al.	1968, Phys. Rev. **170**, 808.
	Safford et al.	1969, J. Chem. Phys. **50**, 4444.
	Harling	1969, J. Chem. Phys. **50**, 5279.
	Franks et al.	1970, Proc. Roy. Soc. **A319**, 189.
	Day and Sinclair	1971, J. Chem. Phys. **55**, 2807.
	Prask et al.	1972, J. Chem. Phys. **56**, 3217.
	Blanckenhagen	1972, Ber. Bunsenges. Phys. Chem. **76**, 891.

Appendix–*contd.*

Compounds	Authors	References
Water of crystallization		
$CuSO_4{\cdot}5H_2O$	Bajorek et al.	1964, IAEA Symp. Bombay **2**, 355.
$BeSO_4{\cdot}4H_2O$	Thaper et al.	1969, Phys. Stat. Sol. **34**, 279.
$BaCl_2{\cdot}2H_2O$	Bajorek et al.	1964, IAEA Symp. Bombay **2**, 355.
$SrCl_2{\cdot}6H_2O$	Bajorek et al.	1964, IAEA Symp. Bombay **2**, 355.
$CuCl_2{\cdot}2H_2O$	Thaper et al.	1969, Phys. Stat. Sol. **34**, 279.
$Li_2SO_4{\cdot}H_2O$	Prask and Boutin	1966, J. Chem. Phys. **45**, 699, 3284.
	Bajorek et al.	1968, IAEA Symp. Copenhagen, **2**, 143.
	Thaper et al.	1969, Phys. Stat. Sol. **34**, 279.
	Mikuli et al.	1972, Rept. Inst. Nucl. Phys. (Kraków).
$LiClO_4{\cdot}3H_2O$	Bajorek et al.	1968, IAEA Symp. Copenhagen **2**, 143.
$CaSO_4{\cdot}2H_2O$	Thaper et al.	1969, Phys. Stat. Sol. **34**, 279.
$NiSO_4{\cdot}6H_2O$	Bajorek et al.	1968, IAEA Symp. Copenhagen **2**, 143.
$NiSO_4{\cdot}7H_2O$	Bajorek et al.	1968, IAEA Symp. Copenhagen **2**, 143.
$KF{\cdot}2H_2O$	Prask and Boutin	1966, J. Chem. Phys. **45**, 699, 3284.
$LiCl{\cdot}H_2O$	Bajorek et al.	1968, IAEA Symp. Copenhagen **2**, 143.
$La_2Mg_3(NO_3)_{12}$ $\cdot 24H_2O$	Bajorek et al.	1968, IAEA Symp. Copenhagen **2**, 143.
$Ba(ClO_3)_2{\cdot}H_2O$	Prask and Boutin	1966, J. Chem. Phys. **45**, 699, 3284.
	Thaper et al.	1969, Phys. Stat. Sol. **34**, 279.
	Thaper et al.	1970, Solid State Commun. **8**, 497.
$K_2C_2O_4{\cdot}H_2O$	Prask and Boutin	1966, J. Chem. Phys. **45**, 699, 3284.
	Thaper et al.	1969, Phys. Stat. Sol. **34**, 279.
	Thaper et al.	1970, Solid State Commun. **8**, 497.
$K_4Fe(CN)_6{\cdot}3H_2O$	Rush et al.	1966, J. Chem. Phys. **45**, 1312.
$Na_2Al_2Si_3O_{10}{\cdot}2H_2O$	Boutin et al.	1963, J. Chem. Phys. **39**, 488.
Hydronium compounds		
$H_3O{\cdot}ClO_4$	Janik et al.	1964, J. Phys. Chem. Sol. **25**, 1091.
	Janik	1965, Acta Physica Polon. **27**, 491.
	Janik et al.	1965, Phys. Stat. Sol. **9**, 905.
	Janik et al.	1973b, Acta Physica Polon. **A43**, 419.
	Janik et al.	1973a, Physica, in print.
$H_3O{\cdot}NO_3$	Janik et al.	1967, Physica **35**, 457.
	Janik et al.	1968, Acta Physica Polon. **33**, 419.
Ammonium compounds		
NH_4Cl	Woods et al.	1960, IAEA Symp. Vienna, 487.
	Rush et al.	1960, Phys. Rev. Letters **5**, 507.
	Palevsky	1962, J. Phys. Soc. Japan Suppl. BII **17**, 367.
	Mikke and Kroh	1962, IAEA Symp. Chalk River **2**, 237.
	Venkataraman et al.	1962, IAEA Symp. Chalk River **2**, 253.

Appendix–*contd.*

Compounds	Authors	References
	Bajorek et al.	1964, IAEA Symp. Bombay **2**, 355.
	Venkataraman et al.	1964, Solid State Commun. **2**, 17.
NH_4Br	Rush et al.	1960, Phys. Rev. Letters **5**, 507.
	Venkataraman et al.	1962, IAEA Symp. Chalk River, **2**, 253.
	Bajorek et al.	1964, IAEA Symp. Bombay **2**, 355.
	Venkataraman et al.	1966, J. Phys. Chem. Sol. **27**, 1103.
NH_4I	Rush et al.	1960, Phys. Rev. Letters **5**, 507.
	Mikke and Kroh	1962, IAEA Symp. Chalk River **2**, 237.
	Bajorek et al.	1964, IAEA Symp. Bombay **2**, 355.
	Rush et al.	1962, J. Chem. Phys. **37**, 234.
	Venkataraman et al.	1966, J. Phys. Chem. Sol. **27**, 1103.
	Kosaly and Solt	1966, Physica **32**, 1571.
NH_4F	Rush et al.	1960, Phys. Rev. Letters **5**, 507.
	Mikke and Kroh	1962, IAEA Symp. Chalk River **2**, 237.
	Bajorek et al.	1964, IAEA Symp. Bombay **2**, 355.
	Rush et al.	1962, J. Chem. Phys. **37**, 234.
$(NH_4)BeF_4$	Rush and Taylor	1964, IAEA Symp. Bombay **2**, 333.
$NH_4 \cdot NO_3$	Mikke and Kroh	1962, IAEA Symp. Chalk River **2**, 237.
	Bajorek et al.	1964, IAEA Symp. Bombay **2**, 355.
NH_4CNS	Bajorek et al.	1964, IAEA Symp. Bombay **2**, 355.
	Leung et al.	1966, unpublished data.
NH_4ClO_4	Janik et al.	1964, J. Phys. Chem. Sol. **25**, 1091.
	Janik	1965, Acta Physica Polon. **27**, 491.
	Janik et al.	1971, Phys. Stat. Sol. **44**, 437.
	Janik et al.	1965, Phys. Stat. Sol. **9**, 905.
NH_4HSO_4	Rush and Taylor	1964, IAEA Symp. Bombay **2**, 333.
	Leung et al.	1966, unpublished data.
NH_4N_3	Boutin et al.	1966, J. Chem. Phys. **45**, 401.
NH_4PF_6	Brajović et al.	1963, J. Phys. Chem. Sol. **24**, 617.
	Janik et al.	1964, J. Phys. Chem. Sol. **25**, 1091.
$(NH_4)_2SO_4$	Bajorek et al.	1964, IAEA Symp. Bombay **2**, 355.
	Rush and Taylor	1964, IAEA Symp. Bombay **2**, 333.
	Leung et al.	1966, unpublished data.
	Kim et al.	1970, Solid State Commun. **8**, 889.
$(NH_4)_2S_2O_8$	Brajović et al.	1963, J. Phys. Chem. Sol. **24**, 617.
	Bajorek et al.	1964, IAEA Symp. Bombay **2**, 355.
$(NH_4)_2Cr_2O_7$	Bajorek et al.	1964, IAEA Symp. Bombay **2**, 355.
	Leung et al.	1966, unpublished data.
$(NH_4)_2CO_3$	Bajorek et al.	1964, IAEA Symp. bombay **2**, 355.
$(NH_4)_2SiF_6$	Schlemper et al.	1966, J. Chem. Phys. **44**, 2499.
$(NH_4)_2SnCl_6$	Venkataraman et al.	1966, J. Phys. Chem. Sol. **27**, 1103.
$(NH_4)_2SnBr_6$	Venkataraman et al.	1966, J. Phys. Chem. Sol. **27**, 1103.

Appendix-*contd.*

Compounds	Authors	References
NH_4SO_3F	Brajović et al.	1963, J. Phys. Chem. Sol. **24**, 617.
NH_4VO_3	Bajorek et al.	1967, Physica **35**, 465.
$(NH_4)CrO_4$	Leung et al.	1966, unpublished data.
Various inorganic compounds		
NH_3	Dasannacharya et al.	1972, IAEA Symp. Grenoble, 477.
PH_4I	Rush	1966, J. Chem. Phys. **44**, 1722.
KH_2PO_4	Palevsky et al.	1962, IAEA Symp. Chalk River **2**, 273.
	Pelah et al.	1964, IAEA Symp. Bombay **2**, 325.
	Steinman and Summerfield	1968, J. Phys. Chem. Sol. **30**, 449.
	Schenk et al.	1968, Phys. Rev. **172**, 576.
	Grimm et al.	1970, Phys. Stat. Sol. **42**, 207.
	Plesser and Stiller	1970, Proceedings of the 2nd Summer School in Hercegnovi (Beograd).
	Arsić-Eskinja et al.	1972, IAEA Symp. Grenoble, 825.
	Antonini et al.	1972, J. de Phys. **33**, C2–83.
KHF_2	Collins et al.	1970, J. Chem. Phys. **52**, 1828.
	Rush et al.	1972, J. Chem. Phys. **56**, 2793.
$NaHF_2$	Rush et al.	1972, J. Chem. Phys. **56**, 2793.
	Ghosh et al.	1972, IAEA Symp. Grenoble, 345.
$CsHCl_2$	Collins et al.	1970, J. Chem. Phys. **52**, 1828.
	Ghosh et al.	1972, IAEA Symp. Grenoble, 345.
$HCrO_2$	Ghosh et al.	1972, IAEA Symp. Grenoble, 345.
$HCoO_2$	Ghosh et al.	1972, IAEA Symp. Grenoble, 345.
$KH(CO_2CCl_3)_2$	Ghosh et al.	1972, IAEA Symp. Grenoble, 345.
$KH(CO_2CF_3)_2$	Collins et al.	1970, J. Chem. Phys. **52**, 1828.
HCl	Boutin and Safford	1964, IAEA Symp. Bombay **2**, 393.
	Trevino et al.	1968, IAEA Symp. Copenhagen **1**, 345.
	Agrawal et al.	1968, Chem. Phys. Letters **2**, 584.
HBr	Boutin and Safford	1964, IAEA Symp. Bombay **2**, 393.
HF	Boutin and Safford	1964, IAEA Symp. Bombay **2**, 393.
Organic acids		
HCOOH	Tubino and Zerbi	1970, J. Chem. Phys. **53**, 1428.
$C_{17}H_{33}COOH$	Larsson and Dahlborg	1964, Physica **30**, 1561.
CH_3COOH	Aldred et al.	1967, Discussions Faraday Soc. **43**, 169.
CF_3COOH	Aldred et al.	1967, Discussions Faraday Soc. **43**, 169.
Alcohols		
CH_3OH	Saunderson and Rainey	1962, IAEA Symp. Chalk River **1**, 413.
	Sampson and Carpenter	1968, IAEA Symp. Copenhagen **1**, 491.
	Aldred et al.	1967, Discussions Faraday Soc. **43**, 169.
	Fischer	1970, Ber. d. Bunsen Ges. f. Phys. Chem. **74**, 696.

Appendix-*contd.*

Compounds	Authors	References
C_2H_5OH	Saunderson and Rainey	1962, IAEA Symp. Chalk River **1**, 413.
	Fischer	1970, Ber. d. Bunsen Ges. f. Phys. Chem. **74**, 696.
$C_3H_5(OH)_3$	Larsson and Dahlborg	1962a, IAEA Symp. Chalk River **1**, 317.
	Larsson	1964, IAEA Symp. Bombay **2**, 3.
	Larsson and Dahlborg	1964, Physica **30**, 1561.
	Alefeld et al.	1969, Naturwiss. **56**, 410.

References

Notes on references

"IAEA Symp. Vienna (1960)" stands for: "Proceedings of the Symposium on Inelastic Scattering of Neutrons in Solids and Liquids, held in Vienna, 1960; International Atomic Energy Agency, Vienna, 1961".

"IAEA Symp. Chalk River (1962)" stands for: "Proceedings of the Symposium on Inelastic scattering of Neutrons in Solids and Liquids, held at Chalk River, 1962; International Atomic Energy Agency, Vienna, 1963".

"IAEA Symp. Bombay (1964)" stands for: "Proceedings of the Symposium on Inelastic Scattering of Neutrons, held in Bombay, 1964; International Atomic Energy Agency, Vienna, 1965".

"IAEA Symp. Copenhagen (1968)" stands for: "Proceedings of a Symposium on Neutron Inelastic Scattering, held in Copenhagen, 1968; International Atomic Energy Agency, Vienna, 1968".

"IAEA Symp. Grenoble (1972)" stands for: "Proceedings of a Symposium on Neutron Inelastic Scattering, held in Grenoble, 1972; International Atomic Energy Agency, Vienna, 1972".

Agrawal, A. K., S. Yip and R. G. Gordon, 1968, Chem. Phys. Letters **2**, 594.
Aldred, B. K., R. C. Eden and J. W. White, 1967, Discussions Faraday Soc. **43**, 169.
Alefeld, B., M. Birr and H. Heidemann, 1969, Naturwiss. **56**, 410.
Anderson, A., H. Gebbie and S. Walmsby, 1964, Mol. Phys. **7**, 401.
Antonini, M., I. Sosnowska and M. Vadacchino, 1972, J. de Phys. **33**, C2–83.
Arnold, G. M. and R. Heastie, 1967, Chem. Phys. Letters **1**, 51.
Arsić-Eskinja, M., H. Grimm and H. Stiller, 1972, IAEA Symp. Grenoble, 825.
Bacon, G. and R. Pease, 1953, Proc. Roy. Soc. **A220**, 397; **A230**, 359.
Bajorek, A., T. A. Machekhina and K. Parliński, 1964, IAEA Symp. Bombay **2**, 355.
Bajorek, A., K. Parliński and M. Sudnik-Hrynkiewicz, 1967, Physica **35**, 465.
Bajorek, A., J. A. Janik, J. M. Janik, I. Natkaniec, K. Parliński, Y. N. Pokotilovsky, M. Sudnik-Hrynkiewicz, V. E. Komarov, R. P. Ozerov and S. P. Solovev, 1968, IAEA Symp. Copenhagen **2**, 143.
Blanckenhagen, P., 1972, Ber. d. Bunsen Ges. f. ph. Ch. **76**, 891.

Boutin, H., G. J. Safford, 1964, IAEA Symp. Bombay **2**, 393.
Boutin, H., G. J. Safford and H. R. Danner, 1963, J. Chem. Phys. **39**, 488.
Boutin, H., S. Trevino and H. Prask, 1966, J. Chem. Phys. **45**, 401.
Brajović, V., H. Boutin, G. J. Safford and H. Palevsky, 1963, J. Phys. Chem. Sol. **24**, 617.
Brugger, R. M., 1968, Nucl. Sci. Eng. **33**, 187.
Burgman, J. O., J. Ściesiński and K. Sköld, 1968, Phys. Rev. **170**, 808.
Collins, M. F., B. C. Haywood and G. C. Stirling, 1970, J. Chem. Phys. **52**, 1828.
Cowley, E. R., 1971, Phys. Rev. **B3**, 2743.
Cribier, D. and B. Jacrot, 1960, IAEA Symp. Vienna, 347.
Dahlborg, U., C. Gräslund and K. E. Larsson, 1972, **59**, 672.
Dasannacharya, B. A., C. L. Thaper and P. S. Goyal, 1972, IAEA Symp. Grenoble, 477.
Day, D. H. and R. N. Sinclair, 1971, J. Chem. Phys. 55, 2807.
Egelstaff, P. A., Ed., 1965, Thermal Neutron Scattering (Academic Press, London, New York).
Egelstaff, P. A., S. J. Cocking, R. Royston and I. Thorson, 1960, IAEA Symp. Vienna, 309.
Egelstaff, P. A., B. C. Haywood and I. M. Thorson, 1962, IAEA Symp. Chalk River **1**, 343.
Eucken, A., 1948, Z. Elektrochem. **52**, 255.
Fischer, C. O., 1970, Ber. d. Bunsen Ges. f. Phys. Chem. **74**, 696.
Forslind, E., 1954, Proc. Swedish Cement and Concrete Res. Inst., 21.
Franks, F., J. Ravenhill, P. A. Egelstaff and P. I. Page, 1970, Proc. Roy. Soc. **A319**, 189.
Ghosh, R. E., T. C. Waddington and F. P. Temme, 1972, IAEA Symp. Grenoble, 345.
Glasstone, S., K. J. Laidler and H. Eyring, 1941, The Theory of Rate Processes (McGraw-Hill, New York).
Goldman, D. T. and F. D. Federighi, 1962, IAEA Symp. Chalk River **1**, 389.
Golikov, V. V., I. Żukowska, F. L. Shapiro and A. Szkatuła, J. A. Janik, 1964, IAEA Symp. Bombay **2**, 201.
Grimm, H., H. Stiller and T. Plesser, 1970, Phys. Stat. Sol. **42**, 207.
Grüner, M., 1969, Thesis (Karlsruhe).
Hall, L., 1948, Phys. Rev. **73**, 775.
Hamilton, W. C. and J. A. Ibers, 1968, Hydrogen Bonding in Solids (W. A. Benjamin, Inc., New York, Amsterdam).
Harling, O. K., 1968, IAEA Symp. Copenhagen **1**, 507.
Harling, O. K., 1969, J. Chem. Phys. **50**, 5279.
Hertz, H. G., 1967, Prog. Nucl. Magn. Res. Spectr. **3**, Eds. J. W. Emsley, J. Feeney and L. H. Sutchliffe (Pergamon Press, Oxford), p. 159.
Hughes, D. J., H. Palevsky, W. Kley and E. Tunkelo, 1960, Phys. Rev. **119**, 872.
Janik, J. M., 1965, Acta Physica Polon. **27**, 491.
Janik, J. A., J. M. Janik, J. Mellor and H. Palevsky, 1964, J. Phys. Chem. Sol. **25**, 1091.
Janik, J. M., J. A. Janik, A. Bajorek and K. Parliński, 1965, Phys. Stat. Sol. **9**, 905.
Janik, J. M., J. A. Janik, A. Bajorek, K. Parliński and M. Sudnik-Hrynkiewicz, 1967, Physica **35**, 457.
Janik, J. A., A. Bajorek, I. Natkaniec, K. Parliński and M. Sudnik-Hrynkiewicz, 1968, Acta Physica Polon. **33**, 419.
Janik, J. A., J. M. Janik and J. Mayer, 1971, Phys. Stat. Sol. **44**, 437.
Janik, J. M., M. Rachwalska and J. A. Janik, 1973a, Physica, **72**, 168.
Janik, J. M., G. Pytasz, M. Rachwalska, J. A. Janik, I. Natkaniec and W. Nawrocik, 1973b, Acta Physica Polon. **A43**, 419.

Kim, H. J., P. S. Goyal, G. Venkataraman, B. A. Dasannacharya and C. L. Thaper, 1970, Solid State Commun. **8**, 889.
Kosaly, G. and G. Solt, 1966, Physica **42**, 1571.
Kottwitz, D. A. and B. R. Leonard, 1962, IAEA Symp. Chalk River, **1**, 359.
Kottwitz, D. A., B. R. Leonard and R. B. Smith, 1962, IAEA Symp. Chalk River **1**, 373.
Larsson, K. E., 1964, IAEA Symp. Bombay **2**, 3.
Larsson, K. E., 1971, Phys. Rev. **A3**, 1006.
Larsson, K. E., 1973, J. Chem. Phys. **59**, 4612.
Larsson, K. E. and U. Dahlborg, 1962a, IAEA Symp. Chalk River **1**, 317.
Larsson, K. E. and U. Dahlborg, 1962b, J. Nucl. Energy **16AB**, 81.
Larsson, K. E. and U. Dahlborg, 1964, Physica **30**, 1561.
Larsson, K. E., S. Holmryd and K. Otnes, 1960, IAEA Symp. Vienna, 329.
Lee, F. S. and G. B. Carpenter, 1959, J. Phys. Chem. **63**, 279.
Leung, P. S., J. J. Rush and T. I. Taylor, 1966, unpublished data.
Mikke, K., 1960, IAEA Symp. Vienna, 351.
Mikke, K. and A. Kroh, 1962, IAEA Symp. Chalk River **2**, 237.
Mikuli, E., J. M. Janik, G. Pytasz, J. A. Janik, J. Ściesiński, E. Ściesińska and A. Mazurkiewicz, 1972, Low Frequency Motions of Water Molecules in Crystalline $Li_2SO_4 \cdot H_2O$, Rept. Inst. Nucl. Phys., No. 791/PS (Kraków).
Miller, S. R., R. Blinc, M. Brenman and J. S. Waugh, 1962, Phys. Rev. **126**, 528.
Mills, R., 1971, Ber. Bunsenges. Phys. Chem. **75**, 195.
Nakahara, Y., 1968, J. Nucl. Sci. Technol. **5**, 635.
Nordman, C. E., 1962, Acta Cryst. **15**, 18.
O'Reilly, D. E., E. M. Peterson and J. M. Williams, 1971, J. Chem. Phys. **54**, 96.
Palevsky, H., K. Otnes and Y. Wakuta, 1962, IAEA Symp. Chalk River **2**, 273.
Parliński, K., 1968, Acta Physica Polon. **34**, 1019.
Parliński, K., 1969, Acta Physica Polon. **35**, 223.
Pelah, I. and J. Imry, 1966, Phys. Letters **21**, 248.
Pelah, I., E. Wiener and J. Imry, 1964, IAEA Symp. Bombay **2**, 325.
Plesser, T. and H. Stiller, 1970, in: Dynamics and Magnetic Properties of Solids and Liquids, Ed. D. M. Jović, Proceedings of the 2nd Summer School in Hercegnovi (Beograd).
Pope, N. K. and R. Nations, 1964, IAEA Symp. Bombay **2**, 141.
Prask, H. J. and H. Boutin, 1964, J. Chem. Phys. **40**, 2670.
Prask, H. J. and H. Boutin, 1966, J. Chem. Phys. **45**, 699, 3284.
Prask, H., H. Boutin and S. Yip, 1968, J. Chem. Phys. **48**, 3367.
Prask, H. J., S. F. Trevino, J. D. Gault and K. W. Logan, 1972, J. Chem. Phys. **56**, 3217.
Rahman, A. and F. H. Stillinger, 1971, J. Chem. Phys. **55**, 3336.
Ranner, N. B., I. D. Datt, A. B. Tovbis and R. P. Ozerov, 1968, Kristallografiya **10**, 914.
Renker, B., 1969, Phys. Letters **30A**, 493.
Rosciszewski, K., 1974, Physica **75**, 268.
Rush, J. J., 1966, J. Chem. Phys. **44**, 1722.
Rush, J. J. and T. I. Taylor, 1964, IAEA Symp. Bombay **2**, 333.
Rush, J. J., T. I. Taylor and W. W. Havens, 1962, J. Chem. Phys. **37**, 234.
Rush, J. J., P. S. Leung and P. I. Taylor, 1966, J. Chem. Phys. **45**, 1312.
Rush, J. J., L. W. Schroeder and A. J. Melveger, 1972, J. Chem. Phys. **56**, 2793.
Safford, G. J., P. S. Leung, A. W. Naumann and P. C. Schaffer, 1969, J. Chem. Phys. **50**, 4444.

Sakamoto, M., B. N. Brockhouse, R. G. Johnson and N. K. Pope, 1962, J. Phys. Soc. Japan, Suppl. BII **17**, 371.

Sampson, T. E. and J. M. Carpenter, 1968, IAEA Symp. Copenhagen **1**, 491.

Saunderson, D. H. and V. S. Rainey, 1962, IAEA Symp. Chalk River **1**, 413.

Schenk, C. H., E. Wiener, B. Weckermann and W. Kley, 1968, Phys. Rev. **172**, 576.

Schlemper, E. O., W. C. Hamilton and J. J. Rush, 1966, J. Chem. Phys. **44**, 2499.

Schofield, P., 1960a, Phys. Rev. Letters **4**, 239.

Schofield, P., 1960b, IAEA Symp. Vienna, 30.

Simpson, J. H. and H. Y. Carr, 1958, Phys. Rev. **111**, 1201.

Singwi, K. S. and A. Sjölander, 1960, Phys. Rev. **119**, 863.

Singwi, K. S., A. Sjölander and A. Rahman, 1962, IAEA Symp. Chalk River **1**, 215.

Sköld, K., 1968, J. Chem. Phys. **49**, 2443.

Sköld, K., E. Pichler and K. E. Larsson, 1964, Atomic Energy Rept., No. AE133 (Stockholm).

Steinman, D. K. and G. C. Summerfield, 1968, J. Phys. Chem. Sol. **30**, 449.

Stiller, H. H. and H. R. Danner, 1960, IAEA Symp. Vienna, 363.

Szkatuła, A. and A. Fuliński, 1967, Physica **36**, 35.

Teh, C. and B. N. Brockhouse, 1971, Phys. Rev. **B3**, 2733.

Thaper, C. L., A. Sequeira, B. A. Dasannacharya and P. K. Iyengar, 1969, Phys. Stat. Sol. **34**, 279.

Thaper, C. L., B. A. Dasannacharya, A. Sequeira and P. K. Iyengar, 1970, Solid State Commun. **8**, 497.

Trevino, S., H. Prask, T. Wall and S. Yip, 1968, IAEA Symp. Copenhagen **1**, 345.

Tubino, R. and G. Zerbi, 1970, J. Chem. Phys. **53**, 1428.

Turchin, W. F., 1963, Myedlennye nejtrony, Moskva.

Venkataraman, G., K. Usha, P. K. Iyengar, P. R. Vijayaraghavan and A. P. Roy, 1962, IAEA Symp. Chalk River **2**, 253.

Venkataraman, G., K. Usha-Deniz, P. K. Iyengar and P. R. Vijayaraghavan, 1964, Solid State Commun. **2**, 17.

Venkataraman, G., K. Usha-Deniz, P. K. Iyengar, A. P. Roy and P. R. Vijayaraghavan, 1966, J. Phys. Chem. Sol. **27**, 1103.

Vineyard, G. H., 1958, Phys. Rev. **110**, 999.

Woods, A. D. B., B. N. Brockhouse, M. Sakamoto and R. N. Sinclair, 1960, IAEA Symp. Vienna, 487.

Yip, S. and K. Osborn, 1963, Phys. Rev. **131**, 2547.

CHAPTER 20

DIELECTRIC PROPERTIES OF HYDROGEN BONDED SYSTEMS

L. SOBCZYK

Institute of Chemistry, University of Wrocław, Poland

H. ENGELHARDT

Universidad del Valle, División del Ciencias, Cali, Colombia, South America

AND

K. BUNZL

Gesellschaft für Strahlen- und Umweltforschung mbH München,
Institut für Biologie, Institut für Strahlenschutz,
D-8042 Neuherberg, West Germany

Contents

20.1. Introduction 939

20.2. Dielectric properties of hydrogen bonded systems in solution (by L. Sobczyk) . 939

- 20.2.1. Solvent influence on the dipole moment and the hydrogen bond 940
- 20.2.2. Dipole moments of complexes with hydrogen bonds in inert solvents . 942
- 20.2.3. Strong hydrogen bond 946
- 20.2.4. Intramolecular hydrogen bonds 953
- 20.2.5. Composition of the formed complexes and their conformation 955
- 20.2.6. Hydrogen bonded systems in a strong electric field 959

20.3. Dielectric polarization and electrical conduction in hydrogen bonded solids (by H. Engelhardt) 963

- 20.3.1. Ice 966
  - 20.3.1.1. Dielectric dispersion 967
  - 20.3.1.2. Molecular dipole moment 969
  - 20.3.1.3. Conduction phenomena 973
  - 20.3.1.4. Proton mobility 979
  - 20.3.1.5. Effect of doping on dielectric properties of ice 982
- 20.3.2. Dielectric properties of clathrates 985
- 20.3.3. Miscellaneous substances 988
- 20.3.4. Experimental methods 995

20.4. Dielectric properties of water adsorbed by solids (by K. Bunzl) 997

- 20.4.1. Water adsorbed by inorganic substances 999
  - 20.4.1.1. Zeolites 999
  - 20.4.1.2. Clay minerals 1003
  - 20.4.1.3. Miscellaneous substances 1006
- 20.4.2. Water adsorbed by organic substances 1010
  - 20.4.2.1. Ion exchange resins 1010
  - 20.4.2.2. Other organic substances 1012
- 20.4.3. Water adsorbed by biopolymers 1012
  - 20.4.3.1. DNA 1012
  - 20.4.3.2. Proteins 1013
  - 20.4.3.3. Miscellaneous substances 1016

References 1017

The hydrogen bond – recent developments in theory and experiments
Eds. P. Schuster et al. © North-Holland Publ. Co., Amsterdam, 1976

20.1. Introduction

Dielectric investigations are becoming increasingly important as a tool for studying the nature of H-bonding. This can be seen from the considerable number of publications devoted to this particular problem. In the present survey, however, which covers the period from about 1966 to 1973, special consideration was given to such topics of this field, which were discussed in earlier reviews only to a limited extent or from other points of view. This restriction seemed appropriate, since within the last few years several books on H-bonding and related studies were published. (Eisenberg et al. [1969], Riehl et al. [1969], Fletcher [1970], Franks [1972], Horne [1972], Whalley et al. [1973] and Hobbs [1974]). All of them contain contributions on dielectric phenomena; some of them are especially devoted to dielectric studies (Hill et al. [1969] and Davies [1970, 1972]). These publications provide thoroughly collected lists of references to original papers. A comprehensive bibliography on the structure and physical properties of liquid water was published by Bell Telephone Laboratories [1969]. A complete list of references covering the recent literature on the physics of ice is regularly attached to each volume of the "Journal of Glaciology".

20.2. Dielectric properties of hydrogen bonded systems in solution (by L. Sobczyk)

The subject of dielectric properties of H-bonded liquids and solutions has previously been reviewed in several papers. The first one, by Magat [1959], is devoted mainly to relaxation phenomena of associated liquids. An extensive supplement is provided by the review of Crossley [1970]. A previous review of Sobczyk [1969] deals, among other things, with general problems of dielectric absorption in H-bonded systems. Selected dielectric phenomena of these systems have been discussed in two reviews by Davies [1969,1970].

Recently Davies [1972] edited an interesting monograph which contains several important chapters dealing with problems of H-bonding. There is a chapter written by Hasted on dielectric properties of water and water solutions and also one by Schwarz on dielectric polarization phenomena in biomolecular systems. The part of this book written by Wyllie is devoted to dielectric relaxation and molecular correlation, and field effects is reviewed by Kielich. Both problems gain more and more importance in any discussion on dielectric properties of H-bonding. Also of particular importance at present are kinetic problems and theoretical descriptions of relaxation phenomena based on the correlation function, which are discussed extensively by Williams [1970, 1972], Anderson [1971] and Schwarz [1972].

This review deals mainly with problems of dipole moments of H-bonded complexes and also with some problems of nonlinear phenomena. These problems, until now, have not been critically reviewed.

20.2.1. Solvent influence on the dipole moment and the hydrogen bond

The increase of dielectric polarization in systems with H-bonds is a well known phenomenon. Earlier experiments on this problem dealt mainly with the so called dioxane effect. It was shown that dipole moments of proton-donor compounds measured in dioxane are distinctly higher as compared with those measured in inert solvents, e.g. cyclohexane, heptane or benzene (its inertness however being open to discussion). Among others, amines (Vassiliev and Syrkin [1941], Few and Smith [1949], Smith [1950, 1953] and Smith and Walshaw [1957]), phenols (Erič et al. [1960], Goode and Ibbitson [1960] and Ibbitson and Sandall [1964]) and alcohols (Ibbitson and Moore [1967]) were also investigated. Recently the dioxane effect was seen in investigations of intra- and intermolecular interactions of more complicated molecular systems (Skulski and Waclawek [1971]). The polarity of H-bond, $\Delta\mu$, defined as the excess dipole moment

$$\overset{\Delta\mu}{\longleftarrow}$$
$$\text{A–H}\cdots\text{B}$$

which is localized along the A–H bond was found to be, based on experimental results, usually in the range 0.3–0.6 D. No correlation could be observed between $\Delta\mu$ and the acidity of the AH component, the ionization potential of component B or the dipole moments of the interacting components. Ibbitson and Sandall [1964] reported that $\Delta\mu$

increases directly with the increase of acidity of the phenol derivative. However, a more exact analysis, taking into account the effect of charge transfer in the phenol derivative itself, revealed (table 20.1) that $\Delta\mu$ is a constant value in the range of experimental error.

TABLE 20.1

Dipole moments of phenols in nonplanar solvents and dioxane, and calculated polarity of H-bonding (in D)

	μ (inert)		μ (dioxane)[a]	$\Delta\mu$[a]	$\Delta\mu$ recalcu-lated
	CCl_4[a]	Benzene[b]			
Phenol	1.43	1.53	1.86	0.42	0.37
p-Cresol	1.44	1.56	1.83	0.35	0.33
p-t-Butylphenol	1.48	—	1.98	—	0.50
Mesitol	1.30	1.40	1.66	—	0.15
p-Chlorophenol	2.25	2.19	2.82	0.54	0.60
p-Bromophenol	2.15	2.19	2.75	0.48	0.54
p-Fluorophenol	—	2.12	2.67	0.43	0.57
s-Trichlorophenol	1.38	1.42	1.88	—	0.40
s-Tribromophenol	1.44	1.42	1.98	—	0.51
p-Nitrophenol	—	5.05	5.37	1.18	0.46
p-Cyanophenol	—	4.93	5.25	0.99	0.45
p-Hydroxybenzaldehyde	—	—	4.28	0.99	—
p-Hydroxyacetophenone	—	—	4.08	0.87	—

[a] Erič et al. [1960] and Ibbitson and Sandall [1964].
[b] Koll et al. [1970].

The increase of dipole moment is not only observed in dioxane solutions but it was demonstrated by Malarski and Sobczyk [1969] that even for weak proton-donors such as the haloforms, phenyl-acetylene, diphenylamine or triphenylcarbinol in weak proton-acceptors such as benzene, an increase of dipole moment caused by a hydrogen interaction (distinctly visible on IR spectra) is observed. Various donors and acceptors were investigated in detail (Lumbroso [1964], Bertin and Lumbroso [1966] and Pigenet and Lumbroso [1966]). Interesting results for HF, HCl, HBr and HJ in various solvents were obtained by Weith et al. [1948] where $\Delta\mu$ is distinctly dependent on the acidity of the donor. A similar situation is observed for HNCS, which is also a strong acid (Detoni et al. [1970]).

For investigations of binary systems donor–solvent, a total complexation of proton-donor with solvent molecules is assumed. This assumption was demonstrated to be correct with IR spectra. Nonpolar solvents with significant basic properties are rather scarce. Furthermore, difficulties arise if efforts are made to discuss more exactly and quantitatively experimental results obtained in polar solvents, especially when the permittivity increment is very small. Thus it seems that investigations of ternary systems (inert solvent–donor–acceptor) are necessary.

20.2.2. Dipole moments of complexes with hydrogen bonds in inert solvents

Determining of molar polarization of complexes with H-bonds in ternary systems requires a knowledge of the formation constants, but so far, with few exceptions, only complexes of 1 : 1 composition have been analyzed.

Simultaneous determination of both the formation constant and the polarization is possible using the method of Cleverdon et al. [1956]. Nevertheless, some authors point out that independent measurements of formation constant, e.g. with IR methods, are necessary. So far the dipole moments of more than 20 complexes with weak H-bonds containing donors like water, alcohols, phenol, pyrrole and acceptors such as pyridine, triethylamine, dioxane, tetrahydrofurane and acetone in various solvents have been determined (see Hawranek and Sobczyk [1971]). The $\Delta\mu$ values change irregularly and do not exceed 0.70 D. Sometimes the calculated $\Delta\mu$ values are negative which is contrary to the basic rule that the $\Delta\mu$ vector is directed from B to AH. This may be caused by different experimental conditions and by the assumed vector scheme (complex configuration), which may not be always well substantiated. This problem will be discussed later. Thus, it appears that there is still a lack of exact and systematic dipole moment measurements of weaker complexes with H-bonding. The necessity of this sort of experiment is self-evident. One attempt in this field is the detailed measurements of dipole moments of a series of complexes of diphenylamine and 2,4,6-trimethylphenol (mesitol) which do not associate under experimental conditions and various proton–acceptors (Hawranek and Sobczyk [1971]). In the method used, the polarization of the complex was determined by extrapolation of the relative increments of permittivity and density to infinite dilution of all components in the solution.

$$\alpha_{\text{solute}} = \lim_{x_A, x_D, x_{AD} \to 0} \frac{1}{\epsilon_1} \left[\left(\frac{\partial \epsilon}{\partial x_A} \right)_{x_D, x_{AD}} + \left(\frac{\partial \epsilon}{\partial x_D} \right)_{x_A, x_{AD}} + \left(\frac{\partial \epsilon}{\partial x_{AD}} \right)_{x_A, x_D} \right]$$

$$= \alpha_A + \alpha_D + \alpha_{AD}, \tag{20.1a}$$

$$\beta_{\text{solute}} = \lim_{x_A, x_D, x_{AD} \to 0} \frac{1}{d_1} \left[\left(\frac{\partial d}{\partial x_A} \right)_{x_D, x_{AD}} + \left(\frac{\partial d}{\partial x_D} \right)_{x_A, x_{AD}} + \left(\frac{\partial d}{\partial x_{AD}} \right)_{x_A, x_D} \right]$$

$$= \beta_A + \beta_D + \beta_{AD}. \tag{20.1b}$$

x_A, x_D, β_A and β_D values were determined independently and thus α_{AD} and β_{AD} are estimated (which were used for calculation of P_{AD} according to Hedestrand [1929]).

$$P^{\infty}_{AD} = \frac{3\epsilon_1 M_1}{(\epsilon_1 + 2)^2 d_1} \alpha_{AD} + \frac{\epsilon_1 - 1}{(\epsilon_1 + 2) d_1} (M_{AD} - \beta_{AD} M_1). \tag{20.2}$$

Respective symbols in this equation have their usual meaning as applied to dielectric problems. It was demonstrated that the above mentioned procedure (which requires independent determination of concentration x_A, x_D and x_{AD}) can be used successfully in investigations of various types of interactions.

In table 20.2, data on polarity of complexes of mesitol and on spectral parameters of these complexes are given. Similarly as in other systems, the approximately constant value of $\Delta\mu$ should be noted. The value of 0.71 D for the piperidine complex, calculated with the assumption of a free rotation about the O–H bond, is not likely to be correct, due to the presence of significant steric hindrance. Perhaps the value of 0.51 D is better calculated for the rigid structure, which corresponds to a minimum of steric hindrance.

It can be demonstrated that the obtained values do not agree with a model which assumes a high contribution of CT (delocalization) effect to the polarity. From this point of view, the CT model of Mulliken, which was recently analyzed by Ratajczak [1972a, b] and Ratajczak and Orville-Thomas [1973] seems to be very suggestive.

In analogy to EDA complexes, the force constant of stretching vibrations of the A–H group in the H-bond is given by the equation

$$k_c = (a^2 + abS)k_0 + (b^2 + abS)k_1 \tag{20.3}$$

where k_0 and k_1 are the force constants of the AH and $(AH)^-$ groups respectively. The last group is the result of a total electron transfer from atom B to the antibonding orbital of the AH group and a, b and S have

TABLE 20.2

Dipole moments and spectral characteristics of mesitol complexes

Proton–acceptor	pK_a (B)	$\Delta\nu_s$ (cm^{-1})	B/B^0	μ (D)	$\Delta\mu$ (D)	$\Delta\mu$ (D) calculated
3,5-Dichloropyridine	0.49	271	19.0	2.58	0.35	2.6
3-Chloropyridine	2.84	330	17.0	3.33	0.25[a)] 0.44[b)]	3.2
Pyridine	5.19	395	21.9	4.01	0.44	3.8
4-Picoline	6.00	419	23.3	4.33	0.36	4.0
2,4,6-Collidine	7.52	481	24.6	3.95	0.43	4.6
Piperidine	11.22	—	—	3.08	0.71[a)] 0.51[c)]	

B – Integrated intensity of the OH stretching vibration band in complex, B^0 – Intensity for the free molecule in CCl_4.

[a)] By assuming free rotation.

[b)] Rigid structure corresponding to the minimum of dipole–dipole interaction energy.

[c)] Rigid structure with minimum steric hindrance.

the usual meaning. From eq. (20.3) the contribution of the dative structure is obtained

$$b_2 + abS = (k_0 - k_c)/(k_0 - k_1) = (\nu_0^2 - \nu_c^2)/(\nu_0^2 - \nu_1^2) \tag{20.4}$$

where ν is the wave number of the corresponding vibration. On the other hand, the dipole moment of a CT complex is given by the following equation

$$\mu_c = (a^2 + abS)\mu_0 + (b^2 + abS)\mu_1 \tag{20.5}$$

where μ_0 and μ_1 are dipole moments of no bond and dative structure respectively.

It follows that

$$|\Delta\boldsymbol{\mu}| = \mu_{ind} + (b^2 + abS)(\mu_1 - \mu_0) \tag{20.6}$$

or including (4)

$$|\Delta\boldsymbol{\mu}| = \mu_{ind} + \frac{\nu_0^2 - \nu_c^2}{\nu_0^2 - \nu_1^2}(\mu_1 - \mu_0) \tag{20.7}$$

if only the inductive (polarization) and the charge-transfer components are taken into account.

As can be seen from table 20.2, $\Delta\boldsymbol{\mu}$ is constant in the range of experimental error and a charge-transfer effect is not detectable. Another treatment on the basis of the CT model was given by Friedrich and Person [1966] who proposed that for a number of CT complexes, $\nu_1 \approx \frac{1}{2}\nu_0$. Also, for many strong bonds, the frequencies of stretching vibrations of A–H decrease by a factor of 2. Thus, equation (20.6) is reduced to:

$$|\Delta\boldsymbol{\mu}| = \mu_{\text{ind}} + \tfrac{8}{3}(\Delta\nu/\nu_0)(\mu_1 - \mu_0). \qquad (20.8)$$

The calculated $|\Delta\boldsymbol{\mu}|$ values listed in table 20.2 were based on this relation and the assumption that $\mu_1 - \mu_0 = 13$ D. It is apparent that there is a discrepancy between the values of $\Delta\boldsymbol{\mu}$ calculated in this manner and the experimental results. This discrepancy is all the more striking if one takes into account the obtained band shifts in IR spectra and their intensity. $\Delta\nu$ and $(\partial\mu/\partial r)$ values calculated from integral intensities B which were very high and distinctly changing directly with the pK, would suggest, in accordance with the CT model, a significant charge transfer which, in fact, is not observed.

Thus, it could be concluded that in weak or medium strong complexes with H-bond, the inductive effect is predominant. The induced moment can, in general, be described by the following equation

$$\boldsymbol{\mu}_{\text{ind}} = \sum_{i=1}^{m_D}\sum_{j=1}^{m_A} \boldsymbol{\alpha}_i f_j + \sum_{p=1}^{m_A}\sum_{q=1}^{n_D} \boldsymbol{\alpha}_p f_q \qquad (20.9)$$

where f_j and f_q are field strengths resulting from atomic multipoles, q and j, and $\boldsymbol{\alpha}_i$ and $\boldsymbol{\alpha}_p$ are tensors of bond polarizability. For calculations of the dipole moment according to the above equation, the electron structure of the donor D and acceptor A as well as the polarizibility of the respective bonds has to be known.

For calculation of μ_{ind}, however, it is possible to confine oneself to a simple relation in which the inductive influence of the dipole of the proton-donor atom (usually the dipole moment of a free electron pair) on A–H is considered (Fauquembergue et al. [1967]).

$$\Delta\mu_{\text{ind}}^{\text{AH}} = \mu\alpha_{\parallel}^{\text{AH}} \frac{\epsilon_a + 2}{3\epsilon_b r^3}(3\cos^2\Theta - 1) \qquad (20.10)$$

where μ is the inducing moment, $\alpha_{\parallel}^{\text{AH}}$ the longitudinal polarizibility of the AH bond, ϵ_b the permittivity of the polarizable system, ϵ_a the medium permittivity between the inducing dipole and the polarizibility center, r the distance between the mid-point of inductive dipole and the polarizibility center and Θ is the angle between $\boldsymbol{\mu}$ and $\boldsymbol{r}$ vectors. This equation is only approximate, because exact values of all parameters are not known. Calculations based on this equation (Sobczyk and Syrkin [1957]) revealed that the μ_{ind} values are in the range 0.2–0.3 D. However it is known (Hanna [1968], Hanna and Williams [1968] and Lippert et al. [1969]) that limitation of the electric field to the dipole field is insufficient and that an important contribution to the field strength comes from the quadrupole moment of the inducing molecule. Thus, it can be ascertained that the calculated induced moment, independent from the approximation used, is the lower limit of the real μ_{ind} value.

These considerations again suggest that the obtained experimental $\Delta\boldsymbol{\mu}$ values which are between 0.3–0.6 D can be totally ascribed to inductive effects. It is obvious that separation of the inductive effect (considered here as an interaction between undisturbed electron systems) and the delocalization effect, is somewhat sophisticated but pertains to contributions with appropriate effects as analyzed by Murrell [1969]. It is proper to add that from ab initio calculation, Kollmann and Allen [1969] obtained dimers with $\Delta\boldsymbol{\mu} = 0.41$ D for water, whereas from charge diagrams for these dimers, Hoyland and Kier [1969] estimated $\Delta\boldsymbol{\mu}$ to be 0.45 D.

20.2.3. Strong hydrogen bond

In recent years much effort was made to investigate the polarization of numerous complexes with H-bonds in which proton-donor and acceptor properties varied over a wide range. The proton-donors were phenols and carboxylic acids. Vector schemes for these complexes are shown in fig. 20.1.

In calculation of $\Delta\boldsymbol{\mu}$, it is usually assumed that the interaction leads only to charge density displacement within the bridge building atoms

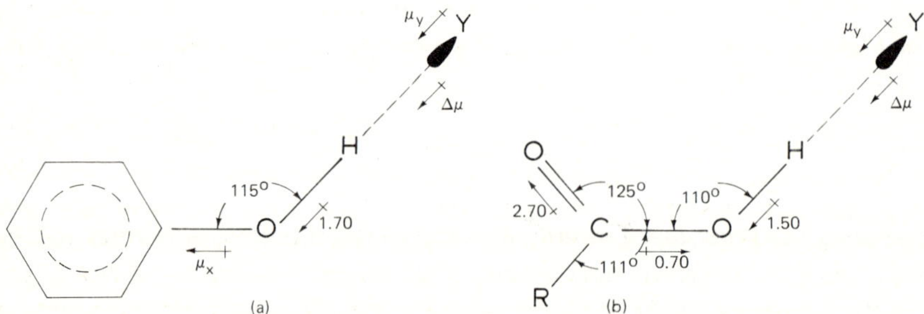

Fig. 20.1. Vector schemes of phenol and carboxylic acid complexes with n-donors.

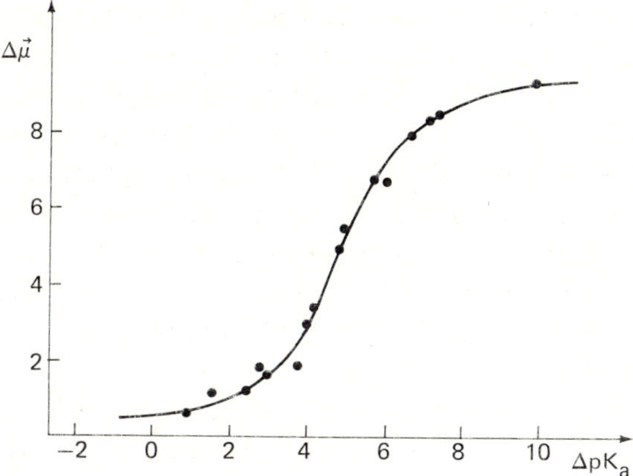

Fig. 20.2. H-bond polarity $\Delta\boldsymbol{\mu}$ related to $\Delta_{pK_a} = pK_{BH^+} - pK_{AH}$ for phenol-triethylamine complexes, according to Ratajczak and Sobczyk [1969].

A···B. In other words, the additional dipole resulting from interaction is directed along the bridge and is treated as the polarity of the H-bond. First systematic investigations were carried out on phenol derivatives–triethylamine systems (Ratajczak and Sobczyk [1969, 1970]) and very characteristic plots were obtained for the function $\Delta\boldsymbol{\mu}(\Delta pK_a)$ shown in fig. 20.2. This type of dependence is very similar to acid-base titration plots and it can be predicted based on the thermodynamical model of proton transfer of Huyskens and Zeegers-Huyskens [1964]. If for the proton a double minimum of potential energy is assumed and both states (i) and (ii) are treated as tautomeric forms of the bridge

$$\underset{\text{(i)}}{\text{A–H} \cdots \text{B}} \rightleftharpoons \underset{\text{(ii)}}{\text{A}^- \cdots \text{H–B}^+} \tag{20.11}$$

then the equilibrium constant of proton transfer, K_{PT}, can be expressed by equation

$$\log K_{PT} = \Delta pK_a + C \tag{20.12}$$

where $\Delta pK_a = pK_{BH^+} - pK_{AH}$, and C is a constant which includes the energy of solvation processes. Experimental values are in agreement with the plot resulting from equation (20.12) only, as in this case, if an empirical parameter $\xi < 1$ is introduced which takes into account the fact that, in reality, states (i) and (ii) are not independent from each other.

$$\log K_{PT} = \xi \Delta pK_a + C'. \tag{20.13}$$

The smaller the coupling between two states the sharper the inflection of the curve and parameter ξ approaches unity. The highest deviation of the curve $\Delta\boldsymbol{\mu}$ (ΔpK_a), %PT (ΔpK_a) is observed for oxygen bases. Also the plot $\Delta\boldsymbol{\mu}$(ΔpK_a) obtained for carboxylic acid + triethylamine systems is almost flat (Sobczyk and Pawełka [1973]). Thus, it can be supposed that the slope of the $\Delta\boldsymbol{\mu}$ curve is dependent on the polarizability of the interacting bridge atoms. Up to this date, experimental results are rather scarce so that no general conclusions can be made.

Changes of $\Delta\boldsymbol{\mu}$ related to ΔpK_a have been found to be much the same for phenol-pyridine derivatives (Hawranek et al. [1972] and Nouwen and Huyskens [1973]) as for phenol-aniline derivatives (Debecker and Huyskens [1971]). In particular, results obtained by Nouwen and Huyskens [1973] shown in fig. 20.3 should be noted. On this picture, small changes of $\Delta\boldsymbol{\mu}$ are visible for high as well as for low ΔpK_a values which would indicate that besides the PT effect, which plays an essential role in the transient area of ΔpK_a, the delocalization also contributes to some extent. It appears that changes in bond length are also important which, for instance, is reflected in permanent energy changes of bridge formation as well as in shifts of stretching vibrations of the OH group (Goldstein et al. [1972a, b] and Huyskens and Hernandez [1973].

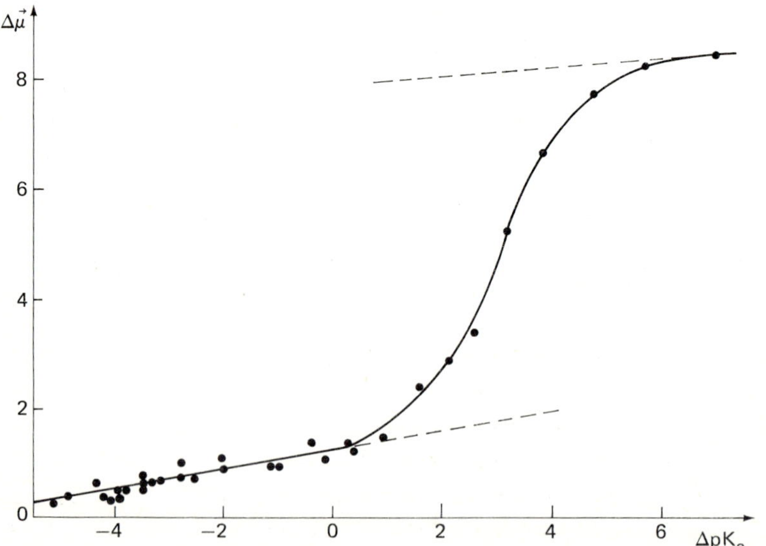

Fig. 20.3. $\Delta\boldsymbol{\mu}$ related to ΔpK_a for phenol-pyridine derivatives complexes, according to Nouwen and Huyskens [1973].

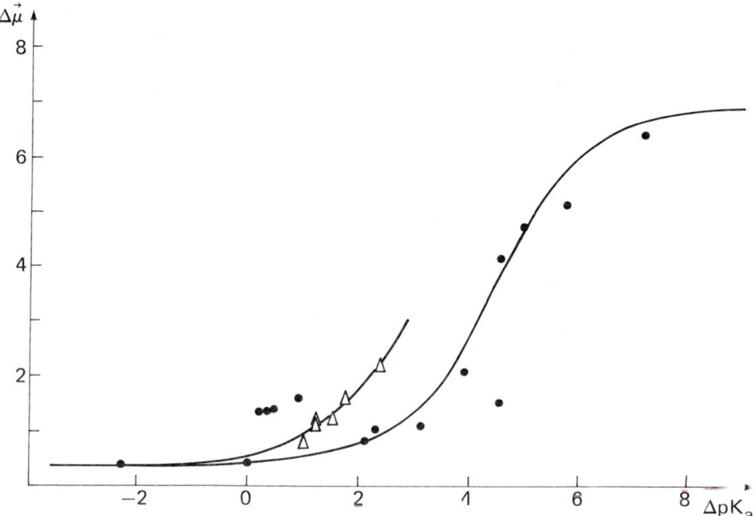

Fig. 20.4. $\Delta\boldsymbol{\mu}$ related to ΔpK_a for carboxylic acid–pyridine derivatives complexes. ●–aliphatic acids, △–benzoic acid derivatives.

Finally, the results obtained for complexes of carboxylic acids and pyridine derivatives should be noticed (Sobczyk and Pawełka [1973]), which were earlier investigated by IR methods (Barrow [1956], Gusakova et al. [1971] and Lindemann and Zundel [1972]) and by UV techniques ($n \rightarrow \pi^*$ transition, Nasielski and Donckt [1963]).

It should be emphasized that results of dipole moment measurements (fig. 20.4) are in good agreement with spectral data, although different experimental conditions were used for the various methods. That is, in the case of dipole moments of strong H-bonded species, strong evidence for the validity of the interpretation is based on the PT effect, particularly for the transient area of ΔpK_a. However, certain perturbations of charge density distribution in donor and acceptor molecules themselves have to be kept in mind. Such phenomena are described by Sobczyk and Syrkin [1956] and Sobczyk and Pawelka [1974].

Phenol-trialkylamine complexes have also been discussed from the point of view of Mulliken charge-transfer theory (Ratajczak [1972b]). If, for a series of complexes of the same type, the assumption is made that the resonance integral, the overlap integral and the Coulomb term are constant, and that there exists a linear dependence between the electron

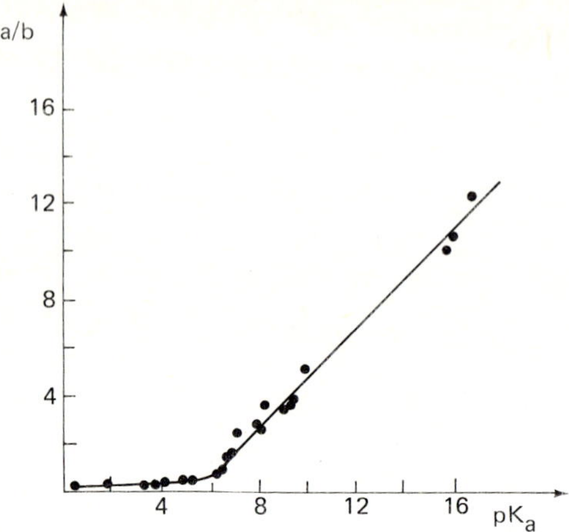

Fig. 20.5. a/b related to pK_a of phenol in complexes with triethylamine, according to Ratajczak [1972a].

affinity of the phenol and its pK_a, then a linear relation between a/b and pK_{AH} has to be expected.

$$a/b = m \times pK_{AH} + n. \tag{20.14}$$

If the dipole moment of the complex μ_c, the moment for total charge transfer μ_1, the induced moment μ_{ind} and the overlap integral are known, then the ratio a/b can be calculated. The results obtained by Ratajczak [1972a] are given in fig. 20.5. Instead of one straight line, two with different slopes were obtained intersecting at about $\Delta pK_a = 6$. This point may be treated as a transition point from "outer" to "inner" (ion pairs) complexes.

It appears however, that the role of the CT effect in the H-bond, as was pointed out by Huyskens and Hernandez [1973], is generally over estimated. Though the CT model may prove to be useful for the construction of semi-empirical correlations, the nature of the interactions is much more like a complex.

Considering again the problem of proton transfer described by eq. (20.11) very interesting results were obtained by Jadżyn and Małecki [1972] who investigated the influence of permittivity of the medium, ϵ, on dipole moments of phenol-triethylamine complexes.

The dipole energy U in a medium of permittivity ϵ is described according to Onsager's theory by eq. (20.15)

$$U = U_0 - \frac{4\pi N}{3V}\mu^2 \frac{\epsilon - 1}{2\epsilon + 1} \tag{20.15}$$

where U_0 is the dipole energy in vacuum, V the molar volume and N is Avogadro's number.

Because there is a great difference in polarity between the two states of the H-bond (i) and (ii) in eq. (20.11), dielectric permittivity should influence the equilibrium constant K_{PT} to a great extent

$$K_{PT} = \exp(-\Delta U/RT) = x_1/x_0 \tag{20.16}$$

where x_1 and x_0 are the mole fractions of the proton and non-proton transferred species respectively. These fractions are calculated based on the additivity principle of dipole polarization,

$$\mu^2 = \mu_0^2 x_0 + \mu_1^2 x_1 \tag{20.17}$$

if the dipole moment for the state with total proton transfer μ_1 is known. The energy difference between the two states of the H-bond is given by

$$\Delta U/RT = \Delta U_0/RT - A(\epsilon - 1)/(2\epsilon + 1) \tag{20.18}$$

where

$$A = \frac{4\pi N}{3RTV}(\mu_1^2 - \mu_0^2).$$

The change of the minima distribution with a change in permittivity is depicted schematically in fig. 20.6. As is evident for certain complexes, an inversion of population of the two minima by changing ϵ of the solvent should be expected. Theoretical expectations based on Onsager's theory were experimentally confirmed by Jadżyn and Małecki [1972] and their results are shown in the diagrams in fig. 20.7. If the μ_1 and μ_0 values are adequately chosen, one can experimentally determine the dipole moment value of the complex in relation to pK_a of phenol as well as the relation to permittivity ϵ of the solvent. Simultaneously it is possible to formulate a generalized relation of the dipole moment of complex from ΔU, which is shown in fig. 20.7 (curve b).

The results of Jadżyn and Małecki, though very suggestive, do not reflect all effects which influence the polarity of the H-bond. Figure 20.8 reveals the dependence of the dipole moment of the 2,4-dinitrophenol-triethylamine complex from the Onsager parameter $(\epsilon - 1)/(2\epsilon + 1)$. As is shown, the dependence is contrary to the expected one. If, however, the G (Allerhand and von Schleyer [1963]) or E_T (Reichardt [1965]) parameter is introduced, which describes the specific donor–acceptor properties of

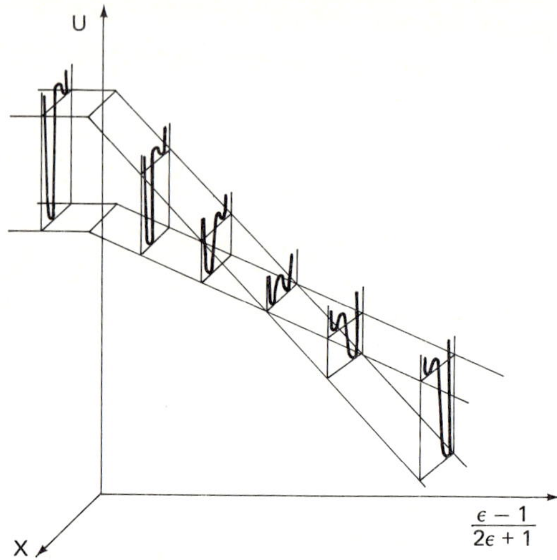

Fig. 20.6. Changes of proton potential energy curve in a hydrogen bridge related to the $(\epsilon - 1)/(2\epsilon + 1)$ parameter. The x-axis is the bridge direction, according to Jadżyn and Małecki [1972].

the solvent, an agreement with expectations is obtained. The type of dependence of dipole moments of complexes with strong H-bonds seems to be from the above parameters, and in the light of very recently obtained results (Pawełka and Sobczyk [1973]), is rather general. This is

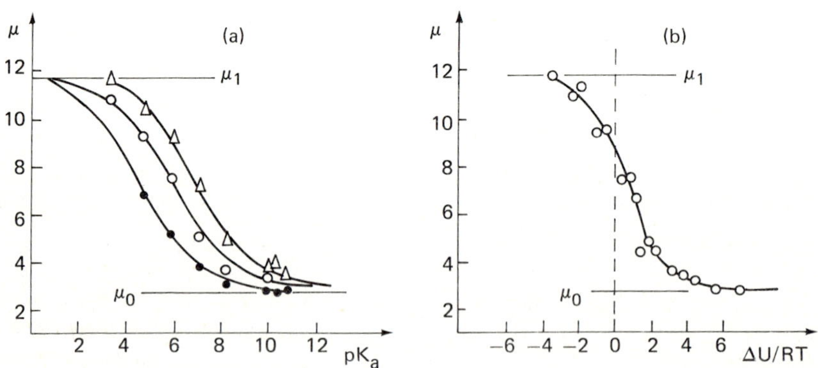

Fig. 20.7. Experimental proof of equation (20.18): a–experimental points for trichloroethylene (△), toluene (○), and cyclohexane (●) compared with theoretical curves μ (pK_a); b–generalized curve μ ($\Delta U/RT$), according to Jadżyn and Małecki [1972].

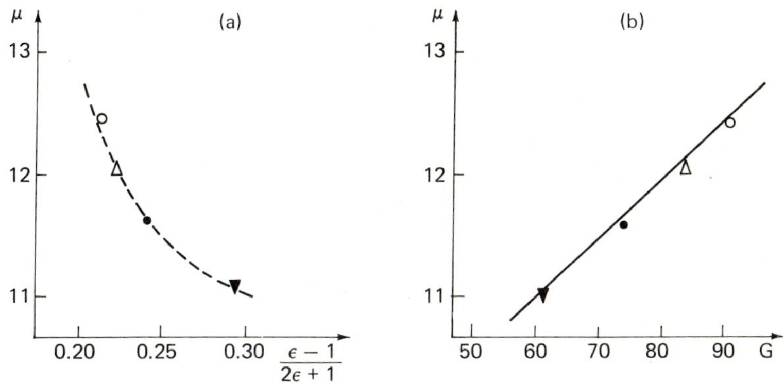

Fig. 20.8. μ related to $(\epsilon - 1)/(2\epsilon + 1)$ and to G-parameter for the 2,4-dinitrophenol-triethylamine complex.

exemplified by previous results which for a series of complexes are shown in fig. 20.9.

Thus generalization of the results of Jadżyn and Małecki [1972] to all complexes and solvents is not possible. It seems therefore that further investigations in this field are necessary.

20.2.4. Intramolecular hydrogen bonds

The number of observations, so far, on charge distribution in intramolecular H-bonds is rather limited. The results obtained by Eda and Ito [1956, 1957] and Koll et al. [1970] indicate that formation of H-bonds originates in a chelate ring which is interacting with the remainder of the molecule. The configuration of the interacting moment vector in ortho-nitrophenol is shown in fig. 20.10 (scheme a). This configuration cannot be explained either with pure classical inductive effect or the CT effect along the H-bond. It is apparent in this case that there is a coupled link of π electrons and the whole chelate ring acts as an electron acceptor.

The configuration of the interacting moment vector in ortho-chlorophenol provides another example. In this case the direction is "normal", which means that the perturbation exists mainly within the bridge (fig. 20.10, scheme b). Also in those cases where the proton–acceptor atom of the intramolecular bridge is isolated from the π electron system by a CH_2 group, no anomalous effects due to changes in charge distribution are observed (Rustanova et al. [1970]). However, it should be added that conclusions on charge distribution in intramolecular

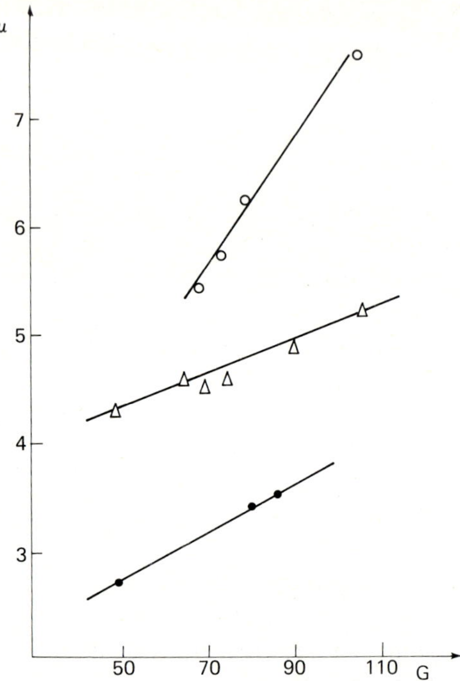

Fig. 20.9. Dependence of dipole moment of hydrogen bonded complexes on G-parameter: ○–dichloroacetic acid with pyridine, △–chloroacetic acid with pyridine, ●–acetic acid with triethylamine.

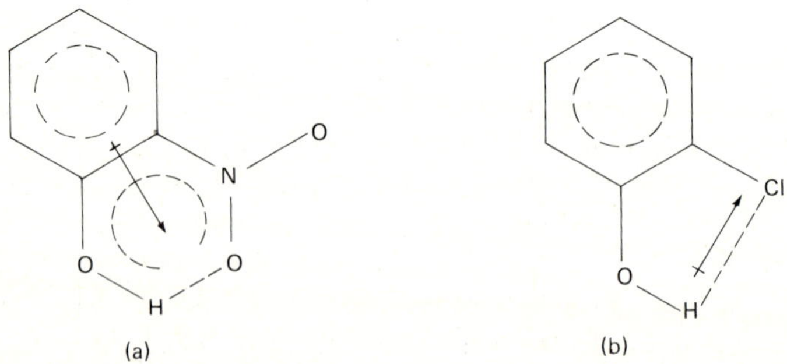

Fig. 20.10. Sense of interacting moment vector for *o*-nitrophenol (a) and *o*-chlorophenol (b).

H-bonds are not without error because it is impossible to take into account all indirect parameters influencing the change of the dipole moment.

20.2.5. Composition of the formed complexes and their conformation

Investigation of the dielectric properties of H-bonded systems should provide some information about the composition and structure of the formed complexes. For this purpose Job's method of continuous variations as well as the method of molar ratios (Sheka [1956a, b], Sheka and Kriss [1956], Guryanova and Goldstein [1962], Ratajczak [1966], Schaarschmidt [1967], Schaarschmidt et al. [1967], Gough and Price [1968] and Goldstein et al. [1972]) can be used. In the first case, changes of ϵ or dielectric polarization of solution mixtures of the interacting components at a constant total concentration are measured, and in the second, the concentration of one component is constant whereas the concentration of the second is changed. In fig. 20.11, typical examples of titration curves (illustrating the usefulness of both methods) are given. Dielectric titration can also be used with good results when the forming complexes differ from each other in relation to the interacting components in polarity. So, for example carboxylic acid-amine complexes with a 1:1 composition as well as with 2:1 can be investigated. The ones with a 2:1 composition are much more polar than the 1:1 complexes and their polarization is much higher than the total polarization of the interacting components. As is evident from dielectric investigations and confirmed by spectral methods, association of an additional acid molecule to the RCOOH-amine complex assists the transition of the proton to the formation of an ion pair (fig. 20.12). Recently, Huyskens [1972] discussed possibilities and limitations of the dielectric titration method for these purposes. He also discussed in detail the influence of H-bond formation on the properties of donor and acceptor groups present in the molecule and also the possibility of formation of composed molecular systems (Clerbaux et al. [1967] and Duterme et al. [1968]).

In general complexation equilibria of primary and secondary amines with carboxylic acids, phenols and alcohols are very complicated because of the multifunctional character of the components when considering their proton-donor–acceptor properties. Only dielectric measurements in connection with other physico-chemical methods, particularly with cryoscopic and calorimetric measurements, afford valuable results (Goldstein et al. [1972a, b], Huyskens [1972]).

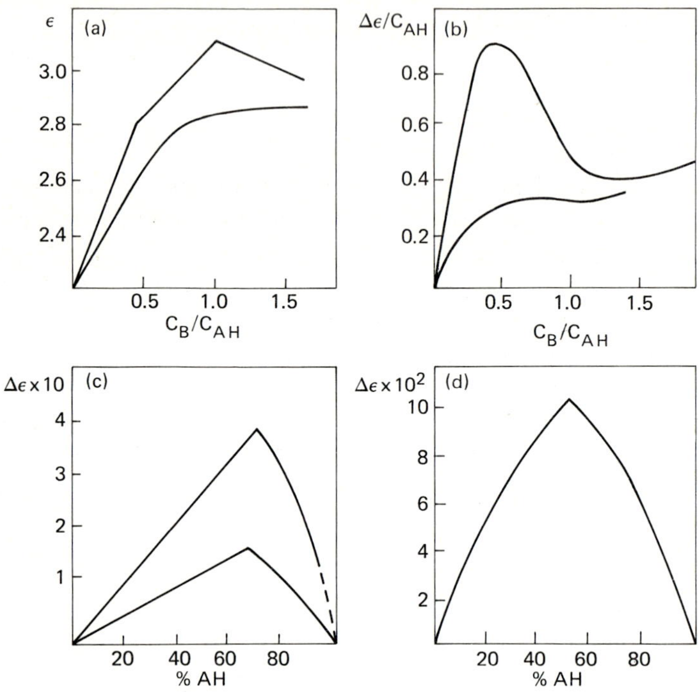

Fig. 20.11. Characteristic dielectric titration curves in benzene: a–trichloroacetic acid–tri-*n*-propylamine and chloroacetic acid–triethylamine (Guryanova and Goldstein [1962]); b–benzoic acid–diisopropylamine and benzoic acid–piperidine (Guryanova [1964]); c–*p*-bromophenol–triethylamine for two various concentrations, and d–*p*-nitrophenol–pyridine (Ratajczak [1966]).

Another aspect of dielectric investigations is the spatial structure of the complexes formed. Dipole moment measurements in these cases can have the same value as in the conformation analysis of simple compounds. The task however is much more difficult and conclusions are less reliable. The

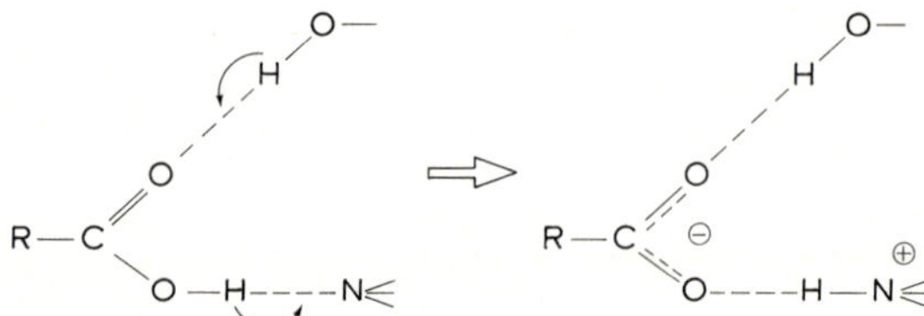

Fig. 20.12. Stimulated proton transfer resulting from attaching of a second acid molecule with a complex.

formation of a H-bond always leads to a change of the dipole moment components, which in general cannot be predicted. However as it was demonstrated for weak H-bonds this change is approximately constant and does not exceed 0.5 D.

Papers worthy of notice are the investigations of Lumbroso [1964], Debecker and Huyskens [1971] and Huyskens and Hernandez [1973]. Lumbroso analyzed the various conformation possibilities of complexation by the pyrrole molecule, in particular the alternatives of bond formation of the n or the π type. Another structural problem of the H-bond are the two possibilities of complexation by oxygen as proton–acceptor, because this atom has two lone electron pairs (Hadži et al. [1967]). A sufficiently meaningful discussion is possible whether the acetonitril molecule is forming the H-bond via the nitrogen atom or with the π electron system (Hawranek et al. [1973]) (see fig. 20.13).

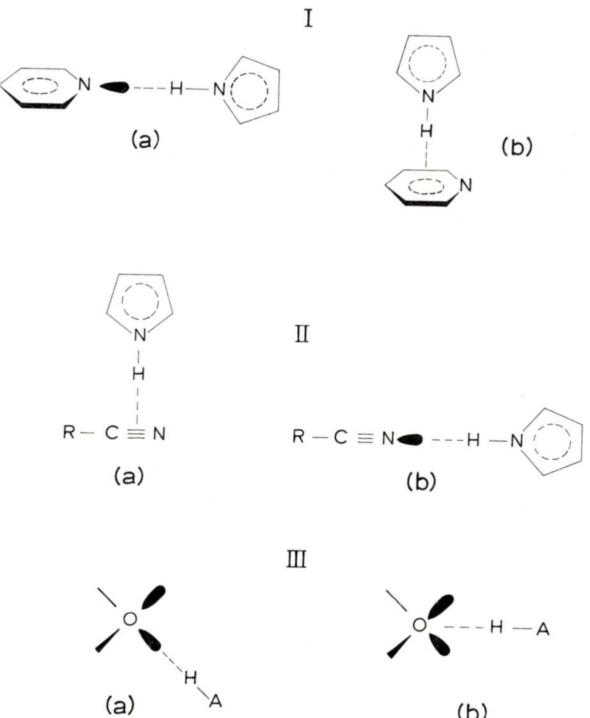

Fig. 20.13. Possible configurations for the pyrrole–pyridine complex (I), pyrrole–nitriles complexes (II), and an A–H acid complex with an oxygen base (III). From dielectric investigations it follows that the a configurations are more stable.

Aniline-phenols systems were used by Debecker and Huyskens [1971] to analyze for the first time the conformation of complexes with H-bonds. These authors measured the dipole moments of a number of complexes and calculated the polarity of the H-bond $\Delta\mu$ for various conformations. The most probable ones are shown in fig. 20.14. Because all investigated complexes are rather weak ones, it could be expected that $\Delta\mu$ will be approximately constant for all of them. Therefore minimalization of value

$$\sigma^2 = \frac{\Sigma(\Delta\mu - \overline{\Delta\mu})^2}{n-1} \tag{20.19}$$

on the basis of the choice of the most probable conformation was assumed. $\Delta\mu$ is the average polarity of H-bond for the given series of complexes and n is the number of investigated complexes. Analysis revealed that state I of fig. 20.14 is the most probable. The H-bond is therefore colinear with the OH bond and the lone electron pair made possible by the free rotation around this bond. However, not all angles of the mutual ring location are obtainable to the same extent. For calculations, a Boltzmann distribution of orientation was assumed, accepting the energy of dipole interaction as the basis.

It appears that for a more exact description of complex structure in solution the repulsion potential as well as the fluctuation of deviations from linearity of the H-bond have to be considered. This problem however, is very difficult and so far no attempts have been made to solve it quantitatively.

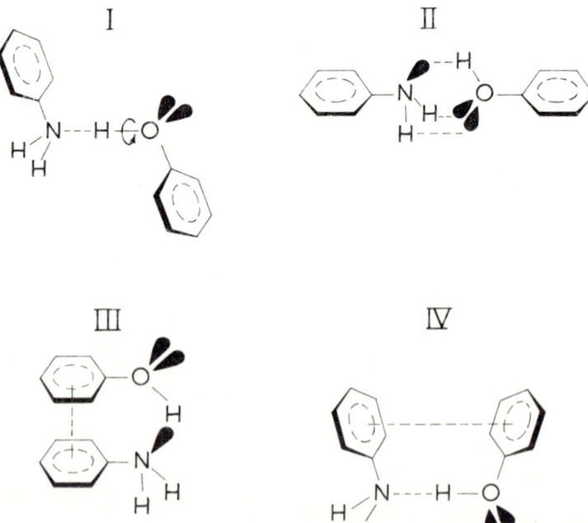

Fig. 20.14. Most probable arrangements of the interacting molecules in phenol-aniline complexes, according to Debecker and Huyskens [1971].

20.2.6. Hydrogen bonded systems in a strong electric field

The amount of information on systems with H-bonds can be considerably extended if permittivity measurements are carried out in strong electric fields (Małecki [1969] and Kielich [1972]). Nonlinear effects (usually called saturation effects–DSE) are observed due to permittivity changes under a field. In the applied range of high electric fields (about 50 kV/cm) the nonlinearity can be limited to the square term $\Delta\epsilon/E^2$. According to Debye's theory this value has to be negative. The anomalous behavior results from various rearrangement (reconstruction) phenomena of ordered dipole clusters under the influence of an electric field. Therefore investigations on nonlinear dielectric effects acquire considerable importance in the case of systems with H-bonds. As it was demonstrated by Małecki [1960, 1962, 1966] extension of dipole polarization measurements of alcohols by investigations of the nonlinear effect, provides completely new information on complicated associated systems. For the description of dielectrics in a strong field, the dipole polarization P^{dip} and the molar saturation constant (introduced by Piekara [1950]) which is usually defined as

$$S^{\mathrm{m}} = \frac{\partial P^{\mathrm{dip}}}{\partial (F^2)} \tag{20.20}$$

where F is the component of the internal field parallel to the external field E, are used. Considering Onsager's model as a base,

$$F = \frac{\epsilon(n^2+2)}{2\epsilon + n^2} E.$$

Frequently the correlation factors of these values, $R^{(\mathrm{P})}$ and $R^{(\mathrm{S})}$, are also calculated.

$$R^{(\mathrm{P})} = \frac{P^{\mathrm{dip}}}{P^{\mathrm{dip}}_{\mathrm{gas}}} = \frac{9kTP^{\mathrm{dip}}}{4\pi N\mu^2} = \frac{\bar{\mu}^2}{\mu^2} \tag{20.21}$$

$$R^{(\mathrm{S})} = \frac{S^{\mathrm{m}}}{S^{\mathrm{m}}_{\mathrm{gas}}} = -\frac{45k^3T^3S^{\mathrm{m}}}{4\pi N\mu^4}. \tag{20.22}$$

For a fairly general case of an associating solute A in a nonassociating solvent S, various groups of composition $S_h A_i$ are formed and according to Małecki

$$R^{(\mathrm{P})} = \sum_{h,i} x_{h,i}\gamma^2_{h,i}R^{(\mathrm{P})}_{h,i} \tag{20.23}$$

$$R^{(\mathrm{S})} = \sum_{h,i} x_{h,i}\gamma^2_{h,i}R^{(\mathrm{S})}_{h,i} \tag{20.24}$$

where $R^{(P)}_{h,i}$ and $R^{(S)}_{h,i}$ are correlation factors of polarization and saturation for the respective molecular species. The coefficients $\gamma_{h,i}$ are defined by the permittivity and by the refraction index of the solution.

$$\gamma_{h,i} = \frac{(n^2_{h,i} + 2)(2\epsilon + n^2_2)}{(n^2_2 + 2)(2\epsilon + n^2_{h,i})} \tag{20.25}$$

whereas $x_{h,i}$ describes the concentration of the (h,i)th complex

$$x_{h,i} = \mathrm{i}\, N_{h,i} / \omega_2 \tag{20.26}$$

where $N_{h,i}$ is the real number of complex molecules in solution and ω_2 the total number of the associating molecules introduced to the solution.

As it was demonstrated by Małecki [1965] and Piekara [1962] in analogy to 1,2-dichloroethane (Piekara et al. [1957]) a reasonable interpretation of experimental facts can be obtained if the assumption is made that the molecular complex can exist in two states differing in energy $\Delta U_{h,i}$ and in dipole moments $\mu^{(1)}_{h,i}$ and $\mu^{(2)}_{h,i}$. A detailed quantitative analysis together with numerical methods for calculations of the main parameters for the respective complexes was made by Małecki [1969] and Małecki and Dopierala [1969a, b]. Though for the quantitative description, the assumption of any molecular model describing the two states in the complex is not incomprehensible as after all, it appears likely that they could correspond to a proton transfer equilibrium. Such a concept was presented for the first time by Piekara [1962].

A typical S^{m}_2 curve for an alcohol in an inert solvent with dependence on concentration is shown in fig. 20.15. A positive "anomalous" effect in the range of lower concentrations is observed as a rule. For pure alcohols this effect is most often negative.

Based on the above mentioned conditions, dipole moments $\mu^{(1)}$ and $\mu^{(2)}$ were determined for complexes of hexanol-1 in dioxane and triethylamine (Małecki [1965]). Surprisingly good agreement between the determined $\mu^{(2)}$ values and the values predicted by Piekara [1962] based on the proton transfer model is obtained. Małecki has determined the fractions of the respective associates over the whole concentration range independently of the dipole moment values. A similar analysis of the structure of solutions was made for other systems (Małecki and Dopierala [1969a, b] and Małecki [1972]).

In fig. 20.16, the dependence of the correlation factor of polarization and saturation for dodecanol-1 in cyclohexane which was determined by Krupkowski and compared with curves calculated by Małecki [1972] is

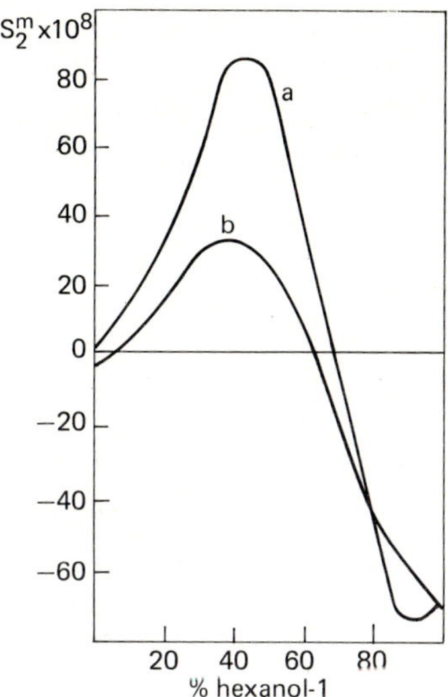

Fig. 20.15. Curves of molar saturation constant S_2^m for hexanol-1 in hexane (a) and benzene (b) in relation to concentration.

shown. Changes of fractions of the appropriate associates, with dependence on alcohol concentration, are given in fig. 20.17.

Another approach to the problem based on the domain theory of associated liquids, taking into account significant structures (Eyring and

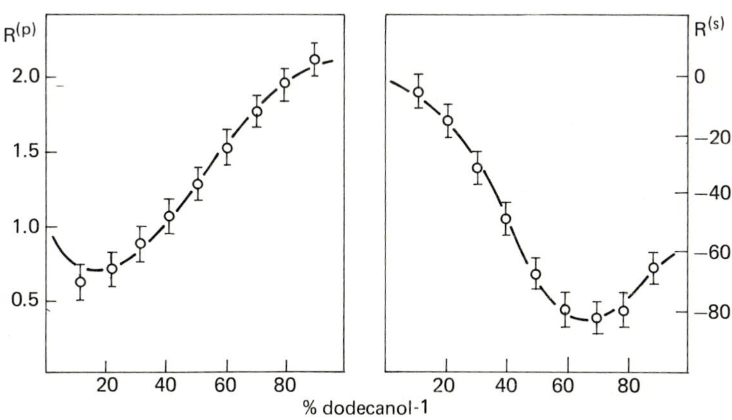

Fig. 20.16. Correlation factors of polarization and saturation for dodecanol-1 in cyclohexane related to concentration. Experimental points as compared with calculated curves (Małecki [1972]).

TABLE 20.3

Dipole moments (in D) of hexanol-1 multimers and hexanol-1 complexes with 1,4-dioxane and triethylamine in normal ($\mu^{(1)}$) and ionized state ($\mu^{(2)}$) (Małecki [1965])

Type of multimer or complex	Symbol	$\mu^{(1)}$	$\mu^{(2)}$	$\mu^{(2)}$ calculated by Piekara [1962]
Monomer	A_1	1.6 ± 0.1	—	—
Dimer	A_2	1.7 ± 0.1	13.0 ± 1.0	13.0
Trimer	A_3	3.0 ± 0.1	8.5 ± 1.0	9.0
Tetramer	A_4	4.9 ± 0.3	7.4 ± 1.0	7.5
Pentamer	A_5	6.2 ± 0.6	8.2 ± 2.0	6.0
Monomer with dioxane	AD	2.0 ± 0.1	4.5 ± 0.1	—
Dimer with dioxane	A_2D	2.7 ± 1.0	6.0 ± 2.0	—
Trimer with dioxane	A_3D	3.5 ± 0.2	7.0 ± 1.0	—
Monomer with triethylamine	AT	2.3 ± 0.1	4.8 ± 1.0	—
Dimer with triethylamine	A_2T	3.0 ± 1.0	6.0 ± 2.0	—
Trimer with triethylamine	A_3T	4.0 ± 0.3	6.5 ± 1.0	—

Ihon [1969]) and a more general Kielich theory of nonlinear effect, was recently presented by Danielewicz-Ferchmin [1972]. From the equation, which was deduced from the Hobbs et al. [1966] formula for the dielectric permittivity of liquids with H-bond, the field effect was calculated and is in good agreement with experiment. The formula of field effect, expressed by $\Delta\epsilon^{\text{dip}}$ has the following form

$$\Delta\epsilon^{\text{dip}} = -\frac{4\pi N}{V}\rho(n,\epsilon)\left(\frac{\Lambda}{V}\Xi + \frac{V-\Lambda}{V}K\right)E^2 \tag{20.27}$$

where V and Λ are the molar volumes of liquid and of solid,

$$\Xi = \mu^2 p^3 \cos^4\vartheta / k^3 T^3, \qquad K = \mu/15k^3T^3$$

where ϑ is the average angle between dipole axis and the direction of the

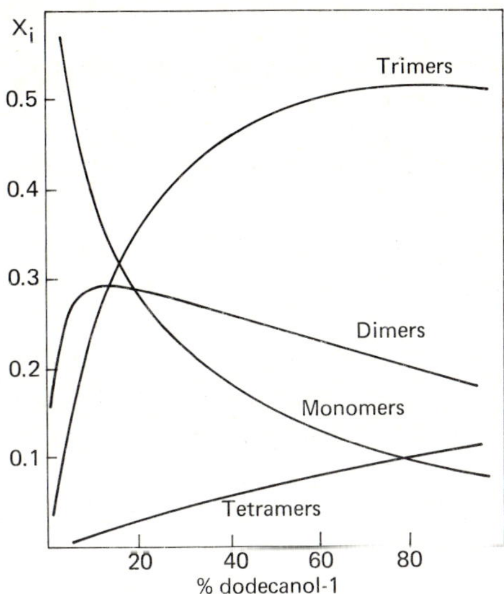

Fig. 20.17. Calculated curves of fractions of the respective multimers in solution related to composition (Małecki [1972]).

maximal polarization of the domain, p the number of orienting molecules and $\rho(n, \epsilon)$ is the function of the light refraction index and permittivity resulting from Onsager's model.

It seems that the proposed theory is worthy of note because no assumption of specified associates is made and it applies to general theories of associated liquids. This problem requires further experimental as well as theoretical investigations.

20.3. Dielectric polarization and electrical conduction in hydrogen bonded solids (by H. Engelhardt)

Faraday [1850] was probably the first to observe properties of a solid H-bonded system which are relevant for its dielectric and electrical behavior, namely, the peculiar properties of the ice surface. The long discussion about the views of Faraday and those of Lord Kelvin (Thomson [1850]) that followed suit is one example of a persistent controversy that has lasted to this day. Kohlrausch and Heydweiler [1894] gave their outstanding example of tenacious scientific effort by establishing the limiting conductivity of water experimentally to $3.84 \times 10^{-8}\,\text{ohm}^{-1}\,\text{cm}^{-1}$ at 18°C, using 36 consecutive vacuum distillations extending over years. Recent developments have increased the efficiency of water purification considerably, so that the same effect can now be achieved in about 100 min (Haller and Duecker [1970]).

During the last 20 years numerous groups around the world emerged to investigate transport properties in ice which is regarded to be the model substance par excellence, and comparatively few workers commenced to study the electrical properties of other H-bonded solids, in the first place ferroelectric crystals as KH_2PO_4. The accumulated results were discussed at several conferences on water and ice, collected in books and reviewed in numerous articles (see introduction). Recently Bruinink [1972] and Glasser [1974] provided extensive surveys on proton conduction in solids.

The presently available experimental results point to a very suggestive correlation between the structures of the H-bonded systems and the dielectric parameters. There are examples of a three dimensional net of interlinked H-bonds as in all of the ice phases, clathrates and some hydrates; examples of the presence of spirals of H-bonds as in acetylene dicarboxylic acid dihydrate; examples of unbroken linear sequences of H_2O molecules as in lithium hydrazinium sulphate; or examples of isolated acid dimers held merely by van der Waals forces to their neighbors as in benzoic, furoic and oxalic acid. The H-bonded system may be disordered as found in ice I, III, IV, V, VI, VII and ordered as in ice II, VIII and IX. The disordered proton configuration has important consequences: zero point entropy and broad IR bands. Together with lattice faults it leads to high dielectric constant and protonic conduction in ice and other H-bonded solids.

Different mechanisms for polarization and conductance occur that involve protonation of H-bonded molecules and/or proton transfer in the H-bond. Several macroscopic properties may be brought about by closely interrelated microscopic mechanisms including cooperative processes where the motion of correlated protons is involved. The precondition for cooperation of protons sitting on two H-bonds of the same molecule is that the donator group (O–H) must be coupled intramolecularly with the acceptor group (O···). In H_2O the same oxygen atom acts as a donator and acceptor at the same time. In other systems these groups may be interlinked by the π electron system, and an intermolecular proton shift is coupled to an intramolecular shift of electron configuration. Many anorganic acids are systems of this kind.

Each H-bond (e.g. O···H–O) has an electrical moment μ_b. In an extended system of H-bonds oriented in one direction the moments add up to a macroscopic moment. Crystals with oriented macroscopic moments are common. In most cases, however, the symmetry of the crystals allows neighboring H-bond moments to be oriented in opposite directions and the unit cell remains non-polar. Ferroelectric crystals, like Rochelle salt, lack such a symmetry center and hence show within some temperature ranges macroscopic electric moments. On applying relatively low electric fields, this polarity can be changed. In ferroelectric Rochelle crystals this change in polarity is equivalent to a shift of charge across the whole crystal and a simultaneous switching of the bridge moments. This process can only occur once. Afterwards, the crystal polarized in field direction is an almost ideal dielectric insulating medium. The displacement current decays with its typical relaxation time. The direct current conductivity is zero, unless a second mechanism is available that reorients the H-bonded chain in such a way that another proton can pass. The situation can be visualized in the following scheme, starting from a favorably oriented chain:

$$\begin{array}{l}
\quad + \qquad\quad \mathrm{H} \qquad\qquad\qquad\quad \mathrm{H} \\
\mathrm{H{-}O{-}H\cdots O{-}H\cdots O{-}H\cdots O{-}H\cdots O{-}H.} \\
\quad \mathrm{O} \qquad\qquad\qquad \mathrm{H} \qquad\qquad\qquad \mathrm{H}
\end{array}$$

After passage of the ionic state the following situation is found:

$$\begin{array}{l}
\qquad\qquad\quad \mathrm{H} \qquad\qquad\qquad\quad \mathrm{H} \qquad\quad + \\
\mathrm{H{-}O\cdots H{-}O\cdots H{-}O\cdots H{-}O\cdots H{-}O{-}H.} \\
\quad \mathrm{H} \qquad\qquad\qquad \mathrm{H} \qquad\qquad\qquad \mathrm{H}
\end{array}$$

The passage of an L-defect

$$\begin{array}{l}
\qquad\qquad\quad \mathrm{H} \qquad\qquad\qquad\quad \mathrm{H} \\
\mathrm{H{-}O\cdots H{-}O\cdots H{-}O\cdots H{-}O \quad O{-}H} \\
\quad \mathrm{H} \qquad\qquad\qquad \mathrm{H} \qquad\qquad \mathrm{H}
\end{array}$$

reorients the chain

$$\begin{array}{l}
\qquad\quad \mathrm{H} \qquad\qquad\qquad \mathrm{H} \\
\mathrm{O{-}H\cdots O{-}H\cdots O{-}H\cdots O{-}H\cdots O{-}H} \\
\mathrm{H} \qquad\qquad\qquad \mathrm{H} \qquad\qquad\qquad \mathrm{H} \quad .
\end{array}$$

The orientational L-defect is complementary to the ionic H_3O^+ state, likewise a D-defect is complementary to an OH^- state. Each mechanism can contribute to dielectric polarization separately. It was recognized by Bjerrum that antipolarization results, if the H_3O^+ state is discharged at the end of its path. Stationary conduction, however, can only occur if both mechanisms play their role alternatively. Macroscopic phenomena, dielectric polarization and direct current conduction can be measured separately, but they must be discussed from the point of view of interlinked molecular processes.

20.3.1. Ice

The concept with four defects (H_3O^+, OH^-, L and D) was originally proposed for ice by Bjerrum [1951]. It was theoretically elaborated by Jaccard [1959] and reviewed by Gränicher [1963], Fletcher [1970] and Jaccard [1972]. Bjerrum L- and D-defects carry an uncompensated local charge and therefore they can compensate, at least incompletely, ionic charges. Onsager and Dupuis [1962] developed a detailed kinetic theory for the propagation of Bjerrum faults. They included the interaction of Bjerrum faults and the recombination kinetics between Bjerrum faults and ionic defects and concluded from a comparison with the Wien effect studies of Eigen and De Maeyer [1956] that most ions in ice may be regarded as associated with single Bjerrum faults.

Most of the conclusions drawn from comparison of theoretical model work and experimental results were nullified in recent years by the extensive work of von Hippel's group and Riehl's group. The experimental situation turned out to be quite complicated as was recently reviewed by Engelhardt [1973].

Dielectric relaxation measurements were widely carried out and reasonable agreement could be obtained by various authors. The considerable efforts in investigating the dc-conduction led to most unsatisfactory discrepancies e.g. the activation energies for the conductivity ranged from 0 to at least 33 kcal $mole^{-1}$ for which a detailed account is given by von Hippel [1971]. A number of reasons for this continuing disagreement have been claimed:

1. Surface phenomena may have obscured bulk properties.
2. The failure to make electrodes that could readily inject or absorb protons; so called ohmic contacts.
3. Aging effects of poorly annealed fresh ice specimens.
4. Inadequate definition of state including impurity and homogeneity.

Improvements of the experimental technique during the last few years seem to clarify the situation at last and a more consistent interpretation is approached. New techniques have added valuable information.

20.3.1.1. *Dielectric dispersion*

The investigation of dielectric properties of H-bonded solids is clearly dominated by ice, notably as single crystals of pure and doped ice Ih, but also studies of most of its other ten phases have been undertaken. The dielectric parameters of ice belong to the best documented quantities. Wörz and Cole [1969] reinvestigated the early work of Auty and Cole [1952] and found in the temperature range 0 to $-40\,°C$ a relaxation time which was closely described by a rate equation

$$\tau = \tau_0 \exp(-E/RT)$$

with activation energy $E = 13.2\ \text{kcal mole}^{-1}$ and a frequency factor $\tau_0 = 5.3 \times 10^{-16}$ s. The temperature dependence of the static dielectric constant ϵ_0 was given by the equation

$$\epsilon_0 - \epsilon_\infty = 20715/(T - 38).$$

Below $-50\,°C$ τ and ϵ_0 deviate from these equations. A new method of improving the dielectric purity of hexagonal ice was applied by Gough and Davidson [1970]. Instead of using zone refined ice as Wörz did, these authors purified their samples frozen from relatively impure water by conversion of ice I to ice II and back to ice I. Their results are comparable to those obtained directly from carefully purified single crystals of ice.

The fact that some faulty ideas established by inadequate experimental procedures and obvious wishful thinking were still perpetuated in ice reviews, caused von Hippel [1971] and von Hippel et al. [1971, 1972, 1973] and his MIT group to carry out a painstaking series of investigations. They decomposed the polarization spectra of ice in the frequency range from 8×10^{-3} to 10^5 Hz into seven discrete components which were assigned to intrinsic dipolar relaxation, space charge effects and imperfections (fig. 20.18). The activation energy for the intrinsic dipolar relaxation is $13.84\ \text{kcal mole}^{-1}$ which is equivalent to the energy of three H-bonds. The relaxation time $\tau_0 = 1.02 \times 10^{-15}$ s increases on replacement of H_2O by D_2O. Ruepp [1973] decomposes his spectra into three components. The Debye component at $T > -50\,°C$ is reproducible and agrees with von Hippel's results. His high frequency component is understood as a mean relaxation time with activation energies between 4–6 kcal mole^{-1}. Its experimental relaxation strength of 0.1–1.0 is 10^{-2} times smaller than the relaxation strength of the Debye type component. Normally present impurities of a concentration of less than $10^{-6}\ \text{mole l}^{-1}$ are very unlikely to

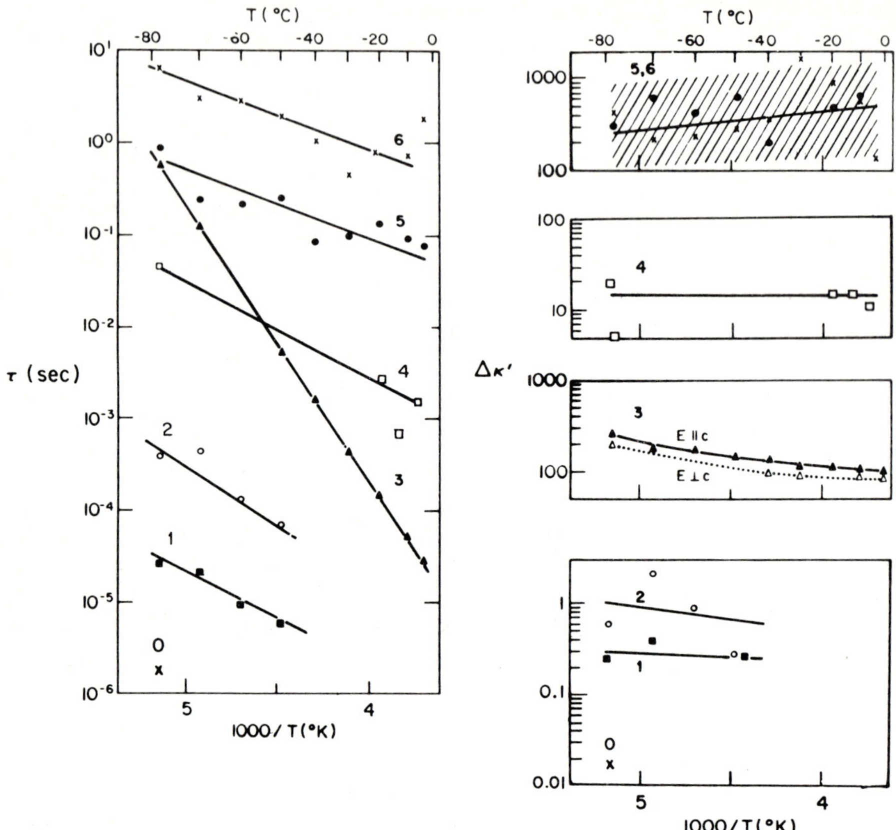

Fig. 20.18. Temperature dependence of discrete spectrum of relaxation times and dielectric strength of ice (von Hippel et al. [1972]).

account for a localized polarization effect of this size. Instead, Ruepp suggests the microstructure of ice that was found by Truby [1955] to be responsible for the high frequency component. Nakamura and Jones [1973] observed high dislocation densities of at least $10^5\,cm^{-2}$ within subgrains. These dislocation densities should be too small to be detected by dielectric methods. However, Noll [1973] using crystals grown by the Bridgman method and Higashi and Teranoto [1972] using Mendenhall Glacier ice with low dislocation densities of $10^4\,cm^{-2}$ as reported by Fukuda and Higashi [1969] found that the low frequency dispersion was influenced by plastic deformation. This dispersion is believed to be caused by ionic defects. Excess charge carriers H_3O^+ in the deformed crystal are assumed to be generated by the following reaction: $2H_2O +$ dislocation jog $\rightarrow H_3O^+ + OH^-$ on jog. The influence of charge dislocations on the dielectric properties of ice was also investigated by Itagaki [1972].

The mechanism of Debye relaxation requires the formation of a small amount of defects in the ice, either rotational (L, D) and/or ionic defects. The distinction between the two types of defects can be done most convincingly by the effect of pressure on the relaxation time (Chan et al. [1965])

$$(\partial \ln \tau/\partial p)_T = \Delta V^{\dagger\dagger}/RT$$

where $\Delta V^{\dagger\dagger}$ is the volume of activation.

The formation of D- and L-defects almost certainly occurs with an increase of volume, because two protons or two lone pairs on a O–O bridge tend to expand the volume even if they push one another from the O–O line. The experimental volume of activation was found by Chan et al. [1965] as 2.9 ± 0.3 cm^3 mole^{-1} at -23.4°C for polycrystalline ice and by Taubenberger et al. [1973] as 3.1 ± 0.2 cm^3 mole^{-1} for single crystals. This provides dielectric conformation for the rotational defects to be responsible for dielectric Debye relaxation in ice. Dielectric relaxation that requires ionic defects would tend to predict a negative volume of activation because the volume change for the self-ionization of water is certainly strongly negative.

20.3.1.2. *Molecular dipole moment*

The dipole moment μ of the solid is determined by measurement of the static dielectric constant ϵ_0 and the approximate formula for the Bernal–Fowler model of ice neglecting contributions of closed rings of dipoles:

$$\epsilon_0 - \epsilon_\infty = \tfrac{4}{3}(N/V)g(\mu^2/RT)$$

where ϵ_0 and ϵ_∞ are the limiting low and high frequency dielectric constants, N = the number of molecules in the volume, V. Improved calculations of the Kirkwood correlation factor g were recently carried out by Nagle [1973] who found for ice Ih parallel to c-axis, $g = 3.0050$ and perpendicular to c-axis, $g = 3.0054$. Cubic ice Ic was also treated by Gobush and Hoeve [1972], who estimated a g value of 2.9958. Using the measurements for ϵ_0 of Wörz and Cole [1969] gives $\mu = 2.73$ debye, which suggests a large enhancement of the dipole moment μ compared to the dipole moment of the water molecule in vapor phase $\mu_0 = 1.83$ debye. The assumption about the dendritic structure, neglecting the formation of closed rings, appears to overestimate the correlation factor g, as shown by an extensive computer analysis by Rahman and Stillinger [1972] who arrive at a value of $g = 2.07$ and 2.11 for ice Ih and ice Ic respectively. Johari and Whalley [1973] employed the general form of the Kirkwood–Fröhlich relation

$$\epsilon_0 - \epsilon_\infty = \frac{4\pi}{3}\frac{N}{V}\left(\frac{3\epsilon_0}{2\epsilon_0 - n^2}\right)\left(\frac{n^2-2}{3}\right)^2 g\frac{\mu_0^2}{RT}$$

to derive the correlation factor g for the high pressure phases of ice. For ice Ih, g is temperature independent and has a value of 3.38 ± 0.08. Ice III and V have values 2.65 (Whalley et al. [1968]) and 2.46 ± 0.02 respectively. Although ice V appears to be partly ordered at low temperature (Hamilton et al. [1969]), it prefers to transform to completely ordered ice II below 227 K and even as low as 133 K as indicated by a decrease in capacitance with time. The dispersion curves of Johari for ice VI are similar to those reported by Wilson et al. [1965]. There is a distribution of relaxation times due to the nonequivalence of water molecules and of their reorientational motion. The static dielectric constant is 145 which is lower than the earlier value of Whalley et al. [1966]. The correlation parameter g increases from 2.96 at 243 K to 3.78 at 133 K. This increase in g with decreasing temperature implies that partial ordering of ice VI is approached. Also Whalley [1969] reported a slow reversible transition of ice VI near 123 K as detected by a change in the high frequency dielectric constant.

Ice VII is particularly interesting because it consists of two interpenetrating cubic ice Ic lattices that are not H-bonded to each other. The X-ray determined density at 25 kbar and $-50\,°C$ is $1.66\ g\ cm^{-3}$, appreciably less than twice that of ice Ic due to longer O–H···O distances of 2.86 Å as compared with 2.76 Å of ice Ic (Kamb [1964]). Johari and Whalley [1973] measured a static dielectric constant of 103, which is close to that of ice I. The g value increases from 1.66 at $-5\,°C$ to 1.82 at $70\,°C$ and is considerably smaller than the value of ice I. There seems to be a substantial correlation of molecules in an antiparallel manner. When ice VII is cooled below $0\,°C$ it transforms to completely ordered ice VIII and a sharp change in the dielectric properties is observed (Whalley et al. [1966]). The two substructures with completely parallel molecular dipoles are oriented antiparallel to one another resulting in a g value which is zero (Kamb and Prakash [1970]).

In the denser high pressure phases the ideal tetrahedral coordination as found in ice Ih, is distorted, allowing the non-nearest neighbors to come closer to one another. This process comes to an end with body-centered cubic ice VII showing an 8-coordinated structure. The repulsion of the close non H-bonded neighbors and the bending of H-bonds leads to an increase of bond length from 2.76 Å in ice Ih to the longest bonds of 2.95 Å in ice VII at 77 K and atmospheric pressure. Since no correlation occurs in ice Ic and since it would be smaller in ice VIII due to longer bonds, the forces causing correlation must therefore originate in the forces between the two sublattices of ice VIII (Kamb [1965]).

The ordering of ice III when cooled through the temperature range from −75 °C to −108 °C was detected by dielectric methods as the limiting low frequency permittivity ϵ_0 gradually changed from 140 to 3.74 and the correlation parameter g from 2.6 to 0. Whalley et al. [1968] designated this new phase, ice IX.

Ice II is the only phase that does not become disordered at high temperatures as shown again by dielectric studies of Wilson et al. [1965]. The dielectric constant of ice II remains at its IR value ϵ_∞ of 3.66 at −30 °C and 2.3 kbar, which is similar to the limiting high frequency values of the other high pressure phases (table 20.4).

The mechanism of dielectric relaxation in the high pressure phases is almost certainly similar to that in ice Ih. The closer packing and the distortion of H-bonds will affect the problem only quantitatively. The volume of activation of about 4.6 cm^3 $mole^{-1}$ for the disordered high pressure phases III, V, VI is in favor of rotational defects. The enthalpies of activation are appreciably smaller than in ice Ih. The dielectric measurements on the ices clearly demonstrate the sensitivity of dielectric

TABLE 20.4

Limiting dielectric constants of ice phases

Ice	P_0 (kbar)	T_0 (°C)	ρ (g cm^{-3})	ϵ_0	ϵ_∞	n^2	g	α (Å^3 $molecule^{-1}$)
I	0	−30	0.917	99	3.1	1.72	3.38	3.21
II	2.3	−30	1.21	3.66	3.66	2.0	0	2.78
III	2.3	−30	1.16	117	4.1	1.96	2.7	3.13
V	5.0	−30	1.26	144	4.6	2.08	2.46	3.10
VI	18.0	−30	1.35	155	5.1	2.19	2.96	3.06
VII	21.5	0	1.56	108	5.1	2.28	1.71	2.65
VIII	21.5	0	1.56	4.65	4.65	2.28	0	2.42
IX	2.3	−100	1.15	3.74	3.74	1.96	0	2.95

properties to structural parameters, particularly order–disorder transitions.

The rate of molecular displacements in ice differs greatly from that in water. The dielectric relaxation time in ice is 10^6 times as long as in water. The activation enthalpy in ice corresponds to the breakage of three H-bonds, but only to the breakage of one bond in water. The limiting high frequency value for the dielectric constant ϵ_∞ is 3.1 in ice, compared to 4.5 in water. The Debye dispersion in ice exhibits a single relaxation time but shows a minor spread in water with a dispersion parameter α of 0.02, less than the value for ice III ($\alpha = 0.04$), ice V (0.025) and ice VI (0.05). In the liquid the concept of Bjerrum L- and D-defects is not applicable, because in water it is much easier to remove the wrongly oriented molecule than in ice. Therefore, the interpretation of dielectric phenomena in ice and in water follows quite different lines. In water a second relaxation time is observed that brings ϵ_∞ from 4.5 down to 1.8, the optical value of n^2. The small enthalpy of activation of less than 2 kcal mole^{-1} indicates that no breakage of H-bonds takes place. Hasted [1961] suggests that this dispersion is caused essentially by the rotation of OH-groups which are abundant in superheated water, exhibiting an increased value of ϵ_∞. From the Raman spectra of Walrafen [1968] who assigns frequencies between 450 cm^{-1} and 780 cm^{-1} to librational modes, Davies [1970] concludes that in water an unusually strong and high frequency Poley absorption is present. In ice the difference between the microwave and visible refractive index, $\epsilon_\infty - n^2 = 0.47$, originates in the lattice vibrations notably the translational band which extends from zero wave number to about 325 cm^{-1}, the librational band centered around 850 cm^{-1} and the fundamental bending and stretching vibrations. Using the Kramer-Kronig relation

$$\Delta n = (2\pi)^{-2} \int k(\nu)\, \nu^{-2}\, \mathrm{d}\nu$$

Bertie et al. [1969] calculated the contributions of the various absorption regions below the visible frequencies to the refractive index. From a total $\epsilon_\infty - n^2 = 0.47$, the far IR region from 17–320 cm^{-1} corresponding to the translational vibrations contributes 74%. The theory of IR spectra of orientationally disordered crystals was developed by Whalley and Bertie [1967] and Whalley and Labbé [1969]. It was recently extended by Klug and Whalley [1973] to include the effect of short range orientational correlation imposed by the ice rules.

The sum of the electronic and IR polarizabilities α_{el} and α_{ir} can be obtained from the Clausius-Mossotti equation using the densities given by Kamb [1968]. As shown in table 20.4, the polarizabilities of the ordered phases II, VIII and IX are appreciably less than those of the disordered phases. Whalley et al. [1968] demonstrated this effect through the transformation III–IX. It is clearly related to the striking differences of the far-IR spectra of disordered and ordered ice crystals. The measurement of the high frequency dielectric constants can conveniently be used to follow order–disorder transformations, particularly in cases where the dielectric relaxation is too long and volume changes are too small to be readily detected.

20.3.1.3. *Conduction phenomena*

Ice has been regarded as the most prominent model substance for the study of intrinsic protonic conduction. In the light of recent measurements this view needs serious corrections. It has already been mentioned that the activation energy of the so called protonic conduction in ice exhibited an unrealistically wide spread with falling tendency as experimental techniques improved. Camp et al. [1969] report on dc-conductivity measurements on ice with zero activation energy and conductivities between 2×10^{-11} and 10^{-8} ohm^{-1} cm^{-1} in the temperature range from -10 to -50°C. The lowest values were obtained from samples grown under conditions such that the water was continuously circulated through a demineralizer. Similarly, a constant conductivity $\sigma_{=}$ from the melting point down to -20°C was obtained by Maidique et al. [1971] and Schneider et al. [1972], when initial currents for H_2O and D_2O ice were plotted as a function of temperature (fig. 20.19a, b). Recently, Ruepp [1973] and Taubenberger [1973] observed aging effects on the low frequency conductivity, its activation energy increasing from 2 kcal mole^{-1} for freshly prepared samples to 7 kcal mole^{-1} for samples which were stored for several months. By these investigations, which were carried out by use of guarded electrodes and short time techniques, it became clear that surface conductivity and space-charge polarization affected previous measurements.

It is now generally accepted that the ice surface exhibits peculiar properties that may obscure bulk properties if not carefully separated from them. The effect of surface on conductivity 20–30°C below the melting point can readily be shown when measurements on single ice crystals are compared with those on polycrystalline samples (fig. 20.20), on samples with cracks (Engelhardt [1964]) or on snow (Kopp [1962]).

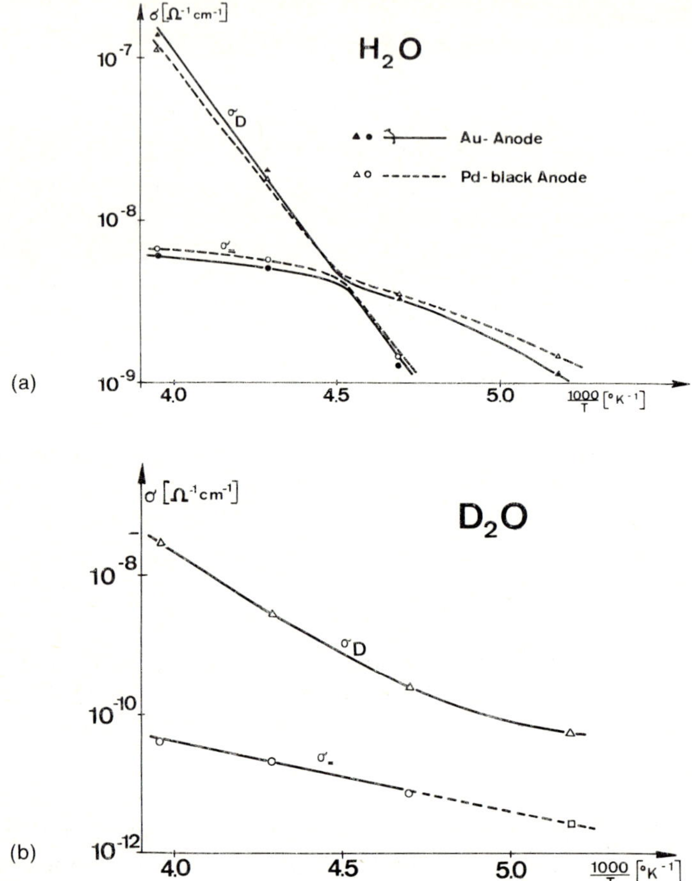

Fig. 20.19a, b. High frequency conductivity σ_D and low frequency conductivity $\sigma_=$ for H_2O and D_2O ice.

Thorough reviews of the ice interface were given by Jellinek [1972] and Fletcher [1973]. The assumption that a transition layer exists on ice explains the high apparent activation energy for the surface conduction of up to 33 kcal mole^{-1} which is due to an enhanced concentration of ions in the surface and a strong change in layer thickness with temperature.

A serious complication arises at the electrodes where discharge is impeded and a pile-up of space-charge in front of the cathode takes place (fig. 20.21). This leads to a time dependent distortion of the applied macroscopic field within the crystal and a genuine steady state dc-

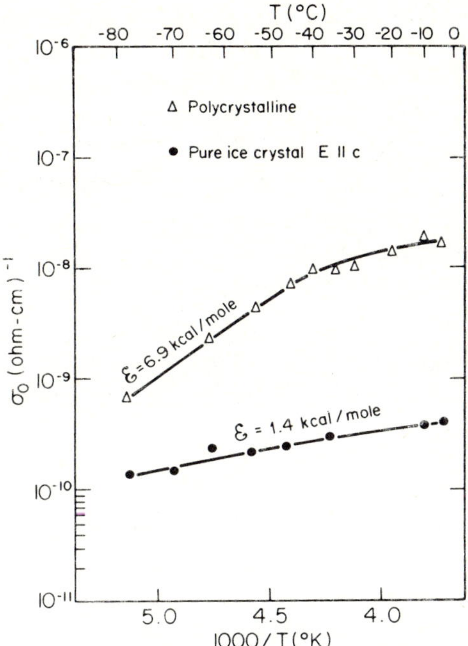

Fig. 20.20. Comparison of polycrystalline and monocrystalline samples of ice. From von Hippel et al. [1973].

conductivity cannot be measured. If the time resolution is increased to the millisecond range, the initial current may be regarded to flow in a field that is still linear and correlated in a simple way to the externally applied voltage. As space charge build-up proceeds, more complicated

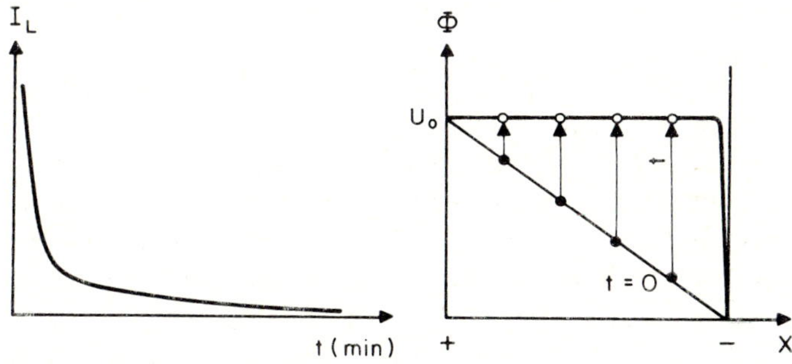

Fig. 20.21. Time dependence of current in pure ice Ih single crystals leading to the development of cathode fall.

current–voltage relations have to be applied Chang and Jaffe [1952] and MacDonald [1953]).

von Hippel et al. [1973] measured the charge carrier transfer through single ice crystals as a function of field strength, sample thickness and time. Figure 20.22 shows a shoulder at 10^{-3} s and a plateau which shift with increasing field strength towards shorter times. The plateau persists only for a limited time, which implies that an exhaustion of charge carrier supply takes place. As the applied voltage increases the plateau current rises superlinearly, presumably because additional charges are released from traps and the impeded discharge at the cathode is reduced by field emission (fig. 20.23). These results clearly show that a true saturation current which is controlled by limited dissociation of intrinsic charge carriers, as claimed by Eigen et al. [1964], does not exist and many theoretical conclusions drawn from it have to be revised. A detailed survey of this important point has recently been presented by von Hippel et al. [1973]. Even the application of water electrodes, which ideally should couple to ice, did not change the general picture. The current

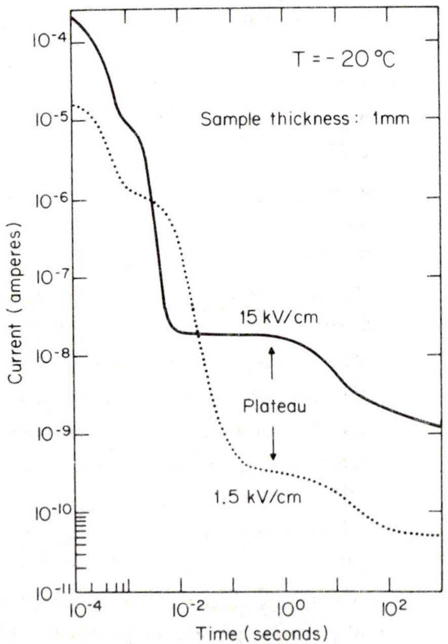

Fig. 20.22. Time dependence of current following step application of voltage. From von Hippel et al. [1973].

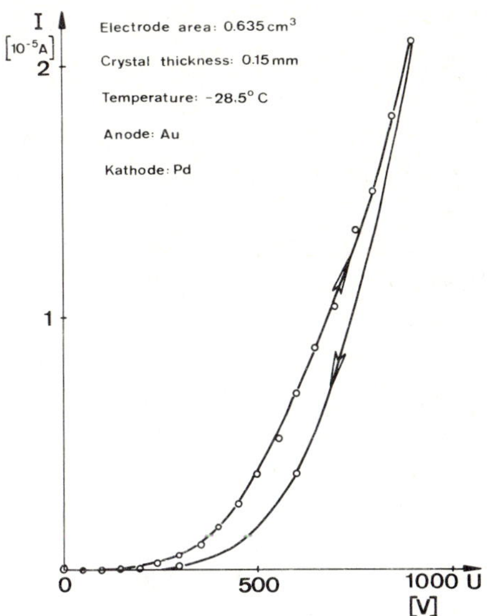

Fig. 20.23. Superlinear current–voltage characteristic for extrinsic charge carriers in ice at high fields.

across single crystals decreased with time indefinitely, whereas, multicrystalline samples showed a much slower decay of current which finally approached a limiting value (fig. 20.24). The latter can be explained by the existence of the transition layer at the ice–ice interfaces that couples more perfectly to the electrodes than to the bulk ice.

The low apparent activation energy (fig. 20.19a, b) for the transient current conductivity of perfect and pure ice Ih single crystals demands an extrinsic origin for the charge carriers. The formation of intrinsic ion pairs would require at least as much energy as in water which has a dissociation energy $E_D = 19.4\ \mathrm{kcal\ mole^{-1}}$ at 25 °C, and probably even more because the immediate shielding of newly generated ions through hydration is strongly suppressed by the requirements of the rigid ice lattice. The effective dielectric constant for the ionic dissociation process may be relatively small and close to the high frequency value of 3.2 and therefore, the contribution of the electrostatic energy to the free energy of dissociation may be appreciable. At lattice faults, e.g. vacancies or impurity sites, however, the charge carrier formation becomes energetically much less expensive.

Lattice faults and impurities of different kinds, the dislocation density

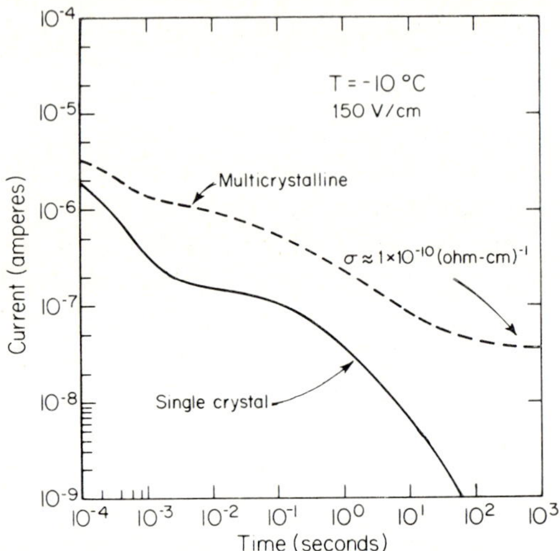

Fig. 20.24. Influence of multicrystallinity on the time dependence of current. From von Hippel et al. [1973].

and mosaic structure, association of defects and trapping centers create a picture of intriguing complexity. This is also revealed in full by thermally stimulated depolarization measurements as shown in fig. 20.25 (Nedetzka and Engelhardt [1972] and Sixou and Jeneveau [1973]). The high tempera-

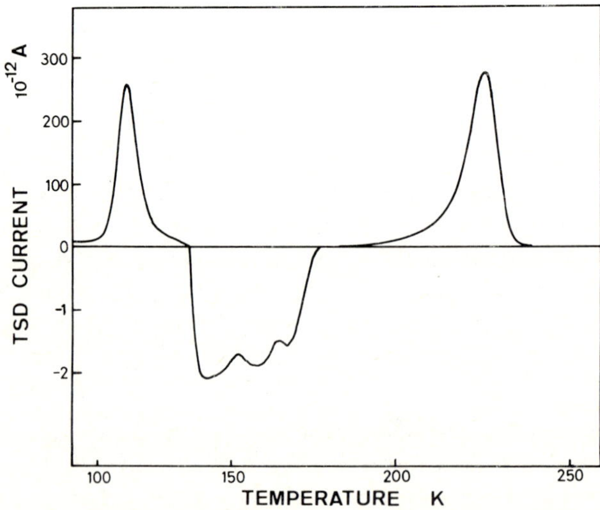

Fig. 20.25. Thermally stimulated depolarization currents.

ture peak that originates in space-charge polarization undergoes remarkable changes by thermal and electrical aging, multicrystallinity and doping. It is supposed that charged dislocations (Itagaki [1972] and Higashi and Teranoto [1973]) and traps (Ruepp [1973]) are the cause. The intrinsic Debye polarization contribution is buried under the intense space-charge peaks and is probably not identifiable.

The low temperature thermally stimulated depolarization (TSD) peak around 110 K in a relatively pure ice specimen has previously been associated with ferroelectricity (Onsager [1967] and Mascarenhas [1969]). Several anomalies observed in this temperature region were claimed to support this hypothesis, e.g. specific heat anomaly (Pick et al. [1971]) and elastic constants (Helmreich [1969]). IR measurements gave no evidence for a change in ordering. In the view of recent results, ferroelectricity in ice Ih does not seem plausible, because the magnitude of the peak is proportional to the polarizing field in a wide range and the effect is dependent on electrodes and follows monoenergetic kinetics. The low temperature experiments, especially those on doped ice, can be explained if H_3O^+ states released from traps produce the TSD current. At least L-faults will be available as traps which were suggested by Engelhardt and Riehl [1966]. The existence of space-charge limited currents with a typical trap filled limit are in support of this view (Engelhardt and Riehl [1965]). Incorporation of other impurities gives rise to a splitting of the low temperature peak into several peaks (Bishop and Glen [1969], Chamberlain and Fletcher [1971] and Nedetzka and Engelhardt [1972]). Very little is known about the origin of four peaks which appear in the intermediate range from 138–186 K.

20.3.1.4. *Proton mobility*

Many attempts have been made to determine a correct value of proton mobility. This quantity is of particular interest as an experimental basis for theoretical model calculations of the motion of proton states (H_3O^+ or $H_5O_2^+$) in a completely interlinked H-bonded lattice. Concerning the molecular mechanism for the proton mobility one should envisage the fact that besides the directed motion of protons along H-bonds in a Grotthus type fashion other more convenient side-paths may be available for the protons, dislocations, interfaces, mosaic grain boundaries, association with other faults and impurity sites. The mobilities of ions, including protons in water, are known from transfer experiments. The value for the H_3O^+ mobility at 0°C is 2×10^{-3} cm^2 V^{-1} s^{-1} which is about 5 to 10 times as much as that for monovalent ions. The effect of pressure and temperature on the equivalent conductance of water measured up to 100 kbar and 1000 °C has been summarized by Holzapfel [1969]. The mobility of protons in water increases very little with temperature. This extra mobility led several workers to advance the concept of quantum mechanical fluctuations of protons within the hydration complex. The rate limiting step for the proton transfer was considered to be the structure diffusion that aligns adjacent H-bridges and allows the proton to leave its complex and to form a new one. Mass spectrographic studies of the hydration series $H_3O^+ + nH_2O$ ($n = 1, \ldots, 8$), however, give no evidence for an exceptional stability of the $H_9O_4^+$ complex. (Kebarle et al. [1967], DePaz et al. [1969] and Cunningham et al. [1972]). The high mobility of the H_3O^+ and OH^- ions in water can also be explained by the smaller disturbance these ions cause in the H-bonded network of water as compared to foreign ions.

The proton mobility in ice has recently been determined by Eckener et al. [1973] who employed the method of measuring the drift velocity of injected protons in a sweep down field (fig. 20.26). At low temperatures where the disturbance from intrinsic Bjerrum faults is negligible the mobility is $10^{-3}\,cm^2\,V^{-1}\,s^{-1}$ and is increasing with falling temperature. An estimate for the mobility of the same order of magnitude was obtained by Maidique et al. [1971] from the measurement of conductivity and of saturation charge of extrinsic charge carriers in the ice sample. The

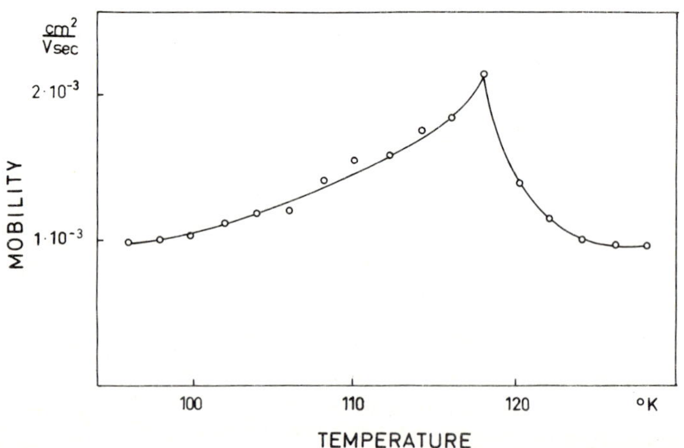

Fig. 20.26. Temperature dependence of mobility of injected protons in ice single crystals. From Eckener [1973].

sample contained about 3×10^9 charges and exhibited a mobility of $5 \times 10^{-3}\,cm^2\,V^{-1}\,s^{-1}$. Camplin and Glen [1973] found that the low frequency conductivity increases linearly with fluoride concentration and is independent of temperature at high temperatures. This indicates full dissociation and it is suggested that the conductivity is governed by extrinsic H_3O^+ states. From this a value for the proton mobility of $3.5 \times 10^{-4}\,cm^2\,V^{-1}\,s^{-1}$ can be estimated. These results have to be compared with the mobilities of other ions, e.g. Li^+, F^-, in ice which are less than $10^{-8}\,cm^2\,V^{-1}\,s^{-1}$. The order of magnitude for the proton mobility in ice and its energy of activation which is close to zero or even negative, suggests a significant deviation from classical hopping transport, and therefore, tunneling has been regarded as an appropriate explanation.

A strong coupling between the proton motion and lattice vibrations proves to be the predominant feature of all theoretical model calculations. Gosar and Pintar [1964] considered a simplified model of the ice lattice, that is, a linear chain of molecules with double well potentials between them. They used the WKB method in order to treat processes that involve at least two phonons. The proton–phonon interaction couples the ground state with the excited states in the double minimum potential well. They obtained a high mobility value of $2.5 \times 10^{-2}\,cm^2\,V^{-1}\,s^{-1}$ at $-10\,°C$. Kim and Schmidt [1967] in a semiclassical theory describe the proton transport in terms of lattice scattering. They treat the mobile proton as a point particle and take the quantized nature of energy transfer between proton and lattice into account. If the effect of translational vibrational modes is included a mobility value of $3.8 \times 10^{-2}\,cm^2\,V^{-1}\,s^{-1}$ is predicted. Two refinements, however, would tend to lower the mobility: the use of an effective dielectric constant lower than the bulk value, and consideration of scattering by water molecules along the path of the mobile proton.

A more refined quantum mechanical treatment of the mobility of structural defects in ice has been developed by Fischer and Hofacker [1969] and Fischer et al. [1969] (see section 6.3.1). They describe the dynamics of structural defects in a linear H-bonded chain and in particular the coupling of the defects to the lattice vibrations. The localized defect proton at a particular lattice site is subject to an inherent potential and interacts with all surrounding protons. The calculation is limited to two states in each double minimum potential well. The effect of the real three dimensional network of H-bonds is taken into account by using O–O distances and oxygen vibrational frequencies from experiment.

The strong proton–phonon coupling creates a new polaron state. One of the consequences is that the width of the ground band shrinks and the defect gets trapped at a certain lattice site. Deviations from the simple activation mechanism are expected because of the strong coupling frequencies which are large compared to kT and because of the long time correlations.

At high temperatures the jump mechanism prevails, at low temperatures the phonon tunneling coupling and the Gosar interband coupling terms determine the mobility. The barrier height of the symmetrical double well potential in which the proton defect propagates, is estimated as 3 kcal $mole^{-1}$.

20.3.1.5. *Effect of doping on dielectric properties of ice*

The usual experimental procedure for gaining deeper insight into the dielectric processes of materials by doping with small amounts of known impurities has been widely used in the investigation of ice. Because the water molecule differs in shape and electronic structure appreciably from most other compounds, the possibility of substitutional incorporation of foreign molecules into the ice lattice is only feasible for a few cases.

Ammonium fluoride is the only molecule that is isostructural, isoelectronic and isoprotonic with two water molecules; the lattice parameters differ only about 4% (Kuriakose and Whalley [1968]). It is very soluble in ice, close to 2.7 mole l^{-1} (Zaromb and Brill [1956]). On the other hand, striking differences of both substances are obvious: the lowest melting point of ice is at −22.2 °C and 2.09 kbar, that of NH_4F is at 213 °C and 2.5 kbar; the proton configuration of ice is fully disordered, in NH_4F, however, fully ordered; water is a molecular liquid, ammonium fluoride is essentially ionic. The segregation of NH_4F into single ice crystals has been discussed in detail by Noll [1969]. Many types of imperfections can be visualized by the following equations for the dissociation equilibria which must be fulfilled: $NH_4F \rightarrow NH_4^+ + F^-$, $NH_4F \rightarrow NH_3 + HF$, $NH_4F \rightarrow NH_2^- + H_2F^+$, $NH_4OH \rightarrow NH_4^+ + OH^-$, $HF \rightarrow H^+ + F^-$, $H_2O \rightarrow H^+ + OH^-$.

Because on freezing, part of the cations are replaced by protons, NH_4F is not stoichiometrically incorporated in ice, that is, the acid component predominates. NH_4^+ and F^- have approximately the same dimensions as H_2O, 1.37 Å and 1.28 Å as compared with 1.38 Å for H_2O.

Extensive work on the electrical properties of ice has been done with the following dopants: NH_4F, HF, LiF, NaF, KF, CsF, HCl, HBr, HI, NH_4OH, LiOH, NaOH, KOH, NH_4Cl, LiCl, NaCl, KCl, RbCl, CsCl, NH_3

and HNO_3. These early investigations have been reviewed by Gross [1968] who also points out discordant results and sources of errors encountered in electrical measurements on doped crystals.

There is still very little known about the actual distribution of the impurities in ice. The impurity segregation during growth is very sensitive to the growth conditions which are different for each laboratory, although a standardizing would be highly desirable. The way dopants are incorporated, substitutionally or interstitially, at dislocations, in clusters or veins has great effects on the dielectric and other properties of the crystals. The segregation coefficient k, the ratio of concentration of the dopant in the crystal and the concentration in the solution from where the crystal is grown, can reach values as low as 10^{-2} to 10^{-3}. Therefore, the concentration in the crystal is not known, and it is practically determined by melting the crystal after completion of the experiment and measuring the constitution of the melt which only gives average concentrations of the dopant in ice. Seidensticker and Longini [1969] and Seidensticker [1972] have calculated the influence of impurities on the population of electrically active defects using a statistical treatment analogous to that used for electronic semiconductors.

In recent years the experimental activities have been concentrated on a few selected doping agents, notably HF (Mounier and Sixou [1969], von Hippel et al. [1972] and Camplin and Glen [1973]), HCl (Young and Salomon [1968]), KCl (Maeno [1972, 1973]), KOH (Kawada [1972]) and on natural ice (Paren and Walker [1971] and Paren [1973]). Among these, ice doped with HF has been studied in greatest detail, because it induces the largest changes in electrical properties of ice. HF is akin to H_2O in its ability to form H-bonds. Substitution of a HF for a H_2O molecule expels one H-bond from the lattice, leading to an L-defect and consequently to a softening of the ice lattice (Jones and Glen [1969] and Nakamura and Jones [1973]). Relatively high concentrations of 2×10^{-3} mole l^{-1} HF in single ice crystals were obtained by Bilgram [1973] who used the Czochralsky method.

Ice grown from KCl solutions was used as a model for natural impure ices like sea ice. It was found that limited amounts of chlorine were incorporated in single ice crystals but no trace of potassium was detected. Unfavorable size factors of HCl and its inability to form H-bonds result in a low solubility limit for substitutional incorporation of less than 1×10^{-4} mole l^{-1}.

The discrete spectra analysis applied by von Hippel and Camplin shows that not only does HF modify the relaxation spectra found for pure ice, but also creates a new spectrum. The inherent intrinsic polarization is destroyed by HF with increasing concentration. The temperature, where the relaxation time deviates from a straight line in a log σ_D versus $1/T$ plot (fig. 20.19a) corresponding to an activation energy of 14 kcal mole^{-1} for intrinsic polarization, shifts to higher temperatures values. The activation energy below this transition temperature reaches 8 kcal mole^{-1} at a concentration of 10^{-4} mole l^{-1} in von Hippel's experiments. Camplin, who decomposes his spectra into three components, obtains 5.8 kcal mole^{-1} for a HF content of 5×10^{-5} mole l^{-1}. A polarization minimum already observed at a concentration of 10^{-8} mole l^{-1} deepens and shifts from -70 to -10°C with increasing HF concentration. The new HF spectrum shows an opposite tendency. The low frequency conductivity $\sigma_=$ increases so strongly with HF doping that field distortion is likely to occur at high temperatures. At low temperatures the activation energy slightly increases with increasing concentration to reach 5 kcal mole^{-1} at a concentration of 10^{-4} mole l^{-1} (von Hippel), compared with 7 kcal mole^{-1} (Camplin). From the behavior of the minimum of polarization it might be concluded that according to previous results and theories (Steinemann [1957] and Jaccard [1959]) a change in majority mechanism, which is responsible for dipolar relaxation, is indicated. Below the temperature T_m where the minimum occurs, the intrinsic Bjerrum defects become the minority mechanism and the extrinsic ionic defects take over the majority mechanism. If this explanation were true, the temperature dependence of σ_D below T_m would be similar to that of $\sigma_=$ above T_m.

There is no unambiguous evidence of this. A low frequency conductivity $\sigma_=$ is independent of temperature above T_m, but σ_D below T_m is not. The existence of traps and associations of defects may possibly account for this behavior.

As expected, the imperfection induced spectra 1 and 2 (fig. 20.18) are also strongly influenced on doping. Time constant and polarization contribution increase rapidly at low HF concentrations and tend to traverse a maximum at about 10^{-6} mole l^{-1}.

The low frequency spectra 4–6 are affected by space charge build-up and field distortion which becomes more pronounced with increasing HF concentration. Multicrystallinity and HF doping have qualitatively the same effect, both influences increase the number of extrinsic charge carriers. Hence, the relaxation time shortens and the polarization contribution grows.

The situation is far from being understood in every detail, but experimental techniques and conceptual interpretations are promising for gaining a deeper insight in future work.

20.3.2. Dielectric properties of clathrates

Clathrates can be thought of as ices in the sense that each water molecule is surrounded by four nearest neighbors in an approximately tetrahedral H-bonded arrangement. (McMullan and Jeffrey [1965] and Mak and McMullan [1965]). Mirror-symmetric bonds, like in cubic ice Ic, enable the formation of nearly planar rings which are built together in three dimensions to form the basic structural unit, a pentagonal dodecahedron. The internal angle of the pentagon is 108° which is close to the ideal tetrahedral angle of 109.5° and therefore favorable for a H-bonded framework. Pentagonal dodecahedra cannot pack to a space filling solid. It is necessary in the regular lattice that other larger polyhedra are included. The cages formed by the polyhedra are large enough (mean free diameter 5–7 Å) to accommodate guest molecules, and in fact, the clathrate structures are stabilized by the presence of molecules of appropriate size. The ratio of the number of water molecules to the number of guest molecules is very high. Structure I clathrates with the general formula $6\,X{\cdot}46H_2O$ are formed by 35 small molecular species X which range in size from argon, nitrogen, ethylene oxid to cyclopropane. Larger guest molecules (some 34 species) like tetrahydrofuran form structure II clathrates with the formula $8\,X{\cdot}136H_2O$ in which, as a rule, only the larger cages are occupied. The smaller cages could be occupied by H_2S if simultaneously present.

The dielectric relaxation of clathrates is determined by the ice framework and the encaged molecules (Davidson et al. [1964], Hawkins and Davidson [1966], Venkateswaran et al. [1967], Gough et al. [1968], Majid et al. [1968], Morris and Davidson [1971] and Gough and Davidson [1973]). The contribution of the guest molecules is markedly dependent upon their nonpolar or polar nature, and their ability to interact with the host lattice by H-bonding or by substitution of water molecules.

The high-frequency dielectric constant for both argon and nitrogen clathrates, are close to that of ice, if one allows for the different densities of water molecules in the empty clathrate lattice and the ice lattice. The contribution to the dielectric constant from the encaged argon and nitrogen is only about 0.1. In contrast, table 20.5 shows much higher $\epsilon_{\infty 1}$ values for the clathrates of dipolar molecules. The permittivity $\epsilon_{\infty 1}$

TABLE 20.5

(a) Host lattice relaxation (−40°C)

Clathrate	μ (D)	τ (μs)	E (kcal mole^{-1})	$\epsilon_{\infty 1}$ (168 K)
Ice Ih	1.84	1500	13.8	3.1
Ar	0	93	5.7	2.85
N_2	0	230	7.9	2.85
SF_6	0	780	12.3	2.9
1,3-dioxolane	1.47	5.4	8.7	4.57
1,2-propylene oxide	2.0	2.0	8.0	5.94
2,5-dihydrofuran	1.54	1.5	7.5	5.03
tetrahydrofuran	1.63	1.0	7.4	5.06
1,3-propylene oxide	1.93	0.48	7.0	5.63
acetone	2.88	0.57	6.5	8.6
cyclobutanone	2.89	0.49	6.5	9.84

(b) Guest molecule relaxation (20 K)

Clathrate	τ (s)	E (kcal mole^{-1})	$\epsilon_{\infty 2}$ (4 K)
1,3-propylene oxide	4.5×10^{-9} [a)]	0.6	3.7
tetrahydrofuran	1.2×10^{-3}	1.06	3.5
acetone	2.1×10^{-4}	1.08	4.0
cyclobutanone	55 [a)]	1.44	3.6

a) Extrapolated value. From Gough and Davidson [1973].

measured on the high frequency side of the ice dispersion region depends on the dipole moment of the encaged molecule. A second absorption found at higher frequencies can therefore be attributed to the rotation of the dipolar guest molecules. The dielectric behavior at 4 K shows a reduced high-frequency dielectric constant $\epsilon_{\infty 2}$, which implies that some orientations within the cage are appreciably more stable than others.

The magnitude of $\epsilon_{\infty 2}$ is due to rotational oscillations about equilibrium orientations to which they are not firmly bound, as evidenced by the low activation energy of about 1 kcal mole^{-1}. The dipole moments of the encaged molecules are indistinguishable from the dipole moments in dilute solutions, because the resultant dipolar field of the water molecules over the surface of the almost spherical ice cage is relatively small (Davidson [1971]).

The static permittivity ϵ_0 of structure II clathrates is independent of the nature of the guest molecules and given by $\epsilon_0 - \epsilon_{\infty 1} = 14900/T$ between 0 and $-80\,°C$. The difference to ice is mainly explained by the smaller density of water molecules in the clathrate lattice. Also the shape of the Debye dispersion associated with the reorientation of water molecules is affected very little by the presence of different guest molecules. A discrete spectra analysis gives two dominant Debye relaxation processes and long time contributions due to ionic conduction and space-charge polarization.

The actual relaxation times and the corresponding activation energies for water dipole reorientation are greatly dependent on the guest molecules. The nonpolar SF_6 clathrate exhibits an activation energy almost as large as for ice Ih, whereas polar guest molecules reduce the activation energy to values only half as great. The relaxation times are relatively slow as in ice I for nonpolar guest molecules such as argon, nitrogen, cyclopropane, but fast in clathrates with polar guest molecules such as ethylene oxide, tetrahydrofuran and acetone.

Thus, H-bonding in clathrates containing small inert guest molecules is very similar to ice Ih. Neither the small tetrahedral distortion nor the planar pentagonal arrangement of water molecules appears to have any significant influence. The polar guest molecules, in contrast, may occasionally form a H-bond with simultaneous injection of an L-defect into the ice structure, thus enhancing the relaxation time by providing a reorientational mechanism for the water molecules. Since there are several non-equivalent water molecule sites, different diffusion paths may account for the two principal relaxation times that have been mentioned.

The dielectric properties of quaternary ammonium salt hydrates have been studied by Koide and Carstens [1972]. In this case, the quaternary ammonium cation serves as a guest ion in the clathrate structure, and the anions are part of the ice structure. The low frequency dielectric constants range from about 100 to 1000. The dielectric data show discontinuities near critical hydration numbers. These inflated values for ϵ_0 together with the relatively high conductivity of these clathrates suggest that the dominant contribution to the dielectric polarization comes from the movement of anions relative to the cations which are entrapped in the clathrate structure. The dipolar orientation mechanism for the water molecules in the ice structure is obscured by the ionic conduction mechanism. Among the quaternary ammonium salt hydrates tetra-*n*-butylammonium hydroxide hydrate stands out. Its conductivity is one order of magnitude higher than in other hydrates of this family, and the relaxation time at 0°C is well below 10^{-8} s. These findings suggest that the high concentration and effective mobility of the hydroxyl ions are the cause.

20.3.3. Miscellaneous substances

Electrical conduction and dielectric polarization in H-bonded solids appears to be caused by protonic states in the first place, as evidenced by numerous electrolysis experiments. It should, however, be emphasized that the evolution of hydrogen gas at the cathode in coulometric experiments only demonstrates the ionic nature of the charge carriers, because of secondary effects at the electrodes, and also non-protonic carriers could lead to a development of hydrogen gas. But isotope effects, the correlation of higher conductivity with crystal directions in which conduction along chains of H-bonds is favored and comparison with non H-bonded solids corroborate the conviction that protons represent the bulk carriers. It goes without saying that the experimental difficulties are similar, as in the case of ice. Very often contact material appears to be the limiting factor. In some cases the chemical and physical state of the crystals under investigation could not be defined satisfactorily, and, therefore, the present review should be regarded as an early stage of a study program which needs continuous efforts.

From a large number of investigations on H-bonded solids we shall consider some typical examples for a more detailed discussion, others can only be mentioned briefly. The ferroelectric properties of H-bonded crystals are expertly discussed by Schmidt in ch. 23. Previous results on organic H-bonded systems can be found in the paper of Pollock and Ubbelohde [1956] and in a systematic study of the semi-conducting properties of biological materials by Eley and Spivey [1960, 1961] and Eley [1962].

A survey on more recent conduction data is given in table 20.6. The electrical conductivity of H-bonded crystals as usual, is given by $\sigma = \sigma_0 \exp(-E/kT)$. Typical values for the apparent activation energy range from 10 to 20 kcal mole^{-1} and room temperature conductivities are of the order of 10^{-6} to 10^{-12} ohm^{-1} cm^{-1}.

Potassium dihydrogen phosphate (KDP). This substance is perhaps the best understood H-bonded solid besides ice, both from an experimental and theoretical point of view. It certainly owes its prominence from the fact that it is ferroelectric with an order–disorder transition at 123 K and at 222 K on deuteration. The early theory on ferroelectric phase

Conductivity data of H-bonded crystals

Crystal	Axis	Temperature (°C)	E [a] (kcal mole^{-1})	σ_0 [a] (ohm cm)$^{-1}$	References
KH_2PO_4	*c*	180	12.65	1.3×10^{-1}	O'Keeffe and Perrino [1967]
	c	180	18.0	1×10^{2}	
	c	180	12.5	6.8×10^{-2}	
	c	180	18.0	1×10^{2}	
	a	298–391	15.9	6.7×10^{1}	
KD_2PO_4	*c*	195–353	13.4	7.9×10^{-1}	Schmidt and Uehling [1962]
KH_2AsO_4	*a*	289–382	13.6	1.1×10^{1}	and Schmidt and Haslett [1964]
$NH_4H_2PO_4$		30–85	14.4	1.54	Murphy [1964]
		85–140	20.4	6.46×10^{3}	
$NH_4H_2PO_4(Ba^{2+})$		25–120	10.1	6.17×10^{2}	
$NH_4H_2PO_4$	*c*	40–150	11.1	1.26×10^{2}	Pollock and Sharan [1969]
$KHSO_4$		366–441	15.2	6.8	Rogers and Ubbelohde [1950]
$(NH_4)_2SO_4$	*b*	225–473	17.5	2.9×10^{1}	Schmidt [1963]
NH_4Cl		39–166	26.6	4.4×10^{5}	Harrington and Staveley [1964]
$NH_4Cl(Cd^{2+})$		0–112	17.3	1.8	
NH_4Cl		130–180	29.0	1.6×10^{7}	Fuller and Patten [1970]
ND_4Cl		130–180	29.0	1.2×10^{7}	
$NH_4Cl(Sr^{2+})$		60–80	3.5		
KHF_2 β-phase		193–247	20.9		Davis and Westrum [1961]
KHF_2 α-phase		100–196	21.7	5×10^{3}	Pollock and Sharan [1967]
KDF_2 α-phase		100–196	23.5	1.6×10^{4}	
$Li(N_2H_5)SO_4$	*c*	20–150	19.6	4.9×10^{5}	Vanderkooy et al. [1964]
	c	0–200	12.5	3.6×10^{-1}	Schmidt et al. [1971]
Borax	$a \sin\beta$	290–320	16.5	4.2	Maričič et al. [1962]
	b	290–320	14.4	6.8×10^{1}	
	c	290–320	12.5	1.4×10^{-3}	
Imidazole	*c*	35–85	39.0	3.0×10^{14}	Kawada et al. [1970]

[a] $\sigma = \sigma_0 \exp(-E/RT)$.

transition of KDP by Slater [1941] was modified by Takagi [1948] and extended by Senko [1961] and Silsbee et al. [1964] to take into account long-range dipole–dipole interactions. The structure of KDP consists of PO_4 tetrahedra which are linked together by H-bonds. The protons reside asymmetrically on the bonds in such a way that two protons belong to each PO_4 tetrahedron. Thus, $H_2PO_4^-$ entities form an open three dimensional H-bonded network filled with K^+ ions to preserve neutrality. There are X and Y bonds which are perpendicular to each other and which are lying approximately perpendicular to the crystalline c-axis. The H-bonds are very short, about 2.5 Å. The conductivities in the direction of the c-axis and perpendicular to the c-axis are very similar and about $1.2 \times 10^{-10}\,\text{ohm}^{-1}\,\text{cm}^{-1}$ at 25 °C (Schmidt and Uehling [1962] and O'Keeffe and Perrino [1967]). The corresponding activation energy is 13 kcal mole^{-1} which is very close to the activation energy of sulphate doped KDP, about 12 kcal mole^{-1}. The high temperature apparent activation energy was found to be 18 kcal mole^{-1}. In analogy to ice two defect types may be postulated: (a) ionic defects formed by reaction $2H_2PO_4^- \rightarrow H_3PO_4 + HOP_4^{2-}$, and (b) L- and D-defects formed by interbond proton transfer. Since incorporation of an HSO_4^- generates one new L-defect and since the conductivity increases linearly with sulphate concentration, the interbond motion of protons as controlling the conductivity is suggested. If ionization of HSO_4^- and ion migration were rate determining, a square root dependence of conductivity on HSO_4^- doping would follow. From deuteron magnetic resonance studies a component of transition probability governed by an activation energy of 1.8 kcal mole^{-1} was assigned to intrabond jumps (Schmidt and Uehling [1962]). The measured activation energy for doped KDP can therefore be attributed to interbond proton jumps among different H-bonds. Hence, in contrast to ice, the rate limiting mechanism for protonic conduction is the transition of protons between different H-bonds. From the apparent activation energy for intrinsic conduction at high temperatures which consists of half the enthalpy of formation of an L-, D-defect pair plus the enthalpy of migration, an enthalpy of formation of about 12 kcal mole^{-1} results. Since the enthalpy of formation and that of migration are equal, the removal of a proton from a H-bond seems to be independent from subsequent steps, be it the formation of a D-defect or the recombination with an existing L-defect. Correspondingly, the attachment of a proton to a H-bond does not seem to come from where the proton stems, from an ordinary H-bond or from a D-defect.

Ammonium dihydrogen phosphate (ADP). It is interesting to compare the results on KDP with ADP, a crystal similar in structure. Again the oxygens sitting at the vertices of the PO_4 groups are linked to oxygens of neighboring phosphate groups by H-bonds. The conduction measurements on single crystals of ADP by Murphy [1964] exhibited an apparent activation energy of 20.4 kcal mole^{-1} at temperatures higher than 85 °C. The similarity of the conductivities of KDP and ADP lends support to the idea that the measured values reflect intrinsic properties of the H-bonded system in these crystals.

On doping ADP with Ba^{2+} ions the activation energy is reduced to 10.1 kcal mole^{-1}. Pollock and Sharan [1969] obtain a similar value of 11 kcal mole^{-1} from ac-conductance measurements.

Ammonium sulphate. In ammonium sulphate N–H···O bonds are formed instead of O–H···O bonds, as in other crystals mentioned above. The ammonium groups are in two different environments of six sulphate ions (Hamilton and Ibers [1968]). The four H-bonds binding one particular N-atom to its surrounding six O-atoms are relatively weak, one is trifurcated and two are bifurcated bonds. A great variety of H···O distances, most of them well above 2.0 Å, are present. The shortest H-bond is 2.9 Å in length; ammonium sulphate undergoes a first-order ferroelectric transition at 223 K.

Schmidt [1963] has measured the electrical conductivity along the crystalline b-axis from −72 to 200 °C. At the ferroelectric transition large conductivity fluctuations occurred. Regardless of how slowly this transition temperature was approached, minute cracks developed because of the first-order transition, that caused the current to increase greatly and low values for the activation energy resulted. When the field was applied to an undamaged crystal, a small decrease in conductivity with time was observed, probably due to electrode polarization. The activation energy was found to be 17.5 kcal mole^{-1}. The low activation energy of about 2.3 kcal mole^{-1} for the rotation of NH_4^+ ions found by Miller et al. [1962], excludes this process from limiting the conductivity. A conduction mechanism similar to that for ammonium chloride discussed in the following section is feasible.

Ammonium chloride. Ammonium chloride like NH_4Br and NH_4I has a sodium chloride type structure at high temperatures. Below 184 °C it has a twofold disordered cubic lattice of caesium chloride type, which undergoes an order–disorder transition near −31 °C. In this crystal, only very weak N–H···Cl bonds are present. Herrington and Staveley [1964], in their extensive comparative study of several ammonium salt crystals, found that the conductivity of ammonium chloride is significantly greater than that of potassium chloride or isomorphous caesium chloride. According to their suggestion a specific three stage mechanism operates in NH_4Cl which is not possible in H-bonds lacking alkali halides: (1) a proton jumps from an ammonium ion to a chloride ion adjacent to a vacancy, in effect producing a NH_3 molecule and a HCl molecule, (2) either the NH_3 or HCl molecule moves into the vacancy, and (3) a reverse proton transfer from HCl to the NH_3 molecule again forming NH_4^+ and Cl^- ions, one of which has exchanged position with the vacancy. Successive processes directed under the influence of the electric field result in a net flow of charge.

A test of this hypothesis has been achieved by Fuller and Patten [1970] who have measured the conductivity of both NH_4Cl and ND_4Cl. If the proton transfer of step (1) is the rate limiting process, then the frequency factor of the moving proton will affect the preexponential factor in the conductivity expression by $\sqrt{2}$. The results were found to be consistent with the proposed model. Further evidence that the reaction $NH_4^+ + Cl^- \rightarrow NH_3 + HCl$ takes place within the crystal to a significant extent, is given by several other anomalous properties of NH_4Cl crystals: (1) additional contributions to the specific heat, (2) anomalous thermal expansion coefficient, and (3) high vapor pressure at room temperature. The measured activation enthalpy for the conductivity in the intrinsic region of about 27 kcal $mole^{-1}$ is similar to the value for isomorphous CsCl of 24 kcal $mole^{-1}$. However, the conductivity of NH_4Cl at 100°C is 10^{-10} ohm^{-1} cm^{-1}, but 4×10^{-12} ohm^{-1} cm^{-1} for CsCl and 8×10^{-15} ohm^{-1} cm^{-1} for KCl.

Fuller and Patten [1970] found, for strontium doped ammonium chloride in the impurity dominated conductivity region, an activation enthalpy of 2.8 kcal $mole^{-1}$, and it is asserted that this value is essentially the activation enthalpy for the proton motion. Thus, an enthalpy of formation of a pair of Schottky defects in NH_4Cl of about 48 kcal $mole^{-1}$ is obtained.

Potassium bifluoride (KHF_2). Just as NH_4Cl is an example of a crystal with the weakest H-bonds, the KHF_2 is an example of a crystal with the strongest H-bonds. There is only one type of H-bond in the crystal and this is very short, 2.26 Å, and symmetrical. The H-bonds formed within the individual $F–H–F^-$ ions are strongly isolated in the crystal, i.e. no H-bonded network exists as in ice. Davis and Westrum [1961] and Pollock and Sharan [1967] have measured the electrical conductivity which was proved to be due to movement of protons alone. At 196 °C, the transition temperature from α to β phase, a sharp discontinuity in conductivity is observed. The activation energies in both phases agree so that the proton transport mechanism is essentially the same in both modifications. The mechanism of conduction of protons across the crystal is by migration of some kind of proton defects, probably interstitial protons. It may be assumed that the migration enthalpy for defects is very small in an open structure such as that of KHF_2. Thus, the observed apparent activation energy of 21.4 kcal mole^{-1} may be attributed to half the enthalpy of formation of a Frenkel-type defect pair.

Lithium hydrazinium sulphate (LiHzS). $Li(N_2H_5)SO_4$ appears to be an interesting case, because it contains nearly linear one dimensional H-bonded N–H···N–H···N–H··· chains. The structure essentially consists of a $LiSO_4^-$ framework containing channels which can accommodate the hydrazinium NH_2–NH_3^+ ions with one hydrogen of the NH_2 group participating in the H-bonded chains running along the c-axis. The unusual electric behavior has been investigated by Vanderkooy et al. [1964] and Schmidt et al. [1971], who found the conductivity along the H-bonded c-axis to be over two hundred times as high as along the a- and b-axis at room temperature. Above 200 °C the conductivity becomes virtually isotropic. Schmidt, however, measured lower absolute conductivities and activation energies than Vanderkooy. The major contribution to the dielectric properties in the direction of the c-axis is supposed to originate from conduction along partially blocked chains of H-bonds which have a very broad distribution of relaxation times. In contrast to ferroelectrics, LiHzS has a large dielectric susceptibility over a wide temperature range. Local extrinsic barriers postulated by Schmidt, account for two observed effects: (1) the conductivity decreases with decreasing frequency, and (2) the interchannel diffusion increases at high temperatures because of thermally induced low barriers, thus reducing anisotropy in conductivity.

Other examples of highly anisotropic conductivity in crystals containing unidimensional chains of H-bonds, as lithium sulphate monohydrate and borax, were studied by Maričić et al. [1961, 1962]. A new method of evaluating the anisotropy of conductivity in crystals has been applied to β-p-nitrophenol, γ-hydrochinone and imidazole by Pigon and Chojnacki [1967]. It consists of measuring the angular distribution of current in a crystal face, using a set of 12 electrodes arranged in a circle and one electrode in the center. The findings of Pigon and Chojnacki [1967] on imidazole which exhibits anisotropy in the (010) plane of about 3.6, led Chojnacki and Golebiewski [1969] and Chojnacki [1970] to attack the mobility problem in this crystal, using the tight binding approximation of the band theory. No satisfactory agreement of the calculated electronic conductivity ratio in the c- and a-axis with experiment was found. The authors admitted that collective motion of protons in H-bridges may be important for charge transfer across the crystal.

Imidazole. This organic crystal deserves closer inspection, because it has been speculated that charge transfer mechanisms mediated in this structure may have biological implications (see Zundel, section 15.9.4). The imidazole molecules form a linear H-bonded chain along the c-axis. The planar rings of two adjacent molecules are twisted by 62°. The N–H · · · N bond length is 2.81 Å at room temperature. Kawada et al. [1970] succeeded in growing transparent single crystals with very low water content by means of a zone refining procedure. The current, passing along the c-axis, led to the evolution of hydrogen gas at the cathode. The coulometric ratio of H-atoms per charges was close to 1. In the a direction the amount of hydrogen involved accounted for only 28% of the total charge passed. Since surface currents in the a direction were 10^5 times greater than the bulk current, the hydrogen collected at the cathode may stem from currents along the surface rather than from the bulk.

The anisotropy of the conductivity measured with guarded electrodes is very high, the ratio of c- to a-axis conductivity reaches 10^3. This implies that Pigon and Chojnacki [1967] basically measured surface currents. The outstanding feature of dc-conductivity in single crystals of imidazole is the movement of protons along directed chains of H-bonds. The following mechanism is apt to explain this property. An ionic state is generated by a translational proton jump in the N–H · · · N bridge. The energy needed for this process is probably not greater than that in liquid, about 10.6 kcal mole^{-1}. Successive jumps along a suitable oriented chain can occur rapidly. If this process was not followed by a reorientation mechanism, the current would cease after the passage of charges along H-bonds which can only be used once. L-defect injection and mobility is most likely. Then, the high apparent activation energy of 39 kcal mole^{-1} has to be assigned to the energy necessary for expansion of the lattice that precedes the rotation of an imidazole molecule. In this way the original H-bond arrangement is restored for a further passage of an ionic defect. The large difference, about 10^6, between the conductivity of liquid and solid imidazole supports the reorientation as being the rate limiting process. In the *a*-direction where no H-bonds are lying, the conduction mechanism is probably electronic and extrinsic at high temperatures with an activation energy of 44 kcal $mole^{-1}$. At low temperatures the activation energy of 25 kcal $mole^{-1}$, similar to values found for surface conduction and in polycrystalline samples is probably due to conduction along grain boundaries and dislocations or leakage currents across the surface not properly shielded from the bulk electrodes.

To demonstrate the significance of proton conduction in H-bonded biological substances, Thomas et al. [1969] selected crystalline isocytosine which contains chains of N–H···N bonds between neighboring pyrimidine rings. It was possible, by using proton injecting electrodes, to draw currents which were greater by a factor of 10^{10} than those associated with so called organic semi-conductors (Eley and Spivey [1961] and Chojnacki [1970]).

Magnesium hydroxide. This substance, like other hydroxides, is a defector from the ranks of H-bonded crystals. The OH^- ions are well separated from one another, so that H-bonding is necessarily very weak and almost non-existent. Yet, it exhibits dielectric properties which are related to H-bonding. Gieseke et al. [1970] measured the dielectric loss dependent on frequency and temperature. The maximum in the loss factor at 180 °C is explained by the motion of protonic states in the water structure formed by the process $OH^- + OH^- = H_2O + O^2$.

20.3.4. Experimental methods

Dielectric measuring and evaluation methods have reached high standards and have been excellently described in several publications. The book of Hill et al. [1969] can be taken as an example and guide for further references. Recent developments on relaxation methods, including time domain reflectometry, have been reviewed by DeMaeyer and Persoons [1974]. In this section only a few supplementary techniques are described.

Thermally stimulated polarization (TSP) *and depolarization* (TSD) offer the advantage of recognizing different mechanisms with long relaxation times separately. The TSP method consists of measuring the polarizing current through a sample in an electric field as it is warmed continuously from low temperatures. Alternatively, in TSD measurements the sample is polarized in an applied electric field as it is cooled to temperatures sufficiently low to prevent depolarization on removal of the electric field. If the crystal is allowed to warm at a constant rate in zero external field, the polarization will be freed at characteristic temperatures, where the temperature dependent relaxation time is of the order of the time for measurement. The build-up and release of polarization causes a current in the circuit containing the sample capacity and the electrometer. Thus, each polarization mechanism corresponds to a TSD (TSP)-peak. From the temperature range, size and shape of the peak, the magnitude of the polarization, its activation energy and relaxation time can be evaluated (Nedetzka and Engelhardt[1972]).

The relaxation time τ for thermally stimulated processes is given by

$$\tau = \frac{h}{4\pi s \nu k T} \exp\left(-\frac{\Delta S}{k}\right) \exp\left(\frac{E}{kT}\right) = \frac{1}{w_0 T} \exp\left(\frac{E}{kT}\right)$$

where h = Planck's constant, s = sterical factor for hindered rotation, ν = transmission factor expressing the fraction of activated dipoles which accomplish the transition, ν is normally close to unity, ΔS = activation entropy and E = activation energy.

The temperature dependence of the depolarization current for dipole reorientation is expressed as

$$j = f w_0 T P_0 \exp\left[-\frac{E}{kT} - \frac{w_0 k (T^3 + T_0^3)}{Eq} \exp\left(-\frac{E}{kT}\right)\right]$$

where f = surface area, P_0 = frozen in polarization, T_0 = initial temperature and q = heating rate. The activation energy E can readily be obtained from the low temperature flank of the peak, if $\log j$ is plotted versus $1/T$. The slope of the straight line gives

$$E = -k \frac{\mathrm{d}\,(\ln j)}{\mathrm{d}\,(1/T)}$$

and

$$w_0 = \frac{q}{kT_\mathrm{m}^3} (E - kT_\mathrm{m}) \exp\left(\frac{E}{kT_\mathrm{m}}\right).$$

In case the TSD peaks overlap so that they cannot be separated satisfactorily, one may proceed as follows. (1) Single peak procedure. On cooling, the polarizing field is applied in a limited temperature range where the peak to be investigated appears on warming. (2) Difference procedure. The peak of interest is obtained as the difference of two TSD-curves. The sample is fully polarized in one measurement and it is shorted within the temperature range of the selected peak in a further heating run. (3) Variation of heating rate. On variation of the heating rate the temperature T_m of maximum peak is shifted. The higher the activation energy the less T_m is shifted.

Since freezing in of polarization mechanisms on cooling depends on previous polarizations, especially in the case of space-charge build-up, a comparison of TSP- and TSD-curves may be useful.

In addition to conductivity measurements, which only yield the product of concentration and mobility of the charge carriers, several techniques were applied, that allows one to determine these quantities separately. *Space-charge limited currents* (Engelhardt and Riehl [1966]) and *transit time measurements* (Eckener et al. [1973]) may provide more information about the mobility which renders insight into the prevailing conduction mechanism. Proton injecting electrodes consisting of a thin layer of palladium powder that has been stored under hydrogen atmosphere, have been successfully applied in these experiments. The protons were injected either under the force of a high field as in space-charge limited currents, or by means of a laser flash which released the protons from the palladium lattice as in the transit time measurements. On the other hand blocking electrodes, teflon foils between sample and electrodes, proton deficient electrodes, or evaporated gold electrodes, were preferred, as in TSD experiments.

Hall effect and *saturation current* measurements carried out on ice e.g., so far failed to yield unambiguous results because of sensitivity limits, surface effects and electrode polarizations.

In H-bonded solids the determination of the nature of charge carriers is very often of primary importance. Schmidt [1965] constructed a simple coulometer for studying the protonic conduction in crystals. Although coulometric measurements alone do not fully prove protonic transport, other checks as isotope effects and anisotropy of conductivity are in accord with the position that protons are very often the only mobile charge carriers in H-bonded crystals.

In summary, dielectric polarization and electrical conduction measurements in H-bonded crystals require very pure and perfect single crystals in defined crystallographic orientations. The crystals have to be grown from carefully purified substances under standardized conditions which avoid inclusion of impurities, grain boundaries and dislocations. Time consuming zone refining procedures are very often indispensable. On doping, the main concern is the way the dopant is incorporated, substitutionally or interstitially, as clusters or in veins. High segregation coefficients may cause inhomogeneous incorporation of foreign molecules.

Axial and radial distributions of impurities were observed depending on growth velocity, direction of growth with respect to crystal orientation, shape of growing crystal face, temperature gradients and convection in the solute. Thermal and electrical prehistory, aging effects as well as mechanical and chemical treatment of crystal faces are parameters which affected the results of dielectric measurements. Finally, surface effects have to be carefully separated from bulk properties by appropriate shielding. Field distortion caused by a pile-up space charge in front of the electrodes can be overcome by short time techniques and low temperature measurements.

Better conclusions from dielectric experiments can be achieved by comparison with other, non electrical, but also defect structure sensitive properties as diffusion, NMR, mechanical, ultrasonic and thermal properties.

20.4. Dielectric properties of water adsorbed by solids (by K. Bunzl)

It is well known that the structure and thus the degree of H-bonding present in bulk water is considerably changed, if the water molecules are adsorbed by a solid substance. Since the dielectric properties of water are strongly dependent on the degree of H-bonding present, dielectric relation spectra yield interesting information on the structure of adsorbed water. If, for example, the adsorbed water will have a structure similar to the one observed in ice, its dielectric relaxation time–compared to the one of bulk water–can be expected to shift by as much as five orders of magnitude. However, even though the determination of the dielectric properties of water in the adsorbed state should be a very sensitive method for detecting small changes of its structure, difficulties arise in the interpretation of experimentally observed relaxation spectra. They are due to the fact that the dielectric adsorption of water containing substances is not only due to the orientation of the water dipoles but can also be caused by the following phenomena: (1) electrode polarization, (2) dielectric adsorption due to the electrical conductivity of the sample, (3) Maxwell–Wagner dispersion due to the inhomogeneity of the sample, consisting in many cases of phases differing in their dielectric constant and conductivity. They will especially be present, if the water is adsorbed by adsorbents carrying ionic groups, as e.g. ion exchangers, where adjacent to the insolating matrix, a concentrated electrolyte solution of high conductivity is formed during the water adsorption process and (4) dielectric dispersion caused by the motion of mobile charges in the adsorbent. Before one interprets, therefore, an observed dielectric dispersion on the basis of changes in the water structure in the adsorbent, one has to examine carefully whether this assignment is indeed justified. In most cases this is a difficult problem.

In the following, the dielectric properties of water containing adsorbents, published since about 1966/67 are reviewed. The various adsorbents studied are classified as inorganic substances (e.g. zeolites, clay minerals, glasses, silica gels, alums or oxides), organic substances (e.g. ion exchange resins) and bio-polymers (e.g. protein powders). For a review of H-bonds in systems of adsorbed molecules, covering especially silica and alumina as adsorbents and also adsorbates other than water, the reader is referred to chapter 27 by H. Knözinger.

20.4.1. Water adsorbed by inorganic substances

20.4.1.1. *Zeolites*

Extensive studies of the dielectric properties of the Linde molecular sieve type 5-A containing various amounts of water were reported by Morris [1969a]. The composition of this zeolite in the hydrated form is $Ca_{4.5}Na_3[(AlO_2)_{12}(SiO_2)_{12}]\cdot 30H_2O$. Since it was impossible to obtain single crystals of the size required for dielectric studies, compacted discs of the crystalline powder were used. Most of the measurements within the range of 5 Hz–148 kHz for partially hydrated samples and within the range of 5 Hz–8.5 GHz for the water saturated material were performed at 22°C. The observed increase in both ϵ' and ϵ'' at low frequencies was attributed to conductivity and electrode polarization. For the completely hydrated sample a large absorption peak was observed at 16 kHz, which moved to lower frequencies when the temperature was decreased. Besides that, a smaller absorption peak at 400 MHz was present, which moved very little with temperature. This latter peak is probably not due to adsorbed water, since the liquid has an activation energy for dipole reorientation of about 5 kcal mole^{-1} and an evenly increased value would be expected for adsorbed water. A further absorption around 10 GHz was also ruled out as originating from the presence of water dipoles and it was concluded that absorptions due to adsorbed water only occur below 100 MHz. If water is desorbed, the 16 kHz peak (absorption I) moves rapidly to lower frequencies, the peak height also falling rapidly. With continuing desorption a further smaller peak (absorption II) appears in the kHz range, which also moves to lower frequencies with continuing desorption, its amplitude remaining essentially constant. This suggests that this peak is not due to adsorbed water but rather to ionic jumping processes. The dielectric isotherm obtained at 148 kHz indicates that the water is adsorbed in three stages: during the initial stage, which extends up to $6H_2O$ per unit cell the water molecules are adsorbed most strongly at the six 8-membered oxygen ring windows and are held there by H-bonds to the lattice oxygen. They would be able to contribute to the total polarization by a rocking movement about the H-bonds. During the second stage, which extends up to $17H_2O$ p.u.c., the rocking movement would be reduced by the presence of neighboring molecules if those water molecules adsorbed above the monolayer also become attached to the cavity walls. In the third region exceeding $17H_2O$ p.u.c. the water molecules are expected to fill the cavity center. Most likely they are H-bonded in this region to these molecules attached to the cavity walls and at most, reorientate about one O–H bond. If one assigns the absorption I as due to the adsorbed water molecules (Morris excludes a Maxwell-Wagner mechanism, since the reproducibility of peak position and height from sample to sample is very good) a relaxation time for the water saturated 5-A·$30H_2O$ p.u.c. of 1.0×10^{-5} s is obtained. Since this value is close to the relaxation time of water molecules in ice at $-0.1\,°C$, one can conclude that the major bulk of adsorbed water has an ice-like character. Morris suggests that the best interpretation can be given by assuming that the whole amount of water can contribute to the high frequency polarization by relaxing some small part of its dipole moment whereas the major part of the dipole moment relaxes as in ice.

In a further paper Morris [1969b] reports dielectric measurements of K^+(3-A), Na^+(4-A) and Ag^+-ion forms of Linde molecular sieve zeolites, also containing various amounts of adsorbed water. The composition of these samples were: 3-A: $K_9Na_3[(AlO_2)_{12}(SiO_2)_{12}]\cdot 27H_2O$ and type 4-A: $Na_{12}[(AlO_2)_{12}(SiO_2)_{12}]\cdot 27H_2O$. The frequency range was 5 Hz to 140 MHz. The effect of adsorbed water on the dielectric properties of 3-A is similar to its effect in 5-A. Below 1 kHz ϵ'' continues to rise, presumably due to conductivity. If water is removed from the water saturated sample, the 200 kHz peak moves rapidly to lower frequencies. As this absorption I moves to lower frequencies, a second dielectric loss peak (absorption II) with constant peak height appears. Similar results are also obtained with the type 4-A and the silver forms obtained from the 4-A and 5-A material. The absorption II is attributed in all cases, as in the 5-A·H_2O system, to cation jumping processes. Absorption I is again most probably due to the relaxation of water molecules. The observed relaxation times at 22°C of the various ion forms saturated with water are: 3-A: 7.9×10^{-7} s, 4-A: 4.9×10^{-7} s and Ag-A: 2.5×10^{-7} s. NMR studies give similar values.

The closeness of these relaxation times to that of water in ice and the fact that absorption I begins to fall in intensity on removal of the first amounts of water from saturation, indicates that all the adsorbed water relaxes the major part of its dipole moment in a manner similar to that of ice and only contributes some small part of its dipole to the polarization at high frequencies by some other process such as a rotation about one O–H bond or the flapping of a molecule attached to the cavity walls. The activation energy of absorption I appears to be constant at 16 kcal mole^{-1} in both 5-A·30H_2O and 4-A at various water contents. This is significantly higher than the activation energy for dipole reorientation in ice (13.4 kcal mole^{-1}) and again reflects the rigidity of the adsorbed water molecules.

There are several papers by Russian scientists, which should be mentioned here. Glazun et al. [1967] investigated the dielectric behavior of the zeolite 0.6Mg0.3NH_4Na-A containing 2.8–51% of the water necessary for complete hydration. The frequency range was 50–10000 Hz at 20–60°C and 100–3.2×10^6 Hz at -100 to $+20$°C. From the values of the activation energy at various water contents it was concluded that the water molecules formed definite structures in the adsorbed phase. In a further paper (Glazun et al. [1966]) they investigated Na-A at a temperature range of 93–213 K from 100 kHz to 10 MHz. They found that the dielectric constant of the system consists of 3 parts: the polarizability of the zeolite, that of the adsorbed water and a relaxation process. The dielectric isotherm at low temperatures showed a minimum at 10% adsorbed water, which is explained by a relaxation process in the dehydrated zeolite, the relaxation of which is suppressed by the adsorbed water. This investigation was extended by Fedorov et al. [1968] to temperatures of $+20$ and -40°C. They observed Maxwell–Wagner losses and found that the activation energy as a function of the amount of water adsorbed passed through a minimum of 58.7×10^{-2} eV at a water content of 10%.

Kopylova et al. [1970] reported measurements on Na-A zeolites at $+50$ and -50°C. At the lower temperature and at very low moisture contents they observed that no correlation exists between the dielectric loss and the amount of water adsorbed. They explained this fact by a possible change in the effective dipole moment of the water molecules. Dubinin et al. [1971] investigated Na-A containing adsorbed water in pressed and unpressed forms. The three observed maxima of dielectric loss as a function of the frequency were explained by field anisotropy in the cavity of the zeolite. The presence of Maxwell–Wagner effects, even in dehydrated Na-A zeolite, was shown at temperatures between 244–360 °C by Glazun and Zhilenkov [1971].

A wide variety of water containing zeolites was investigated by Matron et al. [1971]. They used the following Linde molecular sieves: 4-A: $Na_{12}[(AlO_2)_{12}(SiO_2)_{12}]\cdot 27H_2O$; 3-A: $K_9Na_3[(AlO_2)_{12}(SiO_2)_{12}]\cdot 27H_2O$; 5-A: $Ca_{4.5}Na_3[(AlO_2)_{12}(SiO_2)_{12}]\cdot 30H_2O$; 13-X: $Na_{86}[(AlO_2)_{86}(SiO_2)_{106}]\cdot 264H_2O$; and 10-X: $Ca_{32}Na_{22}[(AlO_2)_{86}(SiO_2)_{106}]\cdot 264H_2O$. The dielectric constant and the dielectric loss were obtained in the frequency range from 300 Hz to 400 kHz between −180 and 210°C. In the dehydrated samples they observed two relaxation processes. The one appearing at low temperatures (respective high frequencies) was called β process and the one at high temperatures (respective low frequencies) α process. Both processes could be described by Cole–Cole formulas. Since they occur already in the dehydrated samples and do not change much with the amount of adsorbed water, they are attributed to the relaxation of ions bound in the zeolites in two different energy levels. A third relaxation process (γ process) is observed only if water is adsorbed by the samples and its intensity increases with the increasing amount of water adsorbed. It is attributed to the water molecules bound to the cations. Since this relaxation process cannot be described by a Cole–Cole formula and is also strongly superimposed by the β process, no activation energies could be calculated. It is interesting to note that one observes for the zeolite 10-X phase transformations of the adsorbed water. The temperatures, at which these transformations occur, decrease with increasing amounts of adsorbed water from +40 to −20 °C.

Chapoton et al. [1970, 1971, 1972] investigated the dielectric properties of A-type zeolites. Since they were mostly interested in the motions of the cations rather than in the structure of the adsorbed water, these papers will not be reviewed here in detail. Lohse et al. [1971a] investigated the zeolite 5-A water system in the frequency range from 30 Hz to 1.2 MHz at −75 to 200 °C as a function of the water content in the NaCa- and the Ca-form of this material. They concluded from their results that water molecules adsorbed in a first range are bound to calcium ions and in a second range to 8-membered oxygen rings. In a third range the hydration shell of the ions is built up further and competed in range four. The system Na-A water (Lohse et al. [1971b]) showed a somewhat different behavior. While the Ca-forms show in the dehydrated form and at small water contents no dielectric dispersion, Na-A exhibits two regions of dispersion which are attributed to the mobility of the Na_2^+ ions (the 4 delocalized Na^+ ions in the large cavities of the zeolite). A corresponding dispersion appears for the Ca-A only at high water contents around 10^2–10^3 Hz. A dispersion due to the adsorbed water was only observed at 10^2 Hz for Ca-A and not for the Na-A. This is explained by assuming that adsorption structures are formed which are stabilized by H-bonds and coordinate bonds to the cations, resulting in a coupling of the cation jumps and the motions of the adsorbed water molecules. This would cause only one region of dispersion for the water molecules and the cations. When dehydrated zeolites of the formula $Na_{11}X$, $Ca_{5.5}X$, $Na_{7.4}Mn_{1.8}X$ and Na_7Cu_2X, where $X = (AlO_2)_{11}(SiO_2)_{13}$, were investigated (Lohse et al. [1970a]), a low frequency dispersion was observed with the exception of the Ca-form. The relaxation time for the Na-form is three orders of magnitude smaller than for those containing divalent ions. The divalent cations are responsible for this relaxation process. If these zeolites are present in the hydrated form (Lohse et al. [1970b]), the dielectric constant within this dispersion region increases considerably with increasing water content for the Na-form, but only very little if the zeolite contains divalent ions. A further dispersion lies at lower frequencies and is attributed to a conductivity dispersion due to the migration of the cations from hole to hole. Dielectric isotherms showed that two regions of the adsorbed water can be distinguished. Within the first region, the dielectric constant increases only to a small extent linearly with increasing water content, while in the second region an exponential increase is observed.

20.4.1.2. *Clay minerals*

Nelson et al. [1969] investigated the dielectric properties of water adsorbed by kaolinite over a temperature range from 35–115 °C and a frequency range from 33 Hz to 20 MHz. They observed a low frequency loss, which rose steadily as the frequency decreased and concluded that most of it is due to some relaxation process, probably arising from surface conductivity of the granules. The high frequency loss observed around 100 kHz moved over many decades to higher frequencies as the water content was increased. The rate of increase, however, decreased as the water content increased. Maxwell–Wagner models were not successful in explaining the observations on the magnitude of this loss. If one calculates the differential enthalpy ΔH^* for the relaxation process using the frequency of maximum loss, one obtains, for high water contents, values between 4–5 kcal mole^{-1}, which are similar to the value observed in bulk water (5.48 kcal mole^{-1}). The authors determined also the differential enthalpy ΔH_{ads} of water adsorption from water adsorption isotherms and found at high water contents $\Delta H \approx 10$ kcal mole^{-1}, while the heat of condensation for water is 9.71 kcal mole^{-1}. These observations indicate that the water molecules are held to the adsorbent largely by H-bonds although the larger values of ΔH and ΔH^* at low water contents suggest that some other binding force then operates. Most probably the origin of the high frequency loss is due to a dipole relaxation process. But even at the highest water contents, when ΔH^* is 4–5 kcal mole^{-1}, the relaxation rate is much less than that for water, f_m being ca. 10^6 instead of 10^{10} Hz. From the magnitude of the loss, one can conclude that adsorption occurs in clusters and that the dipole rotation is cooperative.

The dielectric constant and conductivity of water adsorbed on kaolinite was also measured by Baron [1970] at temperatures between 80–300 K. A dispersion process observed above 200 K had an activation energy of 36 kcal mole^{-1} for dry clay and 27 kcal mole^{-1} for water wet samples. The latter also showed a dispersion near 170 K with activation energies of 15, 11, 11, 11, 13 and 10 kcal mole^{-1} for Na, Li, K, H, Ca and Zn forms respectively of the kaolinite at 60% relative humidity. While the first relaxation process was attributed to dipoles resulting from the replacement of structural Al by divalent ions, the latter relaxation mechanism is attributed to the dipoles of the adsorbed water molecules.

Hoekstra and Doyle [1971] determined the dielectric behavior of water adsorbed by Na-montmorillonite at a frequency of 9.8×10^9 Hz over a temperature range from +20 to −100°C. In a plot of the dielectric loss versus $1/T$ (K^{-1}) a break in the slope is observed at −52°C, when the water content was 0.69 g H_2O/g clay. This indicates a phase transition, which is also observed by differential thermal analysis (Anderson [1967]). The dispersion occurring at about 10^9 Hz is consistent with a water layer of high mobility next to the solid surface. This highly disordered layer would also explain the observed activation energy of ca. 6 kcal mole^{-1} found at microwave frequencies. This microwave absorption need not necessarily indicate dipole rotations. The absorption could also be explained by a rupture of the H-bond between the hydrogen of the water and the oxygen of the clay surface. An observed loss at radio frequencies seemed to be due to free charge carriers and any dispersion superimposed on the large dc-conductivity loss is most likely due to polarization in the ionic double layer, or Maxwell–Wagner effects.

The dielectric constant of intercrystalline films of water in swollen Na-montmorillonite was also determined at moisture contents less than 40% by Deryagin et al. [1970]. They measured in the frequency range from 200 to 700 MHz and found that the dielectric constant of the water adsorbed was lower than that for an equal volume of water. It was attributed to a decrease in the orientation polarization of the adsorbed water molecules due to structuring in the film. K-montmorillonite, studied at low hydration (Mamy and Chaussidon [1967]), from 0.1–20 kHz and from −150 to +40°C, exhibited a sharp increase in the dielectric loss, when a monomolecular water film was completed. Further investigations on the Li, Na, K and Cs-bentonites in the same frequency range (Chaussidon [1966]), indicated that it is reasonable to describe the hydration process as a colloidal mechanism for high hydrations and as a crystal mechanism for low hydrations. The junction system is probably found for water contents, where the cationic field is "efficient" on all water molecules. For nonswelling clays, this water content corresponds probably to a state, where the maximum interlayer spacing detectable by X-rays is present. Dielectric measurements at low temperatures on this material (Mamy and Chaussidon [1967]) gave some indications as to the presence of a characteristic dipole absorption of the adsorbed water molecules.

An interesting study on the dielectric properties of hydrated vermiculite was reported by Schön [1970] and Schön and Weiss [1970, 1973]. Instead of polycrystalline samples, they used single crystals and measured with the electrical field vector perpendicular to the silicate layers, using as divalent interlayer cations Mg^{2+}, Ca^{2+}, Sr^{2+} and Ba^{2+}. The frequency range was 30 Hz to 10^7 Hz and the temperature was usually 25°C. The thickness of the samples was about 10^{-2} cm. With all samples, four different states of hydration have been found. If drying is not complete, a zero state of hydration is established with a basal spacing of 10.3 Å. The first state of hydration is reached at vapor pressures which depend on the nature of the interlayer cation. The basal spacing observed is then about 11.7 Å. The increase corresponds to a monolayer of water. The second state of hydration is reached, when a double layer of adsorbed water is established. In the completely dehydrated vermiculite, the dielectric constant does not change with the frequency and is independent of the nature of the interlayer cation, its mean value being 2.65. If water is adsorbed, ϵ increases with decreasing frequency, the absolute values being influenced by the interlayer cations. During the hydration process no intermediate values of ϵ' are observed. All samples show either the exact value of the lower hydration state or that of the higher state. No dipole or Maxwell–Wagner relaxation was observed in the frequency range covered. From measurements of the high frequency conductivity at different temperatures the activation energies of the first and zero state of hydration were obtained. For both states they were found to be about 2.3 kcal mole^{-1}. The authors believe that this activation energy is exactly the activation energy of diffusion by orientation of structural L-defects, as postulated by Bjerrum for ice. In the first state of hydration their results indicate that only two quick cooperative steps of molecular orientation proceed, starting at the three-bonded water molecules. This would result in a molecular relaxation time of about 10^{-11} s. The authors also explain, why measurements on polycrystalline mica-type-layer-silicates samples yield activation energies as in ice of about 10–14.5 kcal mole^{-1}, rather than the 2.3 kcal mole^{-1} found for single crystals, where the electric field vector is perpendicular to the silicate layer.

20.4.1.3. *Miscellaneous substances*

The dielectric constant of water adsorbed by silica gel was investigated by Nekrasova and Zhilenkov [1968]. They used two types of silica gel, K-2 with a pore diameter of 44 Å and KSK-2 with a pore diameter of 104 Å. The temperature range covered was from -2 to -107°C. The dielectric constant as a function of the water content increased considerably at low water contents but reaches a plateau at higher water contents. The authors concluded that the system investigated cannot be considered as a purely mechanical mixture and that the dielectric constant of the adsorbed water is lower than the one of bulk water. Structuring of the adsorbed water occurs more rapidly in KSK-2 than in K-2.

Water adsorbed by two porous glasses was investigated by Ebert et al. [1971] over a frequency range from 300 Hz to 400 kHz at temperatures between -180 and 200°C. When both glasses are dried at 120°C, no dielectric dispersion is observed. At low water contents of the samples one observes at low temperatures the β_1 relaxation process, which is attributed to the restricted orientation of the water molecules bound to the OH-groups of the glass surface. Also, already at low water contents of the samples, but at higher temperatures, the α process is observed, the intensity of which is independent of the amount of water adsorbed. Calorimetric measurements revealed, that at a transition temperature between -20 and $-5\,°C$ all water froze, except a certain amount corresponding to a bilayer. Because of the freezing the activation energy of the α process increases to values similar to the ones observed in ice (13.2 kcal mole^{-1}). Since the frequency range, at which the α dispersion occurs is also close to the one of ice, the α process is attributed to an ice-like relaxation process. With increasing addition of water, a β_2 relaxation process is observed, which is probably due to more strongly bound water molecules, possibly bound to OH-groups of the glass as well as to the ice-like regions. No essential influence of the different pore size of the two glasses on the dielectric properties was observed.

The dielectric properties of partially dehydrated NH_4Fe, KCr, KAl and NH_4Al sulphate alums were investigated by Hall [1968]. The measurements were performed at 4 MHz from 19–23°C. The change of the dielectric loss with time during the dehydration process showed an initial significant sharp decrease for NH_4Fe and KCr alums, whereas for the NH_4Fe alum a subsequent maximum occurs. NH_4Al and KAl showed only slight decreases in the early stages. Of the 12 molecules of water of crystallization assumed to be originally held by each alum, 9.6, 5.8 and 1.2 were removed from NH_4Fe, KCr and KAl alums, respectively, during this dehydration. During rehydration the loss increases with increasing rehydration pressure, the rate of increase being most marked with NH_4Fe and KCr alums.

In a further paper, the dielectric behavior of water adsorbed by partially dehydrated potassium chrome alum was reported by Hall and Kouvarellis [1968]. The frequency range was 40 Hz to 5 MHz and the temperature 27 and 38°C. If n, the number of moles water taken up per mole of $KCr(SO_4)_2{\cdot}6H_2O$, is zero, no significant dielectric absorption or dispersion is apparent. The six water molecules present at this state of hydration are obviously bound too tight to the lattice to contribute to the dielectric properties within the temperature and frequency range under consideration. As n increases to 0.19, two distinct absorption peaks appear. The high frequency peak B at 4 MHz was attributed to water in equilibrium with water vapor and the low frequency peak A to water of hydration. Cole–Cole plots showed that type A water could be described by a single relaxation time, whereas for type B water there was a definite distribution of relaxation time. The decrease in the distribution factor α with increasing values of n was interpreted as the increasing uniformity of the environment of water molecules as adsorption proceeds. The authors point out that the observed decrease in relaxation time for type A water with increasing n could arise, if the relaxation process involved the cooperative motions of dipoles or charge carriers of neighboring water molecules. In general, the dehydration and rehydration results indicated that only the adsorbed water in equilibrium with the water vapor makes any significant contribution to the dielectric loss. The straight line obtained in a plot of the dielectric loss versus the equilibrium vapor pressure suggests that the adsorbed water is bound by a mechanism similar to that with which water of crystallization is held inside the lattice.

A dielectric relaxation study of water molecules in oxide films on silicon was reported by Dorda and Kaderka [1967]. They measured the dielectric losses from 10^2 to 10^5 Hz for both etched and thermally grown oxide films on silicon under different relative humidities of ambient air. The observed results did not depend on the resistivity and type of silicon nor on the form of the sample. At low humidities the conductivity of the oxide prevails. With increasing humidity relaxation maxima around 8 and 60 kHz appear. The measured relaxation losses indicated dipoles located in the bulk of the oxide layer. Therefore, the authors conclude that water dipoles are not absorbed by the grown oxide even at high humidity. If the oxide is contaminated with sodium, the magnitude of hydration is changed drastically and a series of relaxation maxima located around 0.8, 8 and 60 kHz is observed. Even though the reason for this behavior is not entirely clear, it was suggested that water exists either in different surroundings or is bound in different ways.

The dielectric constant and loss of water adsorbed on a α Fe_2O_3 surface was reported by McCafferty et al. [1970]. They measured at 5, 15, 25 and 35 °C from 70 Hz to 300 kHz. The dielectric isotherms show that water in the first BET monolayer does not cause any change in the dielectric constant. If the second layer is started, the dielectric constant rises sharply and levels off after the formation of about three layers. Relaxation frequencies obtained from Cole–Cole plots increase smoothly from 10 Hz to 10 kHz at BET coverages of three or higher. Activation energies were calculated from the temperature dependence of the relaxation frequency. The following mechanism of hydration is proposed: the first layer of physically adsorbed water is immobile and presumably double H-bonded to the underlying hydroxyl layer. The second layer is mobile, its exact nature is, however, not yet clear. In the next layers the water has an ice-like formation. The activation energy at 25 °C at 2.5 layers where ice-like formation begins, is 16 kcal mole^{-1}. This value is not too different from the 13.3 kcal mole^{-1} observed in bulk ice. The activation energy increases with coverage to about 23 kcal mole^{-1}. The increased interaction due to H-bonding between successive layers is consistent with the gradual approach toward an ice-like film.

The dielectric properties of water sorbed by soda-lime glass, mica, calcinated diatomaceous earth and films of NaCl or $NaNO_3$ were determined by Ida et al. [1968]. Measurements were performed below 0 °C and time effects during adsorption were examined.

Wacrenier et al. [1967] investigated the dielectric behavior of zeolite 13-X, silica gel, activated alumina, clay and sandwater mixtures between 1 Hz to 5 GHz and 8–20 GHz in a temperature range from −75 to +75 °C at various water contents. The samples were surrounded by inert fluids (parrafin oil or nitrogen). The low frequency domains I and II show amplitudes, which are independent of the temperature and the water content, and critical frequencies strongly dependent on these quantities. They are attributed to interfacial polarization according to Maxwell–Wagner occurring around the grains (I) and around the cavities in the porous material (II). The high frequency domains III and IV have critical frequencies almost independent of the water content. The relaxation III is believed to be a superposition of two mechanisms, one existing in the absence of adsorbed water (IIIb), and one depending on the number of water molecules in the pores (IIIa). The domain IIIa has an activation energy of 9 kcal mole^{-1} and is attributed to the live time of the H-bonds between the adsorbed water molecules and the surface i.e. the adsorbed water molecules and the water molecules of constitution. The domain IIIb exists only when the sample contains nothing but hydroxyl water and water of constitution. It is attributed to a partial orientation of the water of constitution. The relaxation domain IV around 12 GHz is attributed to water in the cavities.

The dielectric properties of H_2O and D_2O adsorbed by silica gel were reported by Muroya et al. [1969]. They measured the dielectric loss at −22°C from 50–3000 Hz and at −70 °C from 1–1000 Hz. The relaxation time at −22 °C, which disappears after heating the sample at 200 °C, was attributed to D_2O. The deuterium exchange decreased the relaxation times of the dispersion regions. From the temperature dependence of the relaxation time, the following activation energies were obtained: high temperature dispersion, 2.7 mmole D_2O/g sg: 14.8 kcal mole^{-1}; 2.9 mmole H_2O/g sg: 13.5 kcal mole^{-1}; low temperature dispersion, 0.5 mmole D_2O/g sg: 12.3 kcal mole^{-1}; 0.56 mmole H_2O/g sg: 11.4 kcal mole^{-1}.

Fontaine and Vandorpe [1970] pointed out that changes in the slope and inflection points in dielectric isotherms may not only be caused by changes in the properties of adsorbed water, but also by the presence of low frequency relaxation at the frequency at which the isotherm is measured.

Stuchly [1970] reported the dielectric properties of sand, mashed potato powder and silica gel containing adsorbed water. The measurements, performed at 9.4 GHz indicated two ranges of water bonding. As long as the water content is below a critical value, water has a low permittivity since the absorption forces are large. In a second range, free water with high permittivity is present.

20.4.2. Water adsorbed by organic substances

20.4.2.1. *Ion exchange resins*

The process of freezing of water adsorbed by an ion exchange resin was investigated by dielectric measurements at 10 MHz by Dickel and Bunzl [1966]. If the cation exchange resin was loaded with H^+-ions, a first order phase transformation of the adsorbed water, observable by a sudden decrease of the dielectric constant, was found at −3.4°C if the water content of the sample was 10.4 mole H_2O/eq. ion exchanger. If the water content was 1.85 mole H_2O/eq., no sudden decrease of the dielectric constant was apparent within the temperature range from + 10 to − 30°C. The ion exchanger loaded with potassium ions showed a different behavior. At a water content of 6.14 mole H_2O/eq. ion exchanger, which is almost the saturation water content, a positive temperature coefficient of the dielectric constant is observed at temperatures below − 14°C and a negative coefficient above that temperature. Since positive temperature coefficients are in general observed for solids and negative ones for liquids, one can conclude that at − 14°C a phase transformation of higher order occurs, where no ice is formed, but certain structural rearrangements of the water molecules occur. If the ion exchanger contains less water, the temperature, where the structural rearrangement occurs, is shifted to higher temperatures. The different behavior of the ion exchanger in the H^+-form and in the K^+-form can be understood, if one assumes, that the formation of ice in a hydration shell can occur more easily around a H^+-ion than around a K^+-ion.

The dielectric behavior of water adsorbed by a polystyrene sulfonic acid ion exchange resin was investigated by Urban and Wallace [1968] between 10^2 and 10^5 Hz. They conclude from changes in the slope of the dielectric isotherm that the water adsorption occurs not continuously, but stepwise. This is in agreement with IR measurements by Zundel et al. [1962], Bunzl [1963] and Dickel and Bunzl [1964]. The activation energies observed were for the different cationic forms of the ion exchanger: H^+-form, 1.1 mole H_2O/eq.: 10.8 kcal $mole^{-1}$; for the Na^+-form, 3.8 mole H_2O/eq.: 28.6 kcal $mole^{-1}$ and for the K^+-form, 3.1 mole H_2O/eq.: 31.5 kcal $mole^{-1}$.

A further study on the dielectric behavior of water containing ion exchangers was reported by Gärtner and Sobetzko [1969], who determined the dielectric constant of a water containing weak acid ion exchange resin (polyacrylic acid) at 7 MHz between -50 and $+10$°C. Discontinuities in the dielectric isotherms of the sample in the H^+-form water at contents of $n - 0.5$ and 1.5 mole H_2O/eq. ion exchanger also indicate that the water molecules are adsorbed stepwise. A possible hydration structure is proposed. The dielectric isotherms of the Na^+-form of the ion exchanger showed that the first water molecule is bound much stronger than the ones adsorbed subsequently.

In order to investigate the hydration of the carboxylic acid group, Bunzl and Sansoni [1970] measured the dielectric properties of three ion exchange resins, carrying carboxylic groups of quite different pK values. The frequency range covered was from 30 Hz to 10 MHz. It was observed that in the course of the water absorption the resin with very weak carboxylic acid groups exhibited a discontinuity in the dielectric constant at all frequencies measured, if one water molecule is attached to two carboxylic groups. This can be explained by assuming that two carboxylic groups, which are connected by H-bonds in the dry state, begin to dissociate at the above water content. As a consequence the formation of $H_5O_2^+$ complexes becomes possible. Hydration structures below and above this critical water content are proposed. In the case of the two other ion exchange resins with higher acidity of the carboxylic group, no discontinuity in the dielectric properties during water adsorption originating from a stepwise hydration could be observed. Obviously, the mobility of the proton of the stronger acid groups is increased in these cases, already at very low water contents. The exceptional dielectric behavior of the ion exchanger with very weak carboxylic groups during hydration is also observed in the corresponding water adsorption isotherm, which follow, for this sample, the type 3 in the BET classification. For the other two ion exchangers, BET absorption isotherms of type 2 are observed. (Dickel and Hartmann [1960]).

The dielectric behavior of a cation exchanger in the H^+, Na^+, choline and triethylbenzylammonium form was determined by Frolov et al. [1972]. They measured the dielectric constant and loss as a function of the water content at 22°C from 0.4–100 MHz. A range of dielectric dispersion was observed and was attributed to the motion of opposite ions.

20.4.2.2. *Other organic substances*

All other organic substances, which were investigated in a hydrated state by dielectric measurements in the review period covered, were natural products and are discussed in the following section.

20.4.3. Water adsorbed by biopolymers

20.4.3.1. *DNA*

The dielectric properties of fibers of desoxyribonucleic acid containing 22% water were reported by Brot and Lassier [1966]. They measured ϵ' and ϵ'' between 50 Hz and 25 GHz of this substance at 25°C. Beside the dielectric dispersion observed below 1 Hz described earlier (Brot et al. [1965]), which was attributed to electrode polarization, the authors show the existence of three further relaxation domains: a relaxation domain at a frequency of 90 ± 20 Hz, which was attributed to the migration of ions in the intermolecular space. A relaxation domain at 4 ± 3 kHz, which was attributed to the reorientation of the dipoles of the adsorbed water molecules. The asymmetric distribution of this dispersion is explained by the binding of the water molecules to different types of adsorption sites. A fourth relaxation domain is observed at 14 GHz and is attributed to the presence of "free water", that is, water in adsorption layers, which only interacts with water molecules in other adsorption layers but not with the nucleotides.

A further dielectric study of DNA in the solid phase under different humidities was published by Mesnard and Vasilescu [1966]. The observed dielectric relaxation phenomena showed a very extended distribution of relatively long relaxation times, due to interfacial effects corresponding to ionic movements. The fact that moisture is responsible for the dielectric properties of wet and dry calf thymus DNA was also demonstrated by Hawranek et al. [1967]. They measured at frequencies from about 10^9–10^{10} Hz and found a relaxation time of $(2.8–3.5) \times 10^{-12}$ s.

20.4.3.2. *Proteins*

Dielectric constant and loss angles as well as the water adsorption isotherms at different temperatures were determined for water adsorbed on lyophilized horse hemoglobin pressed to pellets by Brausse et al. [1968]. The frequency applied was within the range of 120 Hz to 30 MHz. The adsorption isotherms indicated a critical hydration, h_c, at approximately 0.12 g H_2O/g protein. Above that value, solution conditions appeared to prevail. The dependence of the dielectric increment on the degree of hydration indicated the following behavior. A slow increase of the increment up to about $h = 0.165$ is followed by a steep increase in the region $h = 0.165$–0.27. Above 0.27 the curve levels off. The relaxation time decreases exponentially with increasing hydration and levels off at a hydration value of about $h = 0.25$–0.30. A distribution of relaxation times is observed. Between $h = 0.168$ and 0.352 an activation energy of 1.38 kcal mole^{-1} was found. The first BET monolayer was calculated to be complete at $h_m = 0.0576$. From the ratio of this quantity to the hydration value $h = 0.0695$ of monomolecular adsorption on polar sites, the fraction of polar groups of the hemoglobin molecule available for hydration was estimated as 0.83%. As possible polarization mechanism the authors discuss: Maxwell–Wagner effects, hindered dipole orientation of the macromolecule and polarization of the water molecule. In a personal communication the authors consider the Maxwell–Wagner effects as the most probable mechanism. The above investigations were continued by Reichle et al. [1970], using the method of thermally stimulated depolarization TSD (see section 20.3.4 for the principle and experimental details of this technique). Three distinct TSD peaks were observed in the temperature range from 85 to 350 K for lyophilized horse hemoglobin. They were denoted 1, 2 and 3 in the same order as they appear when starting at low temperatures. Peak 3, which cannot be observed at degrees of hydration less than 7% and which disappears if insulating electrodes are used, is attributed to electrode polarization, probably caused by the accumulation of hydrogen and oxygen by the polarization near the electrodes. Peak 2, which does not depend on the electrode material but strongly on the degree of hydration is due to an orientational polarization mechanism. The effective dipole moment calculated from susceptibility data is 200 D which suggests that it is the dipole moment of the hemoglobin molecule, which is oriented. Since the susceptibility of the completely dehydrated sample is zero and the activation energy is finite (11.5 kcal mole^{-1}), the authors conclude that the dipole moment of hemoglobin is zero when dehydrated, and is only formed in the presence of water. The activation energy (0.69 kcal mole^{-1} at 25% hydration) corresponds then to the potential barrier between adjacent favorable positions of the molecule. The relaxation time of peak 2 is about 50 s. Peak 1, observed for a process with a relaxation time about 100 s, is attributed to the orientation of polar amino acid residues. The activation energy of this process is 2.3 kcal mole^{-1} for the dehydrated sample and 1.1 kcal mole^{-1} for a 10% degree of hydration. The authors point out the differences in the results obtained by TSD measurements and those obtained by conventional measurements with alternating fields. The characteristics of the susceptibility and the dielectric constant on hydration are different for the two methods. The same is true for the relaxation times. Even though a great part of the discrepancies is due to the difference in the temperatures used in TSD and alternating field measurements, open questions still remain. It is evident, however, that the TSD method is a very interesting tool, capable of yielding additional information on the dielectric properties of solids obtained by conventional methods.

The dielectric properties of feather keratin were investigated by Rehm [1968]. It was observed that the relaxation time decreased by a factor 10^6 compared to free water. The water adsorption occurred at the polarized amino acid fixed ions of the polypeptide chain, which are then surrounded by ca. four water molecules. The observed linear increase of the dielectric adsorption isotherm up to 8% water indicates a constant well defined contribution of each newly adsorbed water molecule to the total polarization. The nature of water bound to wool or horn keratin was studied in dielectric measurements by Algie [1971]. He determined the dielectric loss from 1.2×10^2 to 10^5 Hz within the temperature range from 20 to -120°C. From the temperature dependence of the frequency of maximum dielectric absorption, activation enthalpies ΔH were obtained. For water contents from 0 to 16%, ΔH was about 12 kcal mole^{-1}, which is close to the value of ice. He found no indication for the coexistence of various phases of water with different relaxation times.

The H-bonding of water adsorbed on chicken egg white lysozyme, as deduced from dielectric measurements, was reported by Harvey and Hoekstra [1972]. They measured ϵ' and ϵ'' in the frequency range from 10^7 to 2.5×10^{10} Hz using standing waves and time domain reflectometry methods. The amount of water adsorbed ranged from 0.0 to 0.6 g H_2O/g protein, the temperature interval from -70 to $+60$°C. A break in the dependence of the dielectric constant at 25 GHz was observed near 0.3 g H_2O/g protein, which indicates that at this water content the lysozyme molecule is covered with a single monolayer of water. The dispersion due to this monolayer has a characteristic frequency of 2.5×10^8 Hz. An Arrhenius plot of this dispersion showed that the relaxation time increases with increasing temperature. From the slopes of these curves, the activation enthalpy is seen to be negative. The activation entropy is also negative and the activation energy positive. This can be explained by assuming that the activation process is the breaking of a H-bond between the protein molecule and a water molecule in the primary hydration shell, followed by the formation of better H-bonds between the reorientating water molecule and its neighbors. A second dispersion was observed around 10 GHz. It was attributed to the water molecules in the second layer of hydration. This process is characterized by a distribution of relaxation times ($\alpha = 0.3$). This dispersion persists to about −25 °C. It implies that water in the second layer is also more disordered than in the bulk liquid, but disordering is not as extreme as in the first layer. The results showed that adsorbed water cannot be simply classified as "ice-like" or "not ice-like" because the relaxation time is intermediate between that of ice and water, while the degree of order is not. Molecules of bound water are less mobile than in the bulk liquid, but they are simultaneously less ordered.

Dielectric properties of hydrated protein powders at 9.4 GHz were also reported by Kent [1972]. He investigated white fish meal in the temperature range from 10–90 °C, bovine serum albumen and hemoglobin from 0–40 °C and actomyosin from 0–30 °C. Assuming a proportional dielectric mixture model, he evaluated the dielectric properties in the first and second adsorption BET layer, ϵ_1'' and ϵ_2''. He found that ϵ_1'' is not negligible with respect to ϵ_2''. The relative closeness of ϵ_1'' and ϵ_2'' indicates that both layers are similarly rotationally hindered, possibly the difference being between one or two H-bonds. The second layer is thus much more tightly bound to the surface than was expected. The effect of temperature on the loss factor is, however, still considered to arise partly from the thermal activation of water molecules into less tightly bound sites. At greater coverages, water appears which is characteristically "liquid like" in relation to this dielectric processes. This suggests completion of the hydration shell for the protein and, considering the other data available, implies a shell only two molecules thick.

20.4.3.3. *Miscellaneous substances*

The orientation of sorbed water on chondroitin-4-sulphate, a substance found in animal tissue, was investigated by Lubezky et al. [1967] at 23.5 GHz at various temperatures. This substance has multiple polar groups, such as –OH, –NH, $-COO^-$, $-SO_3^-$ and –C–O–C– capable of binding water molecules by H-bonding. Two dielectric absorption maxima were found around 335 and 309 K, corresponding to potential barriers for molecular orientation of 2.59 kcal $mole^{-1}$ and 2.34 kcal $mole^{-1}$. The potential barrier of 2.59 kcal $mole^{-1}$ was attributed to water H-bonded to –OH type sites, where the binding involves more than one H-bond simultaneously. The broadness of this peak at 335 K was attributed to the binding of a number of different binding sites which are energetically close to each other. Therefore, this absorption band should be due primarily to sorption on free –OH sites and, later, on energetically $-COO^-$ and $-SO_3^-$ and –C–O–C– sites. The absorption peak at 309 K was assigned as arising from water molecules bound preferentially to –NH sites.

The dielectric properties of water adsorbed by potato starch and gelatin were investigated at frequencies from 15 to 100 MHz and from 50 kHz to 10 MHz by Dushchenko and Romanovskii [1969]. The changes in ϵ' and ϵ'' with the water content indicated changes from monomolecular to polymolecular adsorption of the water molecules. The region of dielectric dispersion encompassed a very large frequency range, which was attributed to the polydispersity of the system. Possibilities for the determination of monomolecular, free and intermediate water in intact plant tissues from dielectric measurements are suggested by Sedykh and Ishmukhametpva [1970]. Values for the activation energy and enthalpy of tissue water in wheat leaves, apple pulp and potato tuber are given.

The effect of moisture on the dielectric properties of chrome tanned leather was investigated by Miglyachenko and Krasnobokii [1971]. He observed that for water contents of 45–55%, ϵ' and tan δ have maximum values, which decreased sharply, when the capillary water is removed and decreased less with further dehydration.

A similar behavior was reported by Dushchenko et al. [1971] for the dielectric behavior of water containing cellulose. At low water contents they observed a linear increase of ϵ' and tan δ with the moisture content independent of the frequency (5 kHz to 7.5 MHz). In a second section of this curve, ϵ' and tan δ increased very rapidly with the moisture content and the frequency. This was attributed to a looser contact between cellulose and water molecules.

Water adsorbed by peat was investigated by Gamayunov et al. [1972]. Their results indicate that water sorbed on this substance is weakly structured and that the dipole moment of the water molecules adsorbed corresponds to that of free water molecules.

The dielectric properties of water containing granular and gelatinized potato starch at 1 and 3 GHz (Roebuck et al. [1972]), showed little dependence of ϵ' and ϵ'' on the water content and no frequency dependence below 20% water content.

References

Algie, J. E., 1971, J. Textile Inst. **62**, 696.

Allerhand, A. and P. von Schleyer, 1963, J. Am. Chem. Soc. **85**, 371.

Anderson, D. M., 1967, Nature. **216**, 563.

Anderson, J. E., 1971, Ber. Bunsenges. Phys. Chem. **75**, 294.

Auty, R. P. and R. H. Cole, 1952, J. Chem. Phys. **20**, 1309.

Baron, A., 1970, Acta Univ. Carolinae, Geol. **1**, 45.

Barrow, G. M., 1956, J. Am. Chem. Soc. **78**, 5802.

Bell Telephone Laboratories, 1969, Structure and Physical Properties of Liquid Water, Bibliography 124.

Bertie, J. E., H. J. Labbe and E. Whalley, 1969, J. Chem. Phys. **50**, 4501.

Bertin, D. M. and H. Lumbroso, 1966, C.R. Acad. Sci. Ser. **C263**, 181.

Beukel, A. van den, 1968, Phys. Status Solidi **28**, 565.

Bilgram, J. H., 1973, see Whalley et al. [1973] p. 246.

Bishop, P. G. and J. W. Glen, 1969, see Riehl et al. [1969] p. 231.

Bjerrum, N. 1951, Kgl. Danske, Videnskab, Selskab. Mat. Fys. Medd **27**, 3.

Brausse, G., A. Mayer, T. Nedetzka, P. Schlecht and H. Vogel, 1968, J. Phys. Chem. **72**, 3098.

Brot, C. and B. Lassier, 1966, J. Chim. Phys. **63**, 929.

Brot, C., B. Lassier, A. H. Sharbaugh, S. I. Reynolds and D. N. White, 1965, J. Chem. Phys. **43**, 3603.

Bruinink, J., 1972, J. Appl. Electrochem. **2**, 239.

Bunzl, K., Doctoral Thesis, 1963, University of Munich.

Bunzl, K. and B. Sansoni, 1970, Z. Naturforsch. **25**b, 808.

Camp, P. R., W. Kiszenick and D. Arnold, 1969, see Riehl et al. [1969] p. 450.

Camplin, G. C. and J. W. Glen, 1973, see Whalley et al. [1973] p. 256.

Chamberlain, J. S. and N. H. Fletcher, 1971, Phys. Kondens. Materie **12**, 193.

Chan, R. K., D. W. Davidson and E. Whalley, 1965, J. Chem. Phys. **43**, 2376.

Chang, H. C. and G. Jaffe, 1952, J. Chem. Phys. **20**, 1071.

Chapoton, A., A. Lebrun and G. Ravalitera, 1970, C.R. Acad. Sci., Ser. **C271**, 525.

Chapoton, A., B. Vandorpe and M. Choquet, 1971, C.R. Acad. Sci., Ser. **C272**, 1261.

Chapoton, A., G. Ravalitera, B. Vandorpe, M. Choquet and A. Lebrun, 1972, J. Chim. Phys. **69**, 1191.

Chaussidon, J. 1966, Tech. Rep. Ser. Int, At. Energy Agency **65**, 19.

Chojnacki, H., 1970, Theoret. Chim. Acta **17**, 244.

Chojnacki, H. and A. Gołebiewski, 1969, Molecular Crystals and Liquid Crystals **5**, 317.

Clerbaux, T., P. Duterme, Th. Zeegers-Huyskens and P. Huyskens, 1967, J. Chim. Phys. **64**, 1326.

Cleverdon, D., G. B. Collins and J. W. Smith, 1956, J. Chem. Soc. 4499.

Crossley, J., 1970, Advan. Mol. Relaxation Processes **2**, 69.

Cummins, P. G. and D. A. Dunmur, 1973, J. Phys. Chem. **77**, 423.

Cunningham, A. J., J. D. Payzant and P. Kebarle, 1972, J. Am. Chem. Soc. **94**, 7627.

Danielewicz-Ferchmin, J., 1972, Acta Phys. Polon. **A41**, 627.

Davidson, D. W., 1971, Can. J. Chem. **49**, 1224.

Davidson, D. W., M. Davies and K. Williams, 1964, J. Chem. Phys. **40**, 3449.

Davies, M., 1969, J. Chem. Education **46**, 17.

Davies, M. 1970, Aspects of Recent Dielectric Studies, in: Annual Reports on the Progress of Chemistry, Vol. 67 Chemical Soc. (London) p. 65.

Davies, M. ed., 1972, Dielectric and Related Molecular Processes, Vol. 1 (The Chemical Society, London).

Davis, M. L. and E. F. Westrum, 1961, J. Am. Chem. Soc. **65**, 338.

Debecker, G. and P. Huyskens, 1971, J. Chim. Phys. **68**, 287, 295.

DeMaeyer, L. and A. Persoons, 1974, Electric Field Methods, in: Techniques of Organic Chemistry, A. Wessberger, Vol. 6: Investigation of Rates and Mechanisms of Reactions, Ed. Hammes, G. (John Wiley & Sons, New York) pp. 211–235.

Dengel, O., U. Eckener, H. Plitz and N. Riehl, 1964, Phys. Letters **9**, 291.

DePaz, M., J. J. Leventhal and L. Friedman, 1969, J. Chem. Phys. **51**, 3748.

Deryagin, B. V., N. A. Krylov and V. F. Novik, 1970, Dokl. Akad. Nauk SSSR **193**, 126.

Detoni, S., D. Hadži, R. Smerkolj, J. Hawranek and L. Sobczyk, 1970, J. Chem. Soc. A, 2851.

Dickel, G. and K. Bunzl, 1964, Macromol. Chem. **79**, 54.

Dickel, G. and K. Bunzl, 1966, Z. Physik. Chem. N.F. **51**, 13.

Dickel, G. and J. W. Hartmann, 1960, Z. Physik. Chem. N.F. **23**, 1.

Dorda, G. and M. Kaderka, 1967, Solid State Commun. **5**, 791.

Dubinin, M. M., I. V. Zhilenkov, F. L. Lapinskii and V. M. Fedorov, 1971, Izv. Akad. Nauk SSSR, Ser. Khim. 431.

Dushchenko, V. P. and I. A. Romanovskii, 1969, Elektron. Obrab. Mater **4**, 55.

Dushchenko, V. P., I. A. Romanovskii, Y. V. Olenko and I-I. Poberrzhets, 1971, Inzh. Fiz. Zh. **20**, 642.

Duterme, P., T. Clerbaux, Th. Zeegers-Huyskens and P. Huyskens, 1968, J. Chim. Phys. **65**, 1266.

Ebert, G., W. Matron and F. H. Müller, 1971, Fortschrittsber. Kolloide Polym. **55**, 74.

Eckener, U., D. Helmreich and H. Engelhardt, 1973, see Whalley et al. [1973] p. 242.

Eda, B. and K. Ito, 1956, Bull. Chem. Soc. Japan **29**, 524.

Eda, B. and K. Ito, 1957, Bull. Chem. Soc. Japan **30**, 164.

Eigen, M. and L. De Maeyer, 1956, Z. Elektrochem. **60**, 1037.

Eigen, M., L. De Maeyer and H. Ch. Spatz, 1964, Ber. Bunsenges. Phys. Chem. **68**, 19.
Eisenberg, D. and W. Kauzmann, 1969, The Structure and Properties of Water (Oxford University Press, London)
Eley, D. D., 1962, in: Horizons in Biochemistry (Academic Press, New York) p. 341.
Eley, D. D. and D. I. Spivey, 1960, Trans. Faraday Soc. **56**, 1432.
Eley, D. D. and D. I. Spivey, 1961, Trans. Faraday Soc. **57**, 2280.
Eley, D. D. and D. I. Spivey, 1962, Trans. Faraday Soc. **58**, 405, 411.
Elzbutas, G. and K. Sasnauskas, 1971, Liet. TSR Aukst. Mokyklu Mokslo Darb., Chem. Chem. Technol. **13**, 345.
Engelhardt, H., 1964, Ph.D. Thesis, Physics Department, TU Munich.
Engelhardt, H. 1973, see Whalley et al. [1973] p. 226.
Engelhardt, H. and N. Riehl, 1965, Phys. Letters **14**, 20.
Engelhardt, H. and N. Riehl, 1966, Phys. Kondens. Materie **5**, 73.
Erič, B., E. V. Goode and D. A. Ibbitson, 1960, J. Chem. Soc. 55.
Eyring, H. and M. S. Ihon, 1969, Significant Liquid Structures, (Wiley, New York).
Faraday, M., 1850, Athenaeum p. 640.
Fauquembergue, R., E. Constant and L. Raczy, 1967, C.R. Acad. Sci. Ser. **C264**, 1325.
Fedorov, V. M., I. V. Zhilenkov and B. A. Glazun, 1968, Relaksationnye Yavleniya Tverd. Telakh, Tr. Vses. Nauk. Konf. 4th 1965, 645.
Few, A. V. and J. W. Smith, 1949, J. Chem. Soc. **753**, 2781.
Fischer, S. F. and G. L. Hofacker, 1969, see Riehl et al. [1969] p. 369.
Fischer, S. F., G. L. Hofacker and J. R. Sabin, 1969, Phys. Kondens. Materie **8**, 268.
Fischer, S. F., G. L. Hofacker and M. A. Ratner, 1970, J. Chem. Phys. **52**, 1934.
Fletcher, N. H., 1970, The Chemical Physics of Ice (University Press, Cambridge).
Fletcher, N. H., 1973, see Whalley et al. [1973] p. 132.
Fontaine, J. and B. Vandorpe, 1970, Bull. Soc. Chim. France **3**, 972.
Franks, F. ed. 1972, Water: A Comprehensive Treatise, Vol. 1: The Physics and Physical Chemistry of Water (Plenum Press, New York).
Friedrich, H. B. and W. B. Person, 1966, J. Chem. Phys. **44**, 2161.
Frolov, V. I., B. V. Moskvichev and G. V. Samsonov, 1972, Zh. Fiz. Khim. **46**, 1180.
Fukuda, A. and A. Higashi, 1969, Japan. J. Appl. Phys. **8**, 993.
Fuller, R. G. and F. W. Patten, 1970, J. Phys. Chem. Solids **31**, 1539.
Gärtner, K. and W. Sobetzko, 1969, Z. Physik. Chem. (Leipzig) **240**, 156.
Gamayunov, N. I., A. M. Lych and P. N. Davidovskii, 1972, Inzh. Fiz. Zh. **22**, 795.
Gieseke, W., H. Nägerl and F. Freund, 1970, Naturwissenschaften **10**, 493.
Glasser, L., 1975, Chem. Rev., 75, No. 1.
Glazun, B. A. and I. V. Zhilenkov, 1971, Dielektriki **1**, 28.
Glazun, B. A., V. M. Fedorov, M. M. Dubinin and I. V. Zhilenkov, 1966, Izv. Akad. Nauk SSSR, Ser. Khim. 1297.
Glazun, B. A., M. M. Dubinin, I. V. Zhilenkov and M. F. Rakityanskaya, 1967, Izv. Akad. Nauk SSSR, Ser. Khim. 1193.
Gobush, W. and C. A. Hoeve, 1972, J. Chem. Phys. **57**, 3416.
Goldstein, I. P., E. N. Guryanova and T. I. Perepelkova, 1972a, Zh. Obshch. Khim. **42**, 2091.
Goldstein, I. P., T. I. Perepelkova, E. N. Guryanova and K. A. Kotcheshkov, 1972b, Dokl. Akad. Nauk SSSR, **207**, 636.
Goode, E. V. and D. A. Ibbitson, 1960, J. Chem. Soc. 4265.

Gosar, P., 1963, Nuovo, Cimento **30**, 931.
Gosar, P. and M. Pintar, 1964, Phys. Stat. Solidi **4**, 675.
Gough, S. R., 1972, Can. J. Chem. **50**, 3046.
Gough, S. R. and D. W. Davidson, 1970, J. Chem. Phys. **52**, 5442.
Gough, S. R. and D. W. Davidson, 1973, see Whalley et al. [1973] p. 51.
Gough, S. R. and A. H. Price, 1968, J. Phys. Chem. **72**, 3347.
Gough, S. R., E. Whalley and D. W. Davidson, 1968, Can. J. Phys. **46**, 1673.
Gränicher, H., 1963, Phys. Kondens. Materie **1**, 1.
Gross, G. W., 1968, Advan. Chem. Ser. **73**, 27.
Guryanova, E. N., 1964, in: Vodorodnaya Sviaz, Eds. Sokolov, N. D. and V. M. Czulanovskij (Izd. Nauka, Moskwa) p. 281.
Guryanova, E. N. and I. P. Goldstein, 1962, Zh. Obshch. Khim. **32**, 12.
Gusakova, G. V., G. K. Denisov and A. L. Smolanski, 1971, Zh. Prikl. Spektrosk. **5**, 860.
Hadži, D., H. Ratajczak and L. Sobczyk, 1967, J. Chem. Soc. A, 48.
Hall, P. G., 1968, J. Chem. Soc. A, 66.
Hall, P. G. and G. K. Kouvarellis, 1968, Trans. Faraday Soc. **64**, 1940.
Haller, W. and H. C. Duecker, 1970, J. Res. Nat. Bur. Standards **64A**, 527.
Hamilton, W. C. and J. A. Ibers, 1968, Hydrogen Bonding in Solids (Benjamin, New York).
Hamilton, W. C., B. Kamb, S. J. LaPlaca and A. Prakash, 1969, see Riehl et al. [1969] p. 44.
Hanna, M. W., 1968, J. Am. Chem. Soc. **90**, 285.
Hanna, M. W. and D. E. Williams, 1968, J. Am. Chem. Soc. **90**, 5358.
Harvey, S. C. and P. Hoekstra, 1972, J. Phys. Chem. **76**, 2987.
Hasted, J. B., 1961, in: Progress in Dielectrics, Vol. 3 (Heywood, London).
Hawkins, R. E. and D. W. Davidson, 1966, J. Phys. Chem. **70**, 1889.
Hawranek, J. and L. Sobczyk, 1971, Acta Phys. Polon. **A39**, 651.
Hawranek, J., A. Koll, H. H. Kolodziej and L. Sobczyk, 1967, Proc. Colloq. Ampere (At. Mol. Etud. Radio Elec.) **14**, 527.
Hawranek, J., J. Oszust and L. Sobczyk, 1972, J. Phys. Chem. **76**, 2112.
Hawranek, J., J. Oszust and L. Sobczyk, 1973, unpublished.
Hedestrand, G., 1929, Z. Physik. Chem. **B2**, 428.
Helmreich, D., 1969, see Riehl et al. [1969] p. 231.
Herrington, T. M. and L. A. K. Staveley, 1964, J. Phys. Chem. Solids 25, 921.
Higashi, A. and Y. Teranoto, 1972, International Conference on Physics and Chemistry of Ice, Ottawa 1972.
Hill, N. E., W. E. Vaughan, A. H. Price and M. Davies, 1969, Dielectric Properties and Molecular Behaviour (Van Nostrand Reinhold, London).
Hobbs, M. E., M. S. Ihon and H. Eyring, 1966, Proc. Nat. Acad. Sci. U.S. **56**, 31.
Hobbs, P. V., 1974, Ice Physics (Oxford).
Hoekstra, P. and W. T. Doyle, 1971, J. Colloid Interface Sci. **36**, 513.
Hollins, G. T., 1964, Proc. Phys. Soc. **84**, 1001.
Holzapfel, W. B., 1969, J. Chem. Phys. **50**, 4424.
Horne, R. A., 1972, Water and Aqueous Solution (Wiley-Interscience, New York).
Hoyland, J. R. and L. B. Kier, 1969, Theoret. Chim. Acta **15**, 1.
Huyskens, P., 1972, Ind. Chim. Belge. **37**, 15.
Huyskens, P. and Th. Zeegers-Huyşkens, 1964, J. Chim. Phys. **61**, 81.
Huyskens, P. and G. Hernandez, 1973, Ind. Chim. Belge, **38**, 1237.
Ibbitson, D. A. and L. F. Moore, 1967, J. Chem. Soc. **B76**, 80.

Ibbitson, D. A. and J. P. B. Sandall, 1964, J. Chem. Soc. 4547.

Ida, M., Y. Sakabe and T. Nishimura, 1968, Sci. Rep. Kanazawa Univ. **13**, 91.

Itagaki, K., 1972, International Conference on Physics and Chemistry of Ice, Ottawa 1972.

Jaccard, C., 1959, Helv. Phys. Acta **32**, 89.

Jaccard, C., 1972, Transport Properties of Ice, in: Water and Aqueous Solutions, Ed. Horne, R. A. (Wiley-Interscience, New York) p. 25.

Jadżyn, J. and J. Małecki, 1972, Acta Phys. Polon. **A41**, 599.

Jellinek, H. H. G., 1972, The Ice Interface, in: Water and Aqueous Solutions, Ed. Horne, R. A. (Wiley-Interscience, New York) pp. 65–107.

Johari, G. P. and E. Whalley, 1973, see Whalley et al. [1973] p. 278.

Jones, S. J. and J. W. Glen, 1969, Phil. Mag. **19**, 13.

Kamb, B., 1964, Acta Cryst. **17**, 1437.

Kamb, B., 1965, J. Chem. Phys. **43**, 3917.

Kamb, B., 1968, in: Ice Polymorphism and the Structure of Liquid Water, Eds. Rich, A. and N. Davidson, (Freeman, San Francisco) pp. 508–542.

Kamb, B. and B. L. Davis, 1964, Proc. Nat. Acad. Sci. USA, **52**, 1433.

Kamb, B. and A. Prakash, 1970, Calif. Inst. Techn., Division of Geological Sciences, Contribution Nr. 1882.

Kawada, S., 1972, J. Phys. Soc. Japan **32**, 1442.

Kawada, A., A. R. McGhie and M. M. Labes, 1970, J. Chem. Phys. **52**, 3121.

Kebarle, P., S. K. Searles, A. Zolla, J. Scarborough and M. Arshadi, 1967, J. Am. Chem. Soc. **89**, 6393.

Kent, M., 1972, J. Phys. Appl. Phys. **5**, 394.

Kielich, S., 1958, Acta Phys. Polon. **17**, 239.

Kielich, S., 1972, see Davies [1972] p. 192.

Kim, D. Y. and V. H. Schmidt, 1967, Can. J. Phys. **45**, 1507.

Klug, D. D. and E. Whalley, 1973, see Whalley et al. [1973] p. 93.

Koide, G. T. and E. L. Carstens, 1972, J. Phys. Chem. **76**, 1999.

Kohlrausch, F. and A. Heydweiler, 1894, Wied. Ann. **53**, 209.

Koll, A., H. Ratajczak and L. Sobczyk, 1970, Roczniki Chem. **44**, 825.

Kollmann, P. A. and L. C. Allen, 1969, J. Chem. Phys. **51**, 3286.

Kopp, M., 1962, Z. Angew. Math. Phys. **13**, 431.

Kopylova, V. M., V. M. Fedorov, M. M. Dubinin and I. V. Zhilenkov, 1970, Izv. Akad. Nauk SSSR, Ser. Khim. 977.

Krishnan, P. N., I. Young and R. E. Salomon, 1966, J. Phys. Chem. **70**, 1595.

Kuriakose, A. K. and E. Whalley, 1968, J. Chem. Phys. **48**, 2025.

Lindemann, R. and G. Zundel, 1972, J. Chem. Soc. Faraday Trans. **2**, 979.

Lippert, J. L., M. W. Hanna and P. J. Trotter, 1969, J. Am. Chem. Soc. **91**, 4035.

Lohse, U., H. Stach, M. Hollnagel and W. Schirmer, 1970a, Monatsber. Deut. Akad. Wiss. Berlin **12**, 819.

Lohse, U., H. Stach, M. Hollnagel and W. Schirmer, 1970b, Monatsber. Deut. Akad. Wiss. Berlin **12**, 828.

Lohse, U., H. Stach, M. Hollnagel and W. Schirmer, 1971a, Z. Physik. Chem. (Leipzig), **246**, 91.

Lohse, U., H. Stach, M. Hollnagel and W. Schirmer, 1971b, Z. Physik. Chem. (Leipzig), **247**, 65.

Lubezky, I., F. A. Bettelheim and M. Folman, 1967, Trans. Faraday Soc. **63**, 1794.

Lumbroso, H., 1964, J. Chim. Phys. **61**, 132.
MacDonald, J. R., 1953, Phys. Rev. **92**, 4.
Maeno, N., 1972, J. Appl. Phys. **43**, 312.
Maeno, N., 1973, Can. J. Phys. **51**, 1045.
Magat, M., 1959, Dispersion Dielectrique et Liaison Hydrogene, in: Hydrogen Bonding, the Symposium on Hydrogen Bonding, Ljubljana, 1957, Ed. Hadži, D. (Pergamon Press, London) p. 309.
Maidique, M. A., A. von Hippel and W. B. Westphal, 1971, J. Chem. Phys. **54**, 150.
Majid, Y. A., S. K. Garg and D. W. Davidson, 1968, Can. J. Chem. **46**, 1683.
Mak, T. C. W. and R. K. McMullan, 1965, J. Chem. Phys. **42**, 2732.
Malarski, Z. and L. Sobczyk, 1969, C.R. Acad. Sci. Ser. **C269**, 874.
Małecki, J., 1960/61, Bull. Soc. Sci. Letters, Poznań, **16**, 93.
Małecki, J., 1962, Acta Phys. Polon. **21**, 13.
Małecki, J., 1965, J. Chem. Phys. **43**, 1351.
Małecki, J., 1966, Acta Phys. Polon. **29**, 45.
Małecki, J., 1969, Polaryzacja i nasycenie dielektryczne cieklych ukladow z wiazaniem wodorowym. Roztwory stezone, in: Hydrogen Bonding, collective work, Ed. Sobczyk, L. (PWN, Warszawa) p. 289.
Małecki, J., 1972, Dielectric Discussion Group Meeting, Cambridge.
Małecki, J. and Z. Dopierala, 1969a, Acta Phys. Polon. **36**, 385.
Małecki, J. and Z. Dopierala, 1969b, Acta Phys. Polon. **36**, 401, 409.
Mamy, J. and J. Chaussidon, 1966, Bull. Groupe Franc. Argiles, **17**, 93.
Mamy, J. and J. Chaussidon, 1967, Bull. Groupe Franc. Argiles **19**, 101.
Maričić, S., V. Pravdić and Z. Veksli, 1961, Croat. Chem. Acta **33**, 187.
Maričić, S., V. Pravdić and Z. Veksli, 1962, J. Phys. Chem. Solids **23**, 1651.
Mascarenhas, S., 1969, see Riehl et al. [1969] p. 483.
Matron, W., G. Ebert and F. H. Müller, 1971, Kolloid. Z. Z. Polym. **248**, 986.
McCafferty, E., V. Pravdić and A. C. Zettlemoyer, 1970, Trans. Faraday Soc. **66**, 1720.
McMullan, R. K. and G. A. Jeffrey, 1965, J. Chem. Phys. **42**, 2725.
Mesnard, G. and D. Vasilescu, 1966, C.R. Congr. Nat. Soc. Savantes, Sect Sci. **90**, 93.
Miglyachenko, A. F. and Y. N. Krasnobokii, 1971, Izv. Vyssh. Ucheb. Zaved. Tekhnol. Legk. Prom. **2**, 70.
Miller, S. R., R. Blinc, M. Brenman and J. S. Waugh, 1962, Phys. Rev. **126**, 528.
Morris, B., 1969a, J. Phys. Chem. Solids **30**, 73.
Morris, B., 1969b, J. Phys. Chem. Solids **30**, 103.
Morris, B. and D. W. Davidson, 1971, Can. J. Phys. **49**, 1243.
Mounier, S. and P. Sixou, see Riehl et al. [1969] p. 562.
Muroya, M. and S. Kondo, 1969, Bull. Chem. Soc. Japan **42**, 2724.
Murphy, E. J., 1964, J. Appl. Phys. **35**, 2609.
Murrell, J. N., 1969, Chem. in Britain, **5**, 107.
Nagle, J. F., 1973, see Whalley et al. [1973], p. 175.
Nakamura, T. and S. J. Jones, 1973, see Whalley et al. [1973] p. 365.
Nasielski, J. and E. V. Donckt, 1963, Spectrochim. Acta, **19**, 1989.
Nedetzka, T. and H. Engelhardt, 1972, International Conference on Physics and Chemistry of Ice, Ottawa, 1972.
Nekrasova, E. G. and I. V. Zhilenkov, 1968, Zh. Strukt. Khim. **9**, 406.

Nelson, S. M., H. H. Huang and L. E. Sutton, 1969, Trans. Faraday Soc. **65**, 225.
Noll, G., 1969, see Riehl et al. [1969] p. 113.
Noll, G., 1973, see Whalley et al. [1973] p. 350.
Nouwen, R. and P. Huyskens, 1973, J. Mol. Structure, **16**, 459.
O'Keeffe and C. T. Perrino, 1967, J. Phys. Chem. Solids **28**, 211.
Onsager, L., 1967, Ferroelectricity of Ice, in: Ferroelectricity, Ed. Weller, E. F. (Elsevier, Amsterdam) pp. 17–19.
Onsager, L. and M. Dupuis, 1962, Electrical Properties of Ice, in: Electrolytes, Ed. Pesce, B. (Pergamon Press, New York) pp. 27–46.
Paren, J. G., 1973, see Whalley et al. [1973] p. 262.
Paren, J. G. and J. C. F. Walker, 1971, Nature Phys. Sci. **230**, 77.
Pawełka, Z. And L. Sobczyk, 1973, unpublished.
Pick, M. A., 1969, see Riehl et al. [1969] p. 344.
Pick, M., H. Wenzl and H. Engelhardt, 1971, Z. Naturforschung **26a**, 810.
Piekara, A., 1950, Acta Phys. Polon. **10**, 37, 107.
Piekara, A., 1962, J. Chem. Phys. **36**, 2145.
Piekara, A., A. Chelkowski and S. Kielich, 1957, Z. Physik, Chem. **206**, 375.
Pigenet, C. and H. Lumbroso, 1966, C.R. Acad. Sci. Ser. **C262**, 1221.
Pigoń, K. and H. Chojnacki, 1967, Acta Phys. Polon. **31**, 1061 and 1069.
Pollock, J. M. and M. Sharan, 1967, J. Chem. Phys. **47**, 4064.
Pollock, J. M. and M. Sharan, 1969, J. Chem. Phys. **51**, 3604.
Pollock, J. M. and A. R. Ubbelohde, 1956, Trans. Fraday Soc. **52**, 1112.
Rahman, A. and F. H. Stillinger, 1972, J. Chem. Phys. **57**, 4009.
Ratajczak, H., 1966, Z. Physik. Chem. **231**, 33.
Ratajczak, H., 1972a, J. Phys. Chem. **76**, 3000.
Ratajczak, H., 1972b, J. Phys. Chem. **76**, 3991.
Ratajczak, H. and W. J. Orville-Thomas, 1973, J. Chem. Phys. **58**, 911.
Ratajczak, H. and L. Sobczyk, 1969, J. Chem. Phys. **50**, 556.
Ratajczak, H. and L. Sobczyk, 1970, Bull. Acad. Pol. Sci. Ser. Sci. Chim. **18**, 93.
Rehm, H. J., 1968, Mol. Strukt. Strahlenwirkung, Jahrestag. Deut. Ges. Biophys. Tagungsber. (Ed. Glubrecht, H., Georg Thieme Verlag, Stuttgart) p. 36.
Reichardt, C., 1965, Angew. Chem. **77**, 30.
Reichle, M., T. Nedetzka, A. Mayer and H. Vogel, 1970, J. Phys. Chem. **74**, 2659.
Riehl, N., B. Bullemer and H. Engelhardt, Eds., 1969, Physics of Ice, Munich 1968, (Plenum Press, New York).
Roebuck, B. D., S. A. Goldblith and W. B. Westphal, 1972, J. Food Sci. **37**, 199.
Rogers, E. and A. R. Ubbelohde, 1950, Trans. Faraday Soc. **46**, 1051.
Ruepp, R., 1973, see Whalley et al. [1973] p. 179.
Rustanova, S. N., O. A. Osipov, A. D. Garnovskij, Yu. V. Kolodyashnij and M. A. Salimov, 1970, Khim. Geterotsikl. Soedin. **2**, 180.
Schaarschmidt, K., 1967, Z. Physik. Chem. **235**, 17.
Schaarschmidt, K., L. Ballester and M. Sanfeliz, 1967, Z. Physik. Chem. **236**, 1.
Schmidt, V. H., 1963, J. Chem. Phys, **38**, 2783.
Schmidt, V. H., 1965, J. Sci. Instr. **42**, 889.
Schmidt, V. H. and J. C. Haslett, 1964, Final Progress Report Nat. Sci. Foundation Grant G-17174.

Schmidt, V. H. and E. A. Uehling, 1962, Phys. Rev. **126**, 447.
Schmidt, V. H., J. E. Drumheller and F. L. Howell, 1971, Phys. Rev. **B4**, 4582.
Schneider, K. E., D. Schenk, B. Bullemer and H. Engelhardt, 1972, International Conference on Physics and Chemistry of Ice, Ottawa, 1972.
Schön, G., 1970, Doctoral Thesis, University of Munich.
Schön, G. and A. Weiss, 1970, Reunion Hispano-Belga Miner. Arcilla, An, 36.
Schön, G. and A. Weiss, 1973, Z. Naturforsch. **28**b, 140.
Schwarz, G., 1972, Advan. Mol. Relaxation Processes **3**, 281.
Sedykh, N. V. and N. N. Ishmukhametpva, 1970, Fiziol. Rast. **17**, 945.
Seidensticker, R. G., 1972, J. Chem. Phys. **56**, 2853.
Seidensticker, R. G. and R. L. Longini, 1969, see Riehl et al. [1969] p. 471.
Senko, M. E., 1961, Phys. Rev. **121**, 1599.
Sheka, I. A., 1956a, Zh. Fiz. Khim. **30**, 109.
Sheka, I. A., 1956b, Zh. Obshch. Khim. **26**, 1340.
Sheka, I. A. and E. E. Kriss, 1956, Zh. Neorg. Khim. **4**, 2505.
Silsbee, H. B., E. A. Uehling and V. H. Schmidt, 1964, Phys. Rev. **133**, A 165.
Sixou, P. and A. Jeneveau, 1973, see Whalley et al. [1973] p. 295.
Skulski, L. and W. Waclawek, 1971, Biuletyn WAT, **20**, 45.
Slater, J. C., 1941, J. Chem. Phys. **9**, 16.
Smith, J. W., 1950, J. Chem. Soc. 3532.
Smith, J. W., 1953, J. Chem. Soc. 109.
Smith, J. W. and S. M. Walshaw, 1957, J. Chem. Soc. 3217.
Sobczyk, L., 1969, Absorpcja dielektryczna ukladow z wiazaniem wodorowym, in: Hydrogen Bonding, collective work, Ed. Sobczyk, L. (PWN, Warszawa) p. 307.
Sobczyk, L. and Z. Pawełka, 1973, Roczniki Chem. **47**, 1523.
Sobczyk, L. and Z. Pawełka, 1974, J. Chem. Soc. Faraday Trans. 1 **70**, 832.
Sobczyk, L. and J. K. Syrkin, 1956, Roczniki Chem. **30**, 893.
Sobczyk, L. and J. K. Syrkin, 1957, Roczniki Chem. **31**, 1245.
Steinemann, A., 1957, Helv. Phys. Acta **30**, 581.
Stuchly, S. S., 1970, J. Microwave Power **5**, 62.
Takagi, Y., 1948, J. Phys. Soc. Japan **3**, 271 and 273.
Taubenberger, R., 1973, see Whalley et al. [1973] p. 187.
Taubenberger, R., M. Hubmann, and H. Gränicher, 1973, see Whalley et al. [1973] p. 194.
Thomas, J. M., J. R. N. Evans, T. J. Lewis and P. Secker, 1969, Nature **222**, 375.
Thomson, W., 1850, Proc. Roy. Soc. (Edinburgh) **2**, 267.
Truby, F. K., 1955, J. Appl. Phys. **26**, 1416.
Urban, Z. and R. A. Wallace, 1968, J. Electrochem. Soc. **115**, 276.
Vanderkooy, J., J. D. Cuthbert and H. E. Petch, 1964, Can. J. Phys. **42**, 1871.
Vassiliev, V. G. and Ya. K. Syrkin, 1941, Acta Physicochim. URSS, **14**, 414.
Venkateswaran, A., J. Easterfield and D. W. Davidson, 1967, Can. J. Phys. **45**, 884.
von Hippel, A., 1971, J. Chem. Phys. **54**, 145.
von Hippel, A., D. B. Knoll and W. B. Westphal, 1971, J. Chem. Phys. **54**, 134.
von Hippel, A., R. Mykolajewycz, A. H. Runck and W. B. Westphal, 1972, J. Chem. Phys. **57**, 2560.
von Hippel, A., A. H. Runck and W. B. Westphal, 1973, see Whalley et al. [1973] p. 236.
Wacrenier, J. M., J. Fontaine, A. Chapoton and A. Lebrun, 1967, Rev. Gen. Elec. **76**, 719.

Walrafen, G., 1968, in: Equilibria and Reaction Kinetics, in: Hydrogen-Bonded Solvent Systems, Ed. Covington, A. K. (Taylor and Francis, London) pp. 9–29.

Weissmann, M. and N. V. Cohan, 1965, J. Chem. Phys. **43**, 119.

Weith, A. J., M. E. Hobbs and P. M. Gross, 1948, J. Am. Chem. Soc. **70**, 805.

Whalley, E., 1969, see Riehl et al. [1969] p. 19.

Whalley, E. and J. E. Bertie, 1967, J. Chem. Phys. **46**, 1264.

Whalley, E. and H. J. Labbé, 1969, J. Chem. Phys. **51**, 3120.

Whalley, E., D. W. Davidson, and J. B. R. Heath, 1966, J. Chem. Phys. **45**, 3976.

Whalley, E., J. B. R. Heath, and D. W. Davidson, 1968, J. Chem. Phys. **48**, 2362.

Whalley, E., S. J. Jones and L. W. Gold, Eds., 1973, Physics and Chemistry of Ice, Ottawa 1972, (Royal Society of Canada, Ottawa).

Williams, G., 1970, Advan. Mol. Relaxation Processes, **1**, 409.

Williams, G., 1972, Chem. Rev. **72**, 55.

Wilson, G. J., R. K. Chan, D. W. Davidson and E. Whalley, 1965, J. Chem. Phys. **43**, 2384.

Wong, P. T. T. and D. D. Klug, 1973, see Whalley et al. [1973] p. 87.

Wörz, O. and R. H. Cole, 1969, J. Chem. Phys. **51**, 1546.

Young, I. G. and R. E. Salomon, 1968, J. Chem. Phys. **48**, 1635.

Zaromb, S. and R. Brill, 1956, J. Chem. Phys. **24**, 895.

Zundel, G., H. Moller and G. M. Schwab, 1962, Z. Elektrochem. **66**, 129.

CHAPTER 21

NUCLEAR MAGNETIC RELAXATION IN HYDROGEN BONDED LIQUIDS

H. G. HERTZ AND M. D. ZEIDLER

Institut für Physikalische Chemie und Elektrochemie der Universität, D-75 Karlsruhe, West Germany

Contents

21.1. Basic principles 1029
21.2. Correlation times in non-hydrogen bonded liquids 1036
21.3. A qualitative explanation for the slower motion of hydrogen bonded molecules 1037
21.4. Correlation times for non-aqueous hydrogen bonded liquids 1043
21.5. Water and aqueous mixtures 1049
Note added in proof 1057
References 1058

21.1. Basic principles

The magnetization $\boldsymbol{M}$ of a sample of N nuclear magnetic moments with spin I in the state of thermal equilibrium is a vector parallel to the static magnetic field of magnitude H_0. Identifying the direction of the static magnetic field as the z-direction we have

$$M_z^0 = \frac{N\gamma^2\hbar^2 I(I+1)}{3kT} H_0, \qquad M_x^0 = M_y^0 = 0. \tag{21.1}$$

If by any physical method we produce a nuclear magnetization different from that given by eq. (21.1), then on removing the external stimulus the magnetization moves back to its equilibrium value. This motion consists of a contribution $\gamma \boldsymbol{M} \times \boldsymbol{H}_0$ caused by the torque of the static magnetic field on the magnetization and of a contribution due to the interaction of the nuclear moments with other particles in their molecular environment. The second of the two contributions implies the relaxation process of the magnetization. Especially in liquids the interaction energy of the nuclei with their neighboring particles fluctuates statistically with time, since the relative position of the particles changes continuously due to their thermal motion. Now, qualitatively the rate of approach of the magnetization towards its equilibrium value increases with increase of the spectral intensity of the power spectrum of the fluctuating interaction energy at the resonance frequency $\omega = \gamma H_0$. The power spectrum itself is related to the time constants of the molecular motion or, in a somewhat modified version, to the microdynamic behavior of the liquid and from this stems the possibility of obtaining information concerning the microdynamic behavior of liquids from NMR relaxation studies.

We will not give here any further details of the theory but will simply present the formulae necessary for the discussion to be given in the next sections. Introductions to the theory may be found in a number of textbooks (Andrew [1956], Pople et al. [1959], Abragam [1961], Slichter [1963] and Farrar and Becker [1971]).

If we have a spin system where every molecule carries two like spins and if there exists only a magnetic dipole–dipole interaction between the nuclei–as, for instance, in water–the change with time of the z-component of the magnetization is given by the formula:

$$dM_z/dt = -(1/T_1)(M_z - M_z^0). \tag{21.2}$$

The time constant T_1 of the exponential approach of M_z towards its equilibrium value is called the longitudinal relaxation time. $1/T_1$ is usually called the longitudinal relaxation rate. The description of the change in time of the transversal components of the magnetization towards its equilibrium value is very much facilitated if we observe it in a coordinate system rotating with the resonance frequency about the z-axis, z being the direction of the static magnetic field. Then it may be shown that the following relations hold:

$$dM_x/dt = -(1/T_2)M_x, \qquad dM_y/dt = -(1/T_2)M_y \tag{21.3}$$

if we have again magnetic dipole–dipole interaction between a pair of two like spins. T_2 is called the transversal relaxation time.

With the exception of one H-bonded liquid–glycerol–all systems of interest in this article have sufficiently low viscosity such that–if magnetic dipole–dipole interaction is concerned–$T_1 = T_2$, as may be shown by the theory. In the NMR literature this situation is denoted as "extreme narrowing". However, it is a typical property of molecules forming H-bonds that the lifetime of the X–H proton in the molecule is not very long compared with the time range relevant for the NMR experiments. We may roughly say that $T_2 \leq T_1$ if the residence time of the proton on a given molecule is longer than 10^{-5} s and shorter than 10 s. This effect is due to the electron coupled interaction of the proton with the nuclear magnetic moment of the atom X or to the magnetic interaction with the electron cloud of the molecule (environmental variation of the chemical shift). For exchange faster than this $T_1 = T_2$. Since 10^{-5} s is an extremely long time compared with the times describing the microdynamic process in question here, measurements of T_2 yield no additional information over that obtained by T_1 studies. On the other hand, the reaction kinetics of the proton exchange is not the subject of the present article and nuclear magnetic relaxation studies of the slow exchange of the molecule into and out of the H-bonded state have not been performed yet–to the authors' knowledge. Thus T_2 measurements, which are much more difficult to perform than T_1 measurements, will no longer be mentioned.

The experimental technique used to measure NMR relaxation times will only be sketched very briefly. The relaxation times of interest for the present article cover a range from about 10^{-3} s to about 10^2 s. The experimental procedure is based on the possibility of measuring the magnitude of the nuclear magnetization vector–apart from a constant factor–directly. Of course, if this can be done, its change in time can also be studied. In order to measure the magnitude of the magnetization vector the latter is rotated into a plane perpendicular to the static magnetic field. Here it performs a precessional motion and induces a radiofrequency voltage of the resonance frequency ω in a coil, the axis of which is also perpendicular to the static field and which surrounds the sample containing the spins. The induced r.f. voltage is proportional to the magnitude of the magnetization. The rotation of the nuclear magnetization into the plane perpendicular to the static field is accomplished by the application of a r.f. pulse with frequency ω of appropriate duration (e.g. a "90° pulse") (Slichter [1963], Abragam [1961] and Farrar and Becker [1971]), or by sweeping with the static field through the resonance with proper sweep rate and applying to the sample a steady magnetic r.f. field of frequency ω and of correct amplitude (adiabatic fast passage). Thus, if at $t = 0$ a magnetization zero or $-M_z^0$ is produced (either by one of the methods just mentioned or otherwise), the growth of the magnetization towards equilibrium can be followed at any later time t by rotating it into the plane perpendicular to the static field. The decay of the transverse magnetization is, in principle, directly given by the approach of the r.f. voltage induced by the precessing magnetization towards the value zero. However, certain modifications of the signal appear as a consequence of the inhomogeneity of the static field and the diffusive motion of the particles carrying the spins.

These effects in turn may be utilized to measure the self-diffusion coefficient of a molecule in the liquid. On the following pages we shall report some experimental self-diffusion coefficients, if these are obtained by the nuclear magnetic relaxation method (spin–echo technique).

Short relaxation times $T_2 \leqslant 10^{-3}$ s may simply be obtained from the linewidth of the nuclear magnetic resonance signal

$$1/T_2 = \tfrac{1}{2}\gamma \Delta H_{1/2}.$$

$\Delta H_{1/2}$ is the half-width in gauss. $T_1 = T_2$ when the condition of extreme narrowing is fulfilled.

The relaxation rate $1/T_1$ is composed of two parts, an intramolecular and an intermolecular part:

$$1/T_1 = (1/T_1)_{intra} + (1/T_1)_{inter}. \tag{21.4}$$

There is always an experimental procedure by which it is possible to separate the intramolecular relaxation rate from the intermolecular one. We shall only discuss the intramolecular contribution which originates from the interaction between the two spins within the same molecule.

The intramolecular relaxation rate contains the microdynamic information we are interested in. This is the rotational correlation time τ_c. Since τ_c is the fundamental quantity with which this article is concerned it will be useful to explain the concept of the rotational correlation time a little more thoroughly. It is convenient to consider as an example–which will be described in more detail below–the water molecule in liquid water. To begin with, let us follow the thermal motion of a water molecule in the liquid. It is clear that, as the time goes on, the center of mass of the molecule is displaced from its original position as a consequence of the fluctuating forces exerted by its neighbors. The path of the H_2O molecule is an erratic one as is typical for a diffusive motion. After a mean time τ_t the center of mass is displaced by a length d,

$$\tau_t = d^2/6\,D,$$

where d is the diameter of the molecule and D is its self-diffusion coefficient. With $d \approx 3$ Å and $D = 2.31 \times 10^{-5}$ cm^2/s (Mills [1971]) one finds $\tau_t \sim 7 \times 10^{-12}$ s. Now assume that at $t = 0$ the vector which connects the two water protons within the molecule is parallel to the z-axis of any arbitrary laboratory coordinate system. As we have seen after a time τ_t the fluctuating forces have displaced the molecule by a length d, but at the same time fluctuating torques are acting on the water molecule. It follows that at time τ_t the proton–proton vector will no longer be parallel to the z-axis of our coordinate system, it will have any other direction. Now let us define the reorientation time τ_r of the water molecule as the time after which the proton–proton vector has changed its direction by 180°. We know from dielectric relaxation time measurements that $\tau_r \approx 5 \times 10^{-12}$ s (25 °C) (Pottel and Kaatze [1969]). (More details will be given below.) Actually, the strict definition of the reorientation time is somewhat different, but for the qualitative purpose, which is the intention of this article, the description given suffices. The process we have just envisaged is the rotational diffusion and if we examine the path of the intersection point of the proton–proton vector with a unit sphere we would again obtain an erratic path. The elementary steps of this rotational path may be very small, in this event we speak of microstep diffusion, or they may be larger, implying free paths of 90°, say, in which case we would speak of rotational jump diffusion. The motion we have to assign to the reorientation of the water molecule is closer to the former type (see below) and we may assume that for the reorientational motion of H-bonded molecules the microstep process is generally the better description of the truth.

In the case of microstep rotational diffusion we define the rotational correlation time τ_c:

$$\tau_c = \tfrac{1}{3}\tau_r.$$

As the rotational steps become greater $\tau_c \to \tau_r$, but for the present article it is sufficient to identify the rotational correlation time as one-third of the reorientation time of the molecule considered. For simplicity, in the following we shall use the denotation "correlation time" instead of the full description rotational correlation time.

The interconnection between the intramolecular proton relaxation rate and the correlation time is given by the relation

$$(1/T_1)_{\text{intra}} = \tfrac{3}{2}(\gamma^4\hbar^2/b^6)\tau_c. \tag{21.5}$$

b is the proton–proton separation within the water molecule. Since, in general, b is known from the geometry of the molecule, the measurement of $(1/T_1)_{\text{intra}}$ yields the rotational correlation time τ_c. So far we have considered a molecule like water which contains only one proton vector. Clearly, in general the molecules which are of interest for us contain more than one proton–proton vector. For instance, in methanol we find three vectors connecting the methyl protons and three vectors connecting the OH proton with each of the methyl protons. For each of these proton–proton vectors we may write a formula like eq. (21.5). Then, in a fair approximation the total observed intramolecular relaxation rate is given by the relation

$$1/T_1 = (1/n)\sum_{i<j} 2/T_{ij}. \tag{21.6}$$

n is the number of like spins–here usually protons–in the molecule. T_{ij} is the relaxation time one would calculate for the spin pair formed from spins i and j according to eq. (21.5) and the sum includes all pairs of like spins in the molecule. For each of the rates $1/T_{ij}$ one has the same correlation time τ_c. This implies that the molecule undergoes isotropic rotational diffusion and that no intramolecular reorientations occur. Very often the diffusive motion is an anisotropic one, i.e. the correlation time of a given vector is different from that of another vector in the same molecule. Moreover, there may be internal motions in the molecule, for instance the methyl group may rotate faster than another functional group of the molecule. In the most extreme situation it is no longer meaningful to speak of a correlation time of a molecule, e.g. when the entire molecule is a very flexible system and it is not obvious which part of the molecule represents "the molecule" as a whole. However, on the following pages we shall not go into any details of the more complicated form of motions. The reader should keep in mind that some of the correlation times given below may not concern the rotation of the full molecule or may regard anisotropic rotational behavior. It should be mentioned that a full understanding of the more complicated situation–in particular of the internal motions–is still lacking.

Equation (21.5) applies when the magnetic dipole–dipole interaction between the two protons is the relaxation mechanism. The deuteron has spin $I = 1$ and thus it possesses a nuclear quadrupole moment which interacts with the electric field gradient, produced by the surrounding electron cloud at the position of the nucleus. The nuclear quadrupole interaction represents a very powerful relaxation mechanism (Andrew [1956], Pople et al. [1959], Abragam [1961], Slichter [1963] and Farrar and Becker [1971]) which in most cases is much stronger than the magnetic dipole–dipole mechanism. From the theory one obtains the formula

$$1/T_1 = \tfrac{3}{40}\{(2I+3)/[I^2(2I-1)]\}(1+\tfrac{1}{3}\tilde{\eta}^2)[(eQ/\hbar)\partial^2 V/\partial z'^2]^2\tau_c. \quad (21.7)$$

Here Q is the electric quadrupole moment of the nucleus, $\partial^2 V/\partial z'^2$ is the z-component of the field gradient in a coordinate system fixed in the molecule which is chosen to form the principal axes of the field gradient tensor. The asymmetry parameter, $\tilde{\eta}$, is given by

$$\tilde{\eta} = \left(\frac{\partial^2 V}{\partial x'^2} - \frac{\partial^2 V}{\partial y'^2}\right) \Big/ \frac{\partial^2 V}{\partial z'^2}.$$

τ_c is the same correlation time which appears in eq. (21.5) if the rotational motion is isotropic. For anisotropic motion, in general the direction z' differs from that of the proton–proton vector–if existent–and thus eqs. (21.5) and (21.7) yield different correlation times. Use of eq. (21.7) has the advantage that the experimental relaxation rate is the intramolecular relaxation rate, $1/T_1 = (1/T_1)_{\text{intra}}$, it has the disadvantage that one has to know the quadrupole coupling constant $(eQ/h)\partial^2V/\partial z'^2$ in order to calculate τ_c, which is not always an easy matter (Lucken [1969]). Very often eq. (21.7) has been utilized for the study of the water correlation time in aqueous mixtures where the solvent was D_2O (see below).

We are now prepared to formulate once again the objective of the present article. Many of the chapters of this book treat the H-bond

$$X\text{–}H\cdots X$$

in a liquid. Such a bond always involves a vector defined in a molecular coordinate system which gives the direction of the bond, or more precisely, of the minimum in the orientational part of the intermolecular potentials concerned. Due to the thermal movement in the liquid such a vector very rapidly changes its orientation with respect to the laboratory coordinate system. The present article gives an answer to the question: How long are the times after which such reorientations have occurred? But this leads us immediately to the next question: If the rotational correlation time of the vector pointing originally in the direction of the H-bond is τ_c, then, after a time τ_c, is this direction still the direction of the H-bond between the same two coupled particles? In other words, is the lifetime of the H-bond longer or shorter than the correlation time τ_c? Furthermore, we will have to investigate the question: In what way does the correlation time of a vector fixed in a molecule depend on the fact that this molecule is bound to another one via a H-bond? We shall see that this question is closely connected with the former one, i.e. the influence of the H-bond on the correlation time depends on the lifetime of the molecule in the bonded state.

To follow up this problem the most natural starting point is the consideration of a number of correlation times of those molecules which do not form H-bonds. A presentation of some typical examples for this class of molecules will be given in section 21.2. As already mentioned above, the parameters describing rotational and translational diffusion both reflect the interplay of the thermal energy with the intermolecular forces. Therefore, they usually show the same relative order. This is understandable because the rotational motion is determined by the angular dependence and the translational motion by the radial part of the intermolecular potential. Very often the depths of these potentials do not differ very much. Thus, all facts which have so far been described as regarding rotational diffusion have their analogues in the translational form of motion, which also gives us the justification to quote the self-diffusion coefficients when NMR data are available. Section 21.3 treats the question: Why do H-bonds cause a slowing down of molecular motion?

Thus far we have outlined our general program. Unfortunately, the number of experimental data regarding H-bonded liquids with which the framework formulated could be filled, is relatively small. This is particularly true for non-aqueous H-bonded liquids. We give a review of the experimental data available in section 21.4. The results for aqueous system are summarized in section 21.5.

21.2. Correlation times in non-hydrogen bonded liquids

In table 21.1 we have collected a number of rotational correlation times τ_c for non-H-bonded liquids. In the second column of this table the nucleus whose relaxation has been utilized, is listed. H means proton relaxation, here eqs. (21.5) and (21.6) have been applied. The proton relaxation times given in the table are derived from intramolecular relaxation rates, thus they do not represent direct experimental quantities. We have not included in table 21.1 those molecules whose proton relaxation is caused to an appreciable amount by spin–rotation interaction (Farrar and Becker [1971]). Here the separation of the dipole–dipole contribution from the spin–rotation contribution makes the results somewhat uncertain. For the evaluation of all other (i.e. non-proton) relaxation data in table 21.1 eq. (21.7) has been used. The corresponding quadrupole coupling constants $(eQ/h)\partial^2 V/\partial z'^2$ have not been quoted in table 21.1; they may be recalculated from the relaxation times together with the correlation times listed. The correlation times are effective ones and regard the reorientation of a certain vector in the molecule or they represent an average over several intramolecular vectors. Results of an analysis in terms of rotational diffusion coefficients around various *molecular* axes have not been reproduced, although in a number of cases such an analysis has been performed in the literature. The same is true when the effective correlation times contain contributions due to internal motion within the molecule, like rotation of the methyl group relative to another part of the molecule. In those cases where the reader finds in table 21.1 markedly different correlation times for the same molecule, he may take this as a qualitative indication for the existence of anisotropic or internal motions.

The most important information which we can obtain from table 21.1 is the fact that the correlation times of molecules of small and moderate size which do not form H-bonds range from 0.5 to 2 ps. The shorter times in this range correspond to the smaller molecules, the longer times are found for the larger molecules. If still larger molecules would have been investigated, then the correlation times would turn out to be still longer. Our attention should be focused more towards those molecules which do not contain a halogen atom – or at least not more than one halogen atom – because for these molecules with relatively low molecular mass the effect of the H-bond in molecules with similar composition and mass will become much more pronounced than for those compounds where the H-bond effect is one effect besides strong Van der Waals interaction effects due to the presence of many halogen atoms. Thus we find three molecules in table 21.1 for which τ_c is markedly longer than 2 ps: CCl_3CN, C_6H_5Cl and $VOCl_3$. In all these cases the comparatively long correlation time is surprising and the suspicion that some experimental error is involved cannot quite be ruled out. That dimethyl sulfoxide with its high viscosity and high boiling point has a longer correlation time, is not so surprising. In analogy to the behavior of the rotational motion the translational self-diffusion coefficients range from $\approx 5\times10^{-5}$ down to $\approx 1.5\times10^{-5}$ cm^2/s, where again the high diffusion coefficients correspond to the small molecules and the low ones to the larger molecules*. Here also DMS shows an exceptional behavior.

21.3. A qualitative explanation for the slower motion of hydrogen bonded molecules

A molecule which is H-bonded to its neighbors has a longer rotational correlation time and a smaller self-diffusion coefficient than a non-H-bonded molecule of comparable size and mass. We shall give a qualitative explanation for this behavior for rotational motion; the explanation for translational motion is analogous. Our starting point is the well-known Debye relation (Abragam [1961], Andrew [1956], Pople et al. [1959]).

$$\tau_c = 4\pi a^3 \eta / 3kT. \tag{21.8}$$

a is the radius of the molecule in question. First we assume that this molecule is not H-bonded and that its shape does not deviate too much

* It should be kept in mind that the self-diffusion coefficients listed in table 21.1 are those obtained by the spin–echo technique. In some cases more precise results are available in the literature, which have been measured by other methods. The deviations are within 20%.

TABLE 21.1

Typical correlation times for non-hydrogen bonded molecules in the pure liquid

Substance	Resonance	Temperature (°C)	$T_{1,2}$ (s)	τ_c (ps)	D (cm²/s)[a]	References
BCl_3	^{11}B	24	3.74×10^{-2}	1.05	–	Gillen and Noggle [1970]
	^{35}Cl	25	1.22×10^{-4}	1.11		
CD_3I	D	22	5.27	0.40	3.5×10^{-5}[c]	Gillen et al. [1971] Zeidler [1965]
CD_2Cl_2	D	27	2.9	0.82	4.2×10^{-5}	O'Reilly et al. [1972]
	^{35}Cl	27	4.1×10^{-5}	1.20	–	
$CDCl_3$	D	20	1.35	1.73	2.32×10^{-5}	Huntress [1969]
	^{35}Cl	20	3.1×10^{-5}	1.31		Bender and Zeidler [1971]
CCl_4	^{35}Cl	30	2.2×10^{-5}	1.72	–	Gillen et al. [1971]
CD_3CN	D	25	6.57	0.34	4.7×10^{-5}	Woessner et al. [1968]
	^{14}N	25	4.38×10^{-3}	1.10		Zeidler [1971]
CCl_3CN	^{14}N	27	1.13×10^{-3}	3.65	–	Gillen and Noggle [1970]
	^{35}Cl	25	1.35×10^{-5}	2.70		
CH_3CD_2Br	H	25	31.3[b]	0.67	3.9×10^{-5}	Engelsmann et al. [1974]
CD_3CH_2Br	H	25	50.0[b]	0.84		

$CH_3C{\equiv}CD$	D	−30	1.21	1.29	–	Jonas and DiGennaro [1969]
$CD_3C{\equiv}CH$	D	−30	6.26	0.33	–	Jonas and DiGennaro [1969]
$(CD_3)_2CO$	D	25	5.0	0.47	5.2×10^{-5}	Zeidler [1965]
	^{17}O	25	6.13×10^{-3}	1.1		O'Reilly and Peterson [1971]
						Goldammer and Hertz [1970]
$(CH_3)_2SO$	H	25	5.56[b)]	3.31	0.8×10^{-5}	Zeidler [1965]
$(CH_3)_2{=}CD_2$	D	0	3.03	0.59	–	Assink and Jonas [1970]
$(CH_2)_4O$	H	25	47.6[b)]	0.57	3.4×10^{-5}	Zeidler [1965]
C_5D_5N	D	30	1.17	1.55	–	Zeidler [1965]
	^{14}N	25	1.65×10^{-3}	1.86	1.9×10^{-5}	Kintzinger and Lehn [1971]
C_6H_5Cl	H	24	41.3[b)]	4.0	1.8×10^{-5}	O'Reilly and Peterson [1971]
						Figgins and Rhodes [1969]
C_6H_6	H	25	111[b)]	1.1	2.5×10^{-5}	Zeidler [1965]
C_6H_{12}	H	25	19.6[b)]	1.2	1.7×10^{-5}	Zeidler [1965]
NH_3	H	−71	15.6[b)]	0.66	–	Powles and Rhodes [1967]
$VOCl_3$	^{51}V	25	2.32×10^{-2}	3.67	–	Gillen and Noggle [1970]
	^{35}Cl	25	1.53×10^{-4}	3.11		

a) The diffusion constants were determined for the non-deuterated substances.

b) Intramolecular contribution of the proton relaxation rate.

c) Self-diffusion coefficient measured at 25°C.

from a sphere. Let the temperature be room temperature. Then some typical examples to which eq. (21.8) may refer are liquid ammonia (at its equilibrium pressure), acetone and cyclohexane. For acetone τ_c in eq. (21.8) refers to the CO vector, the CH_3 groups rotate relative to the CO group. The internal motion of C_6H_{12} is so slow that in the present context it may be neglected. $\hat{\eta}$ in eq. (21.8) is often denoted as microviscosity, it is not a well-defined quantity. Here we define $\hat{\eta}$ in terms of eq. (21.8), i.e. it connects the experimental quantity τ_c with a, which the latter may be considered to be a fairly well-defined quantity. We may notice that in general $\hat{\eta}$ will also depend on the size of the molecule, it will increase with increasing molecular radius a. Thus, we obtain the result from eq. (21.8) that τ_c increases as the size of the molecule increases, a fact which we have already realized when discussing table 21.1.

Now imagine that the H-bond is "switched on" for one of our molecules. In effect this means: (i) that the NH_3 molecule is transformed into a H_2O molecule–the slight change in mass and geometry being neglected, (ii) that acetone is transformed into ethanol–again the somewhat greater change in mass and geometry being neglected and (iii) that cyclohexane is transformed into cyclohexanol. Of course, the H-bonds produce associated molecules, i.e. aggregates of molecules. Thus, we may say that now we have "molecules" $(C_2H_5OH)_n$, $n > 1$, and the radius of such a "molecule" is larger than that of the single molecule, or more precisely, if the shape of the aggregates deviates much from a sphere, in at least one direction the extension of the new "molecule" is larger. If we use this statement for eq. (21.8) we may write

$$\tau_c = 4\pi (a^*)^3 \hat{\eta}^* / 3kT, \qquad (21.9)$$

where a^* is the effective radius of the molecule in question, i.e. of H_2O or C_2H_5OH etc. Since in the bound state the actual instantaneous particle is the aggregate of several molecules, $a^* \geqslant a$, where a is the radius of a single "isolated" molecule (in vacuo, say). $\hat{\eta}^*$ is the microviscosity for the H-bonded liquid and since we said that $\hat{\eta}$ increases with increasing particle size, we get $\hat{\eta}^* > \hat{\eta}$. Thus, we arrive at the result that the correlation time of a H-bonded molecule is longer than that of the corresponding molecule without H-bonds.

Next we will ask: In what way is the effective radius connected with the actual radius of the H-bonded aggregate? In this present situation the correlation time is usually of the order 10^{-11} s. Thus, if the aggregate behaves like a stable molecule for a time longer than 10^{-11} s, then with respect to the observation time implied in τ_c it is indistinguishable from an ordinary stable molecule, i.e. it is fully equivalent to a stable molecule. Furthermore, in general we have an association equilibrium involving aggregates composed of varying numbers of particles n. During the process of the relaxation of the nuclear magnetization a given individual molecule for many times is a member of *each* aggregate of type n. Under these conditions, as the theory shows, the effective radius of the molecule, a^*, equals the mean radius of the aggregate:

$$(a^*)^3 = p_0(-4.5)a_0^3 + p_1(-4.5)a_1^3 + p_2(-4.5)a_2^3 + \dots . \tag{21.10}$$

p_1 is the probability that a given molecule is H-bonded to one neighbor, p_2 is the probability that it is bonded to two neighbors and so on.

But as we pointed out the validity of eq. (21.10) is coupled to the condition that $\tau_h \gg \tau_c$, i.e. the lifetime in a given (n-fold) bound state must be longer than the correlation time τ_c. We may write

$$\tau_h = \tau_0 \exp(-\Delta E/RT).$$

$\tau_0 \approx 10^{-14}$ s, thus for $\tau_h \geqslant 10^{-11}$ s we obtain $\Delta E \leqslant -4.5$ kcal/mole. Therefore, we have added the parentheses with (-4.5) at the ps in order to give the correct classification what we mean with the word aggregate. We note that the quantity ΔE regards the potential depth between two molecules in the presence of all the other molecules in the surrounding liquid, i.e., it is an effective intermolecular potential. There may be aggregates for which $\Delta E > -4.5$ kcal/mole. These aggregates contribute to $\hat{\eta}^*$, i.e. they are the cause that $\hat{\eta}^*$ also becomes greater–apart from a certain repulsive contribution–when the H-bonds are formed. We have denoted this effect as the "background effect" (Grüner and Hertz [1972]). The situation just described, $\tau_h \gg \tau_c$, may be called the situation of slow exchange. Slow exchange means that the residence times in the various environments are much longer than the corresponding correlation times, thus we have slow exchange relative to the rotational correlation times, or we may also say slow exchange with respect to the observation times τ_c. If the dynamical details–here the exchange–of a process occur in times longer than the observation times, then dynamical details may be studied. But if we insert eq. (21.10) in eqs. (21.9) and (21.5):

$$(1/T_1)_{\text{intra}} = (2\pi\gamma^4\hbar^2\hat{\eta}^*/kTb^6)[p_0(-4.5)a_0^3 + p_1(-4.5)a_1^3 + \dots], \tag{21.11}$$

then we see that we cannot actually study dynamical details but only one parameter, the mean correlation time or mean radius of the aggregates. In fact, we do not observe the molecular reorientation itself. This would be the case if we studied the dielectric relaxation process. In the measurements to be described here we observe the motion of the nuclear magnetization and thus really our observation time is the nuclear magnetic relaxation time. Relative to this observation time the exchange of a molecule among the various H-bonded states corresponds to the situation of fast exchange (Zimmerman and Brittin [1957])–only a mean value of a dynamic process can be observed.

Next let us consider a liquid with weaker H-bonding for which

$$p_i(-4.5) = 0 \quad \text{for all } i \neq 0.$$

Now we have no aggregates which are long-lived as compared with the rotational correlation time. But still there are aggregates in the liquid, formally association can always be assigned to the molecular structure of a liquid and it is entirely a matter of our convenience which number for ΔE we choose for our classification: bonded aggregate–non-bonded single molecule. Here, for our classification, we select the loosest coupling between two molecules such that the requirement is satisfied that a rotational correlation time or a rotational diffusion coefficient can be ascribed to the resulting aggregate. The definition of an aggregate in this sense implies that the rotational diffusion of all member molecules is the same–or at least that the rotational paths are essentially coupled. We have mentioned that diffusive motion is built up of an appreciable number of small steps. During each of these small steps the instantaneous angular velocity of the member molecules–and thereby of the aggregate–has changed. It follows that the bonding strength between the molecules must fulfill the requirement that the lifetime of the attachment is longer than the correlation time for the angular velocity (Hertz [1970a, 1971a,b]). An estimate yields that for this definition of aggregates $\Delta E \approx -kT$ when the molecules considered are small (Hertz [1971b]). Now the lifetime of a H-bonded aggregate is: $10^{-13} \lesssim \tau_h \lesssim 10^{-12}$ s. Thus the lifetime is smaller, in the limit very much smaller than the rotational correlation time τ_c and we have the situation of fast exchange relative to the correlation time τ_c. Again, τ_c might act as the observation time and if this is the case, we would be unable to get any detailed dynamical information from a study of τ_c. We have fast exchange and thus only the mean value of the dynamical process can be obtained.

It may be shown (Beckert and Pfeifer [1965], Anderson [1967], Hertz [1967b] and Anderson and Fryer [1969]) that under the condition of fast exchange the interconnection between the effective radius and the radii of the aggregates is given by the formula:

$$(a^*)^{-3} = p_0(-0.6)a_0^{-3} + p_1(-0.6)a_1^{-3} + p_2(-0.6)a_2^{-3} + \ldots, \qquad (21.12)$$

thus

$$\tau_c^{-1} = (3kT/4\pi\hat{\eta}^*)[p_0(-0.6)a_0^{-3} + p_1(-0.6)a_1^{-3} + \ldots]. \qquad (21.13)$$

All aggregates with binding energies $\Delta E > -0.6$ kcal/mole contribute to the background effect $\hat{\eta}^*$. Together with a change of the repulsive contribution due to binding, they cause an increase of $\hat{\eta}$. We see from eq. (21.13) that again τ_c becomes longer when H-bonds are present, but the connection between effective radius and aggregate size has changed. Now

$$(1/T_1)_{\text{intra}} = 2\pi\gamma^4\hbar^2\hat{\eta}^*/kTb^6(p_0a_0^{-3} + p_1a_1^{-3} + \ldots). \qquad (21.14)$$

Usually, the direct application of eqs. (21.10), (21.11), (21.12), (21.13) and (21.14) presents the following difficulty: If the monomeric particle is of spherical shape, then in general the higher aggregates will more or less deviate from spherical shape. Thus, each term would have to be corrected for anisotropic motion. However, it is the main purpose of these equations to show that in the case of slow exchange (with respect to the correlation time) the correlation time is mostly influenced by the large aggregates whereas with fast exchange the correlation time is mainly determined by the small aggregates.

The intermediate situation between slow exchange and fast exchange leads to formulae which are fairly complicated even for an exchange between two environments; they cannot be given in closed form for more than two environments (Zimmerman and Brittin [1957], Beckert and Pfeifer [1965], Anderson [1967], Hertz [1967b] and Anderson and Fryer [1969]).

21.4. Correlation times for non-aqueous hydrogen bonded liquids

Correlation times for a number of alcohol and carboxylic acid molecules in the pure liquid are collected in table 21.2. We give a brief discussion of these data. The correlation time for the hydroxyl group of methanol is ≈ 4 ps. The result obtained from the ^{17}O and ^{2}D relaxation need not be the same. This could be due to the incomplete knowledge of the quadrupole coupling constant and possibly to slightly anisotropic reorientation. We used the quadrupole coupling constant for water in both cases (see

TABLE 21.2

Typical correlation times for hydrogen bonded molecules in the pure liquid

Substance	Resonance	Temperature (°C)	T_1 (s)	τ_c (ps)	D (cm²/s)	References
CH_3OH	^{13}C	35	36.5[a)]	0.43	–	Lyerla and Grant [1972]
	^{17}O	25	4.17×10^{-3}	4.3		Versmold and Yoon [1972]
CH_3OD	H	25	21.3[b)]	0.9	2.2×10^{-5}	Goldammer and Zeidler [1969]
	D	25	0.31	4.4[c)]		
CD_3OH	H	25	10.7	–	–	Grüner and Hertz [1972]
	D	25	5.26	0.45		Goldammer and Zeidler [1969]
C_2H_5OH	^{17}O	25	1.54×10^{-3}	11.6	–	Versmold and Yoon [1972]
C_2H_5OD	H	25	9.3[b)]	2.2	1.1×10^{-5}	Goldammer and Zeidler [1969]
	D	25	0.14	9.8[c)]		
CD_3CH_2OH	D	25	0.94	2.6	–	Goldammer and Hertz [1970]
CH_3CD_2OH	D	25	1.16	2.1		Goldammer and Hertz [1970]
C_2D_5OH	H	25	5.9	–	–	Gruner and Hertz [1972]
i-C_3H_7OD	H	25	4.6	4.4	–	Grüner and Hertz [1972]
$(CH_3)_3COD$	H	25	2.15[b)]	7.9	0.26×10^{-5}	Goldammer and Zeidler [1969]
	D	25	2.3×10^{-2}	59.6[c)]		

$(CD_3)_3COH$	D	25	0.17	13.5	–	Goldammer and Zeidler [1969]
$C_3H_5(OD)_3$	H		see text			
HCOOH	^{13}C	35	10.2[a]	4.6	–	Lyerla and Grant [1972] Alger et al. [1971]
$^{13}CH_3COOH$	^{13}C	35	14.9[a]	1.0	–	Lyerla and Grant [1972] Alger et al. [1971]
CH_3COOD	H	25	6.37[b]	3.0	1.3×10^{-5}	Goldammer and Zeidler [1969]
	D	25	0.13	10.5[c]		
CD_3COOH	D	25	1.32	1.8	–	Goldammer and Zeidler [1969]
CD_3CH_2COOH	D	25	0.71	3.3	0.9×10^{-5}[d]	Tutsch [1973] Uedaira
CH_3CD_2COOH	D	25	0.48	4.9	–	Tutsch [1973] Hertz and Tutsch
$CD_3CH_2CH_2COOH$	D	25	0.67	3.5	0.8×10^{-5}[d]	Tutsch [1973] Hertz and Tutsch Uedaira
$CH_3CD_2CH_2COOH$	D	25	0.36	6.5	–	Tutsch [1973] Hertz and Tutsch
$CH_3CH_2CD_2COOH$	D	25	0.25	9.3	–	Tutsch [1973] Hertz and Tutsch

[a] Dipolar contribution of ^{13}C relaxation rate.
[b] Intramolecular contribution of the proton relaxation rate.
[c] Corrected with respect to data given in the literature due to improved knowledge of quadrupole coupling constant.
[d] Self-diffusion coefficients were determined for the non-deuterated substances.

below), thus we did not apply the rather low quadrupole coupling constant which was measured by O'Reilly and Petersen [1971] for solid methanol. The fact that we mention this point may demonstrate that the question of the precise quadrupole coupling constant is not always definitely settled. The reader should be aware of this when using the data given in tables 21.1 and 21.2.

We see that the correlation time of the OH vector in methanol is roughly by a factor of three longer than the correlation times of the CN or CO vector in the non-H-bonded molecules acetonitrile and acetone, respectively. Clearly, this is the effect of H-bonding as has been described in the preceding section. The H-bonded aggregates in alcohols are chain-like, thus it is difficult to assign a definite a_i value to an aggregate built up of i molecules. Since the ratio of the observed correlation times for the bonded and non-bonded molecules is less than 10, we may conclude that the dimensions of those aggregates, which are long-lived as compared with τ_c–eq. (21.10) is valid–are still of the same order of magnitude as the molecular radius of monomeric methanol. From this we may estimate that chain fragments of about four CH_3OH molecules are stable within the time range of the rotational correlation time.

The correlation time of the methyl group is shorter than that of the polar group which is due to internal motion: The methyl group performs a rotation about the CO axis. It is somewhat surprising, however, that the CH_3 correlation time is essentially the same as found for CH_3I, CH_3CN and $(CH_3)_2CO$. The fact that the polar group performs a slow motion should also be reflected in the behavior of the methyl group. We shall see shortly that such a behavior is observed for other molecules, an explanation of the peculiar finding with CH_3OH is still lacking. In formic acid the reorientational movement is also slow when compared with molecules of similar size (see table 21.1). The same is true for the larger alcohol and carboxylic acid molecules listed in table 21.2. For all the latter molecules one observes faster reorientation of the alkyl groups than for the polar part of the same molecule which indicates internal rotation. But now indeed we find the slower motion of the polar group reflected in longer values of the methyl and methylene τ_cs when these are compared with the corresponding figures in table 21.1.

In table 21.2 the proton relaxation times of CD_3OH and C_2D_5OH are given. In the conventional sense these relaxation times are produced by the *intermolecular* magnetic dipole–dipole interaction because there are no other protons in the same molecule. The interactions with the deuterons may be ignored. But we may consider a H-bonded dimer or polymer as a molecular species and then the OH relaxation rate represents an intramolecular relaxation rate with respect to the H-bonded species. Now eq. (21.5) can again be applied. This implies that the lifetime of the dimer, say, is longer than the correlation time τ_c. From the geometry of the H-bonds in the association chain we estimate a proton–proton distance between nearest neighbors $b = 2.3$ Å. With this figure from eq. (21.5) one calculates a correlation time $\tau_c = 7.8 \times 10^{-12}$ s (Göller et al. [1972]) for methanol and $\tau_c = 1.4 \times 10^{-11}$ s for ethanol. The agreement of these times with those derived from the OD and ^{17}O relaxation is satisfactory and, taking account of the influence of next nearest neighbors, could be improved. This shows that at least dimers are stable during times of the order of τ_c, thus, as in methanol, it is very likely that aggregates containing a couple of molecules form firmly attached entities during times of the order of τ_c.

It is interesting to compare our correlation times in methanol and ethanol with the corresponding dielectric relaxation times. In the pure liquid methanol $\tau_r = 5.3 \times 10^{-11}$ s was observed, and for ethanol the main dielectric relaxation time is $\tau_r = 1.70 \times 10^{-10}$ s (23°C) (Saxton et al. [1962], Dalbert et al. [1949], Crossley [1970]). From the NMR data one would expect $\tau \approx 3 \times 4$ ps $= 1.2 \times 10^{-11}$ s and $\tau_r \approx 3 \times 1.1$ ps $= 3.3 \times 10^{-11}$ s for methanol and ethanol, respectively. The fact that the observed (macroscopic) dielectric relaxation times are five times as long as expected from NMR data is very surprising and as yet unexplained.

Glycerol demands a special discussion. Here the network of the H-bonds contains fragments of appreciable size which are long-lived in comparison with the corresponding correlation times. As a consequence, the rotational correlation times are by three orders of magnitude longer than the times we have met so far. This means that the reciprocal correlation time is of the order of magnitude of the nuclear magnetic resonance frequency, thus we no longer have the situation of extreme narrowing. Now it is possible to measure the correlation time directly by studying the frequency dependence of the nuclear magnetic relaxation rate. At 21 °C one finds a distribution of correlation times (Kintzinger and Zeidler [1973]) which, for example, can be described by a Cole–Davidson distribution (Davidson and Cole [1951]) with a limiting correlation time $\tau_0 = 2.12 \times 10^{-9}$ s and a distribution parameter ca. 0.8.

Mixtures of H-bonded liquids with inert solvents offer an interesting possibility to study directly the reaction of the rotational correlation time on the formation and dissociation of the H-bonds. However, so far the number of such investigations is very small. The most extensive study regards mixtures of methanol, ethanol and *i*-propanol with CCl_4, $CHBr_3$ and CS_2 (Grüner and Hertz [1972]). The most important features of these measurements may be described as follows: For the ethyl protons of ethanol over a wide composition range the correlation time shows practically no change as CCl_4 is added to pure ethanol. At alcohol contents $x < 0.2$ (x = mole fraction of alcohol) the dissociation of the H-bonds begins and consequently, with the gradual appearance of more and more monomers, the correlation time decreases. From these experiments one estimates an acceleration of the rotational motion by a factor $2\frac{1}{2}$, but the extrapolation towards $x \to 0$ is not very safe. At low x the decrease of the correlation time may be described by eqs. (21.10) and (21.11) taking into account only monomers and dimers, when the p_is are taken from other physical sources (Grunwald and Coburn [1957]). Thus we have the situation of slow exchange, but at higher alcohol contents the formation of further H-bonds is probably compensated by decreasing lifetimes, thus τ_c remains constant. The qualitative behavior of the self-diffusion coefficient is very similar (Hardt et al. [1959]). Moreover, the composition dependence of the OH relaxation is also qualitatively the same as that of the two aforementioned forms of motion. From this, again we see that the OH $\leftrightarrow$ HO proton–proton vector behaves entirely like an intramolecular vector with respect to the H-bonded aggregates. Were the association effect absent, then the OH relaxation rate should decrease linearly as the alcohol concentration decreases, which is to be expected for an ordinary intermolecular relaxation rate. Model substances like $(C_2H_5)_2CO_2$ (propyl-acetate) and C_2H_5CN for the dimer and monomer, respectively, have been studied and it has been shown that they have correlation times very similar to those of the corresponding alcohol species. Lastly, it has been shown that the correlation times of the alkyl groups and of the OH group go through a maximum as a solvent like CH_3Br is added. The viscosity of this solvent is higher than that of ethanol, thus when the alcohol is diluted with this solvent one observes first an increase of the correlation time of all groups of C_2H_5OH, then, at about $x \leqslant 0.3$ the influence of the dissociation becomes more and more dominant and the correlation time – now corresponding to monomers – decreases. Here we observe a typical superposition of the H-bonding and the background effect with regard to a binding energy of about -4.5 kcal.

In methanol the higher viscosity of CCl_4 as compared with that of CH_3OH is sufficient to cause an increase of the correlation time of both groups as x decreases. Again, τ_c passes through a maximum when the dissociation of the H-bonds becomes effective. From this it may be seen that the background effect, i.e. the effect of the weaker interaction with the large CCl_4 molecules, is appreciable.

In the following we give a brief summary of other studies which have been undertaken with mixtures of H-bonded liquids. Proton T_1 measurements in alcohol–CCl_4 mixtures have been published by Bezrukov and Vuks [1967] which are in qualitative agreement with the data reported by Grüner and Hertz [1972]. The system $CH_3COOH + CCl_4$ has been studied by Anderson [1966]. Here the dissociation of the H-bonds occurs at very low acid concentrations which are out of the range of detectability of the conventional nuclear magnetic relaxation technique. So the relaxation rates of CH_3 and OH only show a weak concentration dependence which roughly parallels the composition dependence of the viscosity. Similar experiments have been performed in our laboratory (Grüner and Hertz [1972]). The qualitative behavior of the concentration dependence of the CH_3 relaxation is in agreement with Anderson's measurements, but Anderson's relaxation rates are by one order of magnitude smaller than ours, which probably is due to the fact that the saturation recovery method applied by Anderson is not reliable.

Chloroform is considered to form weak H-bonds with suitable electron donors. The influence of this H-bond on the reorientation time of CCl_3H and on that of the electron donor molecule has been studied by Anderson [1966] and Huntress [1969]. In some cases the H-bond effect seems to be of the same strength as the background effect corresponding to binding energies $\approx -kT$.

Aniline and phenol in mixtures with different solvents like CCl_4, C_6H_{12} etc. have also been studied (Anderson and Gerritz [1970] and Levy [1972]).

21.5. Water and aqueous mixtures

At temperatures below about 200°C in H_2O the protons are relaxed by magnetic dipole–dipole interaction. Above this temperature spin–rotational interaction contributes increasingly (Hausser [1963], Powles and Smith [1964] and Smith and Powles [1966]). The longitudinal proton relaxation time of H_2O at 25°C is 3.60 s (see for instance, Krynicki

[1966] or Brown [1969] and literature cited therein). The uncertainty of this value is about ±2%. The water molecule is not a strictly stable molecule, the protons are exchanged in times short compared with the relaxation time. Thus, it is not possible to measure the intramolecular relaxation rate separately by the D_2O dilution and extrapolation technique. However, the mean lifetime of a given proton in a water molecule in pure (neutral) water at room temperature is about 10^{-3} s (Loewenstein and Connor [1963]). Since we clearly expect correlation times for the thermal motion of the water molecules much shorter than this, we are allowed to use the formula for the intramolecular relaxation rate of a two-spin system, i.e. eq. (21.5). The intermolecular relaxation rate may be calculated to be $(1/T_1)_{\text{inter}} = 0.096\ \text{s}^{-1}$. The proton–proton distance b is 1.52 Å, thus eqs. (21.4) and (21.5) yield $\tau_c = 2.7 \times 10^{-12}$ s. $(1/T_1)_{\text{inter}}$ for water may also be determined experimentally by an indirect method (Goldammer and Zeidler [1969]). Use of this $(1/T_1)_{\text{inter}}$ yields a rotational correlation time $\tau_c = 2.5 \times 10^{-12}$ s.

The plot of log $1/T_1$ against $1/T$ does not yield a straight line. Thus the "activation energy"* of the molecular processes responsible for the spin relaxation is temperature dependent; it varies from about 5 kcal/mole at 0°C to 3.5±0.2 kcal/mole in the range between 40 and 100°C. It is interesting to note that the "activation energies" derived from corresponding plots for the viscosity, the dielectric relaxation time and the self-diffusion coefficient are very much alike.

In table 21.3 the correlation time τ_c for water is given for different temperatures between 0 and 100°C. The proton relaxation rates as measured by Krynicki [1966] have been used. The intermolecular relaxation rate has been calculated using Torrey's theory [1953] (see e.g. Hertz [1967a, 1973]). The self-diffusion coefficient is taken from Simpson and Carr's paper [1958]. Their results, however, have been corrected by a constant factor so as to yield $D = 2.37 \times 10^{-5}\ \text{cm}^2\ \text{s}^{-1}$ at 25°C (Murday and Cotts [1970]).

Below 25°C the correlation time decreases as the pressure increases. At ≈1500 bar the correlation time goes through a minimum and then increases monotonously with increasing pressure (Hertz and Rädle [1969]). This behavior is in accordance with the viscosity anomaly of water.

For the longitudinal relaxation time of the deuteron in D_2O at 25°C the

* The activation energy E is defined as $\tau = \tau_0 \exp(E/RT)$.

TABLE 21.3
Self-diffusion coefficients and correlation times of the water molecule

°C	$D \times 10^5$ (cm²/s)	$\tau_c \times 10^{12}$ (s)
0	1.08	4.9
5	1.29	4.1
10	1.52	3.6
15	1.76	3.2
20	2.06	2.8
25	2.37	2.5
30	2.74	2.2
35	3.11	2.0
40	3.50	1.9
45	3.92	1.7
50	4.39	1.53
55	4.86	1.42
60	5.37	1.30
65	5.90	1.20
70	6.44	1.11
75	6.98	1.05
80	7.58	0.97
85	8.08	0.91
90	8.63	0.85
95	9.13	0.79
100	9.63	0.74

values 0.45 s (interpolated Woessner [1964]) and 0.44 s ± 5% (Goldammer and Zeidler [1969]) have been published. Such a short relaxation time can only be explained by quadrupole interaction. The quadrupole coupling constant $(eQ/h)\partial^2 V/\partial z'^2$ for the gaseous water molecule is known from microwave spectrum measurements to be 315 ± 0.7 kHz (Posener [1962] and Treacy and Beers [1962]). The coupling constant in ice is about 30% less than this (Waldstein et al. [1964]) due to H-bonding. Let us estimate that in liquid water the coupling constant is 25% less than in the gaseous state, i.e. $(eQ/h)\partial^2 V/\partial z'^2 = 237$ kHz. Thus, from eq. (21.7) with $I = 1$ and $\tilde{\eta} = 0$ we find that $\tau_c = 2.7 \times 10^{-12}$ s. Obviously, the almost exact agreement with the values derived from the proton relaxation time is fortuitous, but it shows that the results obtained are essentially correct.

The correlation time in D_2O should be about 23% longer than in H_2O

because the ratio of the viscosities in D_2O and H_2O is 1.23 (Hardi and Cottington [1949]) (see also Murday and Cotts [1970]). Thus, the most probable value is $\tau_c(D_2O) = 3.1 \times 10^{-12}$ s, which with the experimental $T_1 = 0.44$ s gives a quadrupole coupling constant $(eQ/h)\partial^2V/\partial z'^2 =$ 222 kHz. Hindman et al. [1971] report an activation energy for the deuteron relaxation above 30°C of 3.3 kcal/mole which is also in good agreement with the corresponding result for H_2O.

The linewidth $\Delta H_{1/2}$ of the ^{17}O resonance in water at pH ≈ 12 at 25°C is 82.5 ± 2.5 mgauss (Fister and Hertz [1967]). At this pH value the proton exchange is fast enough to warrant the validity of $T_1 = T_2$. In neutral water $T_2 < T_1$ which is due to electron coupled spin–spin interaction between ^{17}O and the protons (Meiboom [1961]). With $1/T_1 = 1/T_2 = \frac{1}{2}\gamma\Delta H_{1/2}$ we get $T_1 = 6.7 \pm 0.2 \times 10^{-3}$ s. Luz and Meiboom [1963] quote the value $T_1 = 6.3 \times 10^{-3}$ s. Other results given in the literature are slightly shorter (Connick and Stover [1961] and Glasel [1966]). The somewhat greater linewidth corresponding to the shorter T_1 is probably due to modulation and field inhomogeneity effects. T_1 data obtained by pulse experiments (Garret et al. [1967] and Hindman et al. [1971]) are also in agreement with our figure. The nucleus of ^{17}O has spin 5/2 and its quadrupole coupling constant $(eQ/h)\partial^2V/\partial z'^2$ in water is not known exactly; however, the quantities $(eQ/h)\partial^2V/\partial a^2 = 8.1 \pm 0.1$ MHz and $(\partial^2V/\partial b^2 - \partial^2V/\partial c^2)/(\partial^2V/\partial a^2) = 0.7 \pm 0.1$ for the $HD^{17}O$ molecule have been measured by microwave spectroscopy (Stevenson and Townes [1957]). The symbols a, b and c indicate the three principal axes of inertia of the HDO molecule. Assuming that the quadrupole coupling constant is not very much different from the value 8.1 MHz (the a-axis is in the D–O–H plane and deviates by 21°23.5′ from the axis which lies in the HOH plane perpendicular to the C_2 axis) and neglecting the asymmetry parameter $\tilde{\eta}$, we obtain from eq. (21.7) with $I = 5/2$, a value $\tau_c = 2.4 \times 19^{-12}$ s in reasonable agreement with the results found from the ^1H and D resonances.

The ^{17}O quadrupole coupling constant in ice is 6.525 MHz, $\tilde{\eta} = 0.925$ (Edmonds and Zussman [1972]). Connick and Wüthrich [1969] calculated the field gradient in gaseous H_2O in the proper coordinate system from the data given by Stevenson and Townes [1957]: $(eQ/h)\partial^2V/\partial z'^2 = -9.83$ MHz, $\tilde{\eta} = 0.407$. With $\tau_c = 2.5 \times 10^{-12}$ s and the relaxation time quoted above one may derive $\tilde{\eta} = 0.9$, $(eQ/h)\partial^2V/\partial z'^2 = 7.04$ MHz which is in reasonable proportion to the gas and solid values. Summarizing all the data available for τ_c of water,

$$\tau_c = (2.5 \pm 0.2) \times 10^{-12}\ \mathrm{s}$$

at 25 °C may be considered to be the most acceptable value.

Next we turn to the evaluation of the rotational correlation time for the discussion of structural problems in water. Very often the H-bonded aggregates in water have been denoted as "clusters". Let us follow this usage for a moment and let us examine which information regarding the existence of clusters can be extracted from the knowledge of τ_c.

Suppose water is a mixture of clusters, the clusters containing n_i molecules $i = 0, 1, 2, 3 \ldots$; $i = 0$ corresponds to the monomeric water molecule. Let the correlation time of the ith cluster be τ_{c_i}. Then we may rewrite eqs. (21.9), (21.10), and (21.13) as

$$\tau_c = \sum p_i \tau_{c_i}, \qquad \tau_h \gg \tau_{c_i} \tag{21.15}$$

or

$$\tau_c^{-1} = \sum p_i \tau_{c_i}^{-1}, \qquad \tau_h \ll \tau_{c_i}, \tag{21.16}$$

where τ_h is a representative lifetime in a given cluster configuration. We know that for the long-lived octahedral hydration sphere of Mn^{2+} $\tau_c \approx 3 \times 10^{-11}$ s (Bloembergen and Morgan [1961], Pfeifer [1962], Hausser and Noack [1964] and Hertz [1967a, 1973]). Replacing the central Mn^{2+} ion by a water molecule, we have a cluster of seven molecules. Rounding off the number 7 so as to have 10 in order to take account of the partially bound (short-lived) second hydration sphere, we expect $\tau_{c_{10}} \approx 3 \times 10^{-11}$ s. τ_{c_0} is as well known from Hindman's et al. [1971] measurements of dilute solutions of H_2O in inert organic solvents: $\tau_{c_0} \approx 5 \times 10^{-13}$ s. Inserting these numbers into eq. (21.15) with $p_i = 0$ for $i \neq 0, 10$ we find $p_{10} = 0.068$ which is certainly too small. Thus we conclude, if there are larger clusters with $i \geq 10$ in water, the lifetime of these clusters must be very much smaller than an average τ_{c_i}, i.e. $\tau_h \approx 10^{-12}$–10^{-11} s. Putting $\tau_{c_i}^{-1} = 0$ for all clusters present, we obtain from eq. (21.16) $p_0 \approx 0.20$ which is a reasonable figure. If this situation really occurs, then the "free" reorientation is the only contribution to the total molecular reorientation in water. One of the authors (H. G. Hertz) has derived the following formula for τ_c (Hertz [1967a, b; 1973]).

$$\tau_c^{-1} = \tau_{cc}^{-1} + (1 - C_2)/\tau_h,$$

where C_2 is a constant $0 \leq C_3 \leq 1$. τ_{cc} is the average correlation time for all clusters. In the situation just indicated one has to put $C_2 \to 0$ (see Hertz [1967a,b]); then, with $\tau_{cc} \gg \tau_h$ (τ_h = average residence time in a cluster),

$$\tau_c = \tau_h. \tag{21.17}$$

As mentioned above the interesting consequence of eq. (21.17) is the additional validity of $\tau_c = \tau_r$ (Hertz [1967a,b]), where τ_r is the reorientation time. For an ordinary slow diffusion process we would have $\tau_r = 3\tau_c$. Thus, knowledge of τ_r/τ_c would give us the desired information regarding cluster distribution. The dielectric relaxation time τ_d is experimentally well known: 8.2×10^{-12} s at 25°C. However, the macroscopic dielectric relaxation time τ_d is connected with the microscopic dielectric relaxation time or reorientation time τ_r by a factor δ: $\tau_d = \delta\tau_r$. δ is not known exactly; $1 < \delta < 2$ (Cole [1963] and Fatuzzo and Mason [1967]; see also Nee and Zwanzig [1970]). With $\delta = 1.5$ we get $\tau_r = 5 \times 10^{-12}$ s, thus $\tau_r \neq \tau_c$ and the pure free state rotation or rotational jump model seems not to be correct.

It seems worthwhile to point out that the same result is valid for translational motion: Translational jumps displacing the water molecule almost instantaneously after a residence time in a quasi-equilibrium position $\tau_h \geqslant 2 \times 10^{-12}$ s can be ruled out.

In our view the correct model is given by a fast or at least moderately fast exchange of the water molecule among aggregates containing two to five molecules. The residence time in a given state of aggregation is $\leqslant 10^{-12}$ s. The rotational–and translational–diffusion is close to a microstep diffusion, though it probably does not really represent the ideal form of microstep diffusion.

It has been calculated from the correlation times in pure liquid water and from τ_{c0}, the correlation time of monomeric water in an inert solvent, that the probability w_1^+ to find a water molecule (in pure liquid H_2O) attached to another one in the first coordination sphere for a time substantially longer than the decay time of angular velocity correlations, i.e. longer than about 10^{-13} s, is $w_1^+ \approx 0.7$. The corresponding probability for the second coordination sphere is $w_2^+ \approx 0.25$ (Hertz [1970a]).

Furthermore, again from the comparison of $\tau_c = 2.5 \times 10^{-12}$ s in pure water with $\tau_{c0} = 5 \times 10^{-13}$ s for the unbound molecule, it has been estimated that the effective potential depth in liquid H_2O which couples two water molecules together for a time $\geqslant 10^{-12}$ s is $\leqslant -2.3$ kcal/mole (Hertz [1971b]).

It is an interesting feature of the H-bond network in liquid water that it may be decoupled to a certain degree by the presence of inorganic ions with sufficiently large ratio: ion size/ion charge. This is the well-known structure-breaking effect (literature references see below). The mechanism is as follows: The relatively low electric field strength at the surface of these ions competes with the directing forces exerted by the H-bonds. As a consequence, the effective potential causing the preferential orientation of the water molecules will become more shallow, the probability to find water molecules attached to one another decreases and eq. (21.13) tells us that the correlation time will decrease. This in turn, according to eq. (21.5) causes a decrease of the relaxation rate. Indeed, the effect of a decrease of the relaxation rate as the electrolyte concentration increases has been observed for a large number of salts (Hertz [1963, 1967a], Fister and Hertz [1967] and Engel and Hertz [1968]); it is the same effect which manifests itself as a decrease of the viscosity and an increase of the water self-diffusion coefficient (McCall and Douglass [1965] and Endom et al. [1967]). It has been estimated that the effective intermolecular potential depth, which in pure water was given as $\leqslant -2.3$ kcal/mole, in the neighborhood of ions like I^-, Br^-, ClO_4^- etc. is reduced to $\leqslant -1.3$ kcal/mole (Hertz [1971b]). It may be mentioned that even large (complex) anions with charge 2 cause this structure-breaking effect (Engel and Hertz [1968]). The structure-breaking effect is also accompanied by a slowing down of the proton exchange of the water molecules (Hertz and Klute [1970], Hertz et al. [to be published], Yoon [1974]). As the ion radius becomes smaller and the charge higher the electric field strength exerted by the ion becomes greater and, finally, the corresponding directing potential is much stronger than that of the H-bonds. Now we have the ordinary (positive) hydration of ions. Of course, this solvation is not unique for water and not even unique for H-bonded liquids. But it should be kept in mind that in all cases we have an interplay of the electric field of the ion with the forces implicit in the H-bonds. If the latter forces are relatively strong, then the solvation will be weaker. Thus, it is typical for water that here the solvation is the weakest as compared with other solvents (see e.g. Hertz [1970b]). The order in thc hydration sphere is relatively low, the exchange rates of H_2O in the hydration sphere are comparatively high (Hertz [1967a, 1970b]). Nuclear magnetic relaxation studies regarding the internal mobility – i.e. the depth of internal orientational potentials – in the hydration sphere have been performed for Li^+, Al^{3+} (Hertz et al. [1971]) and F^- (Hertz et al. [(1969a]) and it has been shown that there is still rotation of the water molecule about the dipole axis in the case of Li^+. Furthermore, it was possible to prove by nuclear magnetic relaxation measurements that in the hydration sphere of F^- the water molecules are asymmetrically orientated, i.e. there is a H-bond F–HO between the fluoride ion and the water molecule (Hertz and Rädle [1973]), as was predicted on a theoretical basis (Diercksen and Krämer [1970]). The water molecule rotates about the axis F^-–OH (Hertz et al. [1969a]).

Probably the vast majority of solvents, being H-bonded or not, does not show the structure-breaking effect when electrolytes with large ions are dissolved. Two exceptions, however, are known: ethylene glycol and glycerol. Here, like in water, the correlation time decreases when electrolytes like KI, RbI and CsI are dissolved (Engel and Hertz [1968]).

What occurs when inert non-electrolytes are added to water or, in other words, when water is diluted with an inert solvent? In principle, one might expect that the H-bonded network is destroyed, the mean size of the association aggregates is reduced which in turn, according to eq. (21.13), causes a decrease of the correlation time. One practical reason why no direct answer can be given to this question is the very low solubility of really inert substances like hydrocarbons in water–the effects would be so small that their observation would be impossible. Thus we have to consider solvents which are only fairly inert, like ketons, nitriles and aliphatic alcohols and those which have at least one inert part, like the alkyl group (Hertz and Zeidler [1964], Fister and Hertz [1967], Goldammer and Zeidler [1969] and Goldammer and Hertz [1970]). The tetraalkylammonium ions may also be mentioned here (Danyluk and Gore [1964], Hertz and Zeidler [1964] and Hertz et al. [1969b]). For all these and similar substances (Clifford [1965] and Clifford and Pethica [1965]) one observes an increase of the water relaxation rate as their concentration is increased at the *water rich* end of the composition range. In order to interpret this effect we may consider eq. (21.13). This equation is not quite valid here because it does not regard mixtures. Thus we would have to add terms concerning the contribution from solvent-solute aggregates where the coupling occurs via the polar group of the solute. But experience tells us that the increase of $1/T_1$ is stronger when the number of methyl or methylene groups per molecule becomes greater, leaving the polar part unchanged (Hertz and Zeidler [1964]). So we may indeed consider the increase of τ_c to be caused by the inert part of the solute molecule leaving aside the constant contribution of the polar part. Now let us discuss the possible reasons for an increase of τ_c by aid of eq. (21.13). We may imagine three possibilities: (1) p_i, $i \neq 0$, may increase because w_1^+ increases, i.e. the effective potential depth between two water molecules increases. This may be denoted as reinforcement of the structure of the liquid. (2) p_i, $i \neq 0$, increases, but now w_1^+ remains unchanged. p_i increases because the interconnection of p_i and w_1^+ changes due to the fact that the water–water first coordination number in the hydration sphere of the inert solute becomes greater than in pure water. It is somewhat more questionable whether this modification should also be denoted as an increase in the degree of water structure. (3) $\hat{\eta}^*$ increases, i.e. the background effect increases. Now more weakly attractive configurations appear $E \geqslant -kT$, or the repulsive contribution to $\hat{\eta}^*$ becomes greater due to repacking of the water and collisions with the hydrocarbon part of the solute. The third possibility we would not call a "structure increase" effect. A decision as to which of the three effects is the dominant one cannot be given yet. Still, it is general usage to associate with the concept of "hydrophobic" hydration of the alkyl groups the effect of structure reinforcement (Frank and Evans [1945], Frank and Wen [1957] and Hertz [1964]). Furthermore, it is still an open question why the H-bonded network reacts on the presence of non-polar groups in particular rearrangement which is known as the hydrophobic hydration (Franks [1973]), hydration of the second kind (Hertz [1964]), or iceberg formation (Frank and Evans [1945] and Frank and Wen [1957]). This effect–at least to such a high extent–is unique for water. Recently, the orientation of the water molecules in the hydrophobic hydration sphere has been studied by aid of nuclear magnetic relaxation (Hertz and Rädle [1973] and Hertz and Wen [1974]). The measurements yielded the result that the water protons surrounding CH_3 predominantly point towards the bulk water. The chemical shift of the water protons in the hydrophobic hydration sphere suffer almost no change as compared with pure water. There is some controversy in the literature (Zeidler [1973]), and more recently (Oakes [1973]) whether a very small downfield or upfield shift occurs, although it is possible that, depending on the situation, both possibilities are realized.

Note added in proof

The following liquids have been studied in the meantime:

(a) Non-aqueous hydrogen bonded liquids: phenol (Grosescu [1973]; Trepadus et al. [1973]); pyridine/acetone (or benzene, chloroform, cyclohexane, carbon tetrachloride, carbon disulfide) (Tomchuk et al. [1973]);

picoline/carbon tetrachloride (Jurga et al. [1973]); acetic acid (or propionic acid, butyric acid)/carbon tetrachloride (Szcesniak et al. [1973]); ethanol/carbon tetrachloride (Langer et al. [1973]).

(b) Water and aqueous mixtures: water (Hindman and Svirmickas [1973]; Hindman et al. [1973]); water/dioxane (Lee and Jonas [1973]); water/tetraethylammonium bromide (Hertz and Wen [1974]).

References

Abragam, A., 1961, The Principles of Nuclear Magnetism (Clarendon Press, Oxford).

Alger, T. D., D. M. Grant and J. R. Lyerla Jr., 1971, J. Phys. Chem. **75**, 2539.

Anderson, J. E., 1966, Nuova Cimento **45**, 254.

Anderson, J. E., 1967, J. Chem. Phys. **47**, 4879.

Anderson, J. E., 1969, J. Chem. Phys. **51**, 3578.

Anderson, J. E. and P. A. Fryer, 1969, J. Chem. Phys. **50**, 3784.

Anderson, J. E. and W. H. Gerritz, 1970, J. Chem. Phys. **53**, 2584.

Andrew, E. R., 1956, Nuclear Magnetic Resonance, Cambridge University Press.

Assink, R. A. and J. Jonas, 1970, J. Chem. Phys. **53**, 1710.

Beckert, D. and H. Pfeifer, 1965, Ann. Physik 7. Folge **16**, 262.

Bender, H. J. and M. D. Zeidler, 1971, Ber. Bunsenges. Physik. Chem. **75**, 236.

Bezrukov, O. F. and M. F. Vuks, 1967, Ukr. Fiz. Zh. **12**, 160.

Bloembergen, N. and L. O. Morgan, 1961, J. Chem. Phys. **34**, 842.

Brown, R. J. S., 1969, J. Phys. Chem. **73**, 3157.

Clifford, J., 1965, Trans. Faraday Soc. **61**, 1276.

Clifford, J. and B. A. Pethica, 1965, Trans. Faraday Soc. **61**, 182.

Cole, R. H., 1963, in: Magnetic and Dielectric Resonance and Relaxation, Ed. J. Smidt (North-Holland Publ. Co., Amsterdam), p. 96.

Connick, R. E. and E. D. Stover, 1961, J. Phys. Chem. **75**, 2075.

Connick, R. E. and K. Wüthrich, 1969, J. Chem. Phys. **51**, 4506.

Crossley, J., 1970, Advan. Mol. Relax. Proc. **2**, 69.

Dalbert, R., M. Magat and A. Surdut, 1949, Polarisation de la Matière CNRS (Paris), p. 14.

Danyluk, S. S. and E. S. Gore, 1964, Nature **203**, 748.

Davidson, D. W. and R. H. Cole, 1951, J. Chem. Phys. **19**, 1484.

Diercksen, G. H. F. and W. P. Krämer, 1970, Chem. Phys. Letters **5**, 570.

Edmonds, D. T. and A. Zussman, 1972, Phys. Letters **41a**, 167.

Endom, L., H. G. Hertz, B. Thül and M. D. Zeidler, 1967, Ber. Bunsenges. Physik. Chem. **71**, 1008.

Engel, G. and H. G. Hertz, 1968, Ber. Bunsenges. Physik. Chem. **72**, 808.

Engelsmann, K., H. G. Hertz and M. D. Zeidler, 1974, Z. Physik. Chem. N.F. **89**, 134.

Farrar, T. C. and E. D. Becker, 1971, Pulse and Fourier Transform NMR (Academic Press, New York).

Fatuzzo, E. and P. R. Mason, 1967, Proc. Phys. Soc. **90**, 729, 741.

Figgins, R. and M. Rhodes, 1969, Mol. Phys. **17**, 669.

Fister, F. and H. G. Hertz, 1967, Ber. Bunsenges. Physik. Chem. **71**, 1032.

Frank, H. S. and M. W. Evans, 1945, J. Chem. Phys. **13**, 507.
Frank, H. S. and W. Y. Wen, 1957, Discussions Faraday Soc. **24**, 133.
Franks, F., 1973, in: Water, a Comprehensive Treatise, Vol. **2**, Ed. F. Franks (Plenum Press, New York, London), p. 1.
Garret, B. B., A. B. Denison and S. W. Rabideau, 1967, J. Phys. Chem. **71**, 2606.
Gillen, K. T. and J. H. Noggle, 1970, J. Chem. Phys. **53**, 801.
Gillen, K. T., M. Schwartz and J. H. Noggle, 1971, Mol. Phys. **20**, 899.
Gillen, K. T., J. H. Noggle and T. K. Leipert, 1972, Chem. Phys. Letters **17**, 505.
Glasel, J. A., 1966, Proc. Natl. Acad. Sci. **55**, 479.
Goldammer, E. v. and H. G. Hertz, 1970, J. Phys. Chem. **74**, 3734.
Goldammer, E. v. and M. D. Zeidler, 1969, Ber. Bunsenges. Physik. Chem. **73**, 4.
Göller, R., H. G. Hertz and R. Tutsch, 1972, Pure Appl. Chem. **32**, 149.
Grüner, M. and H. G. Hertz, 1972, Advan. Mol. Relax. Proc. **3**, 75.
Grunwald, E. and W. C. Coburn Jr., 1957, J. Am. Chem. Soc. **80**, 1318.
Hardt, A. P., D. K. Anderson, R. Rathburn, B. W. Mahr and A. L. Babb, 1959, J. Phys. Chem. **63**, 2059.
Hardy, R. C. and R. L. Cottington, 1949, J. Chem. Phys. **17**, 509.
Hausser, R., 1963, Z. Naturforsch. **18a**, 1143.
Hausser, R. and F. Noack, 1964, Z. Physik. **182**, 93.
Hertz, H. G., 1963, Ber. Bunsenges. Physik. Chem. **67**, 311.
Hertz, H. G., 1964, Ber. Bunsenges. Physik. Chem. **68**, 907.
Hertz, H. G., 1967a, in: Progress in NMR Spectroscopy, Vol. **3**, Eds. J. W. Emsley, J. Feeney and L. H. Sutcliffe (Pergamon Press, Oxford), p. 159.
Hertz, H. G., 1967b, Ber. Bunsenges. Physik. Chem. **71**, 979.
Hertz, H. G., 1970a, Ber. Bunsenges. Physik. Chem. **74**, 666.
Hertz, H. G., 1970b, Angew. Chem. **82**, 91; Angew. Chem. Intern. Ed. **9**, 124.
Hertz, H. G., 1971a, Ber. Bunsenges. Physik. Chem. **75**, 183.
Hertz, H. G., 1971b, Ber. Bunsenges. Physik. Chem. **75**, 572.
Hertz, H. G., 1973, in: Water, A Comprehensive Treatise, Vol. **3**, Ed. F. Franks (Plenum Press, New York), p. 301.
Hertz, H. G. and R. Klute, 1970, Z. Physik. Chem. N. F. **69**, 101.
Hertz, H. G. and C. Rädle, 1969, Z. Physik. Chem. N. F. **68**, 324.
Hertz, H. G. and C. Rädle, 1973, Ber. Bunsenges. Physik. Chem. **77**, 521.
Hertz, H. G. and R. Tutsch. Ber. Bunsenges. Physik. Chem., to be published.
Hertz, H. G. and W. Y. Wen, 1974, Z. physik. Chem. N.F. **93**, 313.
Hertz, H. G., H. Versmold and C. Yoon, to be published.
Hertz, H. G. and M. D. Zeidler, 1963, Ber. Bunsenges. Physik. Chem. **67**, 774.
Hertz, H. G. and M. D. Zeidler, 1964, Ber. Bunsenges. Physik. Chem. **68**, 821.
Hertz, H. G., G. Keller and H. Versmold, 1969a, Ber. Bunsenges. Physik. Chem. **73**, 549.
Hertz, H. G., B. Lindmann and V. Siepe, 1969b, Ber. Bunsenges. Physik. Chem. **73**, 542.
Hertz, H. G., R. Tutsch and H. Versmold, 1971, Ber. Bunsenges. Physik. Chem. **75**, 1177.
Hindman, J. C., A. Svirmickas and M. Wood, 1968, J. Phys. Chem. **72**, 4118.
Hindman, J. C., A. J. Zielen, A. Svirmickas and M. Wood, 1971, J. Chem. Phys. **54**, 621.
Huntress, W. T. Jr., 1969, J. Phys. Chem. **73**, 103.
Jonas, J. and T. M. DiGennaro, 1969, J. Chem. Phys. **50**, 2392.
Kintzinger, J. P. and J. M. Lehn, 1971, Mol. Phys. **22**, 273.

Kintzinger, J. P. and M. D. Zeidler, 1973, Ber. Bunsenges. Physik. Chem. **77**, 98.
Krynicki, K., 1966, Physica **32**, 167.
Levy, G. C., 1972, J. Magn. Res. **8**, 122.
Loewenstein, A. and T. M. Connor, 1963, Ber. Bunsenges. Physik. Chem. **67**, 280.
Lucken, E. A. C., 1969, Nuclear Quadrupole Coupling Constants (Academic Press, New York).
Luz, Z. and S. Meiboom, 1963, J. Chem. Phys. **39**, 366.
Lyerla, J. R. Jr. and D. M. Grant, 1972, in: Physical Chemistry, Series One, Vol. **4**, Ed. C. A. McDowell (Butterworths, London), p. 155.
McCall, D. W. and D. C. Douglass, 1965, J. Phys. Chem. **59**, 2001.
Meiboom, S., 1961, J. Chem. Phys. **34**, 375.
Mills, R., 1971, Ber. Bunsenges. Physik. Chem. **75**, 195.
Murday, J. S. and R. M. Cotts, 1970, J. Chem. Phys. **53**, 4724.
Nee, T. W. and R. Zwanzig, 1970, J. Chem. Phys. **52**, 6353.
Oakes, J., 1973, J. Chem. Soc. Faraday Trans. II, **69**, 1311.
O'Reilly, D. E. and E. M. Peterson, 1971, J. Chem. Phys. **55**, 2155.
O'Reilly, D. E., E. M. Peterson and E. L. Yasaitis, 1972, J. Chem. Phys. **57**, 890.
Pfeifer, H., 1962, Z. Naturforsch. **17a**, 279.
Pople, J. A., W. G. Schneider and H. J. Bernstein, 1959, High Resolution Nuclear Magnetic Resonance, (McGraw-Hill, New York).
Posener, D. W., 1962, Australian J. Phys. **36**, 1473.
Pottel, R. and U. Kaatze, 1969, Ber. Bunsenges. Physik. Chem. **73**, 437.
Powles, J. G. and M. Rhodes, 1967, Mol. Phys. **12**, 399.
Powles, J. G. and D. W. G. Smith, 1964, Phys. Letters **9**, 239.
Rossegger, P., H. Lischka and P. Schuster, 1972, Theoret. Chim. Acta (Berl.) **24**, 191.
Saxton, J. A., R. A. Bond, G. T. Coats and R. M. Dickinson, 1962, J. Chem. Phys. **37**, 2132.
Simpson, J. H. and H. Y. Carr, 1958, Phys. Rev. **111**, 1201.
Slichter, C. P., 1963, Principles of Magnetic Resonance (Harper & Row).
Smith, D. W. G. and J. G. Powles, 1966, Mol. Phys. **10**, 457.
Stevenson, M. J. and C. H. Townes, 1957, Phys. Rev. **107**, 635.
Stokes, R. H. and R. Mills, 1965, Viscosity of Electrolytes and Related Properties (Pergamon Press, Oxford).
Torrey, H. C., 1953, Phys. Rev. **92**, 962.
Treacy, E. B. and Y. Beers, 1962, J. Chem. Phys. **36**, 1473.
Tutsch, R., 1973, Thesis, Karlsruhe.
Uedaira, H., private communication.
Versmold, H. and C. Yoon, 1972, Ber. Bunsenges. Physik. Chem. **76**, 1164.
Waldstein, P., W. S. Rabideau and J. A. Jackson, 1964, J. Chem. Phys. **41**, 3407.
Woessner, D. E., 1964, J. Chem. Phys. **40**, 2341.
Woessner, D. E., B. S. Snowden Jr. and E. T. Strom, 1968, Mol. Phys. **14**, 265.
Yoon, C., 1974, Thesis, Karlsruhe.
Zeidler, M. D., 1965, Ber. Bunsenges. Physik. Chem. **69**, 659.
Zeidler, M. D., 1971, Ber. Bunsenges. Physik. Chem. **75**, 769.
Zeidler, M. D., 1973, in: Water, A Comprehensive Treatise, Vol. **2**, Ed. F. Franks (Plenum Press, New York), p. 529.
Zimmerman, J. R. and W. E. Brittin, 1957, J. Phys. Chem. **61**, 1328.

Added in proof:

Grosescu, R., 1973, Rev. Roum. Phys. **18**, 133.

Hertz, H. G. and W. Y. Wen, 1974, Z. Physik. Chem., N.F. **93**, 313.

Hindman, J. C. and A. Svirmickas, 1973, J. Phys. Chem. **77**, 2487.

Hindman, J. C., A. Svirmickas and M. Wood, 1973, J. Chem. Phys. **59**, 1517.

Jurga, J., Z. Pajak, K. Jurga and S. Jurga, 1973, Inst. Nucl. Phys. Cracow Rep., No. 819 (PL)-(Pt1), p. 46.

Langer, H., H. G. Hertz and M. D. Zeidler, 1973, Chem. Phys. Letters **23**, 417.

Lee, Y. and J. Jonas, 1973, J. Chem. Phys. **59**, 4845.

Szczesniak, E., Z. Pajak and S. Jurga, 1973, Inst. Nucl. Phys. Cracow Rep. No. 819 (PL)-(Pt1), p. 36.

Tomchuk, E., J. J. Czubryt, E. Bock and N. Chatterjee, 1973, J. Magn. Res. **12**, 20.

Trepadus, V., S. Rapeanu and R. Grosescu, 1973, Rev. Roum. Phys. **18**, 314.

PART E

MATRIX ISOLATION TECHNIQUE

CHAPTER 22

MATRIX-ISOLATION STUDIES OF HYDROGEN BONDING

H. E. HALLAM*

Fachbereich Physikalische Chemie, Philipps-Universität Marburg/Lahn, D-3550 Marburg/Lahn, Biegenstrasse 12, West Germany

* Visiting Professor 1974; permanent address: Department of Chemistry, University College of Swansea, Singleton Park, Swansea SA2 8PP, Wales, U.K.

Contents

22.1. Introduction . . . 1067
22.2. Advantages of matrix-isolation . . . 1068
22.3. Disadvantages of matrix-isolation . . . 1069
22.4. Systems studied . . . 1070
22.5. Hydrogen halides . . . 1071
22.6. Hydrogen cyanide . . . 1083
22.7. Water . . . 1086
22.8. Hydrogen sulphide . . . 1088
22.9. Ammonia . . . 1090
22.10. Hydrazoic acid . . . 1090
22.11. Alcohols . . . 1092
22.12. Thiols . . . 1097
22.13. Carboxylic acids . . . 1098
22.14. Hetero-associated species . . . 1100
22.15. Conclusions . . . 1103
References . . . 1104

The hydrogen bond – recent developments in theory and experiments
Eds. P. Schuster et al. © *North-Holland Publ. Co., Amsterdam, 1976*

22.1. Introduction

Matrix-isolation is a technique for trapping species as isolated entities in an inert solid, or matrix, in order to investigate their properties, usually by spectroscopic methods. A suitable matrix must, at the temperature of the experiment, be a solid which is inert, rigid with respect to diffusion and transparent in the spectral region of interest. The noble gases, nitrogen and carbon monoxide, are ideally suited as matrices both with regard to their inertness and their own spectral simplicity. The technique of trapping in an inert solid matrix at cryogenic temperatures was first developed as a means of studying unstable species such as free radicals and was simultaneously proposed by Norman and Porter [1954] and Whittle et al. [1954]. The subsequent development of the method, particularly in its application to infrared (IR) studies, has been largely pioneered by Pimentel and his associates, who introduced the term "matrix-isolation" and also pointed out its unique potentialities for studying H-bonded species.

The technique involves the rapid condensation of a mixture of the absorbing species (A) and a diluent gas (M) at cryogenic temperatures, usually 4 K (liquid He) or 20 K (liquid H_2). With high matrix: absorber (M/A) ratios the solute may be expected to be "isolated" in the rigid matrix and thus supported in a convenient environment for spectroscopic examination. Under conditions of perfect isolation, the species under study is subject only to solute-matrix interactions. Such conditions are only achieved (and not necessarily so then) at very high M/A ratios, usually greater than 1000. At low M/A ratios interactions between solute molecules become important and molecular aggregates may be formed and trapped in addition to monomers (fig. 22.1). Molecular association will be greatest for solutes capable of forming H-bonds. This chapter reviews the many H-bonding studies which have been made by matrix-isolation spectroscopy. No detailed discussion will be given of either the experimental or the theoretical aspects of the method as major accounts have recently been published by Meyer [1971], covering mainly electronic spectroscopy and

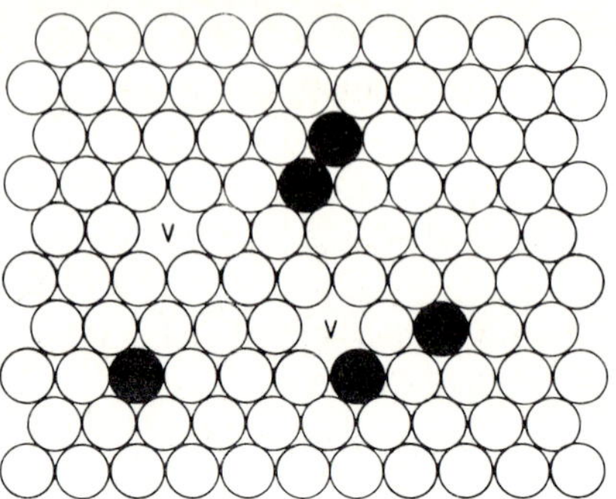

Fig. 22.1. Matrix-trapped dimer and monomers in different environments caused by vacancies and next-nearest neighbours.

by Hallam [1973] on vibrational spectroscopy. Both these texts give detailed accounts of the experimental techniques and provide comprehensive bibliographies of all the original papers and reviews to date of cryogenic spectroscopy.

22.2. Advantages of matrix-isolation

Although the matrix-isolation technique was developed for, and is still widely employed in, the study of unstable species, its advantages as a general tool in vibrational-rotational spectroscopy have recently become recognised (Barnes et al. [1969a]). The basis of its utility lies in the fact that isolation of monomeric solute molecules in an "inert" environment at cryogenic temperatures greatly reduces intermolecular interactions, resulting in a considerable sharpening of solute absorptions compared with other condensed phases. With the exception of a few small hydrides, such as hydrogen halides and ammonia, rotation does not occur in matrices, hence much narrower bands are obtained than in the vapour phase. Thus band half-widths, $\Delta\bar{\nu}_{1/2}$, are typically one-fifth or less than for those of similar concentrations in liquid solutions at ambient temperatures and it is possible to resolve features separated by as little as 1 cm^{-1} and to measure their wave numbers with a precision of about 0.1 cm^{-1}. This is well illustrated by the ability of the method to distinguish features due to energetically-similar molecular conformers (Barnes and Hallam [1970b]).

The same advantages pertain to matrix-trapped multimers, many well-known broad features in ambient-temperature spectroscopy showing quite a dramatic resolution into bands due to "isolated" dimers, trimers, etc. (see for example, figs. 22.8 and 22.10) which can thus be examined individually, and often considerable information obtained with regard to their geometry. Furthermore, the association process can be controlled and followed in a very elegant manner by depositing a low concentration mixture (i.e. high M/A ratio) and allowing the matrix to warm up slightly (say from 20 to 35 K for argon) and thus soften. The trapped monomers then diffuse to form dimers and higher multimers and the process can be monitored spectroscopically, the results being far more informative than their counterpart studies in other phases.

22.3. Disadvantages of matrix-isolation

In all matrix-isolation studies it is assumed that the molecular energy levels are not significantly perturbed by the matrix environment. Perturbations, of course, do occur and are reflected in a wave number shift which is analogous to a solvent shift and, for stretching modes, is usually to lower wave numbers (Hallam [1973]). In noble-gas matrices, however, the wave number shifts are relatively small, typically less than 0.5%. For bonded X–H group vibrational wave numbers such shifts are unimportant but they may be significant for the absorptions due to "free" X–H groups of multimer species.

It is further generally assumed that the quenching process freezes out the vapour-phase equilibrium mixture:

$$n\,\mathrm{A} \rightleftharpoons [\mathrm{A}]_n.$$

Very little work has so far been done to confirm or disprove this assumption but there appear to be clear indications that aggregation is also a function of the transient fluid state prior to solidification. This will be discussed in the later sections covering individual systems.

A more serious disadvantage is the difficulty of making quantitative intensity measurements on solid state systems. This is extenuated in the case of solid matrices since, although accurate concentrations, M/A, can be made up in the gas phase and known quantities sprayed onto the low-temperature support window, there is always uncertainty as to how much has actually been condensed within the aperture of the optical beam. This disadvantage however is largely overcome by the use of growth curves based on an internal standard, usually a vibrational wave number unaffected by H-bonding.

High boiling H-bonded liquids and solids pose experimental problems but these can be overcome by a high-temperature injection system (King [1970]).

22.4. Systems studied

Many of the "classical" H-bonded systems such as water, alcohols and carboxylic acids have been studied in matrix-isolation. In fact water (Van Thiel et al. [1957a]) and methanol (Van Thiel et al. [1957b]) were amongst the first species so examined, primarily because of their H-bonding propensities, as probes for testing the isolating efficiency of the experimental conditions. Both are now classic papers in that they clearly demonstrated the elegance of the technique for distinguishing spectral features due to dimers and trimers and higher multimers. In the excitement of this new technique for examining free radicals and unstable

TABLE 22.1

IR matrix isolation studies of H-bonded species

Monomer	Multimer species			
HX				
X=F, Cl, Br, I	$(HX)_2$	$(HX)_3$	$(HX)_4$	$(HX)_n$
	$(HX{\cdot}DX)$	$(HX)_2DX$	$(HX{\cdot}DX{\cdot}HX)$	
X=CN	$(HCN)_2$	$(HCN)_n$		
X=N_3, NCS	$(HX)_2$			
X=CHO	$(HCHO)_n$			
H_2X				
X=O, S	$(H_2X)_2$	$(H_2X)_3$	$(H_2X)_n$	
NH_3	$(NH_3)_2$			
H_2O_2	$(H_2O_2)_2$?			
ROH				
R=Me, Et, CF_3CH_2, $(CF_3)_2CH$ CF_3CHMe	$(ROH)_2$	$(ROH)_3$	$(ROH)_4$	$(ROH)_n$
RSH				
R=Me, Et	$(RSH)_2$	$(RSH)_4$?	$(RSH)_n$	
RCOOH				
R=H, Me, CF_3	$(RCOOH)_2$	$(RCOOH)_n$		

intermediates the obvious power of the method for studying H-bonded species was neglected for many years although Pimentel had pointed out that an active species includes molecules which are not unstable in the usual sense. Examples of these are molecules which are able to form H-bonds "reacting" to form dimers and multimers. In the case of methanol a pathlength of several metres is required to obtain the vapour spectrum completely free from dimer.

However, the commercial availability since the mid-1960s of mini-cryocoolers has heralded an increase in activity in the field of H-bonding and several important systems have been investigated in detail. Table 22.1 provides a comprehensive listing of the systems studied in matrix-isolation; some of the species listed have only been cursorily examined and only those which have been fairly thoroughly examined from the prime standpoint of molecular association will be discussed in the succeeding sections.

Perhaps one of the most significant applications has been to what, at ambient temperatures, are considered to be weakly H-bonded systems, e.g. hydrogen chloride and hydrogen sulphide. Hetero-associated systems have received some attention and will also be discussed.

22.5. Hydrogen halides

The association characteristics of the hydrogen and deuterium halides have been extensively examined in Ar matrices (Keyser and Robinson [1966], Barnes et al. [1969a], Verstegen et al. [1966], Davies and Hallam [1971] and Barnes et al. [1973a]) with the unfortunate exception of hydrogen fluoride. Although monomeric HF has been well studied in matrices the association behaviour has only been briefly examined (Bowers et al. [1966]) and regrettably these spectra are complicated by the presence of water.

The monomer and multimer spectra of HCl, HBr and HI in solid Ar are shown in fig. 22.2. These have been carried out over a large range of concentrations and growth curves plotted of the multiple spectral features. This allows a qualitative identification of features due to the various multimers assuming that the first formed will be the dimer, followed by the trimer and so on. The interpretations are confirmed by carrying out "warm-up" experiments in which monomers are partly allowed to diffuse to form higher multimers.

The immediate striking feature of these spectra is the wealth of spectral structure (table 22.2). For hydrogen chloride at M/A = 2000, the spectrum obtained is that of the fairly-freely rotating monomer (with well-defined P(1), R(0), R(1) branches and a weak induced Q branch), together with a weak band at 2818 cm^{-1} (A). As the concentration of HCl is increased, this band becomes very intense and a second strong band grows at 2787 cm^{-1} (B). A series of medium intensity bands also appears at 2781 cm^{-1}, 2768 cm^{-1}, 2761 cm^{-1}, 2754 cm^{-1} and 2748 cm^{-1}, of which the 2781 cm^{-1} band grows at a rate similar to band B, the 2768 cm^{-1} and 2761 cm^{-1} band grow at comparable rates (faster than band B), while the 2754 cm^{-1} and

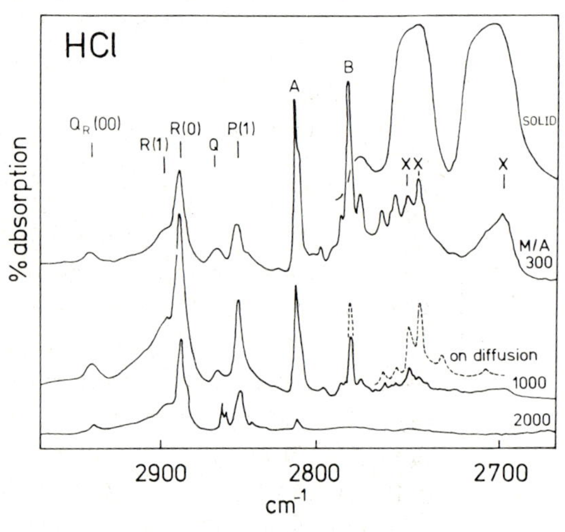

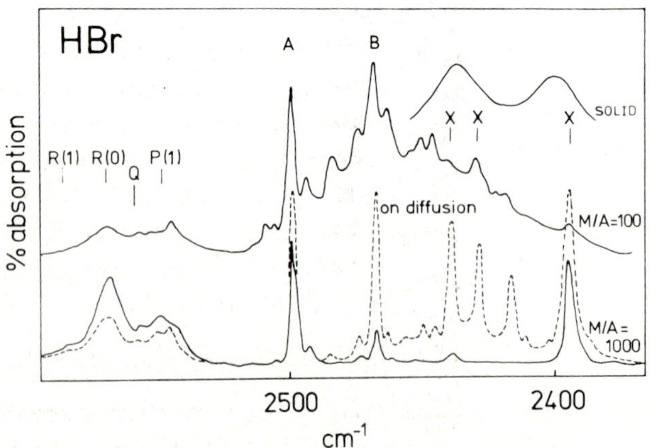

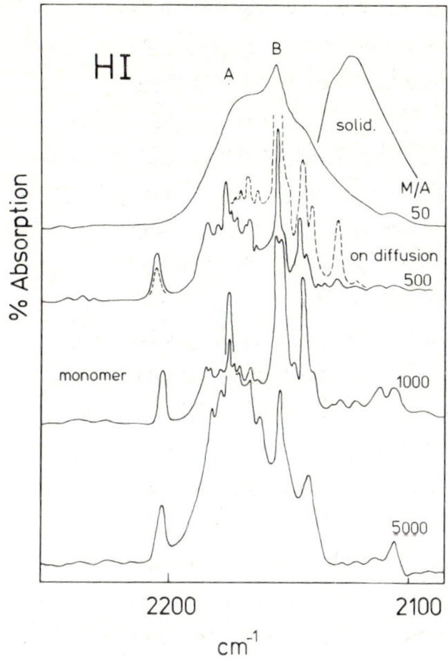

Fig. 22.2. Spectra of fundamental region of hydrogen halides at various concentrations in Ar matrices at 20 K (Barnes et al. [1969a, 1973a]).

2748 cm^{-1} bands rapidly become very strong as the concentration is further increased. In the most concentrated matrix (M/A = 300) a broad absorption also appears at 2701 cm^{-1}. The 2754 cm^{-1}, 2748 cm^{-1} and 2701 cm^{-1} bands (X) are close to the wave numbers found by Savoie and Anderson [1966] for solid hydrogen chloride and are therefore probably due to large aggregates of hydrogen chloride molecules. The behaviour on diffusion is similar to that caused by increasing the concentration of the solute, but the high polymer bands (X) increase in intensity even more rapidly. From these growth rates the other absorptions are assigned as 2818 cm^{-1} dimer, 2787 and 2781 cm^{-1} trimer, 2768 and 2761 cm^{-1} tetramer. The multiplicity of bands have been interpreted by Keyser and Robinson [1966] (KR) in terms of an intermolecular resonance interaction model. The principal assumptions underlying the KR treatment are: (i) that the multimers comprise hydrogen chloride molecules in adjacent substitutional sites, orientated in a similar manner to the pure solid halide and (ii) that intermolecular resonance interactions in the multimer will produce a number of components, split by amounts depending on the intermolecular force constant (f_{12}) of the pure halide. There is some disagreement as to the sign to be taken for f_{12} but the most reasonable value seems to be $-6.5\ \mathrm{Nm}^{-1}$ due to Sayoie and Anderson [1966] which gives a fairly good prediction (Barnes et al. [1969a]) of the bands due to dimer, trimer and tetramer species (table 22.2). The dimer is taken to have a cyclic structure since Katz et al. [1967] have shown in their far IR studies of HCl, DCl and HCl/DCl mixtures in noble-gas matrices that the mixed dimer has only one low frequency H-bond stretching mode ($170\ \mathrm{cm}^{-1}$) intermediate in wave number between the hydrogen chloride dimer ($185\ \mathrm{cm}^{-1}$) and the deuterium chloride dimer ($147\ \mathrm{cm}^{-1}$; all values in xenon). Furthermore no intense low frequency H-bond bending mode is observed, as in the case of the linear hydrogen cyanide dimer (King and Nixon [1968]).

Since considerable discussion arises and controversy still exists (see ch. 11 and later sections in this chapter) concerning the geometry of H-bonded dimers it seems worthwhile at this point to outline the arguments put forward (Katz et al. [1967]) to distinguish spectroscopically between a cyclic or an open-chain configuration, primarily by means of the far IR absorptions. The evidence from the near IR will then be further elaborated in the next section (section 22.6).

For a planar cyclic dimer $(HX)_2$, a molecule containing two equivalent H-atoms, three of the six normal modes will be IR active and are depicted in fig. 22.3.

The antisymmetric H-bond stretching motion, mode 3, will be expected in the near IR ($3000\ \mathrm{cm}^{-1}$ region) whereas modes 1 and 2 will be expected in the far IR ($<$ ca. $200\ \mathrm{cm}^{-1}$). The out-of-plane A_u vibration will be much lower in wave number than the B_u antisymmetric H-bond stretching vibration, which can be described in terms of the out-of-phase in-plane rotations of the two rigid monomers. Both B_u vibrations are expected to have high intensity since they obviously produce a large dipole-moment

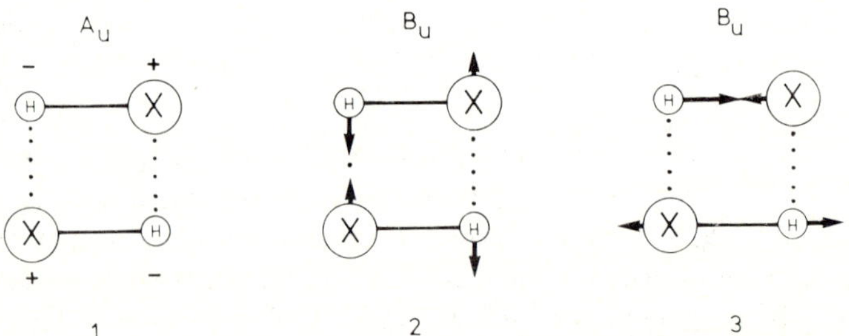

Fig. 22.3. The IR active vibrations of a planar cyclic dimer $(HX)_2$ (point group C_{2h}).

TABLE 22.2

Multimer bands[a] of the hydrogen halides in argon matrices (Barnes et al. [1969a, 1973a])

Multimer	Band[b]	HCl (cm^{-1})	HBr (cm^{-1})	HI (cm^{-1})
—	Q	{2869 2866	~2556	2205.4 (m)
—	—	—	2545.6	2198
—	—	2833		2194
			2507.2	2190
—	—	2830	2503.4	2186.9
—	—	2827		2185.1
Dimer	$\lambda_2+\lambda'$	A 2818.0 (s)	A 2497.8 (s)	2166 (m)
—	—	2804.0	2485	2180.8
Dimer	$\lambda_2-\lambda'$	[2797][c]	2482 (m)	A 2177.5 (m)
—	—	—	—	2174.9
—	—	—	—	2172.4
	—	—	—	~2169
Cyclic trimer (HI)	$\lambda_3+2\lambda'$	—	—	[2137]
Open trimer	$\lambda_3+2^{1/2}\lambda'$	2795	2472.3 (m)	2141
—	—	2791.8	—	—
Cyclic trimer (HCl, HBr)	$\lambda_3+\lambda'$	B 2786.8 (s)	B 2466.4 (s)	—
Open tetramer	$\lambda_4+1.62\lambda'$	[2784]	[2462]	2135
Open trimer	λ_3	2780.5 (m)	2461.4 (m)	B{2151 2155}(s)
Cyclic trimer (HI)	$\lambda_3-\lambda'$	—	—	
Open tetramer	$\lambda_4+0.62\lambda'$	2774	2456.2(?)	2141 (m)
Open trimer	$\lambda_3-2^{1/2}\lambda'$	[2765]	2451.3 (m)	2157
Cyclic tetramer	λ_4	2768.2 (m)	2448.5 (m)	2144.2 (s)
Open tetramer	$\lambda_4-0.62\lambda'$	2760.7 (m)	2444.2 (m)	2151
Cyclic trimer (HCl, HBr)	$\lambda_3-2\lambda'$	[2755]	[2442]	—
—	—	—	—	2135
High polymer	—	X 2753.8 (m)	X 2438.4 (s)	—
Open tetramer	$\lambda_4-1.62\lambda'$	[2750]	[2436]	[2154]
High polymer	—	X 2747.7 (s)	X 2428.6 (s)	X 2128.8 (m)
—	—	2743	—	—
—	—	2737	—	2122
—	—	2733	2416.8 (m)	2113
—	—	2728	—	2108
—	—	2720	—	—
High polymer	—	X 2701 (m)	X 2396.0 (s)	—

[a] The relative intensities are strongly concentration dependent; those of the stronger bands are quoted as a general guide.

[b] For HCl and HBr the sign of λ' should be reversed since f_{12} is negative.

[c] Approximately calculated frequencies of unobserved bands are given in square brackets.

change but the A_u vibration will have low intensity. For a linear dimer all seven modes of vibration are shown in fig. 22.4. There are three Σ vibrations which involve stretches and two doubly-degenerate bending modes of symmetry Π. Modes 1 and 2 represent the in-phase and out-of-phase stretch of the two valence bonds and these are expected to absorb in the near IR; mode 1 will have high intensity but for an exactly co-linear configuration mode 2 will have zero intensity. Mode 3 represents the stretch of the H-bond and will appear in the far IR with low intensity. The Π vibrations involve torsions about the H-bond and can be described in terms of in-phase and out-of-phase rotations of the monomer moieties, i.e. librational modes. They will absorb in the far IR and both will be expected to appear at lower wave numbers than the H-bond stretching mode; mode 4 will be expected to be of low intensity and mode 5 of high intensity. For a strictly co-linear dimer configuration the H-bond stretching vibration (mode 3) will depend on the total mass of the HX moiety and will therefore change very little on deuterium substitution.

Since the HCl dimer is cyclic, Barnes et al. [1969a] assumed that the small multimers (trimers and tetramers) are also mainly cyclic, which seems reasonable behaviour for highly polar molecules in a non-polar medium. Higher multimers in Ar matrices give a similar spectrum to the pure solid HCl and might therefore be expected to have a similar open-chain structure, thus the assumption in the KR treatment of only open-chain multimers is discarded.

A near IR study by Davies and Hallam [1971] of the aggregation band pattern of hydrogen chloride/deuterium chloride mixtures in both the $\bar{\nu}$ (HCl) and the $\bar{\nu}$ (DCl) regions allows a detailed picture of the trimer species to be built up and also provides a more rigorous test of the

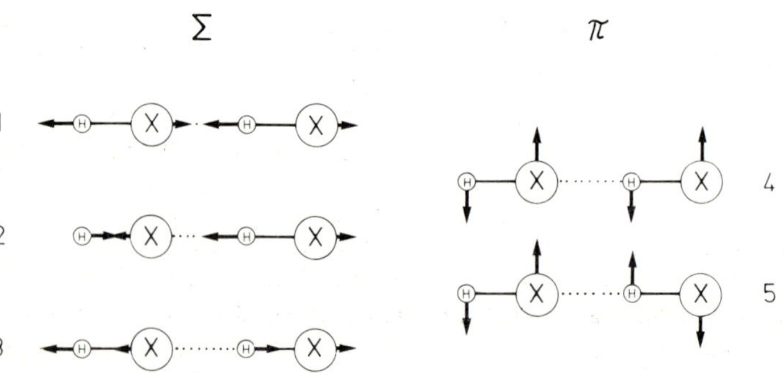

Fig. 22.4. The vibrations of a linear dimer $(HX)_2$ (point group $C_{\infty v}$).

applicability of the KR model. The chloride to argon ratio was kept constant at 1:200 while the percentage deuterium chloride was varied from 25 to 85%. Changes are seen in the spectral features in the deuterium chloride dimer and trimer regions compared with pure deuterium chloride/argon and corresponding changes are observed in the hydrogen chloride region as compared with the spectra (Barnes et al. [1969]) of pure hydrogen chloride/argon. At high deuterium chloride concentrations the band at 2041.3 cm^{-1} in the deuterium chloride dimer region is only a shoulder on the main 2040 cm^{-1} band. At low deuterium chloride concentrations the 2041.3 cm^{-1} feature becomes dominant and the 2040 cm^{-1} band is lost beneath it. Similarly, in the hydrogen chloride dimer region at high hydrogen chloride concentrations, the main band is at 2819.3 cm^{-1} with a shoulder at 2817.6 cm^{-1} whereas at low hydrogen chloride concentrations the latter is the dominant feature. Clearly, therefore, the bands at 2041.3 and 2817.6 cm^{-1} are due to dimeric HCl–DCl species. The KR model predicts a band at the group centre wave number (which is considered a more satisfactory description than the term centre of gravity wave number used by KR) of the dimer λ_2^D for deuterium chloride (2033 cm^{-1}), and λ_2^H for hydrogen chloride (2808 cm^{-1}). The group centres are calculated from the observed value of $\lambda_2 + \lambda'$ for the pure dimer by using the values of f_{12} and $\lambda = 4\pi^2\nu^2$ of Savoie and Anderson [1966].

The trimer regions are depicted in fig. 22.5 from which it can be seen that the mixed trimer species give rise to bands at 2776.8, 2774.4 and 2768.9 cm^{-1} in the hydrogen chloride region and at 2014.2, 2011.3 and 2006.6 cm^{-1} in the deuterium chloride region. These assignments provide a very sensitive test of the KR model since there can be no resonance coupling between bonded hydrogen chloride and deuterium chloride molecules, the splitting into components depending upon the number of isotopically identical molecules in nearest-neighbour contact. However, since the H-bonding capabilities of hydrogen chloride and deuterium chloride are, to a first approximation the same, the group centre of the components will depend on the total number of molecules in the aggregate whatever their isotopic identity. Thus, a cyclic trimer of type $DCl(HCl)_2$ would, with respect to hydrogen chloride, have the group centre of the trimer but the splitting of the dimer, while one of the type $(DCl)_2HCl$ would have the group centre of the trimer and no splitting. With respect to deuterium chloride the latter would have the trimer group centre and the dimer splitting pattern.

In the trimer region of deuterium chloride the dominant features at

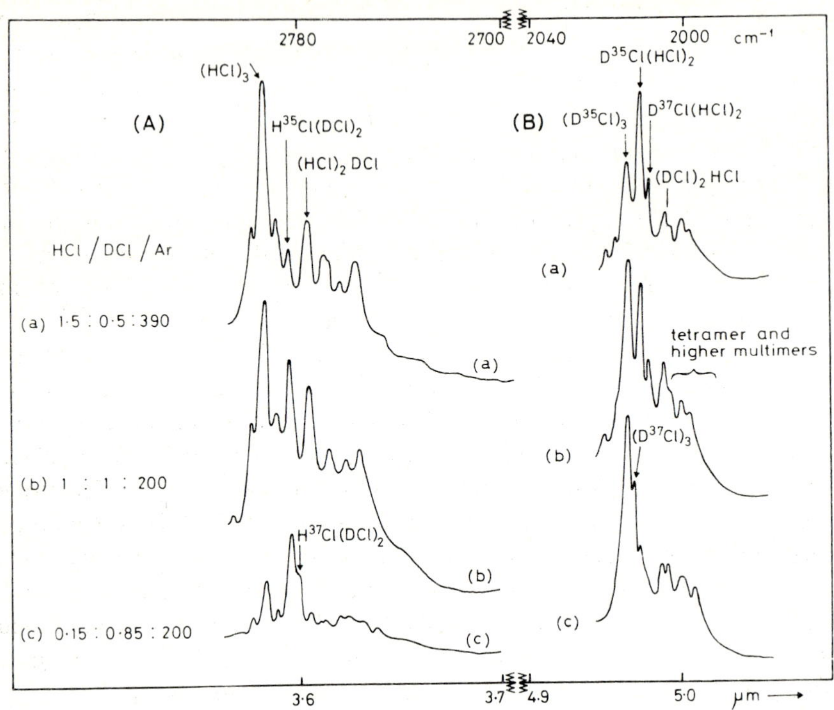

Fig. 22.5. Spectra of mixed HCl/DCl trimer species in (A) HCl fundamental region, (B) DCl fundamental region, at various HCl/DCl ratios in Ar matrices at 20 K (Davies and Hallam [1971]).

2018.6 and 2017 cm^{-1} are assigned to the $\lambda_3^D + \lambda'$ band of $(D^{35}Cl)_3$ and $(D_3^{35}Cl_2^{37}Cl)$ superimposed, and $D_3^{35}Cl^{37}Cl_2$ respectively; these then give calculated values for the group centre wave number of these species of 2012 and 2010 cm^{-1} respectively. The cyclic $(HCl)_2DCl$ trimer should have one band, λ_3^D, at the group centre wave number and two bands are observed at 2014.2 and 2011.3 cm^{-1} which are assigned to the λ_3^D band for this species for $D^{35}Cl$ and $D^{37}Cl$ respectively. The agreement between calculated and observed wave numbers is much better than for the dimer. The variation in intensity of these bands with the deuterium to hydrogen ratio suggests that they are isotopic analogues and that they arise from a trimer with two hydrogen chloride molecules and one deuterium chloride molecule. The $HCl(DCl)_2$ cyclic trimer should have bands corresponding to $\lambda_3^D + \lambda'$ and $\lambda_3^D - \lambda'$ in the deuterium chloride region. The latter should fall around 2002 cm^{-1} and a band is observed at 2006.6 cm^{-1} which increases in intensity as the hydrogen to deuterium ratio goes from 15 to 50% but decreases when it reaches 67%. This suggests that this band arises from a trimer with two deuterium chloride molecules and is assigned to the $\lambda_3^D - \lambda'$ band of the $HCl(DCl)_2$ cyclic dimer. The $\lambda_3^D + \lambda'$ band of this species would then fall about 6 cm^{-1} higher than the group centre at 2018 cm^{-1} and is presumably lost under the main trimer absorption at 2018.6 cm^{-1}.

In the hydrogen chloride trimer region the $\lambda_3^H + \lambda'$ band of the pure $(HCl)_3$ cyclic trimer is at 2786.1 cm^{-1} and this gives the calculated group centre band, λ_3^H, a value of 2776 cm^{-1}. The $(DCl)_2HCl$ cyclic trimer should thus have a band at this wave number and a band is observed at 2776.8 cm^{-1} with an isotopic shoulder at 2774 cm^{-1} which has its largest relative intensity at 15%. At 50 and 67% of hydrogen chloride it is weaker than the 2786.1 cm^{-1} band, so clearly it is due to a trimer containing only one hydrogen chloride molecule. Thus again there appears to be a good correlation between experiment and the KR theory. The $DCl(HCl)_2$ cyclic trimer should have two bands in the hydrogen chloride region, namely the $\lambda_3^H \pm \lambda'$ bands which should be approximately at 2786 and 2766 cm^{-1}. The former band would fall under the most intense band in the region at 2786.1 cm^{-1} and so is lost. A band exhibiting the correct intensity-to-hydrogen-chloride-content relationship is observed at 2786.9 cm^{-1} and is thus assigned to the $\lambda_3^H - \lambda'$ band of the $DCl(HCl)_2$ trimer.

The inadequacy of the KR model to deal with the cyclic dimer is probably due to their configuration which involves two bent H-bonds being far removed from that of the crystalline solid. In the case of the cyclic trimer the configuration involves essentially linear H-bonds.

The pattern in the multimer region of HBr is similar to that of HCl in that two intense bands successively grow at 2498 cm^{-1} (A) and 2466 cm^{-1} (B) with a series of weak to medium intensity bands at lower wave numbers than band B. It should be noted, however, that for low comparable concentrations ($M/A = 1000$) the dimer:monomer ratio is apparently greater for HBr.

For hydrogen iodide, significant differences are found: first, HI is exceptionally difficult to isolate in Ar matrices; at 20 K no monomer absorption could be detected even at $M/A = 5000$ (Barnes et al. [1973a]) although weak monomer bands can be seen (Bowers and Flygare [1966]) when deposited at 12 K. This difficulty may be in part due to the low extinction coefficient of HI monomer, but it is clear that the HI···HI interaction must be very strong so that a high proportion of monomer is obtained only in matrices which isolate most efficiently at 20 K, e.g. CO or CO_2. Secondly, although there is a superficial resemblance to HCl for the major features of band A (at 2178 cm^{-1}), due to dimer, followed by trimer band B (at 2155 cm^{-1}) which increase with increasing concentration at moderate concentrations, the behaviour differs at high concentrations. For HCl at low M/A values several intense bands (X) appear, similar in wave numbers to those of the pure solid, and must arise from small crystallites (large multimers) of HCl embedded in the Ar matrix. These have no counterpart with HI at very low M/A values, all the intensity going into band B and band A, which indicates that the trimer $(HI)_3$, is a greatly preferred configuration–under matrix conditions. It is of related interest to mention that in solid state photochemical kinetic studies of iodoethane at 77 K by Barker and Purnell [1970], the mechanism proposed implies the existence of multimers of hydrogen iodide centring around $(HI)_3$ and $(HI)_4$.

For the low multimers of HBr and HI, dimers to tetramers, the band multiplicity has also been interpreted by Barnes et al. [1969a, 1973a] on the modified KR model. Although the above calculations provide a reasonably satisfactory interpretation of the band splitting patterns (table 22.2) they can be criticised for utilising intermolecular force constants derived from the spectrum of the pure crystalline halide and as was pointed out above for the dimer, the physical situation of the trimer and tetramer is probably different from that of the pure solid.

The reverse approach has thus been adopted by Girardet and Robert [1973b] in a very recent paper which appeared as this review was in preparation. They present a detailed theoretical analysis of the matrix spectra of HCl and DCl, particularly with regard to the structure of the dimers. They first calculate the total interaction potential between a monomer HX molecule and the 12 surrounding atoms of the cell of the fcc lattice of the monatomic matrix. They then proceed to consider the various possible configurations of the two molecular axes of the HX dimers trapped in lattice sites and calculate the molecule–molecule and molecule–atom interaction energies for ranges of intermolecular distance.

In the assumption that the HX molecules are located at the lattice sites, the mean intermolecular distance $\bar{R}$ between the centres of gravity of the two molecules (HX)′ and (HX)″ of a dimer is given by $\bar{R} = a\sqrt{N}$ where $N > 1$ labels the number of cells of parameter a containing one impurity molecule with respect to the localisation of the other molecule. The shortest distance between two such molecules in nearest-neighbour cells is called $\bar{R}_m$ which for HCl/Ar is calculated to be 5.2 Å. The maximum intermolecular distance R_M is calculated to be 15.8 Å $\approx 3a\sqrt{2}$ corresponding to when the molecule–molecule angle dependent interaction may be considered as a weak perturbation with respect to the angle dependent molecule–atoms interaction energy. They evaluate the interaction energy of the two molecules, in terms of the relative orientation $(\Theta_1, \Theta_2, \Phi_1, \Phi_2)$ of the internuclear axes and their relative position $\bar{R}$, and the 24 nearest neighbour atoms pertaining to the cells which surround the molecules. The total interaction energy embraces the interaction potential between each of the trapped molecules and the 12 surrounding molecules of the matrix and the sum of the electrostatic, induction, dispersion and exchange contributions between the two linear dipolar molecules, and depends upon 11 variables. Their calculations indicate the existence of intermediate stages between the monomer and dimer, called "pseudodimers", which exist when the impurity molecules are located *at least* in two different first neighbour cells and *at most* in two second neighbour cells, i.e. in the range of $\bar{R}$ $(\bar{R}_m \leqslant \bar{R} \leqslant \bar{R}_M)$.

They identify three such types of pseudodimer:

(i) When $3a \leqslant \bar{R} \leqslant \bar{R}_M \simeq 3a\sqrt{2}$, the sum of the electrostatic $\bar{R}^{-3}$ dipole–dipole and $\bar{R}^{-4}$ dipole–quadrupole interactions have a similar magnitude to the angle dependent molecule–atoms interaction potential. Thus the electrostatic potential favours a parallel position of the molecular axes, but not necessarily co-linear with the intermolecular axis. They call this configuration pseudodimer PD_1.

(ii) In the second intermolecular distance range, $a\sqrt{3} \leqslant \bar{R} \leqslant 3a$, the dipole–dipole, dipole–quadrupole and quadrupole–quadrupole electrostatic interactions between the two impurities are now greater than the molecule–atoms interaction energy. The more stable orientational equilibrium configuration (pseudodimer PD_2) is obtained for a co-linear and parallel position of the two molecular axes, $(\rightarrow\rightarrow)$.

(iii) The $\bar{R} = a\sqrt{3}$ intermolecular distance corresponds to a particular situation for the orientational equilibrium configuration of the two molecular axes. In the case of HCl, the quadrupole–quadrupole interactions just counterbalance the sum of the dipole–dipole and dipole–quadrupole interactions. Thus for $\bar{R}_m \leqslant \bar{R} \leqslant a\sqrt{3}$, the molecular axes of this pseudodimer, PD_3, are nearly orthogonal $(\rightarrow\uparrow)$, and the centres of gravity of the two molecules are located in two nearest-neighbour cells.

The probability density for finding a molecule that does not have a nearest neighbour located at a given distance, calculated as a function of impurity concentration, shows that the pseudodimers PD_1, PD_2 and PD_3 significantly exist for M/A ratios lower than 2000, 1000 and 200, respectively. The location of their absorption bands is predicted to be in the Q branch monomer region and accounts very satisfactorily for the matrix-induced features observed by Barnes et al. [1969a] and previous workers in this region and variously assigned to HCl–Ar interactions and interacting pairs of HCl molecules.

This detailed study of the equilibrium configurations of the orientation of the two diatomic molecules located at different positions inside a monatomic matrix shows that one type of dimer is to be expected and that it will have a nearly orthogonal configuration of the two molecular axes. In a previous paper, Girardet and Robert [1973a] calculated the theoretical spectra arising from the orientational motions of such dimers, the results of which are in accord with the far IR experimental data. That is, they calculate the possible transitions between the librational levels $|j_1, m_1, j_2, m_2\rangle$ of the dimer, where j_1, m_1, j_2 and m_2 are orientational quantum numbers. In their very recent paper these results are extended to the vibration-orientation (rotation) motions of the dimer and thus calculate the vibrational coupling in these molecular couples trapped in the monatomic matrix, i.e. they calculate the position $\bar{\nu}^{v_1',j_1',m_1',v_2',j_2',m_2'}_{v_1,j_1,m_1,v_2,j_2,m_2}$ and the intensity of the near IR bands.

They calculate the shift, with respect to the $\bar{\nu}$(gas), to low wave numbers (which is characterised by interaction between the dimer and the surrounding atoms and between the molecules of the dimer themselves) and the splitting of each line due to vibrational coupling, which is *only* due to the interaction between the molecules of the dimer. Thus the calculated spectrum of the dimer exhibits two main absorption bands at the wave numbers $+\bar{\nu}^{j_1,m_1,j_2,m_2}_{j_1,m_1,j_2,m_2}$ and $-\bar{\nu}^{j_1,m_1,j_2,m_2}_{j_1,m_1,j_2,m_2}$ and due to the nearly orthogonal configuration of the two molecular axes of the dimer, the two bands will have comparable intensities. For $(HCl)_2$ in Ar these features are calculated to be $+\bar{\nu}^{0,0,0,0}_{0,0,0,0}$ 2817 cm^{-1} and $-\bar{\nu}^{0,0,0,0}_{0,0,0,0}$ 2789 cm^{-1} in striking agreement with the two strongest bands (A and B in fig. 22.2) in medium concentrations in Ar. These bands show lateral features due to orientational transitions and the synthesised spectra of HCl and DCl in Ar and Kr matrices at various concentrations and temperatures which they illustrate, bear an excellent resemblance to the experimental spectra (one example is shown in fig. 22.6).

This treatment of Girardet and Robert thus provides a quantitative

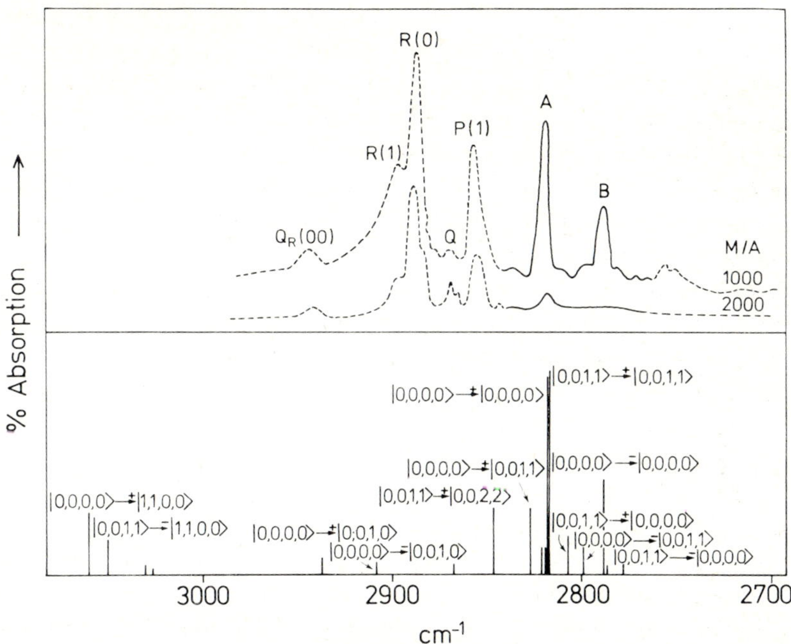

Fig. 22.6. IR spectra of Barnes et al. [1969a] of HCl in an Ar matrix for various concentrations at 20 K and bar spectrum calculated by Girardet and Robert [1973b] for $T = 20$ K and $\bar{R} = 3.72$ Å. In the theoretical spectrum, each line is labeled by four orientational quantum numbers (j_1, m_1, j_2, m_2). The initial vibrational state is $v_1 = v_2 = 0$ and the final one is indexed by the symbol + or − according to the vibrational splitting.

interpretation and their revised assignment of the 2789 cm^{-1} band (for HCl) to the dimer should be adopted (such an alternative assignment had been offered by Verstegen et al. [1966] and Atwood et al. [1967]). It should however be pointed out that the growth curves of the 2817 and 2789 cm^{-1} features do not exactly parallel each other (Barnes et al. [1969a]). This of course may be accounted for by the presence of another feature due to a higher aggregate (trimer?) underlying the 2789 cm^{-1} band. Further calculations on HX trimers and the other hydrogen halides will be awaited with interest.

22.6. Hydrogen cyanide

In their comparative study of the relative efficiencies of isolation of several molecules in various matrices, some preliminary observations of HCN were reported by Becker and Pimentel [1956]. In a N_2 matrix at

M/A = 100 they found HCN exhibited a strong absorption at 3292 cm^{-1} and a weak band at 3206 cm^{-1} with a shoulder at 3174 (all ± 15 cm^{-1}). The weak bands were assumed to arise from dimers and/or multimers. The system has now been studied by King and Nixon [1968] in far more detail with the higher resolution and wider spectral range now available. They have examined HCN and DCN in Ar, N_2 and CO matrices over a wide range of concentrations, at temperatures from 4.5 to 20.5 K from 3400 to 24 cm^{-1}. A multiplicity of both sharp and broad absorption peaks are observed in the near IR associated with each vibrational mode. Growth curves clearly separate these bands into three distinct groups. The optical densities of the first group continually decrease with increasing concentration and are assigned as monomer peaks. With the second group, which are assigned to dimer, the optical densities pass through a maximum at an M/A ratio ca. 500. The third group of peaks increase gradually with increasing concentration, a behaviour attributed to higher multimer species. The spectral pattern varies in the different host "solvent" matrices; a striking difference in the N_2 matrices, compared with those in Ar matrices, is the appearance of many fewer multimer absorptions relative to dimer under the same deposition conditions. Also multiple site trapping in N_2 and in CO causes splitting of the degenerate bending mode.

The N_2 data for monomer and dimer absorptions are shown in table 22.3. In the 3 μm region the dimer has two peaks (3282 and 3204.9 cm^{-1} for HCN, 2616 and 2574.4 cm^{-1} for DCN) due to $\bar{\nu}$ (CH/D); the fact that one of these C–H(D) stretches is almost coincident with that of the monomer while the other is shifted down by 83 cm^{-1} (44 cm^{-1} for C–D) is strong evidence for a H-bonded linear or near-linear dimer configuration. This is supported by the observation of an analogous pair of bands in the $\bar{\nu}$(CN) stretching region and in the hydrogen cyanide bending region; a bent dimer should exhibit four HCN bending modes, the relative wave numbers of which would depend upon the degree of non-linearity. Four bending modes are observed in both N_2 and CO but are attributed to site splitting of degenerate modes, since the monomer $\bar{\nu}_2$ mode is also split in both these matrices. The linear structure is further confirmed by the observation of bands at 1555 and 1480 cm^{-1} attributable to $2\bar{\nu}_2$ (1455 cm^{-1} for HCN monomer) and at 1225 and 1175 cm^{-1} (1155 cm^{-1} monomer) for DCN, the appearance of such an overtone precluding a cyclic structure with a centre of symmetry.

Finally, a linear dimer would have two degenerate low wave number librational modes, corresponding to modes 4 and 5 in fig. 22.4 (see section

TABLE 22.3

Monomer and dimer frequencies (cm^{-1}) of HCN and DCN in N_2 matrices at 20 K (King and Nixon [1968])

	Monomer	Dimer	
		Donor	Acceptor
	N≡C–H	N≡C–H···	N≡C–H
$\bar{\nu}_3$	3287.6	3204.9	3282.0
$\bar{\nu}_1$	2097.3	2092.9	2110.9
$\bar{\nu}_2$	{745.6, 736.0}	{799.0, 794.6}	{757.6, 749.6}
Librational modes corresponding to modes 4 and 5 shown in fig. 30.4		{144, 137}; 96	
	N≡C–D	N≡C–D···	N≡C–D
$\bar{\nu}_3$	2617.8	2574.4	2616.0
$\bar{\nu}_1$	1920.6	1898.4	1927.0
$\bar{\nu}_2$	{594.0, 588.4}	{628.5, 626.2}	{603.0, 598.0}
Librational modes corresponding to modes 4 and 5 shown in fig. 30.4		131; 89	

22.5), and these are observed in the far IR at 96 and 137/144 cm^{-1} for HCN, the latter being split by site effects (89 and 131 cm^{-1} for DCN). The isotope shifts for these two absorptions amount to 5 to 8% which corresponds more closely to the ratio

$$[I(\mathrm{DCN})/I(\mathrm{HCN})]^{1/2} = 1.10$$

than to

$$[M(\mathrm{DCN})/M(\mathrm{HCN})]^{1/2} = 1.02$$

which leads to King and Nixon's assignment to the two librational modes rather than one of them to the H-bond stretch. The H-bond stretching absorption has thus yet to be identified.

The authors have used the one-dimensional model of the H-bond potential function of Lippincott and Schroeder [1955] to calculate wave numbers for the stretching of the H-bond of the dimer and for the torsion about the H-bond. Fair agreement is obtained for the latter, 153 cm^{-1}, with the observed 137/144 cm^{-1} doublet for the HCN dimer and 138 cm^{-1}

calculated, with 131 cm^{-1} observed, for the DCN dimer. For both isotopic dimers the stretching mode is calculated to be 53 ± 6 cm^{-1}; a lattice mode of solid N_2 at 49 cm^{-1} could thus obscure any weak HCN or DCN absorption in this region.

Another feature in the near IR region with Ar matrices is a weak peak at 3219 cm^{-1} in HCN and the corresponding one at 2581.1 cm^{-1} in DCN. From their M/A dependence both appear to be additional dimer peaks and which King and Nixon suggest might represent either a dimer of a different configuration or a second possible site or environment for the dimer in the Ar matrix. With the large number of matrix studies now reported it would appear that multiple trapping sites are a feature more commonly found in N_2 and CO than in Ar matrices (Hallam [1973]), thus the presence of small amounts of dimer of different configuration seems the better interpretation.

22.7. Water

The matrix spectrum of water in a variety of supports has been the subject of many IR cryospectroscopic investigations primarily concerning rotational behaviour (see Hallam [1973] and the review paper by the same author in Luck [1974], where all references may be found). Only three of the investigations, Van Thiel et al. [1957a], Tursi and Nixon [1970a] and Strommen et al. [1973] concern themselves with the detailed association behaviour and the structure of the dimer. In the earlier study Van Thiel et al. found just two absorptions in the OH stretching region (3691 and 3546 cm^{-1}) and one in the bending region (1620 cm^{-1}) definitely attributable to the dimer and they thus proposed a cyclic structure. The possibility that a second weaker band at 1615 cm^{-1} in the bending region might also be due to the dimer led them to speculate further that the cyclic structure might be non-planar and thus without a centre of symmetry. In contrast to these results Tursi and Nixon, with their higher resolution, identify with some confidence *five* fundamentals, and possibly a sixth, both for the H_2O and the D_2O dimer (table 22.4). They thus eliminate a centre of inversion in the dimer structure. The pattern of the wave numbers is significant and bears an interesting comparison with those of matrix-isolated HCN (section 22.6) which, with its single C–H bond provides a simpler spectral interpretation (table 22.3). Two of the dimer fundamentals (3714.4 and 3625.6 cm^{-1} for H_2O are only slightly displaced from the monomer $\bar{\nu}_3$ and $\bar{\nu}_1$ wave numbers and the pair has essentially

TABLE 22.4

Monomer and dimer wave numbers (cm^{-1}) of water (Tursi and Nixon [1970a]) and hydrogen sulphide (Tursi and Nixon [1970b] and Barnes and Howells [1972]) in N_2 matrices at 20 K

	Monomer	Dimer		
		Donor	Acceptor	
	H–O–H	H–O–H···O(H)H		
$\bar{\nu}_3$	3725.7	3697.7	3714.4	
$\bar{\nu}_1$	3632.5	3547.5	3625.6	H_2O
$\bar{\nu}_2$	1569.9	1618.1	(1600.3?)	
$\bar{\nu}_3$	2764.6	2737.6	2756.6	
$\bar{\nu}_1$	2655.0	2599.1	2650.0	D_2O
$\bar{\nu}_2$	1179.2	1193.3	(1181.5?)	
	H–S–H	H–S–H···S(H)–H		
$\bar{\nu}_3$	2632.6	2625.3	2631.1	
$\bar{\nu}_1$	2619.5	2580.3	2617.8	

the same intensity ratio as the monomer $\bar{\nu}_3/\bar{\nu}_1$. This suggests that one unit of the dimer is an only slightly perturbed monomer. Of the remaining two bands in the stretching region, one (3697.7 cm^{-1}) lies between the monomer $\bar{\nu}_3$ and $\bar{\nu}_1$, while the other (3547.5 cm^{-1}) is displaced considerably to a lower wave number. This is thus interpreted as an open-chain structure with a single H-bond for the dimer. The fifth dimer band at 1618.1 cm^{-1} is assigned to the bending mode of the H-bonded unit and a weaker and questionable peak at 1600.3 cm^{-1} is assigned to the corresponding mode of the "free" unit. This accounts for all of the high wave number fundamentals of the dimer. The absorption features of D_2O have an exact correspondence to those of H_2O (table 22.4) and similar features are assigned, but expectedly with less confidence, for HOD. In the latter case the absence of the OH stretch of DOH··· is interpreted as evidence for a stronger deuterium- than H-bond. Force constant calculations are made for the dimer which are reasonable for an open structure with a moderately weak H-bond.

An alternative explanation of the above multiple features has been afforded by Luck (ch. 11) in terms of cyclic multimers and the angle dependence of the H-bonds.

The most recent study briefly reported by Strommen et al. [1973] utilises solid deuterium as a support at 4 K. After a careful concentration study over an M/A range of 83 to 935 they are unable to find more than three bands (3712, 3566 and 1618 cm^{-1}) assignable to the dimer, despite the fact that these bands appear to have greater intensities than the corresponding bands in the N_2 matrices. They thus infer the presence of a cyclic dimer structure in the D_2 matrix. However it is possible to interpret their data in the same manner as Tursi and Nixon if one assumes other (weaker) features, assigned by Strommen et al. to multimer species, at 3704, 3642 and ca. 1605 cm^{-1}, as also due to dimer and comparing the set of wave numbers with those of the monomer (3737, 3645 and 1599 cm^{-1}).

The low wave number modes of water require more detailed study, the only reports to date being that of Miyazawa [1961] for H_2O and D_2O in Ar and N_2 at 20 K in the region 490 to 190 cm^{-1} and some preliminary observations by Mann et al. [1974] also in Ar and N_2, but extended down to ca. 70 cm^{-1}. The results, however, are too sparse to allow a definitive decision as to a cyclic or linear dimer configuration. In the first study Miyazawa finds three bands in N_2 matrices, at 218, 243 and 265 cm^{-1} which are not present for D_2O in the same matrix nor for H_2O in Ar matrices. The 218 cm^{-1} major peak is assigned as primarily due to a librational mode of the monomer and the 243 cm^{-1} band to one of the librational modes of a cyclic structure he assumes from the study of Van Thiel et al. [1957a]. The recent study of Mann et al. [1974] confirms the general observations of H_2O in N_2 but, in contrast to Miyazawa, they find corresponding absorptions for D_2O in N_2 and for H_2O in Ar. The results as yet are mainly at high concentrations ($M/A < 150$) which prevents a complete interpretation. However, it is clear that under all conditions the broad 218 cm^{-1} feature dominates the spectra. It contains structure both to high and low wave numbers and it is apparent that a dimer and multimer librational mode must be superimposed on the monomer mode. Another librational mode may be a weaker feature at lower wave numbers than ca. 150 cm^{-1} and the weak band at 243 cm^{-1} might conceivably be the H-bond stretch.

22.8. Hydrogen sulphide

Matrix-trapped hydrogen sulphide has recently been the subject of two publications concerned with its aggregation behaviour (Tursi and Nixon [1970b] and Barnes and Howells [1972]). These are of considerable interest in light of their comparison with the data of water, especially when one considers the standard comparison of the boiling points of the two compounds, the one being taken as a typical H-bonding molecule and the other having little or no tendency to form H-bonds.

Figure 22.7 shows the concentration dependence in Ar of bands in the SH stretching region and there are clearly distinguishable dimer features even at M/A = 2000. These grow very rapidly with increasing concentration together with peaks due to higher multimers and at comparable concentrations in similar matrices H_2S appears more associated than H_2O! The dimer bands in N_2 (table 22.4) and in CO exhibit a wave number pattern for the SH stretching modes very similar to that for the OH stretches in the H_2O dimer (section 22.7), i.e. three wave numbers falling approximately within the range covered by the two stretching vibration wave numbers of the monomer and one wave number somewhat lower.

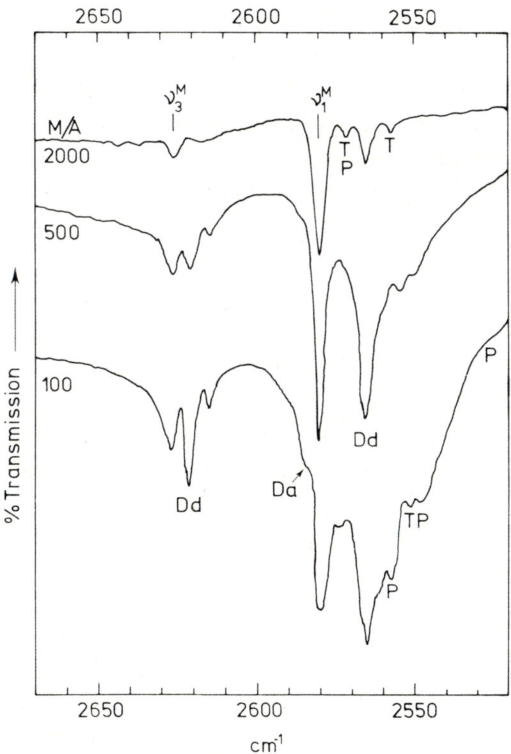

Fig. 22.7. Spectra in the SH stretching region of H_2S in Ar matrices at various M/A ratios and at 20 K (Barnes and Howells [1972]).

Thus the H_2S dimer is also taken to have the same open-chain structure with a single H-bond, the stretching vibration wave numbers at 2631.1 and 2617.8 cm^{-1} being primarily associated with the proton-acceptor unit of the dimer and the ones at 2625.3 and 2580.3 cm^{-1} with the proton donor. At higher concentrations further bands are observed (fig. 22.7) due to larger multimers. The corresponding spectra of D_2S were too contaminated with HDS to allow interpretation.

22.9. Ammonia

Like water, ammonia has been the subject of a number of matrix studies (see Hallam [1973]), the matrix effect on inversion being of special interest in addition to the rotational degree of freedom of the isolated monomer. One of the reports (Pimentel et al. [1962]) includes an examination of the aggregation behaviour. They identify dimer absorptions at 3404, 1004.5 and 987 cm^{-1} and possible absorptions (along with higher multimers) at 3313, 3246 and 3237 cm^{-1}. For the dimer the two highest NH stretching absorptions are assigned as the perturbed E modes, 3404 cm^{-1}, $\bar{\nu}_3'(E)$; 3313 cm^{-1}, $\bar{\nu}_3''(E)$ and the lower two to the perturbed A_1 modes, 3246 cm^{-1}, $\bar{\nu}_1'(A_1)$; 3237 cm^{-1}, $\bar{\nu}_2''(A_1)$ and 987 cm^{-1}, $\bar{\nu}_2''(A_1)$, all with shifts comparable to those for water dimers in N_2 (section 22.7). The appearance of such features rules out any structure with a centre of symmetry, including a cyclic, H-bonded structure analogous to that suggested for water dimers (Van Thiel et al. [1957a]). A symmetrical structure in which one NH_3 pyramid is stacked on the other in C_{3v} symmetry–this is suggested by the crystal structure of solid ammonia where each nitrogen atom acts as a base in three H-bonds–is also considered very unlikely. It is thus proposed that the structure of the dimer is one in which a single H-bond links the two nitrogen atoms.

22.10. Hydrazoic acid

Pimentel et al. [1966] have made a detailed study of the association of HN_3 in a N_2 matrix and a comprehensive assignment offered for the dimer species (table 22.5). There are two dimeric N–H stretching vibration bands, one at 3314 cm^{-1} is shifted only 10 cm^{-1} from the monomer, and the other at 3174 cm^{-1}, is shifted considerably to lower wave numbers (-150 cm^{-1}). This is indicative of an open dimeric structure, either (I) or (II),

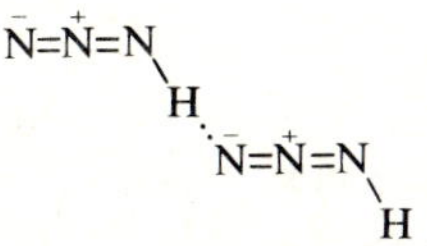

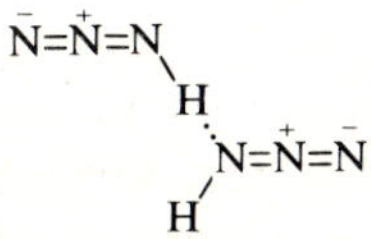

```
N=N=N                    N=N=N
     \                        \
      H                        H
       ·.                        ·.
        N=N=N                      N=N=N
             \                    /
              H                  H
      (I)                      (II)
```

and is inconsistent with the two cyclic structures, (III) and (IV):

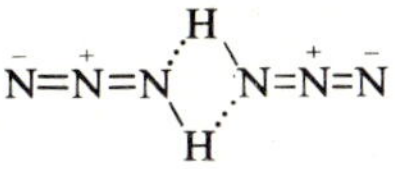

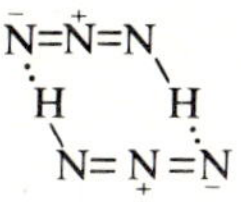

```
         H                     N=N=N
       .·  \                  ·      \
N=N=N       N=N=N            H        H
       \   .·                 \      ·
         H                     N= N =N
       (III)                    (IV)
```

Furthermore, these wave number shifts and their intensities, are similar to those of diols which (Kuhn [1952]) have an intramolecular H-bond,

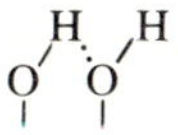

```
  H   H
 / · /
O   O
|   |
```

in which one H-atom is free. The absorption most informative of the dimer structure is the band at 387 cm^{-1} which shifts by a factor of 1.28 to 302 cm^{-1} in DN_3. This band has no monomer counterpart and is thus assigned to the torsional mode of a H-bond. The fact that only two bands are observed, 387 and 302 cm^{-1}, irrespective of the H/D ratio, shows that the $HN_3 \cdots DN_3$ dimer absorbs at the same wave number as $HN_3 \cdots HN_3$ and $DN_3 \cdots DN_3$. This behaviour is consistent only with an open dimeric structure. The authors suggest structure (I) as the most likely because of

TABLE 22.5

Monomer and dimer wave numbers (cm^{-1}) of HN_3 in N_2 matrices at 20 K (Pimentel et al. [1966])

Assignment	Monomer	Dimer	$\Delta\bar{\nu}$
$\bar{\nu}_1$ N–H stretch	3324	3314	−10
		3174	−150
$\bar{\nu}_2$ N–N–N asym stretch	2150	2162	+12
		2143	−7
$\bar{\nu}_3$ N–N–N bend-sym stretch	1273	1279	+6
$\bar{\nu}_4$ H–N–N sym stretch-bend	1168	1177	+9
$\bar{\nu}_6$ N–N–N bend	588 (?)	605 (?)	
$\bar{\nu}_5$ N–N–N bend	527	545 (?)	
$\bar{\nu}_7$ H bond tors		387	

the charge distribution implied by the bonds. They point out, however, that a triple-bonded resonance structure that places negative charge on the first nitrogen atom, makes structure (II) also plausible. This writer favours (II) which better explains the $-10\,cm^{-1}$ shift for $\bar{\nu}NH_{free}$. A comparison of the wave number shift of $150\,cm^{-1}$ with those of other N–H···N bonds of known strength, $NH_3 \cdots NH_3$ and pyrrole/pyridine, suggests a dimer H-bond energy of $9.6\,kJ\,mol^{-1}$.

22.11. Alcohols

The superiority of matrix-isolation spectroscopy is well exemplified in the case of alcohols. The association characteristics of alcohols have been studied, more than any other class of compound, in various phases by IR spectroscopy. In the vapour phase, difficulties arise from the rotational fine structure of breadth, although ill-defined bands can be discerned within the rotational envelope, due to dimers and higher multimers. For example, Inskeep et al. [1958] have reported bands in methanol vapour due to open-chain dimer and cyclic tetramer and have obtained equilibrium constants and values for the enthalpies of association. Barnes et al. [1973b] have used a long pathlength multireflexion cell to examine several other alcohols in the gas phase and also quote ΔH values. However the lack of consistency between these results and those of other workers indicates the difficulties inherent in measuring these band profiles and thus the caution necessary when interpreting multimer structures on the basis of ΔH values alone.

In the pure liquid or solid phases, the OH stretching modes give bands of enormous breadth, due to the presence of large multimeric species. As a result of these difficulties, the association of alcohols has largely been studied in solution in non-polar solvents, which allows some control of the species present by varying the concentration. At high dilution, solute–solute interactions are minimised and the monomer spectrum obtained, whereas at high concentrations spectra approximating to the pure liquid are found (fig. 22.8). On progressive dilution the broad multimer absorption gradually shows indications of a shoulder on the high wave number side, which gives rise to a poorly defined band at ca. $3500\,cm^{-1}$ before the monomer species becomes dominant. The breadth of the bands, however, makes it impossible to locate bands due to multimeric species intermediate between dimer and high polymer.

The advantages afforded by the matrix technique are evident and Van

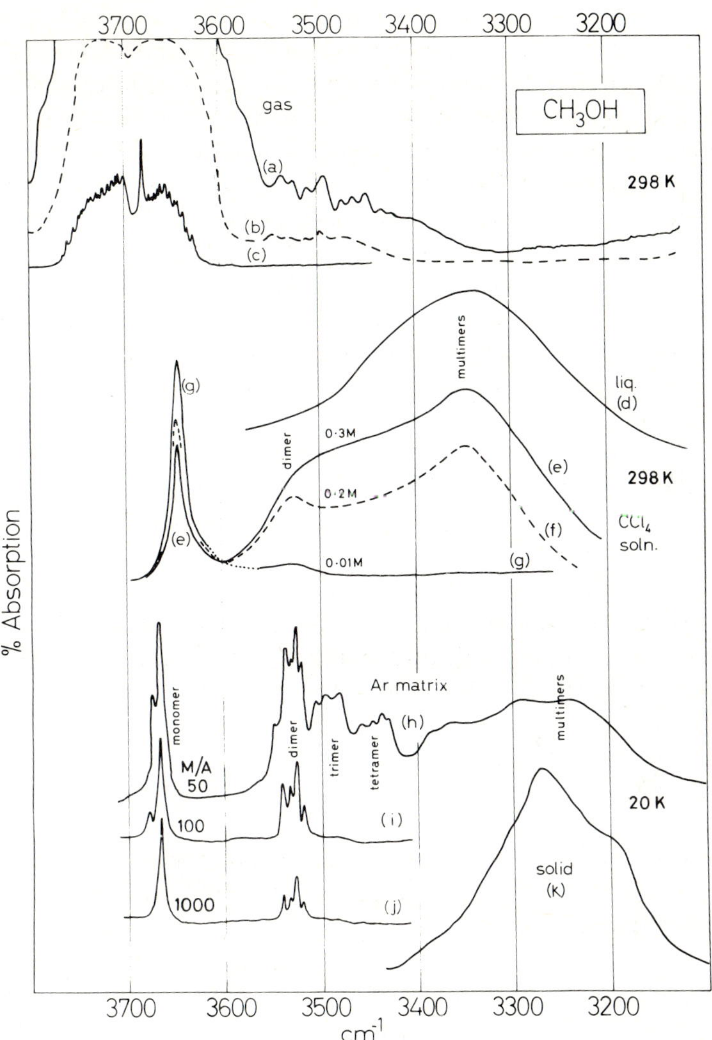

Fig. 22.8. IR spectra of OH stretching region of methanol in various phases.

Thiel et al. [1957b] in one of the earliest matrix studies examined methanol suspended in solid N_2 at 20 K. Their investigation covered the $\bar{\nu}$(OH) and $\bar{\nu}$(CH) spectral region, 3700–2800 cm^{-1}, and by varying the concentration over M/A ratios of 2200 to 10, they were able to identify discrete absorptions due to numerous multimer species. They assigned the bands with low wave number shifts to cyclic dimer and trimer, while an absorption showing a shift close to that of the pure solid was assigned to open-chain tetramer. The system has recently been re-examined by Barnes and Hallam [1970a] in much greater detail with the higher resolution and wider spectral facilities now available. The isotopic species CH_3OH, CH_3OD and CD_3OD were examined between 4000 and 40 cm^{-1} in Ar at 20 K. A similar pattern of bands in the $\bar{\nu}$[OH(D)] stretching region is observed for all three molecules (table 22.6) and the spectra for CH_3OH at various concentrations are shown in fig. 22.8. The improvement in detail over the spectra in all other phases is striking. The OH(D) in-plane bending mode exhibits a corresponding series of bands, shifted to higher wave numbers than the monomer band and terminating in a broad absorption analogous to the solid spectrum.

Despite the wealth of near IR $\bar{\nu}$(OH) studies on alcohols there is still controversy over the structures of the multimer species. The evidence from vapour phase (Inskeep et al. [1958]) and solution (for example,

TABLE 22.6

–OH(D) stretching vibration wave numbers of methanol in Ar at 20 K (Barnes and Hallam [1970a])

Species	CH_3OH	CH_3OD	CD_3OD
"Free" OH dimer / Second site monomer (?)	3678.9	2714.6	2714.6
Monomer	3667.3	2706.1	2706.1
Multimer	3556(vw)	2626(vw)	2626(vw)
Open chain dimer	3541.1	2617.0	2617.0
	3533.5	2611.3	2611.3
	3528.2	2606.6	2606.6
	3519.0	2600.4	2600.4
Open chain trimer	3505	2591	2591
	3495	2584	2584
	3482	2575	2575
Open chain tetramer	3458	~2560(sho)	~2560(sho)
	3446	2545	2545
	3435	~2538(sho)	~2538(sho)
Cyclic tetramer (?)	3379	~2508	~2508
Pentamer (?)	3359	2476	2496
High polymer	3297	2439	2447
	3254	2415	2423
	~3200(sho)	~2393(sho)	~2397(sho)

Bellamy and Pace [1966]) is slightly weighted in favour of the interpretation that small multimers, displaying small wave number shifts compared with the solid, are open-chain, while larger multimers, displaying wave number shifts comparable with the solid, are cyclic. However there is much force in the argument for the reverse interpretation of Luck (ch. 11) based on the angle dependence of the H-bonds.

In the multiplicity of bands resolved for methanol in Ar (fig. 22.8) a band is observed on the high wave number side of the monomer band at 3678.9 cm^{-1}, $\Delta\bar{\nu} = +11.6\ cm^{-1}$, for CH_3OH, whose growth behaviour approximately parallels that of the composite dimer band. Site splitting possibly accounts for this feature but is considered unlikely. A better interpretation is that it arises from the free OH of an open-chain dimer

```
Me        Me
  \      /
   O–H··O
         \
          H
```

which, due to the perturbation its oxygen atom experiences, will absorb at a slightly different wave number from the monomer. Such a shift however would be expected to be to a lower wave number and indeed such a feature does appear in all the other alcohols studied in matrices (fig. 22.9). Moreover, for the OH(D) bending modes of the methanols a corresponding feature is observed, close to the monomer wave number, which in some cases is to *lower* wave number than the monomer, which lends support to the dimer and/or multimer end groups assignment of the 3679 cm^{-1} feature.

Further indications that the dimers predominantly have an open-chain configuration comes from the far IR data (Barnes and Hallam [1970a]) where such a dimer should exhibit H-bond stretching and deformation modes of comparable intensity, as for HCN (King and Nixon [1968]). Two strong bands, at 222 and 116 cm^{-1}, are in fact observed and the low deuteration shift (to 213 cm^{-1}) of the stretching mode lends additional support for an open-chain structure.

The trimer and tetramer which absorb at slightly lower wave numbers than the dimer are also assumed to have open structures. The multiplet structure of each of these absorptions may be due to one or more of the following: (i) multiple trapping sites; (ii) vibrational splitting similar to the hydrogen halides (section 22.5); (iii) other dimer species with different angular orientations of the H-bonds, including the cyclic configuration. It is not possible at this time to give a definitive explanation but with the

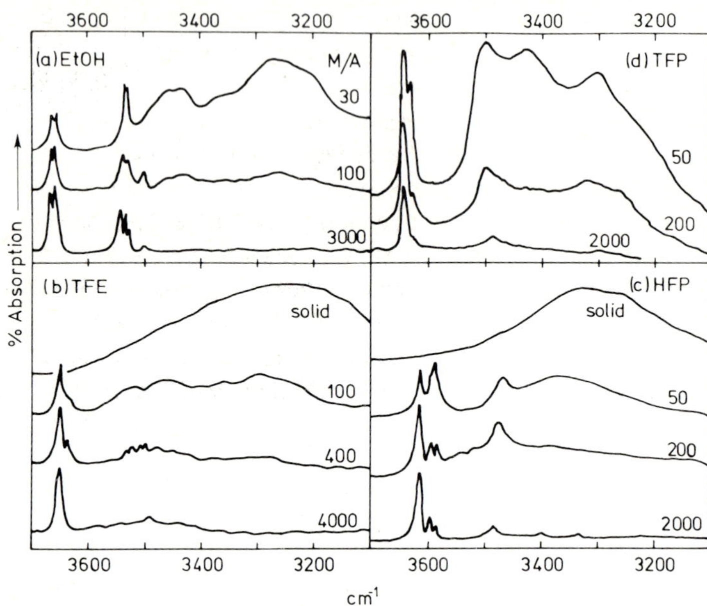

Fig. 22.9. Spectra of the OH stretching region of alcohols at various concentrations in Ar matrices: (a) ethanol (Barnes and Hallam [1970b]); (b) trifluoroethanol (Barnes et al. [1973b]); (c) hexafluoropropanol (Barnes and Murto [1972]); (d) trifluoropropanol (Murto et al. [1974]). (Note that the doublet monomer structure in these cases arises from conformational isomers.)

considerable experience now available on many matrix systems the most likely interpretation is the presence of other species, probably cyclic. Each composite band appears to have a quartet structure and although difficult to follow in detail, the growth behaviour indicates that these are two doublets. A comparison of the CH_3OH spectra of fig. 22.8 with those of Perutz and Turner [1973] recently carried out under a variety of deposition conditions confirms this for the dimer band. Thus the trapping of two types of dimer, linear and cyclic, each band of which is split by vibrational coupling seems the most acceptable explanation.

Ethanol and a number of fluoro-alcohols have also been examined by matrix-isolation spectroscopy. The spectra (fig. 22.9) show similar features to those of methanol though in some cases less well resolved (note that the doublet feature of $\bar{\nu}$(OH) ethanol, observable at very low, as well as high, concentrations, arises from *trans* and *gauche* conformers of this molecule).

22.12. Thiols

Similar studies to methanol and ethanol have been made by Barnes et al. [1972] on methane- and ethane-thiol in Ar at 20 K (fig. 22.10). In the SH

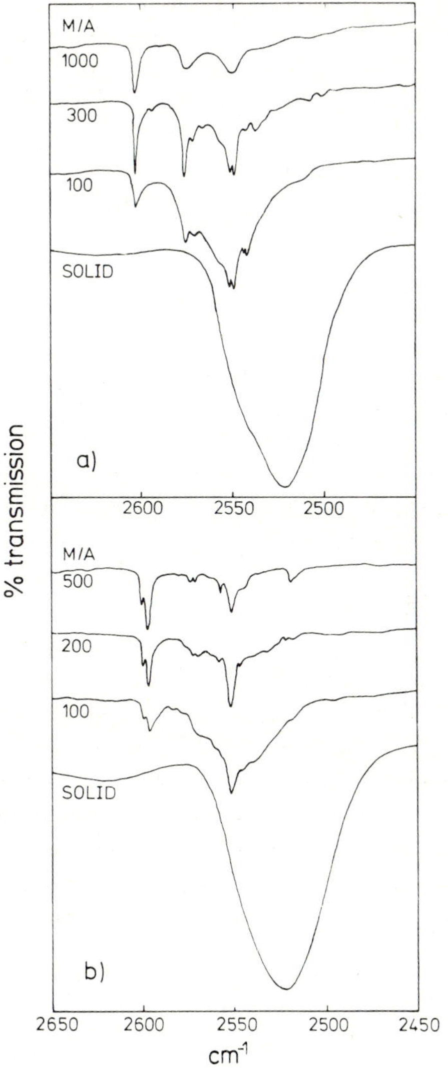

Fig. 22.10. SH stretching region (a) methane-thiol and (b) ethane-thiol, in Ar matrices at various concentrations (Barnes et al. [1972]). (Note that the doublet structure of the monomer band of ethane-thiol arises from conformational isomers.)

stretching region two multimer bands are observed with intensities comparable to that of the monomer even at M/A = 1000. The first feature, at 2576 cm^{-1} for CH_3SH, relative to 2603 cm^{-1} for the monomer is clearly a dimer band and from a comparison with the corresponding shifts for methanol (section 22.11) the authors conclude that the second band, a doublet at 2552 and 2550 cm^{-1}, arises from cyclic tetramer. Since the CSH bond angle is ca. 90° compared with ca. 110° for the COH bond angle of alcohols, it was considered that a cyclic tetramer would have a favourable structure:

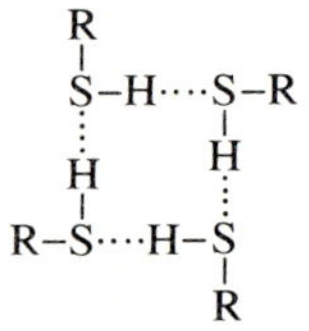

```
     R
     |
     S–H····S–R
     :      |
     H      H
     |      :
   R–S····H–S
            |
            R
```

which would account for the observed stability of the multimer causing the 2551 cm^{-1} absorption compared with the dimer. Such an effect was not, however, observed for H_2S where the dimer is the only important multimer at low concentrations (section 22.8).

Further consideration of these bands, especially for ethane-thiol in which the 2551 cm^{-1} band is considerably more intense than the first band, at 2572 cm^{-1}, suggests that a more rational interpretation would be to assign both features to the dimer, the doublet arising from vibrational coupling.

This interpretation is supported by theoretical calculations (Pecul and Janoschek [1974]) for $(CH_3SH)_2$. They calculate the P.E. and dipole moment surfaces for the SH···S H-bond of a linear dimer by the SCF–MO–LCGO method and hence compute the vibrational spectrum at 20 K. The general appearance of this compares favourably with the observed spectrum of Barnes et al. [1972] although the "origin" of the calculated system is ca. 300 cm^{-1} to lower wave numbers. The two main features are clearly accounted for and also a number of the observed satellites.

22.13. Carboxylic acids

Carboxylic acids form one of the classic examples of the H-bond and have been extensively examined by all structural techniques. It is well known that in the IR the OH stretching vibration appears as an enormously broad and intense absorption for carboxylic acid dimer molecules. There is thus considerable interest in this system and many proposals have been made (sections 6.2.6, 15.6.1, 16.2.2 and ch. 12) to explain this phenomenon, most interpretations being based on combination bands of carboxyl group vibrations and low wave number vibrations. The most quantitative theory has been given by Marechal and Witkowski [1968] who consider the non-adiabatic coupling between the slow O···O stretching coordinate and the fast OH vibration. Their model for the cyclic $(CD_3COOH)_2$ dimer yields absorption lines in excellent agreement with the vapour-phase band contour.

Some of the vapour spectra show detailed structure in the band envelope and it is to be expected that matrix spectra would sharpen these up, especially by the removal of hot band interference. This was first demonstrated by Millikan and Pitzer [1958] who in their study of dimeric and crystalline formic acid briefly report the 3500–2500 cm^{-1} region of a sample trapped in a N_2 matrix at 20 K. They note that the matrix bands are considerably sharper than in the vapour and the most intense bands for the dimer lie near the 3100 cm^{-1} peak observed in the vapour. Miyazawa and Pitzer [1959] have made a detailed comparison of the four isotopic species of formic acid in the vapour phase and in a nitrogen matrix at 20 K, in the region 800–400 cm^{-1}. Their work was undertaken to investigate the internal rotation of formic acid monomer but the torsional vibrations of the terminal OH groups of short-chain multimers of HCOOH were also observed in the matrix at 685 and 694 cm^{-1}.

Like formic acid the IR spectrum of acetic acid has been extensively investigated. However, a new assignment of the fundamental frequencies is proposed as a result of IR studies (Berney et al. [1970]) between 4000 and 400 cm^{-1} of the four species $CH_3COOH(D)$ and $CD_3COOH(D)$ isolated as monomers in Ar and N_2 matrices at 4 K. The interference caused by overlapping absorptions in vapour-phase studies was eliminated and the matrix spectra yield clear and unambiguous monomer absorptions in all spectral regions. At M/A = 1000, weak bands due to acetic acid dimers were observed. These have been the subject of a separate study by Redington and Lin [1971] who have made a detailed examination of the hydroxyl stretching region of $CF_3COOH(D)$ and CH_3COOH matrix-isolated in Ne and Ar at 5 K. Fundamental frequencies for the cyclic dimer ring vibrations are suggested and the broadening phenomenon in the $\bar{\nu}(OH)$ region is accounted for, as being due to combination bands of the low wave number ring vibrations and $\bar{\nu}(OH)$, and excited binary combinations of the COH(D) bending and the CO stretching vibrations.

Grenie et al. [1971] have reported the spectra of CD_3COOH, CH_3COOH and $CH_3C^{18}O^{18}OH$ dimers in an Ar matrix at 20 K. Their results do not agree with Redington and Lin's assignment but partially confirm those of Haurie and Novak [1965]. They point out, however, that it is difficult to correlate clearly the gas and matrix wave numbers in the 3200–2700 cm^{-1} region as the whole absorption is lowered in intensity and wave number, relative to the bands in the 2700–2500 cm^{-1} region, when going from gas to matrix phase. Their spectra are shown in fig. 22.11 which clearly illustrate the sharper features characteristic of matrix spectra.

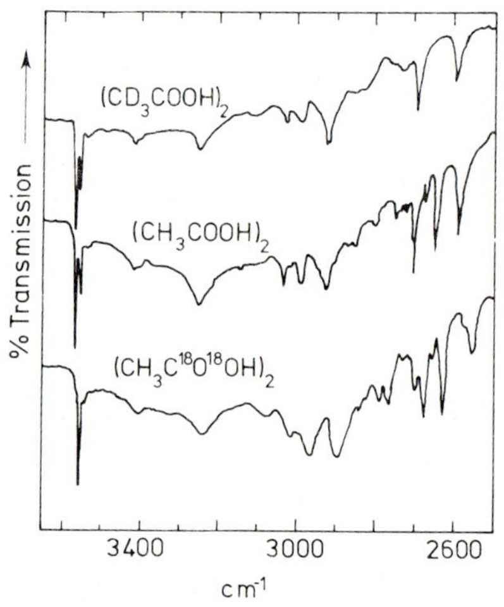

Fig. 22.11. IR spectra in the OH stretching region of three isotopic species of acetic acid in an Ar matrix at 20 K with $M/A \approx 70$ (Grenie et al. [1971]).

22.14. Hetero-associated species

In addition to the self-associated molecules discussed in the previous sections the MI technique has also been utilised to investigate hetero-associated complexes. Work has only recently been commenced in this field but already a number of what one would expect to be very weak interactions, are found to have well-defined characteristics in cryogenic matrices.

Almost all systems involve one of the hydrogen halides as the proton donor and stem from the observations by Bowers and Flygare [1966] and

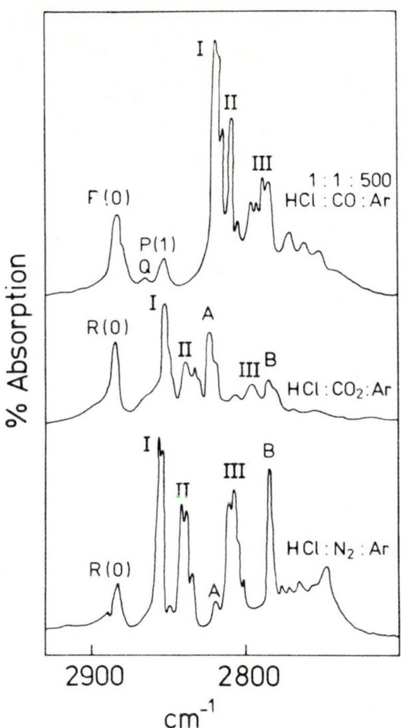

Fig. 22.12. IR spectra of HCl + X mixtures (X = N_2, CO, CO_2) in Ar matrices at 20 K (Barnes et al. [1969b]). Labels identify: (A) $(HCl)_2$ dimers, (I) HCl–X, (II) HCl–X–HCl, (III) $(HCl)_2$–X, species.

Mann et al. [1966] that the presence of impure nitrogen in a matrix of HCl in Ar induced an intense sharp band in the Q branch region of the HCl monomer spectrum at about 2863 cm^{-1}, and a weak band at ca. 2812 cm^{-1}. Barnes et al. [1969b] and Davies and Hallam [1971] have carried out extensive hetero-association studies of binary mixtures of hydrogen chloride, hydrogen bromide or hydrogen iodide with nitrogen, carbon monoxide, carbon dioxide or C_2H_4, (X) and DCl/X in argon matrices at 20 K. Three regions of absorption are induced in the HHal or DHal spectrum due to the dopant molecule X (e.g. fig. 22.12). The most intense induced band (I) falls in the Q branch region of the monomer spectrum and is interpreted as arising from HHal–X interactions. The second induced band(s) (II) are observed at lower wave numbers but still in the monomer region and are assigned to HHal–X–HHal interactions. The third band or group of bands (III) are alongside the dimer band (A) (see section 22.5) which is greatly reduced in intensity; feature III is thus assigned to $(HHal)_2$–X interactions. In the case of carbon monoxide as the dopant molecule, corresponding perturbations are observed in the carbon monoxide spectrum. An intense pair of bands is induced at 2156.2 and 2155 cm^{-1} to higher wave numbers than the pure carbon monoxide monomer band. Since the oxygen sp lone pair has a slightly antibonding character this is interpreted as indicating that (D)HCl is H-bonding with the oxygen atom of carbon monoxide.

Apart from their own intrinsic interest, these results stress the caution necessary to eliminate atmospheric leaks at the injection and from the M.I. apparatus. Minute traces of N_2 impurity cause induced features and several of the matrix studies described in previous sections exhibit these.

A detailed study has also recently been made (Barnes et al. [1973a]) of strong complexes formed by HI with π bonded molecules such as benzene and alkenes, and with iodoalkanes. At HI–alkene(alkane)–Ar ratios of 1 : 1 : 1000, two additional bands are observed in the HI stretching region, e.g. for $HI/C_2H_4(Ar)$, 2177 and 2165 cm^{-1}. The lower wave number band increases in intensity as the proportion of alkene is raised and is thus assigned to a HI interacting with two or more alkene molecules, the main band arising from a 1 : 1 alkene···HI complex. Information concerning the configuration of the 1 : 1 complexes can be derived from the perturbation of the alkene vibrational modes. If the complex is formed by H-bonding to the π electrons along a direction normal to the $\rangle C{=}C\langle$ plane, it would be expected that the C=C stretching mode would be shifted to lower wave number due to weakening of the bond, while the out-of-plane bending motions of the hydrogens would be shifted to higher wave number due to steric hindrance. These effects are observed for ethene (the IR inactive $\bar{\nu}$ (C=C) mode being weakly induced at 1623 cm^{-1}; CH_2 wag, 946 cm^{-1}, mon. 956 cm^{-1}, complex), propene (1650 cm^{-1}, mon. 1645 cm^{-1}, complex; CH_2 wag, 908 cm^{-1}, mon. 917 cm^{-1}, complex) and butenes. No shifts were observed for the in-plane motions. Thus the complexes have an orthogonal configuration:

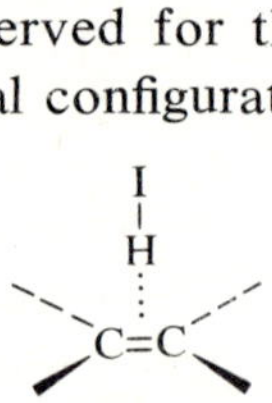

I
|
H
⋮
C=C

The wave number shift of $\bar{\nu}$(HI) of the alkene···HI complex increases with the degree of substitution of the alkene and can be related to the electron donating ability of the alkene; for example there is a correlation between $\bar{\nu}$(HI) and the first ionisation potential of the alkene.

The wave number shifts for the iodoalkane···HI complexes are larger than those of the alkene···HI complexes. The dipole moments of the iodoalkanes are quite large (ca. 2D) and thus the interaction is clearly due to H-bonding: RI···HI, which is confirmed by the shifting to lower wave number of the CI stretching mode of the iodoalkane adduct. Other features are observed which may arise from RI···$(HI)_n$ and/or RI···HI···RI complexes.

22.15. Conclusions

The accumulated experience over a wide variety of systems confirms that matrix-isolation spectra at cryogenic temperatures are far more informative than their counterpart studies in other phases, and fully justifies Pimentel's early predictions of its unique potentialities. In particular it offers spectral evidence of H-bonding in weak complexes, evidence which is difficult to obtain by conventional techniques. Although it is fair to say that we now have a considerable knowledge of the effect of the matrix environments on the trapped species, it should still be stressed that considerable caution is necessary when comparing the spectra, despite their excellent quality, with those of the gas and other phases. The technique certainly freezes out any gas-phase species but there is now overwhelming evidence that the experimental conditions of injection through a capillary nozzle from ca. atmospheric pressure to ca. 1×10^{-5} torr, and the rapid quenching from ambient to cryogenic temperatures itself, leads to higher aggregates. There is considerable evidence from molecular beam deflection and mass spectrometric techniques that molecular clusters are formed when a gas or mixture of gases is expanded through a supersonic nozzle as a result of the drop in local internal temperature (e.g. Golomb et al. [1970]). These combined effects probably account for the apparent increase in aggregation, for comparable concentrations, in the series:

$$HCl < HBr < HI; \qquad H_2O < H_2S; \qquad CH_3OH < CH_3SH,$$

and for the facility, which matrix isolation has, for investigating weak complexes in general. The size, as opposed to the polarity, of the monomer molecule may also play a determining role.

It is also likely that the technique freezes out many configurations of a particular species. Although in the studies reviewed here evidence is presented for the preponderance of a particular configuration, e.g. open-chain or cyclic dimer, in almost all of these cases there are features due to other configurations of the dimer.

There are sometimes striking differences in N_2 matrix studies compared with those in Ar matrices, in the appearance of many fewer multimer absorptions under the same deposition conditions, indicating that N_2 has a greater efficiency of isolation. Studies in a variety of matrices frequently lead to additional information and should be encouraged.

With the developments in cryogenic technology (see chapters in Meyer [1971] and Hallam [1973]) in the last decade, matrix-isolation spectroscopy need no longer be considered as a highly specialised technique. It should now in fact be a routine technique in all vibrational spectroscopy laboratories, particularly those concerned with molecular association. There are many other systems still to be studied and other fruitful areas for exploitation, for example in Raman spectroscopy (despite the experimental difficulties and the generally weak intensity of the H-bond vibrations) and in the overtone regions. Also the far IR demands attention where, as this review shows, there have been very few matrix studies of H-bonding.

References

Atwood, M. R., M. Jean-Louis and H. Vu, 1967, J. Phys. (Paris) **28**, 31.

Barker, P. G. and J. H. Purnell, 1970, Trans. Faraday Soc. **66**, 163.

Barnes, A. J. and H. E. Hallam, 1969, Quart. Rev. Chem. Soc. **23**, 392.

Barnes, A. J. and H. E. Hallam, 1970a, Trans. Faraday Soc. **66**, 1920.

Barnes, A. J. and H. E. Hallam, 1970b, Trans. Faraday Soc. **66**, 1932.

Barnes, A. J. and J. D. R. Howells, 1972, J.C.S., Faraday Trans. II **68**, 729.

Barnes, A. J. and J. Murto, 1972, J.C.S. Faraday Trans. II **68**, 1642.

Barnes, A. J., H. E. Hallam and G. F. Scrimshaw, 1969a, Trans. Faraday Soc. **65**, 3150.

Barnes, A. J., H. E. Hallam and G. F. Scrimshaw, 1969b, Trans. Faraday Soc. **65**, 3172.

Barnes, A. J., H. E. Hallam and J. D. R. Howells, 1972, J.C.S. Faraday Trans. II **68**, 737.

Barnes, A. J., J. B. Davies, H. E. Hallam and J. D. R. Howells, 1973a, J.C.S. Faraday Trans. II **69**, 246.

Barnes, A. J., H. E. Hallam and D. Jones, 1973b, Proc. Roy. Soc. **A355**, 97.

Becker, E. D. and G. C. Pimentel, 1956, J. Chem. Phys. **25**, 224.

Bellamy, L. J. and R. J. Pace, 1966, Spectrochim. Acta **22**, 525.

Berney, C. L., R. L. Redington and K. C. Lin, 1970, J. Chem. Phys. **53**, 1713.

Bowers, M. T. and W. H. Flygare, 1966, J. Chem. Phys. **44**, 1389.

Bowers, M. T., G. I. Kerley and W. H. Flygare, 1966, J. Chem. Phys. **45**, 3399.

Davies, J. B. and H. E. Hallam, 1971, Trans. Faraday Soc. **67**, 3176.
Girardet C. and D. Robert, 1973a, J. Chem. Phys. **58**, 4110.
Girardet C. and D. Robert, 1973b, J. Chem. Phys. **59**, 5020.
Golomb, D., R. E. Good and R. F. Brown, 1970, J. Chem. Phys. **52**, 1545.
Grenie, Y., J.-C. Cornut and J.-C. Lassegues, 1971, J. Chem. Phys. **55**, 5844.
Hallam, H. E. (Ed.), 1973, "Vibrational Spectroscopy of Trapped Species–Infrared and Raman Studies of Matrix-Isolated Molecules, Radicals and Ions" (Wiley, London).
Haurie, M. and A. Novak, 1965, J. Chim. Phys. **62**, 146.
Inskeep, R. G., J. M. Kelliher, P. E. McMahon and B. G. Somers, 1958, J. Chem. Phys. **28**, 1033.
Katz, B., A. Ron and O. Schnepp, 1967, J. Chem. Phys. **47**, 5303.
Keyser, L. F. and G. W. Robinson, 1966, J. Chem. Phys. **44**, 3225.
King, S. T., 1970, J. Chem. Phys. **74**, 2133.
King, C. M. and E. R. Nixon, 1968, J. Chem. Phys. **48**, 1685.
Kuhn, L. P., 1952, J. Am. Chem. Soc. **74**, 2492.
Lippincott, E. R. and R. Schroeder, 1955, J. Chem. Phys. **23**, 1099.
Luck, W. A. P. (Ed.), 1974, "Structure of Water and Aqueous Solutions" (Verlag Chemie, Weinheim).
Mann, D. E., N. Acquista and D. White, 1966, J. Chem. Phys. **44**, 3453.
Mann, B., E. Schmidt, T. Neikes and W. Luck, 1974, Abstracts of paper presented at 73rd meeting of Deutschen Bunsen-Gesellschaft für Physikalische Chemie, Kassel, May 1974.
Marechal, Y. and A. Witkowski, 1968, J. Chem. Phys. **48**, 3697.
Meyer, B., 1971, "Low Temperature Spectroscopy–Optical Properties of Molecules in Matrices, Mixed Crystals, and Frozen Solutions", (Elsevier, New York).
Millikan, R. C. and K. S. Pitzer, 1958, J. Am. Chem. Soc. **80**, 3515.
Miyazawa, T., 1961, Bull. Chem. Soc. Japan **34**, 202.
Miyazawa, T. and K. S. Pitzer, 1959, J. Chem. Phys. **30**, 1076.
Murto, J., A. Kivinen, K. Edelmann and E. Hassinen, 1975, Spectrochim. Acta **31A**, 479.
Norman, I. and G. Porter, 1954, Nature, Lond. **174**, 508.
Pecul, K. and R. Janoschek, 1974, personal communication.
Perutz, R. N. and J. J. Turner, 1973, J.C.S., Faraday Trans. II **69**, 452.
Pimentel, G. C., M. O. Bulanin and M. Van Thiel, 1962, J. Chem. Phys. **36**, 500.
Pimentel, G. C., S. W. Charles and Kj. Rosengren, 1966, J. Chem. Phys. **44**, 3029.
Redington, R. L. and K. C. Lin, 1971, J. Chem. Phys. **54**, 4111.
Savoie, R. and A. Anderson, 1966, J. Chem. Phys. **44**, 548.
Strommen, D. P., D. M. Gruen and R. L. McBeth, 1973, J. Chem. Phys. **58**, 4028.
Tursi, A. J. and E. R. Nixon, 1970a, J. Chem. Phys. **52**, 1521.
Tursi, A. J. and E. R. Nixon, 1970b, J. Chem. Phys. **53**, 518.
Van Thiel, M., E. D. Becker and G. C. Pimentel, 1957a, J. Chem. Phys. **27**, 486.
Van Thiel, M., E. D. Becker and G. C. Pimentel, 1957b, J. Chem. Phys. **27**, 95.
Verstegen, J. M. P. J., H. Goldring, S. Kimel and B. Katz, 1966, J. Chem. Phys. **44**, 3216.
Whittle, E., D. A. Dows and G. C. Pimentel, 1954, J. Chem. Phys. **22**, 1943.

PART F

INVESTIGATIONS ON FERROELECTRICS WITH HYDROGEN BONDS

CHAPTER 23

FERROELECTRIC HYDROGEN BONDED SYSTEMS

V. H. SCHMIDT

Department of Physics, Montana State University, Bozeman, Montana 59715, U.S.A.

Contents

23.1. Introduction 1113
  23.1.1. Significance of ferroelectrics for understanding of hydrogen bonding 1113
  23.1.2. Relation of hydrogen bonded ferroelectrics to ferroelectrics in general 1114
  23.1.3. Pertinent books and review articles 1114
  23.1.4. Outline of topics discussed 1115
23.2. Experimental results for KDP-type crystals 1116
  23.2.1. Introduction 1116
  23.2.2. Structure 1116
  23.2.3. Dielectric behavior 1118
  23.2.4. Specific heat and transition entropy 1120
  23.2.5. Ultrasonic attenuation 1120
  23.2.6. Nuclear magnetic resonance 1120
  23.2.7. Electron paramagnetic resonance 1122
  23.2.8. Neutron scattering 1123
  23.2.9. Light scattering 1124
  23.2.10. Pressure effects 1125
23.3. Theory of KDP-type crystals 1126
  23.3.1. Short-range interactions 1126
  23.3.2. Correlated hydrogen motion 1127
  23.3.3. Coupled mode theories 1128
  23.3.4. Comparison of theory with experiment 1129
  23.3.5. Nature of hydrogen collective motion 1132
23.4. Trihydrogen selenites 1135
  23.4.1. Introduction 1135
  23.4.2. Sodium trihydrogen selenite 1137
  23.4.3. Lithium trihydrogen selenite 1138
  23.4.4. Potassium trihydrogen selenite 1138
  23.4.5. Rubidium trihydrogen selenite 1139
  23.4.6. Cesium trihydrogen selenite 1139
  23.4.7. Discussion 1139
23.5. Antiferroelectric periodates 1141
  23.5.1. Introduction 1141
  23.5.2. Static properties 1141
  23.5.3. Dynamic properties 1141

23.6. Rochelle salt and other ferroelectric tartrates 1142
23.6.1. Introduction 1142
23.6.2. The sodium tartrates 1142
23.6.3. The lithium tartrates 1143
23.7. Ferroelectric sulfates 1144
23.7.1. Introduction 1144
23.7.2. Triglycine sulfate and its isomorphs 1144
23.7.3. Ammonium sulfate and ammonium fluoberyllate 1145
23.7.4. Other ferroelectric sulfates 1145
23.8. Other hydrogen bonded ferroelectrics 1147
23.8.1. Hydrogen halides 1147
23.8.2. Copper formate tetrahydrate 1148
23.8.3. Potassium ferrocyanide trihydrate and isomorphic crystals 1149
23.8.4. Colemanite 1149
23.8.5. Synthetic troegerite 1149
23.8.6. Substituted acetates 1150
23.8.7. Semicarbazide hydrochloride 1150
23.8.8. Thiourea 1150
23.8.9. Two-dimensional perovskites 1151
23.9. $LiN_2H_5SO_4$–one-dimensional conduction mimics ferroelectric behavior . . 1151
23.9.1. Introduction 1151
23.9.2. Dielectric properties 1152
23.10. Discussion 1152
Acknowledgements 1153
Note added in proof 1153
References 1160

The hydrogen bond – recent developments in theory and experiments
Eds. P. Schuster et al. © *North-Holland Publ. Co., Amsterdam, 1976*

23.1. Introduction

23.1.1. Significance of ferroelectrics for understanding of hydrogen bonding

The changes in static properties upon going through the ferroelectric transition have long been of interest in gaining an understanding of H-bonded ferroelectrics. Examples of such changes are structural rearrangements, such as change of H-bond length and of H position within the bond, change from ordered to disordered H arrangements and changes in static dielectric and thermal properties. Structural changes are most commonly detected from X-ray and neutron diffraction and nuclear magnetic resonance (NMR) spectra.

In the past decade, more attention has been focused on dynamic properties such as dielectric relaxation, nuclear spin–lattice relaxation and scattering of neutrons and light. Originally it was thought that only the hydrogens play a key role in those H-bonded ferroelectrics in which the transitions involve an ordering of the H system. Both thermally activated H motion and H tunneling have been considered as the mechanism for dynamic behavior. More recently it has been found that "soft" vibrational modes of the heavy ions are important in the order–disorder transitions of H-bonded ferroelectrics as well as in displacive transitions in other ferroelectrics such as certain perovskites. These soft modes are coupled to the H system in a manner which is yet incompletely understood.

For the person interested primarily in H-bonding, the attraction to study ferroelectrics arises from the rich variety of behavior that can be produced by changing temperature and pressure, or by substitution of different anions or cations in the structure, or most surprisingly by the large changes that frequently result from deuteration. Evidently the H-bond systems in such crystals are quite delicately balanced and small changes in external parameters can cause large changes in certain physical properties.

23.1.2. Relation of hydrogen bonded ferroelectrics to ferroelectrics in general

A ferroelectric is a material which has a spontaneous polarization P_s which can be permanently reversed or changed in direction by applying a (generally large) electric field in the appropriate direction for a relatively short time. For a crystal to be able to possess a polarization in the absence of an applied electric field or stress, it must belong to one of the ten pyroelectric crystal classes. The name "pyroelectric" describes the property that the spontaneous polarization changes as a function of temperature. A ferroelectric then is a pyroelectric in which the direction of polarization can be changed by an external electric field, so this field must be smaller than the field causing dielectric breakdown.

Although a large number of compounds are now known to be ferroelectric, it is still possible to divide most ferroelectrics into two classes, the H-bonded compounds such as Rochelle salt and potassium dihydrogen phosphate and the double oxides such as barium titanate and lithium niobate. In general the H-bonded ferroelectrics have lower transition (Curie) temperatures T_c and smaller values of spontaneous polarization P_s than other ferroelectrics. Also, many H-bonded ferroelectrics have order–disorder transitions, while the other compounds more often have displacive transitions in which ions in positions of high symmetry above T_c move to locations of lower symmetry below T_c. There are however exceptions in both classes of ferroelectrics, and indeed the distinction between the order–disorder and displacive transitions is not drawn now as sharply as it once was.

23.1.3. Pertinent books and review articles

A number of books about ferroelectrics have been written, but unfortunately there are no recent ones which are particularly valuable for understanding the relation between H-bonding and ferroelectricity. The structural aspects of ferroelectricity have been examined with considerable insight by Megaw [1957]. Känzig [1957] has considered both ferroelectrics and antiferroelectrics (in which the polarizations of two sublattices cancel), and emphasizes comparisons of different materials with regard to various properties. Jona and Shirane [1962] subdivide their book in terms of the different families of ferroelectric crystals and give a quite complete account of experimentally measured parameters as well as theory through 1960. A review article by Devonshire [1964] gives some more recent results. Fatuzzo and Merz [1967] organize their book according to phenomena and describe theoretical and experimental results for H-bonded ferroelectrics up to about 1963. They include a discussion of soft modes, as does Burfoot [1967]. His book contains some experimental results, but the theory is in general given without reference to specific ferroelectric materials. A fairly recent book by Grindlay [1970] gives a thorough coverage of the thermodynamic approach to ferroelectricity, but also lacks detailed discussion of theory and experiment relative to specific ferroelectrics. A review of theory and experiment relative to ferroelectric phase transitions has been presented in a series of articles by Nettleton [1970]. More specialized review articles of interest include a description by Cochran [1961] of crystal stability and its relation to ferroelectricity, explanations by Blinc and Žekš [1971] and by Bjorkstam [1973] of how NMR is used to study soft modes and critical phenomena near phase transitions and a discussion by Blinc [1972] of the dynamics of order–disorder ferroelectric and antiferroelectric transitions. Potassium dihydrogen phosphate (KH_2PO_4) and isomorphous crystals are used as examples in each of these articles.

The flavor of ferroelectric research at different times and an idea of "who does what where", can be obtained from the proceedings of various conferences on ferroelectricity. Three international meetings on ferroelectricity have been held, at Prague in 1966 (proceedings edited by Dvořák et al. [1966]), at Kyoto [1969] and at Edinburgh [1973]. In addition there have been European meetings on ferroelectricity at Saarbrücken in 1969 (proceedings edited by Müser and Petersson [1970]) and at Dijon [1971]. Also Weller [1967] has edited the proceedings of a symposium on ferroelectricity held in Warren, Michigan in 1966. A very complete and helpful bibliography of recent articles and books concerning ferroelectricity and related topics provided by Toyoda [1970] appears in most issues of the journal *Ferroelectrics*.

23.1.4. Outline of topics discussed

Rather than begin with a general discussion of ferroelectric phenomena, the behavior of specific materials will be described. Emphasis will be placed on those H-bonded ferroelectrics (and antiferroelectrics) in which the H-bonds appear to have a significant effect on the behavior, but all H-bonded ferroelectrics will be mentioned. The potassium dihydrogen phosphate (KDP) family will be described first and in most detail, as it has been most thoroughly investigated and exhibits perhaps the greatest range of phenomena. Accordingly, section 23.2 deals with experimental results for the KDP family, section 23.3 with theories for KDP-type crystals, section 23.4 with the selenite family, section 23.5 with silver and ammonium periodate, section 23.6 with Rochelle salt and related crystals, section 23.7 with ferroelectric sulfates, section 23.8 with other H-bonded ferroelectrics and section 23.9 with lithium hydrazinium sulfate, which mimics ferroelectric behavior but is now considered as a one-dimensional protonic semiconductor rather than a ferroelectric. A general discussion appears in section 23.10.

Because of space limitations, only certain of the more important or recent experiments and theories can be discussed. References to earlier work can be found in these articles and in the bibliography, review articles and books described in section 23.1.3.

23.2. Experimental results for KDP-type crystals

23.2.1. Introduction

In the KH_2PO_4 family, K can be replaced by Rb, Cs or NH_4, P can be replaced by As and H can be replaced by D. Only the tetragonal structure, which changes to orthorhombic below T_c, is considered here, though other structures exist at high temperature and pressure and KD_2PO_4 (DKDP) and certain other members of the family sometimes grow in a monoclinic modification. Table 23.1 shows the effect of such replacements on Curie point T_c, spontaneous polarization P_s and nature of transition. The ammonium compounds are antiferroelectric, while the others are ferroelectric. Most of the transitions are of first order, but RbDP and DRbDP perhaps exhibit second-order transitions. The near doubling of T_c upon deuteration was the first indication of the importance of H-bonding for the ferroelectric behavior.

23.2.2. Structure

A projection of the structure along the ferroelectric, or c, axis appears in fig. 23.1. Parallel chains consisting of alternating phosphate and potassium ions run along c, but are staggered amounts $c/4$ as indicated. The H-bonds are nearly perpendicular to c. Below T_c the hydrogens are ordered as indicated for the deuterons in fig. 23.1, except in the antiferroelectric ammonium compounds in which each phosphate ion has one hydrogen near a top oxygen and one near a bottom oxygen (top oxygens are shown

TABLE 23.1

Properties of KH_2PO_4-type crystals

Crystal	T_c (K)	T_0 (K)	C (°C)	P_s ($\mu C/cm^2$)	L (J/mol)	dT_c/dp (°C/kbar)	Transitions
KH_2PO_4	122a	122a	2910a	5.10a	46b	− 4.6c	f
KD_2PO_4	220a	218a	4020a	6.21a	440b	− 2.4c	f
RbH_2PO_4	147d						f
RbD_2PO_4	218d						f
CsH_2PO_4	159d						f
$NH_4H_2PO_4$	148e					− 3.40c	af
$ND_4D_2PO_4$	245e					− 1.43c	af
KH_2AsO_4	97d			5.0d	300f		f
KD_2AsO_4	162d				570f		f
RbH_2AsO_4	110d			5.6d			f
RbD_2AsO_4	178d						f
CsH_2AsO_4	143d						f
CsD_2AsO_4	212d						f
$NH_4H_2AsO_4$	216e						af
$ND_4D_2AsO_4$	304e						af

T_c and T_0 are Curie and Curie–Weiss temperatures respectively, C is Curie–Weiss constant, P_s is spontaneous polarization, L is latent heat of transition, dT_c/dp is measured at zero pressure and f and af indicate transitions from a paraelectric to a ferroelectric or antiferroelectric state respectively. The reference key is a = Samara [1973], b = Reese [1969], c = Morosin and Samara [1971], d = Jona and Shirane [1962], e = Känzig [1957] and f = Fairall and Reese [1972].

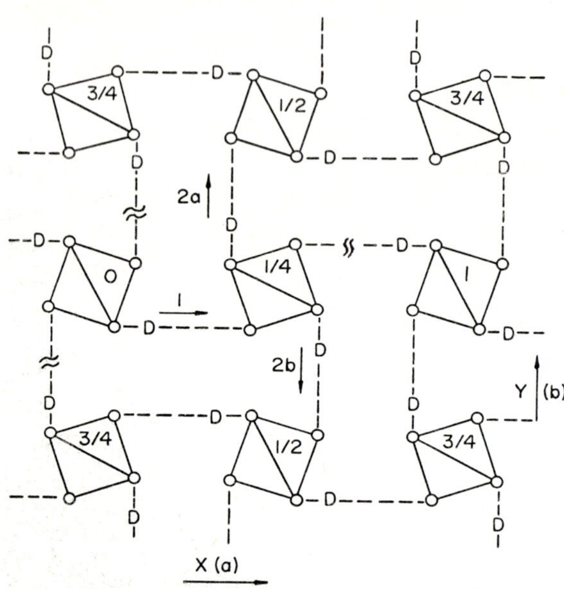

Fig. 23.1. A c-axis projection of the structure of KD_2PO_4, with deuterons (D) shown in positions corresponding to a polarized domain below T_c. See section 23.2.2 for structural description and section 23.3.1 for discussion of effects of deuteron intrabond motions such as indicated by the arrows.

joined by a diagonal line). Relative to their equally spaced positions along the chains in the paraelectric phase above T_c, the phosphorus ions move down and the potassium ions move up when the crystal goes into the ferroelectric phase, for the H configuration shown in fig. 23.1. Also, the square base of the tetragonal cell distorts to a diamond shape below T_c. Above T_c, Nelmes et al. [1972] have shown that both in KDP and DKDP the hydrogens are in disordered positions. Their neutron diffraction results at 294 K for 88% deuterated DKDP show a deuteron site separation of 0.44 ± 0.01 Å, while their reanalysis of the early neutron study of KDP by Bacon and Pease [1955] yielded a site separation of 0.34 ± 0.02 Å. The O–O distances at 293 K found from X-ray measurements by Nakano et al. [1974] are 2.519 ± 0.001 Å for almost completely deuterated DKDP ($T_c = 222$ K) and 2.496 ± 0.001 Å for KDP.

23.2.3. Dielectric behavior

The static dielectric properties for KDP crystals with varying degrees of deuteration have been studied recently by Samara [1973]. For about the first 40 K above T_c the Curie–Weiss law $\epsilon_c = C/(T - T_0)$ is obeyed. Within experimental error, $T_0 = T_c = 122.5$ K for KDP, but measurements of Reese [1969] had shown that the transition is just barely first order. For DKDP (98% deuterated starting solution) $T_0 = 218.3 \pm 0.1$ K while $T_c = 219.8 \pm 0.1$ K, indicating that the transition is farther into the first-order range than for KDP. The Curie constants C are 2910 ± 50 K for KDP and 4020 ± 50 K for DKDP. The spontaneous polarization well below T_c is $6.21 \pm 0.07 \mu C/cm^2$ for DKDP and appears to be approaching $5.10\ \mu C/cm^2$ for KDP. The largest isotope effects were associated with the fields required to move domain walls (which separate domains of opposite polarization). The saturation field E_s required to bring the crystal into a single domain is over 5 kV/cm for KDP and over 32 kV/cm for DKDP. The coercive field E_c which must then be applied in the opposite direction to recreate domains in the previously polarized crystal and bring the net polarization to zero is 160 V/cm in KDP and 3380 V/cm in DKDP.

Although the transition in KDP is of first order, Okada et al. [1973] have shown that it is very close to a critical point at which there is no discontinuity in polarization but the slope of polarization vs. applied field is infinite. The critical temperature is only 0.023 K above T_c and the critical field is 180 V/cm. At a critical point one expects critical fluctuations, and in this case polarization fluctuations result from growth and decay of relatively large polarized regions. The dielectric relaxation rate would depend on the size distribution of these regions. They observed a *single* relaxation time of about 100 seconds. For measurements made at various bias fields and frequencies, but not at the critical point, Hill and Ichiki [1963] reported a spread in relaxation time.

At room temperature, well above T_c both for KDP and DKDP, no clustering of dipoles (critical phenomena) should occur and the interpretation of dielectric relaxation results should be more straightforward. Ludupov [1966] found dielectric relaxation in DKDP at room temperature at 1 GHz (10^9 cyc/s) and above. Kaminow [1965] studied the temperature dependence of dielectric behavior in ADP, KDA and DKDP with varying degrees of deuteration and found strong increase in loss tangent with deuteration at room temperature. Evidently the relaxation process is much slower in DKDP than in KDP.

Perhaps the most striking isotope effect is the observation by Bjorkstam and Oettel [1966] that deuteration decreases the domain wall mobility (velocity/field) by a factor of about 10^6, if measurements are made 10 K below T_c for each crystal. Analysis of domain wall mobility unfortunately involves dislocations and mechanical strain effects, which have been investigated by Bornarel and Lajzerowicz [1972].

23.2.4. Specific heat and transition entropy

Reese [1969] studied the thermal behavior of KDP near T_c and compared the results with earlier work done on DKDP by Reese and May [1968]. The latent heat is 46.2 joules/mole for KDP and 440 joules/mole for DKDP. The discontinuities in spontaneous polarization are 0.31 ± 0.03 and $6.5 \pm 1.0\ \mu C/\mathrm{cm}^2$ for KDP and DKDP respectively. In contrast to the results of Okada et al. [1973], the critical point for KDP appeared to be roughly at 0.01 K above T_c at a bias field of about 300 V/cm. For KH_2AsO_4 (KDA) the latent heat of transition was found by Fairall and Reese [1972] to be 300 joules/mole, while for KD_2AsO_4 (DKDA) the transition latent heat is 570 joules/mole. Evidently KDP is rather unusually close to the borderline between displaying a first-order and second-order transition.

23.2.5. Ultrasonic attenuation

Here again a large isotope effect appears in the dynamic behavior. Above T_c the ultrasonic relaxation time τ_u obeys the Curie–Weiss law $\tau_u = \tau_0/(T - T_c)$. For KDP, Garland and Novotny [1969] found $\tau_0 = 2.4 \times 10^{-11}$ s, while in mostly deuterated DKDP ($T_c = 205.6$ K), Litov and Uehling [1968] obtained $\tau_0 = 2.44 \times 10^{-10}$ s.

23.2.6. Nuclear magnetic resonance

Nuclear magnetic resonance (NMR) can be considered as a microscopic tool for measuring fields and their fluctuations at nuclear sites in a crystal. Nuclei with spin 1/2, such as ^1H and ^{31}P in KH_2PO_4, are sensitive only to magnetic fields and their fluctuations. Nuclei with spin 1 (such as ^2D), spin 3/2 (such as ^{75}As) and larger spin are sensitive also to electric field gradients, offering advantages which will be illustrated for D in DKDP as compared with H in KDP. Only one proton NMR line can be resolved in KDP. The deuteron energy levels for $m_I = 1, 0$ and -1 are split equally by the Zeeman interaction of the nuclear magnetic moment with the applied magnetic field H_0, so the Zeeman interaction alone would yield equal frequencies for the $1, 0$ and $0, -1$ transitions. This degeneracy is lifted by the perturbation caused by the interaction of the deuteron quadrupole moment with the electric field gradient (efg) at its location. The resulting splitting of the two NMR lines depends on the efg tensor and on its orientation relative to H_0, so physically inequivalent deuterons give separate pairs of lines even if they are chemically equivalent, as they are in DKDP. Bjorkstam and Uehling [1959] first observed the deuteron resonance in DKDP and found that the efg tensor undergoes only rather small changes upon going through T_c. Bjorkstam [1967] later was able to make more sensitive measurements, especially below T_c. He observed a 7% increase in the efg strength upon going into the ferroelectric phase and attributed this increase to phosphate ion distortion rather than to a change in the O–D bond length. This conclusion may be incorrect, as Bacon and Pease [1955] observed about a 2% decrease in O–H bond length in KDP below T_c, which for an efg caused by an effective charge at that oxygen nucleus would cause a 6% increase in efg.

Two types of deuteron motion were found to be responsible for deuteron spin-lattice relaxation by Schmidt and Uehling [1962]. One is deuteron diffusion between H-bonds, with the same activation energy of 0.58 eV as found by them for electrical conductivity in that crystal. This is within the range 0.5 to 1.0 eV typically found for protonic conductivity in H-bonded crystals. Blinc and Pirš [1971] studied this process using proton dipolar and rotating-frame spin–lattice relaxation time measurements in KDP, RbDP and CsDP as well as in the trihydrogen selenites $LiH_3(SeO_3)_2$ (LiHSe), NaHSe, KHSe and CsHSe. For both the phosphates and selenites, the activation energy increases with increasing alkali metal ion radius. They concluded that the whole phosphate or selenite group with attached protons rotates in the conduction process. The other possibility is that just the protons diffuse. Suitable relaxation time measurements for ^{17}O in enriched crystals could probably determine which process actually occurs.

Of more significance for ferroelectricity is the intrabond deuteron motion in DKDP, for which Schmidt and Uehling [1962] found an activation energy of about 0.078 eV, with a mean time of about 10^{-11} s between jumps at room temperature. Blinc et al. [1971a] later made more accurate measurements for DKDP, and also investigated DKDA and DCsDA. They observed strong temperature dependence of relaxation rate as T_c was approached from above, which they attributed to critical slowing down. Earlier evidence for critical slowing down of the fluctuating fields causing spin–lattice relaxation had been found by Blinc and Zumer [1968] from a large increase in the ^{31}P relaxation time near T_c in a very pure KDP crystal. In this case the effect of the electric dipole clustering associated with critical behavior is to bring the peak of the frequency spectrum for magnetic field fluctuations resulting mostly from proton intrabond motion down toward the NMR frequency of the protons.

Blinc and Mali [1969] observed a decrease in the ^{87}Rb quadrupolar coupling constant in RbH_2AsO_4 (RbDA) with decreasing temperature above T_c which they attributed to an increase in the Rb^+ ion vibrational amplitude near T_c caused by better coupling with the H system. Thermal contraction would tend to increase the coupling constant, but it is well known that increasing amplitude of vibration tends to decrease quadrupolar coupling constants. An increased vibrational amplitude with decreasing temperature indicates a decrease in vibrational frequency. This then is indirect evidence for a "soft mode" associated with the ferroelectric transition.

In the ADP-type antiferroelectrics there are N–H···O bonds as well as the O–H··· bonds which are similar to those in KDP. The ammonium ion can undergo hindered rotation. A deuteron NMR investigation of $ND_4D_2PO_4$ was made by Genin et al. [1968]. They observed a 10% increase in the quadrupolar coupling constant for the O–D···O deuterons below T_c. They also measured proton relaxation times which have a minimum at the temperature at which the ammonium hindered rotation frequency is equal to the proton NMR frequency. There is a discontinuity in the proton spin–lattice relaxation time at $T_c = 148$ K in a direction which indicates that the barrier against hindered rotation decreases above T_c. It was postulated that there is a corresponding decrease in the ammonium torsional oscillation frequency which accounts for some or all of the excess transition entropy beyond that associated with ordering of the O–H···O hydrogens.

23.2.7. Electron paramagnetic resonance

As KDP-type crystals contain no paramagnetic ions, electron paramagnetic resonance (EPR) can only be performed if paramagnetic centers are introduced by irradiation or doping. EPR spectra then reveal information about the neighborhood of the centers, which may differ from phenomena in undisturbed portions of the crystal. The advantage of EPR is that it can disclose the proton configuration and rate of motion around a particular site. An example is the EPR study of gamma-irradiated KDA and DKDA performed by Blinc et al. [1967]. The irradiation forms AsO_4^{4-} centers, whose EPR spectra are sensitive to the arrangement of the hydrogens in the four adjacent bonds and to the rate of intrabond motion of these hydrogens. As this rate increases with increasing temperature, the EPR spectrum goes from a triplet corresponding to the three possibilities 1, 0 and -1 for the total m_I for the two close protons, to a quintet corresponding to the five total m_I possibilities for the four neighboring protons which the AsO_4^{4-} ion sees as being at the bond centers because of the more rapid motion at high temperature. For DKDA the corresponding high-temperature spectrum has nine weak lines as expected because the deuteron has spin 1, and m_I can vary in integer steps from 4 to -4. The activation energy of 0.08 eV obtained for the deuteron intrabond motion from the temperature dependence of the EPR spectrum is in excellent agreement with the value found by Schmidt and Uehling [1962] from deuteron NMR measurements in DKDP. However, a similar analysis in KDA yields an activation energy of 0.2 eV, which seems surprisingly high.

Dalal and McDowell [1972] performed an EPR and ENDOR (Electron–Nuclear Double Resonance) study of KDA, DKDA and ADA. Well above T_c for each compound they found that for applied magnetic field H_0 along a direction making a 45° angle with the two equivalent tetragonal axes, the EPR spectra of the AsO_4^{4-} center produced by gamma-irradiation indicate only one type of site, as expected. However, at lower temperatures but still well above T_c, the spectra double in a way similar to that shown below T_c to be the result of existence of oppositely polarized domains. In KDA both the ^{75}As and proton spin–lattice relaxation rates were governed by three activation energies in three different temperature regions. At high temperature the proton rate was faster, while in the intermediate range it approached the ^{75}As rate, and at lower temperature but still above T_c the rates were the same. This appears to be quite direct evidence that the transverse optical vibrations of the heavy arsenic and potassium ions become strongly coupled to the proton motion near T_c.

23.2.8. Neutron scattering

Whereas the dielectric, ultrasonic and magnetic resonance methods described heretofore are most sensitive to relatively slow motions up to about 10^{12}/s in frequency, neutron scattering and optical scattering experiments to be described are used to study the nature of lattice vibrations and other fast motions. A study of polarization fluctuations near T_c in DKDP using coherent neutron scattering was made by Paul et al. [1970]. They observed fluctuations which increased in intensity near

T_c, for which the displacements were similar to the static displacements below T_c. Applied electric fields had a large effect on the fluctuations. The fluctuation mode does not have a resonant character and apparently does not interact strongly with optical vibration modes, as no optical phonon changed frequency near T_c.

A somewhat similar study of DKDP was made by Skalyo et al. [1970] by means of neutron triple-axis spectrometry. The results were compared with the ferroelectric mode description for KDP derived by Cochran [1961] on the basis of the neutron diffraction results of Bacon and Pease [1955]. Some striking and unexplained differences relative to the KDP results were found: a large distortional movement of the oxygen tetrahedra occurs, the deuterons have a rather large motion along the ferroelectric c axis perpendicular to the H-bonds and the deuterons move along c in phase with the phosphorus ions, while in KDP the protons move with the potassium ions. The results could not distinguish whether deuteron motion along the bonds originates in vibrations or from tunneling.

23.2.9. Light scattering

Kaminow and Damen [1968] made the first quantitative measurements of the ferroelectric mode in KDP by means of Raman-scattering measurements made both above and below T_c. They were able to fit their results well with a single damped-harmonic-oscillator model. The mode frequency dropped with increasing rapidity as T_c was approached from above. They could offer no explanation for the large and temperature-independent damping constant.

More recent experiments have been analyzed in terms of two coupled modes, each of which has its own frequency, damping and polarization constants, in addition to the coupling strength and phase which are additional adjustable parameters. Such experiments have been carried out by She et al. [1972] for KDP, by Broberg et al. [1972] for ADP and by Katiyar et al. [1971] for CsDA.

Raman scattering involves a shift in the frequency of the incoming light by an amount corresponding to absorption or emission of a relatively high energy quantum of vibrational energy from a phonon or local mode (5 to 270 cm^{-1} in the measurements of She et al. [1972]). Accordingly this method is best suited for studying the higher-frequency transverse optical (TO) modes, specifically (in KDP) the TO mode in which potassium and phosphorus ions vibrate in opposite directions along c, because this mode is coupled to the H system. A complementary method is Brillouin scattering, in which light is diffracted by long-wavelength phonons which act as a three-dimensional diffraction grating. Recently Reese et al. [1973] have employed Brillouin scattering in DKDP to study the soft ferroelectric mode and the soft acoustic shear mode associated with the spontaneous shear that occurs below T_c. Their measurements covered the 0 to 0.8 cm^{-1} range. They concluded that the ferroelectric mode is diffusive (relaxational) rather than oscillatory, and they were able to observe coupling between this mode and the shear mode. Their paper gives numerous references to optical and high-frequency dielectric experiments and to theory as well.

23.2.10. Pressure effects

The hydrogen bonds are the weakest bonds in KDP-type crystals, so it is not surprising that there are marked pressure effects on ferroelectric properties because of the intimate connection between H-bonding and ferroelectricity in these crystals. Samara [1971] discovered that the ferroelectric transitions vanish in KDP and ADP at pressure above 17 and 33 kbar respectively. Morosin and Samara [1971] studied the effect of pressure up to 3 kbar on single-crystal X-ray diffraction in KDP, DKDP, ADP and DADP. The rates of H-bond shortening with increasing pressure were determined. It was estimated that at the 17 kbar pressure at which the KDP transitions disappear, the bond is about 2.44 Å long, which in most H-bonds is characteristic of a single-minimum potential which centers the H in its bond.

Because the soft acoustic shear mode and the K–P transverse optic mode are coupled to the H system in KDP, strong pressure effects are expected on these modes. Fritz [1973] has measured the temperature dependence of the soft acoustic shear mode at atmospheric pressure and at 4.14 kbar ultrasonically. Surprisingly, pressure has little effect on the elastic constant plotted as a function of $T - T_c$. He also measured the room temperature pressure dependence to 20 kbar and found some evidence for a high-pressure phase transition. Peercy [1973] studied the pressure dependence of the soft TO mode up to 9.3 kbar at room temperature. This mode is overdamped at all temperatures for atmospheric pressure and it was not known whether it is an overdamped resonant mode or a purely relaxational mode. This mode becomes underdamped above 6 kbar, so this work has established the resonant nature of the mode.

23.3. Theory of KDP-type crystals

23.3.1. Short-range interactions

The first microscopic theory for KDP, or for any ferroelectric, was produced by Slater [1941]. He assumed that above T_c the hydrogens are in random off-center positions in their bonds, subject to the "ice-rules" which require one H per bond and two hydrogens close to each PO_4 ion. He further assumed that of the six possible configurations which give two "close" protons in the four bonds attached to a given phosphate ion, the two configurations having dipole moment components $\pm\mu$ along the ferroelectric c axis have zero energy, while the other four configurations have zero moment along this axis but have positive energy ϵ_0. The spontaneous polarization P_s, found by minimizing the free energy, goes abruptly from zero to its maximum value at T_c. Because all value of P_s between zero and its maximum value are allowed at T_c, the transition is on the borderline between first and second order. Below T_c all hydrogens are in ordered positions corresponding to one of the two zero-energy configurations (as in fig. 23.1), with the configuration determining the polarity of the domain.

Takagi [1948] modified the Slater theory by introducing the possibility of additional types of groups having one or three hydrogens close to a phosphate ion. These "Takagi groups" were assigned energy ϵ_1. His theory predicts a second order transition with a spontaneous polarization P_s which increases smoothly from zero to its maximum value as temperature decreases below T_c. This agrees better with experiment than Slater's prediction of *no* temperature dependence for P_s below T_c, but does not predict the relatively small discontinuities in P_s resulting from the first order nature of the transitions in most of the KDP-type crystals and does not fit other portions of the P_s vs. T curves in detail.

Senko [1961] modified the Slater–Takagi model by introducing a long-range interaction which adds to the internal energy a negative term $-BP^2$ proportional to the square of the polarization. He obtained best-fit parameters ϵ_0, ϵ_1 and B for KDP and DKDP based on then-existing spontaneous polarization, dielectric constant and specific heat data. Independently, Silsbee et al. [1964] made this same modification to the Slater–Takagi theory and studied its predictions in some detail. They found that if B becomes sufficiently large, the transition goes from second to first order. The best fit for the P_s vs. T curve for DKDP was obtained by using for the Takagi group energy ϵ_1 the value 0.078 eV obtained from NMR measurements by Schmidt and Uehling [1962] for the activation energy for deuteron intrabond jumping.

This relation between Takagi groups and dynamic phenomena in DKDP was further explored by Schmidt [1967]. The creation of a Takagi DPO_4–D_3PO_4 pair from two Slater D_2PO_4 groups results from the intrabond deuteron transfer labeled 1 in fig. 23.1. Further effective transfer of the D_3PO_4 Takagi group can take place by means of intrabond transfers 2a or 2b. Such intrabond transfers associated with Takagi group diffusion are a mechanism for rearranging Slater group configurations; for instance, transfers 1 and 2a convert the central D_2PO_4 from a Slater group of energy zero and dipole moment $-\mu$ along c to a Slater group of energy ϵ_0 with zero moment along c. The equality of the "best-fit" Takagi group energy ϵ_1 with the activation energy for intrabond diffusion implies that the Takagi pair creation requires energy $2\epsilon_1$ while subsequent diffusion of Takagi groups is not thermally activated. The assumption that successive jumps of a given Takagi group are uncorrelated and take place on the average at intervals of 3×10^{-14} s gives good agreement with deuteron spin–lattice relaxation results of Schmidt and Uehling [1962], dielectric relaxation measurements of Ludupov [1966], domain wall mobility measurements of Bjorkstam and Oettel [1966] and ultrasonic attenuation results of Litov and Uehling [1968].

Dynamic phenomena are one to six orders of magnitude faster in KDP than in DKDP, as described in section 23.2. To explain such large isotope effects it appears necessary to invoke correlated hydrogen motions, which are discussed in the following section.

23.3.2. Correlated hydrogen motion

The earlier objections to Slater-type theories were based on their failure to predict static isotope effects such as the shift in T_c, because dynamic effects were not studied until later. Blinc [1960] was the first to propose that proton tunneling occurs between the two minima in the H-bond potential. He also included proton–proton interactions and was able to explain isotope effects such as the near doubling of T_c on the basis of the different tunneling rates of protons and deuterons in a given potential well. De Gennes [1963] converted Blinc's model to a pseudo-spin formalism and discussed consequences of this model particularly for neutron scattering. Using the language of this formalism, Blinc et al. [1972a] have proposed that the vanishing of ferroelectricity in KDP at high pressure results from the "transverse tunneling field" becoming equal to the "longitudinal field" rather than from a change from a double-minimum to a single-minimum well shape as was suggested by Samara [1971].

Blinc and Svetina [1966] developed a model for KDP-type crystals which includes tunneling and also short-range interactions in a cluster approximation. Tokunaga and Matsubara [1966] proposed a somewhat similar model for static properties of KDP-type crystals which essentially considers only the interactions of the hydrogen system. The dipole moments along the ferroelectric *c* axis are considered to originate in K and P ion displacements, but they assumed that this network can adjust adiabatically to the H motions. In the absence of tunneling the model gives results similar to those of Takagi [1948]. Tunneling reduces the transition temperature and can suppress the transition entirely. This result is of interest, relative to the later discovery by Samara [1971] that pressure can eliminate the ferroelectric phase in KDP. Tokunaga [1966] studied the dynamic properties of this model and found dielectric dispersion of the resonance type rather than of the relaxational (Debye) type. He concluded that consideration of the dynamics of the proton system alone is not sufficient for an understanding of the phase transition.

23.3.3. Coupled mode theories

Cochran [1959] originated the important idea that a ferroelectric transition may be accompanied by a "soft mode", that is, a transverse optic (TO) vibrational mode which tends toward zero frequency as the transition temperature is approached. Originally the theory was expected to be useful only for displacive transitions, but was later extended to apply also to modes coupled to a H system which undergoes an order–disorder transition.

Kobayashi [1968] developed the first coupled mode theory for KDP. He considers a system of interacting protons in which a tunneling mode can occur, coupled to a TO vibrational mode involving the K–PO_4 heavy ion system. He assumes these ions are responsible for most of the polarization. He further assumes that the *uncoupled* K–PO_4 mode would not become soft, contrary to the assumption of Tokunaga and Matsubara [1966]. Of the two coupled modes which his theory predicts, one has a high frequency almost independent of temperature, while the frequency of the other becomes small near T_c. In his opinion, KDP-type crystals constitute a "mixed" type of ferroelectric exhibiting order–disorder phenomena of the H system and displacive behavior of the heavy ion system.

Konsin and Kristofel' [1973] have investigated the phase transitions in

KDP and DKDP using a model which contains tunneling, Slater-type proton–proton interactions, phonons and proton–phonon and phonon–phonon interactions. The "bare" K–PO_4 mode has $\omega_0^2 < 0$, which is a quasi-Jahn–Teller effect. This model explains the large isotope effect on T_c while allowing the small observed effect of deuteration on the Curie–Weiss constant. Numerical estimates for parameters in their theory are given.

Tsallis [1972] has added a configuration energy to the theory of Kobayashi [1968] and finds that the order parameter (spontaneous polarization) and the soft mode frequency now vanish at the same temperature, as they must for second-order transition. (The transition in KDP is *almost* of second order.)

23.3.4. Comparison of theory with experiment

Thermal and dielectric properties will be considered first. Careful calorimetric studies have been made by Reese [1969] for KDP, Reese and May [1968] for DKDP and Fairall and Reese [1972] for KDA and DKDA. Results were compared with predictions of the theory of Silsbee et al. [1964]. For DKDP the agreement with theory was good to within 1 K of the transition, both for the specific heat measurements and for the spontaneous polarization results of Hill and Ichiki [1963]. The fit was bad in the last degree below T_c and the transition entropy was not predicted correctly. These failures were attributed to the invalidity of a mean-field theory near T_c. A logarithmic divergence appears to fit the specific heat data best near T_c. For KDP the agreement of calorimetric, dielectric and electrocaloric measurement with theory was poor, presumably because this theory neglects tunneling or other cooperative proton motion. For KDA and DKDA the specific heat and dielectric results near T_c were in excellent agreement with predictions of theory. However, a considerably different value for the parameter ϵ_0 using this same theory was obtained by Blinc and Bjorkstam [1969] from nuclear quadrupole resonance experiments on ^{75}As in KDA, on the assumption that the ^{75}As frequency varies directly as the fraction of the time that the rapidly fluctuating hydrogen configurations around the ^{75}As assume one of the Slater configurations of energy ϵ_0. Blinc and Bjorkstam, as well as Fairall and Reese, have noted very long relaxation times in the arsenates for electric-field-induced nuclear quadrupole resonance spectra changes and for dielectric relaxation respectively, for temperatures a few degrees above T_c. These long relaxation times have been attributed to critical fluctuations, but their occurrence over a relatively large temperature range is not understood.

Litov and Uehling [1968] studied ultrasonic relaxation in DKDP both above and below T_c and compared their results for static parameters with the theory of Silsbee et al. [1964]. Similar measurements were made for KDP by Litov and Garland [1970]. Good agreement with theory was found for both crystals. The polarization relaxation time for DKDP agreed well with a model of Schmidt [1967] based on uncorrelated Takagi group diffusion. The relaxation time for KDP was about ten times shorter, indicating tunneling or other correlated proton motion in this crystal.

A point of comparison between theory and experiment is the order of the transition. While the other KDP-type crystals appear to have first-order transitions, there is still disagreement concerning RbDP. Dielectric constant measurements made above T_c and spontaneous polarization results below T_c obtained by Strukov et al. [1971] showed RbDP to have a second-order transition and KDP a first-order transition. Critical exponents were evaluated for both crystals. On the other hand, Zheludev et al. [1974] report that RbDP has a first-order transition, though it is even closer than KDP to being a second-order transition.

The arsenates tend to exhibit more discontinuous transitions than the corresponding phosphates, according to Blinc et al. [1973a]. For instance, RbDA has a first-order transition. However, the transitions are still close enough to second order to exhibit critical fluctuations. In KDA, Adriaenssens et al. [1972] found additional ^{75}As NMR lines between T_c and $T_c + 15$ K. These lines were similar in angular dependence to those in the polar phase below T_c. The origin of these lines is in polarization fluctuations with long lifetimes compared to 1 ms. Dielectric measurements on these crystals support this hypothesis of long relaxation times.

Blinc et al. [1971a] compared their measurements of deuteron spin–lattice relaxation rate $1/T_1$ in DKDP with predictions of three theories. Schmidt [1967] assumed uncorrelated diffusion of Takagi groups and predicted exponential increase in relaxation rate on approaching T_c from above. Near T_c the temperature dependence was stronger than predicted by this theory which neglected critical polarization fluctuations. Within such small polarized regions few Takagi groups exist, with consequent slowing down of intrabond motion and correspondingly larger Fourier components of the electric-field gradient fluctuations at the relatively low NMR frequency.

In the second theory of Yoshimitsu and Matsubara [1968], the dynamics for a four-deuteron cluster are studied, with effects of other deuterons taken into account by means of a self-consistent electric field. This theory predicts too large an anomaly in T_1 caused by critical polarization fluctuations. Probably the self-consistency condition was applied to too small a cluster.

The third theory is related to one used previously by Blinc et al. [1970a] for $Ca_2Sr(C_2H_5CO_2)_6$ (dicalcium strontium propionate) which is based in turn on the analysis by Suzuki and Kubo [1968] for the dynamic susceptibility of the Ising model. This theory gives quite good agreement with experiment, for arbitrarily chosen values of the "non-interacting dipole correlation time", which are $\tau = 4\times 10^{10} T^{-1}$ s above T_c and $\tau = 16\times 10^{-10} T^{-1}$ s below T_c, for T in K. However no physical interpretation for τ was given. A comparison of its magnitude was made with a value found by Litov and Uehling [1968] from ultrasonic attenuation measurements, but the value compared with, was apparently their τ_w, which has exponential rather than T^{-1} temperature dependence.

The most recent experimental results emphasize the need to consider coupling of the proton system with the TO mode corresponding to the K and P displacements observed below T_c, but the nature of the modes and their coupling is still uncertain. Paul et al. [1970] have analyzed their coherent neutron scattering results in terms of order–disorder and tunneling models and suggest that their observations are of the tunneling mode strongly coupled to a phonon mode. She et al. [1972] analyzed their Raman spectra for KDP with two coupled modes, a TO phonon near 180 cm^{-1} and an overdamped proton mode. The coupling is quite strong (114 cm^{-1}) and temperature-independent. The square of the *uncoupled* proton mode frequency varies linearly with temperature and extrapolates to zero at 30 K. The lower-frequency coupled mode extrapolates to zero frequency at 116 K ($T_c = 122$ K) and the lowering of the frequency by the coupling can be thought of as a level-repulsion effect. Kobayashi [1968] considered two such coupled modes, but predicted that the uncoupled proton mode would have a transition temperature "of order of magnitude 10 K lower" than T_c. Cochran [1969] has made an estimate of this temperature difference which is in better agreement with these results.

Katiyar et al. [1971] compared their Raman spectrum of CsDA with a similar coupled mode model. Attempts to fit their results to the Kobayashi theory achieved the unsettling result that the unpolarizable mode is underdamped while the polarizable mode is overdamped. Kobayashi had assumed that the proton mode is unpolarizable and overdamped, with the underdamped phonon mode giving the observed polarizability.

Scott and Worlock [1973] have analyzed theoretically the temperature dependences of the "tunneling mode", the low-frequency dielectric constant and the ^{75}As quadrupolar coupling constant in KDA and CsDA. The dielectric constant for KDA as measured by Kaminow [1965] obeys a Curie–Weiss law $\epsilon = C/(T - T_c)$ while the tunneling mode obeys $\omega_0^{-2} = C'/(T - \alpha T_c)$, where $\alpha = 0.48$. Accordingly the Lyddane–Sachs–Teller (LST) relation $\epsilon_0(T)/\epsilon_\infty = \omega_{LO}^2/\omega_{TO}^2$ is not obeyed; this failure to obey the LST relation had previously been explained in terms of a frequency-dependent dielectric response by Cowley et al. [1971]. The quadrupolar coupling results of Blinc et al. [1970b] were reinterpreted by integrating the soft mode over the Brillouin zone, using a simple dispersion model. It appears that the NQR results as well as the other results are consistent with a simple soft mode consisting of a "tunneling mode" strongly coupled to a TO mode. Further support for this conclusion comes from later NQR studies by Blinc et al. [1972b], who find similar temperature dependence for D and heavy ion quadrupolar response in KDP-type crystals.

23.3.5. Nature of hydrogen collective motion

The very large isotope effects on dynamic processes in KDP-type crystals described in section 23.2 lead almost inescapably to the conclusion that at least the faster processes in undeuterated crystals must involve collective motion of hydrogens. The nature of this motion depends on the model for the H system. It is now agreed that each H lies in a symmetric double-minimum potential and interacts with neighboring hydrogens. If the neighboring hydrogens are held fixed in off-center positions in their potential wells and their effect is included in the potential of the H in question, this potential will generally be either an asymmetric double well or a well with a single off-center minimum. Uncorrelated (non-collective) motion of a H across a bond center is (except for extremely brief thermally activated excursions) possible only for the symmetric potentials associated with Takagi groups. Such uncorrelated deuteron motion in DKDP has been shown by Schmidt [1967] to provide an explanation for several dynamic phenomena in that crystal, as described in section 23.2. To obtain much faster rates, as observed for KDP for a number of phenomena, it is necessary to assume much higher concentration or mobility of Takagi groups or to look to another mechanism. Much higher concentrations are not consistent with dielectric and thermal measurements in KDP as interpreted by the theory of Silsbee et al. [1964]. Much higher mobilities may, according to Schmidt [1974], result from the 25% smaller mean off-center displacement for protons in KDP compared to deuterons in DKDP. However, the only generally accepted explanation is the assumption of rapid correlation proton tunneling in KDP.

The Hamiltonian for the proton system is now almost invariably written in the pseudo-spin formalism introduced by De Gennes [1963]:

$$H = -\Gamma \sum_i S_{xi} - \sum_{ij} J_{ij} S_{zi} S_{zj}.$$

Here Γ is the tunneling frequency or twice the tunneling integral, J_{ij} is the interproton interaction and the spin up or down states, $S_{zi} = \pm 1/2$, correspond to the proton being at the end of its bond near the top or bottom of a PO_4 group. Even for identical wells the tunneling integral would be much larger for the proton than for the deuteron because of its smaller mass, but in fact the minima are 25% closer together for KDP and the barrier must be correspondingly lower. This Hamiltonian is only an approximation, as de Gennes pointed out. Elliott and Young [1974] have remarked that tunneling becomes most important for weak barriers, but that this is just the situation in which the basic assumption underlying this Hamiltonian–two low-lying levels well separated from the higher energy levels–becomes least valid.

For certain calculations of static properties, for instance the spontaneous polarization and transition temperature as derived by Tokunaga and Matsubara [1966], the inclusion of the tunneling term seems just another way of arriving at the well-known quantum-mechanical result that the proton mean position will be closer to the bond center in an asymmetric well than the deuteron. Here the asymmetry arises because intrabond transfer results in the endothermic reaction $2H_2PO_4^- \rightarrow HPO_4^{2-} + H_3PO_4$. The Slater [1941] model and its modifications have been criticized for not predicting isotope effects in these and other properties, except by means of ad hoc adjustments of the energies ϵ_0 and ϵ_1 for Slater and Takagi groups. But these energies are related to the hydrogen off-center displacements, since for hydrogens at bond centers the various Slater and Takagi configurations would be indistinguishable. Accordingly, isotopic adjustments of ϵ_0 and ϵ_1 seem no more arbitrary than isotope adjustments of the tunneling rate Γ.

Another prediction of the tunneling models is the existence of tunneling modes at certain frequencies, which in the more recent theories are assumed to be coupled to phonon modes. These models are in qualitative agreement with the latest analyses of Raman scattering results which indicate the presence of two strongly coupled resonant modes. The considerable damping indicated by these analyses was not predicted by earlier theories which include tunneling, but results of the Blinc and Žekš [1972] stochastic approach to pseudo-spin motion include such damping. Theories in which the H system response is governed by Takagi group diffusion or by an electric dipole relaxation time predict damping but no resonant behavior for the uncoupled hydrogen mode. Much experimental and theoretical effort is now going into study of the nature of the hydrogen and phonon modes and their coupling and a better understanding of this problem can be expected soon.

While the above tunneling modes which interact with phonons involve simultaneous tunneling of many protons, other tunneling modes have been proposed which only involve a few protons. Reid [1959] misunderstood a c-axis projection of the structure as shown in fig. 23.1 and proposed that there are loops of four phosphate ions connected by H-bonds and that protons can tunnel in phase in these bonds. Schmidt [1966] pointed out that the smallest loops involve six bonds, as shown connecting the six phosphate ions at the right in fig. 23.1, but stated erroneously that tunneling around such a loop could affect dielectric behavior. In fact, if the hydrogens in any closed loop in KDP-type crystals are all arranged so that they can tunnel in phase from, say, clockwise, to counterclockwise positions, their net contribution to the polarization is zero both before and after the tunneling. Cochran [1961] proposed that tunneling takes place for ordered proton sequences which lie in spiral arrangements of phosphate groups; the four phosphates at the upper right in fig. 23.1 are part of such a spiral. Long ordered sequences above T_c would result in too low a transition entropy. Tunneling of short-ordered sequences could take place without cost in energy if the tunneling is accompanied by the effective transfer of a Takagi group from one end of the sequence to the other. Such tunneling would be in competition with thermally activated Takagi group mobility.

Sato [1970] advanced a new proposal for the nature of H-bonding in KDP and DKDP, based on his analysis of infrared spectra. He proposed a broad symmetric single-minimum well for the protons in KDP above T_c which is in conflict with the recent analysis of neutron diffraction results by Nelmes et al. [1972]. For DKDP he proposed that the Slater configurations of energy ϵ_0 do not exist and that above T_c *all* deuterons tunnel in phase between the positive and the negative dipole-moment Slater configurations while the heavy ions vibrate as suggested by Cochran [1961]. His criticism of the evidence advanced by Blinc and Bjorkstam [1969] for the Slater model is unfounded as it is based on a misinterpretation of their eq. (11). In any case, the proposal that all deuterons tunnel in phase seems unphysical because it does not account for the large transition entropy. If one assumes instead that tunneling in phase is restricted to "microdomains" of opposite polarity, then the regions between these microdomains must be equivalent to domain walls. Barkla and Finlayson [1953] showed that domain walls consist mainly of Slater groups of energy ϵ_0, so this interpretation of tunneling is in conflict with Sato's assumption that such Slater groups do not exist. A more reasonable interpretation seems to be that superposition of TO phonons of the type proposed by Cochran will give fluctuating microscopic regions in which the P and K displacements are in the directions favored by "up" or "down" domains and that the hydrogens respond by means of tunneling or Takagi group diffusion and *tend* toward the configurations corresponding to such domains, but, however, Slater groups of all types continue to exist in these regions. These fluctuations at the TO phonon frequencies are to be distinguished from the slow polarization fluctuations described earlier which appear to exist near T_c, for which the polarization within the microscopic regions is presumably much more complete.

To summarize the situation for KDP-type crystals, in spite of the unusually large numbers of studies of these materials and the rapid recent progress, the dynamical processes in these crystals and their relations to ferroelectric behavior and to the H-bond system are still not very well understood (see also section 24.3).

23.4. Trihydrogen selenites

23.4.1. Introduction

The trihydrogen selenites have composition $XH_3(SeO_3)_2$, where X = Li, Na, K, Rb, Cs, or NH_4 and D can be substituted for H. As seen in table 23.2, only in certain instances does deuteration have the large effect on T_c found in KH_2PO_4-type crystals. In sodium trihydrogen selenite, deuteration can actually suppress one of the ferroelectric phases and Shirokov et al. [1969] have shown that pressure can induce a new (δ) phase which appears not to be ferroelectric. The cation has much more effect than in KDP-type crystals, both on the structure and on the dielectric properties. A brief review of the status of these crystals as of 1969 has been given by Shuvalov et al. [1970a].

TABLE 23.2

Properties of selenites and periodates

Crystal	$T_c T_0$ (K)	C (K)	P_s (°C)	ΔS (μC/cm^2)	dT_c/dp (cal/°C mol)	(°C/kbar)	Transitions
$LiH_3(SeO_3)_2$	420[a]g	416[a]g	41000[a]g	15g		−6.0g	f
$LiD_3(SeO_3)_2$	446[a]g	432[a]g	29000[a]g			−5.9g	f
$NaH_3(SeO_3)_2$	194h	193g	460g	6.5j	1.03k	−1.1h	f(β)
$NaH_3(SeO_3)_2$	101h			3.1i	0.5k		f(γ)
$NaD_3(SeO_3)_2$	268i	239i	7900i	6.3i	1.58k	−1.7h	f(γ)
$KH_3(SeO_3)_2$	214h					−5.3h	af?
$KD_3(SeO_3)_2$	297q					−5.4h	af?
$RbH_3(SeO_3)_2$	153n			0.013n		+7.1h	f_i
$RbD_3(SeO_3)_2$	153n			0.000n			f_i?
$CsH_3(SeO_3)_2$	153m	100m	4000m		0.76j	+0.7h	af
$CsD_3(SeO_3)_2$	263m						af
$(NH_4)_2H_3IO_6$	254e				1.40p		af?
$(ND_4)_2D_3IO_6$	266e						af?
$Ag_2H_3IO_6$	230e						af
$Ag_2D_3IO_6$	265e						af

[a] From high-pressure measurements for C; T_c and T_0 extrapolated to 1 bar; crystal melts at 383 K at 1 bar.

Symbols as defined in table 23.1, except ΔS is transition entropy and f_i indicates improper ferroelectric. Numerical values in some cases differ considerably from those obtained by other investigators. The reference key is e = Känzig [1957], g = Samara [1968], h = Shirokov et al. [1971], i = Shuvalov et al. [1970a], j = Makita [1965], k = Makita and Miki [1970], m = Gavrilova-Podol'skaya et al. [1967], n = Shuvalov et al. [1970b], p = Kind [1971] and q = Yagi and Tatsuzaki [1969].

23.4.2. Sodium trihydrogen selenite

The structure of the paraelectric (α) phase of sodium trihydrogen selenite (abbreviated NaHSe) was studied by Kaplan et al. [1970] at room temperature using neutron diffraction. The selenite ions are H-bonded together into parallel sheets which are not connected by H-bonds. The sodium ions lie on interspersed sheets. There are two types of H-bonds. The symmetric O(1)–H(1)$\cdots$O(1′) bonds are 2.602 Å long, while the twice as numerous O(2)–H(2)$\cdots$O(3) bonds are 2.556 Å long. The authors did not specify whether the protons are in single wells or are in disordered positions in wells which in the time average have two minima. The disordered arrangement seems more likely, in view of the recent finding of Nelmes et al. [1972] that protons and deuterons in the shorter bonds in KDP and DKDP occupy disordered positions above T_c.

Recent results on the phase diagram and dielectric properties of $Na(D_xH_{1-x})_3(SeO_3)_2$ have been provided by Shuvalov et al. [1974]. Upon cooling the undeuterated crystal goes from the monoclinic paraelectric phase (α) to a triclinic ferroelectric phase (β) and finally to a monoclinic ferroelectric phase (γ). With increasing deuteration, the crystal goes through the phase sequence $\alpha \to \gamma \to \beta \to \gamma$ upon cooling. Above 30% deuteration the β phase disappears and only the $\alpha \to \gamma$ transition remains. All phase transitions are of first order. There is also a pressure-induced phase (δ) as mentioned before.

Makita and Miki [1970] measured the specific heat of NaHSe and (deuterated) NaDSe and proposed a model for the high-temperature phase based on the axioms: (1) one H per bond in one of two off-center positions; (2) each SeO_3^{2-} ion has either one or two close protons; and (3) selenite ion pairs are chosen in a specific way and required to have three close or connecting protons, that is, the crystal contains only $H_3(SeO_3)_2^-$ ions of a particular configuration and Na^+ ions. While they solved this model only approximately, Abraham and Lieb [1971] solved it exactly and showed that it is not inconsistent with experiment. Nagle and Allen [1971] consider that the choice of selenite ion pairs in axiom (3) is somewhat arbitrary and proposed two other models. The first model requires axiom (3) to hold for *all* neighboring pairs of selenite ions. Perhaps accidentally, the model infinite-temperature entropy is close to the transition entropy for the lower ($\beta \to \gamma$) transition in NaHSe. This revised axiom is in accord with the ordered deuteron configuration in NaDSe in the ferroelectric γ phase inferred by Soda and Chiba [1969a] from deuteron NMR measurements, but not with the deuteron configuration inferred from similar measurements by Blinc et al. [1968]. Knispel [1969] has shown that there are seven other configurations which obey the first two axioms but not the revised axiom (3). The other model of Nagle and Allen [1971] eliminates axiom (3) in the high-temperature limit. This model, unlike their previous one, has sufficient entropy in the disordered state. Some combination of their two models with isotope-dependent interaction strengths could perhaps explain the existence of two- and one-phase transitions in NaHSe and NaDSe respectively.

23.4.3. Lithium trihydrogen selenite

Both the normal (LiHSe) and deuterated (LiDSe) compounds melt before the Curie temperature is reached, but Samara [1968] found that application of pressure brings about a transition in both crystals. The spontaneous polarization of 15 $\mu C/cm^2$ for LiHSe is much larger than for other H-bonded ferroelectrics. The Curie points decrease linearly with pressure and can be extrapolated to 147 ± 1 and $172 \pm 3\,°C$ respectively for zero pressure.

A recent neutron diffraction study by Tellgren and Liminga [1972] shows that the selenite ions are H-bonded together in chains running along *c*. All hydrogens are in ordered off-center positions and all three bonds in a molecular unit are inequivalent, so it would seem that domain reversal would require intrabond transfer of all hydrogens.

23.4.4. Potassium trihydrogen selenite

Neutron diffraction studies of normal and deuterated potassium trihydrogen selenite were carried out at room temperature (above the transition temperature) by Lehmann and Larsen [1971]. The structures are similar, with selenite ions H-bonded together in double chains running along the orthorhombic *c*-axis. The hydrogens in the bonds connecting the two strands of each chain are in disordered positions.

Gorbatyi et al. [1972] have compared results of X-ray studies of NaHSe and KHSe and have related the results to physical properties of these crystals. They speculate that the lack of ferroelectric behavior in the low-temperature phase of KHSe may result from the selenite ions being H-bonded in chains, while in NaHSe they are H-bonded into sheets with correspondingly stronger tendency for three-dimensional order. If the KHSe transition is order–disorder, it must involve at least two-dimensional ordering to produce the sharp transition observed. The possibility of three-dimensional antiferroelectric ordering also exists.

23.4.5. Rubidium trihydrogen selenite

No neutron diffraction results have yet been reported for this crystal, so the locations and possible disorder of the H-bonds are not known with certainty. X-ray studies by Tovbis et al. [1972] and by Tellgren et al. [1973] led both groups to the conclusion that a three-dimensional H-bonded network exists in RbHSe.

Shuvalov et al. [1970b] have shown that while RbHSe is ferroelectric, increasing deuteration reduces the spontaneous polarization and that for more than 90% deuteration the low-temperature phase is no longer ferroelectric and has a different symmetry than the low-temperature phase for RbHSe. Also, unlike other members of the trihydrogen selenite family, RbHSe shows little effect of deuteration on transition temperature. Dvořák [1972] has claimed that both RbHSe and RbDSe must be improper ferroelectrics. In this type of ferroelectric, first proposed by Cross et al. [1968], the transition results from elastic rather than dielectric instability and the dielectric constant above T_c shows little temperature dependence.

23.4.6. Cesium trihydrogen selenite

A preliminary report on a neutron diffraction study of $CsH_3(SeO_3)_2$ by Sato [1972] shows that, as in NaHSe, the selenites are H-bonded together into planes and that the protons undergo an order–disorder transition. Unlike NaHSe this crystal is not ferroelectric. Makita [1965] upon study of dielectric, optical, thermal, crystallographic and pyroelectric properties of CsHSe concluded that the low-temperature phase is probably antiferroelectric. Gavrilova-Podol'skaya et al. [1967] studied dielectric properties of both CsHSe and CsDSe and found that deuteration raises the transition temperature from -120°C to -10°C, which is the largest isotope effect on T_c reported to date.

23.4.7. Discussion

All of the alkali metal trihydrogen selenites show interesting dielectric behavior, ranging from ferroelectric and improper ferroelectric to antiferroelectric and perhaps other types of ordering. There is no discernible trend as one goes through the column of alkali metals, in type of ordering, crystal structure, or magnitude of isotope effect on T_c.

Ammonium trihydrogen selenite is a linear dielectric, while in the KDP family it is lithium and sodium dihydrogen phosphate which show no ferroelectric or antiferroelectric behavior.

There is also no apparent trend in the effect of cation on the pressure dependence of T_c as measured by Samara [1968] for LiHSe and LiDSe and by Shirokov et al. [1971] for other compounds, as shown in table 23.2. Samara explained the negative coefficient for the lithium compounds on the basis of H-bond shortening, leading to lower energy for disordered H configurations. This explanation could also be applied to explain the negative coefficients for the sodium and potassium compounds. The lack of isotope dependence of T_c in RbHSe indicates that H-bonding is not important in its (improper) ferroelectric behavior, so the increase of T_c with pressure is not surprising. However, it is difficult to explain why CsHSe, which has the largest isotope shift of T_c, has a positive albeit small T_c variation with pressure.

In contrast with the KDP family, little theoretical work has been done on the natures of the phase transitions in the trihydrogen selenites, particularly with regard to dynamic phenomena. A difficulty is that the structures are more complex than for KDP. Recently Shuvalov et al. [1973] have proposed two-dimensional ordering in the triclinic β phase of NaHSe and have employed a theory which includes proton dynamics and the cluster approximation which explains some features of the phase transitions in NaHSe.

Regarding critical phenomena in these crystals, Kuroda et al. [1971] and Blinc et al. [1968] obtained evidence for critical slowing down from proton and deuteron spin–lattice relaxation times T_1 in NaHSe and NaDSe respectively. Adriaenssens [1974] showed in NaHSe that this T_1 anomaly disappears at higher NMR frequencies, indicating that a central mode (centered on zero frequency rather than a non-zero "soft phonon" frequency) is responsible. Such a central mode could result from polarization fluctuations in which the polarized regions become larger and decay more slowly as T_c is approached.

In conclusion, the rich variety of behavior in these selenites holds promise that much can be learned about static and dynamic aspects of H-bonding from study of these crystals.

23.5. Antiferroelectric periodates

23.5.1. Introduction

Although they have not been extensively investigated, $(NH_4)_2H_3IO_6$ and $Ag_2H_3IO_6$ are important from the standpoint of H-bonding because both undergo order–disorder transitions of the H-bond system. Early evidence for this, as reviewed by Känzig [1957], is the large transition entropy and the relatively large effect of deuteration on the transition temperature shown in table 23.2.

23.5.2. Static properties

Kind [1971] carried out an NMR investigation to elucidate the nature of the low-temperature phase of ammonium trihydrogen periodate. Because the ^{127}I resonances are too broad to observe directly, he obtained the location of the quadrupolar-split ^{127}I lines by observing dips in the proton spin–lattice relaxation time T_1 as a function of crystal orientation. The high-temperature electric-field gradient (efg) tensor for ^{127}I is axially symmetric, consistent with dynamic disorder of the six nearest protons. The low-temperature results are consistent with either the non-polar $R\bar{3}c$ or the antiferroelectric R32 structures. Both structures have ordered H-positions in only half of the H-bonds. Low-temperature structure studies, deuteron NMR, or other methods must be used to determine whether this crystal becomes antiferroelectric.

Similar experiments were carried out by Roos and Kind [1974] on silver trihydrogen periodate. They found that the proton system goes from complete disorder to complete order as the crystal goes from the room-temperature hexagonal $R\bar{3}$ phase to the low-temperature triclinic $P\bar{1}$ phase, which is antiferroelectric. Both the ammonium and the silver trihydrogen periodate have the same room-temperature structure and both exhibit *two* phase transitions closely spaced in temperature. It would be interesting to learn the structures and degrees of H order for these unknown intermediate phases.

23.5.3. Dynamic properties

Blinc et al. [1973b] employed a clever NMR technique to determine the rate of proton intrabond motion in both silver and ammonium trihydrogen periodate. The ^{127}I spin–lattice relaxation time T_{1I} is governed by efg fluctuations caused by motion of the protons surrounding these nuclei. The ^{127}I resonance is not directly observable, but in the dipolar and the rotating reference frames the protons and ^{127}I nuclei are strongly coupled and the measured proton relaxation times T_{1D} and $T_{1\rho}$ in these respective frames are determined by the ^{127}I spin–lattice relaxation time. A proton intrabond jump time of order 10^{-10} s at room temperature was inferred, which is similar to values for KDP-type crystals. Critical slowing down of the intrabond jumping was noted near T_c. The spontaneous polarizations of the antiferroelectric sublattices were inferred from the T_{1I} vs. temperature curves, which in turn were inferred from the temperature dependences of T_{1D} and $T_{1\rho}$. In view of these interesting phenomena, it would seem worthwhile to study static and dynamic properties of the deuterated analogs of these crystals.

23.6. Rochelle salt and other ferroelectric tartrates

23.6.1. Introduction

The ferroelectric tartrates all have orthorhombic high-temperature phases, but otherwise exhibit considerable individuality. Rochelle salt was the first ferroelectric discovered, but only recently has its behavior become fairly well understood on a microscopic basis. H-bonding appears to have less connection with ferroelectric behavior in the tartrates than in the phosphates, selenites and periodates.

23.6.2. The sodium tartrates

Rochelle salt (RS) has composition $NaKC_4H_4O_6 \cdot 4H_2O$, while its isomorph ammonium Rochelle salt (ARS) contains NH_4 in place of K. Rochelle salt displays the Curie–Weiss behavior common to most ferroelectrics, but has *two* transitions at 24 °C and −18 °C, between which temperatures it is ferroelectric. Deuteration moves these transitions to 35 °C and −21 °C.

It is now believed that the hydroxyl radicals in the tartrate ions make the major contribution to spontaneous polarization in RS, resulting from two possible positions for the hydrogens which are believed to lie in intrinsically asymmetric double-minimum potential wells. Mitsui [1958] extended earlier theories based on such wells and was able to account for many properties of RS. Sandy and Jones [1968] made extensive dielectric measurements at microwave frequencies up to 13 GHz which were consistent with Mitsui's theory. However, Abe and Tokumaru [1971] on the basis of dielectric measurements made between liquid helium and room temperatures concluded that either the well asymmetry is smaller than previously believed, or else tunneling between the minima in these wells must be considered.

Later extensions of the theory of RS were made by Takagi and Ishibashi [1972] who assumed four rather than two inequivalent dipoles and by Žekš et al. [1971] who assumed two sublattices with asymmetric wells. For appropriate coupling constants between hydroxyl dipoles on the same, and on opposite, sublattices, they obtained three distinct temperature regions. At low temperature each sublattice is mostly ordered and the polarizations cancel. The high-temperature phase is similar, but with less order. The ferroelectric intermediate phase arises because in this temperature range either sublattice can "overpower" the other, resulting in incomplete cancellation of their polarizations. The equilibrium properties of both RS and deuterated RS could be explained by assuming proton tunneling but no deuteron tunneling. The temperature dependence of the sublattice polarizations is in agreement with subsequent NMR measurements of the electric-field gradients at the ^{23}Na sites obtained by Fitzgerald and Casabella [1973].

Unlike RS, ammonium Rochelle salt has only one Curie point, at 109 K. It shows little temperature dependence of dielectric constant above T_c, and its spontaneous polarization cannot be reversed by application of electric field, but only by a suitable mechanical stress. This behavior is explained by Sawada and Takagi [1972] who classify ARS as an improper ferroelectric, while Aizu [1972] considers ARS to be a "faint" ferroelectric and draws a distinction between "faint" and "improper".

23.6.3. The lithium tartrates

The members of this group have formula $LiXC_4H_4O_6{\cdot}H_2O$ and are isomorphic in the high-temperature phase. For X = Tl, the Curie point has the unusually low value of 10 K which does not change upon deuteration, but the Curie–Weiss temperature T_0 drops from 10.8 K to 1.8 K upon deuteration while the Curie–Weiss constant increases from 1666 K to 1966 K.

For X = NH_4, the Curie temperature is 98 K and the ferroelectric axis is along *b* rather than *a* as in the thallium compound. El Saffar et al. [1972] performed an NMR study of this compound (LAT) and found from quadrupolar splitting of the deuteron resonance that ND_4^+ ion distortion accounts for 60% of the spontaneous polarization. The temperature dependence of the proton spin–lattice relaxation time T_1 at low temperature indicates coherent rotation of the ammonium ions involving quantum-mechanical tunneling. There is also evidence that the hydroxyl protons move between several inequivalent sites, as in Rochelle salt.

23.7. Ferroelectric sulfates

23.7.1. Introduction

The ferroelectric sulfates are grouped together mainly for convenience, as they vary widely in chemical complexity, structure and ferroelectric behavior. H-bonding is not strongly related to ferroelectric properties in these compounds, as evidenced by the small effect of deuteration on T_c. This family is composed of triglycine sulfate (TGS) and related crystals which have been widely studied, ammonium sulfate and other simple sulfates, various alums and guanidinium aluminum sulfate hexahydrate (GASH) and isomorphous compounds.

23.7.2. Triglycine sulfate and its isomorphs

Considerable study has been made of triglycine sulfate, or $(NH_2CH_2COOH)_3 \cdot H_2SO_4$, because it is ferroelectric at room temperature ($T_c = 49°C$), is easily grown and has reasonably large spontaneous polarization. Its isomorphs are formed by substituting selenate (SeO_4^{2-}) or fluoberyllate (BeF_4^{2-}) ions for the sulfate ions to form compounds designated as TGSe and TGFB respectively. All of these compounds and their deuterated analogs are quite similar. From the theoretical point of view they are interesting because they exhibit second-order phase transitions, so that critical phenomena can be studied near T_c without the complication encountered with KH_2PO_4, that its transition is slightly within the first-order region. Studies of critical behavior made by Sawada et al. [1971] for TGS and by Mercado and Gonzalo [1973] for TGFB show that both crystals are well described by mean-field theory, for which clustering effects are ignored.

To clear up the question of whether the glycine I ion in TGS undergoes a displacive or an order–disorder transition, Blinc et al. [1971b] studied the temperature dependence of the electric-field gradient at the nitrogen sites by means of pulsed proton-^{14}N double resonance in the rotating frame. They found strong evidence for an order–disorder transition involving flipping of the glycine I ion. This flipping is probably associated with proton intrabond motion in the H-bond joining the glycine II and glycine III ions which in the time average are equivalent above T_c. Below

T_c the glycine I ordering makes these bonds asymmetric in the time average, with resulting ordering of the protons in these bonds. This hydrogen ordering appears to be more an effect than a cause of the transition, because deuteration only raises T_c by about 11 °C. Similar conclusions concerning the transition in TGFB were reached by Blinc et al. [1971c] on the basis of ^{9}Be quadrupolar-perturbed NMR in that crystal. A recent neutron diffraction study by Kay and Kleinberg [1973] of the phase transition in TGS tends to support the order–disorder view of the transition.

23.7.3. Ammonium sulfate and ammonium fluoberyllate

Both of these compounds, $(NH_4)_2SO_4$ and $(NH_4)_2BeF_4$, are orthorhombic both above and below T_c, yet are not isomorphic either in the ferroelectric or paraelectric phases. Both exhibit first-order phase transitions and both show little effect of deuteration upon T_c. Ammonium sulfate (AS) has about twice as much transition entropy as ammonium fluoberyllate (AFB). Both crystals show the lack of temperature dependence of dielectric constant above T_c characteristic of improper ferroelectrics. Kobayashi et al. [1972] have presented a phenomenological theory of AS as an improper ferroelectric, while Strukov et al. [1973a] characterize AFB as an improper ferroelectric and present thermal evidence for additional phase transitions near T_c.

O'Reilly and Tsang [1967] on the basis of NMR studies of AS and AFB and the deuterated crystals DAS and DAFB concluded that an order–disorder transition of the ammonium ions is responsible for the ferroelectric transition. On the basis of the transition entropies and additional deuteron NMR studies in DAFB, Kydon et al. [1969] suggested that only half of the ammonium ions order in AFB while all order in AS. Evidence against an order–disorder transition has been presented by Hamilton [1969] and Torrie et al. [1972] report that results of their Raman and infrared studies of AS support neither the order–disorder model nor the view that the H-bonds change significantly at T_c. Evidently the nature of the ferroelectric phase transitions in these crystals is still not very well understood.

23.7.4. Other ferroelectric sulfates

Ammonium cadmium sulfate, $(NH_4)_2Cd_2(SO_4)_3$, exhibits virtually no temperature dependence in its dielectric behavior above T_c and is one of the original materials named as a possible improper ferroelectric when

Cross et al. [1968] first defined this term. Aizu [1972] proposed a specific mechanism for the improper transition, which involves three degenerate conjugate pairs of lattice vibrations which slow down with decreasing temperature and "condense" at T_c.

The mineral lecontite, $NaNH_4SO_4 \cdot 2H_2O$ becomes ferroelectric at 101 K and also has a dielectric anomaly at 92 K. The corresponding selenate compound shows similar behavior and deuteration has little effect on T_c. Genin and O'Reilly [1969] performed proton, deuteron and ^{23}Na NMR experiments in lecontite and concluded that the water and ammonium groups are not directly involved in the transition. From the change in activation energy for the NH_4^+ ion hindered rotation at T_c, they concluded that the transition is displacive. However, results of dielectric measurements by Aleksandrov et al. [1971] in both compounds and deuteron NMR studies of $NaND_4SO_4 \cdot 2D_2O$ by Aleksandrova et al. [1971] brought these workers to the conclusion that the ammonium groups order to produce the spontaneous polarization as in ammonium sulfate.

Ammonium bisulfate shares with Rochelle salt the property of being ferroelectric only between two temperatures, 270 K and 154 K in the case of NH_4HSO_4. Nelmes [1972] found from X-ray measurements that the ferroelectricity is associated with sulfate ion ordering and is performing a neutron diffraction study of ND_4DSO_4 to study the hydrogen behavior. Trontelj and Rebič [1972] studied deuteron T_1 in 90% deuterated polycrystalline ND_4DSO_4 and found fast ND_4^+ ion hindered rotation in all three phases, with discontinuities in T_1 at the transitions at 265 ± 10 K and 174 ± 5 K.

In the course of investigation in which they discovered a number of ferroelectrics, Pepinsky and Vedam [1960] determined that although rubidium bisulfate, $RbHSO_4$, is isomorphous with ammonium bisulfate at room temperature, it has only one phase transition, at 258 K. The transition seems clearly of second order, as evidenced by a rather slow rise of spontaneous polarization P_s with decreasing temperature below T_c. This second-order nature seems in conflict with the report by Silvidi et al. [1969] that proton T_1 in a polycrystalline sample shows thermal hysteresis extending over a range of about 40°C below the originally reported T_c. Silvidi and coworkers suggest that proton tunneling sets in *below* T_c.

The alums constitute a rather large ferroelectric family, in which deuteration has little effect on T_c. Recent work has been done with methylammonium aluminum sulfate dodecahydrate,

$CH_3NH_3Al(SO_4)_2\cdot 12H_2O$ or MASD for short and with ferric ammonium sulfate dodecahydrate, $NH_4Fe(SO_4)_2\cdot 12H_2O$, normally abbreviated as FAS. Makita and Sumita [1971] have studied dielectric properties of MASD from 10^6 to 10^9 Hz and find a single dielectric relaxation time and Curie–Weiss temperature dependence. They attribute this behavior to the $CH_3NH_3^+$ dipoles which are disordered above T_c and undergo random 180° flips. The ferroelectric behavior in FAS has been attributed to Fe^{3+} ion displacement by Montano et al. [1971] on the basis of single crystal Mössbauer effect measurements.

The prototype of another fairly large ferroelectric family is guanidinium aluminum sulfate (GASH), with formula $C(NH_2)_3Al(SO_4)_2\cdot 6H_2O$, where Al can be replaced by Ga, Cr, or V, S by Se and H by D. These crystals decompose near 200 °C before the ferroelectric Curie point is reached. An ^{14}N quadrupole resonance study of GASH performed by Oja [1969] contains references to previous NMR work, but unfortunately NMR in this case has been unable to provide understanding of the nature of ferroelectric ordering in this crystal.

23.8. Other hydrogen bonded ferroelectrics

23.8.1. Hydrogen halides

The hydrogen halides contain the simplest molecular units of any ferroelectric, yet their behavior is quite complex. For instance, X-ray studies by Simon [1971] showed that HBr has two high-temperature face-centered cubic phases and two low-temperature orthorhombic phases, with transitions at 89.8, 113.6 and 116.9 K. Similar behavior is shown by HCl and HI, while HF exhibits no transitions. Although H-bonding is of basic importance in the ferroelectric transitions, deuteration only raises the transition temperature by 3 to 7 °C.

An X-ray and neutron diffraction experiment by Niimura et al. [1972] resulted in the seeming paradox that the cubic to orthorhombic transition at 120K is accompanied by little change in specific heat, dielectric constant and spontaneous polarization, while at 98 K where large anomalies occur in these parameters no structural change detectable by the diffraction experiments occurs. They proposed that zigzag HCl chains form at 120 K and order at 98 K. Dynamic behavior of such chains had already been postulated by O'Reilly [1970] for HCl and by Kadaba and

O'Reilly [1970] for HBr and DBr on the basis of ^{35}Cl and ^{79}Br and ^{81}Br nuclear quadrupole resonance (NQR) results. They proposed not only 180° flips of chain segments, but also 90° flips in which every other molecule in a zigzag chain leaves that chain and becomes part of an adjoining chain. Such interchain transfers must take place cooperatively, and "quasi-spin-wave modes" were proposed as the dynamic mechanism.

A statistical theory was developed by Yi and Gavrielides [1971] to explain previous neutron diffraction, NQR, dielectric and infrared absorption results. They assumed that each chain has two possible orientations and found that this two-dimensional Ising model predicts the temperature dependence of the spontaneous polarization better than does the molecular field approximation.

Tsang and Shaw [1971] attempted to explain ferroelectric behavior in these halides by means of an empirical intermolecular potential. They used two arbitrary parameters and obtained qualitative agreement with experiment.

23.8.2. Copper formate tetrahydrate

This crystal, $Cu(HCOO)_2 \cdot 4H_2O$, exhibits an antiferroelectric transition at 234 K which is raised 12°C by deuteration and also makes transitions to a partly-ordered magnetic state at 50 K and to an antiferromagnetic state at 17 K. The structure contains parallel layers of water molecules, a sort of "two-dimensional ice". Before a neutron study had been made, Okada [1967] proposed a model in which the water molecules are partly disordered above T_c, solved it approximately and obtained reasonable agreement with the measured transition entropy. Nagle [1969] later solved this model exactly. Ishibashi et al. [1973] extended this model to take into account anisotropy within the water layers and interactions between layers by which alternately ordered ferroelectric layers build up an antiferroelectric crystal. Their theory predicts a first-order transition as observed.

A deuteron NMR study of $Cu(DCOO)_2 \cdot 4D_2O$ by Soda and Chiba [1969b] revealed separate NMR lines below T_c for different D_2O orientations, with line broadening above T_c indicative of hindered rotation of water molecules. A neutron diffraction study of the undeuterated compound by Kay and Kleinberg [1972] confirmed the ordering of water molecules.

23.8.3. Potassium ferrocyanide trihydrate and isomorphic crystals

In $K_4Fe(CN)_6 \cdot 3H_2O$, Fe can be replaced by Ru, Os, or Mn with little change in ferroelectric properties. Deuteration also has little effect, raising T_c by about 6°C. The iron compound exhibits complicated twinning, which together with the ferroelectric behavior has been studied by Krasnikova et al. [1972] using proton NMR. Crystals of the ruthenium compound can be grown free of such twinning, and NMR results of Habuda et al. [1970] in such crystals show that the spontaneous polarization results from one type of water molecule, which orders below T_c. Krasnikova and Polandov [1970] found that pressure increases T_c and makes comparisons with pressure effects in other H-bonded ferroelectrics.

23.8.4. Colemanite

Another ferroelectric containing water molecules is colemanite, $CaB_3O_4(OH)_3 \cdot H_2O$, with T_c at 270 K. Proton spin–lattice relaxation times T_1, $T_{1\rho}$ and T_{1D} were measured by Watton et al. [1973] in the laboratory, rotating, and dipolar reference frames respectively, each measurement being most sensitive to proton motions of a different rate. These measurements disclosed water molecule 180° flips having no dielectric effect and also a "jump mode" by which water and hydroxyl protons exchange roles. This jump mode is thermally activated well above T_c, but its frequency goes critically toward zero near T_c because of coupling to the structure. As in KDP-type crystals, the H motions "trigger" the transition but contribute little to the spontaneous polarization.

23.8.5. Synthetic troegerite

Unusual ferroelectric behavior was recently reported by De Benyacar et al. [1973] in the uranyl compound $HUO_2AsO_4 \cdot 4H_2O$. Near 20°C the crystal becomes antiferroelectric, but upon cooling in an ac field the crystal becomes ferroelectric at −20°C as evidenced by appearance of hysteresis loops. Upon subsequent heating, the crystal remains ferroelectric up to 30°C. A later structural description by De Benyacar and De Abeledo [1974] does not mention any ferroelectric behavior.

23.8.6. Substituted acetates

Properties of a number of ferroelectric crystals in this group have been reviewed by Jona and Shirane [1962], including TGS, TGSe and TGFB which were discussed in section 23.7.2 and the non-H-bonded dicalcium strontium propionate mentioned in section 23.3.4. Blinc et al. [1970c] measured proton T_1 and linewidth in diglycine nitrate, $(NH_2CH_2COOH)_2HNO_3$, trisarcosine calcium chloride, $(CH_3NHCH_2COOH)_3CaCl_2$ and its deuterated analog. They found evidence for flipping of the glycine and sarcosine units which appears to freeze out below T_c. The diglycine nitrate structure and glycine flipping are reminiscent of TGS.

Yamamoto et al. [1970] measured ^{35}Cl quadrupolar resonance in normal and deuterated $NH_4H(ClCH_2COO)_2$ and postulated on the basis of the 11°C increase in T_c upon deuteration that H-ordering is responsible for the transition. They also considered a displacive transition possible because the C–Cl directions change below T_c.

23.8.7. Semicarbazide hydrochloride

Rocaries and Boldrini [1972] reported a new ferroelectric $H_2NCONHNH_2{\cdot}HCl$ with dielectric constant maxima of about 2000 at temperatures of −230, 19 and 21°C. A Curie–Weiss law is obeyed above 21°C. This is a molecular crystal with some ionic character and has a H-bond network which may be three-dimensional. They expect that other ferroelectrics will be discovered in this crystal family.

23.8.8. Thiourea

Thiourea, $SC(NH_2)_2$, might be considered a borderline case between a H-bonded and a molecular crystal. Jona and Shirane [1962] stated that H-bonding is not expected in thiourea, while O'Reilly et al. [1971] mention N–H···S bonds in their report on proton T_1 and $T_{1\rho}$ measurements and deuteron NMR spectra in normal and deuterated thiourea. There are five phases of thiourea, the ferroelectric phases being phase I which exists below 169 K and phase III which exists between 176 and 190 K. Deuteration increases the transition temperatures by 11 to 16°C. The NMR results showed that the thiourea molecule undergoes 180° flips and molecular diffusion becomes evident above 385 K. Two ferroelectric modes were also found in the high-temperature phase, which are believed to correspond to flips between molecular orientations for the two domains in phase I (for one mode) and phase III (the other mode).

Gesi [1969] showed that for applied pressures up to 7 kbar all transition temperatures decrease with increasing pressure. He discovered two new phases which are not ferroelectric. He presented a table of pressure effects on various ferroelectrics.

23.8.9. Two-dimensional perovskites

The three-dimensional perovskites, of which $BaTiO_3$ is the prototype, are the best-known family of non-H-bonded ferroelectrics. Arend et al. [1974] have suggested on the basis of optical and thermal data that $(C_nH_{2n+1}NH_3)_2MnCl_4$, where $n = 1$, 2 or 3 may be antiferroelectric between $-50\,°C$ and 151.3 °C (for $n = 2$). The "two-dimensional perovskite" designation arises because the Cl^- octahedra surrounding each manganese ion join corners in a nearly square planar array. These planes have been studied as nearly two-dimensional magnetic systems by Boesch et al. [1971]. An X-ray study by Peterson and Willett [1972) of the $n = 2$ compound has not revealed the nature of the N–H···Cl bonding, but the structure seems favorable for H order–disorder transitions, even though deuteration has almost no effect on the transition temperatures. A number of other crystals with similar planar structures exist.

23.9. $LiN_2H_5SO_4$ – one-dimensional conduction mimics ferroelectric behavior

23.9.1. Introduction

Lithium hydrazinium sulfate (abbreviated LiHzS) was considered ferroelectric after Pepinsky et al. [1958] observed hysteresis loops for large electric fields applied along the c-axis. The loops degenerated above 80 °C because of high electrical conductivity, but absence of dielectric anomalies below the decomposition temperature above 200 °C indicated that the crystal was ferroelectric up to its decomposition temperature. Vanderkooy et al. [1964] found that the conductivity is protonic, and is over 100 times greater along c than along a or b. The crystal structure is orthorhombic, with sulfate ions pointing nearly along c and Li^+ ions surrounding channels aligned with c which contain the hydrazinium ions. Neutron diffraction studies by Padmanabhan and Balasubramanian [1967] showed that these $N_2H_5^+$ ions are bonded together by N–H···N–H··· chains running along c.

23.9.2. Dielectric properties

Whereas most ferroelectrics have a large dielectric constant only near T_c, Schmidt et al. [1971] found that in LiHzS ϵ_c is large (up to 10^6) and lossy over a very wide temperature and frequency range. Studies of etching and pyroelectric behavior after application of fields larger than the reported coercive field (field necessary to reverse domains) indicated that electric fields cannot reverse the structure, so that LiHzS is not ferroelectric. Anomalous neutron scattering measurements by Anderson and Brown [1972] in $^6LiN_2H_5SO_4$ confirmed its non-ferroelectric nature.

One explanation considered for the dielectric behavior was reversal of some of the N–H dipoles in the N–H···N–H··· chains, considering these chains as biased Ising chains, with the non-centric structure supplying the effective bias field. Schmidt [1971] showed that the ac susceptibility for such intrinsically biased chains should depend strongly on the sign of an applied dc field. No such strong dependence was found.

Next considered by Schmidt et al. [1971] was a model in which the protonic conduction along the channels parallel to c is interrupted by barriers of random height resulting from structural defects. Fixed barrier height between channels was assumed. This model accounts quite well for the dielectric behavior. Schmidt and Parker [1972] were able to show that a similar model is able to predict hysteresis loops similar in form to those observed in LiHzS. The nature of the defects and the microscopic mechanism for conductivity are not yet known.

Similar anomalous dielectric behavior might be expected in other crystals containing H-bonded chains.

23.10. Discussion

A bewildering array of structures and phenomena has been presented. One might ask what central theme ties the subject of H-bonded ferroelectrics together and gives it a unique character.

The answer that comes to mind is that the interest in ferroelectrics is focused on the phase transitions. Most of the H-bonded ferroelectrics undergo order–disorder transitions, involving the H-bond network. What is unique about order–disorder transitions of ferroelectrics?

Other order–disorder phase transitions fall into two classes. In one, exemplified by gas–liquid and liquid–solid transitions, the entire system goes from a chaotic to a less chaotic state. In the other class, an essentially fixed crystal lattice is the framework within which the order–disorder transition occurs, examples being magnetic and superconducting transitions.

Ferroelectric order–disorder transitions fall into an intermediate class which is more complex and displays greater richness of phenomena, because the ordering elements interact quite strongly with the so-called fixed lattice. Accordingly, research in H-bonded ferroelectrics consists primarily in observation and explanation of the static and dynamic aspects of these interactions between the protons or other ordering elements and the lattice.

Acknowledgements

This work was supported in part by the National Science Foundation. The author is grateful also to many colleagues throughout the world for informative discussions and for preprints and reprints of their work.

Note added in proof

This Note reports primarily on developments published between the date of submission of the manuscript for this chapter, and April, 1975. Items are organized below under the same headings used in the text.

23.1. Introduction

23.1.3. *Pertinent books and review articles.* Smolenski and Krainik [1972] have written *Ferroelectrics and Antiferroelectrics*, which has been translated from Russian into German. Mitsui et al. [1973a] have written *An Introduction to the Physics of Ferroelectrics.* Zheludev [1971] has discussed symmetry of ferroelectric and antiferroelectric crystals, including symmetry considerations of domain structure. He does not refer to the considerable body of work published by Aizu on this topic. Subbarao [1973] has compiled a table of spontaneous polarization P_s, Curie temperature T_c, and other transformation temperatures for ferroelectrics and antiferroelectrics known through 1971. Bonera et al. [1972] described the NMR relaxation method of studying critical phenomena at structural phase transitions. The theory of coherent neutron scattering by H-bonded ferroelectrics at low temperature has been reviewed by Stamenković [1972]. Comparison of KD_2PO_4 results with several models was made.

Finally Mitsui et al. [1973b] have reviewed critical phenomena in ferroelectric phase transitions. They find that the static indices agree with mean-field theory except very near to critical points.

23.2. Experimental results for KDP-type crystals

23.2.3. *Dielectric behavior.* Based on the large critical field E_{cr} of 7100 V/cm for KD_2PO_4 (DKDP) measured by Sidnenko and Gladkii [1973a], at which the ferroelectric transition near 223 K becomes second-order, the much smaller E_{cr} of 180 V/cm reported by Okada et al. [1973] for the KH_2PO_4 (KDP) transition near 123 K, and the rate of drop of transition temperature with pressure reported for KDP by Samara [1971], Schmidt [1974] has suggested that KDP may exhibit a tricritical point at a pressure of order 5 kbar, at which the transition becomes second-order for zero applied field. Eberhard and Horn [1975] question this conclusion on the basis of 1000 Hz dielectric measurements in the presence of small dc fields which by extrapolation indicate an E_{cr} of roughly 6500 V/cm for KDP, and cite also X-ray dilatometric results of Kobayashi et al. [1971] which indicate E_{cr} near 8500 V/cm.

23.2.4. *Specific heat and transition entropy.* Fairall and Reese [1974], continuing their series of calorimetric and dielectric measurements on KDP-type crystals, find that RbDA has a first-order transition with latent heat of 255 ± 5 J/mole and Curie constant $C = 223$ K. They found poorer agreement with the theory of Silsbee et al. [1964] (SUS theory) than was found for KDA, and noted further that their Slater energy ϵ_0 disagreed markedly with that found by Blinc and Bjorkstam [1969] from NMR relaxation results.

Strukov et al. [1974] found first-order transition heats of 54 ± 2 and 117 ± 2 cal/mole for CsDA and DCsDA respectively, and noted that tunneling seems less important in CsDA than in KDP and RbDP.

23.2.6. *Nuclear magnetic resonance.* Nuclear quadrupole resonance (NQR) studies of ^{75}As in KDA and CsDA and their deuterated analogs by Blinc et al. [1973c] demonstrated that the arsenate groups fluctuate among the various Slater configurations above T_c. Existence of sharp ^{75}As lines above T_c which correspond to the crystal symmetry below T_c showed that long-lived polarized clusters giving the "central mode" exist above T_c. They also noted relaxation caused by the "soft" optical mode near T_c.

Similar studies by Blinc et al. [1973d] of ^{75}As in ADA showed that Nagamiya's [1952] proposal of O–H$\cdots$O hydrogen ordering in the antiferroelectric ordered phase is correct. They further found a positive value for the Slater energy ϵ_0, which using the Slater model would predict a *ferroelectric* ordered state. They concluded that long-range dipole–dipole forces in the a–b plane are responsible for the antiferroelectric ordering.

23.2.7. *Electron paramagnetic resonance.* Further confirmation of Nagamiya's [1952] ordering scheme for ADA came from ENDOR (electron–nuclear double resonance) experiments by Gaillard et al. [1974]. Ishibashi et al. [1974] have developed a theory of the antiferroelectric phase transition in these ammonium-based crystals which explains the antiferroelectric ordering even for a positive Slater energy ϵ_0.

Kawano [1974] has used the quintet to triplet transition in the EPR spectrum of SeO_4^{3-} radicals in ADP doped with K_2SeO_4 to find the activation energy for the correlation time for hydrogen configurations around that radical, and has tabulated similar results for related crystals. These activation energies (in eV) are 0.19 ± 0.03 for ADP, 0.11 ± 0.03 for KDP, 0.21 ± 0.03 and 0.12 for KDA, and $0.33 + 0.03$ and 0.25 for ADA.

23.2.8. *Neutron scattering.* Dietrich et al. [1974] studied neutron scattering from DCsDA and found that, unlike DKDP, no ferroelectric critical scattering appears. They found evidence for a complex structure in the ferroelectric phase, with periodicities of about 30, 15, 10, and 5 lattice spacings.

23.2.9. *Light scattering.* The low-frequency laser Raman spectra of KDP, DKDP, and RbDP were studied by Blinc et al. [1973e]. In KDP and RbDP they observed underdamped "tunneling" modes well below T_c, which decrease in frequency, broaden and disappear as T_c is approached. The results were explained using equations similar to the Bloch NMR equations. In low-frequency Raman studies of KDP and RbDP and partly deuterated analogs, Mavrin et al. [1974] noted bands at 80 and 60 cm^{-1} in KDP and RbDP respectively, and at 10 cm^{-1} in 85% deuterated crystals of both types. Both bands were present in the partly deuterated crystals. They attributed these bands to proton and deuteron tunneling.

Lowndes et al. [1974] investigated Raman scattering by the paraelectric and ferroelectric phases of KDA, RbDA, and CsDA and their deuterated isomorphs. They found evidence for collective proton and deuteron motions in the paraelectric phases. The marked breadth of many features disappeared below T_c, so this breadth is probably related to hydrogen disorder. They found only partial agreement with Kobayashi's [1968] theoretical predictions, but their results fit into the framework of Cowley and Coombs' [1973] theory.

Brillouin scattering studies of the elastic anomaly associated with the

KDP transition were performed by Brody and Cummins [1974] between 113 and 291 K and for bias fields up to 3,937 V/cm. Their values found for the shear elastic constants fit quite well with predictions based on SUS theory (Silsbee et al. [1964]), but with different parameters than those which best fit the spontaneous polarization data. Further Raman and Brillouin scattering studies by Lagakos and Cummins [1974] gave indication of a "central peak" too narrow to be resolved. They analyzed their Raman results in terms of coupled optic, acoustic, and ferroelectric soft modes. They further analyzed both the Brillouin and Raman results in terms of the relaxing self-energy model of Cowley and Coombs [1973]. They found that the soft mode extrapolates to zero frequency at the Curie–Weiss temperature T_0. In similar experiments with CsDA, however, Lagakos and Cummins [1975] found no evidence for a central peak. They studied CsDA because Katiyar et al. [1971] reported that the ferroelectric soft mode extrapolates to zero frequency at 69 K, far below the ferroelectric transition temperature of 143 K. According to relaxing self-energy theories of Cowley and Coombs [1973] and of Young and Elliott [1974], such a large temperature difference should be accompanied by a large central peak. Lagakos and Cummins showed that this temperature difference is instead related to the large difference between the free and clamped Curie temperatures, which in turn are caused by an unusually small $C_{66}^{P,T}$ elastic constant.

23.2.10. *Pressure effects.* The pressure and temperature dependences of dielectric properties and Raman spectra of RbDP were investigated by Peercy and Samara [1973]. They found that $\mathrm{d}T_c/\mathrm{d}p = -6.2$ K/kbar at low pressure, and that T_c approaches 0 K (ferroelectricity vanishes) at 15.2 kbar. The Raman investigation of the soft mode yielded results inconsistent with the Kobayashi [1968] model. In KDP, Peercy [1975] has found from Raman spectra below T_c at pressures up to 4 kbar that there are two modes which decrease in frequency with increasing pressure because pressure lowers the transition temperature. Other modes show the usual increase of frequency with pressure. He identifies these two modes as the soft mode and the optic mode, and his results show that these modes remain coupled in the ferroelectric phase as well as in the paraelectric phase.

23.3. Theory of KDP-type crystals

23.3.3. *Coupled mode theories.* The pressure and temperature dependences of the coupled proton-optic mode for KDP and RbDP obtained from Raman scattering studies were used by Peercy [1974] to obtain microscopic parameters for the Kobayashi [1968] model and their pressure derivatives. Using these parameters, quite good predictions were obtained for the temperature and pressure dependences of dielectric constant, Curie constant, spontaneous polarization, and pressure required for vanishing of the ferroelectric phase.

Young and Elliott [1974] extended the pseudo-spin model to include interactions of the pseudo-spins describing the proton positions with pairs of phonons. Their theory predicts a very narrow central peak, as observed by Lagakos and Cummins [1974]. Two other treatments of strong proton–phonon coupling, both of which predict observed isotope effects, have been put forward by Pak [1973] and by Godzik and Blumen [1974].

23.3.4. *Comparison of theory with experiment.* Sidnenko and Gladkii [1973b] measured polarization in KDP as a function of temperature and electric field, and from these data determined coefficients in the Landau phenomenological theory. They predicted a jump in spontaneous polarization at T_c of $1.8 \pm 0.4\ \mu C/cm^2$, in good accord with measured values. Strukov et al. [1973b] found that both RbDP and DRbDP exhibit second-order transitions, and found the Slater and Takagi group energy parameters. They determined tunneling to be present in RbDP but not in DRbDP. Their energy and tunneling parameters do not agree very well with parameters found recently by Fairall and Reese [1975] for RbDP, in a compilation including seven other members of the KDP family. This compilation is based on the Vaks and Zinenko [1973] small-tunneling approximation incorporated into the Silsbee et al. [1964] (SUS) theory. Another method of obtaining these parameters, by fitting transverse dielectric susceptibility curves, was developed by Havlin et al. [1974a]. Their tunneling parameters for KDP and KDA do not agree closely with those found by Fairall and Reese. Measurements by Havlin et al. [1974b] showed a transverse susceptibility peak at 132 K, ten degrees above T_c. They extended the model of Blinc [1960] and De Gennes [1963] using a cluster approximation, and found that the Slater and Takagi energies giving best fit were quite different above T_c than below T_c.

23.4. Trihydrogen selenites

23.4.2. *Sodium trihydrogen selenite.* Hydrogen ordering in the $Na(H_{1-x}D_x)_3(SeO_3)_2$ systems has been studied by Raman scattering and NMR methods. The Raman studies of Torrie and Knispel [1973] indicated hydrogen disorder in the α phase, partial disorder in the β phase, and order in the γ phase. The NMR studies of Stepišnik et al. [1973] indicated deuteron order in the β and γ phases and disorder in the α phase. Evidence for an overdamped soft mode was found.

On the theoretical side, Ishibashi and Takagi [1973] have modified the configurational entropy calculation of Makita and Miki [1970]. A cluster approximation approach by Vaks and Zein [1974] predicted features of the phase transitions, including appearance of the β phase. Holakovský [1973] proposed the NaHSe $\alpha \rightarrow \beta$ transition as an example of a triggered phase transition (TFT), in which the ferroelectric transition is triggered by another phase transition parameter.

23.6. Rochelle salt and other ferroelectric tartrates

23.6.2. *The sodium tartrates.* Sailer and Unruh [1975] studied order parameter fluctuations in Rochelle salt by means of Brillouin scattering, and found piezoelectric coupling of the soft mode to the polarization fluctuations. Takagi et al. [1975] applied the idea of irreducible susceptibilities to Rochelle salt, and were able to clarify concepts of the Mitsui [1958] theory and of the two-sublattice model.

Sato et al. [1974] from ESR results on Cr^{3+}-doped ammonium Rochelle salt confirmed the improper ferroelectric designation for this crystal, and showed Aizu's [1971, 1972b] soft mode proposals to be incorrect.

23.6.3. *The lithium tartrates.* Abe and Matsuda [1974] calculated the asymmetric well potential for O–H rotation in lithium ammonium tartrate, based on which they offered an explanation for ferroelectric behavior in this crystal.

23.7. Ferroelectric sulfates

23.7.2. *Triglycine sulfate and its isomorphs.* Deuteron NMR studies by Stepišnik and Slak [1975] in deuterated TGS confirmed that disordered hydrogen intrabond motion occurs between glycines II and III, but cast doubt on the hypothesis of critical flipping of glycine I.

Leonidova et al. [1974] analyzed thermodynamically the behavior of TGSe near the triple point at a pressure of 6 kbar, at which a first-order transition to the high-pressure phase occurs, while the ferroelectric transition is second-order. In deuterated TGSe, Peshikov [1972] on the basis of dielectric and hysteresis measurements reports a first-order ferroelectric transition at zero pressure which he states is related to the high-pressure first-order transition in TGSe.

23.7.3. *Ammonium sulfate and ammonium fluoberyllate.* A far-infrared study of ammonium sulfate (AS) was conducted by Petzelt et al. [1974]. They classify this crystal as a "pseudoproper ferroelectric", a type of improper ferroelectric with a phase transition parameter having the same symmetry as the polarization, but in which the soft-mode-related displacements contribute only a small amount to the static dielectric constant. They believe that the phase transition in AS occurs when the NH_4 oscillations slow down and couple linearly to a translational mode. Jain and Bist [1974] report that ammonium and sulfate ion distortion accompanies the phase transition, with the sulfate ions playing a major role in triggering the transition.

23.7.4. *Other ferroelectric sulfates.* Kruglik et al. [1973] have done an X-ray study of the complex phase transition which occurs in $NaNH_4SeO_4{\cdot}2H_2O$ at 180 K, and have correlated the structure with previous NMR and dielectric results. A strong effect of H-bonding on the ferroelectric transition is indicated by the negative pressure coefficient of T_c, -4.1 ± 0.2 K/kbar, reported by Gesi et al. [1973]. In MASD, by way of contrast, they found a positive coefficient of 2.5 ± 0.1 K/kbar.

Aizu [1974] has developed an explanation of the two phase transitions in ammonium bisulfate in terms of successive instabilities of vibrational modes. Deuteron NMR results of Kasahara and Tatsuzaki [1974] in $RbDSO_4$ show that the deuteron stays in an acentric H-bond position even above T_c, so the ferroelectric transition does not involve hydrogen order. The positive pressure coefficients of T_c found by Gesi and Ozawa [1975], of 12.0 ± 0.1 K/kbar for $RbHSO_4$ and 12.7 ± 0.2 K/kbar for $RbDSO_4$, corroborate this conclusion. They found a new phase to occur at 7.5 kbar, similar to behavior in NH_4HSO_4.

23.8. Other hydrogen bonded ferroelectrics

Properties of a new H-bonded ferroelectric, $PbHPO_4$ (lead monetite) have been reported by Negran et al. [1974]. The normal and deuterated crystals undergo second-order ferroelectric transitions at 37 and 179°C respectively, the largest isotope shift yet reported. Their spontaneous polarizations are 0.72 and 2.1 $\mu C/cm^2$ respectively. The crystal is monoclinic below T_c and the polar axis lies in the $a-c$ plane. The phosphate tetrahedra are linked by H-bonds to form chains parallel to the c-axis. Ordering of these H-bonds presumably causes the ferroelectric transition. Shultenite, $PbHAsO_4$, is presumed to have the same structure, and $BaHPO_4$ also may have an H-bonded chain, so these crystals may turn out to be ferroelectric also.

23.8.1. *Hydrogen halides.* From the small-signal ac dielectric response of HBr with a dc bias applied, Cichanowski and Cole [1973] found effects of both domain wall motion and intradomain dipole reorientation, but were unable to establish the dipole reorientation mechanism.

23.8.2. *Copper formate tetrahydrate.* Allen [1974] has proposed a dimer model for the antiferroelectric phase transition in $Cu(HCOO)_2 \cdot 4H_2O$. Kameyama et al. [1973] have performed an ultrasonic study of elastic properties in this crystal, and noted a large ultrasonic absorption at T_c accompanied by a discontinuous change in sound velocity. From pulse NMR studies of T_1 and $T_{1\rho}$ of protons in $Cu(CHOO)_2 \cdot 4D_2O$, Bonera et al. [1974] conclude that the relaxation is caused by paramagnetic excitons, and not by critical slowing down of water molecule motions as was proposed by Žumer and Pirš [1974]. Gesi and Ozawa [1973] found dT_c/dp of 4.36 ± 0.02 K/kbar, indicating lack of influence of H-bonding.

23.8.4. *Colemanite.* In an ^{11}B NMR investigation of critical behavior in colemanite, Theveneau and Papon [1974] found logarithmic dependence of T_1^{-1} in the critical region. Brosowski et al. [1974] studied proton T_1 in colemanite, TGS, and potassium ferrocyanide trihydrate, and in colemanite did not find the change in activation energy reported by Watton et al. [1973], who based extended conclusions upon this change.

23.8.6. *Substituted acetates.* Chihara et al. [1973] performed a nuclear quadrupole resonance (NQR) study of $NH_4H(ClCH_2COO)_2$ and $ND_4D(ClCH_2COO)_2$, but were unable to decide between a displacive transition and a transition involving hydrogen order. From an X-ray structural study of glycine silver nitrate, Guha [1972] concluded that ferroelectricity results from silver ion displacement rather than from H-bonding.

References

Abe, R. and Y. Tokumaru, 1971, J. Phys. Soc. Japan **31**, 1748.

Abraham, D. B. and E. H. Lieb, 1971, J. Chem. Phys. **54**, 1446.

Adriaenssens, G. J., 1974, Ferroelectrics **7**, 123.

Adriaenssens, G. J., J. L. Bjorkstam and J. Aikins, 1972, J. Magnetic Resonance **7**, 99.

Aizu, K., 1971, J. Phys. Soc. Japan **31**, 1521.

Aizu, K., 1972a, J. Phys. Soc. Japan **32**, 135, 1033.

Aleksandrov, K. S., I. P. Aleksandrova, L. I. Zherebtsova, M. P. Zaitseva and A. T. Anistratov, 1971, Ferroelectrics **2**, 1.

Aleksandrova, I. P., V. I. Yuzvak and V. N. Shcherbakov, 1971, Bull. Acad. Sci. USSR, Phys. Ser. **35**, 1644.

Anderson, M. R. and I. D. Brown, 1972, Acta Cryst. **A28**, 663.

Arend, H., R. Hofmann and J. Felsche, 1974, Ferroelectrics **8**, 413.
Bacon, G. E. and R. S. Pease, 1955, Proc. Roy. Soc. **A230**, 359.
Barkla, H. M. and D. M. Finlayson, 1953, Phil. Mag. **44**, 109.
Bjorkstam, J. L., 1967, Phys. Rev. **153**, 599.
Bjorkstam, J. L., 1973, NMR Studies of Collective Atomic Motion near Ferroelectric Phase Transitions, in: Advances in Magnetic Resonance (Academic Press, New York).
Bjorkstam, J. L. and R. E. Oettel, 1966, Proc. International Meeting on Ferroelectricity, Prague, 1966, vol. 2, p. 91.
Bjorkstam, J. L. and E. A. Uehling, 1959, Phys. Rev. **114**, 961.
Blinc, R., 1960, J. Phys. Chem. Solids **13**, 204.
Blinc, R., 1971, NMR and NQR Studies of Critical Effects in Structural Phase Transitions, in: Structural Phase Transitions and Soft Modes. Eds. Samuelsen, E. J., E. Andersen and J. Feder (Universitetsforlaget, Oslo).
Blinc, R. and B. Žekš, 1972, Adv. Phys. **21**, 693.
Blinc, R. and J. L. Bjorkstam, 1969, Phys. Rev. Letters **23**, 788.
Blinc, R. and M. Mali, 1969, Phys. Rev. **179**, 552.
Blinc, R. and J. Pirš, 1971, J. Chem. Phys. **54**, 1535.
Blinc, R. and S. Svetina, 1966, Phys. Rev. **147**, 423 and 430.
Blinc, R. and S. Žumer, 1968, Phys. Rev. Letters **21**, 1004.
Blinc, R., P. Cevc and M. Schara, 1967, Phys. Rev. **159**, 411.
Blinc, R., J. Stepišnik and I. Zupančič, 1968, Phys. Rev. **176**, 732.
Blinc, R., S. Žumer and G. Lahajnar, 1970a, Phys. Rev. **B1**, 4456.
Blinc, R., D. E. O'Reilly and E. M. Peterson, 1970b, Phys. Rev. **B1**, 1953.
Blinc, R., M. Jamsek–Vilfan and G. Lahajnar, 1970c, J. Chem. Phys. **52**, 6407.
Blinc, R., J. Stepišnik, M. Jamsek–Vilfan and S. Žumer, 1971a, J. Chem. Phys. **54**, 187.
Blinc, R., M. Mali, R. Osredkar, A. Prelesnik, I. Zupančič and L. Ehrenberg, 1971b, J. Chem. Phys. **55**, 4843.
Blinc, R., J. Slak and J. Stepišnik, 1971c, J. Chem. Phys. **55**, 4848.
Blinc, R., S. Svetina and B. Žekš, 1972a, Solid State Commun. **10**, 387.
Blinc, R., M. Mali, J. Slak, J. Stepišnik and S. Žumer, 1972b, J. Chem. Phys. **56**, 3566.
Blinc, R., M. Burgar and A. Levstik, 1973a, Solid State Commun. **12**, 573.
Blinc, R., J. Pirš and S. Žumer, 1973b, Phys. Rev. **B8**, 15.
Boesch, H., U. Schmocker, F. Waldner, K. Emerson and J. E. Drumheller, 1971, Phys. Letters **36A**, 461.
Bornarel, J. and J. Lajzerowicz, 1972, Ferroelectrics **4**, 177.
Broberg, T. W., C. Y. She, L. S. Wall and D. F. Edwards, 1972, Phys. Rev. **B6**, 3332.
Burfoot, J. C., 1967, Ferroelectrics (Van Nostrand, Princeton).
Cochran, W., 1959, Phys. Rev. Letters **3**, 412.
Cochran, W., 1961, Adv. Phys. **10**, 401.
Cochran, W., 1969, Adv. Phys. **18**, 157.
Cowley, R. A., G. J. Coombs, R. S. Katiyar, J. F. Ryan and J. F. Scott, 1971, J. Phys. **C4**, L203.
Cross, L. E., A. Fousková and S. E. Cummins, 1968, Phys. Rev. Letters **21**, 812.
Dalal, N. S. and C. A. McDowell, 1972, Phys. Rev. **B5**, 1074.
De Benyacar, M. A. R. and M. J. De Abeledo, 1974, Am. Mineral. **59**, 763.
De Benyacar, M. A. R., M. J. De Abeledo and H. L. De Dussel, 1973, Third International Meeting on Ferroelectricity, Abstracts (unpublished).

De Gennes, P. G., 1963, Solid State Commun. **1**, 132.

Devonshire, A. F., 1964, Some Recent Work on Ferroelectrics, in: Reports on Progress in Physics, vol. 27, p. 1. Ed., Strickland, A. C., (Institute of Physics and Physical Society, London).

Dijon, 1971, Proceedings of the Second European Meeting on Ferroelectricity, 1972, J. de Physique **33**, C2.

Dvořák, V., 1972, Phys. Stat. Solidi **b51**, K129.

Dvořák, V., A. Fousková and P. Glogar, Eds., 1966, Proceedings of the International Meeting on Ferroelectricity, Prague, 1966, vols. 1 and 2 (Institute of Physics of the Czechoslovak Academy of Science, Prague).

Edinburgh, 1973, Proceedings of the Third International Meeting on Ferroelectricity, 1974, Ferroelectrics **7** and **8**.

Elliott, R. J. and A. P. Young, 1974, Ferroelectrics **7**, 23.

El Saffar, Z. M., D. E. O'Reilly, E. M. Peterson and C. Flick, 1972, J. Chem. Phys. **57**, 2372.

Fairall, C. W. and W. Reese, 1972, Phys. Rev. **B6**, 193.

Fatuzzo, E. and W. J. Merz, 1967, Ferroelectricity (North-Holland, Amsterdam).

Fitzgerald, M. E. and P. A. Casabella, 1973, Phys. Rev. **B7**, 2193.

Fritz, I. J., 1973, Ferroelectrics **5**, 17.

Garland, C. W. and D. B. Novotny, 1969, Phys. Rev. **177**, 971.

Gavrilova-Podol'skaya, G. V., M. L. Afanas'ev, A. L. Yudin and A. G. Lundin, 1967, Bull. Acad. Sci. USSR, Phys. Ser. **31**, 1126.

Genin, D. J. and D. E. O'Reilly, 1969, J. Chem. Phys. **50**, 2842.

Genin, D. J., D. E. O'Reilly and T. Tsang, 1968, Phys. Rev. **167**, 445.

Gesi, K., 1969, J. Phys. Soc. Japan **26**, 107.

Gorbatyi, L. V., V. I. Ponomarev and D. M. Kheiker, 1972, Soviet Phys.-Cryst. **16**, 781.

Grindlay, J., 1970, An Introduction to the Phenomenological Theory of Ferroelectricity (Pergamon Press, New York).

Habuda, S. P., A. G. Lundin and É. P. Zeer, 1970, Ferroelectrics **1**, 71.

Hamilton, W. C., 1969, J. Chem. Phys. **50**, 2275.

Hill, R. M. and S. K. Ichiki, 1963, Phys. Rev. **130**, 150.

Ishibashi, Y., S. Ohya and Y. Takagi, 1973, J. Phys. Soc. Japan **34**, 888.

Jona, F. and G. Shirane, 1962, Ferroelectric Crystals (Macmillan, New York).

Kadaba, P. K. and D. E. O'Reilly, 1970, J. Chem. Phys. **52**, 2403.

Kaminow, I. P., 1965, Phys. Rev. **138**, A1539.

Kaminow, I. P. and T. C. Damen, 1968, Phys. Rev. Letters **20**, 1105.

Känzig, W., 1957, Ferroelectrics and Antiferroelectrics, in: Solid State Physics, vol. 4, Eds. Seitz, F. and D. Turnbull (Academic Press, New York).

Kaplan, S. F., M. I. Kay and B. Morosin, 1970, Ferroelectrics **1**, 31.

Katiyar, R. S., J. F. Ryan and J. F. Scott, 1971, Phys. Rev. **B4**, 2635.

Kay, M. I. and R. Kleinberg, 1972, Ferroelectrics **4**, 147.

Kay, M. I. and R. Kleinberg, 1973, Ferroelectrics **5**, 45.

Kind, R., 1971, Phys. Kond. Materie **13**, 217.

Knispel, R. R., 1969, Ph.D. Thesis, Montana State University (unpublished).

Kobayashi, K. K., 1968, J. Phys. Soc. Japan **24**, 497.

Kobayashi, J., Y. Enemoto and Y. Sato, 1972, Phys. Stat. Solidi **b50**, 335.

Konsin, P. I. and N. N. Kristofel', 1973, Soviet Phys.-Solid State **14**, 2484.

Krasnikova, A. Ya. and I. N. Polandov, 1970, Soviet Phys.-Solid State **11**, 1421.
Krasnikova, A. Ya., É. P. Zeer and V. A. Koptsik, 1972, Soviet Phys.-Cryst. **17**, 287.
Kuroda, N., Y. Tabata and A. Kawamori, 1971, J. Phys. Soc. Japan **31**, 609.
Kydon, D. W., H. E. Petch and M. Pintar, 1969, J. Chem. Phys. **51**, 487.
Kyoto, 1969, Proceedings of the Second International Meeting on Ferroelectricity, 1970, J. Phys. Soc. Japan **48** Suppl.
Lehmann, M. S. and F. K. Larsen, 1971, Acta Chem. Scand. **25**, 3859.
Litov, E. and C. W. Garland, 1970, Phys. Rev. **B2**, 4597.
Litov, E. and E. A. Uehling, 1968, Phys. Rev. Letters **21**, 809.
Ludupov, Ts.-Zh., 1966, Soviet Phys.-Cryst. **11**, 416.
Makita, Y., 1965, J. Phys. Soc. Japan **20**, 1567.
Makita, Y. and H. Miki, 1970, J. Phys. Soc. Japan **28**, 1221.
Makita, Y. and M. Sumita, 1971, J. Phys. Soc. Japan **31**, 792.
Megaw, H. D., 1957, Ferroelectricity in Crystals (Methuen, London).
Mercado, A. and J. A. Gonzalo, 1973, Phys. Rev. **B7**, 3074.
Mitsui, T., 1958, Phys. Rev. **111**, 1259.
Montano, P. A., H. Shechter and A. Biran, 1971, Solid State Commun. **9**, 2029.
Morosin, B. and G. A. Samara, 1971, Ferroelectrics **3**, 49.
Müser, H. E. and J. Petersson, Eds., 1970, Proceedings of the European Meeting on Ferroelectricity, Saarbrücken, 1969 (Wissenschastliche Verlag., Stuttgart).
Nagle, J. F., 1969, Phys. Rev. **186**, 594.
Nagle, J. F. and G. R. Allen, 1971, J. Chem. Phys. **55**, 2708.
Nakano, J., Y. Shiozaki, and E. Nakamura, 1974, Ferroelectrics **8**, 483.
Nelmes, R. J., 1972, Ferroelectrics **4**, 133.
Nelmes, R. J., V. R. Eiriksson and K. D. Rouse, 1972, Solid State Commun. **11**, 1261.
Nettleton, R. E., 1970, Ferroelectrics **1**, 3, 87 and 207, and **2**, 5 and 77.
Niimura, N., K. Shimaoka, H. Motegi and S. Hoshino, 1972, J. Phys. Soc. Japan **32**, 1019.
Oja, T., 1969, Phys. Letters **30A**, 343.
Okada, K., 1967, Phys. Rev. **164**, 683.
Okada, K., H. Sugié and K. Kan'no, 1973, Phys. Letters **44A**, 59.
O'Reilly, D. E., 1970, J. Chem. Phys. **52**, 2396.
O'Reilly, D. E. and T. Tsang, 1967, J. Chem. Phys. **46**, 1291, 1301.
O'Reilly, D. E., E. M. Peterson and Z. M. El Saffar, 1971, J. Chem. Phys. **54**, 1304.
Padmanabhan, V. M. and R. Balasubramanian, 1967, Acta Cryst. **22**, 532.
Paul, G. L., W. Cochran, W. L. J. Buyers and R. A. Cowley, 1970, Phys. Rev. **B2**, 4603.
Peercy, P. S., 1973, Phys. Rev. Letters **31**, 379.
Pepinsky, R. and K. Vedam, 1960, Phys. Rev. **117**, 1502.
Pepinsky, R., K. Vedam, Y. Okaya and S. Hoshino, 1958, Phys. Rev. **111**, 1467.
Peterson, E. R. and R. D. Willett, 1972, J. Chem. Phys. **56**, 1879.
Reese, W., 1969, Phys. Rev. **181**, 905.
Reese, W. and L. F. May, 1968, Phys. Rev. **167**, 504.
Reese, R. L., I. J. Fritz and H. Z. Cummins, 1973, Phys. Rev. **B7**, 4165.
Reid, C., 1959, J. Chem. Phys. **30**, 182.
Rocaries, C. and P. Boldrini, 1972, Appl. Phys. Letters **20**, 49.
Roos, J. and R. Kind, 1974, Ferroelectrics **8**, 553.
Samara, G. A., 1968, Phys. Rev. **173**, 605.

Samara, G. A., 1971, Phys. Rev. Letters **27**, 103.
Samara, G. A., 1973, Ferroelectrics **5**, 25.
Sandy, F. and R. V. Jones, 1968, Phys. Rev. **168**, 481.
Sato, S., 1972, J. Phys. Soc. Japan **32**, 1670.
Sato, Y., 1970, J. Chem. Phys. **53**, 887; Phys. Letters **33A**, 156.
Sawada, A. and Y. Takagi, 1972, J. Phys. Soc. Japan **33**, 1071.
Sawada, A., Y. Ishibashi and Y. Takagi, 1971, J. Phys. Soc. Japan **31**, 823.
Schmidt, V. H., 1966, Proc. International Meeting on Ferroelectricity, Prague, 1966, vol. 2, p. 97.
Schmidt, V. H., 1967, Phys. Rev. **164**, 749; **173**, 630 (1968). Because of an error of a factor of 2 in the theory on p. 754, the pre-exponential times T_0 on that page and in fig. 5 should be half as large as given.
Schmidt, V. H., 1971, J. Math. Phys. **12**, 992.
Schmidt, V. H., 1974, Ferroelectrics **7**, 199.
Schmidt, V. H. and R. S. Parker, 1972, J. de Physique **33**, C2–109.
Schmidt, V. H. and E. A. Uehling, 1962, Phys. Rev. **126**, 447.
Schmidt, V. H., J. E. Drumheller and F. L. Howell, 1971, Phys. Rev. **B4**, 4582.
Scott, J. F. and J. M. Worlock, 1973, Solid State Commun. **12**, 67.
Senko, M. E., 1961, Phys. Rev. **121**, 1599.
She, C. Y., T. W. Broberg, L. S. Wall and D. F. Edwards, 1972, Phys. Rev. **B6**, 1847.
Shirokov, A. M., L. A. Shuvalov and N. R. Ivanov, 1969, Phys. Letters **29A**, 599.
Shirokov, A. M., A. I. Baranov and L. A. Shuvalov, 1971, Bull. Acad. Sci. USSR, Phys. Ser. **35**, 1727.
Shuvalov, L. A., N. V. Gordeyeva, N. R. Ivanov, L. F. Kirpichnikova, A. M. Shirokov and N. M. Schagina, 1970a, J. Phys. Soc. Japan **28**, Suppl., 75.
Shuvalov, L. A., N. R. Ivanov, L. F. Kirpichnikova and N. V. Gordeyeva, 1970b, Phys. Letters **33A**, 490.
Shuvalov, L. A., V. G. Vaks, N. R. Ivanov and N. E. Zein, 1974, Ferroelectrics **8**, 409.
Silsbee, H. B., E. A. Uehling and V. H. Schmidt, 1964, Phys. Rev. **133**, A165.
Silvidi, A. A., A. J. Falzone and J. L. Roupe, 1969, Solid State Commun. **7**, 359.
Simon, A., 1971, J. Appl. Cryst. **3**, 138.
Skalyo, Jr., J., B. C. Frazer and G. Shirane, 1970, Phys. Rev. **B1**, 278.
Slater, J. C., 1941, J. Chem. Phys. **9**, 16.
Soda, G. and T. Chiba, 1969a, J. Phys. Soc. Japan **26**, 723.
Soda, G. and T. Chiba, 1969b, J. Phys. Soc. Japan **26**, 249.
Strukov, B. A., M. A. Korzhuev and V. A. Koptsik, 1971, Bull. Acad. Sci. USSR, Phys. Ser. **35**, 1678.
Strukov, B. A., T. L. Skomorokhova, V. A. Koptsik, A. A. Boika and A. N. Izrailenko, 1973a, Soviet Phys.-Cryst. **18**, 86.
Suzuki, M. and R. Kubo, 1968, J. Phys. Soc. Japan **24**, 51.
Takagi, Y., 1948, J. Phys. Soc. Japan **3**, 273.
Takagi, Y. and Y. Ishibashi, 1972, J. Phys. Soc. Japan **33**, 1381.
Tellgren, R. and R. Liminga, 1972, J. Solid State Chem. **4**, 255.
Tellgren, R., D. Ahmad and R. Liminga, 1973, J. Solid State Chem. **6**, 250.
Tokunaga, M., 1966, Progr. Theor. Phys. **36**, 857.
Tokunaga, M. and T. Matsubara, 1966, Progr. Theor. Phys. **35**, 581.

Torrie, B. H., C. C. Lin, O. S. Binbrek and A. Anderson, 1972, J. Phys. Chem. Solids **33**, 697.
Tovbis, A. B., T. S. Davydova and V. I. Simonov, 1972, Soviet Phys.-Cryst. **17**, 81.
Toyoda, K., 1970, Ferroelectrics **1**, 43 and subsequent issues.
Trontelj, Z. and M. Rebič, 1972, Solid State Commun. **11**, 1337.
Tsallis, C., 1972, J. de Physique **33**, 1121.
Tsang, T. and E. L. Shaw, 1971, J. Chem. Phys. **55**, 2337.
Vanderkooy, J., J. D. Cuthbert and H. E. Petch, 1964, Can. J. Phys. **42**, 1871.
Watton, A., H. E. Petch and M. M. Pintar, 1973, Can. J. Phys. **51**, 1005.
Weller, E. F., 1967, Ed., Ferroelectricity (Elsevier, New York).
Yagi, T. and I. Tatsuzaki, 1969, J. Phys. Soc. Japan **26**, 865.
Yamamoto, T., N. Nakamura and H. Chihara, 1970, J. Phys. Soc. Japan **28**, Suppl., 112.
Yi, P.-N. and A. T. Gavrielides, 1971, J. Chem. Phys. **54**, 3777.
Yoshimitsu, K. and T. Matsubara, 1968, Progr. Theor. Phys. Suppl., Extra Number, 109.
Žekš, B., G. C. Shukla and R. Blinc, 1971, Phys. Rev. **B3**, 2306.
Zheludev, I. S., V. V. Gladkii, E. V. Sidnenko and V. K. Magataev, 1974, Ferroelectrics **8**, 567.

Added in proof:
Abe, R. and M. Matsuda, 1974, J. Phys. Soc. Japan **37**, 437.
Aizu, K., 1972b, J. Phys. Soc. Japan **33**, 629.
Aizu, K., 1974, J. Phys. Soc. Japan **36**, 937.
Allen, G. R., 1974, J. Chem. Phys. **60**, 3299.
Blinc, R., M. Mali, J. Pirš and S. Žumer, 1973c, J. Chem. Phys. **58**, 2262.
Blinc, R., M. Mali, R. Osredkar, R. Parker, J. Seliger and S. Žumer, 1973d, J. Chem. Phys. **59**, 2947.
Blinc, R., B. Lavrenčič, I. Levstek, V. Smolej and B. Žekš, 1973e, Phys. Stat. Solidi **b60**, 255.
Bonera, G., F. Borsa and A. Rigamonti, 1972, Riv. Nuovo Cimento **2**, 325.
Bonera, G., C. I. Massara, E. R. Mognaschi and A. Rigamonti, 1974, Solid State Commun. **15**, 19.
Brody, E. M. and H. Z. Cummins, 1974, Phys. Rev. **B9**, 179.
Brosowski, G., W. Buchheit, D. Müller and J. Petersson, 1974, Phys. Stat. Solidi **b62**, 93.
Chihara, H., A. Inaba, N. Nakamura and T. Yamamoto, 1973, J. Phys. Soc. Japan **35**, 1480.
Cichanowski, S. W. and R. W. Cole, 1973, J. Chem. Phys. **59**, 2420.
Cowley, R. A. and G. J. Coombs, 1973, J. Phys. **C6**, 121 and 143.
Dietrich, O. W., R. A. Cowley and S. M. Shapiro, 1974, J. Phys. **C7**, L239.
Eberhard, J. W. and P. M. Horn, 1975, Bull. Am. Phys. Soc. **20**, 285.
Fairall, C. W. and W. Reese, 1974, Phys. Rev. **B10**, 882.
Fairall, C. W. and W. Reese, 1975, Phys. Rev. **B11**, 2066.
Gaillard, J., P. Bloux and B. Lamotte, 1974, Solid State Commun. **14**, 417.
Gesi, K. and K. Ozawa, 1973, J. Phys. Soc. Japan **35**, 943.
Gesi, K. and K. Ozawa, 1974, J. Phys. Soc. Japan **38**, 459.
Gesi, K., K. Ozawa, Y. Makita and T. Osaka, 1973, J. Phys. Soc. Japan **35**, 1562.
Godzik, K. D. and A. Blumen, 1974, Phys. Stat. Solidi **b66**, 569.
Guha, S., 1972, Indian J. Phys. **46**, 255.
Havlin, S., E. Litov and E. A. Uehling, 1974a, Phys. Rev. **B9**, 1024.
Havlin, S., E. Litov and H. Sompolinsky, 1974b, Phys. Letters **49A**, 33.
Holakovský, J., 1973, Phys. Stat. Solidi **b56**, 615.

Ishibashi, Y. and Y. Takagi, 1973, J. Phys. Soc. Japan **35**, 814.

Ishibashi, Y., S. Ohya and Y. Takagi, 1974, J. Phys. Soc. Japan **37**, 1035.

Jain, Y. S. and H. D. Bist, 1974, Solid State Commun. **15**, 1229.

Kameyama, H., Y. Ishibashi and Y. Takagi, 1973, J. Phys. Soc. Japan **35**, 1450.

Kasahara, M. and I. Tatsuzaki, 1974, J. Phys. Soc. Japan **37**, 167.

Kawano, T., 1974, J. Phys. Soc. Japan **37**, 848.

Kobayashi, J., Y. Uesu and Y. Enomoto, 1971, Phys. Stat. Solidi **b45**, 293.

Kruglik, A. I., V. I. Simonov and V. I. Yuzvak, 1973, Soviet Phys.-Cryst. **18**, 177.

Lagakos, N. and H. Z. Cummins, 1974, Phys. Rev. **B10**, 1063.

Lagakos, N. and H. Z. Cummins, 1975, Phys. Rev. Letters **34**, 883.

Leonidova, G. G., V. P. Mylov and V. M. Varikash, 1974, Soviet Phys.-Solid State **16**, 2138.

Lowndes, R. P., N. E. Tornberg and R. C. Leung, 1974, Phys. Rev. **B10**, 911.

Mavrin, B. N., Kh. E. Sterin, A. V. Mishchenko and L. N. Rashkovich, 1974, Soviet Phys.-Solid State **15**, 2468.

Mitsui, T., I. Tatsuzaki and E. Nakamura, 1973a, An Introduction to the Physics of Ferroelectrics (Gordon and Breach, New York).

Mitsui, T., E. Nakamura and M. Tokunaga, 1973b, Ferroelectrics **5**, 185.

Nagamiya, T., 1952, Prog. Theor. Phys. Japan **7**, 275.

Negran, T. J., A. M. Glass, C. S. Brickenkamp, R. D. Rosenstein, R. K. Osterheld and R. Susott, 1974, Ferroelectrics **6**, 179.

Pak, K. N., 1973, Phys. Stat. Solidi **b60**, 233.

Peercy, P. S., 1974, Phys. Rev. **B9**, 4868.

Peercy, P. S., 1975, Solid State Commun. **16**, 439.

Peercy, P. S. and G. A. Samara, 1973, Phys. Rev. **B8**, 2033.

Peshikov, E. V., 1972, Soviet Phys.-Solid State **14**, 1377.

Petzelt, J., J. Grigas and I. Mayerová, 1974, Ferroelectrics **6**, 225.

Sailor, E. and H. G. Unruh, 1975, Solid State Commun. **16**, 615.

Sato, K., A. Sawada, Y. Takagi and Y. Ishibashi, 1974, J. Phys. Soc. Japan **36**, 616.

Schmidt, V. H., 1974, Bull. Am. Phys. Soc. **19**, 649.

Sidnenko, E. V. and V. V. Gladkii, 1973a, Soviet Phys.-Cryst. **17**, 861.

Sidnenko, E. V. and V. V. Gladkii, 1973b, Soviet Phys.-Cryst. **18**, 83.

Smolenski, G. A. and N. N. Krainik, 1972, Ferroelektrika und Antiferroelektrika (Teubner, Leipzig).

Stamenković, S., 1972, J. Low-Temp. Phys. **9**, 475 and 485.

Stepišnik, J. and J. Slak, 1975, J. Chem. Phys. **62**, 34.

Stepišnik, J., J. Slak, R. Blinc, L. A. Shuvalov, N. R. Ivanov and N. M. Schagina, 1973, Solid State Commun. **13**, 1053.

Strukov, B. A., A. Baddur, V. N. Zinenko, A. V. Mishchenko and V. A. Koptsik, 1973b, Soviet Phys.-Solid State **15**, 939.

Strukov, B. A., A. Baddur, V. I. Zinenko, V. K. Mikhailov and V. A. Koptsik, 1974, Soviet Phys.-Solid State **16**, 1347.

Subbarao, E. C., 1973, Ferroelectrics **5**, 267.

Takagi, Y., Y. Ishibashi and A. Sawada, 1975, J. Phys. Soc. Japan **38**, 497.

Theveneau, H. and P. Papon, 1974, Phys. Letters **49A**, 218.

Torrie, B. H. and R. R. Knispel, 1973, Ferroelectrics **5**, 53.
Vaks, V. G. and N. E. Zein, 1974, Ferroelectrics **6**, 251.
Vaks, V. G. and V. I. Zinenko, 1973, Soviet Phys.-JETP **37**, 330.
Young, A. P. and R. J. Elliott, 1974, J. Phys. **C7**, 2721.
Zheludev, I. S., 1971, Ferroelectricity and Symmetry, in: Solid State Physics, vol. 26, Eds. Seitz, F. and D. Turnbull (Academic Press, New York).
Žumer, S. and J. Pirš, 1974, Ferroelectrics **7**, 119.

CHAPTER 24

COHERENT NEUTRON SCATTERING FOR OBSERVATIONS ON COLLECTIVE PROTON MOTIONS IN SYSTEMS OF HYDROGEN BONDS

H. STILLER

Institut für Festkörperforschung der Kernforschungsanlage, D-517 Jülich, West Germany

Contents

24.1. Introduction 1171
24.2. Coherent neutron scattering from solids 1172
24.3. The problem of collective motions of hydrogen bonded protons 1178
24.4. Experiments and comparison to theory 1186
Acknowledgements 1195
References 1195

The hydrogen bond – recent developments in theory and experiments
Eds. P. Schuster et al. © *North-Holland Publ. Co., Amsterdam, 1976*

24.1. Introduction

The collective behavior of protons (or deuterons) in systems of H-bonds is one of the most fascinating aspects of the H-bonding problem. Collective motions of protons appear to play an essential role in the mechanisms of phase transformations in H-bonded ferroelectrics as well as in the mechanisms of conformation changes in certain macromolecules, perhaps, for instance, also in DNA replication.

Unfortunately, a direct experimental observation of the collective nature of proton motions is difficult, for it obviously requires a microscopic determination of pair correlations in space and time. Most microscopic probes presently available determine either the instantaneous correlations in space, as with X-rays, or a space-averaged behavior of the atoms in time, as with optical spectroscopy, or motions of individual atoms, as for instance with NMR, but not their phase relations to the motions of other atoms. The one exception to these alternatives is coherent neutron scattering. The reason lies in the fact that for neutrons the wave length λ is related to the energy E by

$$\lambda = h/mv = h/\sqrt{2mE}, \tag{24.1}$$

if m is the neutron mass, and v its velocity. Consequently, neutrons of a wave length comparable to the distances between neighboring atoms, say 0.5–10 Å, have energies which are comparable to the kinetic energies of atoms in condensed matter, i.e. 10^{-1}–10^{-4} eV (10^3 to 1 cm^{-1}). Thus one can measure simultaneously the interferences of waves scattered from neighboring atoms, as with X-rays, and energy exchanges between the waves and the atoms, as with Raman scattering. It has been shown by Van Hove [1954] that from such a measurement one obtains a function $G(\boldsymbol{r}-\boldsymbol{r}_0, t-t_0)$ which classically represents the probability of finding an atom at $\boldsymbol{r}$ at time t, if at an earlier time t_0 an atom had been at $\boldsymbol{r}_0$;

$$G(\boldsymbol{r}-\boldsymbol{r}_0, t-t_0) = \frac{1}{N}\left\langle \sum_{j,l=1}^{N} \int \mathrm{d}\boldsymbol{r}'\, \delta(\boldsymbol{r}-\boldsymbol{r}_0+\boldsymbol{r}_j(t_0)-\boldsymbol{r}')\, \delta(\boldsymbol{r}'-\boldsymbol{r}_l(t))\right\rangle.$$

That is to say, one obtains information on the average behavior of the atoms relative to each other in space and time. As for all measurements, the average is a thermal assembly average and, eventually, a quantum-mechanical average. In the next section this general statement shall be illustrated by recalling a few formulae which reveal the information content of neutron radiation scattered from atomic nuclei in a crystal. These formulae will be needed to understand the results of experiments described in section 24.4.

Unfortunately, for studies on the collective motions of protons in H-bonded systems, coherent neutron scattering has been used in the past in only a very few cases, for reasons which will be outlined at the end of section 24.2. However, it may be hoped that with the advent of high flux research reactors and with the development of new experimental techniques more cases will become accessible. In section 24.3 we will outline the problem of collective proton motions for the case of KH_2PO_4 which is especially favorable; we will also outline a theoretical model which has been suggested as a phenomenological solution to the problem. In section 24.4 neutron scattering experiments will be described which partly confirm and partly contradict the model.

24.2. Coherent neutron scattering from solids

Equation (24.1) holds for all kinds of particles, of course. In practice, in scattering experiments with wave lengths between 0.5 and 10 Å and energies of the order of $k_B T$, only neutrons can be used, because other particles of such wave lengths and energies cannot be obtained in sufficient numbers and will not penetrate sufficiently deep into matter.

In a scattering experiment neutrons of wave vector $\boldsymbol{k}_0$ impinge upon a sample, and we then measure scattered neutrons of wave vectors $\boldsymbol{k}$. By definition $|\boldsymbol{k}_0| = 2\pi/\lambda_0$, $|\boldsymbol{k}| = 2\pi/\lambda$, so that with eq. (24.1)

$$|\boldsymbol{k}_0| = \sqrt{2mE_0}/\hbar, \qquad |\boldsymbol{k}| = \sqrt{2mE}/\hbar.$$

We determine wave vector changes

$$\boldsymbol{\kappa} = \boldsymbol{k}_0 - \boldsymbol{k}, \qquad \kappa^2 = k_0^2 + k^2 - 2k_0 k \cos\theta \tag{24.2}$$

(where θ is the scattering angle) and energy changes

$$\pm\hbar\omega = E_0 - E = (\hbar^2/2m)(k_0^2 - k^2). \tag{24.3}$$

Figure 24.1 shows such an experiment schematically. With a mono-

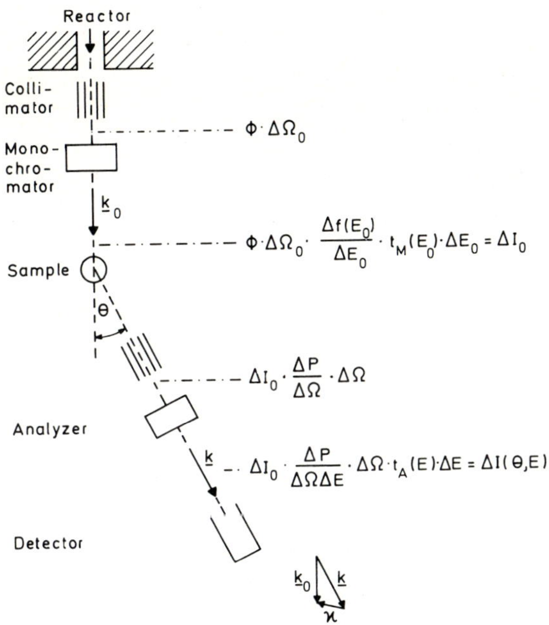

Fig. 24.1. Principle of a neutron spectroscopic measurement. The formulae are explained in the text. $\Delta P/\Delta\Omega\Delta E = Nd\ \mathrm{d}^2\sigma/\mathrm{d}\Omega\,\mathrm{d}E$ is the probability for scattering into solid angle $\mathrm{d}\Omega$ and energy interval $\mathrm{d}E$.

chromator neutrons of a desired $\boldsymbol{k}_0$ are selected from a reactor beam and are then scattered by the sample. We measure the flux of scattered neutrons simultaneously (a) as a function of the scattering angle and (b), with an analyzer, as a function of the energies E of the scattered neutrons, $I(\theta, E)$. From this function distributions $I(\boldsymbol{\kappa}, E)$ can be obtained with eqs. (24.2) and (24.3). As indicated in the figure these distributions are proportional to the cross-section $\mathrm{d}^2\sigma/\mathrm{d}\Omega\,\mathrm{d}E$ for scattering into solid angle $\mathrm{d}\Omega$ and energy interval $\mathrm{d}E$, which is the quantity from which one obtains the information on the space–time behavior of the atoms in the scattering system. In first Born approximation the cross-section may be written (Marshall and Lovesey [1971])

$$\frac{\mathrm{d}^2\sigma}{\mathrm{d}\Omega\,\mathrm{d}E} = \frac{1}{\hbar}\frac{\mathrm{d}^2\sigma}{\mathrm{d}\Omega\,\mathrm{d}\omega} = \frac{1}{\hbar}\frac{k}{k_0}S(\boldsymbol{\kappa}, \omega), \tag{24.4}$$

where $S(\boldsymbol{\kappa}, \omega)$ is called the scattering function. Its Fourier transform is the function $G(\boldsymbol{r} - \boldsymbol{r}_0, t - t_0)$ mentioned in the introduction.

Monochromator and analyzer both serve the same purpose, i.e. to select neutrons of a certain wave length or velocity out of a polychromatic beam. This can be done either by using the particle nature of the neutron through a velocity selection, or by using the wave nature of the neutron through a wave length selection. The two possibilities are illustrated in figs. 24.2 and 24.3. For more details on experimental techniques the reader is referred to the literature (Brockhouse et al. [1964]).

If the system of scattering atoms is a crystal, with the atoms performing harmonic oscillations around their equilibrium sites, the coherent scattering function $S(\boldsymbol{\kappa}, \omega)$ is of the general form (Marshall and Lovesey [1971])

$$S(\boldsymbol{\kappa}, \omega) = |F(\boldsymbol{\kappa})|^2 \delta(\omega) \sum_{\boldsymbol{\tau}} \delta(\boldsymbol{\kappa} - 2\pi\boldsymbol{\tau}) + \sum_{s,\boldsymbol{q}} |F_{s,\boldsymbol{q}}(\boldsymbol{\kappa})|^2 \frac{\langle n(\omega) + \frac{1}{2} \pm \frac{1}{2}\rangle}{\omega} \delta(\omega \mp \omega_{s,\boldsymbol{q}}) \sum_{\boldsymbol{\tau}} \delta(\boldsymbol{\kappa} + \boldsymbol{q} - 2\pi\boldsymbol{\tau}). \tag{24.5}$$

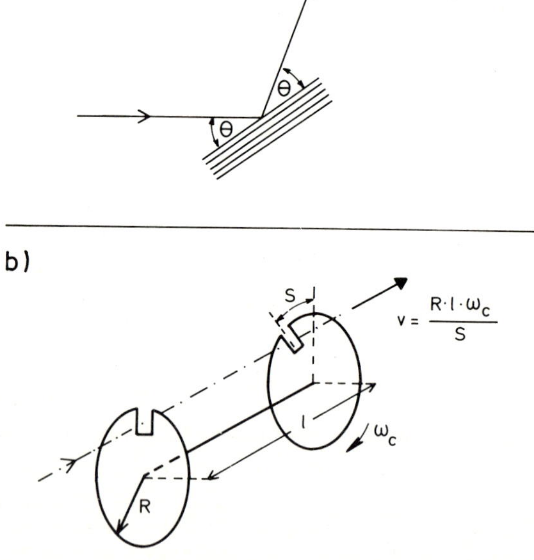

Fig. 24.2. (a) Selection of neutrons with wave length λ out of a polychromatic beam by Bragg reflection from a single crystal. d is the spacing of the reflecting crystal planes. (b) Selection of neutrons of velocity v out of a polychromatic beam with the help of two rotating wheels.

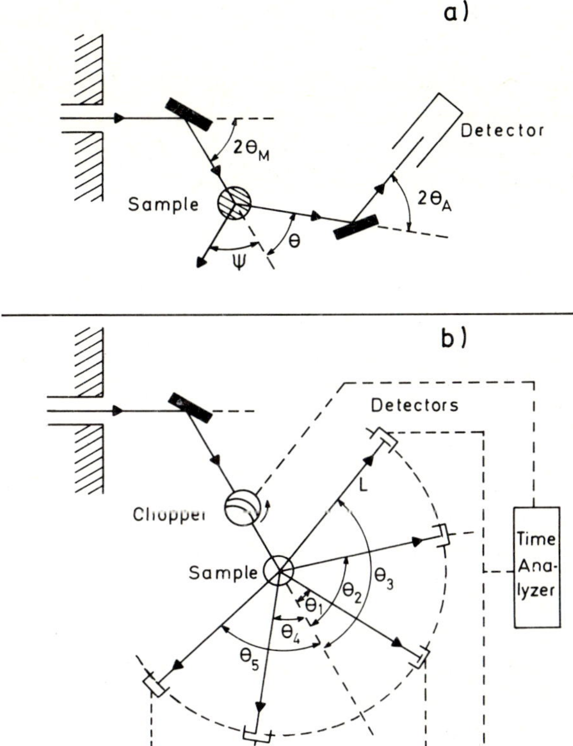

Fig. 24.3. Principles of two commonly used instruments: (a) A double-crystal spectrometer. The angles θ_M, θ and θ_A as well as the orientation ψ of a single crystal sample can be varied simultaneously and automatically. (b) A time-of-flight spectrometer. The distribution of the velocities v of scattered neutrons is determined by measuring the times $\tau = L/v$ which the neutrons need to travel through a distance L between the sample and the detectors. For this purpose the incident beam is pulsed with a chopper. Distances L between 1 and 10 m are used.

The first term represents Bragg scattering and is used for structure determinations. $2\pi\boldsymbol{\tau}$ is a reciprocal lattice vector. $\boldsymbol{\kappa} = 2\pi\boldsymbol{\tau}$ is the well-known Laue interference condition. $F(\boldsymbol{\kappa})$ is the structure factor

$$F(\boldsymbol{\kappa}) = \frac{1}{V^{1/2}} \sum_j b_j \exp(\mathrm{i}\,\boldsymbol{\kappa} \cdot \boldsymbol{R}_j) \exp[-W_j(\boldsymbol{\kappa})],$$

with b_j the coherent scattering amplitude, $\boldsymbol{R}_j$ the equilibrium position, and $W_j(\boldsymbol{\kappa})$ the Debye–Waller exponent of atom j. The summation is over the atoms in the unit cell of volume V. The Debye–Waller factor

$\exp[-W_j(\boldsymbol{\kappa})]$ represents a diffraction from the space filled by the mean amplitude of the thermal motion of the nucleus of atom j.

The second term of eq. (24.5) is the one which interests us here. It represents inelastic interference scattering due to single excitation or deexcitation of collective atomic motions having frequencies $\omega_s(\boldsymbol{q})$. The first δ-function is the conservation of energy

$$\pm\hbar\omega = E_0 - E = \hbar\omega_s(\boldsymbol{q}).$$

The upper sign refers to energy loss ($E < E_0$) of the neutron, the lower sign to energy gain. $\boldsymbol{q}$ is the wave vector of the excited or deexcited collective atomic motion, and s is an index specifying the various possible modes of collective atomic motions.

The second δ-function in the second term of eq. (24.5) is the inelastic interference condition: the neutrons are diffracted by density waves of wave vector $\boldsymbol{q}$ originating from the collective atomic displacements which propagate in phase through the crystal. The waves retain the periodicity of the lattice. The quantity in brackets $\langle \ldots \rangle$ is the thermal population of the initial state. $F_{s,q}(\boldsymbol{\kappa})$ is the so-called dynamic structure factor

$$F_{s,q}(\boldsymbol{\kappa}) = \frac{1}{V^{1/2}}\sum_j b_j\left(\frac{\hbar}{M_j}\right)^{1/2} \exp[\mathrm{i}(\boldsymbol{\kappa}+\boldsymbol{q})\cdot\boldsymbol{R}_j]\exp[-W_j(\boldsymbol{\kappa})](\boldsymbol{\kappa}\cdot\boldsymbol{e}_{j,s,q}), \tag{24.6}$$

where M_j is the mass of atom j and $\boldsymbol{e}_{j,s,q}$ is a unit vector in the direction of the displacement of atom j in mode s at wave vector $\boldsymbol{q}$. One sees that with neutrons, through the constructive interference of inelastically scattered waves, we can single out each one of the $3N$ collective excitations characterized by ω_s and $\boldsymbol{q}$. Each interference maximum is determined by three equations

$$\omega = \omega_s \tag{24.7a}$$

$$\omega_s = \omega_s(\boldsymbol{q}) \tag{24.7b}$$

and

$$\boldsymbol{\kappa} + \boldsymbol{q} = 2\pi\boldsymbol{\tau} \tag{24.7c}$$

for three variables, namely ω_s and $\boldsymbol{q}$ and one experimental quantity, for instance $|\boldsymbol{k}|$. Figure 24.4 illustrates this interference scattering. From the intensities of the interference maxima we can determine $|F_{s,q}(\boldsymbol{\kappa})|^2$ and hence the dynamic displacements

$$\left(\frac{\hbar}{M_j\omega_{s,q}}\right)^{1/2} \boldsymbol{e}_{j,s,q}$$

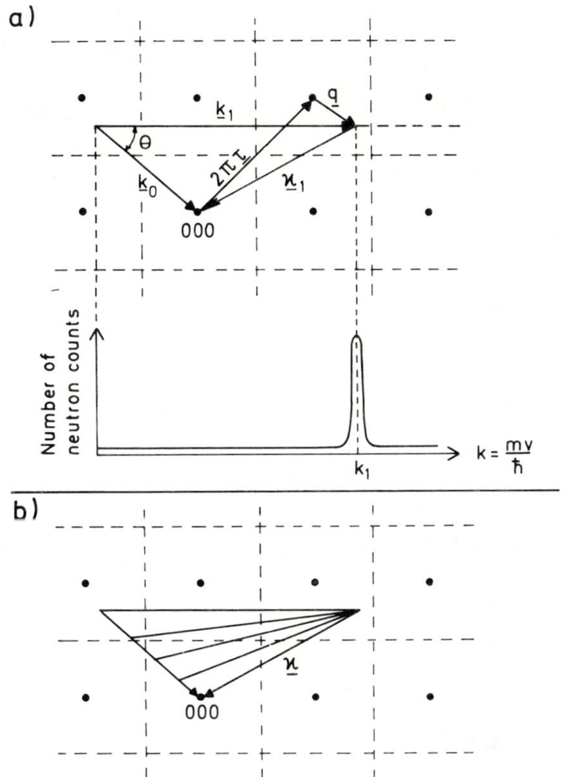

Fig. 24.4. Example of an experiment with inelastic interference scattering. The dashed grid shall represent the boundaries of Brillouin zones in a reciprocal lattice plane. (a) Neutrons incident with wave vector $\boldsymbol{k}_0$ are, after scattering, observed under scattering angle θ. With the analyzer the number of scattered neutrons is measured as a function of their velocities. Because of the eqs. (24.7) this function will have maxima at certain values of $|\boldsymbol{k}|$. Let k_1 be such a value. Then $\boldsymbol{\kappa}_1$ is found by eq. (24.2), $\boldsymbol{q}$ by eq. (24.7c) and ω_s by eqs. (24.7a) and (24.3). (b) Measurement with a double-crystal spectrometer. $\boldsymbol{\kappa}$ is kept fixed by simultaneous variation of $k_0 = \pi(d_M \sin \theta_M)^{-1}$, of θ and of $k = \pi(d_A \sin \theta_A)^{-1}$.

of the atoms in mode s at $\boldsymbol{q}$, in much the same way as the relative equilibrium positions are determined with X-ray diffraction. Coherent inelastic neutron scattering is a kind of combination of X-ray diffraction and Raman spectroscopy.

In addition to the coherent scattering, as for a harmonic solid given by eq. (24.5), certain kinds of atoms give rise to incoherent scattering, owing to a random orientation of nuclear spins or to the presence of different isotopes (Marshall and Lovesey [1971]). With this scattering waves scattered from different nuclei do not interfere with each other and eq. (24.7c) no longer holds. With this scattering we thus obtain information on the average space–time behavior only of individual atoms and not about correlations among them. Such information is described in ch. 19 by Janik. Here we are concerned with coherent scattering only, for which incoherent scattering, if present, is nothing but a background.

Measurements are subject to limitations for practical reasons. These are indicated in fig. 24.1. The formulae in the figure give the differential flux at various points in the neutron path: ϕ is the thermal neutron flux at the surface of the reactor core; $\Delta\Omega_0$ is the solid angle seen by the monochromator; $\Delta\Omega$ is the one seen by the analyzer; $t_M(E_0)$ and $t_A(E)$ are the transmissions of the monochromator and the analyzer, respectively; $f(E_0)$ is the distribution of neutron energies in the reactor beam; $d(\theta)$ is the distance travelled by the neutron through the sample; N is the density of nuclei in the sample; and ΔE_0 and ΔE are the energy widths transmitted by the monochromator and analyzer, respectively. They and the solid angles determine the resolution in ω and $\boldsymbol{\kappa}$. One sees how they affect the flux in the detector. One must realize that at usual research reactors $\Phi\Delta\Omega_0$, the flux before monochromatization is of the order $10^8\ \mathrm{cm}^{-2}\ \mathrm{s}^{-1}$. This is to be compared to a flux of some 10^{17} monochromatic photons $\mathrm{cm}^{-2}\ \mathrm{s}^{-1}$ obtained from a Laser. Consequently, in most experiments the resolution in ω, for instance, cannot be made better than some $10^{11}\ \mathrm{s}^{-1}$. With special techniques (Alefeld et al. [1969] and Mezei [1973]) and at high flux reactors $10^8\ \mathrm{s}^{-1}$ may be reached in special cases. With respect to motions of protons in H-bonds this limiting resolution in ω means that proton motions cannot be observed if their frequencies are lower than $10^8\ \mathrm{s}^{-1}$. Also the concentration of such protons (or deuterons) must be rather large; and large sample sizes are necessary. As pointed out in ch. 19 by Janik, the neutron proton scattering is predominantly incoherent. Hence, for studies on collective motions it will be advantageous to use deuterated compounds.

24.3. The problem of collective motions of hydrogen bonded protons

A two-dimensional grid of H-bonds is shown in fig. 24.5: (a) in an ordered state and (b) in a disordered state. However, the disorder is not complete, the proton arrangement still obeys the well-known ice rules (Pauling [1935]) or Slater rules (Slater [1941]), i.e. there is only one proton in each bond and there are only two protons close to each cross-point of the grid.

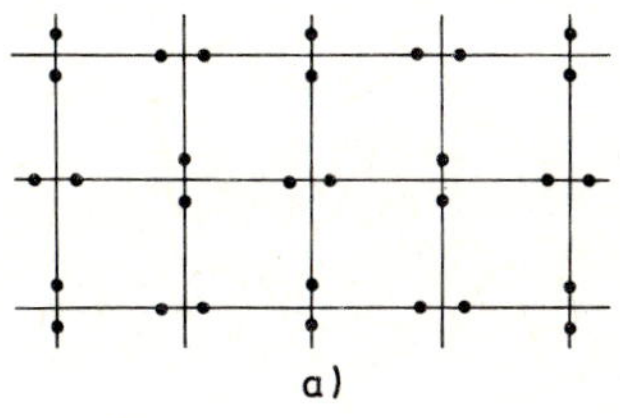

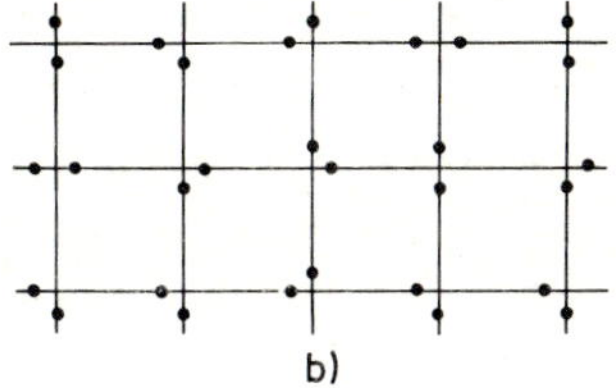

Fig. 24.5. Two-dimensional arrangement of protons H-bonded between the cross-points of a grid: (a) in an ordered state, (b) in a state disordered within Slater's rules.

The best known examples for disorder limited by such rules are ice and KH_2PO_4. Values of the zero point entropy of ice and the total entropy change at the ferroelectric phase transformation in KH_2PO_4 calculated with the assumption of these rules (Takagi [1948]) are in very good agreement with experiment. Similar, though eventually more complicated rules presumably apply to many systems of H-bonds.

Of the two most studied examples, ice and KH_2PO_4, the latter is simpler and, because of the phase transformation and its particularities, also more interesting. Figure 24.6 shows a projection of the crystal structure onto the lattice planes perpendicular to the axis of polarization for the ferroelectric phase. One sees from this projection that the proton arrangement corresponds to the ordered state shown in fig. 24.5a. Upon heating the sample, two things happen at the phase transition ($T_c = 122$ K in KH_2PO_4, $T_c = 213$ K in KD_2PO_4). In the first place, the P ions at the centers of the oxygen tetrahedra and the K ions between the tetrahedra (neither is shown in fig. 24.6) are displaced relative to each other along the ferroelectric axis. These displacements remove the major part of the polarization (Frazer and Pepinski [1953] and Bacon and Pease [1955]). Secondly, the proton system goes into a disordered state as shown in fig. 24.5b. Experiments with incoherent neutron scattering (Grimm et al. [1970]), as described by Janik in ch. 19, have shown that the proton system in fact goes into a disordered state and not into an arrangement in which the equilibrium sites were in the centers of the H-bonds which

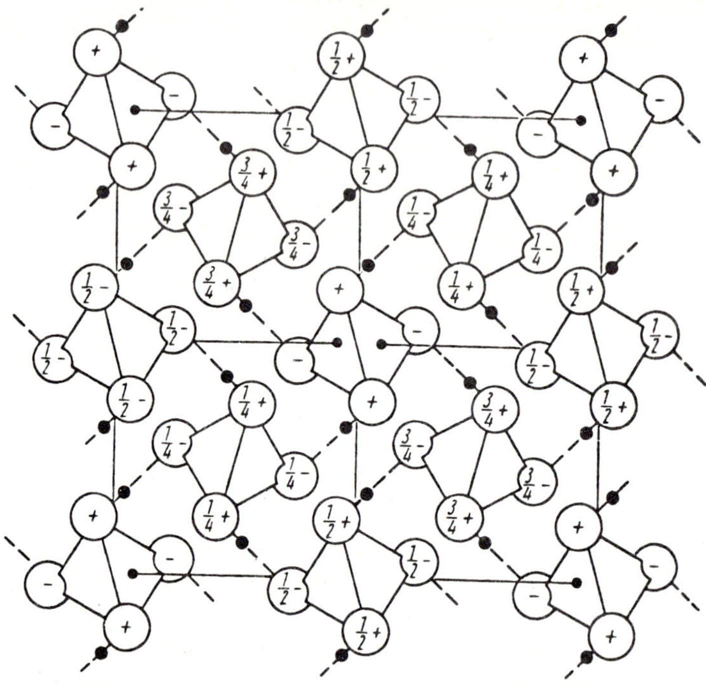

Fig. 24.6. Projection of the ferroelectric KH_2PO_4 lattice onto x–y planes, i.e. planes perpendicular to the axis of polarization. The dashed lines indicate the H-bonds, the solid points the proton equilibrium sites and the open circles the oxygen positions. The K ions lie along the polarization axis between the oxygen tetrahedra. They and the P ions, in the centers of the tetrahedra, are not shown.

could not be excluded by diffraction studies (Bacon and Pease [1953]). Thus we have a displacive phase transformation, primarily with respect to the K and P ions, coupled with an order–disorder transformation for the proton system. Such a coupling of two, or even more, types of collective processes may be the origin of quite many transformations in H-bonded systems. How can such a coupling take place? The process is little understood; but the ferroelectric transition in KH_2PO_4 seems to be the simplest example. In what follows, we will describe a phenomenological model first worked out by de Gennes [1963], Tokunaga and Matsubara [1966], Kobayashi [1968] and Cochran [1969] (see also ch. 23 section 3).

Without going into the long history of the model, its primary assumptions are: (a) that in the disordered state the distribution of the protons is dynamic – by tunneling, each proton fluctuates between two available sites

(Blinc [1960] and Blinc and Ribaric [1963]), and (b) that in order to yet fulfill Slater's rules these fluctuations are correlated. The first assumption has been confirmed again by incoherent neutron scattering (Grimm et al. [1970]), as well as by NMR measurements (Blinc [1973]). With respect to the second assumption, the correlations in principle may be either of short range, as shown in fig. 24.7, or of long range, i.e. the protons may move in "tunneling modes" propagating through the entire system. The first

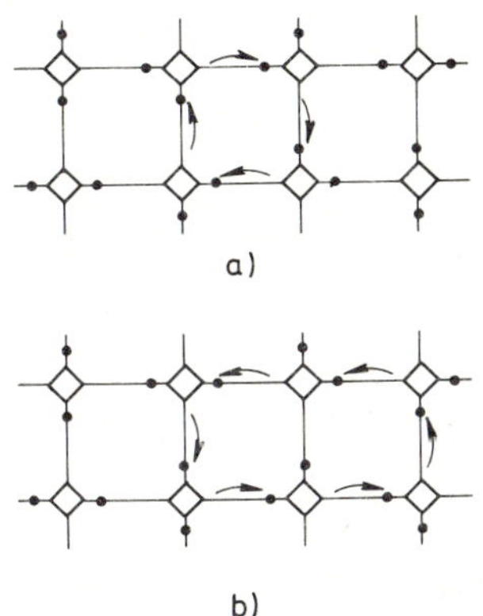

Fig. 24.7. Short range tunneling correlations. If, through simultaneous tunneling of four protons, a lattice defect is created in one cell (a), correlations may be such that further tunneling transitions constitute a migration of the defect (b).

possibility has been studied in the first instance by Villain and Aubry [1969], the second possibility by Tokunaga and Matsubara [1966]. Since there are four H-bonds in the unit cell of the crystal we have four different tunneling modes as shown in fig. 24.8. One sees that they lead to all the configurations allowed by Slater's rules.

We assume that in the disordered state, the double-minimum potential for each proton is symmetric and that in this potential states higher than the lowest asymmetric and symmetric ones are not excited, either thermally or by the scattering. The latter assumption is certainly valid for thermal neutron scattering in KH_2PO_4 and KD_2PO_4. The Hamiltonian for the model then may be written

$$H_p = -\hbar\omega_t \sum_j \sigma_j^x - \frac{1}{2} \sum_{j,l} J_{jl} \sigma_j^z \sigma_l^z. \tag{24.8}$$

Here pseudo-spin operators are introduced, defined as shown in fig. 24.9 in analogy to the Pauli spin matrices, such that the eigenvalue $\bar{\sigma}^z$ of σ^z is

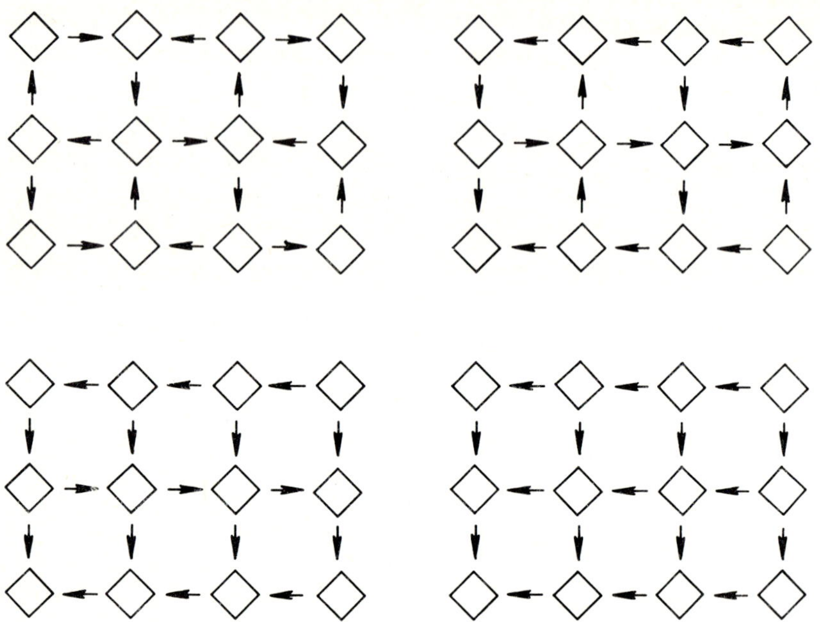

Fig. 24.8. Four "tunneling modes". The protons are tunneling in phase throughout the crystal. Condensation of the first mode leads to the ordered arrangement of fig. 24.5a.

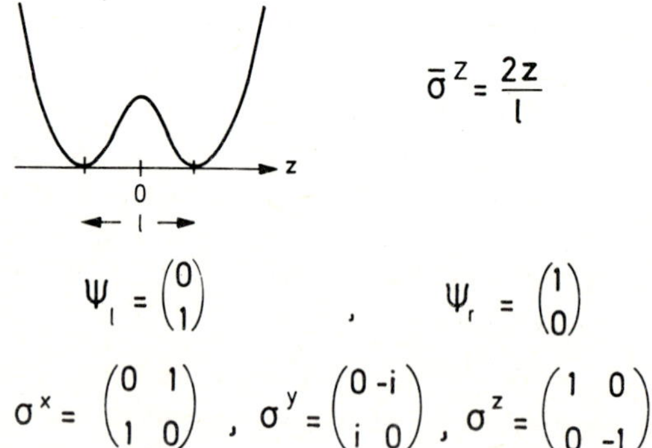

Fig. 24.9. Definition of pseudo-spin operators. ψ_l, ψ_r, respectively, are the proton ground states in the left and in the right well of the double-minimum potential. For KH_2PO_4 and KD_2PO_4 higher states are known to have such high energies that they need not be considered here.

± 1 for the proton on the right or on the left equilibrium site, respectively (de Gennes [1963]). The operator σ_j^z localizes proton j on one site. The operator σ_j^x shifts proton j from one site to the other. The meaning of σ_j^y is not as obvious but also not important; it can be shown to represent the local current of the proton transfers. ω_t is the tunneling frequency. In what follows, we first consider long-range correlations, because with them a coupling to the displacements of the other atoms is more easily visualized.

In an excellent recent review on this model by Blinc and Žekš [1972] the proton motions resulting from the Hamiltonian (24.8) are considered in random phase approximation. That is, we imagine a field $E_j(t)$ acting on proton j, such that

$$H_{\mathrm{p}}^{\mathrm{eff}} = -\hbar\omega_t \sum_j \sigma_j^x - \frac{1}{2}\sum_{j,l} J_{jl}\sigma_j^z\sigma_l^z - 2\mu \sum_j E_j(t)\sigma_j^z;$$

and the averages over operator products are replaced by products of averages

$$\langle\sigma_j^\alpha\sigma_l^\beta\rangle = \langle\sigma_j^\alpha\rangle\langle\sigma_l^\beta\rangle \qquad \text{for } j \neq l,$$

with $\alpha, \beta = x, y, z$.

The Heisenberg equations of motion are then

$$\frac{\mathrm{d}\langle\boldsymbol{\sigma}_j\rangle}{\mathrm{d}t} = \langle\boldsymbol{\sigma}_j\rangle \times \boldsymbol{H}_j^{\mathrm{p}}(t), \tag{24.9}$$

with $\boldsymbol{H}_j^{\mathrm{p}}(t) = -\partial\langle H(t)\rangle/\partial\langle\boldsymbol{\sigma}_j(t)\rangle$. With

$$E_j(t) = E_j \exp(\mathrm{i}\,\omega t) \quad \text{and} \quad \langle\boldsymbol{\sigma}_j(t)\rangle = \langle\boldsymbol{\sigma}_j\rangle + \delta\langle\boldsymbol{\sigma}_j\rangle \exp(\mathrm{i}\,\omega t),$$

so that the equations of motion can be linearized, the solutions of eq. (24.9) are, for one H-bond per cell, three modes with frequencies

$$\omega_1(\boldsymbol{q}) = 0, \qquad \omega_{2,3}^2(\boldsymbol{q}) = \omega_t[\omega_t - \hbar^{-1}J(\boldsymbol{q})\langle\sigma^x\rangle]. \tag{24.10}$$

Here $\boldsymbol{q}$ represents the wave vector of the mode and

$$J(\boldsymbol{q}) = \sum_{j,l} J_{jl} \exp[\mathrm{i}\,\boldsymbol{q}(\boldsymbol{R}_j - \boldsymbol{R}_l)], \tag{24.11}$$

with $\boldsymbol{R}_j$ and $\boldsymbol{R}_l$ the equilibrium sites of protons j and l. (For details of the mathematics the reader is referred to the review mentioned above.) ω_1 is the frequency of a motion in the direction of the molecular field, i.e. a longitudinal motion. ω_2 and ω_3 are the frequencies of transverse modes, i.e. of free "precessions" of the pseudo spins around the molecular field, or, in other words, of tunneling modes, as they involve the x-component of $\boldsymbol{\sigma}$. With four H-bonds per cell (Nettleton [1971, 1972]) we get three further modes with frequency zero and six further modes with a dispersion of the form (24.10).

As $\langle\sigma^x\rangle$ is the probability for a transition between two states which differ by an energy $\hbar\omega_t$, it must be in equilibrium

$$\frac{\exp(\hbar\omega_t/2k_BT)-\exp(-\hbar\omega_t/2k_BT)}{\exp(\hbar\omega_t/2k_BT)+\exp(-\hbar\omega_t/2k_BT)}=\tanh\left(\frac{\hbar\omega_t}{2k_BT}\right).$$

Thus eq. (24.10) may also be written

$$\omega^2_{2,3}(\boldsymbol{q})=\omega_t\left[\omega_t-\frac{1}{\hbar}J(\boldsymbol{q})\tanh\left(\frac{\hbar\omega_t}{2k_BT}\right)\right]. \qquad (24.12)$$

One sees that for $k_BT\gg\hbar\omega_t$ the protons become independent, $\omega_{2,3}=\omega_t$. On the other hand, $\omega_{2,3}$ approaches zero as

$$J(\boldsymbol{q})\tanh(\hbar\omega_t/2k_BT)\rightarrow\hbar\omega_t. \qquad (24.13)$$

At a temperature T_c, given by eq. (24.13), the correlated tunneling condenses out. Figure 24.10 represents a schematic picture of the average tunneling motion of a proton in the disordered phase.

Next, we introduce in the same phenomenological way, a coupling to the other atoms, with a Hamiltonian

$$H_{PL}=-\sum_{j,\nu}\sigma^z_j\boldsymbol{I}_{j,\nu}\cdot\boldsymbol{u}_\nu. \qquad (24.14)$$

This description was first suggested by Kobayashi [1968] and Cochran [1969]. $\boldsymbol{u}_\nu$ is the displacement of atom ν due to lattice vibrations. The coupling to the σ^z can be imagined, for instance, as an interaction between the electric dipole moment of an O–H···O configuration and the electric field created by atomic displacements in a polar mode of the lattice. In the harmonic approximation

$$\boldsymbol{u}_\nu=\frac{1}{\sqrt{NM_\nu}}\sum_{s,\boldsymbol{q}}Q_{s,\boldsymbol{q}}\,\boldsymbol{e}_{\nu,s,\boldsymbol{q}}\exp(\mathrm{i}\,\boldsymbol{q}\cdot\boldsymbol{R}_\nu).$$

Here $\boldsymbol{q}$ is the wave vector of a collective atomic motion in the lattice. The index s specifies the mode as in eq. (24.5). $\boldsymbol{e}_{\nu,s,\boldsymbol{q}}$ is, as in eq. (24.6), a unit vector in the direction of the displacement of atom ν in mode s at $\boldsymbol{q}$. M_ν is the mass, and $\boldsymbol{R}_\nu$ the equilibrium site of atom ν. The Hamiltonian for the lattice vibrations is then

$$H_L=\frac{1}{2}\sum_{s,\boldsymbol{q}}\{|\dot{Q}_{s,\boldsymbol{q}}|^2+\omega^2_{s,\boldsymbol{q}}|Q_{s,\boldsymbol{q}}|^2\}$$

and the total Hamiltonian is now $H=H_p+H_L+H_{PL}$.

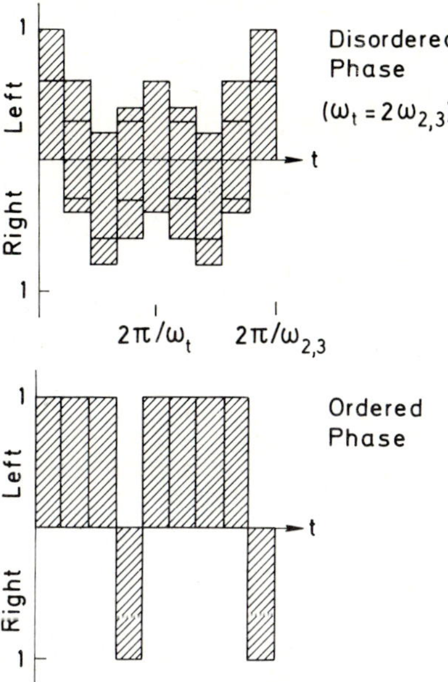

Fig. 24.10. Time development of the density distribution of a proton onto two equilibrium sites, left and right; if at $t = 0$ it is on the left side. Notice how the correlated motion with frequency $\omega_{2,3}$ renders the effective single-particle potential asymmetric, if $\omega_t/\omega_{2,3}$ is not an odd number. In contrast to the behavior of spins in a ferromagnet the propagating mode $\omega_{2,3}(\boldsymbol{q})$ here exists in the disordered and not in the ordered state.

For this model the motions of the protons and the other atoms have again been considered in random phase approximation and by linearizing the equations of motion. For the mathematical procedure the reader is referred once more to the review article by Blinc and Žekš [1972]. One finds in this approximation that the tunneling mode mixes with an optic lattice mode of the same symmetry (B_2). Two mixed modes are created with frequencies

$$\omega_{\pm}^2(\boldsymbol{q}) = \tfrac{1}{2}[\omega_{\mathrm{B}}^2(\boldsymbol{q}) + \omega_{2,3}^2(\boldsymbol{q})] \pm \frac{1}{2}\left[[\omega_{\mathrm{B}}^2(\boldsymbol{q}) - \omega_{2,3}^2(\boldsymbol{q})]^2 + \frac{2N\omega_t}{\hbar}|I_q|^2 \tanh\left(\frac{\hbar\omega_t}{2k_{\mathrm{B}}T}\right)\right]^{1/2} \quad (24.15)$$

with

$$\boldsymbol{I}_q = \sum_{\nu} \frac{1}{\sqrt{NM_\nu}} \boldsymbol{I}_{j,\nu} \cdot \boldsymbol{e}_{\nu,q} \exp[\mathrm{i}\,\boldsymbol{q} \cdot (\boldsymbol{R}_\nu - \boldsymbol{R}_j)], \quad (24.16)$$

$\omega_{\mathrm{B}}(\boldsymbol{q})$ is the frequency of the B_2 lattice mode. The frequency of the minus-mode, $\omega_-(\boldsymbol{q})$, approaches zero as

$$\omega_{\mathrm{B}}^2(\boldsymbol{q}) + \omega_{\mathrm{t}}^2 \to \frac{1}{2\hbar\omega_{\mathrm{t}}}(N|I_q|^2 + \tfrac{1}{2}J_q) \tanh\left(\frac{\hbar\omega_{\mathrm{t}}}{2k_{\mathrm{B}}T}\right), \qquad (24.17)$$

where eq. (24.12) has been used for $\omega_{2,3}^2$.

Thus the model predicts the mechanism of the phase transformation in all details. The tunneling mode and a B_2 lattice mode mix into two modes. One of these, the so-called minus-mode, becomes soft as a critical temperature T_c, given by eq. (24.17), is approached from the high temperature side. The critical temperature depends on $\boldsymbol{q}$, but from the symmetry of the tunneling and the B_2 lattice modes one easily finds that $\boldsymbol{I}_q$ and J_q are largest for $q = 0$. Hence, the transition actually takes place in the center of the Brillouin zone. Figure 24.11, taken from Kobayashi [1968] and Cochran [1969], shows the essential atomic displacements in the minus-mode. As ω_- approaches zero the protons are trapped on ordered sites and the other atoms involved are shifted into new equilibrium positions, thus producing the new lattice structure. The shift of the K and the P ions produces the main part of the lattice polarization; at the same time the oxygen tetrahedra are slightly distorted. Whether the model is true must be tested by experiments.

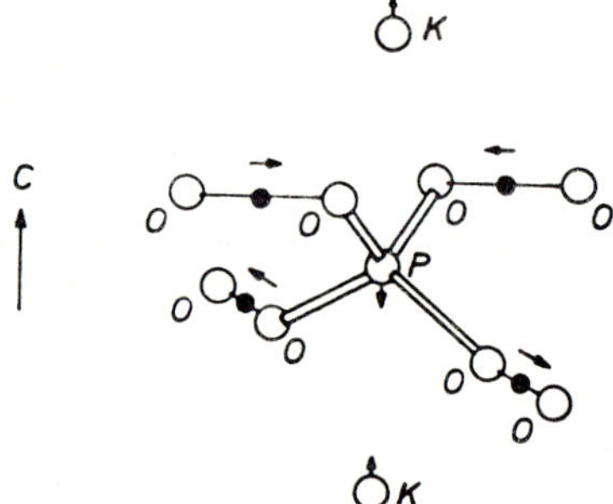

Fig. 24.11. Displacements of protons (or deuterons) and of the K and P ions in the minus-mode. The oxygen displacements are not shown. From Kobayashi [1968] and Cochran [1969].

24.4. Experiments and comparison to theory

Experimentally in KH_2PO_4 a frequency decreasing to zero as $T \to T_c$ was first found by Raman scattering (Kaminow and Damen [1968]). With further Raman and infrared absorption measurements (Sugawara and

Nakamura [1970] and Coignac and Poulet [1971]), it also could be established that the mode to which the observed temperature-dependent frequency belongs is of B_2 symmetry as predicted by the model, and that, again as predicted, a second mode of the same symmetry exists with a frequency which does not decrease but slightly increases with decreasing temperature. However, unpredicted by the model, the mode which becomes soft was found to be strongly damped, in fact overdamped, i.e. concentrated around $\omega = 0$, even at temperatures far above T_c.

As stated in the preceding section the model predicts that the minus-mode should become soft first at $q = 0$. For this reason neutron scattering experiments on KD_2PO_4 (Buyers et al. [1968] and Skalyo et al. [1970]) and on KH_2PO_4 (Arsić-Eskinja et al. [1972]) were both done in the vicinities of the centers of various Brillouin zones. Figure 24.12 shows an observed

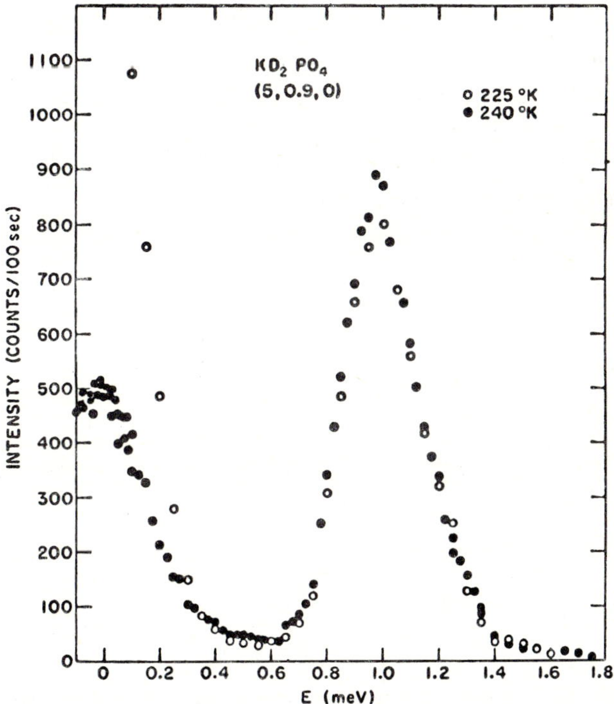

Fig. 24.12. Part of the energy distribution of neutrons scattered from KD_2PO_4 at two temperatures, 12 and 27 degrees above T_c. The measurements are done with $\boldsymbol{\kappa}$ = const. = (5, 0.9, 0). The peak at about 1 meV originates from deexcitation of an acoustic phonon. The distribution around $E = 0$ is due to deexcitation of the minus-mode. From Skalyo et al. [1970].

intensity distribution as a function of ω near $q=0$ in KD_2PO_4. The distribution around $\omega=0$ has a similar shape as observed in Raman scattering: the mode is overdamped. The second intensity peak originates in an acoustic phonon.

Further experiments were then done with a resolution in ω such that the measurements integrated over the ω-distribution of the overdamped mode. That is to say, a quantity

$$I_-^{(1)}(\boldsymbol{\kappa}) = \int R(\boldsymbol{\kappa}-\boldsymbol{\kappa}')S_-^{(1)}(\boldsymbol{\kappa}')\,\mathrm{d}\boldsymbol{\kappa}'$$

was measured, where $R(\boldsymbol{\kappa}-\boldsymbol{\kappa}')$ is the instrumental resolution and

$$S_-^{(1)}(\boldsymbol{\kappa}) = \int S_-^{(1)}(\boldsymbol{\kappa},\omega)\,\mathrm{d}\omega.$$

Here, the second term of eq. (24.5) is called $S^{(1)}(\boldsymbol{\kappa},\omega)$ and written as

$$S^{(1)}(\boldsymbol{\kappa},\omega) = \sum_s S_s^{(1)}(\boldsymbol{\kappa},\omega).$$

For the mode $s=-$ (the minus-mode) we take the damping into account by replacing (Cowley [1963])

$$\frac{\delta(\omega-\omega_-(q))}{\omega} \to \frac{\omega\Gamma_-}{[\omega^2-\omega_-^2(q)]^2+\omega^2\Gamma_-^2}. \tag{24.18}$$

The damping $\Gamma_- \gg 2\omega_-$ is assumed independent of ω and $\boldsymbol{q}$. With

$$\langle n(\omega)+1\rangle \approx \frac{k_BT}{\hbar\omega}$$

we obtain

$$S_-^{(1)}(\boldsymbol{\kappa},\omega) = \frac{k_BT}{\hbar}\sum_{\boldsymbol{q}} |F_{-,\boldsymbol{q}}(\boldsymbol{\kappa})|^2 \frac{\Gamma_-}{[\omega^2-\omega_-^2(\boldsymbol{q})]^2+\omega^2\Gamma_-^2}\sum_{\boldsymbol{\tau}}\delta(\boldsymbol{\kappa}+q-2\pi\boldsymbol{\tau}),$$

and hence

$$S_-^{(1)}(\boldsymbol{\kappa}) = \sum_q |F_{-,q}(\boldsymbol{\kappa})|^2 \frac{k_BT}{\hbar\omega_-^2(q)}\sum_\tau \delta(\boldsymbol{\kappa}+q-2\pi\tau). \tag{24.19}$$

This quantity was measured for various Brillouin zones and in various directions of reciprocal space as well as in dependence of temperature as $\omega_-(\boldsymbol{q}) = \omega_-(\boldsymbol{q}, T)$. Figure 24.13 shows two examples. In the centers of the distributions, at $q=0$, there are, of course, Bragg reflections; the minus-mode intensity must be interpolated into them. In this respect KH_2PO_4

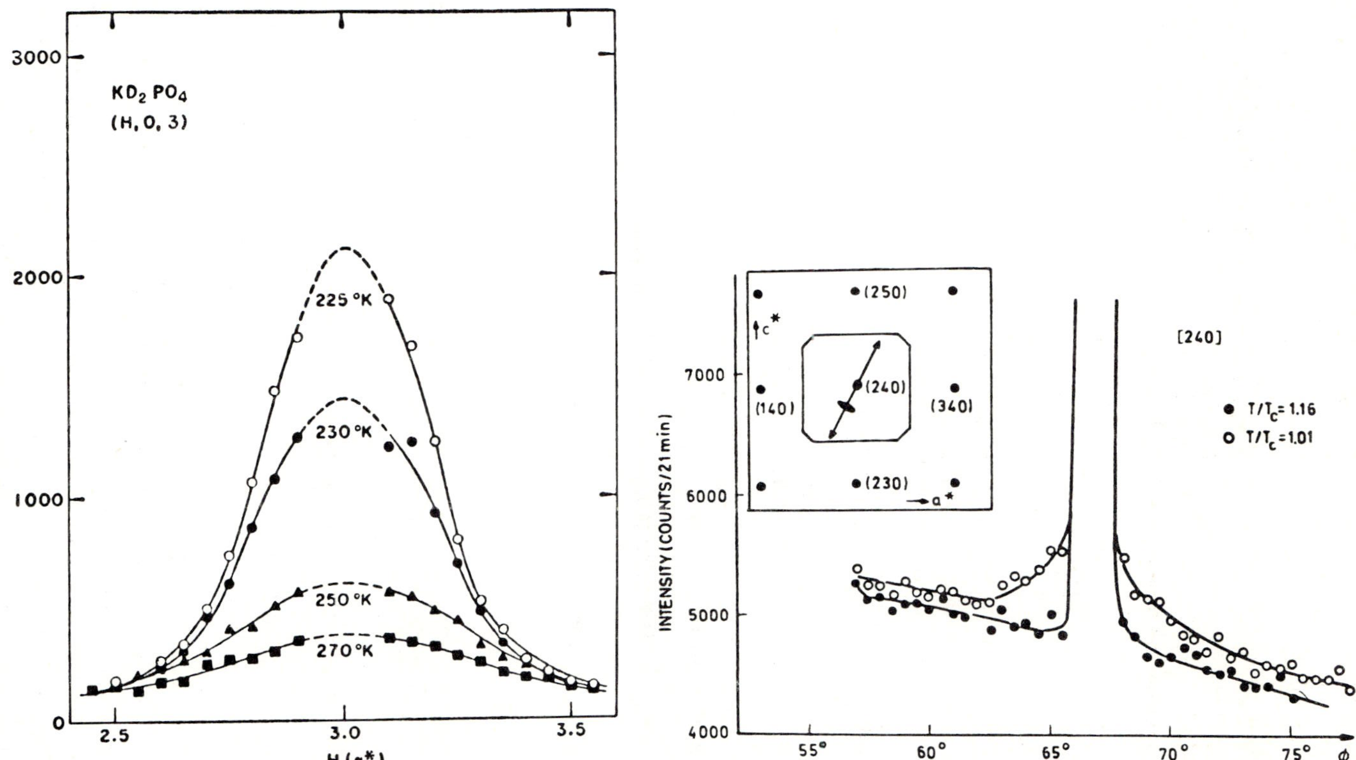

Fig. 24.13. $I_{-}^{(1)}(\boldsymbol{\kappa})$: (a) measured in KD_2PO_4 around (303), from Skalyo et al. [1970], (b) measured in KH_2PO_4 around (240), from Arsić-Eskinja et al. [1972]. The insert in (b) indicates the particular experimental path in reciprocal space used for this experiment. The small ellipsoid represents a projection of the instrumental resolution function onto the reciprocal lattice plane.

and KD_2PO_4 are not very well suited for neutron measurements. Yet obviously we now obtain from the measured quantity (24.19) information on the space behavior, on the displacements and the phase relations of the atoms in the minus-mode, information unobtainable by other means.

The dynamic structure factor $F_{q,s=-}(\boldsymbol{\kappa})$, eq. (24.6), was calculated with the atomic displacements predicted from the model for the minus-mode. For the protons, which according to the model are involved with the mode merely by their correlated tunneling, the centers of motion $\boldsymbol{R}_j$ were taken as the centers of the H-bonds. For their amplitude $(\hbar/M\omega_{s,q})^{1/2}$ was replaced by l, the distance between the two potential minima as determined by incoherent neutron scattering (Grimm et al. [1970]). In the same way the proton Debye–Waller factors were taken to be anisotropic for KH_2PO_4, as determined by incoherent scattering. For the other atoms the Debye–Waller factors were assumed isotropic, so that $W_j(\boldsymbol{\kappa}) = \kappa^2\langle u_j\rangle^2$ if $\langle u_j\rangle^2$ is the mean square amplitude. Values were taken from diffraction measurements (Bacon and Pease [1953]). For KD_2PO_4 all amplitudes were used as fitting parameters (Skalyo et al. [1970]). Figure 24.14 shows as an example the results of such calculations for KH_2PO_4. A comparison of these calculations to the experimental observations constitutes a detailed test of the model. For both KH_2PO_4 and KD_2PO_4, the results of calculations, as shown in fig. 24.14, were found to agree very well with the experiments, with, it seems, only one minor difference between KH_2PO_4 and KD_2PO_4. In KH_2PO_4 the z-component of the proton displacements, originating from a slight inclination of the H-bonds with respect to the x–y planes (Nehnes et al. [1972]) is found to be in phase with the displacements of the K ions, as expected from the static displacements in the phase transition, while the z-component of the deuterium displacements in KD_2PO_4 unexpectedly seems to be in phase with the P ions.

Thus a soft mode with the correlated atomic displacements predicted by the tunneling–lattice coupling model in fact exists, in KH_2PO_4 essentially in the same way as in KD_2PO_4. However, two observations are in disagreement with the model. The first is shown in fig. 24.15. In addition to a comparison of relative intensities of the soft mode for various Brillouin zones, as calculated in fig. 24.14, one may, of course, measure in each zone the dependence on $\boldsymbol{q}$; it is due to the extension of the soft mode in $\boldsymbol{q}$-space. Two examples are shown in fig. 24.15: contours of equal counting rates around the point (303) in the [010] zone and the point (510) in the [001] zone of KD_2PO_4. Around (303) the contours have the shape expected qualitatively from the symmetry of the mode. However, around (510) the experiment reveals something entirely unexpected, without any

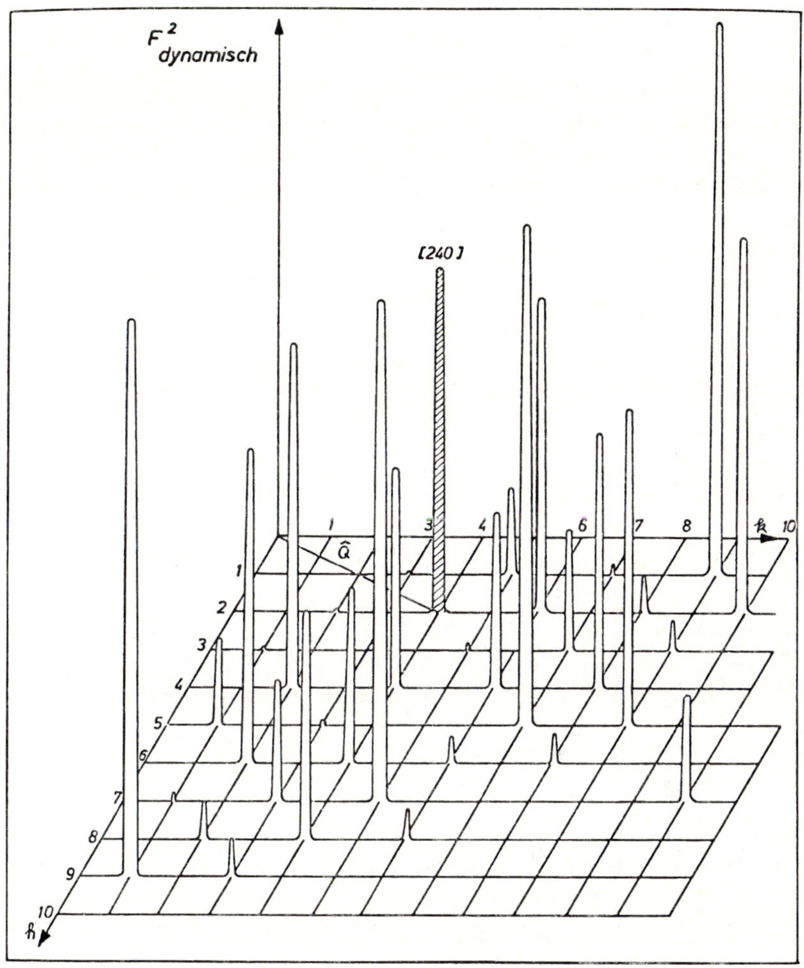

Fig. 24.14. Relative strengths of $S_-^{(1)}$ calculated for KH_2PO_4 from the model of Kobayashi for various Brillouin zones in the reciprocal lattice plane of the measurement shown in fig. 24.13b. From Arsić-Eskinja [Thesis, 1972].

apparent symmetry.

Secondly, as shown in ch. 19 by Janik, for KH_2PO_4 the elongation of the proton cloud, which originates through tunneling between two sites, was found by incoherent neutron scattering to disappear at the phase transition temperature, 122 K (Arsić-Eskinja et al. [1972]). However, the elongation can be observed in these experiments only as long as the tunneling frequency is larger than the experimental resolution, because

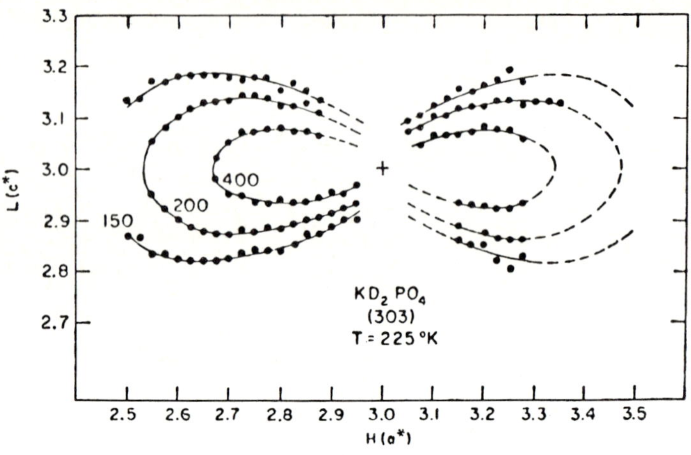

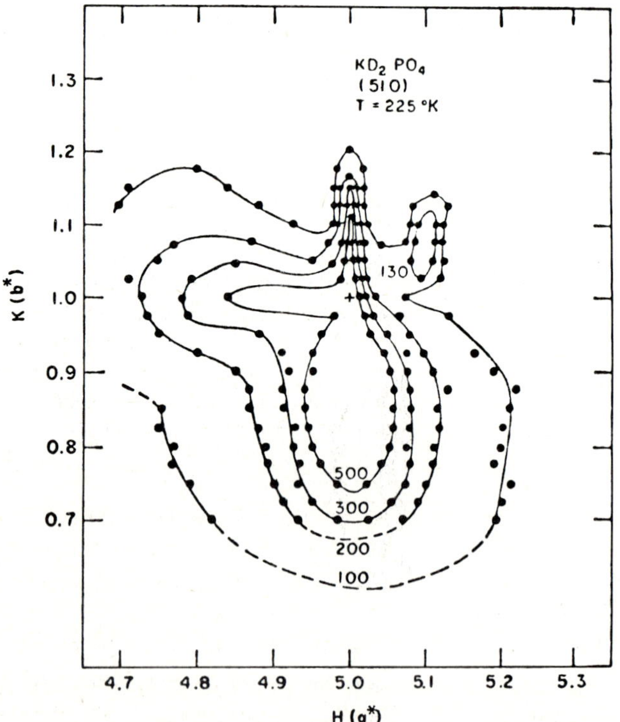

Fig. 24.15. Intensity contours of $S_{-}^{(1)}(\boldsymbol{\kappa})$ in KD_2PO_4: (a) in the [010] zone around (303), (b) in the [001] zone around (510). From Skalyo et al. [1970].

only then does the neutron wave packet travel over an individual proton for a time which is longer than the mean time the proton stays at one site. Therefore the disappearance of the elongation should have been found at 126 K, since ω_-, as observed in Raman as well as in the neutron experiments, becomes in fact smaller than the resolution at this temperature. In the frame of the model it must be concluded that, at least in the region $0 \leqslant T - T_c \leqslant 4$ K, only a small part of the protons move with ω_-, and most of them fluctuate with the temperature-independent ω_t only (cf. fig. 24.10).

These two observations are not understood. They add to the already mentioned unpredicted overdamping of the mode. Since the model takes the dipole moment of a H-bond as the origin of the proton–lattice interaction it also fails, of course, to predict the large shift of T_c upon deuteration. This shift recently has been obtained (Godzik and Blumen [1974]) through a renormalization of the tunneling frequency and the hydrogen interaction J_{jl} via a coupling also of the x-components of the $\boldsymbol{\sigma}_j$. However, this problem shall not be further discussed here, as this article is concerned with the neutron scattering experiments only.

A few possible improvements of the model can immediately be visualized. (i) The equations of motion may be set-up with non-linear terms to account for the damping. (ii) An interaction of the minus-mode with a transverse acoustic mode of the same symmetry, as observed by Brillouin scattering (Brody and Cummins [1968]), may be taken into account. It will, however, be of importance only very near T_c, where the frequency of the minus mode for small $\boldsymbol{q}$ approaches the frequency of the acoustic mode. (iii) The lattice vibrations may modify the tunneling frequency and also the interaction between protons, so that terms

$$-\hbar \sum_{j,l,\nu} \sigma_j^x (\text{grad}_l\, \omega_t) \cdot \boldsymbol{u}_\nu \quad \text{and} \quad -\frac{1}{2} \sum_{j,l,m,\nu} (\text{grad}_j\, J_{lm} \sigma_l^z \sigma_m^z) \cdot \boldsymbol{u}_\nu$$

must be added to the Hamiltonian H_{PL} (Blinc and Žekš [1972]). However, such improved approximations render the mathematical treatment so difficult that to the author's knowledge they have not been attempted as yet. They also are unlikely to provide an answer to what appears to be the principal question: If on approaching T_c fewer and fewer protons take part in correlated tunneling, as indicated by the incoherent neutron scattering experiments, how can tunneling modes propagate?

Two entirely different approaches have been suggested. The first one is the so-called stochastic model by Blinc and Žekš [1972]. Extending the Villain–Aubry concept [1969] to include the lattice dynamics they proposed that the tunneling motions may not be as directly coupled to the lattice as is suggested by eq. (24.14), but rather become stochastically affected by the other atoms. The expectation value for the position of proton j is written, therefore, as

$$\langle\sigma_j^z(t)\rangle = \langle\sigma_j^z\rangle + \delta\langle\sigma_j^z\rangle\,\mathrm{e}^{\mathrm{i}\omega t} + \Delta\langle\sigma_j^z(t)\rangle,$$

where the second term is, as before, the response to the molecular field $E_j \exp(\mathrm{i}\,\omega t)$ and the third term is a stochastic fluctuation. The problem is then treated by linear response theory. In addition to the local field term $-2\mu\Sigma_j E_j(t)\sigma_j^z$, a term $\Delta H_j(t)\sigma_j^z$ representing a field due to spontaneous pseudo-spin fluctuations in the σ^z direction, is added to the Hamiltonian (24.8). In this Hamiltonian products of operators $\sigma_j^z\sigma_l^z$ are replaced by $\langle\sigma_j^z(t)\rangle\sigma_j^z$ and $\Delta H_j(t)$ is written

$$\Delta H_j(t) = \sum_l J_{jl}\,\Delta\langle\sigma_j^z(t)\rangle,$$

so that $H = \Sigma_j H_j$. In this way the problem is reduced to the classical problem of a spin in a constant field $\boldsymbol{H} = (\hbar\omega_t, 0, J\langle\sigma_j^z\rangle)$ and a random local field $\boldsymbol{H}_{\mathrm{loc}}(t) = (0, 0, \Delta H(t))$. The model can in fact reproduce the Raman line shapes quite well. However, it must be remembered here that the intensity evaluations in the neutron scattering experiments unquestionably show that the scattering intensity around $\omega = 0$ is determined by a dynamic structure factor and, hence, that defined phase relations exist between the proton motions and the motions of the other atoms. Perhaps a combination of the two models would come closer to reality.

The second approach is the so-called pseudo-polaron model of Fischer et al. [1970]. An extended version of the model is described in this volume (ch. 6 section 3.1). The coupling between H-bonded protons and the lattice is assumed to originate not, as in Kobayashi's model, from the dipole-moment of an O–H···O configuration, but rather from the differences in populations of symmetric and antisymmetric states within the local double-minimum potential. By itself such a model, of course, will not lead to a phase transformation. However, if combined with the dipole interaction, the model may be able to describe some of the features which the Kobayashi model fails to predict, for instance the mass-dependence of T_c. Unfortunately, to the author's knowledge dynamic atomic displacements have not been calculated as yet on the basis of such more general models.

The problem is unsolved in this respect. It has been outlined here primarily in order to demonstrate the possibilities of the application of coherent neutron scattering to the problem of correlated atomic motions in systems of H-bonds. The list of references, therefore, is also far from complete.

Acknowledgements

The author is indebted to Drs. H. Grimm and M. Arsić-Eskinja for many helpful discussions.

References

Alefeld, B., M. Birr and H. Heidemann, 1969, Naturwiss. **56**, 410.
Arsić-Eskinja, M., 1972, Thesis Techn. Hochschule Aachen.
Arsić-Eskinja, M., H. Grimm and H. Stiller, 1972, Neutron Inelastic Scattering, IAEA Symp. Grenoble, 825.
Bacon, G. and R. Pease, 1953, Proc. Roy. Soc. **A220**, 397.
Bacon, G. and R. Pease, 1955, Proc. Roy. Soc. **A230**, 359.
Blinc, R., 1960, J. Phys. Chem. Solids **13**, 204.
Blinc, R., 1973, private communication.
Blinc, R. and M. Ribaric, 1963, Phys. Rev. **130**, 1816.
Blinc, R. and B. Žekš, 1972, Advan. Phys. **21**, 693.
Brockhouse, B., S. Hautecler and H. Stiller, 1964, in: Interaction of Radiation and Solids, Ed. R. Strumane (North Holland), p. 580.
Brody, E. and H. Cummings, 1968, Phys. Rev. **L21**, 1263.
Buyers, W., R. Cowley, G. Paul and W. Cochran, 1968, Neutron Inelastic Scattering, IAEA Symp. Vienna, **1**, 267.
Cochran, W., 1969, Advan. Phys. **18**, 157.
Coignac, J. and H. Poulet, 1971, J. Phys. **32**, 679.
Cowley, R., 1963, Advan. Phys. **12**, 421.
Fischer, S., G. Hofacker and M. Ratner, 1970, J. Chem. Phys. **52**, 1934.
Frazer, B. and R. Pepinsky, 1953, Acta Cryst. **6**, 273.
Gennes, P. de, 1963, Solid State Commun. **1**, 132.
Godzik, K. and A. Blumen, 1974, Phys. Stat. Sol. (b) **66**, 569.
Grimm, H., H. Stiller and Th. Plesser, 1970, Phys. Stat. Sol. **42**, 207.
Kaminow, I. and T. Damen, 1968, Phys. Rev. **L20**, 1105.
Kobayashi, K., 1968, J. Phys. Soc. Japan **24**, 497.
Marshall, W. and S. Lovesey, 1971, Theory of Thermal Neutron Scattering, Oxford.
Mezei, F., 1973, Z. Physik **255**, 146.
Nehnes, R., V. Ericksson and K. Rouse, 1972, Solid State Commun. **11**, 1261.
Nettleton, R., 1971, Z. Physik **248**, 101.
Nettleton, R., 1972, J. Phys. Soc. Japan **33**, 641.
Pauling, L., 1935, J. Am. Chem. Soc. **57**, 2680.
Skalyo, J., B. Frazer and G. Shirane, 1970, Phys. Rev. **B1**, 278.
Slater, J., 1941, J. Chem. Phys. **9**, 16.
Sugawara, F. and T. Nakamura, 1970, J. Phys. Soc. Japan **28**, 158.
Takagi, Y., 1948, J. Phys. Soc. Japan **3**, 271.
Tokunaga, M. and T. Matsubara, 1966, Progr. Theoret. Phys., Osaka **35**, 581.
Van Hove, L. 1954, Phys. Rev. **95**, 249.
Villain, J. and S. Aubry, 1969, Phys. Stat. Sol. **33**, 337.

PART G

THERMODYNAMICS OF HYDROGEN BONDED SYSTEMS

CHAPTER 25

SPECTROSCOPIC AND CALORIMETRIC STUDIES OF HYDROGEN BONDING

A. DEAN SHERRY

Institute for Chemical Sciences, University of Texas at Dallas, P.O. Box 688, Richardson, Texas 75080, U.S.A.

Contents

25.1. Introduction 1201
25.2. Calorimetry 1202
25.2.1. Inert solvent method 1204
25.2.2. Pure base method 1205
25.2.3. Polar solvent method 1206
25.3. Infrared techniques 1207
25.4. Nuclear magnetic resonance techniques 1208
25.5. Linear enthalpy–spectroscopic relations 1209
25.6. The double scale enthalpy equation 1216
25.7. Single scale enthalpy equations 1220
25.8. Conclusions 1222
References 1223

The hydrogen bond – recent developments in theory and experiments
Eds. P. Schuster et al. © *North-Holland Publ. Co., Amsterdam, 1976*

25.1. Introduction

Since introduction of the original concept of a H-bond (Moore and Winmill [1912]), a great deal of experimental effort has been directed towards understanding the energetics of H-bond formation. The volumes by Hadži [1959] and Pimentel and McClellan [1960] give representative data of these earlier experimental approaches. The prediction of a linear relationship between the change in the X–H stretching wave numbers upon formation of a X–H···Y H-bond and the thermodynamic strength of the resulting H-bond (Badger and Bauer [1937, 1940]) has led to controversy regarding the validity and generality of such an approach. Much of this controversy resulted from the large errors often involved in a spectrophotometric determination of a small enthalpy. Linear ΔH versus $\Delta\bar{\nu}$ relationships, however, were observed using the van't Hoff method for the determination of enthalpies for phenol reacting with a series of oxygen donors (Joesten and Drago [1962], Gramstad [1963] and West et al. [1964]), phenol and nitriles (Gramstad and Sandstrom [1969]), phenol and thioamides (Gramstad and Sandstrom [1969]), the acids phenol, pentachlorophenol, α-naphthol, and methanol with a series of organophosphorus donors (Aksnes and Gramstad [1960] and Gramstad [1961]), and various substituted phenols with *n*-heptylfluoride (Jones and Watkinson [1964]). It is clear from these reports that one linear ΔH versus $\Delta\bar{\nu}$ relationship does not hold for every acid–base pair but is different for each acid and class of donors studied. Many researchers have misinterpreted the degree of generality of such relationships and have therefore argued against their existence (Becker [1961], West et al. [1962], Powell and West [1964], Ghersetti and Lusa [1965], Barakat et al. [1966], Yoshida and Osawa [1966], Kito and Jarboe [1967], Hallam [1969] and Arnett et al. [1970]).

With the development of solution calorimetric techniques, we have in recent years been able to test relationships between accurately measured thermodynamic functions and spectroscopic shifts. The more reliable calorimetric enthalpies strongly support the validity of such linear relationships within a given "class" of donor molecules. A promising theoretical potential function even predicts a linear ΔH versus $\Delta\bar{\nu}$ relation for data in the 3–12 kcal/mole region with curvature below 3 kcal/mole to the origin (Lippincott and Schroeder [1955]). The double scale enthalpy equation (Drago and Wayland [1965]) allows not only an accurate means of calculating enthalpies which are difficult to measure, but also provides an experimental insight into the relative electrostatic and covalent contributions in a given H-bond. It is the intent of this article to discuss recent enthalpy-spectroscopic shift data and the information they contain regarding the energetics of H-bond formation.

25.2. Calorimetry

Gas phase enthalpies and spectroscopic shifts on a wide variety of H-bonding acids and electron donors would be most beneficial in extracting the important contributions to the strength of a H-bond. Unfortunately, limitations in volatility of reactants and products, laborious experimental procedures involving gas pressure changes (Brown and Gerstein [1950]) and large experimental uncertainties in gas phase spectroscopic techniques (Long and Strong [1965]) have resulted in only isolated studies of one or two acid–base systems. No definitive gas phase study has been completed on one acid with a variety of Lewis donors or one donor with a variety of acids. Various workers have shown, however, that solution enthalpies measured in non-coordinating, inert solvents, such as carbon tetrachloride and the aliphatic hydrocarbons, approximate the gas phase values (Tamres and Searles [1962], Goodenow and Tamres [1965], Epley and Drago [1967] and Sacks et al. [1968]). It is expected that solvents with negligible dipole moments and small polarizabilities will weakly solvate all species in solution and contribute little to the total energy of bond formation.

Let us further examine this cancellation in terms of a Hess Law cycle:

$$\begin{array}{ccccc} A_{(g)} & + & B_{(g)} & \xrightarrow{\Delta H_{(g)}} & AB_{(g)} \\ \downarrow{\scriptstyle \Delta H_A} & & \downarrow{\scriptstyle \Delta H_B} & & \downarrow{\scriptstyle \Delta H_{AB}} \\ A_{(s)} & + & B_{(s)} & \xrightarrow{\Delta H_{(s)}} & AB_{(s)}. \end{array}$$

For the solution reaction enthalpy to approximate the gas phase enthalpy, ΔH_A plus ΔH_B must equal ΔH_{AB}. This may be reasonable in view of the fact that measured dipole moments of H-bonded adducts are larger than the individual dipoles of the acid and base (Kimura and Fujishiro [1961]). It has been argued that the transfer energies do not necessarily cancel but are rather proportional by a constant α (Christian et al. [1966], Christian and Grundnes [1968] and Mollendal et al. [1969]).

$$\Delta G^0_{AB} = \alpha(\Delta G^0_A + \Delta G^0_B). \tag{25.1}$$

The standard free energies in the above equation represent changes for transfer of A, B and AB from gas phase to solvent or from one solvent to another and α is a constant related to the fraction of the free energy of solvation of the monomers that is retained by the molecular pair after the complex is formed. From association and distribution constants, an α value of 0.71 has been calculated for the pyridine–H_2O acid–base reaction in cyclohexane, carbon tetrachloride, toluene, benzene and 1,2-dichloroethane. This suggests that some solvent molecules are "squeezed out" when the adduct is formed (Christian et al. [1966]). A similar equation can be written in terms of transfer enthalpies, ΔH_{AB}, ΔH_A and ΔH_B. However, in view of the fact that limited adduct solvation transfer energies are available, the assumption that ΔH_{AB}, ΔH_A and ΔH_B are also related by the parameter, α, has not been critically tested. Recently reported transfer enthalpies for a few *m*-fluorophenol–base systems suggests that these energies may be solvent dependent (Nozari and Drago [1972]). So, without further data, the assumption of near cancellation of ΔH_{AB}, ΔH_A and ΔH_B transfer energies remains for H-bond enthalpies measured in inert non-coordinating solvents.

Simple, inexpensive solution calorimeter designs have been presented (Arnett et al. [1965] and Sherry and Purcell [1970a]). They principally consist of a temperature-sensing element such as a thermistor, a standard reference resistor, potentiometer and recorder, a calibration heater, a timing device, an isolated, silvered Dewar flask, a mechanical or magnetic stirrer and a sample injection port. The common procedure involves injection of a small volume of an acid as a neat liquid or as a concentrated solution into a solution of the base and the measured heat is corrected for the heat of solution or dilution of the acid. Calibration of the recorder chart paper with the calorimeter heater before and after each injection allows an accurate determination of the heat of reaction, ΔH_s. The precision of this method is better than ± 0.1 kcal/mole while the estimated accuracy of the enthalpies is ± 0.2 kcal/mole.

25.2.1. Inert solvent method

The equilibrium between a H-bonding acid, A, and a Lewis base molecule, B, in an inert solvent may be represented by:

$$A_{(s)} + B_{(s)} \rightleftharpoons AB_{(s)}$$

$$K_a = \frac{[AB]}{[A][B]} = \frac{[AB]}{([A]_0 - [AB])([B]_0 - [AB])}. \tag{25.2}$$

To allow for incomplete formation of the adduct in calculating $\Delta H_{(s)}$, the adduct concentration is expressed as

$$[AB] = \frac{Q'}{v\Delta H_{(s)}} \tag{25.3}$$

where v is the final volume of the solution after injection of the acid, Q' is the measured heat in calories corrected for the heat of solution or dilution of the acid, and $\Delta H_{(s)}$ is the desired enthalpy of solution. If large base concentrations are used such that $[B]_0 - [AB] \approx [B]_0$, the above equation may be rearranged to a form of the Scott equation used in spectrophotometric studies (Scott [1956]).

$$\frac{[A]_0[B]_0 v}{Q'} = \frac{[B]_0}{\Delta H_{(s)}} + \frac{1}{K_f \Delta H_{(s)}}. \tag{25.4}$$

The above graphical method is unfortunately often complicated by base aggregation at concentrations larger than 0.3–0.5 M (Purcell et al. [1969a]). Such aggregation could lead to an enhanced solvation of both free acid and adduct and give misleading reaction enthalpies (Drago and Epley [1969] and Purcell et al. [1969a]).

Aggregation of *N,N*-dimethylacetamide in cyclohexane results in larger phenol-DMA reaction enthalpies in this solvent than in carbon tetrachloride (Drago and Epley [1969]). These authors suggest the use of carbon tetrachloride as an inert solvent for all polar oxygen donors. Conversely, hexane or cyclohexane should be used as solvents for amines, pyridine, and sulfur donors to avoid the well demonstrated specific interactions between carbon tetrachloride and these donors (Morcom and Travers [1966], Vogel and Drago [1970] and Sherry and Purcell [1970b, 1972b]).

The recommended procedure for calorimetric determination of H-bond enthalpies in an inert solvent is to measure the heats of solution of all acids and bases to detect the concentrations at which aggregation becomes important, measure heats of reaction, Q', at a variety of acid and base concentrations below that limit found for aggregation, and simultaneously determine the reaction enthalpy, $\Delta H_{(s)}$, and the association constant, K_a, with the following equation (Bolles and Drago [1965]).

$$K_a^{-1} = \frac{Q'}{v\,\Delta H} + \frac{[A]_0[B]_0 v\,\Delta H}{Q'} - ([A]_0 + [B]_0). \qquad (25.5)$$

Substitution of the measured quantities, Q' and v, for each experiment at concentrations $[A]_0$ and $[B]_0$ results in a set of equations containing two unknowns. A discussion of the advantages of this equation over previous methods (Benesi and Hildebrand [1949], Keefer and Andrews [1952] and Scott [1956]) and a computer program for the determination of this set of equations with two unknowns has recently been outlined (R. S. Drago, personal communication).

25.2.2. Pure base method

To eliminate the need for accurately determining K_a at high dilution in an inert solvent, Arnett et al. [1970] have proposed the injection of an acid into pure base as solvent. Under these conditions, complete complexation of the acid is assured and the resulting heat is simply a summation of the heat due to H-bond formation, the heat of dissociation of the acid, and the heat of solvation of the acid and the adduct by the base molecules. Arnett corrects for the heat of solvation by subtracting (a) the heat evolved when a model compound (i.e., anisole as a model compound for phenol) is injected into the same pure base solvent and (b) the difference in the heats of solution of both model compound and acid in an inert solvent. A comparison of some reaction enthalpies of *p*-fluorophenol determined by the pure base method and the inert solvent method show reasonable agreement (± 0.2 kcal mole^{-1}) with the less polar ether donors, but poor agreement with the stronger, more polar oxygen and nitrogen donors (Arnett et al. [1970]). However, even the fortuitous agreement of the weak donors strongly depends upon the choice of model compound and inert solvent. Duer and Bertrand have proposed a more promising pure base method which compares the heat of solution of an acid and its model compound in two different pure bases (Duer and Bertrand [1970]). A cancellation of enthalpies should give accurate heats of transfer of an acid from one base to another. Unfortunately it assumes that the solvation of a given base molecule in itself, for example pyridine in pyridine, is equal to the solvation of a different base molecule in itself, for example a diethyl ether molecule in diethyl ether. Therefore, it would be unreliable for donors which undergo extra self-association because of such effects as intermolecular H-bonding (Sherry and Purcell [1970b]). So it appears, in general, that calorimetric reactions determined by a pure base method have the advantage of complete complexation but have a greater disadvantage of unusual solvation effects by polar donor molecules which can lead to erroneous enthalpy values.

25.2.3. Polar solvent method

H-bond enthalpies should be calorimetrically measured in a non-coordinating inert solvent for donors which are sufficiently soluble in these solvents. In order to advance our current knowledge of H-bond strengths, it will become necessary to include Lewis donors which are only slightly soluble in inert solvents, such as sulfides, selenides, phosphines, and arsines. Drago and colleagues have suggested a convenient elimination of solvent procedure (ESP) for determination of calorimetric enthalpies in weakly polar solvents (Drago et al. [1972], Nozari and Drago [1972] and Nozari et al. [1973]). The method is essentially based upon the displacement procedure recommended for the pure base method by Duer and Bertrand [1970]. An examination of the reaction enthalpies of *m*-fluorophenol with ethyl acetate and dimethyl sulfoxide in the solvents, carbon tetrachloride, benzene and *o*-dichlorobenzene suggests that the displacement enthalpy, ΔH_3, is a constant, independent of solvent.

$$\begin{aligned} A_{(s)} + \mathrm{EtOAc}_{(s)} &\rightarrow A\cdot\mathrm{EtOAC}_{(s)} \qquad \Delta H_1 \\ A_{(s)} + \mathrm{DMSO}_{(s)} &\rightarrow A\cdot\mathrm{DMSO}_{(s)} \qquad \Delta H_2 \\ A\cdot\mathrm{DMSO}_{(s)} + \mathrm{EtOAc}_{(s)} &\rightarrow A\cdot\mathrm{EtOAc}_{(s)} + \mathrm{DMSO}_{(s)} \qquad \Delta H_3 = \Delta H_1 - \Delta H_2. \end{aligned} \tag{25.6}$$

Thus, the enthalpies for the reaction of A with a new series of donors which are insoluble in the inert solvents may be measured in a more polar solvent and corrected by ΔH_3 to obtain the corresponding gas phase or inert solvent enthalpy. Solvation enthalpy constants have been reported for *m*-fluorophenol in *o*-dichlorobenzene (0.5 kcal mole^{-1}) and in benzene (1.2 kcal mole^{-1}).

$$-\Delta H_{\text{gas phase}} = -\Delta H_{\text{inert solvent}} = -\Delta H_{\text{polar solvent}} + \text{constant}. \tag{25.7}$$

This equality suggests that rather than a cancellation of ΔH_A, ΔH_B and ΔH_{AB} transfer energies from gas phase to an inert solvent, the sum of the transfer energies from an inert solvent to a polar solvent is a constant.

When the donors are varied with a given acid, this implies that the difference in transfer energies, $\Delta H_B - \Delta H_{AB}$, is a constant dependent only upon the acid and the polar solvent. Care must be taken in the selection of polar solvents as the cancellation will be inaccurate if the solvents undergo specific interactions with the donors or the adducts (Nozari and Drago [1972]). The same ESP technique has been successful in the study of a variety of Lewis acids with a constant donor in polar solvents (Nozari et al. [1973]). The ESP technique at this point seems experimentally much more promising than Christian's α parameter or Duer and Bertrand's pure base method. In comparing these three techniques, we are really comparing three different approaches toward the cancellation of acid, base, and adduct solvation energies. The fact that the ESP technique seems most general again argues for the use of non-polar, inert media whenever possible. This will insure the most reliable calorimetrically determined solution enthalpies for H-bond reactions.

25.3. Infrared techniques

The measurement of a H-bond wave number shift is a simple procedure compared to a calorimetric enthalpy determination. An infrared spectrum of a small amount of acid and base dissolved in an appropriate solvent will show both a free $\nu_{X\text{-}H}$ stretching vibration band and a H-bonded $\nu_{X\text{-}H\cdots Y}$ stretching vibration band due to incomplete adduct formation. Both bands are solvent dependent showing lower wave numbers in the more polar solvents. The difference between these two wave numbers, $\Delta\bar{\nu}_{X\text{-}H}$, in an inert solvent has been shown to approximate the equivalent gas phase wave number shift (Allerhand and Schleyer [1963]). Furthermore, unlike the free $\nu_{X\text{-}H}$ stretching vibration band, the H-bonded band, $\nu_{X\text{-}H\cdots Y}$, is often extremely dependent upon the base concentration. Thus the wave number shifts should be measured at the lowest possible concentration of acid and base which shows both free and bonded stretching vibrations in an inert solvent. Extrapolation of the observed $\Delta\bar{\nu}$'s to infinite base dilution should be performed whenever a base concentration dependence is observed. Wave number shifts should never be measured using the pure base as solvent.

Acceptable solvents include hexane, cyclohexane, carbon tetrachloride, and carbon disulfide. Extrapolation to infinite base dilution can be avoided by using a long path length infrared cell (2.5 cm) and carbon tetrachloride or carbon disulfide as solvent (Purcell et al. [1969b]). These solvents have no absorption in the normal H-bonded stretching vibration band region while the long path length allows acquisition of data at low concentrations of both acid and base ($\sim 10^{-3}$ M). Short path length cells, cancellation of C–H vibrations, and extrapolation to infinite base dilution are often necessary when one of the hexanes is used as solvent. The precision of most wave number shift measurements is limited to ± 5–$10\ \text{cm}^{-1}$ because of the large band width of the $\nu_{\text{X-H}\cdots\text{Y}}$ stretching frequencies.

25.4. Nuclear magnetic resonance techniques

When a H-bond is formed between an acid X–H and a base, Y, the magnetic resonance signal of the proton is generally shifted downfield from the free acid resonance peak position. These shifts are a summation of a decrease in the proton shielding resulting from X–H bond elongation, an increase in the proton shielding resulting from Y to X–H charge transfer in the formation of the H-bond, and variable shielding effects resulting from anisotropic magnetic currents existing in the Y donor molecule. The magnitude of the donor anisotropy term can usually be approximated (Narasimhan and Rodgers [1959], Howard et al. [1963], Eyman and Drago [1966] and Slejko and Drago [1973a]) leaving a shift which is proportional to the polarization of X–H bond electron density by the donor atom, Y. Recent reports advocate studying a variety of H-bonding acids with one Lewis donor to assure a constant neighbor anisotropy contribution to the H-bond chemical shifts (Slejko and Drago [1973a, b]).

For the equilibrium, $A + B \rightleftarrows AB$, the observed proton chemical shift will be an average of both free and complexed proton resonances.

$$\delta_{\text{obs.}} = \frac{[AB]}{[A]+[AB]}\,\delta_{\text{complex}} + \frac{[A]}{[A]+[AB]}\,\delta_{\text{free}} \tag{25.8}$$

$$\Delta = \delta_{\text{complex}} - \delta_{\text{free}}. \tag{25.9}$$

The free proton chemical shift of the monomeric acid, δ_{free}, is easily determined from a dilution study of the acid in an inert solvent. The complex chemical shift, δ_{complex}, is most commonly determined by lowering the temperature to a point where all of the acid is in the complex form. The assurance of completely complexed acid requires high base concentrations (2.0–4.0 M) and a lowering of the temperature to $\sim -60\,°\text{C}$. The most common solvent for these temperature studies has been methylene chloride. Unfortunately, the effect of solvent upon a H-bonded proton chemical shift may be as important as it is in determining the calorimetric enthalpy of bond formation. Gramstad and Becker [1970] have shown that the chemical shift of the phenol OH proton extrapolated to infinite dilution is 0.2 ppm further downfield in carbon tetrachloride and methylene chloride than in cyclohexane. Thus, the presence of potential donor atoms in the normally inert solvent, carbon tetrachloride, may lead to a non-gas-phase H-bonded chemical shift. If it is assumed that complex chemical shift will be the same in carbon tetrachloride, methylene chloride, and cyclohexane, the small difference in δ_{free} may be subtracted to give the correct gas-phase shift due to H-bonding. A method for simultaneous determination of Δ and K similar to the above calorimetric procedure alleviates the need for low temperatures and therefore allows the use of the inert solvents, hexane and cyclohexane (Slejko et al. [1972] and Slejko and Drago [1973b]). Experimentally, this procedure requires more sample preparation but has the advantage that high concentrations of base are not needed to assure complete complexation. The effects of solvation of the adduct by large base aggregates which could lead to unusual donor-anisotropy effects is thereby eliminated. This procedure is potentially the best method for determining gas-phase H-bond chemical shifts providing all precautions concerning experimental conditions and data analysis are observed (Drago [1973]).

25.5. Linear enthalpy-spectroscopic shift relations

Several linear enthalpy–spectroscopic shift relationships have been observed using calorimetrically determined enthalpies. The acid and base systems studied and their linear equations are listed in table 25.1. These equations, together with a spectroscopic shift for any new base within the same class of donor studied, will predict the enthalpy of adduct formation in an inert solvent within ± 0.2 kcal/mole. The importance of accurate enthalpy determinations is exemplified by comparing the linear ΔH versus $\Delta\bar{\nu}$ equations found for phenol reacting with a series of oxygen and nitrogen donors when ΔH is determined by the van't Hoff method (Joesten and Drago [1962]) versus the calorimetric method (Epley and Drago [1967]). The ΔH values determined by the spectrophotometric method for the weaker acid–base systems were consistently smaller than those determined by the calorimetric method. A slope of 0.016 and an intercept of 0.63 found in the spectrophotometric study is considerably different from that reported in the calorimetric study (see table 25.1). Another calorimet-

TABLE 25.1

Linear enthalpy-spectroscopic shift relations

Acid	Base	Equation
ΔH versus $\Delta \bar{\nu}$		
Perfluoro-*t*-butanol[a]	Oxygen and nitrogen donors	$\Delta H_{PFTB}(\pm 0.2) = 0.0106(\pm 0.0004)\Delta \bar{\nu}_{OH} + 5.0$
1,1,1,3,3,3-Hexafluoro-2-propanol[b]	Oxygen and nitrogen donors	$\Delta H_{HFIP}(\pm 0.2) = 0.0114(\pm 0.0008)\Delta \bar{\nu}_{OH} + 4.3$
Phenol[c]	Oxygen and nitrogen donors	$\Delta H_{p}(\pm 0.2) = 0.0105(\pm 0.0007)\Delta \bar{\nu}_{OH} + 3.0$
2,2,2-Trifluoroethanol[d]	Oxygen and nitrogen donors	$\Delta H_{TFE}(\pm 0.1) = 0.0121(\pm 0.0005)\Delta \bar{\nu}_{OH} + 2.7$
Pyrrole[e]	Oxygen and nitrogen donors	$\Delta H_{pyrrole}(\pm 0.1) = 0.0115(\pm 0.0007)\Delta \bar{\nu}_{NH} + 2.0$
t-Butyl alcohol[f]	Oxygen and nitrogen donors	$\Delta H_{t\text{-butanol}}(\pm 0.1) = 0.0108(\pm 0.0003)\Delta \bar{\nu}_{OH} + 1.6$
1,1,1,3,3,3-Hexafluoro-2-propanol[g]	Sulfur donors	$\Delta H_{HFIP}(\pm 0.2) = 0.0287(\pm 0.0020)\Delta \bar{\nu}_{OH} - 3.5$
2,2,2-Trifluoroethanol[h]	Sulfur donors	$\Delta H_{TFE}(\pm 0.2) = 0.0319(\pm 0.0020)\Delta \bar{\nu}_{OH} - 2.4$
ΔH versus Δ		
1,1,1,3,3,3-Hexafluoro-2-propanol[i]	Oxygen and nitrogen donors	$\Delta H_{HFIP}(\pm 0.3) = 0.89(\pm 0.05)\Delta + 4.3$
2,2,2-Trifluoroethanol[j]	Oxygen and nitrogen donors	$\Delta H_{TFE}(\pm 0.3) = 0.98(\pm 0.06)\Delta + 2.3$
Phenol[k]	Oxygen and nitrogen donors	$\Delta H_{p}(\pm 0.4) = 0.98(\pm 0.04)\Delta + 2.1$

a) Sherry and Purcell [1972a]. The average difference between the calorimetric enthalpies and those calculated with the new E and C parameters (which include a contribution to intramolecular bonding) is 1.1 kcal/mole (see Guidry and Drago [1973]). The intercept of this equation has been accordingly increased by 1.1 kcal/mole.

b,i) Purcell et al. [1969a]. The intercept has been corrected by 0.7 kcal/mole (see footnote a).

c,k) Calculated from data in Epley and Drago [1967].

d,j) Sherry and Purcell [1970a].

e) Nozari and Drago [1970].

f) Drago et al. [1970].

g) Sherry and Purcell [1972b]. The intercept has been corrected by 0.7 kcal/mole (see footnote a).

h) Sherry and Purcell [1972b].

ric study of a series of substituted phenols with similar oxygen and nitrogen donors resulted in the same ΔH versus $\Delta\bar{\nu}$ equation as that established for phenol (Drago and Epley [1969]). This suggests that the *p*-tert-butyl, *p*-chloro, and *m*-trifluoromethyl substituents have little effect upon the ΔH versus $\Delta\bar{\nu}$ relationship of phenol. Further studies of a series of fluoro-substituted aliphatic alcohols (Purcell et al. [1969a] and Sherry and Purcell [1970a, 1972a]) with the same class of donors resulted again in linear ΔH versus $\Delta\bar{\nu}$ plots with similar slopes but different intercepts. The incorporation of two weak acids, pyrrole and *t*-butanol, into this series was possible only through the use of the double scale enthalpy equation (Drago [1973]).

$$-\Delta H = E_A E_B + C_A C_B. \tag{25.10}$$

In this equation, E and C represent electrostatic and covalent contributions to the enthalpy from the acid, A, and base, B. By measuring the enthalpies of adduct formation of these two acids with three donors whose E_B and C_B parameters are known, the acid E_A and C_A parameters were solved and used to estimate the enthalpies of formation of the weaker adducts. The agreement between experimentally determined and calculated enthalpies is reportedly within 2% (Nozari and Drago [1970] and Drago et al. [1970]).

The ΔH versus $\Delta\bar{\nu}$ equations established with oxygen and nitrogen donors seem to have two regions of enthalpy and wave number shift response. Below the linear portion of the curves in the experimentally untested weak interaction regions, ΔH increases more rapidly than $\Delta\bar{\nu}$. The ratio of response increases with acid strength thereby producing the largest intercept along the enthalpy axis for the strongest acid. The parallel between intercept and acid strength (as measured by ΔH or $\Delta\bar{\nu}$) would suggest that the intercept is related to the ground state acid dipole. As shown in fig. 25.1, an excellent correlation exists between the acid electrostatic parameters (E_a) and the ΔH versus $\Delta\bar{\nu}$ intercepts. This suggests that the weak interaction region (assumed curved portion through the origin) is dominated by an electrostatic attraction between the acid and base. This dipole–dipole attraction could lead to moderate enthalpy values with only a small infrared OH or NH wave number shift. As the donor strength increases, charge transfer from the base to the acid becomes important and the data fall on the linear portion of the curve. Thus, the more electronegative R–OH substituents (a) place a greater positive charge on the H-atom which in turn leads to a larger intercept and (b) more efficiently withdraw the electron density transferred to the

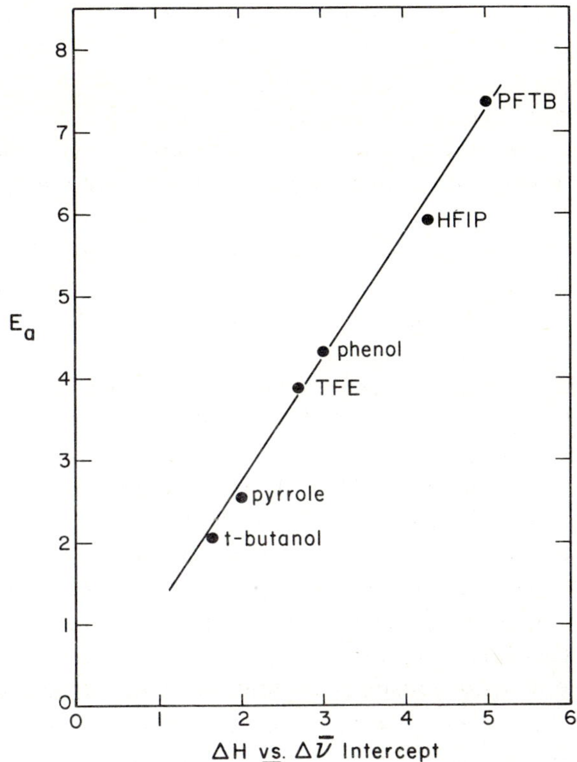

Fig. 25.1. The acid electrostatic parameters, E_a, for a series of H-bonding acids versus the corresponding intercepts of their enthalpy-wave number linear relationships.

H-atom which slides the data points along the ΔH versus $\Delta\bar{\nu}$ line. As evidenced by the indistinguishable slopes of the equations in table 25.1, electron withdrawal apparently causes a similar response in ΔH and $\Delta\bar{\nu}$ for every acid studied. This model is in contrast to an earlier model which proposed that charge transfer effects are also important in the weak interaction region (Drago et al. [1970]).

The positive intercepts of the ΔH versus Δ equations also suggest that the dipole–dipole attraction in the weak interaction region leads to moderate enthalpy values with only small changes in the chemical shift of the H-atom. This would imply that charge transfer to the H-atom is not important in this region. However, as the donor strength increases, charge transfer again becomes important and the chemical shift of the H-atom becomes more responsive. Of the two spectroscopic indices of acid strength, the chemical shift of the H-atom should be most sensitive to charge transfer effects. This is reflected in a smaller slope of the ΔH versus Δ equation for hexafluoroisopropanol rather than for trifluoroethanol or phenol while the corresponding slopes of the ΔH versus $\Delta\bar{\nu}$ equations are indistinguishable. The hexafluoroisopropyl group is, of course, more efficient at removing electron density from the H-atom than the trifluoroethyl or phenyl group (Sherry and Purcell [1970a]).

The ΔH versus $\Delta\bar{\nu}$ equations found for the reactions of hexafluoroisopropanol and trifluoroethanol with six sulfur donors are also recorded in table 25.1. It should be noted that the range of enthalpies and OH wave number shifts from the weakest to the strongest sulfur donor is approximately one-half the range covered by the oxygen donor analogs. Thus, the acidities of the relatively "hard" fluoroalcohols are masked when the "hard" oxygen donor atoms are replaced by the weaker, "soft" sulfur atoms (Sherry and Purcell [1972b]). The terms "hard" and "soft" refer here to the acid–base concepts of Pearson [1968]. The negative intercepts of the sulfur donor equations reflect the small contribution of dipole–dipole attractions to the enthalpy in the weak acid–base region. The enhanced importance of covalency is reflected in the immediate sensitivity of the OH wave number shift as one moves away from the origin. Charge transfer effects apparently dominate the H-bond strength (as measured by ΔH or $\Delta\bar{\nu}$) in the inaccessible weak interaction region and in the linear portion of the curve.

One semiempirical model of H-bond formation successfully predicts a linear ΔH versus $\Delta\bar{\nu}$ relationship with curvature in the weak interaction region toward the origin (Lippincott and Schroeder [1955]). According to the assumptions of this model, the enthalpy of H-bond formation may be divided into three terms: an endothermic OH bond breaking term (δE_{OH}), an exothermic term due to HB bond formation and OB coulombic attractions resulting from B to OH charge transfer (E_{HB}), and a $\mathrm{O}\cdots\mathrm{B}$ van der Waals repulsion term (V_{OB}). The role of the two endothermic terms is to moderate the rate of increase in the exothermic term as the $\mathrm{O}\cdots\mathrm{B}$ distance decreases (Purcell et al. [1969a]). A comparison of three pairs of analogous oxygen and sulfur donors is found in table 25.2. In every pair, the OH wave number shifts make little distinction between the oxygen and sulfur donors while the enthalpy is always largest for the oxygen donor. Obviously, ΔH is more sensitive to a change in the donor atom. Since the endothermic terms of the Lippincott–Schroeder Model ($\delta E_{\mathrm{OH}} + V_{\mathrm{OB}}$) are linearly related (Purcell and Drago [1967]) to the

TABLE 25.2

A comparison of ΔH, $\Delta\bar{\nu}$, and Δ for similar oxygen and sulfur donors

Donor	ΔH_{HFIP} ±0.2 kcal/mole	$\Delta\bar{\nu}_{HFIP}$ ±10 cm^{-1}	Δ_{HFIP} ±0.1 ppm	ΔH_{phenol} ±0.2 kcal/mole	$\Delta\bar{\nu}_{phenol}$ ±10 cm^{-1}	Δ_{phenol} ±0.1 ppm
Diethyl ether	7.7[a)]	357[a)]	4.3[a)]	6.0[c)]	279[a)]	3.7[e)]
Diethyl sulfide	6.4[b)]	348[b)]	3.8[a)]	4.6[d)]	256[d)]	2.9[e)]
Tetrahydrofuran	8.5[a)]	386[a)]	–	6.0[c)]	285[e)]	–
Tetrahydrothiophene	6.8[b)]	361[b)]	–	4.9[d)]	274[d)]	–
N,N-Dimethylacetamide	9.4[a)]	428[a)]	–	–	–	–
N,N-Dimethylthioacetamide	8.5[b)]	423[b)]	–	–	–	–

a) Purcell et al. [1969a]. The enthalpies have been corrected for intramolecular H-bonding (see footnote a, table 25.1).

b) Sherry and Purcell [1972b]. The enthalpies have been corrected for intramolecular H-bonding (see footnote a, table 25.1).

c) Drago et al. [1971].

d) Vogel and Drago [1970].

e) Eyman and Drago [1966].

exothermic term (E_{HB}), the decreased enthalpy for the sulfur donor reaction in each pair may be directly ascribed to larger van der Waals repulsions in these reactions. The increased van der Waals repulsions in the surfur donor reactions is also reflected in the larger slopes of the ΔH versus $\Delta\bar{\nu}$ equations (Sherry and Purcell [1972b]). The limited chemical shift data available for these donor pairs indicate a greater deshielding of the acid proton when the atom is oxygen. This may be a result of more effective charge transfer in the sulfur donor adducts. However, if a paramagnetic correction of 0.5 ppm is added for magnetic anisotropy of the sulfur atom (as recently suggested in Purcell et al. [1969b]), the difference in chemical shifts would be nearly leveled.

25.6. The double scale enthalpy equation

The linear enthalpy–spectroscopic shift equations discussed in the previous section are very convenient for estimating an enthalpy of reaction from an experimentally simple spectroscopic shift measurement. The equations do not however provide a great deal of direct information about the process of H-bond formation. Fortunately, all of the H-bond reactions discussed in this chapter fit the double scale enthalpy equation (Drago and Wayland [1965] and Drago et al. [1971]). This empirical correlation (see eq. (25.10)) allows the determination of an electrostatic and covalent parameter for each acid and base studied, thereby providing some insight into the processes which determine the magnitude of an enthalpy of adduct formation. The original set of E and C parameters were determined using the enthalpies of reaction of iodine with a series of alkylamines. The E_a and C_a were set equal to 1.0 for iodine while $E_b = b\mu$ (where b is a constant and μ is the ground state dipole moment) and $C_b = aR_D$ (where a is a constant and R_D is the total distortion polarization) for the alkylamines. All other E and C parameters are then determined relative to this original basis set. As pointed out by Drago and coworkers, the double scale enthalpy equation has a definite advantage over the Hard and Soft Acid–Base Model (Pearson [1968]) for it not only predicts the relative hardness or softness of an acid or base through the C/E ratios but it also predicts the strength of any acid–base pair whose parameters are known (Drago et al. [1971]). An elegant discussion of acid–base interactions and the advantages of double scale enthalpy relationships has recently been published (Drago [1973]). Table 25.3 lists the E and C parameters of the H-bonding acids and several donors used in these studies.

TABLE 25.3

E and C parameters[a] for H-bonding acids and several bases

Acid	C_a	E_a	C/E
Perfluoro-*t*-butanol (PFTB)	0.731	7.34	0.100
Phenol	0.442	4.33	0.102
Hexafluoroisopropanol (HFIP)	0.623	5.93	0.105
Trifluoroethanol (TFE)	0.451	3.88	0.116
Pyrrole	0.295	2.54	0.116
Tert-butyl alcohol	0.300	2.04	0.147
Base	C_b	E_b	C/E
Acetonitrile	1.34	0.886	1.51
Ethyl acetate	1.74	0.975	1.79
N,N-Dimethylacetamide (DMA)	2.58	1.32	1.95
N,N Dimethylformamide (DMF)	2.48	1.23	2.02
Dimethyl sulfoxide (DMSO)	2.85	1.34	2.13
Hexamethylphosphoramide (HMPA)	3.55	1.52	2.34
Acetone	2.33	0.987	2.36
Diethyl ether	3.25	0.963	3.38
Tetrahydrofuran (THF)	4.27	0.978	4.37
Pyridine	6.40	1.17	5.48
Triethylamine	11.09	0.991	11.20
Quinuclidine[b]	13.20	0.704	18.75
Diethyl sulfide	7.40	0.339	21.80

[a] From tables in Drago [1973].
[b] Slejko and Drago [1973b].

The E_a parameters of the H-bonding acids reflect the order of acid strength toward oxygen and nitrogen donors as discussed in section 25.5. This indicates that the electrostatic contribution (E_aE_b) to the enthalpy dominates in these reactions. These H-bonding acids would be classified as "hard" acids when compared with a $C/E = 1.0$ for the typically "soft" acid, I_2. Interesting inversions in acid strength, however, can take place. For example, toward all oxygen and nitrogen donors ("hard" donors), phenol is always a stronger acid than trifluoroethanol as measured by the enthalpy, OH wave number shift, or chemical shift. However, toward a typical "soft" donor like diethyl sulfide, trifluoroethanol produces a larger enthalpy of reaction than does phenol (Sherry and Purcell [1972b]). The OH wave number shifts and chemical shifts however still rank phenol as the stronger acid. This suggests that the electrostatic attractions continue to dominate the charge transfer effects in determining the wave number shifts and chemical shifts of these H-bond reactions.

The twelve oxygen and nitrogen donors are listed in order of increasing C/E ratios, from the "hardest" donor, acetonitrile, to the "softest" donor, quinuclidine. Even though the H-bonding acids are considered "hard", the strength of acid–base interaction in this series always increases from acetonitrile to quinuclidine. When placed in increasing order of donor strength toward H-bonding acids, all bases follow the C/E ratios except DMF, DMA, DMSO and HMPA which always lie between THF and pyridine. The increased strength of these donors (over that predicted by the C/E ratio) results from an extra large electrostatic interaction. A closer examination reveals resonance structures for these four donors which would place a greater negative charge on the donor atom and hence promote a greater electrostatic contribution.

$$(CH_3)_2N{-}C(=O)CH_3 \rightleftharpoons (CH_3)_2\overset{\oplus}{N}{=}C(O^{\ominus})CH_3$$

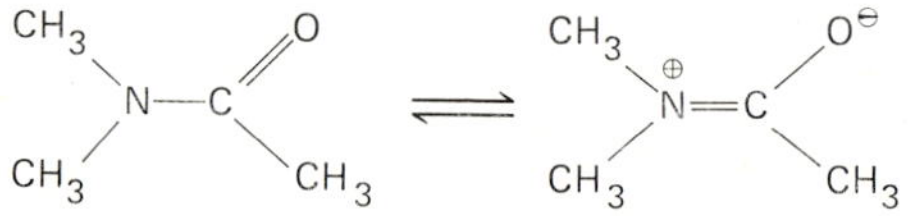

A plot of the trifluoroethanol H-bonding enthalpies versus the donor C_b parameters is shown in fig. 25.2. The solid line has a slope equal to the C_a parameter of trifluoroethanol ($C_a = 0.451$) and its intercept is equal to an electrostatic contribution (E_aE_b) of 3.8 kcal/mole. The electrostatic contribution to the reaction enthalpies with acetonitrile, ethylacetate, acetone, diethyl ether, tetrahydrofuran and triethylamine, fit the reference line well (their E_aE_b is very nearly constant at 3.8 kcal/mole). The difference between this constant electrostatic contribution and the measured enthalpy is the covalent contribution for these six donors. The total electrostatic contribution to a H-bond energy can arise from two terms (Lippincott and Schroeder [1955]). These are the dipole–dipole attractions in the weak interaction region before charge transfer becomes important (Structure I) and the attractions resulting from the transfer of charge from the donor to the acceptor (Structure II).

$$\text{R–X–}\overset{\delta+}{\text{H}} \qquad \overset{\delta-}{\text{Y}} \qquad\qquad \text{R–}\overset{\delta-}{\text{X}}\text{–H}\cdots\cdots\overset{\delta+}{\text{Y}}$$

$$\text{I} \qquad\qquad\qquad\qquad \text{II}$$

Hence, an increased charge migration also increases the electrostatic contribution to the total bond energy. The constant electrostatic contribu-

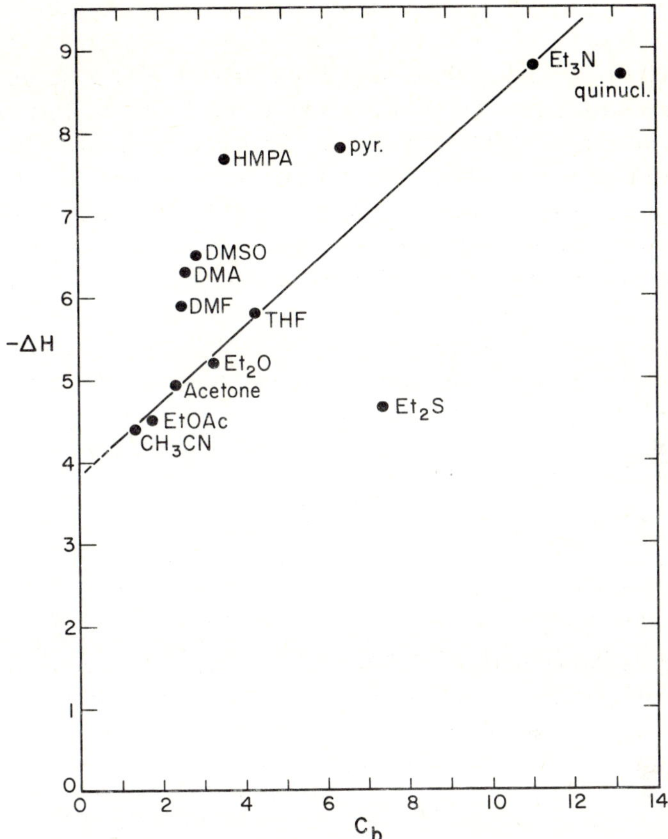

Fig. 25.2. Calorimetric solution enthalpies for the H-bonding acid, 2,2,2-trifluoroethanol versus the base covalent parameters, C_b.

tion of the six reference donors therefore represent a constant sum of the above two electrostatic terms. A total electrostatic contribution of 3.8 kcal/mole to the acetonitrile-trifluoroethanol enthalpy is largely due to a dipole–dipole attraction as in Structure I (dipole moment = 3.97 μ) whereas that same total for the triethylamine-trifluoroethanol reaction primarily results from electrostatic attractions resulting from charge transfer as in Structure II (dipole moment = 0.64 μ). The four remaining reference donors have intermediate contributions from both resonance structures. Those points lying above the solid line represent donors which have larger electrostatic contributions than the six reference donors, while the two points below the line represent donors with smaller electrostatic contributions (see table 25.3). It should be noted that all donors which lie on the solid line or above it adhere to the ΔH versus $\Delta\bar{\nu}$ and ΔH versus Δ relationships of trifluoroethanol. The nitrogen donor, quinuclidine, and the sulfur donor, diethyl sulfide, however, do not adhere to these same relationships (Sherry and Purcell [1970b] and Slejko and Drago [1973b]). Again, this may reflect the high sensitivity of the spectroscopic shifts to the total electrostatic contributions, i.e., these donors fall off the enthalpy-spectroscopic shift lines because of their unusually low electrostatic contributions and not because of their unusually large covalent contributions.

A comparison of C_b parameters with donor atom hybridization in table 25.4 shows another interesting trend. As the p character of the donor lone pair orbital increases, the C_b parameter increases, suggesting a greater tendency to transfer charge to the acceptor. Thus, the order of nitrogen donor strength (acetonitrile < pyridine < triethylamine) toward H-bonding acids results primarily from the greater charge transfer capabilities of a donor hybrid orbital containing the largest proportion of p character. The larger C_b parameters of the cyclic donors, tetrahydrofuran and quinuclidine, (over diethyl ether and triethylamine, respectively) suggests that internal bond strain increases the electron donor orbital p character.

25.7. Single scale enthalpy equations

Many two-parameter equations have been successfully used to correlate infrared OH stretching vibration wave numbers (Pimentel and McClellan [1960]) or solution enthalpies (Drago and Epley [1969], Sherry and Purcell [1970a] and Vogel and Drago [1970]) with acid substituent constants. For example, the ratio of enthalpies, $\Delta H_x/\Delta H_{TFE}$ (where ΔH_x is the reaction enthalpy of any other OH acid), was nearly constant for all of the oxygen and nitrogen donors used in establishing the linear ΔH versus $\Delta\bar{\nu}$ relationships (Sherry and Purcell [1970a, 1972b]). This ratio, defined as α_A, could therefore be used to estimate the enthalpy of reaction via the two-parameter equation, $-\Delta H = \alpha_A\beta_B$, where β_B is a constant representative only of the strength of each donor. A plot of the acid parameters, α_A, versus the Taft polar substituent constants (σ^*) for each R–OH substituent shows excellent linearity for all alcohols included, except phenol and di-tert-butylcarbinol. The positive deviation of phenol and the negative deviation of di-tert-butylcarbinol were suggestive of π reso-

TABLE 25.4

Correlation of hybridization of donor atoms with C_b parameters

Oxygen donors	Donor atom hybridization	C_b
$(CH_3)_2N{-}C(=O){-}H$	sp^2	2.48
$CH_3{-}C(=O){-}CH_3$	sp^2	2.33
Et–O–Et	sp^3	3.25
(tetrahydrofuran ring, O)	sp^3	4.27

Nitrogen donors	Donor atom hybridization	C_b
$CH_3{-}C{\equiv}N{:}$	sp	1.34
(pyridine ring) N:	sp^2	6.40
$Et_3N{:}$	sp^3	11.09
(quinuclidine cage) N:	sp^3	13.20

nance enhancement of acidity and steric crowding, respectively. The enthalpy of reaction of any new alcohol whose σ^* is known with any donor whose β_B parameter is known could easily be calculated using the two-parameter equation (Sherry and Purcell [1970a]). Such equations, however, are limited to donors with similar C/E ratios (i.e., they will not predict inversions in acid strength) and are therefore generally less informative than a double scale relationship. A comparison of the E and C parameters and their relationship to other Hammett equations and the Hard and Soft Acid–Base Model has been published (Drago et al. [1971]). The versatility of the four-parameter E and C equation makes it by far the best model for predicting enthalpies of adduct formation in poorly solvating media.

25.8. Conclusions

An examination of the current electronic theories of H-bonding (Puranik and Kumar [1963], Bratož [1967] and Schuster [1970]) reveals the importance of charge transfer, electrostatic attractions and short range repulsion effects in explaining spectroscopic wave number shifts and intensity changes and in determining the strength of a H-bond. The incorporation of H-bond reaction enthalpies determined in a non-polar inert solvent into the double scale E and C correlation provides an estimate of these relative electrostatic and covalent contributions. All H-bonding acids included in this correlation are considered "hard" based upon their small C/E ratios. Thus, the energy of the bond formed between a "hard" oxygen or nitrogen donor and a H-bonding acid is primarily electrostatic in nature. The importance of covalency increases and in fact often dominates when the donor atom is a "soft" sulfur atom. Although the E and C equation is only empirical and has never been proven, it does seem intuitively correct when compared to existing theoretical arguments. For example, Schuster has found a near linear relationship between the calculated energies of weak H-bonds and the amount of charge transferred from donor to acceptor (Schuster [1970]). The order of calculated covalency in the donor series ($R_2CO < R_2O < R_3N$) and acceptor series ($N\text{–}H < O\text{–}H < F\text{–}H$) is the same as that predicted on the basis of C_a and C_b parameters. The electrostatic parameters, E_a and E_b, are obviously derived from both acid–base dipole–dipole attractions and adduct attractions resulting from base to acid charge transfer. A fairly linear correlation between the E_a parameters and pK_a's or OH substituent constants suggest that the ground state dipole attractions dominate the acid electrostatic contributions to the H-bond energy. No correlation, however, is found between the donor E_b parameters and their ground state dipole moments. This would suggest that the excited state electrostatic contributions weigh heavily in determining the total donor electrostatic contribution.

The importance of accurate solution calorimetric or gas phase enthalpy determinations in studying the H-bond process cannot be overemphasized. Linear enthalpy-spectroscopic shift relationships appear to be general within a given class of Lewis donors, i.e., "hard" oxygen and nitrogen donors and "soft" sulfur donors. When comparing the intercepts and slopes of these equations, one is actually comparing the relative electrostatic and covalent contributions to the enthalpies and spectroscopic shifts with increasing donor strength. The similar slopes of these equations within a given class of donors suggests that the bond breaking energy (as reflected by the magnitude of $\Delta\bar{\nu}$) is a linear function of the enthalpy of these reactions. Further calorimetric and spectroscopic data from a wider variety of donors and acceptors is needed to develop a complete understanding of the mechanisms involved in the formation of a H-bond.

References

Aksnes, G. and T. Gramstad, 1960, Acta. Chem. Scand. **14**, 1485.

Allerhand, A. and P. V. R. Schleyer, 1963, J. Am. Chem. Soc. **85**, 371.

Arnett, E. M., W. G. Bentrude, J. J. Burker and P. McC. Duggleby, 1965, J. Am. Chem. Soc. **87**, 1541.

Arnett, E. M., L. Joris, E. Mitchell, T. S. S. R. Murty, T. M. Gorrie and P. V. R. Schleyer, 1970, J. Am. Chem. Soc. **92**, 2365.

Badger, R. M. and S. H. Bauer, 1937, J. Chem. Phys. **5**, 839.

Badger, R. M. and S. H. Bauer, 1940, J. Chem. Phys. **8**, 288.

Barakat, T. M., M. J. Nelson, S. M. Nelson and A. D. E. Pullin, 1966, Trans. Faraday Soc. **62**, 2674.

Becker, E. D., 1961, Spectrochim. Acta **17**, 436.

Benesi, H. A. and J. H. Hildebrand, 1949, J. Am. Chem. Soc. **71**, 2703.

Bolles, T. F. and R. S. Drago, 1965, J. Am. Chem. Soc. **87**, 5015.

Bratož, S., 1967, Adv. Quant. Chem. **3**, 209.

Brown, H. C. and M. Gerstein, 1950, J. Am. Chem. Soc. **72**, 2923.

Christian, S. D. and J. Grundnes, 1968, Acta Chem. Scand. **22**, 1702.

Christian, S. D., J. R. Johnson, H. E. Affsprung and P. J. Kilpatrick, 1966, J. Phys. Chem. **70**, 3376.

Drago, R. S., 1973, Structure and Bonding, Vol. 15, Springer-Verlag, Heidelberg, pp. 73–139.

Drago, R. S. and T. D. Epley, 1969, J. Am. Chem. Soc. **91**, 2883.

Drago, R. S. and B. B. Wayland, 1965, J. Am. Chem. Soc. **87**, 3571.

Drago, R. S., N. O'Bryan and G. C. Vogel, 1970, J. Am. Chem. Soc. **92**, 3924.

Drago, R. S., G. C. Vogel and T. E. Needham, 1971, J. Am. Chem. Soc. **93**, 6014.

Drago, R. S., M. S. Nozari and G. C. Vogel, 1972, J. Am. Chem. Soc. **94**, 90.

Duer, W. C. and G. L. Bertrand, 1970, J. Am. Chem. Soc. **92**, 2587.

Epley, T. D. and R. S. Drago, 1967, J. Am. Chem. Soc. **89**, 5770.

Eyman, D. P. and R. S. Drago, 1966, J. Am. Chem. Soc. **88**, 1617.

Ghersetti, S. and A. Lusa, 1965, Spectrochim. Acta **21**, 1067.

Goodenow, J. M. and M. Tamres, 1965, J. Chem. Phys. **43**, 3393.

Gramstad, T., 1961, Acta. Chem. Scand. **15**, 1337.
Gramstad, T., 1963, Spectrochim. Acta **19**, 497.
Gramstad, T. and E. D. Becker, 1970, J. Mol. Struct. **5**, 253.
Gramstad, T., and J. Sandstrom, 1969, Spectrochim. Acta **25A**, 31.
Guidry, R. M. and R. S. Drago, 1973, J. Am. Chem. Soc. **95**, 759.
Hadži, D., 1959, Hydrogen Bonding, Pergamon Press, Inc., New York, N.Y.
Hallam, H. E., 1969, J. Mol. Struct. **3**, 43.
Howard, B. B., C. F. Jumper and M. T. Emerson, 1963, J. Mol. Spectrosc. **10**, 117.
Joesten, M. D. and R. S. Drago, 1962, J. Am. Chem. Soc. **84**, 3817.
Jones, D. A. K. and J. G. Watkinson, 1964, J. Chem. Soc. 2367.
Keefer, R. M. and L. J. Andrews, 1952, J. Am. Chem. Soc. **74**, 1891.
Kimura, K. and R. Fujishiro, 1961, Bull. Chem. Soc. Japan **34**, 304.
Kito, T. and C. H. Jarboe, 1967, J. Org. Chem. **32**, 407.
Lippincott, E. R. and R. Schroeder, 1955, J. Chem. Phys. **23**, 1099.
Long, F. T. and R. L. Strong, 1965, J. Am. Chem. Soc. **87**, 2345.
Mollendal, H., J. Grundnes and E. Augdahl, 1969, Acta Chem. Scand. **23**, 3525.
Moore, T. S. and T. F. Winmill, 1912, J. Chem. Soc. **101**, 1635.
Morcom, K. W. and D. N. Travers, 1966, Trans. Faraday Soc. **62**, 2063.
Narasimhan, P. T. and M. T. Rodgers, 1959, J. Phys. Chem. **63**, 1388.
Nozari, M. S. and R. S. Drago, 1970, J. Am. Chem. Soc. **92**, 7086.
Nozari, M. S. and R. S. Drago, 1972, J. Am. Chem. Soc. **94**, 6877.
Nozari, M. S., C. D. Jensen and R. S. Drago, 1973, J. Am. Chem. Soc. **95**, 3162.
Pearson, R. G., 1968, J. Chem. Educ. **45**, 581, 643.
Pimentel, G. and A. McClellan, 1960, The Hydrogen Bond, W. H. Freeman and Co., San Francisco.
Puranik, P. G. and V. Kumar, 1963, Proc. Indian Acta. Sci. **58**, 29.
Purcell, K. F. and R. S. Drago, 1967, J. Am. Chem. Soc. **89**, 2874.
Purcell, K. F., J. A. Stikeleather and S. D. Brunk, 1969a, J. Am. Chem. Soc. **91**, 4019.
Purcell, K. F., J. A. Stikeleather and S. D. Brunk, 1969b, J. Mol. Spectrosc. **32**, 202.
Sacks, L. J., R. S. Drago and D. P. Eyman, 1968, Inorg. Chem. **7**, 1484.
Schuster, P., 1970, Theoret. Chim. Acta. **19**, 212.
Scott, R. L., 1956, Rec. Trav. Chim. **75**, 787.
Sherry, A. D. and K. F. Purcell, 1970a, J. Phys. Chem. **74**, 3535.
Sherry, A. D. and K. F. Purcell, 1970b, J. Am. Chem. Soc. **92**, 6386.
Sherry, A. D. and K. F. Purcell, 1972a, J. Am. Chem. Soc. **94**, 1853.
Sherry, A. D. and K. F. Purcell, 1972b, J. Am. Chem. Soc. **94**, 1848.
Slejko, F. L. and R. S. Drago, 1973a, J. Am. Chem. Soc. **95**, 6935.
Slejko, F. L. and R. S. Drago, 1973b, Inorg. Chem. **12**, 176.
Slejko, F. L., R. S. Drago and D. G. Brown, 1972, J. Am. Chem. Soc. **94**, 9210.
Tamres, M. and S. Searles, Jr., 1962, J. Phys. Chem. **66**, 1099.
Vogel, G. C. and R. S. Drago, 1970, J. Am. Chem. Soc. **92**, 5347.
West, R., D. L. Powell, L. S. Whatley, M. K. T. Lee and P. V. R. Schleyer, 1962, J. Am. Chem. Soc. **84**, 3221.
West, R., D. L. Powell, M. K. T. Lee and L. S. Whatley, 1964, J. Am. Chem. Soc. **86**, 3227.
Yoshida, Z. and E. Osawa, 1966, J. Am. Chem. Soc. **88**, 4019.

CHAPTER 26

VAPOR PRESSURE STUDIES ON HYDROGEN AND DEUTERIUM BONDING

H. WOLFF

Physikalisch-Chemisches Institut, Universität Heidelberg, D-69 Heidelberg, West Germany

Contents

26.1. Introduction 1227
26.2. Activity coefficients and results derived from their representation according to Wilson 1227
26.3. Results derived from the theory of ideal associated solutions 1229
26.4. Results derived from the vapor pressure ratios 1240
26.5. Explanations 1243
26.6. Supplements and confirmations 1251
26.7. Summary 1257
References 1257

The hydrogen bond – recent developments in theory and experiments
Eds. P. Schuster et al.

26.1. Introduction

In addition to spectroscopic, dielectric and other investigations, vapor pressure measurements can be used for the study of H-bonding (Prigogine and Defay [1954] and Prausnitz [1969a]). In such investigations it is of particular interest to study the behavior of the simplest aliphatic amines, alcohols and thiols as well as that of their various deuterated forms. The present report deals with the results of investigations in which the activity coefficients were determined from measurements of the vapor pressure isotherms and in which these activity coefficients were then used to obtain information on the hydrogen and deuterium bonding of these compounds in solutions, mainly in *n*-hexane. The observations and calculations for some other compounds in the undiluted state or in mixtures with appropriate solvents supplement and confirm the results.

26.2. Activity coefficients and results derived from their representation according to Wilson

26.2.1. The activity coefficients of a binary mixture are defined as

$$f_1 = \frac{p_1}{x_1 R_1 P_1} \tag{26.1}$$

and

$$f_2 = \frac{p_2}{x_2 R_2 P_2}, \tag{26.2}$$

where p_1 and p_2, x_1 and x_2, and R_1 and R_2 are the partial pressures, the mole fractions, and the corrections for non-ideal gas of the respective components of the mixtures, and where P_1 and P_2 are the pressures of these components in the undiluted state.

Formerly, the activity coefficients were derived mostly from dynamic measurements, i.e. the measurements in a circulating distillation apparatus. The measurements in such an apparatus are, however, affected by many uncertainties, and the determination of the total pressure and the composition of the liquid and the gas at equilibrium is a somewhat tedious method, particularly when the measurements are performed over a greater range of temperatures.

Now the determination of activity coefficients by static measurements is facilitated by an evaluation method introduced by Barker [1953]. In applying this method, one measures solely the total vapor pressure of the liquid. If the volume of the vapor phase is very small, evaporation in reaching the equilibrium does not affect the composition of the liquid, and the composition of the vapor is obtained by calculation. Applying a least-square fit approximation which can be programmed for computer calculation, expressions for the activity coefficients are obtained which give them as function of the mole fractions of the mixtures. When the activity coefficients are represented by means of the equations of Redlich and Kister [1948], the expressions are

$$\ln f_1 = Ax_2^2 - Bx_2^2(1-4x_1) + Cx_2^2(1-8x_1+12x_1^2) + \dots \tag{26.3}$$

and

$$\ln f_2 = Ax_1^2 + Bx_1^2(1-4x_2) + Cx_1^2(1-8x_2+12x_2^2) + \dots, \tag{26.4}$$

where A, B, C, etc. are the constants determined by least squares.

For strongly non-ideal mixtures, as for many solutions of H-bonded substances, a representation of the activity coefficients suggested by Wilson [1964], and successfully applied by other authors (Gölles and Wolfbauer [1968], Prausnitz [1969a, b], Krug et al. [1971], Wolff et al. [1975] and Wolff and Götz [1976]), is more suitable. These equations are

$$\ln f_1 = -\ln(x_1 + \Lambda_{12}x_2) + x_2\left(\frac{\Lambda_{12}}{x_1+\Lambda_{12}x_2} - \frac{\Lambda_{21}}{\Lambda_{21}x_1+x_2}\right) \tag{26.5}$$

and

$$\ln f_2 = -\ln(x_2 + \Lambda_{21}x_1) - x_1\left(\frac{\Lambda_{12}}{x_1+\Lambda_{12}x_2} - \frac{\Lambda_{21}}{\Lambda_{21}x_1+x_2}\right), \tag{26.6}$$

where Λ_{12} and Λ_{21} are given by the relations

$$\Lambda_{12} = \frac{V_2}{V_1}\exp\left(-\frac{\lambda_{12}-\lambda_{11}}{RT}\right) \tag{26.7}$$

and

$$\Lambda_{21} = \frac{V_1}{V_2}\exp\left(-\frac{\lambda_{12}-\lambda_{22}}{RT}\right). \tag{26.8}$$

V_1 and V_2 are the mole volumes of the two components of the mixture, and $\lambda_{12}-\lambda_{11}$ as well as $\lambda_{12}-\lambda_{22}$ are constants, likewise to be determined by least squares.

26.2.2. In a physical interpretation of the λ's, λ_{11} and λ_{22} can be conceived as energy parameters reflecting the intermolecular forces between the like molecules 1 and 2, whereas λ_{12} is associated with the interaction between the unlike molecules 1 and 2. The sign of these parameters is negative by definition. Positive values of $\lambda_{12}-\lambda_{11}$ and $\lambda_{12}-\lambda_{22}$, therefore, indicate a greater interaction between the like than between the unlike molecules, the interaction between the like molecules being stronger with greater positive values of $\lambda_{12}-\lambda_{11}$ and $\lambda_{12}-\lambda_{22}$.

If on the basis of these statements the data (I and II) for solutions of amines, alcohols and thiols in *n*-hexane are compared, one observes for all the mixtures investigated that $\lambda_{12}-\lambda_{11}$ is positive and greater than $\lambda_{12}-\lambda_{22}$ (table 26.1). That means, that as expected the interaction between the H-bonding molecules is greater than the interaction between the H-bonding and the solvent molecules. In addition, $\lambda_{12}-\lambda_{11}$ decreases in the sequence: methanols, ethanols, methylamines, dimethylamines, methanethiols and diethylamines; that is, $\lambda_{12}-\lambda_{11}$ reflects the strength and degree of association which, from other observations, is also known for these compounds. The degree of association is also reflected by the temperature dependence of the values, showing an increase of $\lambda_{12}-\lambda_{11}$ with decrease of temperature (Krug et al. [1971], Wolff et al. [1975] and Wolff and Götz [1976]).

Now comparing the variously deuterated forms, it is most conspicuous that $\lambda_{12}-\lambda_{11}$ of the compounds deuterated in the associating group is greater than $\lambda_{12}-\lambda_{11}$ of the corresponding non-deuterated compound. Assuming λ_{12} to be nearly equal or at least only slightly different for the corresponding compounds, this observation suggests that the association energy and the association degree are somewhat greater for the OD, ND and ND_2 compounds than for their OH, NH and NH_2 analogues (Wolff et al. [1975] and Wolff and Götz [1976]).

26.3. Results derived from the theory of ideal associated solutions

26.3.1. Figure 26.1 shows the vapor pressure isotherms for mixtures of methylamine with *n*-hexane and for mixtures of trimethylamine with *n*-hexane. The straight line corresponds to Raoult's law, the curve is the

TABLE 26.1

Wilson parameters $\lambda_{12}-\lambda_{11}$ and $\lambda_{12}-\lambda_{22}$ of *n*-hexane solutions of amines, alcohols and thiols, differing in the deuteration of the associating group[a)]

Temperature (°C)	I. Compound non-deuterated in the associating group			II. Compound deuterated in the associating group		
	Compound	$\lambda_{12}-\lambda_{11}$	$\lambda_{12}-\lambda_{22}$	Compound	$\lambda_{12}-\lambda_{11}$	$\lambda_{12}-\lambda_{22}$
+20	CH_3NH_2	929	268	CH_3ND_2	973	263
−20		1118	420		1160	453
+20	CD_3NH_2	941	249	CD_3ND_2	972	268
−20		1118	439		1159	483
+20	$C_2H_5NH_2$	690	85	$C_2H_5ND_2$	729	81
−20		836	133		864	159
+20	$(CH_3)_2NH$	571	−47	$(CH_3)_2ND$	552	4
−20		670	+55		696	61
+20	$(CD_3)_2NH$	526	6	$(CD_3)_2ND$	565	−13
−20		657	72		689	69
+40	$CH_3C_2H_5NH$	346	−7	$CH_3C_2H_5ND$	371	−22
+20		381	+22		415	−1
+60	$(C_2H_5)_2NH$	238	−61	$(C_2H_5)_2ND$	240	−62
+20		242	+21		331	−73
+60	CH_3OH	2597	690	CH_3OD	2636	658
+40		2726	759		2774	726
+60	CD_3OH	2604	736	CD_3OD	2637	707
+40		2751	798		2777	746
+60	C_2H_5OH	2123	324	C_2H_5OD	2160	312
+40		2242	351		2285	346
+60	C_2D_5OH	2162	329	C_2D_5OD	2175	323
+40		2290	362		2317	350
+20	CH_3SH	514	130	CH_3SD	508	147
−20		589	222		578	248

[a)] Wolff and Dill [1976], Wolff and Götz [1976], Wolff and Schiller [1976] and Wolff et al. [1975].

result of our measurements (Wolff and Würtz [1968]). The measured curve for the system with methylamine, i.e. the system with H-bonding, shows positive pressure deviations from the straight line. Trimethylamine has no group capable of H-bonding, and the measured points nearly coincide with the straight line. These observations make it clear that the positive pressure deviations result mainly from the association of the H-bonding component of the system. The same explanation seems justified for the positive deviations of other systems from a self-

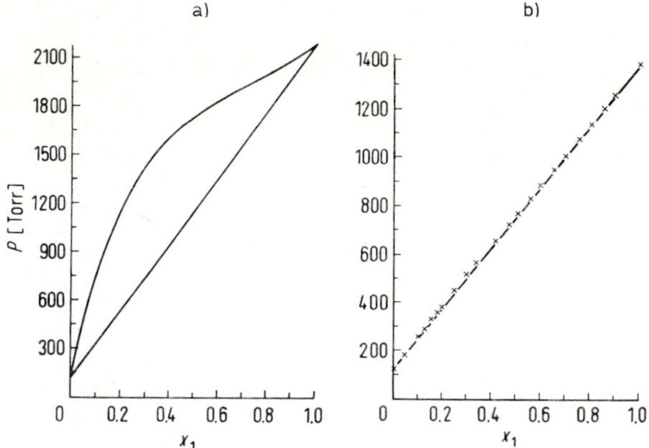

Fig. 26.1. Vapor pressure isotherms (a) of the mixtures of methylamine and *n*-hexane and (b) of trimethylamine and *n*-hexane at +20 °C. p = pressure, x_1 = mole fraction of amine.

associating compound and an inert solvent in which the self-associating component is similar in chemical nature, size and shape of the molecules to the solvent, apart from the H-bonding group (Prigogine and Defay [1954]). The explanation, therefore, can be assumed to be valid also for the much stronger positive deviations of the solutions of alcohols in *n*-hexane (fig. 26.2, Wolff and Höppel [1968a]) and for the weaker deviations of the solutions of thiols in this solvent.

26.3.2. An equilibrium of monomers with tetramers, as proposed by some authors (Feeney and Sutcliffe [1961, 1962] and Bystrow and Lezina [1964]), cannot explain the changes with temperature, observed for the vapor pressure isotherms of amine solutions (Wolff and Wolff [1968]). Calculations of association data from the excess functions based on the Mecke–Kempter equations likewise make some difficulties (Kehiaian [1966], Kehiaian et al. [1967] and Wolff and Wolff [1968]). Therefore, it seems reasonable to assume equilibria of monomers with dimers, dimers with trimers, etc. (continuous association) where the equilibrium constants of dimerization, trimerization, etc. may be different if necessary. The same assumption seems admissible for hexane solutions of the only weakly associated thiols. For the alcohols, regarding the variety of association models discussed (table 26.2), a decision in favor of a definite model is difficult. But, as for amines and thiols, the evaluation of measurements can be based on the model of continuous association admitting different equilibrium constants. This model makes the fewest

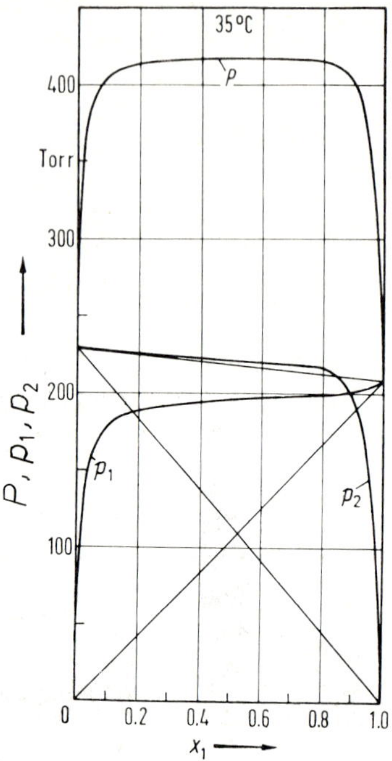

Fig. 26.2. Vapor pressure isotherms and partial pressure curves of the mixture of methanol and *n*-hexane at +35 °C. P, p_1 and p_2 = total pressure, partial pressure of methanol and partial pressure of *n*-hexane, x_1 = mole fraction of methanol.

specific assumptions, and no definite evidence seems to exist for a different model. In comparing the results for deuteroisomeric compounds, the model may also be of minor importance.

One can then conclude, as suggested by Prigogine and Defay [1954], that the binary solutions behave ideally if they are regarded as mixtures consisting of the solvent and the different forms of the associating component. Thus two expressions for the Gibbs energy of the solution are obtained, the one viewing the mixture as a non-ideal binary system

$$dG = \mu_A \, dn_A + \mu_B \, dn_B \tag{26.9}$$

(μ_A and μ_B = chemical potentials of the components of the binary system, n_A and n_B = moles of each component), and the other expression viewing the mixture as an ideal multi-component system

$$dG = \sum_{\nu} \mu_{A_\nu} \, dn_A + \mu_{B_1} \, dn_B \tag{26.10}$$

TABLE 20.2

Models of alcohol association

Assumed association equilibrium	System	Method[a]	References
Monomer-trimer(cyclic)	t-butanol/CCl_4	NMR	Storek and Kriegsmann [1968]
Monomer-tetramer(cyclic + linear)	t-butanol/decane, 1-octanol/decane	IR	Fletcher and Heller [1967]
Monomer-dimer(linear)-trimer or tetramer(cyclic)	Heptanol-1/CCl_4	DM	Bordewijk et al. [1973]
Monomer-trimer(linear)-polymer(octamer or hexamer, cyclic)	Methanol/hexadecane, t-butanol/hexadecane	NMR, IR VP, PVT	Tucker et al. [1969] and Tucker and Becker [1973]
Monomer-dimer-trimer- . . . polymer (Chain assoc., $K_{1,2} = K_{2,3} = K_{3,4} \ldots$)	Methanol/CCl_4, ethanol/CCl_4, etc.	IR	Mecke [1948]
	Alcohols/hydrocarbons	VP	Redlich and Kister [1947]
	Alcohols/CCl_4	VP, IR	Prigogine and Defay [1954]
Monomer-dimer(linear)-trimer- . . . polymer	Methanol/solvents	IR	Bellamy and Pace [1966]
Monomer-dimer(linear)-trimer(linear)- . . . polymer(cyclic)	Alcohols/hydrocarbons, etc.	DM	Ibbitson and Moore [1967a, b]
Monomer-dimer(cyclic)- . . . polymer	Methanol/CCl_4, etc.	IR	Liddell and Becker [1957]
Monomer-dimer(cyclic)-trimer(cyclic)- . . . polymer(linear)	Methanol/matrix	IR	Van Ness et al. [1967]
	Ethanol/hexane	DM	Oster [1946]
Monomer-dimer(cyclic)- . . . polymer ($K_{1,2} \neq K_{2,3} = K_{3,4} = K_{3,5} \ldots$)	Ethanol/heptane, ethanol/toluene	VP, IR	Van Ness et al. [1967] and Haskell et al. [1968]
Monomer-dimer-trimer- . . . polymer ($K_{1,2} \neq K_{2,3} \neq K_{3,4} \ldots$)	Methanol/hexane, methanol/CCl_4, etc.	VP	Wolff and Höppel [1968a, c]
Monomer-dimer-trimer- . . . polymer + physical interaction ($K_{1,2} = K_{2,3} \ldots$)	Alcohols/hydrocarbons	VP	Renon and Prausnitz [1967] and Prausnitz [1969a]

[a] NMR = nuclear magnetic resonance measurements, IR = infrared measurements, DM = dielectric measurements, VP = vapor pressure measurements, PVT = PVT measurements.

(μ_{A_ν} and μ_{B_1} = chemical potentials of the components of the multi-component system, n_{A_ν} and n_B = moles of these components). The definitions for the various chemical potentials, showing their dependence on the activity coefficients f_1 and f_2 and on the stoichiometric and true mole fractions are

$$\mu_A = \mu_A^0 + RT \ln f_1 x_1, \tag{26.11}$$

$$\mu_B = \mu_B^0 + RT \ln f_2 x_2, \tag{26.12}$$

$$\mu_{A_\nu} = \mu_{A_\nu}^0 + RT \ln x_{A_\nu} \tag{26.13}$$

and

$$\mu_{B_1} = \mu_{B_1}^0 + RT \ln x_{B_1}. \tag{26.14}$$

Stoichiometric and true mole fractions are

$$x_1 = \frac{n_A}{n_A + n_B}, \tag{26.15}$$

$$x_2 = \frac{n_B}{n_A + n_B}, \tag{26.16}$$

$$x_{A_\nu} = \frac{n_{A_\nu}}{\sum_\nu n_{A_\nu} + n_B} \tag{26.17}$$

and

$$x_{B_1} = \frac{n_B}{\sum_\nu n_{A_\nu} + n_B}. \tag{26.18}$$

Appropriate transformation and solution of eqs. (26.9–26.18) yields the quantities characterizing association (Prigogine and Defay [1954] and Wolff and Höpfner [1962]).

The fractions of molecules of the associating component which exist as monomers, dimers, trimers, etc. are found to be

$$\beta_1 = f_1/(f_2 \lim_{x_1 \to 0} f_1), \tag{26.19}$$

$$\beta_2 = 2K_{1,2} x_1 f_2 \beta_1^2, \tag{26.20}$$

$$\beta_3 = 3K_{1,2} K_{2,3} x_1^2 f_2^2 \beta_1^3, \qquad \text{etc.} \tag{26.21}$$

The mole fraction equilibrium constants of dimerization, trimerization, etc. are

$$K_{1,2} = \frac{1 - \beta_1}{2x_1 f_2 \beta_1^2}, \tag{26.22}$$

$$K_{2,3} = \frac{\frac{1}{3} - \frac{1}{3}\beta_1 - \frac{2}{3}K_{1,2}x_1 f_2 \beta_1^2}{K_{1,2}x_1^2 f_2^2 \beta_1^3}, \qquad \text{etc.} \tag{26.23}$$

If the activity coefficients are given by the three-constant Redlich–Kister equations, the equilibrium constants can be derived from the constants of these equations

$$K_{1,2} = A - 3B + 5C \tag{26.24}$$

and

$$K_{2,3} = 2\left[\frac{B - 4C}{A - 3B + 5C} + (A - 3B + 5C)\right]. \tag{26.25}$$

From the constants of the Wilson equations the equilibrium constants follow as

$$K_{1,2} = \frac{2 - \Lambda_{12} - \Lambda_{12}\Lambda_{21}^2}{2\Lambda_{12}} \tag{26.26}$$

and

$$K_{2,3} = \frac{\frac{9}{\Lambda_{12}} - 6 + \Lambda_{12} - 12\Lambda_{21}^2 + 3\Lambda_{12}\Lambda_{21}^2 + 2\Lambda_{12}\Lambda_{21}^3 + 3\Lambda_{12}\Lambda_{21}^4}{6 - 3\Lambda_{12} - 3\Lambda_{12}\Lambda_{21}^2} \tag{26.27}$$

(Wolff et al. [1975]). Multiplying the mole fraction constants by the mole volume of the solvent, the concentration equilibrium constants K^c are obtained. The association energies follow from the equilibrium constants according to van't Hoff. The average degree of association results as

$$\bar{n} = x_1 f_1 / (\beta_1 \lim_{x_1 \to 0} f_1 - x_2 f_1). \tag{26.28}$$

26.3.3. Looking at the results obtained with eqs. (26.19 – 26.28) for the compounds non-deuterated in the associating group (table 26.3), one sees that, in agreement with the much more pronounced positive deviations, the association constants of the alcohols are 10–20 times larger than those of the corresponding amines and than those of the thiol. Conversely, the fractions of the molecules which exist as monomers in the solutions are largest for the amines and the thiol and smallest for the alcohols. The association energies of the alcohols, being about 5 kcal/mole, are 2–3 times larger than those of the amines and the thiol. The constants of dimerization and trimerization are found to be different for all the systems investigated*.

Constants and energies have also been determined by infrared (IR)

* As in the work by Haskell et al. [1968], $K_{1,2} \neq K_{2,3} = K_{3,4}$, etc. can be assumed. From the thus determined association constants of the two-constant model the activity coefficients of this model can be calculated. It is interesting to note that they agree with the regularly calculated activity coefficients within about 10% over the whole concentration range. This is true for the amines as well as for the alcohols (Wolf et al. [1976]).

TABLE 26.3

Association constants $K_{1,2} = x_2/x_1^2$ ($K_{1,2}^c = c_2/c_1^2$) and $K_{2,3} = x_3/x_2x_1$ and association energy $\Delta U_{1,2}$ for dilute hexane solutions of amines[a], alcohols[b] and methanethiol[c], non-deuterated in the associating groups[d]. Vaporization energy $\Delta U_v = RT^2(\mathrm{d}\ln P_1/\mathrm{d}T) - RT$ of the same compounds at the undiluted state

Compound	Temperature (°C)	$K_{1,2}$	($K_{1,2}^c$) (l/mole)	$\Delta U_{1,2}$ (cal/mole)	$K_{2,3}$	ΔU_v (cal/mole)
CH_3NH_2	+20	1.5_2	(0.19_9)	−2090	2.5_6	5770
	−20	2.6_7	(0.33_1)		4.7_2	
CD_3NH_2	+20	1.6_5	(0.21_6)	−1830	2.7_9	5760
	−20	2.7_2	(0.33_7)		4.81	
$C_2H_5NH_2$	+20	1.1_7	(0.15_3)	−2030	2.0_6	6410
	−20	2.02	(0.25_0)		3.6_2	
$(CH_3)_2NH$	+20	0.9_1	(0.11_9)	−1930	1.4_8	6150
	−20	1.5_3	(0.19_0)		2.6_6	
$(CD_3)_2NH$	+20	0.8_5	(0.11_1)	−2020	1.3_8	6100
	−20	1.4_7	(0.18_2)		2.5_7	
$CH_3C_2H_5NH$	+40	0.5_4	(0.07_3)	−2340	0.8_2	6430
	+20	0.7_0	(0.09_1)		1.1_1	
$(C_2H_5)_2NH$	+60	0.2_6	(0.03_6)	−2090	–	6850
	+20	0.4_0	(0.05_3)		–	
CH_3OH	+60	$15._0$	(2.0_8)	−4930	$23._0$	8390
	+40	$24._2$	(3.2_5)		$36._8$	
CD_3OH	+60	$15._2$	(2.1_1)	−5250	$23._3$	8380
	+40	$25._2$	(3.3_9)		$38._3$	
C_2H_5OH	+60	$10._4$	(1.4_4)	−4360	$16._0$	9340
	+40	$15._8$	(2.1_2)		$24._2$	
C_2D_5OH	+60	$11._0$	(1.5_2)	−4530	$17._0$	9400
	+40	$17._1$	(2.3_0)		$26._1$	
CH_3SH	+20	0.6_8	(0.08_9)	−1700	–	5710
	−20	1.0_8	(0.13_4)		–	

[a] Wolff et al. [1964], Wolff and Höpfner [1967a], Wolff and Würtz [1970b] and Wolff and Schiller [1976].

[b] Wolff and Götz [1976] and Wolff et al. [1975].

[c] Wolff and Dill [1976].

[d] The calculation of association constants for alcohols is based on Wilson type representations of the activity coefficients. The calculations for amines and methanethiol is basing on Redlich–Kister equations with three constants, the calculation for diethylamine on equations with only two constants.

spectroscopy. Where this has been possible, as with solutions of dimethylamine and diethylamine and of methylamine and its homologues, the constants or energies thus determined agree reasonably well with the values from vapor pressure measurements (Wolff and Schmidt [1964], Wolff and Würtz [1970b] and Wolff and Gamer [1972a, b]). Likewise, the correspondence is reasonable to the results of other measurements, e.g. in the case of the alcohols to the results of calorimetric measurements by Savini et al. [1965]).

For primary and secondary amines the association energy is found to be the same. This is in agreement with Raman and IR investigations which revealed that usually only one of the two H-atoms of the amino group forms a H-bond (Wolff and Staschewski [1962], Wolff and Horn [1967, 1968], and Wolff and Mathias [1973]).

Finally, the calculations do not distinguish between cyclic and linear aggregates. Thus when both occur simultaneously, the calculated values represent an average.

26.3.4. If one now compares the findings for the non-deuterated amines and alcohols and the non-deuterated thiol with the results for the deuterated compounds (table 26.4 and fig. 26.3), one observes a behavior

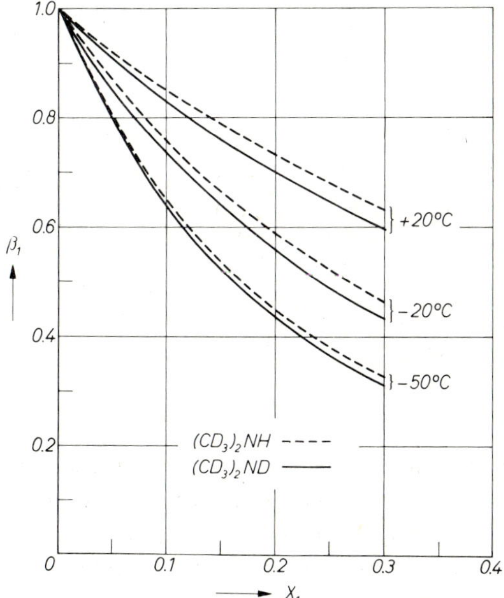

Fig. 26.3. Fractions β_1 of the molecules of hexadeuterodimethylamine and *N*-deuterohexadeuterodimethylamine, in *n*-hexane solutions at +20, −20 and −50°C existing as monomers. x_1 = mole fraction of amine.

TABLE 26.4

Association constants $K_{1,2} = x_2/x_1^2$ ($K_{1,2}^c = c_2/c_1^2$) and $K_{2,3} = x_3/x_2x_1$ and association energy $\Delta U_{1,2}$ for dilute hexane solutions of amines[a] alcohols[b] and methanethiol[c], deuterated in the associating groups[d]. Vaporization energy $\Delta U_v = RT^2(\mathrm{d}\ln P_1/\mathrm{d}T) - RT$ of the same compounds at the undiluted state

Compound	Temperature (°C)	$K_{1,2}$	($K_{1,2}^c$) (l/mole)	$\Delta U_{1,2}$ (cal/mole)	$K_{2,3}$	ΔU_v (cal/mole)
CH_3ND_2	+20	1.6_0	(0.20_9)	−2140	2.7_5	5880
	−20	2.8_6	(0.35_5)		5.0_7	
CD_3ND_2	+20	1.6_0	(0.20_0)	−2150	2.7_5	5870
	−20	2.8_5	(0.35_3)		5.0_6	
$C_2H_5ND_2$	+20	1.2_8	(0.16_7)	−2130	2.2_5	6520
	−20	2.2_7	(0.28_1)		4.0_2	
$(CH_3)_2ND$	+20	0.9_6	(0.12_5)	−1950	1.5_3	6190
	−20	1.6_3	(0.20_2)		2.8_4	
$(CD_3)_2ND$	+20	1.0_0	(0.13_1)	−2170	1.5_9	6170
	−20	1.6_5	(0.20_5)		2.8_6	
$CH_3C_2H_5ND$	+40	0.5_8	(0.07_9)	−2000	0.8_8	6480
	+20	0.7_3	(0.09_5)		1.2_1	
$(C_2H_5)_2ND$	+60	0.2_6	(0.03_6)	−2870	–	6880
	+20	0.4_7	(0.06_2)		–	
CH_3OD	+60	$15._9$	(2.2_0)	−5130	$24._4$	8520
	+40	$26._2$	(3.5_2)		$39._7$	
CD_3OD	+60	$16._0$	(2.2_1)	−5150	$24._5$	8470
	+40	$26._3$	(3.5_4)		$39._9$	
C_2H_5OD	+60	$11._0$	(1.5_2)	−4490	$17._0$	9440
	+40	$17._0$	(2.2_9)		$25._9$	
C_2D_5OD	+60	$11._3$	(1.5_7)	−4790	$17._3$	9420
	+40	$17._9$	(2.4_1)		$27._3$	
CH_3SD	+20	0.7_1	(0.09_3)	−1110	–	5720
	−20	0.9_5	(0.11_8)		–	

[a] Wolff and Höpfner [1965, 1967a], Wolff and Würtz [1970b] and Wolff and Schiller [1976].
[b] Wolff and Götz [1976] and Wolff et al. [1975].
[c] Wolff and Dill [1976].
[d] See footnote d) of Table 26.3.

which agrees with that derived from the differences in the Wilson constant $\lambda_{12} - \lambda_{11}$, and which also follows from the fact that the activity coefficients f_1 of the OD, ND or ND_2 compounds are greater than the coefficients of the non-deuterated analogues* (table 26.5). All data for the

* The activity coefficients of undiluted alcohols are unity by definition. The greater values of the activity coefficients of the OD, ND or ND_2 compounds, therefore, are observed only with dilution.

compounds deuterated in the associating group (with exception of the data for the only weakly associated thiols and diethylamines) indicate a somewhat enhanced association. This is true for the association constants and energies (thus also for the vaporization energies), as well as for the computed distribution of monomers and associates and for the computed average degree of association. Solely the solvation energy, which is a

TABLE 26.5

Activity coefficients f_1 and f_2 of variously deuterated methylamines and methanols (1) in solutions with n-hexane (2) at + 20 and + 60°C, respectively

(a) methylamine-n-hexane (20°C)

Mole fraction amine	CH_3NH_2		CH_3ND_2		CD_3NH_2		CD_3ND_2	
	f_1	f_2	f_1	f_2	f_1	f_2	f_1	f_2
0	4.0_0	1	4.2_6	1	4.1_5	1	4.2_8	1
0.01	3.8_8	1.00_0	4.0_1	1.00_0	3.8_9	1.00_0	4.0_2	1.00_0
0.05	3.4_7	1.00_4	3.6_7	1.00_4	3.5_6	1.00_4	3.6_8	1.00_4
0.10	3.0_6	1.01_4	3.2_0	1.01_5	3.1_0	1.01_5	3.2_1	1.01_5
0.15	2.7_5	1.03_1	2.8_4	1.03_3	2.7_5	1.03_3	2.8_5	1.03_3
0.20	2.4_6	1.05_3	2.5_3	1.05_7	2.4_7	1.05_7	2.5_5	1.05_7
0.30	2.0_5	1.1_2	2.1_0	1.1_3	2.0_5	1.1_2	2.1_1	1.1_3
0.40	1.7_6	1.2_2	1.7_8	1.2_3	1.7_5	1.2_2	1.7_9	1.2_3
0.50	1.5_3	1.3_6	1.5_4	1.3_9	1.5_2	1.3_7	1.5_4	1.3_9
0.75	1.1_{47}	2.2_4	1.1_{47}	2.3_0	1.1_{46}	2.2_4	1.1_{47}	2.3_1
1	1	6.3_5	1	6.5_0	1	6.3_3	1	6.5_2

(b) methanol-n-hexane (60°C)

Mole fraction alcohol	CH_3OH		CH_3OD		CD_3OH		CD_3OD	
	f_1	f_2	f_1	f_2	f_1	f_2	f_1	f_2
0	$37._9$	1	$39._9$	1	$38._5$	1	$40._3$	1
0.01	$28._9$	1.00_1	$30._0$	1.00_1	$29._3$	1.00_1	$30._3$	1.00_1
0.05	$14._0$	1.02_3	$14._2$	1.02_3	$14._1$	1.02_3	$14._3$	1.02_3
0.10	8.2_1	1.06_7	8.2_3	1.06_8	8.2_8	1.06_7	8.3_0	1.06_3
0.15	5.7_4	1.1_2	5.7_3	1.1_2	5.7_8	1.1_2	5.7_8	1.1_{24}
0.20	4.3_9	1.18_7	4.3_8	1.19_0	4.4_2	1.18_8	4.4_1	1.19_0
0.30	2.9_8	1.34_9	2.9_7	1.35_3	3.0_{00}	1.35_0	2.9_9	1.35_3
0.40	2.2_5	1.5_7	2.2_4	1.5_7	2.2_7	1.5_7	2.2_6	1.5_7
0.50	1.8_0	1.8_7	1.8_0	1.8_8	1.8_2	1.8_8	1.8_2	1.8_8
0.75	1.2_3	3.6_2	1.2_3	3.6_2	1.2_3	3.6_4	1.2_3	3.6_4
1	1	$23._6$	1	$22._5$	1	$26._3$	1	$24._3$

quantity associated with monomers, appears to be smaller for the compounds deuterated in the associating group than for the non-deuterated analogues (Wolff and Würtz [1970b]).

26.4. Results derived from the vapor pressure ratios

26.4.1. Thus far, mainly the data based on the theory of ideal associated solutions have been considered. More striking differences between the behavior of the non-deuterated and the deuterated compounds are observed when the partial pressures in solutions are compared. Accordingly, the ratio of these pressures is considered. Assuming the gas imperfections to be equal for the deuteroisomers, the ratio of the partial pressures of the heavy and the light compound in the same solvent at the mole fraction x_1 follows from eq. (26.1) as

$$\frac{p_1}{p'_1} = \frac{f_1 P_1}{f'_1 P'_1}, \tag{26.29}$$

where the prime refers to the lighter compound. That is to say, the partial pressure ratio can be calculated from the vapor pressures of the two deuteroisomers at the undiluted state and from the activity coefficients which have already been used for the calculation of the association constants, association energies, etc. (Wolff and Höpfner [1965]).

26.4.2. Figure 26.4 shows the results of the calculation for solutions of variously deuterated methylamines in n-hexane. The figure shows, as function of the amine mole fraction x_1, the partial pressure ratio for methylamines with the deuteration difference in the alkyl group, the amino group, and finally with the deuteration difference in both groups (Wolff and Höpfner [1967a]).

When the partial pressure quotients of the compounds, having the deuteration difference only in the methyl group, are considered (fig. 26.4a and b), there is hardly any change in the ratios with dilution. The deviations of the curves from a nearly horizontal course must be ascribed to the insufficiencies of the representations of the activity coefficients (Redlich–Kister equations with three constants). One speaks of a normal vapor pressure isotope effect when p/p' is smaller than one, and one speaks of an inverse effect when p/p' is greater than one. Using this terminology, one states that the vapor pressure isotope effect of the pairs with the deuteration difference in the alkyl groups is inverse in the whole concentration range.

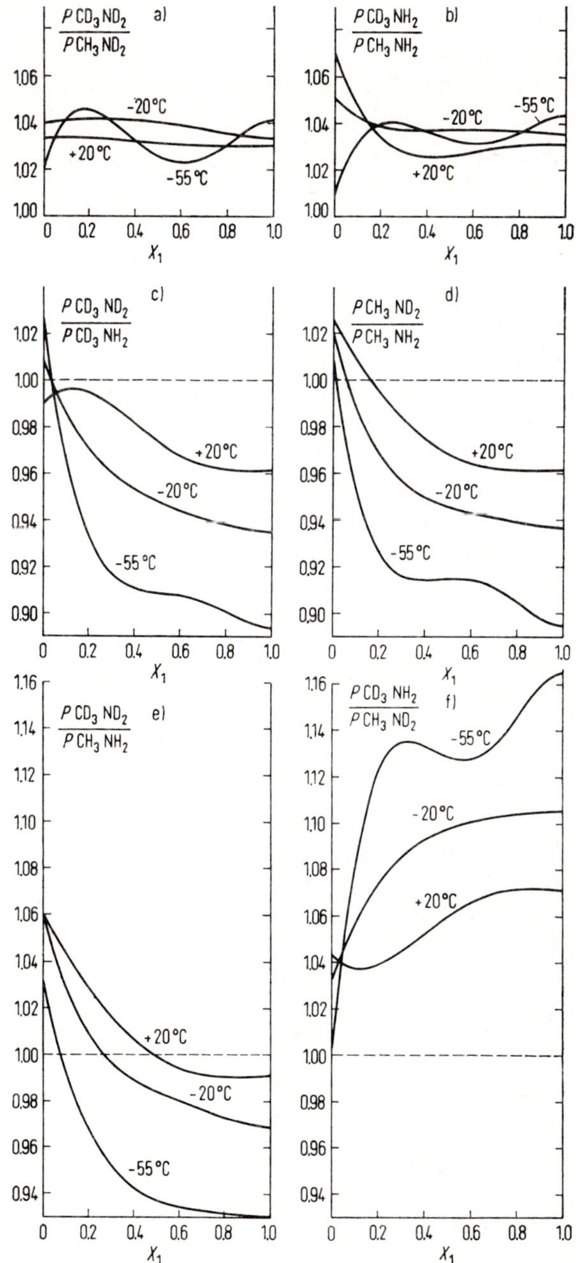

Fig. 26.4. Changes of vapor pressure ratios of variously deuterated methylamines on dilution with *n*-hexane at temperatures from +20 to −55°C. x_1 = mole fraction of methylamine.

When the vapor pressure ratios of the methylamines showing the deuteration difference in the amino group are examined (fig. 26.4c and d), a change from an isotope effect, which is normal in the undiluted state, to an effect which is inverse at high dilution is distinctly seen. For the pairs with the deuteration difference in both groups (fig. 26.4e and f) a superposition of the effects of the amino and the methyl group deuteration is seen. The oscillations of the curves in fig. 26.4c and d as well as in fig. 26.4d and e are again due to the errors of the calculated p/p'.

All curves with a deuteration difference in the amino group (fig. 26.4c–f) show, in addition, the transition from the normal to the inverse effect or at least a hint to that transition, when the temperature is increased. This is more distinctly seen in fig. 26.5, the dotted lines of which show the temperature dependence of the vapor pressure ratios of the undiluted methylamines for which fig. 26.4 has shown the concentration dependence of the ratios. Again, one observes an inverse effect of the

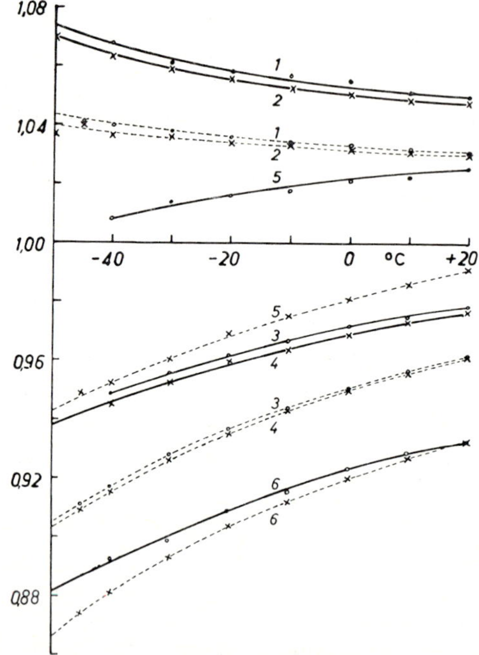

Fig. 26.5. Changes of vapor pressure ratios of variously deuterated methylamines (----) and dimethylamines (———) in the undiluted state for temperatures from +20 to −50°C. (1) $P_{CD_3NH_2}/P_{CH_3NH_2}$ and $P_{(CD_3)_2NH}/P_{(CH_3)_2NH}$, (2) $P_{CD_3ND_2}/P_{CH_3ND_2}$ and $P_{(CD_3)_2ND}/P_{(CH_3)_2ND}$, (3) $P_{CH_3ND_2}/P_{CH_3NH_2}$ and $P_{(CH_3)_2ND}/P_{(CH_3)_2NH}$, (4) $P_{CD_3ND_2}/P_{CD_3NH_2}$ and $P_{(CD_3)_2ND}/P_{(CD_3)_2NH}$, (5) $P_{CD_3ND_2}/P_{CH_3NH_2}$ and $P_{(CD_3)_2ND}/P_{(CH_3)_2NH}$, (6) $P_{CH_3ND_2}/P_{CD_3NH_2}$ and $P_{(CH_3)_2ND}/P_{(CD_3)_2NH}$.

compounds with the deuteration difference in the alkyl groups; this inverse effect is only slightly dependent on temperature. Furthermore, one again observes a normal effect of the pairs with the deuteration difference in the amino groups; this effect tends to change into the inverse effect with increase of temperature. Finally, the pairs with the deuteration difference in both groups again show the behavior which derives from the superposition of the single effects.

The vapor pressure ratios of other variously deuterated amines (Wolff and Höpfner [1965] and Wolff and Würtz [1970a]) and the ratios of the variously deuterated alcohols (Wolff and Höppel [1968b], Wolff et al. [1975] and Wolff and Götz [1976]) show some differences, compared with the curves for the methylamines, but fundamentally the situation is the same. Due to the larger number of deuterium atoms, the inverse effect of the dimethylamines [fig. 26.5 (solid lines) and fig. 26.6] is considerably greater than for the methylamines. The inverse effect of the variously deuterated methanols and ethanols (figs. 26.7 and 26.8) is unexpectedly small. For the alcohols with the difference in the hydroxyl groups the transition from the normal to the inverse effect with dilution seems to occur at concentrations so low that the calculated vapor pressure ratios are noticeably affected by the errors of their determination. But within the limits of error the curves conform to the curves for the methyl- and the dimethylamines, and at the pair C_2D_5OD/C_2H_5OH a cross-over is already seen for the total pressure curves (fig. 26.9). Therefore, the evidence of the transition cannot be doubted. For the very weakly associated diethylamines and methanethiols no distinct vapor pressure isotope effect is seen (Wolff and Dill [1976] and Wolff and Schiller [1976]).

If the temperature is not too low, undiluted compounds usually show the inverse effect if they exist as monomers, whereas they show the normal effect if they are associated (table 26.6, Rabinovich [1970]). The normal effect, however, is not observed if the associating groups of the two compounds of the pair are the same. With regard to these results and to the fact that the transition from the normal to the inverse effect occurs with dilution of associated molecules, the explanation of the transition by the change from the associated molecules to the monomers is unquestionable (Wolff and Höpfner [1965]).

26.5. Explanations

26.5.1. A more detailed explanation of the characteristic results follows from the relation which statistical mechanics gives for the vapor pressure

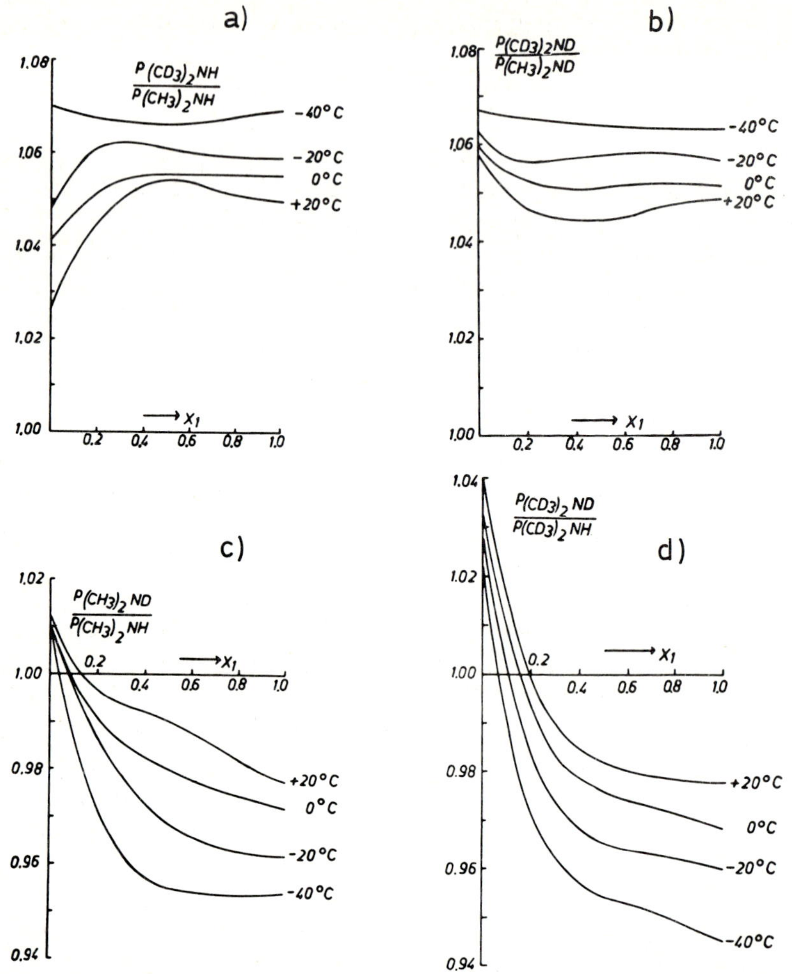

Fig. 26.6. Changes of vapor pressure ratios of variously deuterated dimethylamines on dilution with *n*-hexane at temperatures from +20 to −40°C. x_1 = mole fraction of dimethylamine.

ratio. In its general form this relation has been formulated by Bigeleisen [1961] and Jancso and Van Hook [1974]. Neglected corrections for the isotope effect on the mole volume of the condensed phase, for the influence of gas imperfection and for non-classical rotation in the gas phase, the equation can be reduced to a form in which the ratio depends only on the inter- and the intramolecular vibrations of the pair. Assuming harmonic vibrations Bigeleisen's equation can then be formulated as

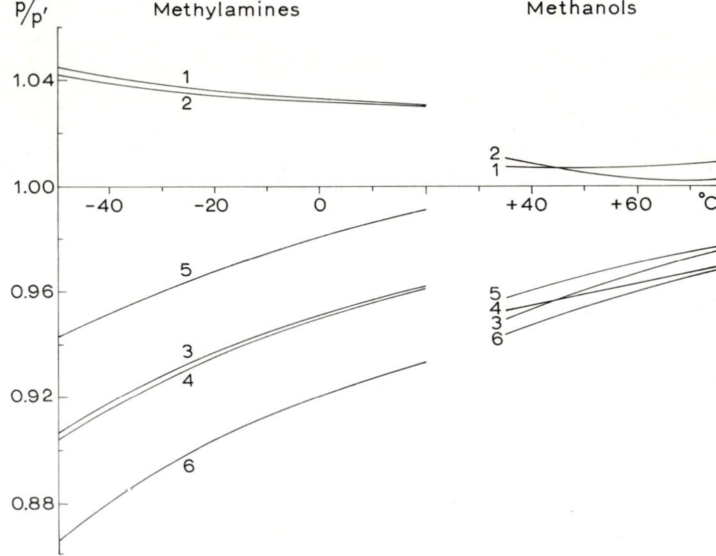

Fig. 26.7. Changes of vapor pressure ratios of variously deuterated methanols and methylamines in the undiluted state for temperatures from +35 to +75°C and from −50 to +20°C. (1) P_{CD_3OH}/P_{CH_3OH} and $P_{CD_3NH_2}/P_{CH_3NH_2}$, (2) P_{CD_3OD}/P_{CH_3OD} and $P_{CD_3ND_2}/P_{CH_3ND_2}$, (3) P_{CH_3OD}/P_{CH_3OH} and $P_{CH_3ND_2}/P_{CH_3NH_2}$, (4) P_{CD_3OD}/P_{CD_3OH} and $P_{CD_3ND_2}/P_{CD_3NH_2}$, (5) P_{CD_3OD}/P_{CH_3OH} and $P_{CD_3ND_2}/P_{CH_3NH_2}$, (6) P_{CH_3OD}/P_{CD_3OH} and $P_{CH_3ND_2}/P_{CD_3NH_2}$.

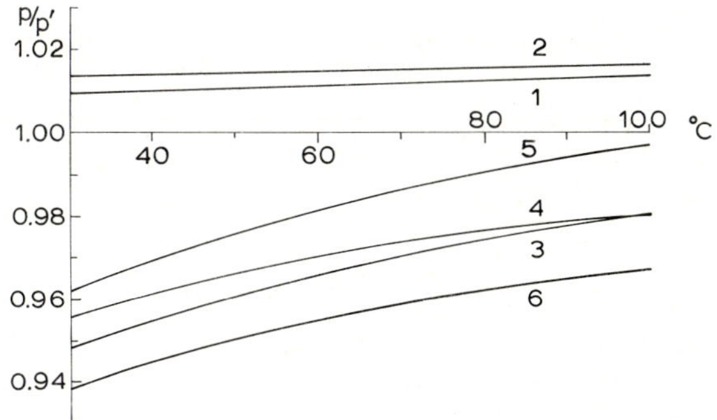

Fig. 26.8. Changes of vapor pressure ratios of variously deuterated ethanols in the undiluted state for temperatures from +30 to +100°C. (1) $P_{C_2D_5OH}/P_{C_2H_5OH}$, (2) $P_{C_2D_5OD}/P_{C_2H_5OD}$, (3) $P_{C_2H_5OD}/P_{C_2H_5OH}$, (4) $P_{C_2D_5OD}/P_{C_2D_5OH}$, (5) $P_{C_2D_5OD}/P_{C_2H_5OH}$, (6) $P_{C_2H_5OD}/P_{C_2D_5OH}$.

TABLE 26.6

Pairs of undiluted compounds showing the vapor pressure isotope effect

I. Pairs of non-associated compounds showing the normal effect ($P_{heavy} < P_{light}$)	Ref.	II. Pairs of compounds showing the inverse effect ($P_{heavy} > P_{light}$)	Ref.	III. Pairs of associated compounds showing the normal effect ($P_{heavy} < P_{light}$)	Ref.
H_2/D_2	a	CH_4/CH_3D ⎫		$^{14}N^{16}O/^{15}N^{18}O$	o
$^{20}Ne/^{22}Ne$	b	CH_4/CH_2D_2 ⎬ $T > 98$ K	e	H_2O/D_2O ($T < 225$ °C)	n
$^{14}N_2/^{15}N_2$	c	CH_4/CHD_3		NH_3/ND_3	p
$^{16}O_2/^{18}O_2$	d	CH_4/CD_4 ⎭			
CH_4/CH_3D ⎫		C_2H_6/CH_2DCH_2D ⎫		CH_3OH/CH_3OD ⎫	
CH_4/CH_2D_2 ⎬ $T < 74$ K	e	C_2H_6/CH_2DCHD_2 ⎬	f	CD_3OH/CD_3OD ⎭	i
CH_4/CHD_3		$C_2H_6CD_3CH_3$ ⎭			
CH_4/CD_4 ⎭		C_2H_4/C_2H_3D ⎫		C_2H_5OH/C_2H_5OD ⎫	
		$C_2H_4/C_2H_2D_2$ ⎭	g	C_2D_5OH/C_2D_5OD ⎭	k
		C_6H_6/C_6D_6 ⎫		CH_3NH_2/CH_3ND_2 ⎫	
		C_6H_{12}/C_6D_{12} ⎭	h	$C_2H_5NH_2/C_2H_5ND_2$ ⎭	l
		CH_3OH/CD_3OH ⎫		$(CH_3)_2NH/(CH_3)_2ND$ ⎫	
		CH_3OD/CD_3OD ⎭	i	$(CD_3)_2NH/(CD_3)_2ND$ ⎭	m
		C_2H_5OH/C_2D_5OH ⎫			
		C_2H_5OD/C_2D_5OD ⎭	k		
		CH_3NH_2/CD_3NH_2 ⎫			
		CH_3ND_2/CD_3ND_2 ⎭	l		
		$(CH_3)_2NH/(CD_3)_2NH$ ⎫			
		$(CH_3)_2ND/(CD_3)_2ND$ ⎭	m		
		H_2O/D_2O ($T > 225$ °C)	n		

References: **a** = Woolley et al. [1948]; **b** = Bigeleisen and Roth [1961]; **c** = Clusius and Schleich [1958]; **d** = Clusius et al. [1961]; **e** = Armstrong et al. [1955]; **f** = Van Hook [1964]; **g** = Bigeleisen et al. [1963] and Stern et al. [1963]; **h** = Davies and Schiessler [1953]; **i** = Beersman and Jungers [1947] and Rabinovich [1970]; **k** = Rabinovich [1970] and Wolff and Götz [1976]; **l** = Wolff and Höpfner [1967a]; **m** = Wolff and Würtz [1970a]; **n** = Landolt-Börnstein [1960]; **o** = Clusius et al. [1959a, b]; **p** = Kirshenbaum and Urey [1942], Groth et al.

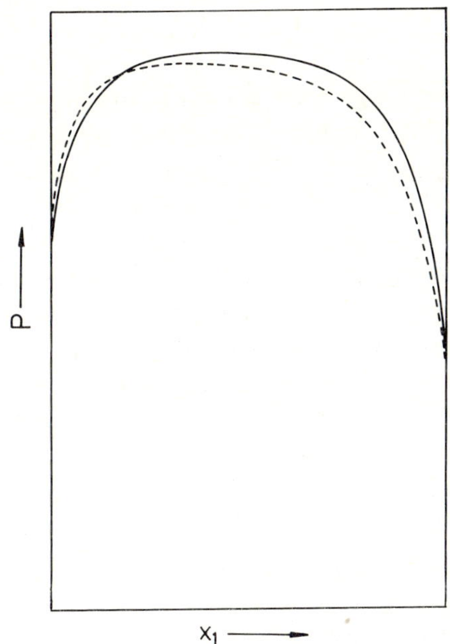

Fig. 26.9. Schematic representation of vapor pressure isotherms of binary mixtures of C_2D_5OD and *n*-hexane (----) and of C_2H_5OH and *n*-hexane (——). P = total pressure, x_1 = mole fraction of alcohol.

$$\frac{p}{p'} = \prod_{i=1}^{3N} q_i = \prod_{i=1}^{6} \left(\frac{\Theta'_{i_{\text{cond}}}}{\Theta_{i_{\text{cond}}}} \frac{\sinh(\Theta_{i_{\text{cond}}}/2T)}{\sinh(\Theta'_{i_{\text{cond}}}/2T)} \right) \times \prod_{i=7}^{3N} \left(\frac{\Theta'_{i_{\text{cond}}} \Theta_{i_{\text{gas}}}}{\Theta_{i_{\text{cond}}} \Theta'_{i_{\text{gas}}}} \frac{\sinh(\Theta'_{i_{\text{gas}}}/2T) \sinh(\Theta_{i_{\text{cond}}}/2T)}{\sinh(\Theta_{i_{\text{gas}}}/2T) \sinh(\Theta'_{i_{\text{cond}}}/2T)} \right) \tag{26.30}$$

and if, in addition, the temperature dependence of the intramolecular vibrations is ignored, as

$$\frac{p}{p'} = \prod_{i=1}^{3N} q_i = \prod_{i=1}^{6} \left(\frac{\Theta'_{i_{\text{cond}}}}{\Theta_{i_{\text{cond}}}} \frac{\sinh(\Theta_{i_{\text{cond}}}/2T)}{\sinh(\Theta'_{i_{\text{cond}}}/2T)} \right) \times \exp\left\{ \frac{1}{2T} \sum_{i=7}^{3N} [(\Theta'_{i_{\text{gas}}} - \Theta'_{i_{\text{cond}}}) - (\Theta_{i_{\text{gas}}} - \Theta_{i_{\text{cond}}})] \right\}. \tag{26.31}$$

q_i is the factor with which a single vibration enters the ratio, and Θ_i is the characteristic temperature of the vibrations, with i from 1 to 3 for the translations, from 4 to 6 for the librations, and from 7 to $3N$ (N = number of atoms in the molecule) for the intramolecular vibrations in the gaseous and the condensed state. Prime again refers to the lighter isotopic compound.

26.5.2. In applying Bigeleisen's equation to the explanation of the observations, the intermolecular vibrations of the pairs must be assumed to be different. The influence of this difference on the vapor pressure ratio is seen in fig. 26.10. It gives the factor q_i, which results for a single vibration as function of T/Θ^H, for different ratios of Θ^H/Θ^D.

For the rather low-frequency translations, T/Θ^H is comparatively large; therefore, q_i lies on the right side of the diagram. Θ^H/Θ^D for the translations is given by the square root of the reciprocal mass ratio of the two compounds, i.e. by values which are only slightly greater than one; therefore, q_i lies not only on the right side of the diagram but also on one of the upper curves. That means, factor q_i is approximately one, and the vapor pressure ratio is nearly uninfluenced by the translations.

For the higher-frequency librations T/Θ^H is comparatively small; therefore, q_i lies in the middle or on the left of the diagram. Besides, Θ^H/Θ^D, which is given for the librations by the square root of the reciprocal moments of inertia of the two compounds, is essentially greater than one; accordingly, q_i lies not only in the middle or on the left of the diagram but also on one of the lower curves. This means, q_i of the librations is essentially smaller than one, and the vapor pressure ratio is strongly decreased under the influence of the intermolecular vibrations in general (Wolff and Höpfner [1967b]).

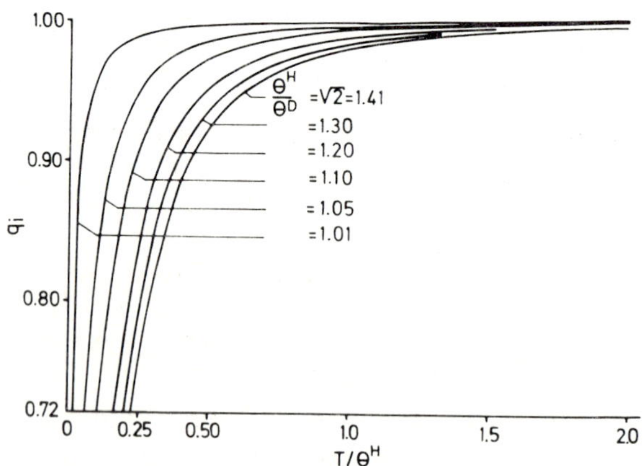

Fig. 26.10. Factor q_i of a single intermolecular vibration as a function of T/Θ^H for ratios Θ^H/Θ^D from 1.01 to $\sqrt{2}$.

Considering the influence of the intramolecular vibrations of the associated molecules on the vapor pressure ratio, it can be stated that the stretching frequencies of the associating groups (amino groups, hydroxyl groups, etc.) are strongly shifted to lower wave numbers due to the association, and that the shifts of the vibrations with H-atoms are greater than those of the vibrations with D-atoms. Consequently, the ratio is usually strongly decreased under the influence of the intramolecular vibrations. However, considering the influence of the inter- and the intramolecular vibrations together, it must be stated for substances which form H-bonds that the decrease under the influence of the intermolecular vibrations is mostly greater than the increase of the ratio under the influence of the intramolecular vibrations. Therefore, the observed effect of the associated molecules is usually normal.

Explaining that the effect of the monomers is inverse, one has to consider that the intramolecular vibrations of the monomers are shifted only by small amounts to lower wave numbers due to condensation or solvation. Therefore, only a small increase of the vapor pressure ratio results. And the intermolecular vibrations of the monomers, the librations included, are assumed to lie at essentially smaller wave numbers than the corresponding vibrations of the associated molecules. They can be assumed to lie in the range of an approximately classical behavior of the vibrations where both, the vibrations of the deuterated and those of the non-deuterated compound, are nearly fully excited and where, therefore, the factor with which the vibrations enter the vapor pressure ratio, is approximately one. Consequently, mainly the small inverse effect of the intramolecular vibrations is observed. The transition to this state explains the changes of the vapor pressure ratios with dilution (Wolff and Höppel [1968b] and Wolff and Würtz [1970a]).

26.5.3. The differences in the data based on the theory of ideal associated solutions for the compounds with the deuteration difference in the associating groups can be explained by the same principles as the differences in the vapor pressure ratios. The association constant and the association energy of the deuterated compounds are found to be greater than the corresponding values of the non-deuterated molecules because of the non-classical behavior of the intermolecular vibrations of the associated molecules and the predominating influence of these vibrations. The solvation energy, which is associated with the monomers, is found to be smaller because of the approximately classical behavior of the intermolecular vibrations of these monomers, and because of the predominating influence of the intramolecular vibrations under these circumstances (Wolff and Würtz [1970b]).

26.5.4. In interpreting the behavior of the vapor pressure ratios so far, only the pairs with the deuteration difference in the associating groups have been considered. Considering also the behavior of the pairs with the deuteration difference in the alkyl groups, obviously the inter- and the intramolecular vibrations related to the associating groups are nearly the same. That means, their vapor pressure isotope effect is inverse, independent of the concentration, and only slightly dependent on the temperature. For the pairs with differences in the associating as well as in the alkyl groups, both mechanisms are effective. From this, an enhancement of the normal effect due to the difference of the associating groups results in one case, and a reduction in the other (Wolff and Höpfner [1967a] and Wolff and Würtz [1970a]).

26.5.5. Taking into account the equilibrium between the molecules with free groups and the molecules with bonded groups, the vapor pressure ratio is given by the two-state function

$$\frac{p}{p'} = \left(\frac{p}{p'}\right)_{\text{bonded}}^{1-x} \left(\frac{p}{p'}\right)_{\text{free}}^{x}, \tag{26.32}$$

where x is the mole fraction of the molecules with free groups and where $(p/p')_{\text{bonded}}$ and $(p/p')_{\text{free}}$ are the ratios of the molecules with both the bonded and the free groups to be calculated from eq. (26.30). From this it follows that for the pairs with the deuteration difference in the associating groups p/p' increases with temperature, not only due to the vibrational excitation according to eq. (26.30), but also due to the augmentation of the molecules with free groups, according to the change in the association equilibrium.

From eq. (26.32) the values of $(p/p')_{\text{bonded}}$ and $(p/p')_{\text{free}}$ may be calculated, provided sufficiently accurate values of p/p' are known for some mole fractions. One can then estimate the mole fraction x of the free groups in the undiluted state if one reinserts the calculated values in the equation and inserts the value for the ratio of the undiluted compounds. In this way it was found that undiluted dimethylamine contains 60–40% molecules with free amino groups between +20 and −20°C (Wolff and Würtz [1970b]), and that undiluted methylamine contains 40–30% molecules with free groups at the same temperatures (Wolff

[1971]). For dimethylamine a similar result was obtained by IR measurements in the range of the first harmonics of the NH stretching vibration (Wolff and Gamer [1972a]). Likewise, the result for methylamine agrees with other observations on its association, e.g. the determination of association constants about 50% higher than those for the secondary analogue (tables 26.3 and 26.4).

26.6. Supplements and confirmations

26.6.1. If the vibrations are available, the vapor pressure ratios of the monomers and the associates can be calculated. Figure 26.11 shows, for solid and liquid H_2O and D_2O, the calculated vapor pressure ratios as well as the factors of the inter- and the intramolecular vibrations from which the ratios have been calculated. Because of the simplifications, the curves can claim only a restricted accuracy. But they make clear that the vapor pressure isotope effect of water and ice results from superposition of a normal effect of the intermolecular vibrations, mainly the librations, and

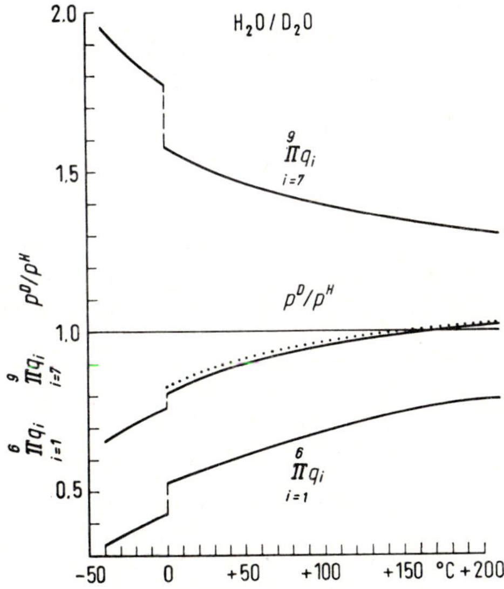

Fig. 26.11. Vapor pressure ratios p^D/p^H of H_2O/D_2O ice and water and partial factors $\Pi_{i=1}^{6} q_i$ and $\Pi_{i=7}^{9} q_i$ of the inter- and the intramolecular vibrations from which the ratios have been calculated. The ratios calculated assuming equilibrium between free and bonded groups are shown with dotted lines.

of an inverse effect of the intramolecular vibrations (Wolff and Wolff [1969] and Wolff [1969a]).

Figure 26.12 compares, for H_2O and D_2O as well as for H_2O and HDO, the results of calculations with the measurements. To avoid errors due to Fermi resonance, the calculations have been performed by using the stretching frequencies of HDO instead of those of H_2O and D_2O. Only by this procedure the approximate agreement between the calculations (curves) and the measurements (symbols) is obtained, particularly for the solids. The procedure implies that the H-bonds of ice and water are equal in their strengths.

For the liquids two lines are seen. The solid lines are calculated, taking into account only the wave numbers of the associated molecules. The dotted lines are calculated, assuming the fractions of molecules with free groups derived from dielectric measurements by Haggis et al. [1952] (about 10% at room temperature). The ratios thus derived deviate from both the measurements and the values, calculated without the assumption

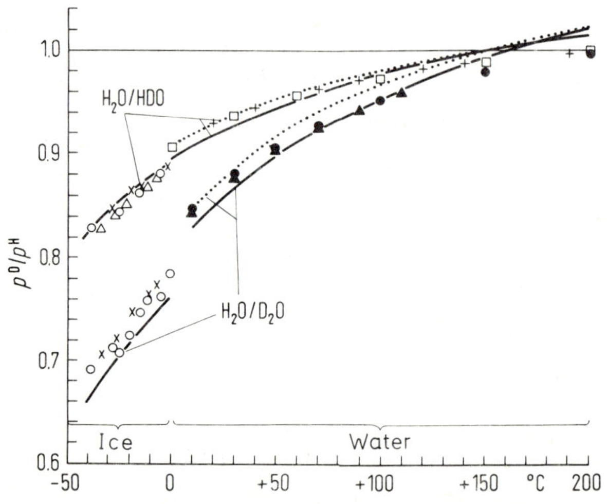

Fig. 26.12. Calculated and measured vapor pressure ratios p^D/p^H of H_2O/D_2O and H_2O/HDO in the solid and liquid state. ——— = ratios calculated without the assumption of equilibrium, ---- = ratios calculated with the assumption of equilibrium between free and bonded groups as indicated in the text. Measured values: H_2O/HDO ice (Matsuo et al. [1964], Kiss et al. [1966] and Merlivat and Nief [1967]), H_2O/HDO water (Kirshenbaum [1951], Landolt-Börnstein 6. Aufl. II, 2a and Merlivat et al. [1963]), H_2O/D_2O ice (Matsuo et al. [1964] and Kiss et al. [1966]), H_2O/D_2O water (Kirshenbaum [1951], Landolt-Börnstein 6. Aufl. II, 2a and Jones [1968]).

of equilibrium, only by amounts which lie within the limits of error. But the ratios which are calculated using considerably higher values for the mole fractions of free groups, as postulated by some theories of water structure (Eucken [1948]), show deviations from the measurements which are larger than the errors of the calculations. These theories, therefore, can be rejected.

Obviously, the influence of the free groups on the vapor pressure ratio is minor at temperatures below the cross-over where p/p' equals one. But at the highest temperatures, particularly near the critical point, this influence and that of the downward shift of the librations and the upward shift of the stretching frequencies with increasing temperature, must be assumed predominant. The two influences reflect that the concepts of both models of water structure, the mixture and the continuum model, are justified to some extent simultaneously.

Majoube [1971] recalculated the vapor pressure ratios of liquid water, inserting three different librational wave numbers instead of the value for the unresolved overall band and taking into account the temperature dependence of the wave numbers. But, the value which was used for the highest wave number libration is not accurately known, and the value for the temperature dependence of the librational wave numbers, in the meantime, was substantially corrected (Walrafen [1972]); the presence of molecules with free groups, as demonstrated by Worley and Klotz [1966], Luck [1967], Luck and Ditter [1969] and Walrafen [1972], was neglected. Therefore, this calculation likewise is of limited accuracy. Calculations by Jones [1968] and Van Hook [1968] used wave numbers which, according to the present state of investigations, are in part apparently erroneous (Wolff [1969b] and Majoube [1971]).

26.6.2. Figure 26.13 shows the measured and the calculated vapor pressure ratios of solid and liquid ammonia and deuteroammonia, and the ratios of highly diluted solutions of these compounds in propane. In the solid state the calculated as well as the measured vapor pressure isotope effect is normal, as expected for the H-bonded state of compounds (Wolff et al. [1971]). The same holds true for the liquid state. At high dilutions the effect is inverse as would be expected in the monomeric state. The spectroscopic investigations by Datta and Barrow [1965, 1968] and by Corset et al. [1966] revealed a considerable degree of rotational freedom for the ammonia molecules dissolved in saturated hydrocarbons. Therefore, this inverse effect is calculated assuming a classical behavior of the

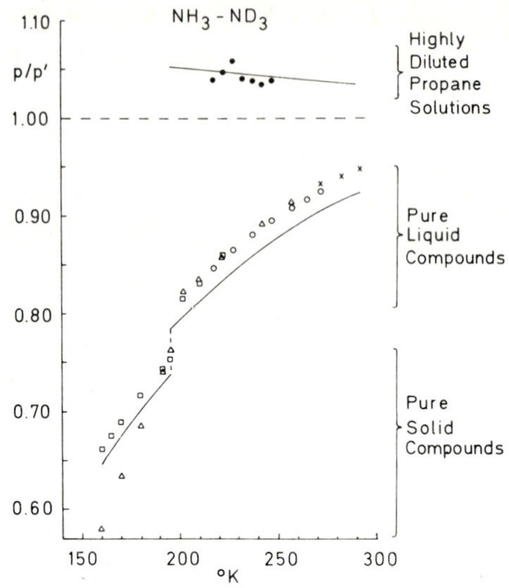

Fig. 26.13. Calculated and measured vapor pressure ratios of solid and liquid ammonia and deuteroammonia, and of highly diluted solutions of these compounds in propane. ——— = calculated ratios, ● ○ □ △ × = measured, or from the measured values extrapolated ratios, □ = Kirshenbaum and Urey [1942], △ = Kiss et al. [1962], × = Groth et al. [1956], ○ and ● = Wolff and Höpfner [1969].

intermolecular vibrations of monomers. The approximate agreement of the calculated ratios with the measured values seems to confirm that such an assumption is justified. The calculations for the solid and the liquid state imply again that the H- or D-bonds are equal in their strengths.*

* Calculating the inverse effect of the highly diluted ammonia solutions alone from the intramolecular vibrations, the wave-numbers measured by Benedict and Plyler [1957] and by Garing et al. [1959] were used for gaseous NH_3 ($\nu_1 = 3336\ cm^{-1}$, $\nu_2 = 950\ cm^{-1}$, $\nu_3(e) = 3444\ cm^{-1}$, and $\nu_4(e) = 1628\ cm^{-1}$). The values measured by Datta and Barrow [1965b] for pentane solutions were employed for the dissolved compound ($\nu_1 = 3320\ cm^{-1}$, $\nu_2 = 961\ cm^{-1}$, and $\nu_4 = 1620\ cm^{-1}$). A value of 3428 cm^{-1}, calculated from the Howard–Wilson equations (Herzberg [1945]), was used for ν_3, since the value of 3400 cm^{-1} observed by Datta and Barrow is not associated with monomers. The solvent shifts $\Delta\nu = \nu_{gas} - \nu_{cond}$ of the deuterated molecule were calculated from the ratios $\Delta\nu(NH_3)/\Delta\nu(ND_3)$ by inserting the values for $\Delta\nu(NH_3)$. The ratios $\Delta\nu(NH_3)/\Delta\nu(ND_3)$ resulted from the Howard–Wilson equations by appropriate differentiation and division.

Liquid ammonia contains a temperature-dependent equilibrium of triple-, double-, single- and non-H-bonded molecules (Corset and Lascombe [1967]). Replacing the vibrations of three single-bonded molecules by those of one triple-bonded molecule and of two non-bonded

26.6.3. The normal vapor pressure isotope effect and its transition to an inverse effect with dilution in an inert solvent have been explained from the transition of the intermolecular vibrations of self-associates to the intermolecular vibrations of monomers. In this respect it is of interest to consider the effect of dilution of a self-associated compound with a solvent with which mixed associates are formed. Figure 26.14 shows the vapor pressure ratios which have been calculated from data given by Rabinovich [1970] for solutions of light and heavy water with pyridine. There is no transition to the inverse effect, which can be understood in light of the preceding explanations from the fact that with dilution the intermolecular vibrations of the self-associates are replaced by the

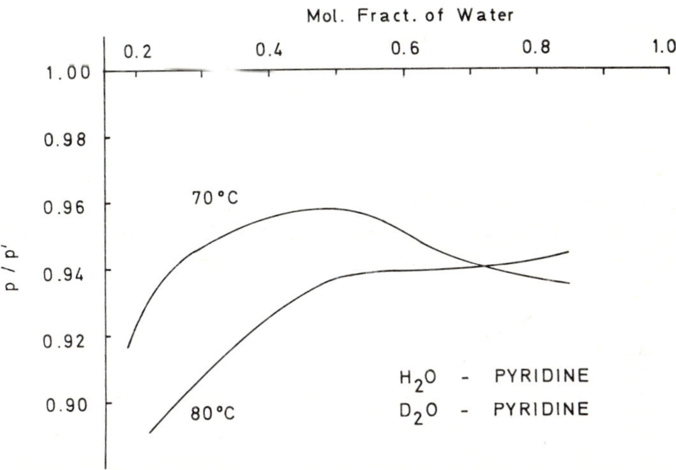

Fig. 26.14. Changes of vapor pressure ratios of H_2O and D_2O on dilution with pyridine at temperatures of +70 and +80°C, calculated from measurements of Rabinovich [1970].

---

molecules, and replacing the vibrations of three double-bonded molecules by those of two triple-bonded molecules and of one non-bonded molecule, the ratios can be calculated from eq. (26.32). Near the melting point the liquid contains mainly triple-bonded molecules, the wavenumbers of which only slightly differ from those of the solid, consisting too of triple-bonded molecules (Plint et al. [1954], Corset and Lascombe [1967], Kruh and Petz [1964], Pauling [1962], and Olovsson and Templeton [1959]). Therefore, $(p/p')_{\text{bonded}}$ could be calculated from the vibrations already used for the calculation of the ratios of the solid. The values determined for the highly diluted solutions served as $(p/p')_{\text{free}}$. x was calculated from Eyring and Marchi's expression $(V - V_s)/V$ where V_s and V are the mole volumes of the solid at the melting point and of the liquid at the temperature under consideration (Eyring and Marchi [1963], Marchi and Eyring [1964]).

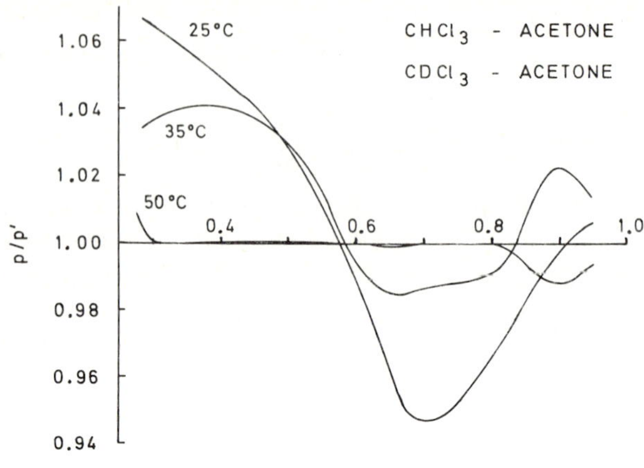

Fig. 26.15. Changes of vapor pressure ratios of chloroform and deuterochloroform on dilution with acetone at temperatures of +25, +35 and +50°C, calculated from measurements of Rabinovich [1970].

intermolecular vibrations of the mixed associates, and not by those of the monomers.

26.6.4. Likewise, it is of interest to consider compounds which do not self-associate but do form mixed associates. Figure 26.15 shows the vapor pressure ratios which we determined for solutions of chloroform and deuterochloroform with acetone, again from data given by Rabinovich [1970]. In this case the vapor pressure isotope effect is inverse at low as well as at high mole fractions, where the monomers exist. But the effect is normal in the middle range where the mixed associates are formed, i.e. where classical behavior of the intermolecular vibrations of monomers is replaced by non-classical behavior of the intermolecular vibrations of the mixed associates.

26.6.5. With compounds which are associated in the gaseous as well as in the liquid state, the inter- and intramolecular vibrations occur in both phases and are nearly the same. Therefore, the vibrations cancel each other in the vapor pressure ratio, and the vapor pressure ratio of the associates is determined only by small differences in these vibrations, due to unspecific interactions or due to dissociation of the associates into the monomers. Rabinovich [1970] has reported that an inverse effect is observed for undiluted hydrogen and deuterium fluoride as well as for the undiluted carboxylic acids and the corresponding deuterated acids.

26.7. Summary

In summary one can state that the distribution of the monomers and the associates, the association constants, the association energies, etc. can be derived from vapor pressure measurements. In investigating the various deuterated forms of the associating compound the differences of these quantities can also be studied. The process of association can be followed by changes of the vapor pressure ratios. These ratios are closely correlated to the changes of the inter- and the intramolecular vibrations of the compounds under consideration, and in appropriate cases the ratios may be calculated from these vibrations. Thus, the study of the normal as well as of the deuterated forms of the substances forming H-bonds permits the examination of current concepts of association.

References

Armstrong, G. T., F. G. Brickwedde and R. B. Scott, 1955, J. Res. Natl. Bur. Std. **55**, 39.
Barker, J. A., 1953, Australian J. Chem. **6**, 207.
Beersman, J. and J. C. Jungers, 1947, Bull. Soc. Chim. Belg. **56**, 72.
Bellamy, L. J. and R. J. Pace, 1966, Spectrochim. Acta **22**, 525.
Benedict, W. S. and E. K. Plyler, 1957, Can. J. Phys. **35**, 1235.
Bigeleisen, J., 1961, J. Chem. Phys. **34**, 1485.
Bigeleisen, J. and E. Roth, 1961, J. Chem. Phys. **35**, 68.
Bigeleisen, J., S. V. Ribnikar and W. A. Van Hook, 1963, J. Chem. Phys. **38**, 489.
Bordewijk, P., M. Kunst and A. Rip, 1973, J. Phys. Chem. **77**, 548.
Bystrov, V. F. and V. P. Lezina, 1964, Optics and Spectroscopy **16**, 430.
Clusius, K. and K. Schleich, 1958, Helv. Chim. Acta **41**, 1342.
Clusius, K., M. Vecchi, A. Fischer and U. Piesberger, 1959a, Helv. Chim. Acta **42**, 1975.
Clusius, K., K. Schleich and M. Vecchi, 1959b, Helv. Chim. Acta **42**, 2654.
Clusius, K., F. Endtinger and K. Schleich, 1961, Helv. Chim. Acta **44**, 98.
Corset, J., P. V. Huong and J. Lascombe, 1966, C.r. Acad. Sci. Paris, Sér. **C262**, 959.
Corset, J. and J. Lascombe, 1967, J. Chim. Phys. **64**, 665.
Datta, P. and G. M. Barrow, 1965, J. Chem. Phys. **43**, 2137.
Datta, P. and G. M. Barrow, 1965b, J. Am. Chem. Soc. **87**, 3053.
Datta, P. and G. M. Barrow, 1968, J. Chem. Phys. **48**, 4662.
Davies, Jr., R. T. and R. W. Schiessler, 1953, J. Phys. Chem. **57**, 966.
Eucken, A., 1948, Z. Elektrochem. Angew. Physik. Chem. **52**, 255.
Eyring, H. and R. P. Marchi, 1963, J. Chem. Educat. **40**, 562.
Feeney, J. and L. H. Sutcliffe, 1961, Proc. Chem. Soc. (London) **1961**, 118.

Feeney, J. and L. H. Sutcliffe, 1962, J. Chem. Soc. (London) **1962**, 1123.
Fletcher, A. N. and C. A. Heller, 1967, J. Phys. Chem. **71**, 3742.
Garing, J. S., H. H. Nielsen, and K. N. Rao, 1959, J. Mol. Spectroscopy **3**, 496.
Gölles, F. and O. Wolfbauer, 1968, Mh. Chem. **99**, 1814.
Groth, W., H. Ihle and A. Murrenhoff, 1956, Angew. Chem. **68**, 605.
Haggis, G., J. B. Hasted and J. Buchanan, 1952, J. Chem. Phys. **20**, 1452.
Haskell, R. W., H. B. Hollinger and H. C. Van Ness, 1968, J. Phys. Chem. **72**, 4534.
Herzberg, G., 1945, Infrared and Raman Spectra of Polyatomic Molecules, Van Nostrand, New York, 1945, p. 187.
Ibbitson, D. A. and L. F. Moore, 1967a, J. Chem. Soc. B, 76.
Ibbitson, D. A. and L. F. Moore, 1967b, J. Chem. Soc. B, 80.
Jancso, G. and W. A. Van Hook, 1974, Chem. Rev. **44**, 689.
Jones, W. M., 1968, J. Chem. Phys. **48**, 207.
Kehiaian, H., 1966, Bull. Acad. Polon. Sci., Sér. Sci. Chim. **14**, 703.
Kehiaian, K., K. Sosnkowska-Kehiaian and A. Treszczanowicz, 1967, Thermodynamics of n-Amine + n-Alkane Mixtures, Kl. Schäfer, Thermodynamik-Symposium Heidelberg, 1967 (AZ-Werbung + Weberdruck, Heikelberg).
Kirshenbaum, I., 1951, Physical Properties of Heavy Water (McGraw-Hill Book Co., New York).
Kirshenbaum, I. and H. C. Urey, 1942, J. Chem. Phys. **10**, 706.
Kiss, I., L. Matus and I. Opauszki, 1962, Kernenergie **5**, 329.
Kiss, I., H. Jakly and H. Illy, 1966, Acta Chim. Acad. Sci. Hung. **47**, 379.
Krug, Kl., D. Haberland and H.-J. Bittrich, 1971, Chem. Technol. **23**, 410.
Kruh, R. F. and J. I. Petz, 1964, J. Chem. Physics **41**, 890.
Landolt-Börnstein, 1960, 6. Aufl. II, 2a.
Liddell, U. and E. D. Becker, 1957, Spectrochim. Acta **10**, 70.
Luck, W. A. P., 1967, Discussions Faraday Soc. **43**, 115.
Luck, W. A. P. and W. Ditter, 1969, Z. Naturforsch. **24B**, 482.
Majoube, M., 1971, J. Chim. Phys. **68**, 1423.
Marchi, R. P. and H. P. Eyring, 1964, J. Phys. Chem. **68**, 221.
Matsuo, S., H. Kuniyoshi and Y. Miyake, 1964, Science **145**, 1454.
Mecke, R., 1948, Z. Elektrochem. **52**, 269.
Merlivat, L. and G. Nief, 1967, Tellus **19**, 122.
Merlivat, L., R. Botter and G. Nief, 1963, J. Chim. Phys. **60**, 56.
Oster, G., 1949, J. Am. Chem. Soc. **68**, 2036.
Olovsson, I. and D. H. Templeton, 1959, Acta crystallogr. **12**, 832.
Pauling, L., 1962, Die Natur der chemischen Bindung, Verlag Chemie, Weinheim/Bergstrasse.
Plint, C. A., R. M. B. Small, and H. L. Welsh, 1954, Can. J. Phys. **32**, 653.
Prausnitz, J. M., 1969a, Molecular Thermodynamics of Fluid-Phase Equilibria (Prentice Hall, Englewood Cliffs, N.J.).
Prausnitz, J. M., 1969b, The Oil and Gas Journal, June 2, p. 61.
Prigogine, I. and R. Defay, 1954, Chemical Thermodynamics (Longmans Green and Co., London), pp. 409–436.
Rabinovich, I. B., 1970, Influence of Isotopy on the Physicochemical Properties of Liquids (Consultants Bureau, New York).

Redlich, O. and A. Kister, 1947, J. Chem. Phys. **15**, 848.
Redlich, O. and A. Kister, 1948, Ind. Eng. Chem. **40**, 345.
Renon, H. and J. M. Prausnitz, 1967, Chem. Eng. Sci. **22**, 299, Errata 1891.
Savini, C. G., D. R. Winterhalter and H. C. Van Ness, 1965, J. Chem. Eng. Data **10**, 168.
Stern, M. J., W. A. Van Hook and M. Wolfsberg, 1963, J. Chem. Phys. **39**, 3179.
Storek, W. and H. Kriegsmann, 1968, Ber. Bunsenges. Physik. Chem. **72**, 706.
Tucker, E. E. and E. D. Becker, 1973, J. Phys. Chem. **77**, 1783.
Tucker, E. E., S. B. Farnham and S. D. Christian, 1969, J. Phys. Chem. **73**, 3820.
Van Hook, W. A., 1964, J. Chem. Phys. **40**, 3727.
Van Hook, W. A., 1968, J. Phys. Chem. **72**, 1234.
Van Ness, H. C., J. Van Winkle, H. H. Richtol and H. B. Hollinger, 1967, J. Phys. Chem. **71**, 1483.
Walrafen, G. E., 1972, Raman and Infrared Spectral Investigations of Water Structure, in: Water, Vol. **1**, Ed. Felix Franks (Plenum Press, New York), p. 151.
Wilson, G. M., 1964, J. Am. Chem. Soc. **86**, 127.
Wolff, H., 1969a, Ber. Bunsenges. Physik. Chem. **73**, 399.
Wolff, H., 1969b, The Vapor Pressure Isotope Effect of Ice and its Isomers, in: Physics of Ice, Eds. N. Riehl, B. Bullemer and H. Engelhardt (Plenum Press, New York), pp. 305–319.
Wolff, H., 1971, J. Phys. Chem. **75**, 160.
Wolff, H. and L. Dill, 1976, *in preparation.*
Wolff, H. and G. Gamer, 1972a, J. Phys. Chem. **76**, 871.
Wolff, H. and G. Gamer, 1972b, Spectrochim. Acta **28A**, 2121.
Wolff, H. and R. Götz, 1976, *in preparation.*
Wolff, H. and A. Höpfner, 1962, Z. Elektrochem., Ber. Bunsenges. Physik. Chem. **66**, 149.
Wolff, H. and A. Höpfner, 1965, Ber. Bunsenges. Physik. Chem. **69**, 710.
Wolff, H. and A. Höpfner, 1967a, Ber. Bunsenges. Physik. Chem. **71**, 461.
Wolff, H. and A. Höpfner, 1967b, Ber. Bunsenges. Physik. Chem. **71**, 730.
Wolff, H. and A. Höpfner, 1969, Ber. Bunsenges. Physik. Chem. **73**, 480.
Wolff, H. and H.-E. Höppel, 1966, Ber. Bunsenges. Physik. Chem. **70**, 874.
Wolff, H. and H.-E. Höppel, 1968a, Ber. Bunsenges. Physik. Chem. **72**, 710.
Wolff, H. and H.-E. Höppel, 1968b, Ber. Bunsenges. Physik. Chem. **72**, 722.
Wolff, H. and H.-E. Höppel, 1968c, Ber. Bunsenges. Physik. Chem. **72**, 1173.
Wolff, H. and D. Horn, 1967, Ber. Bunsenges. Physik. Chem. **71**, 467.
Wolff, H. and D. Horn, 1968, Ber. Bunsenges. Physik. Chem. **72**, 419.
Wolff, H. and D. Mathias, 1973, J. Phys. Chem. **77**, 2081.
Wolff, H. and O. Schiller, 1976, *in preparation.*
Wolff, H. and U. Schmidt, 1964, Ber. Bunsenges. Physik. Chem. **68**, 579.
Wolff, H. and D. Staschewski, 1962, Z. Elektrochem., Ber. Bunsenges. Physik. Chem. **66**, 140.
Wolff, H. and E. Wolff, 1968, Ber. Bunsenges. Physik. Chem. **72**, 98.
Wolff, H. and E. Wolff, 1969, Ber. Bunsenges. Physik. Chem. **73**, 393.
Wolff, H. and R. Würtz, 1968, Ber. Bunsenges. Physik. Chem. **72**, 101.
Wolff, H. and R. Würtz, 1969, Z. Physik. Chem. (Neue Folge) **67**, 115.
Wolff, H. and R. Würtz, 1970a, J. Phys. Chem. **74**, 1600.
Wolff, H. and R. Würtz, 1970b. Z. Physik. Chem. (Neue Folge) **69**, 67.

Wolff, H., A. Höpfner and H.-M. Höpfner, 1964, Ber. Bunsenges. Physik. Chem. **68**, 410.
Wolff, H., H.-G. Rollar and E. Wolff, 1971, J. Chem. Phys. **55**, 1373.
Wolff, H., O. Bauer, R. Götz, H. Landeck, O. Schiller and L. Schimpf, 1975, J. Phys. Chem., *submitted.*
Wolff, H., H. Landeck and R. Götz, 1976, in preparation.
Wooley, H. W., R. B. Scott and F. G. Brickwedde, 1948, J. Res. Natl. Bur. Std. **41**, 379.
Worley, J. D. and I. M. Klotz, 1966, J. Chem. Phys. **45**, 2868.

PART H

HYDROGEN BONDING AND ADSORPTION ON SURFACES

CHAPTER 27

HYDROGEN BONDS IN SYSTEMS OF ADSORBED MOLECULES

HELMUT KNÖZINGER

Physikalisch-Chemisches Institut, Universität München, Sophienstr. 11, D-8 München 2, West Germany

Contents

27.1. Introduction 1265
27.2. Silica and related surfaces 1266
- 27.2.1. Characterization of surfaces 1266
- 27.2.2. Specific adsorption interactions 1273
  - 27.2.2.1. Perturbation of surface silanol groups 1274
  - 27.2.2.2. Correlations 1284
  - 27.2.2.3. Further spectroscopic information on hydrogen bonding . . 1290
- 27.2.3. Adsorption of benzene 1291
- 27.2.4. Adsorption of ammonia 1295
- 27.2.5. Adsorption of water 1303
- 27.2.6. Adsorption of alcohols 1318

27.3. Alumina 1329
- 27.3.1. Characterization of alumina surfaces 1329
- 27.3.2. Adsorption of water 1333
- 27.3.3. Adsorption of alcohols 1337

27.4. Review of recent literature 1341
Notes added in proof 1351
References 1353

The hydrogen bond – recent developments in theory and experiments
Eds. P. Schuster et al.

27.1. Introduction

Many solid surfaces carry potential H-bond donor and acceptor groups so that adsorption interactions via H-bonds are principally possible. Untreated metal oxides, in particular, terminate usually with a hydroxylated surface due to the contact with moist air and after heat treatment, oxygen ions are created. Such surfaces can therefore take over both the H-bond donor and the H-bond acceptor function depending on the degree of hydroxylation and on the function of the adsorbate molecule. The most prominent adsorbate molecule in this respect is, of course, water which is of extreme importance in a wide variety of surface processes. Thus H-bonding interactions play a dominant role in all drying processes, i.e. in selective adsorptions. The selectivity of adsorption from aqueous solutions may be strongly influenced by alterations of the structure of liquid water near solid surfaces. The use of adsorbents in the prevention of water pollution and the regeneration of waste water is thus intimately related to H-bonding at and near solid surfaces. The fixation of pharmaceuticals to carriers such as silica and their liberation in the organism involves H-bond interactions. In catalysis–heterogeneous as well as enzymatic–H-bonded species can be the precursors of catalytically important intermediates. Even C–H···X H-bonds between a C–H proton in an "activated" molecule and some surface acceptor site X, may play a crucial role during the course of a heterogeneously catalyzed reaction. The "activation" of a reactant molecule may be brought about by the fluctuations of protons in H-bonded systems on a catalyst surface. Again adsorbed water can effectively interfere in these processes. Out of a variety of possible examples the possibility is discussed that petroleum genesis occurred by decomposition reactions of organic species interacting with clay minerals contained in sediments. Moreover, the presence of water certainly is an important factor in geochemical transformations. Finally biological systems may be mentioned here, since e.g. the hydration of polypeptides or the bonding between an enzyme and a substrate may be described as an adsorption process on the respective "surfaces".

It cannot be the task of this contribution to review the unlimited number of systems and processes, in which H-bonding occurs on solid surfaces. Instead the H-bonding interactions on silica surfaces will be discussed in some detail, since it is this type of adsorbent to which most emphasis has been paid regarding its H-bonding properties. Adsorption onto alumina as an example of an acidic oxide surface follows in section 27.3 and in section 27.4 a literature survey is given regarding recent research work devoted to H-bonding interactions on surfaces of other pure and mixed oxides, zeolites and clay minerals.

27.2. Silica and related surfaces

27.2.1. Characterization of surfaces

The various forms of silica can be considered as polycondensation products of orthosilicic acid. Although there is some evidence for a structural order, they are generally defined as "amorphous". The primary particles of these materials consist of a more or less irregular three-dimensional network of SiO_4-tetrahedra, in which the valence of the surface silicon atoms is satisfied with hydroxyl groups. These surface OH groups were first detected by Kiselev [1936]. Thermal treatment of hydroxylated silica surfaces leads to a condensation of surface hydroxyl groups with the formation of water and surface siloxane bonds. There is a strong $d\pi$–$p\pi$ bonding between silicon and oxygen in these siloxane bonds. Consequently, as shown by Huggins [1961], the oxygen loses much of its basic properties and shows only a very weak tendency to work as a H-bond acceptor. Highly dehydroxylated silica surfaces therefore are hydrophobic in nature (Boehm [1966a, b, 1968]). The chemical nature and the adsorptive properties of silica surfaces are thus predominantly due to the surface hydroxyl groups (silanol groups). The details of specific adsorption interactions and of H-bonding in particular will be governed by the silanol density and their relative distribution, which strongly depend on the temperature of pretreatment.

The surface *silanol groups* have been extensively characterized by *infrared spectroscopy* (Little [1966], Hair [1967] and Kiselev and Lygin [1972]). Some typical spectra of the OH fundamental stretching region are shown in fig. 27.1. A sharp band is observed at 3750 cm^{-1} which is assigned as the OH stretching vibration of unperturbed free surface silanol groups. Hair and Hertl [1969] argued that this band could be resolved into three components. This was, however, disregarded by Hockey [1970], who demonstrates that the splitting is an artifact due to residual atmospheric water vapor in the spectrometer. The discussion on this subject is still going on and different opinions have been put forward by various research groups very recently. Thus, Van Cauwelaert et al. [1972, 1973] believe in a possible resolution of more than one OH stretching band of surface hydroxyl groups, whereas Morrow and Cody [1973] and Klier et al. [1973] disregard the existence of multiple OH stretching bands.

After evacuation at moderate temperatures the broad absorption with center around 3500 cm^{-1} is decreased and bands appear near 3680 and 3550 cm^{-1}. Both are reduced in intensity by a more severe heat treatment (400 °C and higher). The 3680 cm^{-1} band had been attributed to neighboring silanol groups which are perturbed through mutual H-bonding (McDonald [1958] and Sidorov [1960]). More recently, however, this band was ascribed to inaccessible internal, or bulk (intraglobular) OH groups (Armistead et al. [1969] and Kubelkova et al. [1969]), whereas the band at 3550 cm^{-1} is assumed to be due to mutually interacting surface silanol groups (Davydov et al. [1964a] and Armistead et al. [1969]). The deformation vibration corresponding to free silanols has been observed at 870 cm^{-1} by Benesi and Jones [1959] and a very broad absorption in the far IR region at 150 cm^{-1} was connected with intermolecular vibrations of mutually interacting neighboring OH groups by Brodskii et al. [1970].

A band around 950 cm^{-1} was assigned as the Si–O stretching vibration of surface silanol groups by Hino and Sato [1971] since its position responded to the substitution of ^{16}O surface groups by ^{18}O and it was sensitive to adsorption via H-bonds onto the surface silanol groups. The overtone band of the unperturbed silanol groups was observed at 7326 cm^{-1} by Yaroslavsky [1950] and Anderson and Wickersheim [1964] and at 7420 cm^{-1} by Jeziorowski et al. [1973].[1)]* A band at 4550 cm^{-1} was assigned as a combination band of the OH stretching fundamental with the Si–OH deformation at 870 cm^{-1} (Anderson and Wickersheim [1964]). On deuterium exchange of the surface silanol groups the band of the free groups at 3750 cm^{-1} moves to 2760 cm^{-1} (Benesi and Jones [1959] and Peri [1966a]), which is a clear indication for the assignment of the 3750 cm^{-1} band as an OH stretching vibration. In addition Peri [1966a] observed weak P and R branches for both surface OH and OD groups

* See Notes added in proof.

which were separated by 200 and 140 cm^{-1} respectively. This indicates that the free Si–OH groups are non-linear. The Si–O–H bond angle could be estimated to be 113°.

Some *NMR data of surface silanol groups* have also been reported. Egorov et al. [1966] claim to resolve three absorption signals on completely hydroxylated surfaces. From the second moments of these bands they conclude that there exist free silanols (IR band at 3750 cm^{-1}) and two types of interacting groups with distances of 2.5–2.6 Å. These interacting groups are only distinct due to differing numbers of nearest neighbors. O'Reilly et al. [1958] and Hall et al. [1963] found a Lorentzian line shape of 0.31 G line width, which was temperature independent, with a corresponding spin–spin relaxation time of 1.8×10^{-4} s on a partially hydroxylated silica containing 2.6 OH/100 $Å^2$. A chemical shift of $-(0.3 \pm 0.2)$ ppm, relative to water, was observed suggesting that most of the

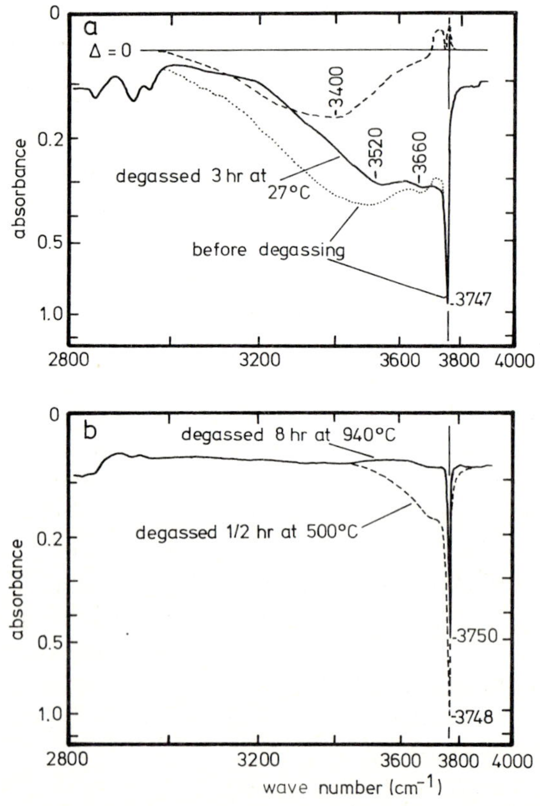

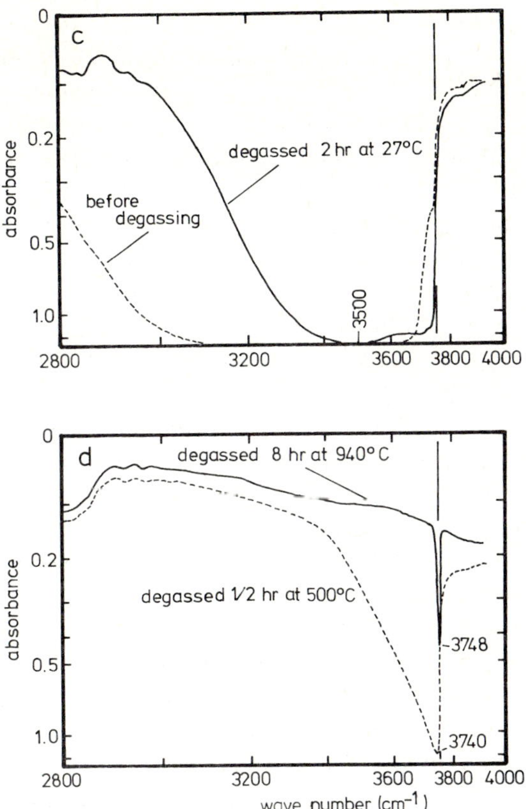

Fig. 27.1. IR spectra of cabosil (a and b) and of a pure silica gel (c and d). Broken line in (a): difference between solid line and dotted line. (Reproduced by permission of the American Chemical Society from McDonald [1958]).

protons were bound as rather isolated silanol groups in this case. Spin–spin relaxation times have been observed for surface silanol groups to fall in the range of $1.5–2.0 \times 10^{-4}$ s (Kamiyoshi [1959], Woessner [1963] and Kvilividse [1964]).

Hall et al. [1963] concluded from their NMR work that the isolated silanol groups were chemically comparable to alcoholic hydroxyl groups. A pK_a value of 7.1 ± 0.5 is reported for the OH surface groups (Dugger et al. [1964] and Hair and Hertl [1970]).

The *number of surface silanol groups* per unit surface area has been determined by various physical and chemical methods. Thus, Borello et al. [1967a] adopting an extinction coefficient $\epsilon = (35 \pm 1)$l cm^{-1} mole^{-1} for the OH fundamental stretching band at 3750 cm^{-1} calculated the number of surface silanols from IR spectroscopic data. Bermudez [1970] and Morariu and Mills [1972a] were able to obtain this quantity from NMR line widths. By far the most determinations of surface hydroxyl densities have been made purely chemically through reaction of the silanol groups with adequate reagents (Boehm [1966a, b, 1968]). The results of all methods coincide very well and lead to an average value of 4.6 ± 0.2 OH/100 Å^2 for a fully hydroxylated silica surface. The number of silanol groups per unit surface area depends on the temperature of vacuum treatment; it is, however, relatively insensitive to the previous history of the silica sample and to its disperseness, as mentioned by Davydov et al. [1964b] and Zhuravlev and Kiselev [1970]. The temperature dependence of the number of silanols per unit surface area is shown in figure 27.2 for various silica samples, where the data from more than 20 published papers from various laboratories applying different experimental methods fall between the limits shown by the two parallel curves. Thus, the average number of 4.6 OH/100 Å^2 of a fully hydroxylated surface drops to an average number of about 1 OH/100 Å^2 after vacuum treatment of the silica at 800°C.

4.6 OH/100 Å^2 very nearly correspond to the number of silicon atoms per 100 Å^2 in the rhombohedral face of β tridymite. De Boer and Vleeskens [1958] were therefore led to the conclusion that this crystallographic plane would form the surface layer of the silica crystallites. In conclusion, the surface silanol groups should all be free and unperturbed

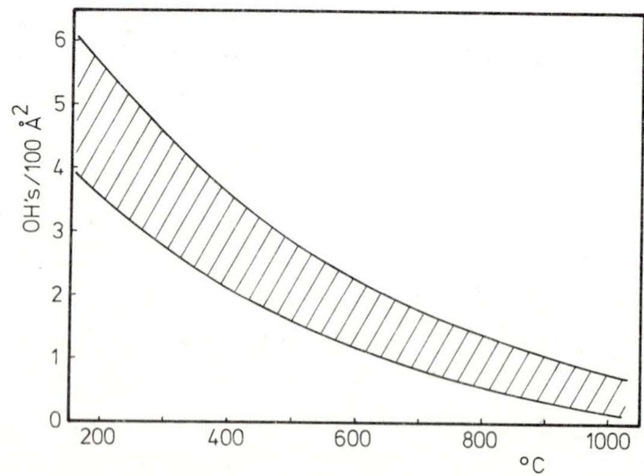

Fig. 27.2. Number of surface hydroxyl groups of silica as a function of degassing temperature.

since their corresponding oxygen atoms would be 5 Å apart, so that mutual interactions cannot occur due to this relatively large distance. This cannot be true, however, since IR as well as NMR information shows that perturbed silanol groups are present at least on fully hydroxylated surfaces. Similarly chemical methods proved the existence of pairs of OH groups (Synder and Ward [1966]). Principally, three different configurations of surface silanol groups may be distinguished, i.e. isolated (structure (I)), geminal (structure (III)) in which one silicon atom carries two OH groups and vicinal (structure (II)) where two (or more) OH groups are located at neighboring silicon atoms:

$$\mathrm{Si{-}O{-}H} \xrightarrow{MX_3} \mathrm{Si{-}O{-}MX_2} \xrightarrow{H_2O} \mathrm{Si{-}O{-}H} + \mathrm{M(OH)_3} \qquad \text{(I)}$$

$$\mathrm{Si{-}O{-}H\cdots O(H){-}Si} \xrightarrow{MX_3} \mathrm{(Si{-}O)_2M{-}X} \xrightarrow{H_2O} \mathrm{(Si{-}O)_2M{-}OH} \qquad \text{(II)}$$

$$\mathrm{HO{-}Si{-}OH} \xrightarrow{MX_3} \mathrm{Si(O)_2M{-}X} \xrightarrow{H_2O} \mathrm{Si(O)_2M{-}OH} \qquad \text{(III)}$$

these silanol groups may react with trivalent halides MX_3 to form surface compounds of different composition and geometry. Thus, Bermudez [1971] very convincingly proved the presence of all three hydroxyl configurations by very careful analyses of the IR spectra of the surface compounds formed on reaction with BCl_3 and of their hydrolysis products. The relative distribution of the different types of silanol groups is of fundamental importance for the detailed structure of H-bonded systems on hydroxylated silica surfaces and it is the reason for an energetic heterogeneity of the surfaces which shows up in a dependence of heats of adsorption on surface coverage (Kiselev [1964a]). A clear distinction between these OH configurations cannot be made from IR data since geminal silanols appear to absorb at 3750 cm^{-1} as do the isolated silanols. Therefore considerable effort has been made to determine the relative distributions of the various hydroxyl configurations through the kinetics or the stoichiometry of surface reactions with chlorosilanes (Snyder and Ward [1966], Armistead and Hockey [1967], Hair and Hertl [1969], Hertl and Hair [1969] and Tertykh et al. [1973]), metal halides (Boehm et al. [1963], Hambleton and Hockey [1966], Peri and Hensley [1968], Camara et al. [1968], Armistead et al. [1969] and Bermudez [1971]), organometalics (Fripiat and Uytterhoeven [1962] and Peglar et al. [1971]), diborane (Shapiro and Weiss [1953]; Fripiat and Van Tongelen [1965]), alkyl halides (Van Cauwclaert et al. [1970]) and of deuterium exchange reactions (Fripiat et al. [1962]). Though paired groups certainly exist, there is still no systematic agreement on the extent of pairing as a function of the pretreatment temperature. It appears to be tested, however, that even after high temperature pretreatment in vacuo, when the infrared spectra only show free surface hydroxyls, there is a considerable contribution of paired geminal silanols (Hair and Hertl [1969], Hertl and Hair [1969] and Bermudez [1971]). Armistead et al. [1969] believe that after heat treatment at 400°C only free hydroxyls are retained on a silica surface, which give rise to the IR band of unperturbed OH groups at 3750 cm^{-1}. These groups may either be isolated or geminal. Hair and Hertl [1969] and Bermudez [1971] showed that 60% of the 1.7 OH/100 $Å^2$ retained after heat treatment at 800°C are single OH groups and the remaining 40% are geminal.

Peri and Hensley [1968] proposed a *model for the silica surface which corresponds to a surface resembling an idealized* (100)-*face of* β cristobalite. In this model in a fully hydroxylated surface, each silicon atom carries two hydroxyl groups. The high percentages of 85% paired OH's/100 $Å^2$ after heat treatment at 600°C are explained by this model. As pointed out by Armistead et al. [1969], however, the assumption of only one crystallographic face forming the surface of small silica particles (approximate diameter of primary particles 10 Å, aggregates of approximately 100 Å diameter) seems to be unrealistic in general. These authors therefore prefer a picture in which the surface is made up by various planes, which allow the formation of isolated and paired hydroxyls. Thus, a (0001)-tridymite face would expose silicon atoms 5 Å apart which could carry isolated and non-interacting OH groups since the Si–O bond is orthogonal to the surface plane. In a (100)-β cristobalite face on the other hand, the surface silicon atoms occur in pairs and carry two OH groups, each with the Si–O axis at a low angle (approximately 35°) to the surface plane. Vicinal and geminal configurations could be formed on such a face. The most realistic picture seems therefore to be one which assumes a surface that is formed by various crystallographic faces, the relative contributions of which may vary for different preparations and treatments. The relative distribution of isolated and geminal silanol groups on highly dehydroxylated surfaces may thus also vary for different silica samples and should, therefore, be determined for each individual case.

Silica surfaces can be modified by halides such as PCl_3, BCl_3, $GeCl_4$ etc. (Fink et al. [1970]) to form POH, BOH, GeOH etc. groups in the surface. These groups are characterized by IR bands at 3665, 3707 and 3682 cm^{-1}, respectively, of the corresponding free hydroxyl stretching fundamental (Fink et al. [1971, 1972]). The acidities of these hydroxyl groups and correspondingly their interaction energy with H-bond acceptor molecules increases in the series

$$\mathrm{BOH} < \mathrm{SiOH} \leq \mathrm{GeOH} < \mathrm{POH}$$

as concluded from the respective OH wave number shifts due to the adsorption of benzene and pyridine (Fink et al. [1972]). This acidity sequence is the same as that proposed by Hair and Hertl [1970, 1973]. The BOH groups which are formed on silica surfaces on reaction with BCl_3 (Camara et al. [1968] and Bermudez [1971]) and their properties are of particular interest, since porous Vycor glass shows a band around 3700 cm^{-1} which has been assigned as a BOH stretching fundamental (Low and Ramasubramanian [1965, 1966] and Low et al. [1967a]). This material has been used frequently in adsorption studies.

27.2.2. Specific adsorption interactions

The *energy of interaction* between adsorbent, i.e. silica surface and adsorbate depends on the nature of the surface, in particular on its degree of hydroxylation and on the electronic structure of the adsorbate molecules. Thus, differential heats of adsorption of apolar adsorbates (e.g. n-hexane, carbon tetrachloride) on hydroxylated and highly dehydroxylated silica surfaces are identical and rapidly approach the heat of liquefaction with increasing surface coverage. On the contrary, the differential heats of adsorption of molecules containing π electron systems or lone pair electrons are usually appreciably higher on hydroxylated surfaces than on dehydroxylated surfaces (Kiselev [1964a]). This observation is a clear indication for specific interactions of the adsorbate with the surface silanol groups, which can be understood as H-bonding, whereas only non-specific interactions can occur with dehydroxylated parts of the surface in agreement with the almost inert character of siloxane bonds (see section 27.2.1). Differential heats of adsorption fall with increasing coverage due to the geometrical and energetic heterogeneity of the silanol adsorption sites. Davydov et al. [1963a, b, 1965] concluded from the changes in the IR spectra of the hydroxyl groups on silica samples of varying hydroxyl density due to the adsorption of molecules of different electronic structure, that the free hydroxyl groups are the preferred adsorption sites. A 1:1 adsorption is likely to occur for molecules containing a single lone pair of electrons, whereas molecules such as ethers with two sets of lone-pair electrons may form two H-bonds on, for example, geminal hydroxyl groups (Davydov et al. [1963a] and Fripiat [1964]) on surfaces of sufficiently high hydroxyl content. The H-bonded vicinal silanol groups seemed to participate in interactions with the adsorbate to only a lesser extent (Galkin et al. [1964]). Van Cauwelaert et al. [1971] on the contrary, have shown that ammonia and primary amines interact with the H-bonded hydroxyl groups. Such an interaction will be possible in cases where the H-bond between neighboring surface hydroxyls is replaced by a stronger H-bond between a hydroxyl group and the adsorbed molecule, e.g. an amine. Steric hindrance appeared to influence these interactions strongly as evidenced by the lower tendency of secondary and tertiary amines to interact with H-bonded hydroxyl groups.[2)]

Differential heats of adsorption are a measure of the total energy of interaction between the surface and an adsorbed molecule including specific and non-specific interactions at a given coverage. Dzhigit et al. [1961] and Davydov et al. [1962] tried to separate the specific contributions ΔH_{OH} to the differential heat of adsorption of a given adsorbate molecule by taking the difference between the differential heat of adsorption of that molecule on a hydroxylated and a strongly dehydroxylated (temperature treatment at 900 °C) surface at 50% coverage:

$$\Delta H(\text{hydr.}) - \Delta H(\text{dehydr.}) = \Delta H_{OH}. \qquad (27.1)$$

ΔH_{OH} should approximately represent the energy of specific interaction of the corresponding molecule with the surface silanol groups. Typical values are 1.9, 6.2 and 5.4 kcal/mole for benzene, acetone and pyridine, respectively. (Davydov et al. [1970]).

27.2.2.1. *Perturbation of surface silanol groups*

The interpretation of the specific interactions between surface silanol groups and adsorbate molecules such as H-bonding is clearly proved by the changes in the IR spectra of the surface hydroxyls and by the influence on the internal vibrations of the adsorbed molecules.

Particularly, the intensity of the stretching fundamental of free hydroxyl groups at 3750 cm^{-1} decreases on adsorption, while intense and broad absorption bands appear at lower wave numbers. Such changes in the OH stretching region clearly demonstrate the formation of H-bonds (see ch. 12). The wave number shift $\Delta\bar{\nu}_{OH}$ to lower values on H-bonding is considered as a measure of the bond strength, i.e. of the energy of specific interaction. Table 27.1 summarizes the observations obtained for a variety of systems in different laboratories.[3)] The wave number shifts are given as well as heats of adsorption ΔH_{ads} and ionization potentials I of the adsorbate. Basila [1961] and Galkin et al. [1964] have shown that the shift $\Delta\bar{\nu}_{OH}$ is a function of the surface coverage, the dependence being most pronounced below 50% coverage.

As $\Delta\bar{\nu}_{OH}$ changes very slowly between 50 and 100% coverage, the values given in table 27.1 generally refer to this range. Discrepancies in $\Delta\bar{\nu}_{OH}$ values from different sources for comparable systems may be due to some differences in coverage, to different pretreatment temperatures (Galkin et al. [1964]), to different sample temperatures (sample temperatures cannot be measured accurately because of the heating effect of the IR beam) and due to inaccuracies in the location of the maximum absorption of the broad bands of perturbed hydroxyl groups. A further difficulty can be brought about by the superposition of other bands such as C–H stretching fundamentals. For wave number shifts $\Delta\bar{\nu}_{OH} > 100\ cm^{-1}$ the error in $\Delta\bar{\nu}_{OH}$ should be less than 10% in general. Wave number shifts in table 27.1 correspond to silanol Si–OH groups. On modified silica surfaces containing other types of hydroxyl groups and on porous vycor glass with B–OH groups the respective wave number shifts due to interaction with these groups are also observed.

The adsorbate molecules in table 27.1 are ordered into three groups, that is to say, apolar molecules and rare gases, π electron donor molecules and lone-pair electron donor molecules. Obviously the wave number shifts strongly depend on these gross differences of the electronic structure. Thus, $\Delta\bar{\nu}_{OH}$ is smallest for the apolar molecules ($\Delta\bar{\nu}_{OH} < 100\ cm^{-1}$); it falls into the range of $100 \leq \Delta\bar{\nu}_{OH} \leq 200\ cm^{-1}$ for the group of π electron donors and reaches values as high as $1000\ cm^{-1}$ for the molecules containing lone-pair electrons ($200 \leq \Delta\bar{\nu}_{OH} \leq 1000\ cm^{-1}$). Evidently the energy of specific H-bonding interactions increases in this order.

TABLE 27.1
Wave number shifts of the OH-stretching fundamental band at 3750 cm^{-1} on adsorption

Adsorbate	$\Delta\bar{\nu}_{OH}$ (cm^{-1})	ΔH_{ads} (kcal/mole)	I (eV)	References
I. Non-polar				
1. Argon	8			McDonald [1957]
2. Krypton	16			McDonald [1957]
3. Xenon	19			McDonald [1957]
4. Nitrogen	24			McDonald [1957]
5. Oxygen	12			McDonald [1957]
6. Methane	32			McDonald [1957]
	45	8.8		Davydov et al. [1962]
7. *n*-Pentane	30	5.7[d)]	10.58	Hertl and Hair [1968]
	40			Elkington and Curthoys [1968]
8. *n*-Hexane	45	9.0[a)]	10.54	Kiselev and Lygin [1964]
	30		10.43	Hertl and Hair [1968]
	45	8.8		Davydov et al. [1970]
9. *n*-Heptane	45		10.34	Hertl and Hair [1968]
10. Cyclohexane	45	6.4[d)]	10.06	Hertl and Hair [1968]
	27		9.88	Cusumano and Low [1971]
II. Electron donors				
11. Benzene	106[b)]		9.25	Basila [1961]
	110[c)]	10.2		Davydov et al. [1962]
	110		9.23	Galkin et al. [1964]
	110	11[a)]	9.25	Kiselev and Lygin [1964]
	120			Zecchina et al. [1968]
	120	5.6[d)]	9.2	Hertl and Hair [1968]
	125			Elkington and Curthoys [1968]
	115–130			Cusumano and Low [1970a]
	110	10.4		Davydov et al. [1970]
	115			Fink et al. [1972]
12. Toluene	120			Sidorov [1954]
	120[b)]		8.82	Basila [1961]
	130[c)]	11.8		Davydov et al. [1962]
	143		8.82	Galkin et al. [1964]
	130	12.5[a)]	8.82	Kiselev and Lygin [1964]
	127		8.81	Hertl and Hair [1968]
	130			Elkington and Curthoys [1968]
	145			Cusumano and Low [1971]
13. *p*-Xylene	139[b)]		8.45	Basila [1961]
	146[c)]	13.4		Davydov et al. [1962]
	156		8.46	Galkin et al. [1964]

TABLE 27.1–(*contd.*)

Adsorbate	$\Delta\bar{\nu}_{OH}$ (cm^{-1})	ΔH_{ads} (kcal/mole)	I (eV)	References
	145	13.7[a)]	8.44	Kiselev and Lygin [1964]
	155	6.2[d)]	8.45–8.60	Hertl and Hair [1968]
	160		8.45	Cusumano and Low [1971]
14. Mesitylene	166[b)]		8.39	Basila [1961]
	160[c)]	14.0		Davydov et al. [1962]
	169		8.39	Galkin et al. [1964]
	160	14.2[a)]	8.41	Kiselev and Lygin [1964]
	166		8.39	Cusumano and Low [1971]
15. Ethylbenzene	120			Sidorov [1954]
16. *t*-Butylbenzene	150		8.69	Cusumano and Low [1971]
17. Fluorobenzene	59		9.19	Cusumano and Low [1971]
18. Anisole	150		8.20	Cusumano and Low [1971]
19. CF_3 Benzene	33		9.68	Cusumano and Low [1971]
20. Phenylamine	150		7.70	Cusumano and Low [1971]
21. Nitrobenzene	140			Abramov et al. [1964a]
	140	15.2[b)]		Kiselev and Lygin [1964]
	140		10.1	Cusumano and Low [1971]
22. Furan	120			Elkington and Curthoys [1968]
	140	9.3		Davydov et al. [1970]
23. Thiophene	130			Elkington and Curthoys [1968]
III. Lone-pair electron donors				
24. Ammonia	820			Folman and Yates [1958a]
	750			Cant and Little [1964]
	750			Low et al. [1967b]
	675	8.9[d)]	10.16	Hertl and Hair [1968]
	675		10.16	Cusumano and Low [1971]
	750		10.16	Van Cauwelaert et al. [1971]
25. Pyridine	850			Sidorov [1956]
	765	10.8[d)]	9.7	Hertl and Hair [1968]
	760	17.1		Davydov et al. [1970]
	765		9.7	Cusumano and Low [1971]
	770			Fink et al. [1972]
	735			Knözinger [1974]
26. 2-methyl-py.	785			Knözinger [1974]
27. 4-methyl-py.	775			Knözinger [1974]
28. 2-ethyl-py.	775			Knözinger [1974]
29. 3-ethyl-py.	775			Knözinger [1974]
30. 4-ethyl-py.	785			Knözinger [1974]
31. 2-*t*-butyl-py.	745			Knözinger [1974]

TABLE 27.1–(*contd.*)

Adsorbate	$\Delta\bar{\nu}_{OH}$ (cm^{-1})	ΔH_{ads} (kcal/mole)	I (eV)	References
32. 4-*t*-butyl-py.	805			Knözinger [1974]
33. 2,6-dimethyl-py.	805			Knözinger [1974]
34. 2,6-di-isopropyl-py.	825			Knözinger [1974]
35. 2,4,6-trimethyl-py.	855			Knözinger [1974]
36. 2,6-di-*t*-butyl-py.	100			Knözinger [1974]
37. 2-fluoro-py.	485			Knözinger [1974]
38. 2-chloro-py.	525			Knözinger [1974]
39. 3-chloro-py.	645			Knözinger [1974]
40. 4-cyano-py.	525			Knözinger [1974]
41. 2-methoxy-py.	615			Knözinger [1974]
42. Methylamine	840		8.97	Van Cauwelaert et al. [1971]
43. Ethylamine	840		8.86	Van Cauwelaert et al. [1971]
44. *n*-Propylamine	850		8.78	Van Cauwelaert et al. [1971]
45. *t*-Butylamine	860		8.71	Van Cauwelaert et al. [1971]
46. Dimethylamine	900		8.24	Van Cauwelaert et al. [1971]
47. Diethylamine	990			Basila [1961]
	890		8.01	Van Cauwelaert et al. [1971]
48. Di-*n*-propylamine	940		7.84	Van Cauwelaert et al. [1971]
49. Di-*t*-butylamine	940		7.69	Van Cauwelaert et al. [1971]
50. Trimethylamine	990			Davydov et al. [1965]
	915		7.82	Van Cauwelaert et al. [1971]
51. Triethylamine	975	11.0[d)]	7.50	Hertl and Hair [1968]
	990	20.0		Davydov et al. [1970]
	930		7.50	Van Cauwelaert et al. [1971]
52. Phenylamine	550			Abramov et al. [1964a]
	550		7.70	Cusumano and Low [1971]
	550	19.5[b)]	7.69	Kiselev and Lygin [1964]
	550			Low and Hasegawa [1968]
	550			Low and Subba Rao [1969]
53. Pyrolidine	900			Elkington and Curthoys [1968]
54. Pyrole	325			Elkington and Curthoys [1968]
55. Hydrogen cyanide	180			Kozyrovski and Folman [1964]
	190			Low et al. [1968]
56. Acetonitrile	300			Roev et al. [1958]
	305			Davydov et al. [1965]
	320			Elkington and Curthoys [1968]
	305	12.7		Davydov et al. [1970]
	290			Knözinger [1973]
57. Benzylnitrile	280			Knözinger [1973]
58. Acrylnitrile	280			Elkington and Curthoys [1968]
59. 4-cyanopyridine	210			Knözinger [1974]

TABLE 27.1–(*contd.*)

Adsorbate	$\Delta\bar{\nu}_{OH}$ (cm^{-1})	ΔH_{ads} (kcal/mole)	I (eV)	References
60. Dimethylether	430			McDonald [1958]
	420			Elkington and Curthoys [1968]
61. Diethylether	450			Davydov et al. [1963a]
	450	14.9[a)]	9.72	Kiselev and Lygin [1964]
	445			Elkington and Curthoys [1968]
	450	14.8		Davydov et al. [1970]
62. Di-*n*-propylether	470			Elkington and Curthoys [1968]
63. Ethylvinylether	320			Elkington and Curthoys [1968]
64. Chlorodimethylether	350			Elkington and Curthoys [1968]
65. Phenylmethylether	340			Elkington and Curthoys [1968]
	350		8.2	Cusumano and Low [1971]
66. 2-methoxypyridine	250			Knözinger [1974]
67. Tetrahydrofuran	470			Davydov et al. [1965]
	475			Elkington and Curthoys [1968]
68. Tetrahydropyran	490			Elkington and Curthoys [1968]
69. 2,3-Dihydropyran	430			Elkington and Curthoys [1968]
70. Dioxane	430			Elkington and Curthoys [1968]
	390	16.0		Davydov et al. [1970]
71. $CH_3Si(OCH_3)_3$	363	6.0[d)]		Hertl and Hair [1968]
72. $(CH_3)_2Si(OCH_3)_2$	403	6.7[d)]		Hertl and Hair [1968]
73. $(CH_3)_3SiOCH_3$	477	8.3[d)]		Hertl and Hair [1968]
74. Acetone	370			Sidorov [1954]
	330			Folman and Yates [1958a]
	330			Davydov et al. [1965]
	340			Young and Sheppard [1967b]
	395 (25°C)	12[d)]		Hertl and Hair [1968]
	345 (100°C)			Hertl and Hair [1968]
	360			Elkington and Curthoys [1968]
	330	15.2		Davydov et al. [1970]
75. Cyclopentanone	340			Davydov et al. [1965]
	340	17.3		Davydov et al. [1970]
76. Acetaldehyde	290			Sidorov [1954]
	280	10.5[d)]		Hertl and Hair [1968]
	290			Elkington and Curthoys [1968]
77. Methylethylketone	330			Elkington and Curthoys [1968]
78. Benzaldehyde	290			Sidorov [1954]
	290		9.51	Cusumano and Low [1971]
79. Acetophenone	320			Elkington and Curthoys [1968]
80. Methylacetate	270			Elkington and Curthoys [1968]

TABLE 27.1–(*contd.*)

Adsorbate	$\Delta\bar{\nu}_{OH}$ (cm^{-1})	ΔH_{ads} (kcal/mole)	I (eV)	References
81. Ethylacetate	280			Davydov et al. [1965]
	280	16.1		Davydov et al. [1970]
	270			Elkington and Curthoys [1968]
82. Phenylacetate	250			Elkington and Curthoys [1968]
83. CCl_4	40		11.47	Basila [1961]
	45	6.0[d]	11.47	Hertl and Hair [1968]
84. $CHCl_3$	48		11.42	Basila [1961]
	45	5.6[d]	11.42	Hertl and Hair [1968]
85. CH_2Cl_2	72		11.35	Basila [1961]
86. CH_3Cl	110			Folman and Yates [1958a]
	106		11.28	Basila [1961]
87. CH_3I	125	6.0[d]	9.1	Hertl and Hair [1968]
88. CH_3Br	120		10.5	Van Cauwelaert et al. [1970]
89. $SiCl_4$	25	5.4[d]	11.6	Hertl and Hair [1968]
90. CH_3SiCl_3	57	6.5[d]	11.17	Hertl and Hair [1968]
91. $(CH_3)_2SiCl_2$	90	7.8[d]	10.50	Hertl and Hair [1968]
92. $(CH_3)_3SiCl$	135	8.7[d]	10.07	Hertl and Hair [1968]
93. CH_3NO_2	160			Davydov et al. [1965]
	160	11.9		Davydov et al. [1970]
94. SO_2	115			Folman and Yates [1958a]
95. CS_2	60	6.9[d]	10.4	Hertl and Hair [1968]

[a] 25% coverage.
[b] Zero coverage (GC determination).
[c] 50% coverage.
[d] Determined from isotherms based on the coverages which were measured from relative decrease in intensity of free hydroxyl stretching band on adsorption.

Some typical systems mentioned in table 27.1 should be discussed in more detail. Among the *apolar adsorbates*, oxygen adsorption produces a wave number shift of only 12 cm^{-1}, which is comparable to those observed for rare gases. Nitrogen, on the contrary, leads to a wave number shift of 24 cm^{-1}. Furthermore, as observed by McDonald [1957], the band of the free surface hydroxyl groups vanishes when the surface coverage by nitrogen is still considerably less than the monolayer capacity determined from adsorption isotherms. In agreement with this result Kiselev and Khrapova [1957] could show that the adsorption capacity for nitrogen continuously decreased with increasing surface dehydroxylation. Accordingly, nitrogen interacts specifically with the surface hydroxyl groups and this phenomenon was explained as being due to the quadrupole moment of nitrogen. Frohnsdorf and Kington [1959] have shown, on the basis of a reasonable model of the surface hydroxyl environment, that the observed wave number shift can be understood in terms of a proton–quadrupole interaction provided the effective protonic charge is not less than about 0.2×10^{-10} e.s.u.

As pointed out by Kiselev [1964b] and later by Davydov et al. [1970], the *specific molecular interactions of adsorbate molecules with hydroxyl groups* and the corresponding wavenumber shifts $\Delta\bar{\nu}_{OH}$ are not governed by the total dipole moment of polar molecules but rather by the local distribution of electrons at the periphery of the given adsorbate molecules. On this basis the relative wave number shifts of various adsorption systems can be interpreted qualitatively. Thus, in the furan molecule (no. 22 in table 27.1) the electron density in the lone-pair orbitals of the oxygen is distributed by conjugation throughout the ring and the molecules show aromatic properties. Therefore the silanol groups form $O\text{–}H\cdots\pi$ complexes with π electrons of the furan molecules and the observed wave number shifts of 120–140 cm^{-1} are very close to those observed for benzene (no. 11 in table 27.1). On the contrary, tetrahydrofuran (no. 67 in table 27.1) behaves as a cyclic ether, the oxygen lone-pair electrons being the H-bond acceptors and the corresponding wave number shift of 470 cm^{-1} is identical to that of diethylether (no. 64 in table 27.1). The nitrogen lone-pair electrons in pyridine (no. 25 in table 27.1) are also involved in conjugation with the π bonds in the aromatic ring, so that the electron density in the sp^2 hybrid lone-pair orbital is less than in the sp^3 lone-pair orbital at the nitrogen in aliphatic amines (nos. 42–51 in table 27.1). Correspondingly, the wave number shifts $\Delta\bar{\nu}_{OH}$ are smaller on pyridine (735–770 cm^{-1}) than on amine adsorption (840–990 cm^{-1}).

A dual or *multiple interaction* may occur between suitably located hydroxyl groups and adsorbate molecules which contain an aromatic nucleus and lone-pair electrons in substituents or side chains. Thus, wave number shifts of 150 and 550 cm^{-1} for aniline (nos. 20 and 52 in table 27.1) adsorption and of 150 and 340 cm^{-1} for anisole (nos. 18 and 65 in table 27.1) adsorption are observed. The lower shifts are close to those obtained for benzene adsorption, whereas the higher shifts approach those for amine and ether adsorption. A dual site adsorption might be adopted for these cases for surfaces of sufficiently high hydroxyl density, where the two H-bonds will mutually influence the respective bond energies and wave number shifts. For anisole this situation may be demonstrated by the following model structure:

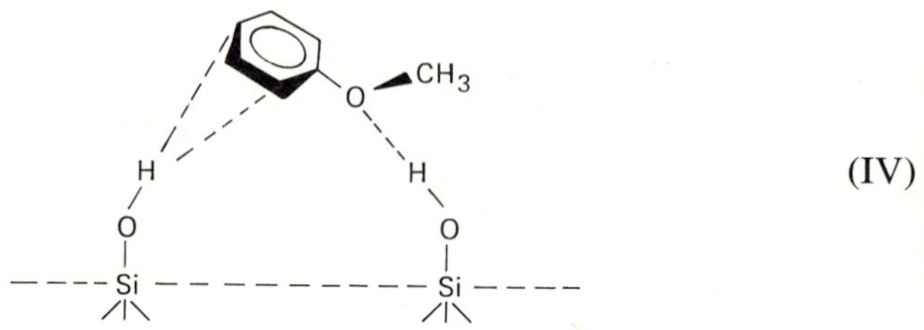

(IV)

For 2-methoxy-pyridine (nos. 41 and 66 in table 27.1) wave number shifts of 250 and 615 cm^{-1} are observed which are typical for ether and pyridine adsorption, respectively. A dual interaction with the lone-pair electrons at the nitrogen and oxygen, taking over the H-bond acceptor functions seems to be an adequate model. For geometric reasons two individual surface species must be formed on adsorption of 4-cyano-pyridine (nos. 40 and 59 in table 27.1) to explain the two wave number shifts of 210 and 525 cm^{-1} shown in fig. 27.3. The shift of 525 cm^{-1} corresponds to an adsorption via the lone electron pairs of the ring nitrogen, the shift being lower than for pyridine due to the electron withdrawing power of the cyano group. A second adsorption through the nitrogen of the cyano group leads to the shift of 210 cm^{-1} which is comparable to those observed for nitrile adsorption.

As already mentioned, the *wave number shifts of the stretching vibration of surface hydroxyl groups on adsorption is a function of the surface coverage* (McDonald [1957], Basila [1961] and Galkin et al. [1964]). Usually the wave number shifts in gas–solid interactions increase remarkably at coverages below 50% and vary only slowly at coverages between 50 and 100%. An example is given in fig. 27.4 for the adsorption of alkyl benzenes. McDonald [1957] who observed such a dependence of $\Delta\bar{\nu}_{OH}$ on the coverage during nitrogen adsorption, explained this phenomenon as being due to an interaction of one nitrogen molecule with several surface hydroxyls at low coverages followed by a redistribution at higher coverages, as a result of which a 1 : 1 adsorption should be formed. Consequently, this redistribution should be accompanied by an increase in the interaction energy per hydroxyl group which leads to an increase in the wave number shift towards lower wave numbers. As pointed out by Galkin et al. [1964] an analogous interpretation cannot hold for the adsorption of

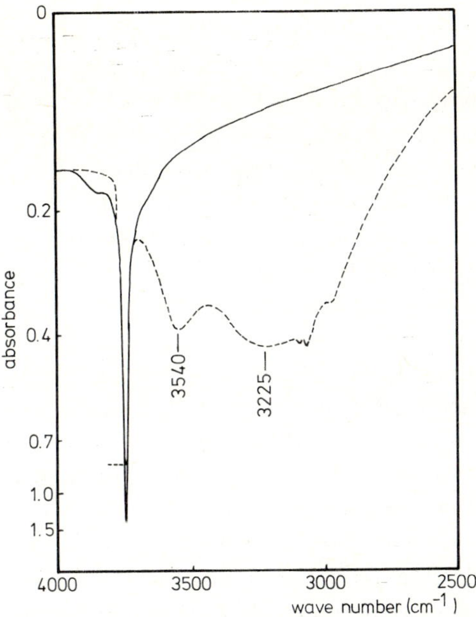

Fig. 27.3. Transmission IR spectrum of 4-cyano-pyridine adsorbed on silica "Standard" (Degussa). Solid line: silica background after degassing for 2 h at 700°C; broken line: after adsorption of 4-cyano-pyridine at ambient temperature.

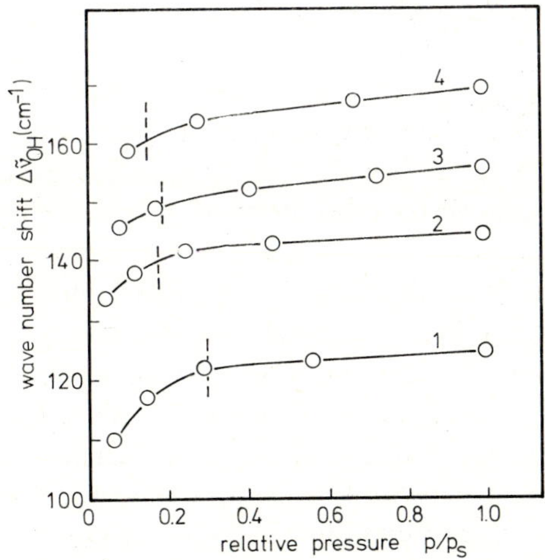

Fig. 27.4. Dependence of the wave number shift of aerosil hydroxyl groups on surface coverage upon adsorption of benzene(1), toluene(2), *p*-xylene(3) and mesitylene(4). Vertical dotted lines indicate the completion of a statistical monolayer. (Reproduced by permission of the Chemical Society, London, from Galkin et al. [1964]).

aromatic molecules, since their molecular size is large enough to cover several surface hydroxyls. This is true, however, only for adsorption on surfaces with high hydroxyl density, whereas the wave number shifts are also a function of the coverage after temperature treatment at 700 °C when predominantly isolated and some geminal hydroxyls are retained on the silica surface (see section 27.2.1). Galkin et al. [1964] explain the phenomenon by assuming a closer interaction or reorientation of molecules with growing coverage. The fact that the differential heat of adsorption decreases while the wave number shift and, therefore, the strength of specific interactions increases, is assumed to be due to the differences in total and specific interaction energies. It should be mentioned in this connection that Joris and Raguè-Schleyer [1968] observed an analogous dependence of the wave number shift $\Delta\bar{\nu}_{OH}$ of methanol on the concentration of the H-bond acceptor molecule, when studying the system methanol/substituted pyridines in solution.[4)] A general explanation of the phenomenon appears to be still unavailable.

In the search for quantitative relations between wave number shifts $\Delta\bar{\nu}_{OH}$ and molecular properties of the adsorbate, the coverage dependence of $\Delta\bar{\nu}_{OH}$ must be taken into account and comparable conditions have to be chosen for different adsorbates particularly when the shifts are small ($\Delta\bar{\nu}_{OH} < 200\ cm^{-1}$) and the coverages are low ($< 50\%$). For large wave number shifts and coverages higher than 50%, the dependence of $\Delta\bar{\nu}_{OH}$ on coverage appears to be negligible and to fall within the limits of accuracy in the determination of $\Delta\bar{\nu}_{OH}$.

27.2.2.2. *Correlations* (see also chapter 25)

Various attempts have been made to correlate the wave number shifts of the stretching fundamental of the surface hydroxyl groups with molecular properties of the adsorbate. McDonald [1957] found a smooth relationship between wave number shifts and the polarizabilities of rare gases, whereas nitrogen, oxygen and methane did not follow this correlation. The sequence of dipole moments for various adsorbates is also not in accord with their wave number displacements (Kiselev [1964b] and Davydov et al. [1970]). Plots of $\Delta\bar{\nu}_{OH}$ vs. the Kirkwood function $(\epsilon - 1)/(2\epsilon + 1)$, which were derived by West and Edwards [1937] and by Bauer and Magat [1938], with ϵ being the macroscopic dielectric constant of the surrounding medium, were also unsuccessful. This failure is probably due to the fact that the macroscopic dielectric constant of polar media can be largely different from the local dielectric constant in the immediate environment of an adsorbing molecule (Little [1966]). This effect has been taken into account by White et al. [1967] while considering the perturbation of OH groups in zeolites.

A number of *empirical correlations* have been reported in the literature. Kiselev and coworkers (Davydov et al. [1962, 1970], Galkin et al. [1964] and Kiselev [1965]) obtained a linear correlation between $\Delta\bar{\nu}_{OH}$ and the approximate energy of the specific interaction with hydroxyl groups ΔH_{OH} as defined by eq. (27.1). As shown in fig. 27.5 this relation holds nicely up to approximately $\Delta H_{OH} = 6.5$ kcal/mole. Anderson [1965] showed that his value of $\Delta\bar{\nu}_{OH} = 200\ \text{cm}^{-1}$ for water adsorption also fits Kiselev's correlation, though originally only molecules were taken into account which did not show self association. Later, Kiselev [1971] and Curthoys et al. [1974] reported a linear correlation between ΔH_{OH} and the change in the square root of the integrated intensity ($A^{1/2} - A_o^{1/2}$), where A refers to the band of the OH stretching mode in the H-bonded complex and A_0 to the free OH stretching band.

Noller et al. [1972] obtained a fairly good linear relation between $\Delta\bar{\nu}_{OH}$ and the pK_a values of 14 chemically very different adsorbate molecules, which they ascribe to the parallelism of pK_a values and the electron pair donor strength which is a measure of the H-bond acceptor strength.[5)]

Basila [1961] found two separate non-linear relationships between $\Delta\bar{\nu}_{OH}$

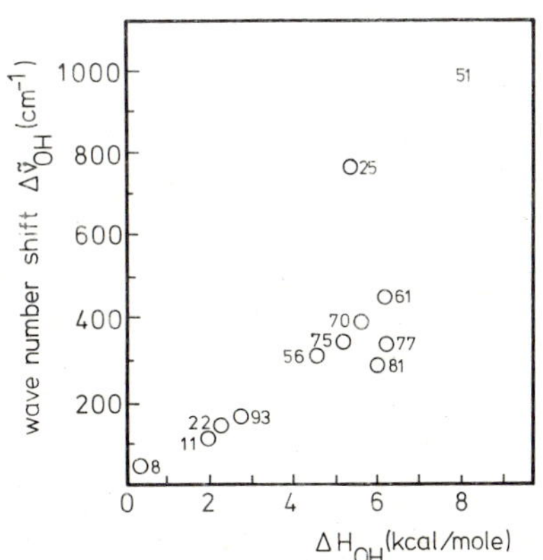

Fig. 27.5. Relation between wave number shift $\Delta\bar{\nu}_{OH}$ of the stretching vibration of surface silanol groups and ΔH_{OH}, the difference in the heats of adsorption on hydroxylated and strongly dehydroxylated silica surfaces at 50% coverage. The numbers in the diagram correspond to those of table 27.1. (Reproduced by permission of the Chemical Society, London, from Davydov et al. [1970]).

and the ionization potentials I of the series of chloromethanes and methylbenzenes, respectively. Similarly, Hertl and Hair [1968] gave a relation

$$\Delta\bar{\nu}_{OH} = \frac{10.9 - I(\mathrm{eV})}{0.015} \tag{27.2}$$

with the ionization potential, which was found to be valid for methyl benzenes, saturated hydrocarbons and chloromethanes, for which they found identical heats of adsorption of (6.0 ± 0.4) kcal/mole from spectroscopic isotherms. For those molecules, for which the heat of adsorption depends on the wave number shift and on the location of the lone-pair electrons responsible for the H-bonding,the same authors found the empirical equation

$$\Delta H_{ads}(\mathrm{kcal/mole}) = 0.455(\Delta\bar{\nu}_{OH})^{1/2} + \mathrm{constant}. \tag{27.3}$$

The constant is equal to 3.0 when the lone-pair electrons are located in p orbitals and -2.3 when the lone-pair electrons are located in sp^2 or sp^3 hybrid orbitals. Some non-linear correlations between $\Delta\bar{\nu}_{OH}$ and the Hammett parameter σ_p for the adsorption of substituted benzenes was reported by Cusumano and Low [1971]. A linear correlation, with the Hammett parameter was obtained for the adsorption of substituted pyridines by Knözinger [1974] (see fig. 27.6) and with the Taft parameter of aliphatic amines by Van Cauwelaert et al. [1971]. Analogous correlations have recently been reported by Rouxhet [1974] for a series of alkyl benzenes and alkyl bromides. As pointed out by Cusumano and Low [1971] such correlations between $\Delta\bar{\nu}_{OH}$ and ionization potential or Hammett and Taft parameters suggest a charge-transfer mechanism of H-bonding. Electrostatic interaction, dispersion forces, exchange repulsion and delocalization energy have to be taken into account as contributions to the energy of H-bonding. On this basis an equation was derived which relates the ionization potential I of the adsorbate molecule and the Mulliken–Puranik parameter $(\Delta\bar{\nu}_{OH}/\bar{\nu}_0)^{-1/2}$ where $\bar{\nu}_0$ is the wave number of the unperturbed hydroxyl stretching fundamental (Mulliken [1952], Puranik and Kumar [1963a, b]):

$$I = [(\tfrac{5}{4})^{1/2} cS\lambda^{1/2}] \left(\frac{\Delta\bar{\nu}_{OH}}{\bar{\nu}_0}\right)^{-1/2} + E_A + W. \tag{27.4}$$

λ is the polarity of the acceptor orbital, S is the overlap integral between donor orbital and the molecular orbital of the acceptor molecule, E_A is the electron affinity of the acceptor molecule and W is the net attraction

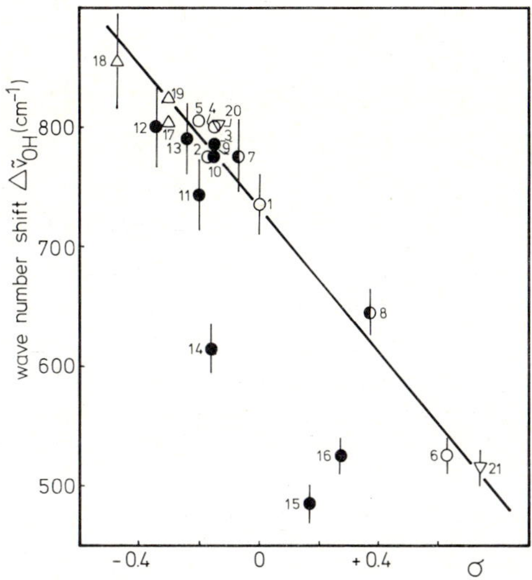

Fig. 27.6. Relation between wave number shift $\Delta\bar{\nu}_{OH}$ of the stretching vibration of surface silanol groups (degassing temperature 700 °C, 2 h) and Hammett's polar substituent parameters of substituted pyridines (adsorbed at room temperature, 70% coverage): ○ para-, ◐ meta-, ● ortho-substitution, △ 2,6-di-substitution, ▽ other di-substituted compounds: (1) pyridine; (2) 4-methyl-; (3) 4-ethyl-; (4) 4-isopropyl-; (5) 4-*t*-butyl-; (6) 4-cyano-; (7) 3-ethyl-; (8) 3-chloro-; (9) 2-methyl-; (10) 2-ethyl-; (11) 2-*t*-butyl-; (12) 2,4-dimethyl-; (13) 2,5-dimethyl-; (14) 2-methoxy-; (15) 2-fluoro-; (16) 2-chloro-; (17) 2,6-dimethyl-; (18) 2,4,6-trimethyl-; (19) 2,6-di-isopropyl-; (20) 3,5-dimethyl-; (21) 3,5-dichloro. (Reproduced with permission from Knözinger [1974]).

energy between acceptor and donor. From eq. (27.4), linear plots of the ionization potential vs. the Mulliken–Puranik parameter can be expected for the adsorption of series of chemically similar donor molecules having similar donor orbitals, so that the overlap integral S is approximately constant in the series. Cusumano and Low [1971] and Van Cauwelaert et al. [1970, 1971] were in fact able to test successfully eq. (27.4) for the series of alkanes, chloromethanes, fluoro-aromatics, alkyl aromatics, oxy-aromatics, ethers, aliphatic amines and methyl chlorosilanes (fig. 27.7). Cusumano and Low [1971] could additionally derive an equation – also based on the charge-transfer model – relating the integrated absorption intensity of the stretching fundamental of perturbed hydroxyl groups to the reciprocal Mulliken–Puranik parameter. As this relationship was also shown to hold, the *charge-transfer model*, at present, seems to be the most adequate description of the specific adsorption interaction on silica surfaces.

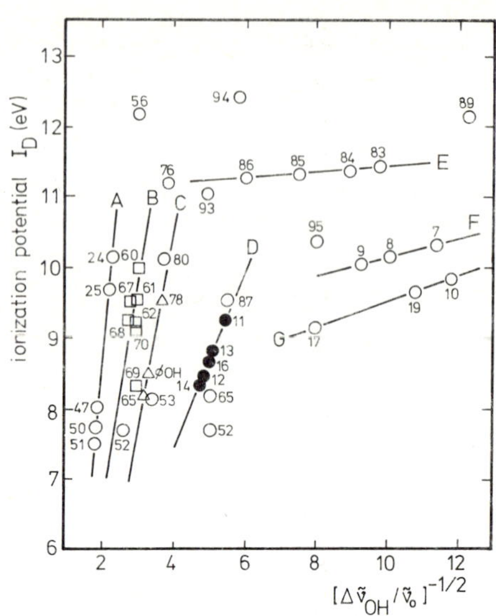

Fig. 27.7. Application of charge transfer theory to the hydroxyl wave number shift, for a variety of molecules interacting with hydroxyls at room temperature at a surface coverage of 50%. (A) amines; (B) ethers; (C) oxy-aromatics; (D) CH_3 aromatics; (E) chloromethanes; (F) alkanes; (G) fluoro-aromatics. The numbers in the diagram correspond to those given in table 27.1. (Reproduced with permission from Cusumano and Low [1971]).

It should be pointed out that the charge-transfer model describes only one aspect of H-bonding. For H-bonding in solution, the wave number shifts of two H-bond donors can be correlated by plotting the relative wave number shift $\Delta\bar{\nu}/\bar{\nu}_0$ of one donor with respect to that for another (Bellamy et al. [1958]). Such graphs give linear relationships and very different H-bond acceptors fit the same straight line when the proton bearing atoms of the donors are identical (Bellamy and Pace [1969]). Rouxhet and Sempels [1974] succeeded in relating the relative wave number shifts of the stretching vibration of surface silanol groups using *p*-fluorophenol, phenol and methanol as references. The slopes of such plots with phenol or *p*-fluorophenol as reference could be successfully correlated with the pK_a values of the H-bond donors for liquid systems and the value for silica also fell on the same straight line if a pK_a value of 6.8 as determined by Schindler and Kamber [1968] was adopted as characterizing the surface hydroxyl groups. This value agrees excellently with that obtained by Hair and Hertl [1970] applying the same method as Rouxhet and Sempels [1974].

In their study on adsorption of aliphatic amines, Van Cauwelaert et al. [1971] observed deviations from the linear correlation for diethylamine and the tertiary amines, which they explain by the assumption of a strong influence on the H-bonding interaction by *steric restrictions* between large substituent groups of the adsorbate and the silica surface. In the pyridine series studied by Knözinger [1974] deviations from the straight line relating $\Delta\bar{\nu}_{OH}$ with Hammett's inductive parameter (see fig. 27.6) are observed only for the tertiary butyl group in orthoposition in 2-*tert*.butyl pyridine whereas the point for 2.6-diisopropyl pyridine falls onto the straight line within the limits of accuracy. The steric hindrance of a single tertiary butyl group therefore seems to be appreciably stronger than that of two isopropyl groups in orthoposition. In the case of 2.6-di-*tert*.butyl pyridine the nitrogen lone-pair electrons are shielded so effectively that H-bonding via the nitrogen becomes impossible. This compound is thus adsorbed via an $OH\cdots\pi$ bond analogous to the aromatic molecules producing a wave number shift of only $100\ cm^{-1}$ which compares very well to that observed on benzene adsorption (see fig. 27.8). There are

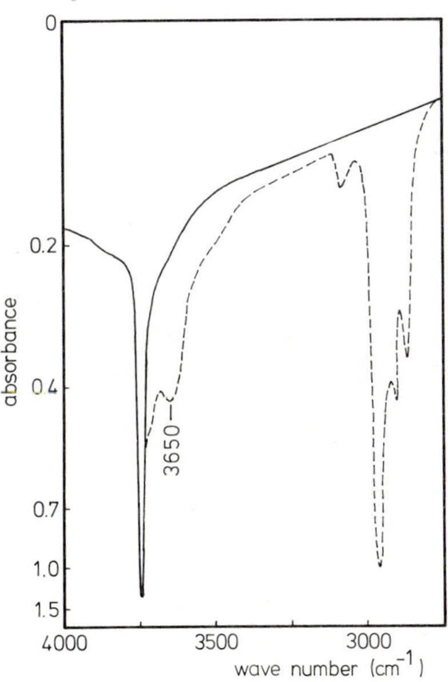

Fig. 27.8. Transmission IR spectrum of 2,6-di-*tert*.-butyl pyridine adsorbed on silica "Standard" (Degussa). Solid line: silica background after degassing for 2 h at 700°C; broken line: after adsorption of 2,6-di-*tert*.butyl pyridine at ambient temperature. (Reproduced with permission from Knözinger [1974]).

three other ortho-substituted pyridine compounds which do not fit the straight line in fig. 27.7, i.e. 2-methoxy-pyridine and 2-fluoro- and 2-chloro-pyridine. Steric restrictions may contribute to the observed deviations for the methoxy and the chloro compound. However, as pointed out in section 27.2.2.1, these compounds can interact simultaneously with two suitably located OH groups through the nitrogen and the ortho substituent. The bond strengths of the two H-bonds will be mutually influenced which leads to a weakening of the OH···N bond and a corresponding decrease in the wave number shift.

27.2.2.3. *Further spectroscopic information on hydrogen-bonding*

The IR fundamental region has been mostly dealt with up to this point. Other spectroscopic information is also available to characterize the specific H-bonding interactions between adsorbate molecules and surface hydroxyl groups.

The influence of the adsorption of a large number of molecules on the *overtone bands* of surface OH groups on silica and porous glass surfaces has been studied by Kurbatov and Neuimin [1949], Yaroslavsky and Terenin [1949], Yaroslavsky and Karyakin [1952], Filimonov [1956], and Filimonov and Terenin [1956]. In agreement with observations in the fundamental region, the intensity of the hydroxyl stretching overtone at 7320 cm^{-1} decreased on adsorption, the overtones of internal vibrations of the adsorbed molecules, however, were observed only at relatively high coverages. Typically the displaced bands of H-bonded hydroxyl groups–particularly those which were perturbed by strong H-bond acceptor molecules–were not observable or were only very weak. This phenomenon was interpreted as being due to a reduction of the anharmonicity of the O–H oscillator on H-bonding. A re-interpretation based on the influences of mechanical and electrical anharmonicity on the intensity of X–H stretching vibrations is more adequate as shown by Di Paolo et al. [1972] (see also ch. 13).

Changes in the IR (Kiselev and Lygin [1960, 1961a, b], Abramov et al. [1963, 1964b] and Galkin et al. [1964]) and electronic (Abramov et al. [1963] and Kiselev and Lygin [1964]) spectra on adsorption will be discussed in the next sections on benzene, ammonia and water adsorption.

The analysis of the *IR band shapes* determined by an unresolved rotational structure enabled Sheppard and Yates [1956] and Sheppard et al. [1960] to conclude that methane and bromomethane had rotational freedom in the adsorbed state only around a C_3 axis (the C–Br axis in the case of bromomethane) which is oriented vertically to the surface. Measurements of NMR (Freude [1971]) *relaxation times* and of *self diffusion coefficients* in the adsorbed layer of *n*-hexane, which is adsorbed by non-specific dispersion forces only, showed that only very few molecules at coverages below 1% are localized at adsorption centers with residence times of up to 10^{-8} s at 25 °C. The local self diffusion coefficients are nearly independent of the surface coverage. Diethylether on the contrary, which specifically interacts with surface OH groups through H-bonding, is localized with residence times of 10^{-8} s already at 50% coverage and the self diffusion coefficient decreases by a factor of 100 with decreasing coverage. This relatively strong localization of ether on silica surfaces may be compared with the conclusion drawn by Kiselev and Lygin [1962] from calorimetric adsorption studies, that two H-bonds are formed between the sets of oxygen lonc-pair electrons of the adsorbate and two neighboring hydroxyl groups, according to the model structure

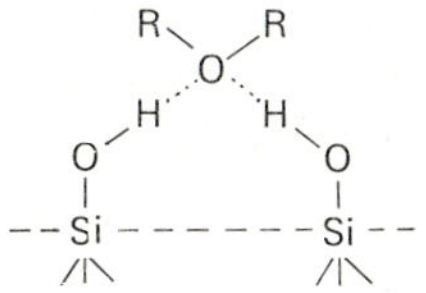

```
           R     R
            \   /
              O
            ..  ..
          H'      'H
         /          \
        O            O
        |            |
   -- Si --------- Si --
       /|\          /|\
```
(V)

Beckert [1971] showed by a relaxation analysis that intermolecular dipolar interactions and the interaction with the surface hydroxyl groups are responsible for the NMR relaxation of methane adsorbed on silica. A rapid reorientation of the adsorbed molecules with a correlation time less than 10^{-12} s at 106 K occurred. Translational motion of adsorbed methane can be described by correlation times in the range of 10^{-9}–10^{-11} s. As with *n*-hexane, only a very small fraction is in a localized state at surface OH groups with correlation times in the order of 10^{-7}–10^{-8} s at 106 K. Further properties of H-bonded systems on silica surfaces will be dealt with in the subsequent subsections.

27.2.3. Adsorption of benzene

Benzene is taken as an example to discuss some additional properties of adsorbed π electron donor molecules. As concluded from heats of adsorption and from the influence of benzene on the OH stretching fundamental and overtone bands, benzene and other adsorbate molecules, in which the aromatic character dominates the molecular properties, interact specifically with surface hydroxyl groups through formation of OH$\cdots\pi$ bonds. These interactions not only decrease the force constant of the OH stretching vibrations but also influence the *intramolecular vibrations* of the adsorbate in a typical manner. Galkin et al. [1964] and Cusumano and Low [1970a] have shown that CH stretching and CC stretching wave numbers are only slightly displaced on adsorption onto silica surfaces by a few wave numbers as compared to their position in the liquid state. The out-of-plane CH deformation vibration, however, undergoes larger shifts of 10 cm^{-1} (Galkin et al. [1964]). These shifts are a characteristic indication for the perturbation of the π electron system on adsorption, since the out-of-plane CH deformation vibration is particularly sensitive to electron redistributions of the π electron system. Accordingly, the changes in the position of these bands should strongly depend on the strength of interaction. It therefore approaches the wave number of the respective vibration in liquid benzene at 675 cm^{-1} with increasing coverage and that of gaseous benzene at 671 cm^{-1} on highly dehydroxylated, methylated or fluorided surfaces, where only dispersion forces are acting on the adsorbed molecule. The relative integrated absorption of the same vibration mode very sensitively shows completely analogous trends as the band position, if the integrated absorption of the CH stretching modes is taken as a standard (Galkin et al. [1964]). It is assumed that their integrated intensities change relatively little on adsorption since they scarcely respond to changes in the π electron density.

Compatible with these changes in the IR spectra of the adsorbate are changes of the adsorbate bands in the *electronic spectra* (Abramov et al. [1963]). Ron et al. [1962] have shown that the electronic spectra of adsorbed benzene largely resemble the features of molecular benzene. Additionally, the usually forbidden $0 \rightarrow 0$ transition occurs weakly due to the perturbation of the benzene molecule by the surface and a wave number shift of 140–170 cm^{-1} towards lower energy, relative to the gas phase, is observed. Some combination bands have also been found, which were attributed to vibrational levels of the excited state and which are forbidden in the free molecule. As soon as the symmetry plane in the molecular ring is removed , these transitions are allowed and an adsorption is assumed with the molecular plane parallel to the surface so that the symmetry of the molecule is reduced as necessary from D_{6h} to C_{6v} due to its asymmetric interaction with the surface. Thorp and Woolf [1967, 1968] attributed a reduced electrical polarizability of adsorbed benzene to an electrical distortion by adsorption, which was assumed to render a further charge separation by an externally applied field more difficult. This adsorption mode finds some support by the results of high-resolution NMR studies of mesitylene adsorbed on silica (Karagounis [1964] and Pickett and Rogers [1967]). Identical line width of bands of the methyl and ring protons suggests that these are located at approximately equal distances from the surface, i.e. the ring plane must be oriented parallel to the surface.

The vibrational structure of the electronic spectra of adsorbed benzene obtained by Ron et al. [1962] furthermore showed a transition at 135 cm^{-1} which was believed to belong most probably to a vibration of the molecule relative to the surface. This value agrees fairly well with that of 190 cm^{-1}, which can be estimated from an equation given by Hill [1946] if the binding energy is assumed to be 10 kcal/mole. In Hill's equation the wave number of the vibration of an adsorbed molecule vertical to the surface is shown to be proportional to the square root of the binding energy.

A 1 : 1 interaction between surface silanol groups and adsorbed benzene molecules at low coverages on surfaces that do not contain more than 2 OH/100 Å^2, was proposed by Zecchina et al. [1968]. $\Delta\bar{\nu}_{OH}$ and the width of the band of H-bonded surface silanols remains constant and independent of the coverage under such conditions. The 1 : 1 interaction model, however, becomes inadequate for surfaces which contain more densely packed hydroxyl groups. Interaction of more than one silanol group then becomes possible with one benzene molecule and as a consequence, $\Delta\bar{\nu}_{OH}$ becomes a function of the surface coverage with benzene at a given pretreatment temperature and a function of the latter parameter for a fixed surface coverage (Galkin et al. [1964]).

The most reliable model for the specific interaction of adsorbed benzene on the time scale of IR spectroscopy, and of aromatic molecules in general, on silica surfaces of low hydroxyl content, as concluded from the presented material, is an $OH \cdots \pi$ bond with the molecular plane of the molecule being oriented approximately "parallel" to the adsorbent surface. Relying on this picture, statistical mechanical model calculations, particularly of the *adsorption entropy*, can be carried out and the results can be compared with experimental values. This will give information about the translational and rotational freedom of the benzene molecules in the adsorbed phase. High heats of adsorption and high entropy losses on adsorption of benzene on porous glass surfaces (degassed at 750 °C) at low coverages (less than 20%) suggest a relatively high degree of localization of the adsorbate, with residence times of the order of magnitude of 100 s as estimated from the Frenkel equation (Cusumano and Low [1970b]). With increasing coverage the adsorbate becomes mobile. For coverages above 50% the model calculations suggest a two-dimensional translational freedom with one additional degree of freedom of rotation around the C_6 axis which is perpendicular to the adsorbent surface and to the molecular plane. The residence times of mobile molecules at an adsorption site have been estimated to be of the order of magnitude of 10^{-8} s, i.e. less by a factor of 10^{10} as compared to the localized state. Similar conclusions have been reached by Ron et al. [1962].

More direct information about the *dynamic behavior* of H-bonded benzene on silica surfaces can be obtained from NMR *relaxation studies.* A detailed relaxation analysis at 3/4 of a statistical monolayer of benzene on a silica of high hydroxyl density (temperature treatment at 210 °C) was carried out by Woessner [1966]. From the temperature dependence of the intramolecular proton–proton interaction, a rapid anisotropic rotation of the benzene molecules around their hexad axis, which wobbles at a much slower rate, was shown to occur even at 77 K in agreement with the thermodynamic considerations. Two T_1 minima were observed, or at least indicated, at a temperature below 125 K and around room temperature. The low-temperature minimum which was associated with a long transverse relaxation time was assumed to belong to almost freely reorienting benzene molecules. The high-temperature minimum on the other hand was ascribed to less freely moving molecules which are more strongly held by specific interactions. Resing [1967–1968] on reinterpreting Woessner's data preferred an alternative explanation for the high-temperature T_1 minimum. A surface diffusion process was proposed with the assumption of a Gaussian distribution of barrier heights. Diffusional motions of adsorbed benzene have been shown to occur by gravimetric studies (Haul and Boddenberg [1968, 1969] and Boddenberg et al. [1969] and by NMR spinecho experiments (Haul and Boddenberg [1968], Freude [1969] and Boddenberg et al. [1969, 1972]). The local surface self-diffusion coefficients are found to be of the order of magnitude of 10^{-5} cm^2/s and to depend on the surface coverage in a characteristic fashion. At coverages between 1/2 and 2 statistical monolayers activation energies of surface self-diffusion of 3.5 kcal/mole corresponding to 1/3 of the isosteric heat of adsorption have been reported for the temperature range between −15 and +62 °C by Boddenberg et al. [1972]. Pfeifer and Staudte [1968] give additional evidence for translational mobility of adsorbed benzene molecules, when they observed a decrease of the line width of surface OH groups on adsorption of increasing amounts of C_6D_6 up to 1/3 of a statistical monolayer. This phenomenon was explained by assuming the superimposition of a statistical motion upon the OH groups caused by the translational diffusion of the adsorbed C_6D_6 molecules. The onset of translational motions appears to occur at about −120 °C and leads to a uniform relaxation behavior at higher temperatures due to spin exchange (Haul and Boddenberg [1968]). Freude et al. [1966] have extended relaxation measurements to high temperatures. They conclude that the correlation times for benzene molecules held to specific adsorption sites are determined by two dynamic processes which can be described as surface diffusion and desorption. The respective correlation times are approximately equal at 360 °C, whereas at room temperature the correlation time for desorption is lower by a factor of 100 as compared to that of surface diffusion.

27.2.4. Adsorption of ammonia

Ammonia is taken as an example which shows up the typical behaviour of lone-pair electron donor molecules adsorbed on silica surfaces. The large wave number shifts $\Delta\bar{\nu}_{OH}$ of the stretching fundamental of surface silanol groups towards lower wave numbers (see table 27.1) on adsorption of ammonia suggest a specific interaction of ammonia molecules with these surface groups. The corresponding asymmetric and symmetric NH stretching vibrations of the H-bonded ammonia molecules on porous vycor glass and silica gel were observed at 3400 and 3320 cm^{-1}, respectively, by Little and Mathieu [1960] and by Cant and Little [1964, 1965] (see fig. 27.9). The symmetric deformation mode, which is usually obscured by the absorption of the adsorbent and which appears at 938 and 968 cm^{-1} (inversion doubling) for gaseous ammonia, was determined from combination and overtone bands and located at 1050 cm^{-1} by Cant and Little [1967]. The shift to higher wave numbers was considered due to a reduction of anharmonicity by the H-bonding onto surface silanol groups. The asymmetric deformation mode can hardly be observed, since Si–O combination and overtone bands appear in the same region. A band at 1600 cm^{-1} has, however, been found on a silica surface pretreated at 200 °C and assigned to this vibration mode by Fripiat et al. [1970]. Similar spectra in the OH and NH stretching region have been obtained by

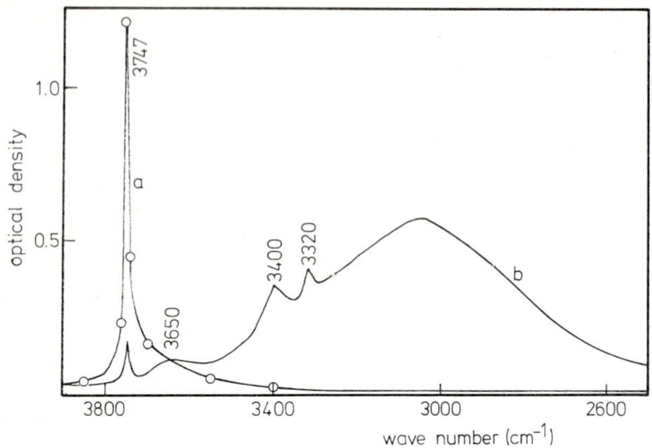

Fig. 27.9. (a) Transmission IR spectrum of cabosil after evacuation for 3 h at 450°C; (b) Spectrum 1 h after admission of 60 mm Hg of ammonia at room temperature. The circles represent the resulting optical densities after evacuation at room temperature for 1 h. (Reproduced by permission of the National Research Council of Canada from Cant and Little [1965]).

Sidorov [1956], Folman and Yates [1958a, 1959] and Kiselev et al. [1964] and the wave number shifts of the normal modes of ammonia on H-bonding have been calculated theoretically by Abramov et al. [1964b].

The broad absorption band around 3100 cm^{-1} (fig. 27.9) was assigned to once free hydroxyl groups which are perturbed due to H-bonding with ammonia molecules. Since intensity considerations showed that the ammonia bands at 3400 and 3320 cm^{-1} were sufficiently intense to account for all ammonia interacting with hydroxyl groups, an alternative adsorption structure with surface oxygen acting as H-bond acceptor and the ammonia molecule as the donor was disregarded (Cant and Little [1964]). The preferred adsorption structure is therefore the following:

$$\begin{array}{l} \quad\quad\quad\quad\quad\;\; \diagup H \\ \quad\quad\quad\;\; \cdots N - H \\ \quad\quad\;\; \diagup H \quad\;\; \diagdown H \\ O \\ | \\ Si \\ \diagup | \diagdown \end{array} \qquad\qquad (VI)$$

H-bonding of ammonia onto silica surfaces leads to *anomalous length changes of the adsorbent.* Whereas normally rigid porous materials expand when molecules are adsorbed onto their surfaces due to the lowering of their surface free energy, the adsorption of ammonia onto hydroxylated silica surfaces causes a contraction (Folman and Yates [1958a, b, c]). A model was assumed as an explanation of this phenomenon, in which vicinal silanol groups interact through H-bonds and ammonia is adsorbed via a H-bond with a free OH group:

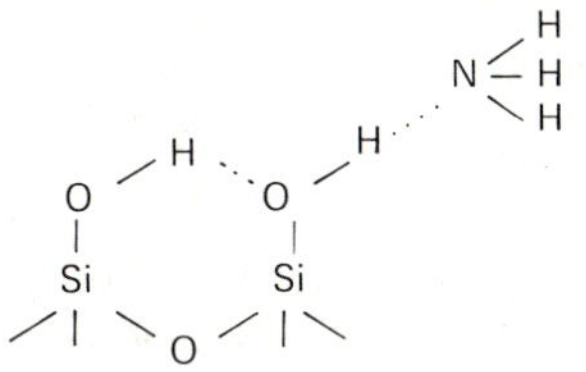

(VII)

The formation of the O–H···N bond on ammonia adsorption increases the polarity of both O–H bonds and consequently the strength of the H-bonding between the surface OH groups is increased, which may lead to a contraction of the adsorbent. This picture was confirmed by the observation, that the contraction clearly increased with increasing base strength of the adsorbate and was absent on methylated or strongly dehydroxylated surfaces.

More strongly bound species than H-bonded ammonia were already detected by Sidorov [1956] by IR spectroscopy. The formation of NH_2 groups on ammonia adsorption has been shown to occur by Peri [1966a] on silica and by Low and Ramasubramanian [1966] and by Low et al. [1967b] on porous vycor glass. These results found some confirmation by second moment calculations and NMR wideline experiments (Bonardet and Fraissard [1973]). Moreover, not only on porous glass (Chapman and Hair [1965], Low and Ramasubramanian [1966] and Low et al. [1967b]), and mixed oxides (Cant and Little [1967]), but also on silica gel (Sidorov [1956]) a *donor-acceptor complex* between adsorbed ammonia and electron deficient surface sites was formed (asymmetric and symmetric NH stretching bands at 3365 and 3280 cm^{-1}). These electron deficient sites on silica may be connected with small impurities of foreign oxides or they may be due to coordinatively unsaturated Si atoms in the silica surface. As pointed out by Egorov et al. [1966], the degree of coordinative unsaturation of surface Si atoms increases with increasing numbers of OH groups connected with the respective Si atom. Accordingly, the decrease in the adsorption capacity with decreasing hydroxyl density was assumed to be due to the decrease in coordinative unsaturation. This conclusion, however, is not unambiguous since the decrease in adsorption capacity could equally well be due to a reduced number of H-bonded ammonia. However, differential heats of adsorption between 18 and 15 kcal/mole have been measured in the initial ranges of coverage (Bliznakov and Polikarova [1966]). These values are too high to be explained by H-bonding and therefore suggest also a second, more strongly held species.

Coordinately, as well as H-bonded ammonia also shows up in the dielectric absorption, as demonstrated by Fiat et al. [1961], Feldman and Folman [1964] and by Lubezky et al. [1965]. Adsorption of ammonia on porous vycor glass containing a moderate concentration of surface OH groups, caused two absorption bands in the temperature range 250–350 K at a frequency of 23.6 Mc/s. The potential barriers for reorientation of the ammonia molecules in the two adsorption states were calculated to be 2.56 and 1.98 kcal/mole, respectively, the latter value probably being due to H-bonded ammonia.

Coordination of ammonia molecules to electron deficient sites is of some importance for H-bonding of ammonia, since in the coordinated molecule the N–H bonds will be strongly polarized and thus may provide a much stronger H-bond donor ability than the surface silanol groups. Kvlividze et al. [1969] concluded from the observation of a stepwise change of the second moment and of the NMR line width, that ammonia forms clusters through intermolecular H-bonding interactions around a central adsorption site, which is assumed to be a coordinately unsaturated Si atom.

Reuben et al. [1966] report on the *proton longitudinal and transversal relaxation times* in ammonia adsorbed on porous vycor glass. Both relaxation times depend strongly on the degree of surface coverage. This phenomenon was attributed to a distribution of correlation times which belong to diffusional and rotational motions of the adsorbed molecules. Dipolar interactions between the protons of ammonia and a proton of the nearest OH group were assumed to be influenced by a diffusional motion of adsorbed ammonia jumping from one OH group to another one located nearby. Correlation times between 1×10^{-9} and 4×10^{-7} s were estimated on the basis of this model. An intramolecular relaxation mechanism due to rotational freedom of the adsorbed molecule could not be excluded, though it does not fit the estimated correlation times equally well. Furthermore, an average correlation time for rotational motion of 6×10^{-12} s had been found from dielectric loss measurements (Feldman and

Folman [1964] and Lubezky et al. [1965]). This value is lower by two to three orders of magnitude than the lowest value of the NMR correlation time.

The formation of *protonated ammonia* (NH_4^+) on silica surfaces had been invoked from the observation of a usually weak absorption at 3450 cm^{-1} in the IR spectra (Felden [1959], Chapman and Hair [1963] and Boyle et al. [1965]). This assignment is, however, not unambiguous due to the proximity of the NH_2 stretching band as pointed out by Low et al. [1967b]. Thus, with only IR data of the region of OH and NH stretching vibrations available, a clear decision could not be made, whether NH_4^+ is formed on silica surfaces. Fripiat et al. [1970] observed two bands at 1450 and 1500 cm^{-1} in addition to the asymmetric deformation of NH_3 at 1600 cm^{-1} on a silica gel which had been pretreated at 200°C. From the dependence of the intensities of these three bands on the surface coverage the following assignment was put forward: the band at 1450 cm^{-1} is due to the asymmetric deformation of the protonated species NH_4^+, which is formed by a proton transfer either from a surface silanol group or from a chemisorbed water molecule; the band at 1500 cm^{-1} was assigned to the asymmetric deformation vibration of a NH_4^+ species which is H-bonded to an ammonia molecule, $H_3NH^+ \cdots NH_3$. The probability of the formation of this species increases with the surface coverage, whereas NH_3 H-bonded to silanol groups and NH_4^+ species are present simultaneously, even at the lowest coverages. The ratio of NH_4^+/NH_3 at a statistical monolayer was estimated roughly to be 0.3. The formation of the associated species $H_3NH^+ \cdots NH_3$ suggests the possibility of a proton transfer along the $NH \cdots N$ axis. This view found confirmation from the results of high resolution NMR experiments performed by Bonardet and Fraissard [1973]. The chemical shift with respect to gaseous ammonia was a function of the hydroxyl group density of the silica surfaces. The high values of the chemical shift were taken as evidence for a contribution of a proton exchange between OH groups, NH_3 and NH_4^+ in addition to H-bonding of ammonia to surface silanol groups. Moreover, an exponential increase of the surface electrical conductivity on adsorption of ammonia on silica was observed by Soffer and Folman [1966] and though not unambiguously verified, a protonic conduction mechanism was considered a possible explanation.

A detailed picture of the *dynamic behavior*, including *proton transfer processes*, of adsorbed ammonia on a highly hydroxylated silica surface (pretreatment at 200°C leading to 2.5 OH/100 Å^2) was developed by

Fripiat et al. [1970]. Surface conductivity was observed to follow an exponential relationship with the number of charge carriers and a quantitative relation could be deduced adopting the following model: charge carriers have to jump over potential barriers in order to move from one equilibrium position to the next. Assuming a protonic conduction mechanism and taking the NH_4^+/NH_3 ratio of 0.3 as estimated from the IR data (see above), good quantitative agreement was obtained. Two domains of dipolar absorptions were found at a frequency of 1592 cps. The first region between -120 and $-60\,°C$ has been assigned to a Maxwell–Wagner effect of dielectric heterogeneity. The second region between -180 and $-130\,°C$ showed broader and weaker absorption bands, the intensities of which varied strongly with the applied frequency. It was assumed that this absorption was probably due to a rotation of the ammonia molecules with a correlation time $\tau_r = 2 \times 10^{-10}\ e^{2900/RT}$ s. This molecular rotation was shown to represent the main contribution to the transversal NMR relaxation time T_2, while a protonic translational motion contributed mostly to the spin–lattice relaxation time T_1. The respective correlation time was 1.2×10^{-9} s for a statistical monolayer at 25 °C. The following model was put forward on the basis of these results: proton transfer from a surface silanol or a strongly held residual water molecule (acidic surface site AH) to an ammonia molecule leads to the formation of NH_4^+ species which may subsequently render a rapid proton transfer possible along the $^+NH \cdots N$ axis in a $H_3NH^+ \cdots NH_3$ species. The following steps should be involved:

$$AH \rightleftharpoons A^- + H^+ \tag{a}$$

$$NH_3 + H^+ \rightleftharpoons NH_4^+ \tag{b}$$

$$NH_4^+ + NH_3 \rightarrow NH_3 + NH_4^+ \tag{c}$$

and reasonable values for the equilibrium constant of step (a) could be estimated from the experimentally obtained ratios of $[NH_4^+]/[NH_3] = 0.3$ and $[H^+]/[NH_3] \approx 10^{-2}$. The main contribution to the electrical conductivity is due to “free” protons, the rate limiting process for the proton transfer being the reorientation of adsorbed species. Thus, the electrical conductivity is determined by two dynamic processes, the proton tunneling along the $^+NH \cdots N$ axis and the rotational reorientation of molecules. At a statistical monolayer the proton diffusion coefficient is found to be $0.5 \times 10^{-11}\ cm^2\ s^{-1}$ with an activation energy of approximately 8 kcal/mole. This value is equal to the sum of the activation energies of both the rotation of adsorbed species (approximately 2.9 kcal/mole) and the proton exchange process (4 to 6 kcal/mole, depending on the temperature) as determined from the temperature dependence of the respective NMR correlation times. Questions related to the lifetime of the NH_4^+ species were recently considered by Fripiat [1973].

Another phenomenon is observed in the IR spectra of ammonia adsorbed on silica and mixed silica–oxide adsorbents, which is to be related to the proton mobility in the adsorbed layer. As shown in fig. 27.10 a *continuous absorption* extends towards low wave numbers till the cut-off due to the strong absorption of the adsorbent itself (Cant and Little [1964, 1968]). Such continuous absorptions on adsorption have first been reported by Roev [1960] and Basila [1962] and explained as being due to a photodesorption process. This explanation is certainly incorrect, since the effect should then increase with decreasing bond strength, while the optical density of the continuum in fact increases with increasing proton affinity of the adsorbate (see below). Continuous absorptions have also been detected for adsorption of aniline on porous glass by Low and Subba Rao [1969], of pyridines on silica by Knözinger [1974] and of a wide variety of H-bond acceptor molecules on silica by Noller et al. [1972]. The optical density of the continuum is a function of the surface coverage and increases at a given wave number and coverage with increasing proton affinity of the adsorbate. Noller et al. [1972] have shown

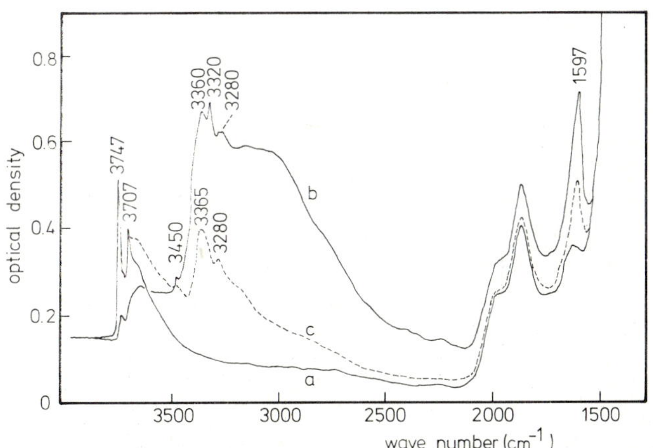

Fig. 27.10. (a) Transmission IR spectrum of boric oxide-silica mixture evacuated at 450°C; (b) after admission of 50 mm Hg of ammonia; (c) after evacuation at 20°C. (Reproduced by permission of the National Research Council of Canada from Cant and Little [1967]).

that the optical density of the continuum parallels the pK_a values of the adsorbate. For a series of *para*- and *meta*-substituted pyridines a roughly linear correlation was found between the optical density A_{1950} (taken at 1950 cm^{-1} and 70% coverage) as shown in fig. 27.11, while *ortho*-substituted pyridines did not fit the correlation. Cant and Little [1964] and Low and Subba Rao [1969] already related the continuous absorption with a high proton mobility in the H-bonded species formed. In line with the model of proton transfer processes in the ammonia-silica-system as proposed by Fripiat et al. [1970], the formation of $NH^+ \cdots N$ (and $OH^+ \cdots O$) species in the adsorbed phase can be assumed. The tunneling probability in these more or less symmetrical and very strong H-bonds is high, the jump frequency being of the order of magnitude of 10^{-9} s^{-1} as estimated by Fripiat et al. [1970] for the ammonia/silica system at 25 °C and for a statistical monolayer. These H-bonds seem to be highly polarizable (see ch. 15). Due to the field inhomogeneities on the surface, which also must be responsible for the distribution of correlation times for the different dynamic processes that were found by Fripiat et al. [1970], a continuous distribution of energy levels is created for the protons and a continuous IR absorption is the consequence (Zundel [1970]). This model simply explains the dependence of the optical densities of the continuum on surface coverage and proton affinity of the adsorbate by the higher probability for the formation of the $NH^+ \cdots N$ (or $OH^+ \cdots O$) species with an increase of these two parameters. A strong

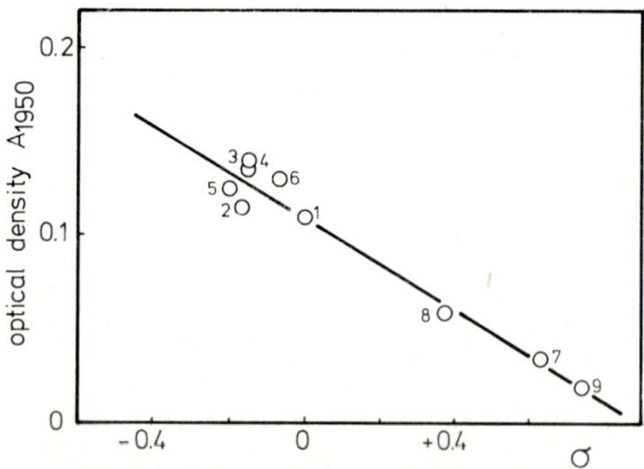

Fig. 27.11. Correlation of the optical density in the range of the continuous absorption at 1950 cm^{-1} and at 70% coverage with substituted pyridines with Hammett's polar substituent parameter: (1) pyridine; (2) 4-methyl-; (3) 4-ethyl-; (4) 4-isopropyl-; (5) 4-*tert*.butyl; (6) 3-ethyl-; (7) 4-cyano-; (8) 3-chloro-; (9) 3,5-dichloro-. (Reproduced with permission from Knözinger [1974]).

influence of steric restrictions on the optical density of the continuum as observed for ortho-substituted pyridines by Knözinger [1974] may be explained by a distortion of the potential well along the $NH^{+} \cdots N$ axis.

27.2.5. Adsorption of water

With the exception of ammonia and a few other adsorbates only H-bond acceptor molecules have been treated thus far. One of the most important molecules in surface chemistry is water, which develops H-bond donor and acceptor properties due to its two protons and two sets of lone-pair electrons. Water therefore has a pronounced tendency to interact with ionic surfaces by various types of forces and shows the ability for self-association and organization of long-range order even on solid surfaces, which makes up the "propriétés curieuses de l'eau adsorbée" as expressed by Fripiat [1971a]. The state of adsorbed water on solid oxide surfaces in general and on silica in particular is intimately related to the properties, concentration and configuration of surface hydroxyl groups. These parameters are governing the formation of short-range and long-range order and the possible nucleation of ice-like structures on the silica surface at a given coverage. The molecular motions of water in adsorbed layers and the degree of dissociation likewise depend on the nature of the underlying surface.

Sidorov [1960] concluded from his *IR data*, that water does not adsorb on free hydroxyl groups, since the band of the OH stretching fundamental of free silanol groups at 3750 cm^{-1} was relatively unaffected by the water adsorption. Davydov et al. [1964b] on the contrary showed that all hydroxyl groups are participating in interactions with adsorbed water layers. Sidorov [1960] observed two new bands in the OH stretching region which appeared on adsorption of water vapor at 3665 and 3450 cm^{-1} and which were superimposed on the bands assigned to H-bonded surface hydroxyl groups. The band at 3665 cm^{-1} was assigned to water molecules adsorbed on non-hydroxylic sites. Though Kiselev and Lygin [1962] disregard any evidence for adsorption sites other than hydroxyl groups, later IR spectroscopic measurements by Kiselev [1971], Fink and Camara [1972] and by Kunath [1972] render a coordination of water onto exceptionally coordinated silicon atoms possible under certain conditions. These molecules seem to be held very strongly and they are extremely polarized. The second band observed at 3450 cm^{-1} was assigned to the water molecules, H-bonded to initially mutually interacting silanol groups. Kubelkovà and Jirů [1972] came to the same conclusion very recently. The importance of mutually interacting surface silanol groups as adsorption sites has been pointed out by Van Cauwelaert et al. [1971] regarding the adsorption of ammonia and aliphatic amines (see section 27.2.2).

In agreement with their explanation of the anomalous contraction of porous glass specimens on ammonia adsorption (see section 27.2.4 and structure), Folman and Yates [1958c] explained the absence of any contraction in the porous glass/water system by the fact that water seemed not to interact with free hydroxyl groups.

It should be mentioned in this connection, that a small effect on the intensity of the $3750\,cm^{-1}$ band of free silanol groups could also be explained by the formation of water clusters with only a few anchoring H-bonds with silanol groups, even far below a statistical monolayer (Klier et al. [1973]). An analogous phenomenon was observed on adsorption of alcohols by Jeziorowski et al. [1973] (see section 27.2.6).

Zhdanov [1958] and Kiselev and Lygin [1959, 1962] consider OH pairs on fully or highly hydroxylated surfaces as the preferred adsorption sites. The pair sites may be geminal or vicinal OH groups favoring structure (VIII) or (IX), respectively:

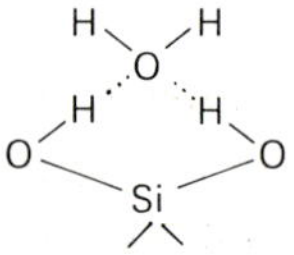

(VIII) 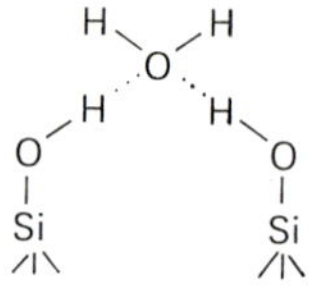 (IX)

Accordingly, Nair and Thorp [1965a] observed a H_2O/OH ratio of 1:2 and Fripiat et al. [1962] and Fripiat [1964] also report some evidence in favor of structures (VIII) or (IX) on hydroxylated silica surfaces. In contrast, Morimoto et al. [1971] found a 1:1 ratio on a less hydroxylated silica surface containing only 2.7 OH/100 $Å^2$, which would correspond to structure (X):

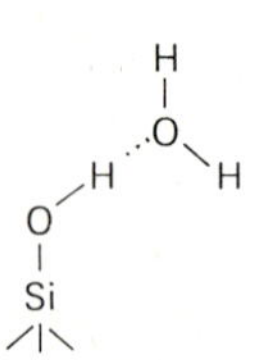

(X)

This structure, which was also favored by Klier et al. [1973], will progressively become more important with decreasing hydroxyl content and decreasing number of pair sites, while structure (XI) will be relatively unimportant in the case of silica because of the low basicity of surface

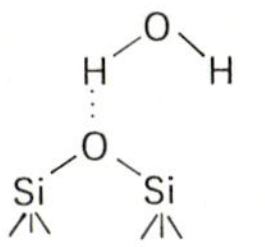

O
H H
⋮
O
Si Si

(XI)

oxygen atoms on silica (see section 27.2.1). The relative distribution of species (VIII), (IX) and (X) on a given surface will additionally depend on the acidity of the OH groups (Morariu and Mills [1972a]) and on the water coverage. However, general agreement appears to exist now that OH pairs, if present, are the preferred sites for adsorption of oxygen containing molecules with two sets of lone-pair electrons (see also structure (V) section 27.2.2). The strong *localization of water molecules* in the first layer as concluded from entropy calculations by Fripiat et al. [1965] and as verified also by NMR correlation times of individually adsorbed water molecules by Michel [1967], may be due to the dual site adsorption via two H-bonds per molecule.

The O–H···O distances and bond angles in species (VIII) and (IX) and thus the specific energy of adsorption must strongly depend on the local topography (relative orientation of Si–O bonds) of the surface. A wide distribution of interaction energies is therefore to be expected. Kiselev [1971] ascribes an asymmetry of the combination band near 5300 cm^{-1} (see below) towards lower wave numbers to such a distribution of energies of H-bonding.

The existence of *individually adsorbed water molecules* has been assumed by Tempelhoff et al. [1972] for adsorption of an amount of water corresponding to the number of residual OH groups on a silica surface, heat treated at 400 °C. They observed the torsional mode of the adsorbed water at 345 cm^{-1} in a neutron time-of-flight spectrum. This wave number is considerably lower than the corresponding values for liquid water (425 cm^{-1}) and ice (454 cm^{-1}) and it was therefore assigned to a molecule of lower coordination number. From the quasi-elastic peak a correlation time of 2×10^{-10} s (lower limit) was estimated, which is higher than that of liquid water at 20 °C by about two orders of magnitude*.

* The correlation time 2×10^{-10} s is the same as that obtained by Michel [1967] from NMR relaxation measurements. Michel ascribed this value to water in clusters (see below).

At increasing coverages a *liquid-like phase* (*water clusters*) appears to be formed as shown by slow neutron inelastic scattering (Boutin and Prask [1964]) and NMR relaxation studies (Michel [1967]). Kiselev and Lygin [1960] distinguished water molecules adsorbed in a monolayer on a hydroxylated surface from capillary condensed water by means of the differences of the respective IR fundamental spectra. This distinction could be verified by measurements of heat capacities of the adsorbed phases (Berezin et al. [1963]).

The details of adsorption of individual water molecules and cluster formation do not clearly show up in the *IR fundamental region* due to the overlap of the bands of surface hydroxyl groups and of the adsorbed water. The *overtone region* is most useful to follow the various stages of water adsorption in conjunction with the fundamental region. The combination band of the bending and stretching modes of a water molecule, which occurs around 5300 cm^{-1} and which was first observed by Wirzing [1963] for the water/silica system, is of considerable interest; this band does not contain any contribution from surface groups and is therefore characteristic for the adsorbed water. Anderson and Wickersheim [1964] observed a decrease in intensity of the bands of free silanol groups at 3750 and 7326 cm^{-1} on adsorption of water and the appearance of a new combination band at 4420 cm^{-1} (see section 27.2.1) of silanols which they assigned as those groups interacting with the adsorbed water molecules. A wave number shift of the fundamental stretching mode of 200 cm^{-1} was estimated from this combination band (see also section 27.2.2). This wave number shift suggests an energy of interaction of the primary adsorption bond of roughly 2–3 kcal/mole. At low levels of water adsorption three sets of bands develop, namely at 3660 and 3540 cm^{-1} in the OH fundamental region, at 7090 and 6850 cm^{-1} in the corresponding overtone region and at 5290 and 5180 cm^{-1}. As the intensity of all bands behaves similarly as the coverage is varied, it is assumed that all bands belong to the same surface species. From the occurence of two OH stretching fundamentals split by roughly 100 cm^{-1}, the presence of a pair of weaker absorptions 1600 cm^{-1} higher in position than the fundamentals, the general profile of the bands and the growth of the bands with increasing water content, Anderson and Wickersheim [1964] conclude that the surface species is molecular water. The high wave numbers and the sharpness of the bands suggest this is "monomeric" or individually adsorbed water. With increasing amounts of water adsorbed, a broad absorption grows with center at 3400 cm^{-1} and a shoulder at 3320 cm^{-1}. This band strongly resembles that of liquid water and is therefore assigned to the formation of a liquid-like H-bonded network on the surface of silica. Because of the comparably low relative integrated intensities of the overtone bands of such polymeric species (see ch. 13), the corresponding absorption could not be detected in the overtone region. A further important observation of Anderson and Wickersheim [1964] is that the fundamental bands of the water molecules belonging to the network already occur even at relatively low water contents and that the spectrum of the isolated molecules persists to fairly high coverages. This suggests that both species exist simultaneously in a rather broad range of coverages, i.e. adsorption takes place either on silanol groups or on H_2O molecules with comparable probability. A complete monomolecular layer is not really formed on the surface. For this reason it should be extremely difficult to make any clear-cut distinction between preferred primary adsorption sites (see discussion on single and pair sites above).

Recently further studies in the overtone region have been reported. Kiselev [1971] observed the combination band of the water molecule at a fixed wave number of 5265 cm^{-1} for relative pressures of 0.008 up to saturation. A second band could not be resolved, but the band was asymmetric towards lower wave numbers. This asymmetry was explained as being due to a distribution of energies of the H-bonded species. *Diffuse reflectance spectra* of the silica water system have been reported by Zettlemoyer and Klier [1971] and by Klier et al. [1973]. These authors observed the combination band at 5271 cm^{-1} and they mention that this wave number is only 20 cm^{-1} lower than that of water in the critical state and is of the order of perturbation of the OH mode in water that is acting as a H-bond acceptor. The band is therefore assigned to an isolated water molecule adsorbed via a single H-bond (structure (X)) on a nonporous silica heat treated at 650°C. The corresponding wave number for liquid water at 25°C is 5181 cm^{-1} and that of ice at 0°C is 5051 cm^{-1}. In the adsorbed state no change in position of the 5271 cm^{-1} band was observed on cooling the sample to −60°C. A two-dimensional liquefaction or solidification was therefore ruled out when starting from a surface hydrated by species (X). On increasing the amount of adsorbed water the 5271 cm^{-1} band grows and simultaneously a shoulder between 5100 and

5200 cm^{-1} gradually develops. Contrary to the interpretation of Anderson and Wickersheim [1964]*, this band is assigned to mutually interacting H-bonded water molecules in clusters:

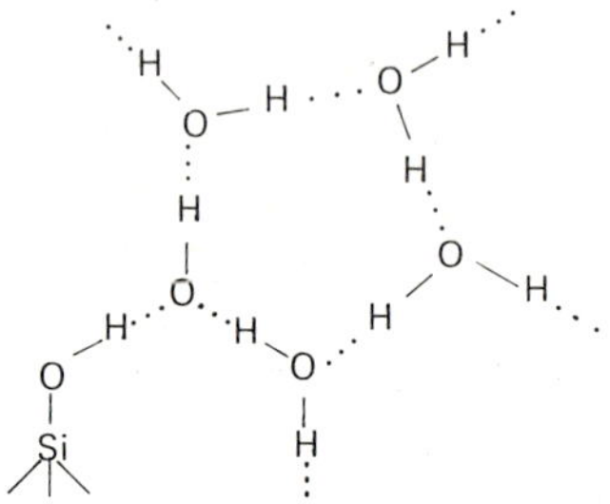

(XII)

This cluster formation is assumed to occur simultaneously with the adsorption of isolated water molecules in agreement with Anderson's and Wickersheim's view. The initial linear portions of the quantitative relations between the intensities at maximum of the combination band of the water molecule and of the overtone band of the free silanol groups and the amount of adsorbed water (fig. 27.12) proves the formation of species (X) during the initial stages of adsorption. The following deviations from the linear relation are due to the simultaneous cluster formation. Finally, on further increasing the surface coverage the low wave number shoulder becomes the dominant band and the position of the water combination band shifts to lower wave numbers and approaches that of liquid water. This indicates a growth of the cluster sizes on the expense of individually adsorbed water molecules.

A band at 225 cm^{-1} and a second absorption at 170 cm^{-1} at high coverages was observed by Lorenzelli and Bonino [1965] and Bertoluzza et al. [1966] in the *far IR region.* The band at 225 cm^{-1} was also reported by Brodskii et al. [1973]. Bertoluzza et al. [1966] assigned the band at 225 cm^{-1} as the O···O stretching vibration in a Si–OH···OH_2 H-bond

* Contrary to Anderson and Wickersheim [1964], Kiselev [1971], Zettlemoyer and Klier [1971] and Klier et al. [1973] do not distinguish between the $(\bar{\nu}_2+\bar{\nu}_3)$ and $(\bar{\nu}_2+\bar{\nu}_1)$ combination bands of water. It should be noted that the $(\bar{\nu}_2+\bar{\nu}_3)$ combination band of isolated water molecules at 5282 cm^{-1} shifts to 5181 cm^{-1} in liquid water. The $(\bar{\nu}_2+\bar{\nu}_1)$ combination band of isolated water was observed near 5190 cm^{-1} in CCl_4 solution and its relative intensity was only about 1:10 as compared to the $(\bar{\nu}_2+\bar{\nu}_3)$ combination band (Luck and Ditter [1969]). For this reason it is extremely difficult to unambiguously assign a weak absorption between 5100 and 5200 cm^{-1} to either the $(\bar{\nu}_2+\bar{\nu}_3)$ band of liquid-like water or the $(\bar{\nu}_2+\bar{\nu}_1)$ band of isolated water molecules.

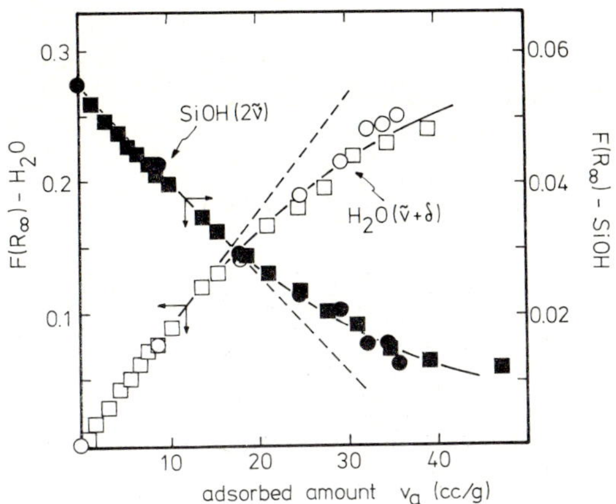

Fig. 27.12. Relation between intensities at band maxima (Kubelka–Munk function $F(R_\infty)$) of the SiOH ($2\bar{\nu}$) band (right hand scale, full symbols), of the $H_2O(\bar{\nu}+\delta)$ band (left hand scale, open symbols) and the amount of water adsorbed at 25°C on Na–HiSil (degassed at 650°C). (Reproduced by permission of the American Chemical Society from Klier et al. [1973]).

between a surface silanol group and an adsorbed water molecule. The absorption at 170 cm^{-1} corresponds to the respective vibration in liquid water and may therefore be assigned as the O···O stretching vibration in liquid-like islands or clusters which form at higher coverages.

The observed IR bands of adsorbed water are summarized in table 27.2 and their assignments are given. The most probable *picture of the water adsorption on the time scale of IR spectroscopy* is in conclusion the following: individual water molecules are adsorbed on surface silanol groups at low coverages forming species (VIII), (IX) or (X), depending on the properties and density of the OH groups. As the coverages increase, the tendency of water towards self-association leads to a simultaneous cluster formation which prevents a monolayer formation. On approaching the saturation vapour pressure a liquid-like film, several molecular diameters thick, may be formed on open surfaces and capillary condensation occurs. The structure of the film may be identical with that of liquid water at large distances from the surface, it is, however, different from that of liquid water and is determined by the topography of the underlying adsorbent surface at shorter distances, as shown by Tyler et al. [1971] by heat of immersion measurements. Moreover, alterations in the structure and properties of adsorbed water layers by silica surfaces as compared to bulk liquid water clearly show up in the heat capacity

TABLE 27.2

IR bands of water adsorbed on silica and their assignments

Adsorption conditions	$\bar{\nu}$ (cm^{-1})	Assignment	Species	References
No molecular water present	3750	$\bar{\nu}_{OH}$	Si–O–H	Anderson and Wickersheim [1964]
	4550	$\bar{\nu}_{OH}+\delta_{OH}$		Anderson and Wickersheim [1964]
	7326	$2\bar{\nu}_{OH}$		Anderson and Wickersheim [1964]
Low levels of adsorption	(3550)	$\bar{\nu}_{OH}$	Si–OH···O	Anderson and Wickersheim [1964]
	4420	$\bar{\nu}_{OH}+\delta_{OH}$		Anderson and Wickersheim [1964]
	3540	$\bar{\nu}_1$	VIII, IX *or* X	Anderson and Wickersheim [1964]
	3660	$\bar{\nu}_3$		Anderson and Wickersheim [1964]
	5180	$\bar{\nu}_2+\bar{\nu}_1$		Anderson and Wickersheim [1964]
	5290	$\bar{\nu}_2+\bar{\nu}_3$		Anderson and Wickersheim [1964]
	6850	$2\bar{\nu}_1$		Anderson and Wickersheim [1964]
	7090	$2\bar{\nu}_3$		Anderson and Wickersheim [1964]
	5265	$\bar{\nu}_2+\bar{\nu}_{1,3}$	X	Kiselev [1971]
	5271	$\bar{\nu}_2+\bar{\nu}_{1,3}$		Zettlemoyer and Klier [1971] and Klier et al. [1973]
High levels of adsorption	3400	$\bar{\nu}_1,\ \bar{\nu}_3$	XII	Anderson and Wickersheim [1964]
	5100–5200	$\bar{\nu}_2+\bar{\nu}_{1,3}$	XII	Zettlemoyer and Klier [1971] and Klier et al. [1973]
	5271	$\bar{\nu}_2+\bar{\nu}_{1,3}$	X	Zettlemoyer and Klier [1971] and Klier et al. [1973]
Near saturation pressure	5235	$\bar{\nu}_2+\bar{\nu}_{1,3}$	liquid-like water	Zettlemoyer and Klier [1971] and Klier et al. [1973]

measurements carried out by Plooster and Gitlin [1971]. These authors used non-porous silica powders, in which effects due to narrow pores are completely absent. With films of 5 to 7.5 nm thickness, a *phase transition* occurred entirely below 0 °C and the heat changes were below that of the ice–water transition. Only at increasingly larger thickness of the adsorbed layers was a second phase transition observed at 0 °C. In complete agreement, Klier et al. [1973] conclude that freezing on a nonporous silica surface requires at least 6–10 monolayers and that heterogeneous nucleation of ice on silicas is a three-dimensional phenomenon in which long-range order can only be attained in large-size prenucleation clusters. The reduced wave number of the torsional mode of adsorbed water molecules as compared to bulk liquid water and ice (Tempelhoff et al. [1972]) is a further indication for the different structure of the adsorbed films or clusters.

The NMR data obtained by Kvlividze et al. [1962] and Kvlividze [1964] are in agreement with the described model of the water adsorption insofar as these authors also propose the formation of patches of liquid-like water clusters (H_2O II) which coexist with individually adsorbed water molecules (H_2O I). A multiphase behavior was generally observed for the water/silica system (Zimmerman et al. [1956], Zimmerman and Brittin [1957], Zimmerman and Lasater [1958] and Woessner and Zimmerman [1963]). The two environmental states were characterized by different nuclear spin relaxation times. Evidence for a transfer of protons between the two states, the average residence time of a proton in a given state being approximately 1 ms at a statistical monolayer, was obtained by Zimmerman and Brittin [1957] and Zimmerman and Lasater [1958]. Woessner [1963] and Woessner and Zimmerman [1963] studied the relaxation behavior of adsorbed water on silica at 3/4 of a statistical monolayer in the temperature range between −80 and +100 °C. The relaxation times of the two distinct environmental states *a* and *b* were strongly influenced by the temperature-dependent rates of proton exchange between the two phases. The activation energy as determined from temperature dependence of the average residence time of a proton in state *a* was approximately 5 kcal/mole. It was therefore concluded that the proton exchange was connected with the rupture of a H-bond.

Michel [1967] studied the *relaxation behavior of adsorbed water* at 3/4 of a statistical monolayer in the temperature range between −100 and +80 °C using a high purity silica heat treated at 300 °C. He found a two-phase behavior for the transversal relaxation, however, only a one-phase behavior of the longitudinal relaxation was found. The apparent transverse relaxation times T_{2b} of state *b* with the lower mobility were independent of the temperature and corresponded fairly well to the relaxation time of the free surface hydroxyl groups before water adsorption. State *b* was therefore attributed to these surface groups. The apparent relaxation time T_{2a} of state *a* showed a maximum as a function of the temperature, which was explained by a temperature-dependent proton exchange between phases *a* and *b*. The high ratios obtained for T_{1a}/T_{2a} at the T_1 minimum are not in accord with the case of an isotropic motion with only one correlation time. Two sub-states *a*,1 and *a*,2 have therefore been adopted between which a rapid exchange takes place. Some of the results obtained by Michel [1967] are summarized in table 27.3. A frequency of 16 MHz was used. τ_c^i is the correlation time for state *i*, V_c^i is the activation energy of the corresponding proton exchange process, M_2^i is the second moment and p_i is the residence probability in state *i*. The two sub-states *a*,1 and *a*,2 are attributed to adsorbed molecular water in different environments. Sub-state *a*,2 represents islands of water clusters H_2O II which are made up by approximately 95% of the adsorbed molecules at 0 °C and 3/4 of a statistical monolayer. The correlation time of 2.7×10^{-10} s which is lower by two orders of magnitude than that of liquid water at 25 °C, demonstrates the reduced mobility of the water molecules in these clusters. The individually adsorbed molecules of sub-state *a*,1 with a correlation time of only 2.3×10^{-8} s are

TABLE 27.3
NMR relaxation data of the silica/water system (Michel [1967]

	State			
	a,1	*a*,2	*b*	H_2O
τ_c^i (0 °C) (s)	2.3×10^{-8}	2.7×10^{-10}	2×10^{-3}	2.5×10^{-12} (25 °C)
V_c^i (kcal/mole)	5.4–5.8	4.5–4.8	4	3.8
M_2^i (G^2)	(27)	12.4–13	—	—
p_i	$\leq 0.05\times0.7$	$\geq 0.95\times0.7$	0.3	
Assignment	Individually adsorbed water molecules H_2O I	Water clusters H_2O II	Surface OH	Liquid water

even more hindered in mobility. These results are in excellent agreement with the model developed from the IR data. The structure of the water clusters seem to depend strongly on the nature and properties of the silica surface (Michel [1966, 1967] and Morariu and Mills [1972a]) and depending on their size, it will deviate more or less from the structure of liquid water.

Considering the *correlation functions* of water in the liquid state, in adsorbed multilayers and submonolayers on non-porous silica, Klier [1973] recently concluded, that the rotational motion of water molecules is strongly perturbed. The perturbation seemed to be strongest in monolayers, less in multilayers and still less in bulk liquid water. This is a further indication for structural differences of multilayers or clusters of water adsorbed on silica surfaces as compared to bulk liquid water.

Recently Morariu and Mills [1973] measured the line width for water adsorbed in two silicas of differing acidity as a function of temperature and coverage. The results were compatible with a surface diffusion model proposed by Resing [1967–1968]. The model is based on the process of surface diffusion and it is assumed that relaxation occurs only by the translational jump mechanism.

The proton mobility in water/silica systems is also responsible for the increase in *surface electrical conductivity* on adsorption of water which has been observed by a variety of research groups (Freyman and Freyman [1954], Cook et al. [1954], Le Bot and Le Montagner [1955], Kurosaki et al. [1955], Van Beek [1960], Baldwin and Morrow [1962], Dransfeld et al. [1962], Simkovich [1963], Fripiat et al. [1965], Soffer and Folman [1966] and Anderson and Parks [1968]). An exponential dependence of the specific conductivity on the surface coverage is usually found empirically (e.g. Fripiat et al. [1965] and Soffer and Folman [1966]):

$$\sigma = \sigma_0 \, e^{c\Theta}, \tag{27.5}$$

where c is a temperature-independent constant. An analogous relation was found for the real part of the dielectric constant and for the tangent of the loss angle by Fripiat et al. [1965]. The phenomenon of surface conductivity had been explained by assuming a promotion of the mobility of impurity cations (e.g. Na^+) by the water adsorption (Soffer and Folman [1966]). Comparing the conductivities of a high purity silica gel and a sodium containing porous glass as well as a Na-montmorillonite, Fripiat et al. [1965] concluded that the conduction mechanism was protonic. Moreover, Anderson and Parks [1968] observed a change in conductivity when H_2O was replaced by D_2O, the magnitude of which suggests that the charge carriers are protons. Dielectric dispersion studies also demonstrated that the main contribution in the dispersion region was the absorption of electrical energy by free charge carriers. A model was therefore adopted by Fripiat et al. [1965] which suggests charge carriers (protons, charge e) jumping over potential wells U in order to move from one equilibrium position to the next. The specific conductivity is then given by eq. (27.6):

$$\sigma = n_p \nu (e^2 a^2 / kT) \exp (4\pi e^2 l n_p / kTA\rho), \tag{27.6}$$

where n_p is the number of charge carriers per unit volume, ν is the number of "jumping attempts" per second, a is the distance between two equilibrium positions, l is the potential difference between surface and "conduction level", ρ is the density of the sample and A its surface area. The protons originate from the dissociation of adsorbed water molecules. The degree of dissociation was estimated from eqs. (27.5) and (27.6) to be of the order of magnitude of 10^{-2} at 20°C and at a statistical monolayer. This value is higher by a factor of 10^6 than that of liquid water at 25°C. A strong enhancement of the water dissociation by surface electrical fields was therefore assumed by Fripiat et al. [1965] and one might argue that the water molecules strongly held by non-hydroxylic sites (see above) may be the most potential proton donor molecules.

Anderson and Parks [1968] provided a model in which they assume an increase of the dielectric constant on adsorption of water, which in turn reduces the dissociation energy for protons and hence increases the degree of dissociation and the surface conductivity. According to Fripiat et al. [1965] the proton migration proceeds following a chain of exchange reactions of the type

$$H_3O^+ + H_2O \rightleftharpoons H_2O + H_3O^+ \tag{XIII}$$

just as in liquid water (Conway et al. [1956]). If the adsorbed water layer is mostly localized so that the molecules have only rotational but no translational freedom, the proton transfer over large distances is possible only if an adequate distribution of the molecules exists in the two-dimensional adsorbed phase. This must be the case at 35°C at coverages below a statistical monolayer on hydroxylated silica surfaces as verified by entropy calculations carried out by Fripiat et al. [1965]. Figure 27.13 shows that the experimental entropy loss comes close to the theoretical values for an immobile film at coverages $\Theta < 1$ and approaches those for a

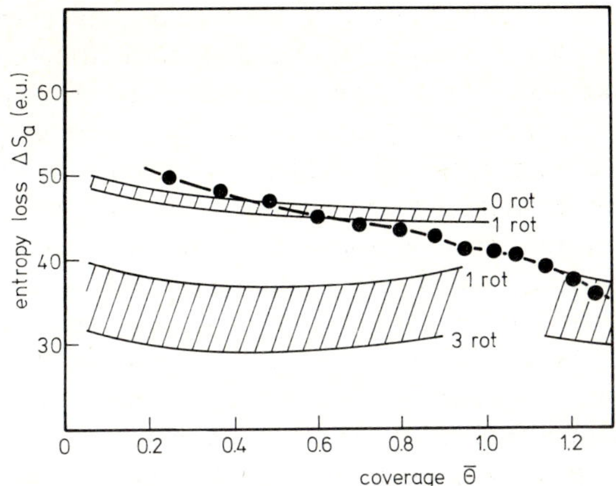

Fig. 27.13. Entropy loss ΔS_a (e.u.) of water molecules adsorbed on Aerogel at 35.5°C as a function of the averaged coverage $\bar{\Theta}$: black dots, experimental values; upper shaded band, theoretical entropy loss for an immobile film; the upper and lower limits of this band correspond to the loss of three and two degrees of rotational freedom, respectively; lower shaded band, theoretical entropy loss for a mobile film characterized by one (upper limit) and three (lower limit) degrees of rotational freedom. (Reproduced by permission of the American Chemical Society from Fripiat et al. [1965]).

mobile film around an average statistical monolayer. Consequently, a favorable distribution of adsorbed molecules is no longer required, since a long-range proton transfer is now possible due to the two translational degrees of freedom of the adsorbed molecules.

Dielectric measurements have been carried out for the silica or porous glass/water system by various research groups (Rolland and Bernard [1951], Freyman and Freyman [1954], Kämpf and Kohlschütter [1958], Nair and Thorp [1965a, b], Kondo and Mūroya [1969], Ebert et al. [1971] and Nekrasova and Zhilenkov [1973]). The minds of the various workers do not seem to be closed with respect to the interpretation of the data. Two dispersion regions have usually been found. The first region which occurs near room temperature at frequencies between 10^2 and 10^5 cps was attributed to the relaxation of water molecules, the mobility of which was restricted due to the adsorption interaction (Kämpf and Kohlschütter [1958] and Kondo and Mūroya [1969]). The same dispersion region, however, was explained by Nair and Thorp [1956a, b] on the basis of the Maxwell–Wagner theory for heterogeneous dielectrics. Some evidence for a relaxation process at −100°C of adsorbed molecules with hindered orientational mobility was reported by Ebert et al. [1971] on porous glass surfaces. A molecular description of a second relaxation process at higher temperatures could not be given.[6)] It seems that dielectric methods applied to the study of adsorbate systems provide much more relevant information of the properties of capillary condensed water than of the direct adsorbate–surface interaction. Due to the reduced saturation vapor pressure in narrow pores, these will be filled by water even at relatively low pressures. The structure of this capillary condensed water will depend on the pore diameter and it will deviate even more from the structure of multilayers on an open surface and liquid water, the narrower the pores are. Nekrasova and Zhilenkov [1973] claim to observe two Debye-type relaxation regions besides a Maxwell–Wagner type of dispersion. The first region of Debye relaxation was observed at all coverages in the frequency range 10^2 to 3×10^9 cps. The activation energies increased with coverage from 9.2 to 15.8 kcal/mole, which is an indication for energetic and structural inhomogeneity of the adsorbed water layers. Molecules in the first layers or in narrow pores seem to be relatively easily oriented whereas conditions change in subsequent layers (wider pores), leading to some kind of multiphase behavior. Increasing temperature apparently levels out these structural differences due to the higher molecular mobility and the greater frequency of transfers between layers.

The second relaxation region was observed by Nekrasova and Zhilenkov [1973] in the centimeter range between 8.3×10^9 and 15×10^9 cps at higher water coverages only. Activation energies of orientation have been found to be approximately 3.7–4.6 kcal/mole. This relaxation region has been assigned to a phase transition between "liquid-like" and "ice-like" water. Both phases coexist in certain temperature ranges. The coexistence of the phases and the absence of the phase transition at coverages below two or three layers of adsorbed water was taken as an indication of the inability of the first water layers to freeze. This may be due to the structuring effect caused by the adsorbent surface, since the nucleation of ice is largely controlled by the adsorption properties of the adsorbent surface in the absence of a good crystallographic match between adsorbent and ice (Basset et al. [1970]). Exact analogous conclusions have been reached from calorimetric measurements of the phase transitions of adsorbed water layers by Antoniou [1964], Berezin et al. [1970] and Ebert et al. [1971] and from NMR studies on the properties of adsorbed water layers by Anderson [1967], Derouane [1969a] and Kvlividze [1971]. The NMR linewidth measurements of Morariu and Mills [1972b] also show that at least one monolayer of water next to the surface does not freeze, although its mobility characteristics may change with temperature. Freezing occurs at temperatures below 0 °C if the surface coverage is higher than 1.5 as evidenced by an abrupt increase in the NMR line width. Moreover, the temperature at which freezing begins decreases as the coverage is lowered.

The *freezing of water in porous systems* is of great practical interest regarding *frost damages.* Sidebottom and Litvan [1971] and Litvan [1972a] studied the behavior of a silica glass/water system during temperature cycles between +5 and −40 °C. Due to the differences of the saturation pressures of liquid-like undercooled water in narrow pores (p_l^0) and of solid ice ($p_s^0 < p_l^0$), water migrates from the larger voids and pores to the outer surface, giving rise to a high-temperature freezing between 0 and −8 °C. Finally water migrates from the small pores to the larger pores, leading to a low-temperature freezing around −20 °C. Taking transport phenomena (permeability) into account, Litvan [1972b] applied this model to explain the mechanism of frost action in hardened cement paste.

On talking about capillary condensed water, the so-called "*anomalous*" *water* should at least shortly be mentioned. Deryagin and Fredyakin [1962] and Deryagin et al. [1967] first reported on the anomalous physico-chemical properties of a water which was formed on condensation of the vapor onto the walls of fused quartz capillaries. From spectroscopic studies Bellamy et al. [1969], Lippincott et al. [1969] and Page et al. [1970] concluded that the anomalous water was some kind of polymeric water. Many objections have since been raised against the existence of a water modification of the observed anomalous properties. A thesis has been put forward recently by Prigogine and Fripiat [1971a, b] as an explanation of the observed properties of the so-called "anomalous" water, which is essentially based on the surface properties of silicious surfaces. As already pointed out above, a highly organized and well-ordered phase of strongly adsorbed water may form on a silica surface provided a favorable distribution of surface hydroxyl groups exists. Silica and possibly other constituents of glass may become soluble due to the exceptional properties of the adsorbed water phase, particularly its high degree of dissociation (Fripiat et al. [1965]). The quantity of silica dissolved is determined by the initial conditions, i.e. the activity of the surface. Prigogine and Fripiat [1971a] conclude that there is no evidence for the existence of a polymerized water of anomalous properties, instead they propose that these properties are due to a hydrogel which is formed in the capillaries by means of a dissolution of some silica within the first water layers.

This interpretation found some confirmation when Davis [1972a, b] measured the ESCA spectra of polywater and polyheavywater. His conclusion is that polywater is "an ionic honey, a mixture of large amounts of inorganic salts with organic carboxylates and just enough water to keep the material as a semiliquid honey". Deryagin and Churaev [1973] very recently explained the unusual properties of polywater similarly, by impurities of colloidal silicic acid with dissolved Na-, Si-, C-, O-, K-, Cl- and S-containing compounds. Polywater could not be prepared on specially cleaned and polished quartz surfaces.

27.2.6. Adsorption of alcohols

On adsorption of alcohols on silica surfaces, surface reactions occur and consequently adsorption is not completely reversible. These surface reactions have been studied extensively by Beliakova and Kiselev [1950], Sidorov [1956], McDonald [1958], Boehm [1966a], Borello et al. [1967a], Low and Harano [1968], Bavarez and Bastick [1969], Kubelkovà et al. [1969], Kunath [1971] and Mertens and Fripiat [1973]. An esterification of surface hydroxyl groups according to the scheme:

$$\underset{\diagup|\diagdown}{\overset{\displaystyle OH}{\underset{|}{Si}}} + ROH \longrightarrow \underset{\diagup|\diagdown}{\overset{\displaystyle OR}{\underset{|}{Si}}} + H_2O \qquad \text{(XIV)}$$

has been proposed by Sidorov [1956] and McDonald [1958], whereas Beliakova and Kiselev [1950] have shown that a surface alkylation proceeded through an opening of strained siloxane bonds:

$$-\underset{|}{Si}\overset{O^-}{-O-}\underset{|}{Si^+}- \;+ROH \longrightarrow -\underset{|}{\overset{OR}{\overset{|}{Si}}}-O-\underset{|}{\overset{OH}{\overset{|}{Si}}}- \qquad \text{(XV)}$$

It is now certain that both reactions occur simultaneously (Boehm [1966a], Low and Harano [1968], Bavarez and Bastick [1969], Kubelkovà et al. [1969], Kunath [1971] and Mertens and Fripiat [1973]). Low and Harano [1968] have shown that reaction (XV) is the preferred route at 30 °C on silicas outgassed at 400 °C and higher temperatures. Mertens and Fripiat [1973] conclude from their kinetic measurements between 150 and 190 °C that the esterification (XIV) is favored at lower temperatures. The latter authors propose a protonated $CH_3OH_2^+$ species as an important intermediate in the alkoxylation reaction.

H-bonding interactions occur additionally on alcohol adsorption onto silica surfaces. The electronic structure of alcohols is very similar to that of water and one therefore expects similar behavior with respect to the primary adsorption interaction. Thus, depending on the degree of hydroxylation, adsorption structures analogous to the water structures (VIII), (IX) and (X) might be formed. The direct interaction of the alcoholic hydroxyl group with the silica surface has in fact been demonstrated by high-resolution NMR measurements. The width of the lines of hydroxyl protons was larger than that of methylene and methyl protons of ethanol adsorbed on silica (De La Hardrouyère [1960] and Geschke and Pfeifer [1966]) and of methanol on silica (Geschke [1969] and Volkov et al. [1972]). This observation was explained by the closer proximity of the hydroxyl group to the silica surface. Karagounis and Gutbrod [1967] could not observe the OH signal of adsorbed methanol at room temperature. Only at temperatures below $-20\,°C$ did the OH signal appear. This phenomenon was interpreted by the slowing down of proton exchange processes, which again proves the proximity between the alcoholic OII group and the surface.

Sewell and Morgan [1969] observed a drop in the heat of adsorption of methanol on silica and soda-lime silica glass at low coverages from approximately 10 kcal/mole at low degassing temperatures (high hydroxyl content) of the surface to approximately 4.5 kcal/mole at high degassing temperatures (low hydroxyl content) of the surface. They interpreted this drop as an indication of a transition from a dual-site adsorption on hydroxyl-rich surfaces to a single-site adsorption on surfaces which are poor in hydroxyl density. The respective species (XVI), (XVII) and (XVIII) correspond to those proposed for water adsorption, namely

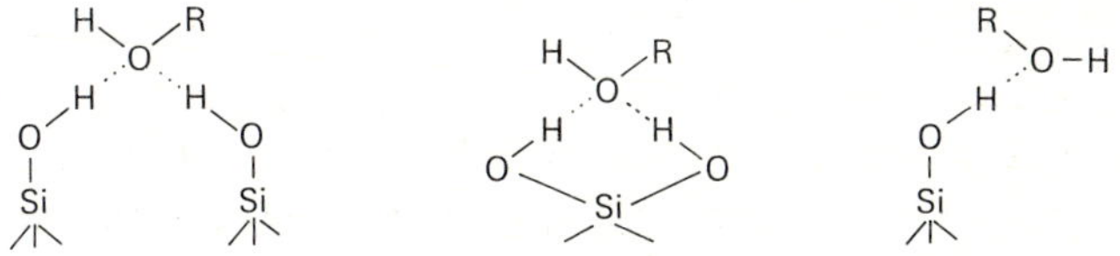

(XVII) (XVI) (XVIII)

(VIII), (IX) and (X). This view is confirmed by the findings of Borello et al. [1967b] that for outgassing temperatures of the silica of 500 and 650 °C two silanol groups are perturbed for every adsorbed methanol molecule at low coverages ($\Theta \to 0$), whereas for outgassing temperatures of 800 and 950 °C only one silanol group is perturbed for every adsorbed methanol molecule at low coverages. Also consistent with the assumption of surface hydroxyl groups being the main adsorption sites (at least at low coverages) is the decrease of the adsorption capacity for methanol as the hydroxyl content of the surface is reduced by degassing at increasingly higher temperatures (Borello et al. [1967b]). Hence, as in the case of water adsorption, the number and the arrangement of surface hydroxyl groups governs the nature of the adsorbed species formed at low coverages. A maximum of 3 H-bonds can be formed per alcohol molecule.

Alcohols like water have a tendency towards *self-association* (Fletcher [1971] and Sandorfy [1972]), which might even be enhanced cooperatively through the H-bonding to a surface OH group (Zundel [1969], p. 75 and Hankins et al. [1970]). One should therefore expect polymeric aggregates of alcohol in addition to the individually adsorbed molecules, at least at higher coverages.

IR spectroscopy has revealed direct evidence for H-bonding interactions between adsorbed alcohols and silica surfaces. Thus, Borello et al. [1963–64, 1967a, b], Low and Harano [1968], Kubelkovà [1969] and Cruz et al. [1973] studied the methanol/silica system, Jeziorowski et al. [1973] the ethanol and *tert.*butanol adsorption and Davydov et al. [1963b] the *tert.*butanol adsorption. Borello et al. [1963–64, 1967b] observe a decrease in intensity of the 3750 cm^{-1} stretching fundamental of free surface hydroxyl groups and the simultaneous growth of a broad absorption between 3200 and 3600 cm^{-1} (fig. 27.14). Additionally a weak band occurs at 3625 cm^{-1} which vanishes with increasing coverage. The broad absorption shows some structure, particularly a shoulder near 3500 cm^{-1} and the maximum absorption of the main band shifts from 3380 cm^{-1} at low coverages to 3337 cm^{-1} at coverages above a statistical monolayer. Because of the disordered topology of the silica surface with regions of higher and lower hydroxyl density, Borello et al. [1967b] assume the existence of a wide variety of different adsorbed species which bring about a wide spectrum of adsorption energies in agreement with experiment. In regions of high hydroxyl density, interactions are suggested with three H-bonds for every adsorbed alcohol molecule, while in regions of low hydroxyl density only one H-bond per adsorbed alcohol molecule will be formed. The energies of the H-bonds formed and the corresponding wave number shifts are thus dependent on various factors such as distance between sites and bond distortion from linearity. According to these considerations, Borello et al. [1967b] assigned the band around 3500 cm^{-1}, which becomes more easily detected on surfaces of high OH content, to adsorbed species of methanol that are held by more than one H-bond (e.g.

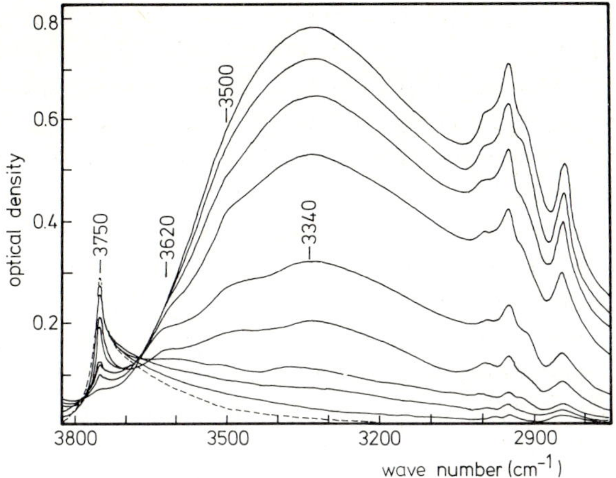

Fig. 27.14. Transmission IR spectra of methanol at 22°C adsorbed on Aerosil (degassed at 350°C). Relative pressures, starting from the uppermost spectrum, are 0.096, 0.057, 0.038 and 0.019. The last five spectra were recorded at pressures lower than 1 mm Hg. (Reproduced by permission of the American Chemical Society from Borello et al. [1967b]).

species (XVI) and (XVII)). It is suggested that as the corresponding sites are progressively saturated at increasing coverages, a new phase is reached in which each methanol molecule is adsorbed by only one H-bond. These single H-bonds are assumed to be formed between a methanol molecule and either an isolated silanol group (species (XVII)) or a previously adsorbed methanol molecule (cooperative effect). The broad absorption near 3340 cm^{-1} is attributed to these two types of single H-bonds*. The maximum position of this band shifts towards the position of the liquid methanol band and it is the only one present at coverages higher than a statistical monolayer. Under such conditions the polymeric species aggregate to form a liquid-like film and the heat of adsorption approaches the heat of liquefaction. Finally, the relatively weak band at 3625 cm^{-1} that is detected at low coverages, is ascribed to the terminal free hydroxyl group of methanol molecules in polymeric aggregates or of single molecules bonded to surface hydroxyl groups. At 24°C and 33% coverage 12% of the alcohol hydroxyl groups with respect to the total

* According to Sandorfy [1972] (see also ch. 13) the large width of bands of H-bonded systems is intimately related to a polymeric character of the aggregate (see also discussion below of the silica-*tert.*butanol system).

amount of alcohol adsorbed have been estimated to be free (Borello et al. [1963–64]).

Adsorption isotherms, heats of adsorption and IR and dielectric data have been discussed by Cruz et al. [1973]. These authors propose a *model for the physical adsorption of methanol on a silica surface* heat-treated at 110–120 °C, which suggests the formation of H-bonds between silanols and alcohol hydroxyls and also between adsorbed species which become adsorption sites due to cooperative effects. Both processes lead to the formation of clusters of various sizes; the "degree of polymerization" is assumed to be probably quite high. The clusters are assumed to be anchored randomly onto the surface by certain methanol molecules (type S–H bonds) through an acceptor H-bond with a surface silanol group, while the other molecules of the cluster form linear chains of H-bonds (type L–H bonds):

```
                                 type L
         R       R       R     /R        R
         |       |       |    ↙ |        |
     ···O—H ···O—H ···O—H ···O—H ···O—H···
         :                      :
         H                      H ← type S
         |                      |
     ---O----------------------O---------- surface
   —Si´                  —Si´
   ⁄|                    ⁄|
```
(XIX)

*Tert.*butanol has the advantage over methanol and ethanol that it does not undergo surface reactions with silica and that adsorption is therefore completely reversible. This system has been studied recently in the temperature range between 50 and 180 °C in the *IR fundamental* and *overtone* region by Jeziorowski et al. [1973]. Overtone spectra had only been reported by Filimonov [1956] and by Filimonov and Terenin [1956] before. The heat of adsorption as measured by Jeziorowski et al. [1973] was 14 kcal/mole on a silica degassed at 500 °C (~2 OH/100 Å^2) and remained fairly constant in a range of coverages between 0.4 to 3 molecules adsorbed per 100 Å^2. Figure 27.15 shows some typical spectra of the OH stretching fundamental region and in fig. 27.16 the difference between these spectra and the silica background is plotted in absorbance units. The band at 3680 cm^{-1} belongs to intraglobular OH groups. On adsorption of *tert.*butanol, the intensity of the 3745 cm^{-1} band of free surface silanols is reduced while a band at 3605 cm^{-1} and a broad absorption with center at 3330 cm^{-1} appears. These two bands are removed simultaneously on desorption and the intensity of the band of free silanols is restored. The difference spectra in fig. 27.16 give additional

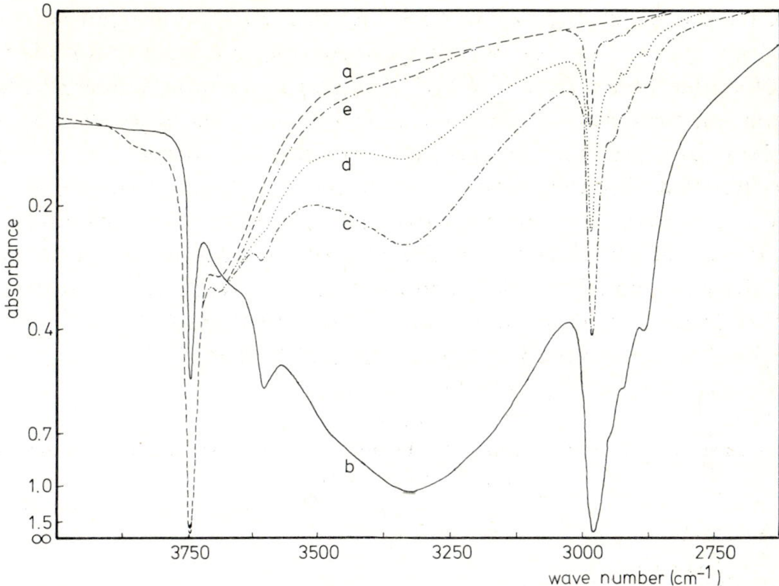

Fig. 27.15. Transmission IR spectra of *tert.*butanol at 55°C adsorbed on silica "Standard" (Degussa): (a) background of silica after degassing at 500°C; (b) after adsorption of 30 mm Hg *tert.*butanol vapor, coverage 3.1 molecules/100 Å^2; (c) coverage 0.7 molecules/100 Å^2; (d) coverage 0.35 molecules/100 Å^2; (e) coverage 0.18 molecules/100 Å^2. (Reproduced by permission of the Chemical Society, London, from Jeziorowski et al. [1973]).

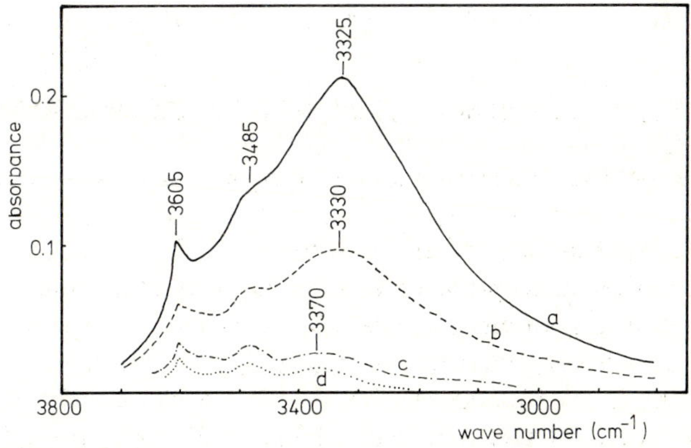

Fig. 27.16. Difference between spectra shown in fig. 27.15 and the silica background (the CH_3 stretching have been omitted): (a) coverage 0.7 molecules/100 Å^2; (b) coverage 0.35 molecules/100 Å^2; (c) coverage 0.21 molecules/100 Å^2; (d) coverage 0.18 molecules/100 Å^2. (Reproduced by permission of the Chemical Society, London, from Jeziorowski et al. [1973]).

information in that they show that the main absorption band seems to shift to higher wave numbers with decreasing coverage. Moreover an additional narrower band, the relative intensity of which increases with decreasing coverage, is detected at 3485 cm^{-1}. This absorption should correspond to the band at 3500 cm^{-1} observed by Borello et al. [1967b]. The observed IR bands together with the overtones are summarized in table 27.4. The most striking feature of the overtone spectra is the occurrence of a band at 6820 cm^{-1}, which is assigned as the overtone of the 3485 cm^{-1} band. The relative intensity of this overtone with respect to its fundamental is significantly higher than that of the 6410 cm^{-1} band as compared to the respective fundamental at 3330 cm^{-1}.

TABLE 27.4

IR fundamental and overtone bands of *tert.*butanol adsorbed on silica and their assignments (Jeziorowski et al. [1973]).

Fundamentals		Overtones			
$\bar{\nu}_{01}$ (cm^{-1})	$b_{1/2}$ (cm^{-1})	$\bar{\nu}_{02}$ (cm^{-1})	$\frac{1}{2}\bar{\nu}_{02}$ (cm^{-1})	X_{12} (cm^{-1})	Assignment
3745	15	7420	3710	35	"Free" surface silanols
3605	30	7130(7062)	3565	40	"Unperturbed" alcohol hydroxyls
3485	70–80	6820(6750)[a)]	3410	75	Anchoring surface silanols
3330(3370)	350	6410(6370)	3205	125	H-bonded alcohol hydroxyls in clusters

The values in brackets are those for liquid *tert.*butanol as obtained by Bourderon and Sandorfy [1973].

[a)] This overtone of liquid *tert.*butanol belongs to oligomeric species.

An appreciable number of OH groups (more than 10%) remain unperturbed even after adsorption of a number of alcohol molecules that is higher than the total number of silanol groups present (see spectrum b in fig. 27.15). Moreover, even at coverages below a statistical monolayer there are about 20–30% alcohol molecules, with respect to the total amount adsorbed, the OH groups of which remain unperturbed. The band at 3605 cm^{-1} is assigned to the respective stretching fundamental. These observations suggest a model similar to that proposed by Cruz et al. [1973], i.e., individually adsorbed molecules and molecules forming clusters may be present at the same time, the clusters being formed even before completion of a statistical monolayer and some silanol groups remaining unperturbed even after completion of a statistical monolayer.

An assignment of the IR bands (table 27.4) has been given by Jeziorowski et al. [1973] which is different from that proposed by Borello et al. [1967b]. The assignment is based on recent observations on liquid alcohols and dilute solutions (Sandorfy [1972] and Bourderon and Sandorfy [1973]). For such systems very broad bands with high relative band shifts are observed, e.g. at $-40°C$ *tert.*butanol shows a band at 3321 cm^{-1}, which corresponds to a relative band shift $\Delta\bar{\nu}/\bar{\nu}$ of 0.083. The band width was about 140 cm^{-1}, the anharmonicity constant ($X_{12} = \bar{\nu}_{01} - \bar{\nu}_{02}/2$) was 166 cm^{-1}. The band width and anharmonicity constant increased with increasing temperature. This band could clearly be assigned to a "polymer" species, that is, to a species in which the oxygen atom is both proton donor and acceptor and it is not a member of a small oligomer. Sandorfy [1972] has shown that the observed high values of wave number shifts, band widths and anharmonicity constants can only occur in such "polymer" species and that a disordered medium (distribution of $O \cdots O$ distances) is needed for this. Consequently, narrow bands at wave numbers intermediate between these free groups and the polymer species have been assigned to oligomeric species, such as e.g. open dimers:

$$\mathrm{R{\setminus}O{-}H \cdots R{\setminus}O{-}H} \tag{XX}$$

The corresponding overtones for polymer and oligomer species of *tert.*butanol have been found at approximately 6750 and 6370 cm^{-1}, respectively, by Bourderon and Sandorfy [1973], in the liquid phase. Relying on these results, the bands observed in the *tert.*butanol/SiO_2 system are assigned as follows: the band at 3605 cm^{-1} represents the free OH stretching wave number of alcohol molecules, whereas the broad band between 3300 and 3400 cm^{-1} (band width $b_{1/2} = 350\ cm^{-1}$) is characteristic for "polymeric" species. The wave number increase with decreasing coverage may probably indicate a decrease of the average size of the respective clusters. The band at 3485 cm^{-1} is assigned as the OH stretching wave number of bonded silanols, which form single, linear H-bonds with alcohol molecules. The relatively low half width of only 70–80 cm^{-1} may also be an indication for the assignment of this band to single H-bonds. These H-bonds may be understood as the S-bond type described by Cruz et al. [1973] as the anchoring bridges of the clusters to the surface (see above). A H-bond energy of these single H-bonds of

6–7 kcal/mole may be estimated from the corresponding wave number shift of 260 cm^{-1}. These assignments are confirmed by the observed overtone spectra (table 27.4), in particular by the fact that the relative intensity of the first overtone of the oligomer band at 6820 cm^{-1} with respect to the fundamental at 3485 cm^{-1} is significantly higher than the corresponding relative intensity of the polymer overtone at 6410 cm^{-1} with respect to its fundamental at 3330 cm^{-1}.

Luck and Ditter [1967–68] first observed that the relative intensities of OH stretching bands of free groups and of groups of oligomers are higher than those of polymeric species in the overtone region as compared to the fundamentals. This phenomenon has recently been interpreted by Di Paolo et al. [1972] on the grounds of mechanical and electrical anharmonicity. Thus, one should be able to observe the bands of anchoring H-bonds, which may be compared to dimeric species in the liquid, much easier in the overtone region. Positions and band widths of the various OH stretching fundamentals and the relative wave number shifts are fairly well comparable to the values reported for the liquid. The increase of the anharmonicity constants of "polymeric" species as compared to free hydroxyls and "dimeric" species is also observed for the adsorbed phase. From these spectroscopic results the formation of a polymeric aggregate with randomly distributed anchoring H-bonds to surface silanols–as proposed by Cruz et al. [1973] for the adsorption of methanol–also seems to be the most probable situation in the case of *tert.*butanol.

Some qualitative information regarding changes of the *average size of the polymeric aggregates* with temperature and coverage can also be obtained from the spectra. The high relative intensity of the OH stretching overtone of free alcoholic hydroxyls even at low coverages suggests an increasing number of free groups with decreasing coverage. The average size of the clusters should then decrease simultaneously, in agreement with the shift to higher frequencies of the "polymer" fundamental band. Accordingly, the integral absorption of the alcoholic OH fundamental deviates from a linear correlation with the integral absorption of the "polymer" fundamental, indicating the above mentioned decrease of the relative number of free groups with increasing coverage. Analogously, the ratio of the absorbances at 3330 and 3605 cm^{-1} decreases from 5.2 to 4 with decreasing adsorbed amount in the covered range. Assuming an extinction coefficient of approximately 50 l $mole^{-1}$ cm^{-1} for the band of unperturbed alcoholic hydroxyls at 3605 cm^{-1}, one estimates an increase of the percentage of free hydroxyls from approximately 21 to 28%, when the coverage decreases from 0.7 to 0.2 molecules/100 Å^2 at a temperature of 55 °C. These values agree quite well with that of 12% as given by Borello et al. [1963–64] for methanol at 24 °C and a surface coverage of 33%, if their higher coverages and lower temperatures are kept in mind.

The relative number of *anchoring H-bonds* also increases with decreasing coverage as shown by the increasing relative intensity of the 3485 cm^{-1} band (fig. 27.16). It may be proposed that adsorption structures of type (XVIII) with only one single alcohol molecule bonded to a single surface silanol may be the dominating species at sufficiently high temperatures and low coverages. These considerations do not exclude the formation of H-bonded species with bridging H-bonds on surfaces of higher silanol density.

The *isosteric heat of adsorption* of *tert.*butanol of 14 kcal/mole has been found to be fairly independent of the surface coverage (Jeziorowski et al. [1973]). Since, in the respective range of coverages (i.e. between 0.4 and 3 molecules/100 Å^2) most of the adsorbed molecules are members of the polymeric aggregates, this result may become understandable. The absolute value of 14 kcal/mole – though it exceeds the heat of liquefaction by about 4 kcal/mole – is the direct consequence of the fact that each alcohol molecule within the clusters forms at least two H-bonds.[7)]

The *dynamic behavior* of methanol molecules in the clusters formed on silica surfaces has been studied by pulsed NMR. Fiat et al. [1967], while measuring the relaxation times in the systems CD_3OH and CH_3OH on porous vycor glass and CH_3OD on deuterated porous vycor glass at 22 °C and at coverages between 0.1 and 1, came to the conclusion that the spin–lattice relaxation time was mainly determined by molecular diffusion processes. The corresponding correlation times were of the order of magnitude of 10^{-10} to 10^{-7} s. The surface diffusion coefficients were of the order of magnitude of 10^{-8} to 10^{-7} cm^2/s in the range of coverages studied. Diffusional motion was therefore strongly restricted as compared to liquid methanol for which the self-diffusion coefficient is about two to three orders of magnitude higher (O'Reilly and Peterson [1971]).

Cruz et al. [1972a] studied the dependence of *relaxation times* in wide ranges of temperatures and coverages in the *methanol/silica system*. The silica used had an average pore radius of 17.5 Å, their diameter thus being about four times larger than the molecular diameter of methanol. This material was degassed at 110 °C before adsorption. A single phase behavior was observed for both longitudinal and transversal relaxation times for CH_3OD on a deuterated surface. In agreement with the results of Fiat et al. [1967], a molecular diffusion was found to contribute mainly to the relaxation process. The activation energies of this diffusional motion was approximately half the heat of adsorption. A broad gaussian distribution of correlation times demonstrated the energetic heterogeneity of the surface. Heat of adsorption and activation energy of diffusion passed through a maximum near a statistical monolayer. This was considered as an indication that a more ordered molecular arrangement is built up on the surface under these conditions. It was also shown that the fraction of nuclei in the translational mobile phase increases with temperature at a given coverage and that it shifts towards lower temperatures as the coverage is increased. The methyl group of the alcohol is rotating very rapidly around the triad axis. A two-phase behavior was observed by Cruz et al. [1972a] for the spin–spin relaxation time of CD_3OH on hydroxylated silica and the long T_2 was attributed to the diffusion motion and the short T_2 was assumed to be due to a proton exchange process between hydroxylic protons of CD_3OH and surface silanols or between methanol molecules. The correlation times for the exchange were always higher than for the diffusion; the activation energies, however, were found to be approximately the same for both processes. This was expected since both processes, diffusion and proton exchange, demand the rupture of H-bonds. From the comparison of the correlation times of the exchange process on the surface with those reported for exchange in liquid basic or acid methanol, it was suggested that, most probably a proton was transferred from the silica surface into an adsorbed cluster via an anchoring H-bond (see structure (XIX)) and that it undergoes rapid exchange between the cluster molecules according to (XXI) before it comes back to the surface. Dielectric absorption data also confirmed the existence of protonated methanol (Cruz et al. [1972b]), the lifetime of which was discussed recently by Fripiat [1973].[8)]

$$CH_3OH_2^+ + CH_3OH \rightleftharpoons CH_3OH + CH_3OH_2^+ \qquad \text{(XXI)}$$

Process (XXI) may also deliver an explanation for the increased *surface electrical conductivity* on methanol adsorption on a silica surface. Levy and Folman [1963] could describe their experimental data applying eq. (27.5) and they also postulated the formation of $CH_3OH_2^+$. However, they suggest that these species could also be formed through an autoprotolysis of methanol and they already believed that some clustering was occurring far below monolayer coverage.

As shown by Cruz et al. [1972b], the *mobility of adsorbed methanol molecules* on chemically very similar surfaces strongly depends on the porosity. A comparison was made by these authors of the dynamic behavior of methanol adsorbed on the silica described above and on an aerogel containing pores of molecular dimensions. Applying a method described by Ross and Olivier [1964] the equations of state for the adsorbed methanol were determined for the two adsorbents at 20 °C. With the narrow pore silica the best fit of the experimental adsorption data was obtained applying the Fowler–Guggenheim equation which describes the behavior of an immobile film with lateral interaction between adsorbate molecules. The Hill–de Boer equation on the other hand was needed to fit the adsorption data on the wide-pore silica. This equation describes the state of a mobile adsorbed layer with lateral interactions between adsorbate molecules. Thus, the mobility of adsorbed methanol at 20 °C was clearly more restricted in the narrow pores of molecular dimensions. This result agreed excellently with NMR relaxation data obtained by the same authors. The surface diffusion coefficient was calculated to be less than 10^{-11} cm^2 sec^{-1} in the narrow-pore sample and between 2×10^{-7} and 3×10^{-6} cm^2 sec^{-1} at coverages between 0.6 and 2 statistical monolayers in the wide-pore adsorbent. These figures attribute some quantitative meaning to the terms "mobile" and "immobile" adsorption.

27.3. Alumina

The metal-oxygen bond of many metal oxides is more ionic in character than that of silica. The H-bond donor and acceptor properties of these oxides are therefore distinct from those of silica. Alumina has probably been studied most frequently among these acid oxides and will therefore be discussed in some detail as an example.

27.3.1. Characterization of alumina surfaces

Alumina Al_2O_3 occurs in a wide variety of crystallographic modifications. Among the transition aluminas the η and γ forms are of particular interest for adsorption and catalysis since they develop fairly high specific surface areas (usually between 100 and 200 m^2/g). Both modifications are built up by a defect lattice of the spinel type and they are distinct by their defect character and the distribution of the cations in tetrahedral and octahedral sites (Leonard et al. [1967]). The water content of γ Al_2O_3 is greater than that of the η form whereas the latter form develops the higher surface acidity after thermal treatment. It is still not clear how structural differences of the bulk reflect themselves in the surface properties of different modifications. According to Leonard et al. [1967] one has to assume that the less organized the starting material is, the more defective and the more energetic the surface will be.

The δ modification has also been used frequently in adsorption studies. The δ phase is crystallographically similar to the γ phase (Ginsberg et al. [1957]).

Alumina surfaces which were previously exposed to water vapor (or moist air), are terminated by a monolayer of hydroxyl groups, each occupying about 8 $Å^2$ in the surface (Kipling and Peacall [1957]). The presence of hydroxyl groups in the surface has been shown by IR spectroscopy and deuterium exchange (Peri and Hannan [1960], Peri [1965a] and Carter et al. [1965]) and by chemical methods (Boehm [1966a]). Regarding γ alumina Peri and Hannan [1960] observed three resolved OH stretching bands at 3698, 3737 and 3795 cm^{-1} after degassing at 650 °C, which were assigned as "isolated" non-interacting OH groups. With higher resolution two additional bands at 3733 and 3750 cm^{-1} were observed (Peri [1965a]). At degassing temperatures below approximately 500 °C, broad unresolved bands are obtained in the OH stretching region due to mutual interactions of the hydroxyl groups. The isolated surface OH groups are progressively removed from the surface at increasing degassing temperatures but with different rates and even at 800 °C about 2% of the total hydroxyl content is retained on the surface (Peri [1965a]). Based on his weight loss, rehydration and IR studies and applying a statistical (Monte Carlo) method for the simulation of the dehydroxylation process, Peri [1965b] developed a *model for the* γ Al_2O_3 *surface.* The (100)-plane is assumed to be exposed predominantly with an aluminium ion in an octahedral site lying immediately below each surface hydroxyl group. The removal of water on heating occurs through the condensation of two immediately neighboring hydroxyl groups. This process leaves an oxide ion in the outermost layer and an exposed aluminium ion in the next lower layer. Assuming a random removal of hydroxyl pairs, a regular surface lattice with no two oxide ions adjacent can be maintained until a stage where about 67% of the monolayer has been removed. Further pairs of adjacent hydroxyls can only be condensed to eliminate water with the creation of defects, which comprise adjacent aluminium and adjacent oxide ions. This process can be continued until 90.4% of the surface hydroxyl content has been eliminated. At this stage no hydroxyls exist on adjacent sites and further dehydroxylation is possible only by surface migration.

The five observed OH stretching bands were attributed to hydroxyl groups in distinct surface environments (Peri [1965b]). Thus, the high wave number band at 3800 cm^{-1} will be characteristic for an OH group with four oxide nearest neighbors (A-type), whereas the low wave number band at 3700 cm^{-1} points to an OH group which contains no oxide ion as nearest neighbor, but 4 aluminium ions (C-type). The other three hydroxyl groups which give rise to the intermediate wave numbers at 3733, 3744 and 3780 cm^{-1} are located at one, two and three nearest oxide neighbors, respectively (E-, B-, and D-type).

These environmental differences lead to different *acidic and H-bond donor properties of the various OH groups.* One would expect the C-type hydroxyls to be the most acidic and the A-type the least acidic ones, according to the proposed model. However, the exchange rate with D_2 of the A-type hydroxyls is appreciably higher than of the other hydroxyls (Carter et al. [1965] and Dunken and Fink [1966]). Dunken and Fink [1966] therefore propose the reverse acidity sequence, where the A-type hydroxyls (3800 cm^{-1}) are the most acidic and have four aluminium ions and no oxide ions as nearest neighbors. The C-type hydroxyls (3700 cm^{-1}) on the other hand are assumed to be immediately adjacent to four oxide ions with the OH force constant and the proton reactivity (acidity) being lowered due to the H-bond interactions with the neighboring oxide ions (compare also the low H-bond donor function of mutually interacting vicinal silanol groups, section 27.2.2). Thus, the assignment of the observed OH stretching bands still seems to be ambiguous and this situation unfortunately renders a discussion of the H-bond donor properties and the interpretation of H-bonded systems of alumina surfaces rather difficult.

Very similar OH stretching bands have been found for other modifications of alumina as shown in table 27.5. Since equal numbers of OH stretching bands and nicely comparable positions are obtained, one may conclude that the respective surface OH groups reside in comparable local environments on the surfaces of the three modifications of alumina.

A totally different interpretation of multiple wave numbers in the OH stretching region has been suggested by Hallam [1969]. He points to the possibility that multiplet wave numbers in the OH stretching region may arise from combination sum and difference modes of low-frequency

TABLE 27.5

Comparison of the OH(OD) stretching fundamental wave numbers (cm^{-1}) for γ, η and δ Al_2O_3

References	Peri [1965a] Peri and Hannan [1960]	Dunken and Fink [1966]	Carter et al. [1965]	Knözinger and Stolz [1971]	Meye [1972]
Oxide	γ Al_2O_3	γ Al_2O_3	η Al_2O_3	δ Al_2O_3	δ Al_2O_3
	3700	3700	3710	3670	3680
	3733			3685	
$\bar{\nu}$(OH)	3744	3745	3740	3727	3730
	3780	3760		3775	
	3800	3785	3785	3790	3795
	2733	2725	2730	—	2715
$\bar{\nu}$(OD)	2759	2760	2755	—	2755
	2803	2790	2790	—	2800

vibrations. The number of distinct types of OH groups present in the surface is then less than the number of bands observed in the IR spectra. In this case, the central band and the corresponding sum and difference mode belonging to a certain OH group should respond to exchange or adsorption processes in a predictable manner. Inspection of spectra during D_2 exchange experiments (e.g. Carter et al. [1965] and Dunken and Fink [1966]) seem not to show behavior consistent with Hallam's interpretation, at least for the aluminas used in their studies.

The model proposed by Peri [1965b], though certainly very helpful, is an extremely idealized one, particularly since it assumes the predominant appearance of the (100)-face as the terminating plane. The occurrence of various exposed crystal planes, which would offer different local environments for isolated OH groups, appears to be likely for polydisperse materials such as the transitional aluminas. Furthermore, it seems possible that in the surface layers a structure is stabilized which is distinct from the bulk structure. Evidence for such phenomena has been accumulated by French and Samorjai [1970] for the (0001)-face of α alumina by means of low-energy electron diffraction. These authors also showed that the surface layer is oxygen deficient and that the aluminium ions are in a reduced valence state. The surface structure was assumed to have a composition corresponding to Al_2O (or AlO). Weller and Montagna [1971] have shown that also η and γ Al_2O_3 may exist in a non-stoichiometric composition and they also believe in the presence of surface aluminium ions in a reduced valence state.

The particular H-bonding properties of surface OH groups and oxygen ions will certainly depend on the structure and composition of the surface layer in their immediate environments. These details unfortunately are still not clear. However, OH groups and oxygen ions will certainly be present in a partially dehydroxylated alumina surface and they will act as potential H-bond donors and acceptors, respectively. There are also incompletely coordinated aluminium ions (possibly in different valence states) exposed in such surfaces, which form Lewis-acid sites. In many cases the interactions of adsorbate molecules with these aprotonic acid sites are much more energetic than H-bonding interactions. Consequently, these only play an important role at higher coverages after saturation of the Lewis-acid sites.

27.3.2. Adsorption of water

The exposure of partially dehydroxylated alumina surfaces to a water vapor atmosphere will lead to a rehydroxylation and partly to coordination of water molecules to exposed aluminium ions during the first stages (Peri [1965a] and Della Gatta et al. [1973]). The corresponding heat evolution amounts to as much as 100 kcal/mole at the lowest coverages (Cornelius et al. [1955], Venable et al. [1965], Spannheimer and Knözinger [1966] and Hendriksen et al. [1972]) indicating the strong chemical interaction between surface and water molecules. Two peaks at 648 and 360 cm^{-1} in a neutron time-of-flight spectrum of water adsorbed on γ Al_2O_3, have been assigned as torsion and translation vibrations of an H_2O molecule coordinated to an exposed aluminium ion by Boutin and Prask [1964], since similar peaks are observed in the spectrum of $AlCl_3 \cdot 6H_2O$. This assignment may be questioned, since the γ Al_2O_3 sample had been heat-treated at only 150°C, so that only very few or no aluminium ions should have been exposed.

The coordinately bonded water molecules may act as H-bond donors (Clement et al. [1973]) due to the increased acidity which is brought about by the strong polarization of the water molecule (Zundel [1969]) (section 15.6.8.1):

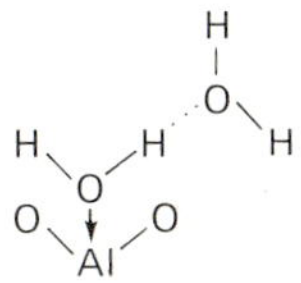

```
            H
            |
         ..O
H     H'     \
 \   /         H
   O
O  ↓  O
 \ Al /
```
(XXII)

Antipina et al. [1972] and Kirina et al. [1973] argue that such coordinately bound water molecules are able to protonate pyridine, though in an earlier study by Knözinger and Stolz [1971] a transformation of Lewis sites into Brönsted sites on water adsorption had been ruled out.

Since alumina surfaces are usually completely rehydroxylated (Hendriksen et al. [1972] and Della Gatta et al. [1973]) at temperatures above approximately 50 °C and the OH content amounts to 70–80% even after heat treatment at 200 °C (Peri [1965a]), only very few surface oxygen ions are available as H-bond acceptor sites after adsorption of water at temperatures between approximately 50 and 200 °C. At temperatures below 50 °C rehydroxylation does not readily occur (Peri [1965a]) and water will probably preferentially adsorb through coordination bonds. Under such conditions (strongly dehydroxylated surface, water adsorption below 50 °C) water may be strongly H-bonded onto the surface, the oxygen ions being the acceptor sites:

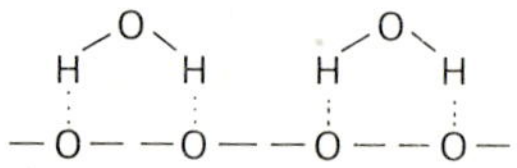

(XXIII) 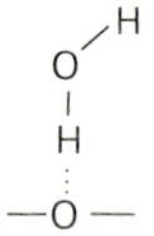 (XXIV)

Structure (XXIII) was proposed by De Boer et al. [1963] to explain the molecular adsorption of water by very strong H-bonds.

Most adsorption studies of water have, however, been made on nearly completely hydroxylated surfaces, which do not expose oxygen ions. The surface OH groups, which are densely packed (approximately 8 Å^2/OH group, corresponding to 12.5 OH/100 Å^2), will then take over the H-bond donor function. Water molecules will be adsorbed through one or two passive H-bonds as shown in structures (XXV) and (XXVI):

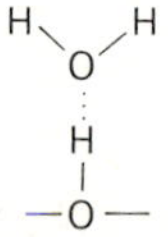

(XXV) 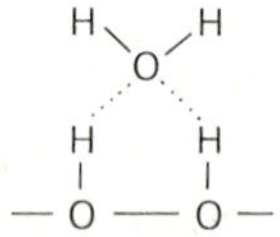 (XXVI)

Structure (XXVI) will be preferred due to the dense packing of OH groups with approximate interhydroxyl distances of 3.2 Å as estimated for the (100)-face according to Peri's [1965a] model. Accordingly, a

H_2O/OH ratio of 1:2 was reported by Morimoto et al. [1971] for the monolayer formation on a γ Al_2O_3 and Della Gatta et al. [1973] also propose structure (XXVI) as the most probable one relying on differential heats of adsorption.

Unfortunately, the *IR spectra of adsorbed water* cannot contribute much to our knowledge in the OH stretching region, since only very broad and unresolved bands are observed which contain contributions from both surface OH groups and molecular water. There is one paper on *Raman scattering of adsorbed water* (Careri et al. [1972]) from which, again, no detailed information on the adsorption structures can be gained. Overtone spectra which probably would provide the required information have not been reported so far. However, the presence of molecular water in the adsorbed phase can be detected by means of the $\bar{\nu}_2$ deformation mode of molecular water around 1600 cm^{-1} (Glemser and Hartert [1958]). Knözinger and Ress [1968] observed a band at 1630 cm^{-1} in the temperature range between 80 and 168°C using a δ Al_2O_3. This band shifted to 1600 cm^{-1} with decreasing coverage. Van Thiel et al. [1957] applying the N_2 matrix isolation technique for studies on water, assigned a band at 1633 cm^{-1} to polymeric water and a band at 1600 cm^{-1} to monomeric water. Stevenson [1965] showed that the molecular vibrations of molecules which participate in H-bonds only as acceptors through their lone-pair electrons, do not change much. For these reasons, the two water deformation bands observed for the adsorbed system at 1630 and 1600 cm^{-1} were assigned as water clusters at higher coverages and as "monomeric" molecules corresponding to structure (XXVI) at lower coverages, respectively. Similar observations were reported by Yates [1961] for the water/TiO_2 system. The occurrence of the band at 1630 cm^{-1} suggests a cluster formation, which is probably enhanced by cooperative effects, before a monolayer is completed (see also section 27.2.5).

Some support for the preferred formation of structure (XXVI) comes also from the *dielectric behavior of adsorbed water layers* on hydroxylated aluminas. Ebert and Langhammer [1961] observed various linear portions in dielectric isotherms of water adsorbed on γ Al_2O_3 in the temperature range between -60 and $+20$°C in the frequency range between 0.1 and 10 Mc/s. The slope of subsequent linear portions increased with increasing coverage and the first change in slope coincided with the monolayer capacity. The slow increase of the dielectric constant with coverage in the monolayer region was explained as being due to a strong restriction of the orientational freedom in an external field of the molecules within the first adsorption layer. With increasing coverage the molecules obviously become increasingly free to orientate in an external field. Similar observations were reported by Baldwin and Morrow [1962] studying the frequency range between 50 cps and 100 kc/s at temperatures between 0 and 86°C. Finally, Dransfeld et al. [1962] followed the temperature dependence of the relaxation rate of capillary condensed water in γ Al_2O_3 between 180 K and room temperature at frequencies between 100 kc/s and 20 Mc/s. They find only one relaxation process which they attributed to a transition of a super-cooled liquid into a "glassy" solid state. Capillary condensed water in aluminas apparently does not freeze to form an ice-like structure, it merely solidifies. The restriction of the orientational freedom of molecules within the first water layer on a hydroxylated alumina surface is compatible with structure (XXVI), if the immobilization of these molecules is assumed to be due to the formation of two H-bonds per molecule. This explanation was put forward by McCafferty and Zettlemoyer [1971], when they obtained analogous results for the H_2O/α Fe_2O_3 system.

It should be mentioned, however, that the interpretation of dielectric data obtained for the water/alumina system does not appear to be completely resolved. Thus, Hoekstra and Doyle [1971] concluded that the dielectric loss at radio frequencies appears to be due to free charge carriers and that any dispersion superimposed on a dc conductivity loss is most likely due to Maxwell–Wagner effects. A dc conductivity was also observed by Clement et al. [1973] at higher temperatures (>100°C). Kondo et al. [1973] working at 30 cps through 30 Mc/s in the temperature range between +120 and −196°C observed three relaxations. The one occurring at the highest temperatures (>0°C) was attributed to loosely adsorbed molecular water, the other two, occurring at lower temperatures, were explained as being due to two types of hydroxyl groups. Applying approximately the same frequency range, Umeya and Kanno [1973] interpreted their results by a polarization of the surface layer formed by an agglomeration of adsorbed ions around non-conducting regions.

Pulsed NMR investigations of water adsorbed on γ Al_2O_3 have been reported for water coverages corresponding to one or more statistical monolayers (Winkler [1961] and Reball and Winkler [1964]). The behavior of the proton relaxation times of the adsorbed water was studied as a function of the coverage. The transversal relaxation time T_2 showed a single-phase behavior at the lower coverages, while a two-phase behavior was observed at higher coverages. The two phases were attributed by Winkler [1961] to water in micropores and capillary condensed water in wider pores. Reball and Winkler [1964] suggested a fairly rapid proton exchange between the phases. The water molecules adsorbed in the first monolayer seemed to be relatively immobile and a corresponding correlation time for the proton–proton interaction of 10^{-7} s was calculated by Winkler [1961]. This value comes closer to that of ice than of liquid water. Winkler [1961] and Reball and Winkler [1964] explained the behavior of the longitudinal relaxation time T_1 by interactions with paramagnetic impurities, whereas Michel [1966] came to the conclusion that both T_1 and T_2 were determined by this kind of interaction.

In conclusion, the adsorption of water onto a hydroxylated alumina surface leads to a relatively immobile first layer corresponding to species (XXVI). Multilayer or cluster formation most probably begins before the completion of a statistical monolayer. The structure of the clusters and of capillary condensed water is distinct from that of bulk liquid water. Normal freezing and nucleation of ice does not occur, but temperature decrease leads to a solidification comparable to a glass transition. Because of the stronger ionic character of an alumina surface as compared to silica, one may also expect a very high degree of dissociation for water adsorbed on alumina surfaces, though detailed experimental information is still lacking.

27.3.3. Adsorption of alcohols

Various modes of interaction are possible in principal between alcohol molecules and alumina surfaces. As in the case of water adsorption, coordinative and dissociative adsorptions may occur besides H-bonding on partially hydroxylated surfaces. However, contrary to the chemisorption of water which leads to a complete rehydroxylation of the surface, the dissociative adsorption of alcohols with formation of surface alkoxide groups and new surface OH groups (Greenler [1962], Treibmann and Simon [1966], Kagel [1967], Arai et al. [1967], Deo and Dalla Lana [1969], Knözinger et al. [1968], Fink [1969], Knözinger and Stolz [1970], Deo et al. [1971] and Chuang and Dalla Lana [1972]) does not cover the alumina surface completely, even at low temperatures. One can therefore expect that accessible basic oxygen ions may act as H-bond acceptor sites besides the OH donor groups and it will be an interesting question with regard to the surface properties of alumina as compared to those of silica, which of the two possible structures (XXVII) or (XXVIII) is formed preferentially:

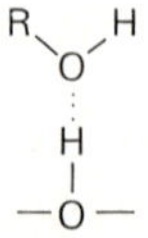

(XXVII) 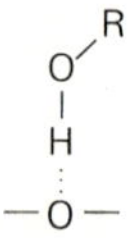 (XXVIII)

Barto et al. [1966] observed strong *monolayer formation* on adsorption of C_1–C_4 *n*-aliphatic alcohols on two alumina samples of widely differing surface areas (γ and α modification, respectively) at 20 °C. Completely analogous results were obtained for the adsorption of C_1–C_5 and C_{12} *n*-aliphatic alcohols on a flat superficially oxidized aluminium foil (Blake and Wade [1971] and Blake et al. [1971]) and also for some branched-chain alcohols (Blake and Wade [1972]). A tendency to *multilayer formation* was only detected at relative pressures around 0.9, provided the carbon chain contained at least 3–4 carbon atoms. The surfaces containing monolayers of alcohol (C_3 and higher) were shown to be effectively autophobic. Steps were observed at relative pressures between approximately 0.9 and 1.0, the inflection points of which correspond to the completion of the first and third statistical monolayer (rather than the first and second). It was therefore concluded by Blake and Wade [1972] in analogy to similar results obtained by Avgul et al. [1961] for the adsorption of *tert.*butanol on Graphon, that the second and third monolayers are formed concertedly such that the molecules in the second layer are H-bonded to those of the third layer. Only dispersion forces will then hold these two layers onto the first monolayer. From estimations of the effective molecular cross-sections in the first monolayer, Blake and Wade [1972] concluded that the carbon chains of the alcohol molecules should be aligned approximately vertical to the surface. In an earlier work, Smith and Thorp [1963] proposed an adsorption structure of cyclohexanol with the six-membered ring lying flat on the surface.

Adsorption isotherms showing sharp knees at low relative pressures and a fairly well developed saturation character, have also been reported for temperatures between 58 and 120 °C for the adsorption of ethanol on a δ Al_2O_3 (Jeziorowski et al. [1973]). The isosteric heats of adsorption between 18 and 11 kcal/mole indicated strong interactions of alcohol and surface in the first monolayer, the high heats of adsorption and their fall to lower values probably being due to the simultaneous adsorption through coordination bonds and strong H-bonds.

The *autophobicity of the first monolayer of alcohols* (C_3 and higher) clearly shows that the alcoholic hydroxyl group directly interacts with the alumina surface. This conclusion has been confirmed by high-resolution NMR data (De La Hardrouyère [1960], Geschke and Pfeifer [1966] and Derouane [1969b]), which show a preferential broadening of the band corresponding to the alcoholic hydroxyl group. This indicates that the adsorption onto the surface occurs through this group.

The *IR spectroscopic work* on alcohol adsorption on alumina surfaces has mainly been devoted to the study of surface reactions and H-bonding has been discussed only occasionally. Surface OH groups have been considered as donor groups in most cases (Kagel [1967], Deo and Dalla Lana [1969], Řeřicha and Kochloefl [1969], Deo et al. [1971] and Chuang and Dalla Lana [1972]), though the acceptor function of the surface had also been taken into account (Greenler [1962], Kagel [1967] and Knözinger et al. [1967]). Greenler [1962], in particular, pointed to the fact that on methanol adsorption the OH bending vibration is observed around 1420 cm^{-1}, this region being characteristic for alcohol OH groups acting as H-bond donors. Kagel [1967] mentioned that no absorption could be detected above 3600 cm^{-1} which would have indicated a freely vibrating alcohol OH group. Both these observations invoke a possible preference of species (XXVIII).

The temperature ranges in which alcohol adsorption can be studied are limited because of the high catalytic activity of aluminas towards dehydration. Especially the most reactive tertiary alcohols start decomposing over alumina already around 50 °C. *Ethanol adsorption* in the temperature range 58–120 °C has recently been studied on a δ Al_2O_3 pretreated at 500 °C by Jeziorowski et al. [1973]. This surface exposes OH groups and oxygen ions as potential H-bond donors and acceptors, but also incompletely coordinated aluminium ions with acidic properties (see section 27.3.1). Besides possible H-bonded structures such as (XXVII) and (XXVIII) a coordinative adsorption between the alcohol hydroxyl oxygen and a Lewis site (structure (XXIX) must therefore be taken into account

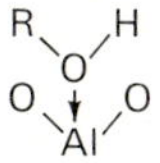

R   H
 \ /
  O
O ↓ O
 \Al/   (XXIX)

for the interpretation of the IR spectra. Figure 27.17 shows the difference between the spectra of the adsorbate and the background (spectra b and c). Spectrum a shows the OH stretching region of the alumina before

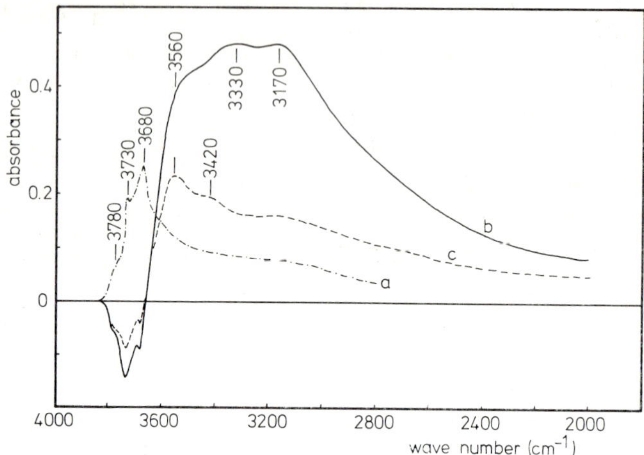

Fig. 27.17. (a) Background transmission IR spectrum of δ Al_2O_3, degassed at 500°C; (b) difference between transmission IR spectrum of ethanol adsorbed on δ Al_2O_3 (at 58°C, 44 mm Hg) and background; (c) after desorption 45 min. at 58°C. (Reproduced by permission of the Chemical Society, London, from Jeziorowski et al. [1973]).

adsorption of the alcohol, indicating three bands of free surface OH groups (see section 27.3.1). The negative absorbances in spectra b and c in this region are due to the surface OH groups, which are perturbed on adsorption. The following observations can be extracted from the IR spectra:

(a) No band attributable to the stretching vibration of unperturbed alcoholic hydroxyls ($\bar{\nu} > 3600\ cm^{-1}$) can be detected. The number of adsorbed alcohol molecules with free hydroxyls must therefore be negligibly small.

(b) A comparison of intensities of the free surface OH groups and of those perturbed on alcohol adsorption (negative absorbance in spectra b and c of fig. 27.17) indicate, that the band at $3780\ cm^{-1}$ is nearly completely reduced in intensity even after desorption of the reversibly adsorbed species. The band at $3730\ cm^{-1}$ is still appreciably perturbed, whereas the band at $3680\ cm^{-1}$ characteristic for the most acidic OH's (see section 27.3.1) is only slightly reduced in intensity on adsorption and is nearly completely restored after removal of the reversibly adsorbed species.

(c) An intense band occurs at $3560\ cm^{-1}$ after adsorption. This band cannot be removed on desorption and can therefore not be related to reversibly adsorbed species.

(d) In the presence of the alcohol vapour, two absorption centers are observed at 3330 and 3170 cm^{-1} which indicate the presence of H-bonded species. The surface hydroxyls should not dominantly be the donors in these H-bonds, particularly, since the most acidic OH groups are only slightly perturbed even in the presence of the vapour, while the less acidic groups remain strongly perturbed even after desorption. The band of free alcoholic hydroxyls on the other hand are not observed. The most probable H-bonded species will therefore be one in which a surface oxygen (basic site) takes over the acceptor function and the alcoholic hydroxyl group is the donor (species (XXVIII)). The high displacement of the OH stretching band of up to 450 cm^{-1} as compared to 280 cm^{-1} in the liquid, suggests relatively strong H-bonding due to the high basicity of the surface oxygens.

(e) On desorption, the bands at 3330 and 3170 cm^{-1} are removed and a shoulder appears around 3420 cm^{-1}. Furthermore, a very diffuse absorption without any structure remains detectable down to 2000 cm^{-1}.

The band at 3560 cm^{-1}, which does not vanish on desorption, indicates the formation of new OH groups during the adsorption of ethanol. These groups very probably result from the dissociative adsorption of the alcohol on acid-base pair sites, which leads to the formation of an alkoxide-like species and a hydroxyl group. OH groups initially present–mainly those with typical stretching wave numbers at 3780 and 3730 cm^{-1}–that may be perturbed by the alkoxide species or because of the increased hydroxyl density, may also contribute to the absorption at 3560 cm^{-1}. Because of surface heterogeneity not all of the acid–base pair sites will be able to dissociate the OH bond of an alcoholic hydroxyl group completely, but lead only to a polarization of the bond. A wide spectrum of energetically distinct species with varying degrees of polarization of the OH bonds will exist, the main surface bond being a coordination bond of the alcoholic hydroxyl oxygen and an incompletely coordinated aluminium ion. The shoulder around 3420 cm^{-1} and the diffuse structureless absorption towards lower wave numbers may be attributed to such species. They can be represented by a model structure (XXX):

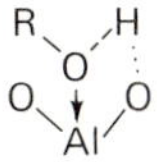

```
R   H
 \ ,'
  O  :
O  ↓  O
 \ Al/
```
(XXX)

Some of these species may be reversibly adsorbed, as indicated, by the higher diffuse absorption of spectrum b in fig. 27.17. The binding energy of such species can be expected to be higher than that of a H-bond. The high isosteric heats of adsorption at low coverages thus become understandable. Fermi resonance between some OH stretching vibrations and the CH_3 and CH_2 stretching fundamentals and bending overtones may also contribute to the diffuse absorption.

Species similar to structure (XXX) have also been proposed for the adsorption of alcohols on TiO_2 (Knözinger [1970] and Jackson and Parfitt [1972]), MgO (Kagel and Greenler [1968] and Tench et al. [1971]) and GeO (McManus et al. [1969]). These species more or less resemble a surface alkoxide species depending on the degree of loosening of the alcoholic O–H bond.

In conclusion, the oxygen ions of an alumina surface seem to be effective H-bond acceptor sites, whereas the OH groups appear to be relatively poor H-bond donors, at least in cases where they have to compete with oxygen ions for the adsorbate. It is interesting to note in this connection, that *ammonia adsorbed on* γ *Al_2O_3* was proposed to be H-bonded either onto a surface OH group or a surface oxygen ion (Peri [1965c] and Dunken and Fink [1967]). In this respect, alumina surfaces behave completely differently as compared to silica surfaces, where the oxygen ions have lost their H-bond acceptor properties due to the low ionic character of the Si–O bond and the possible $d\pi$–$p\pi$ contributions (see section 27.2.1). The Al–O bond in alumina is more ionic (63% against 50% as calculated from Pauling's electronegativity values) which leads to the comparably high basicity and enhanced H-bond acceptor function of the surface oxygen ions. One may expect similar behavior for other ionic metal oxides such as e.g. titanium dioxide.

27.4. Review of recent literature

In the following literature review publications have been included which deal with H-bonded systems on solid surfaces that have not been discussed in the preceding sections. Parts I–V of table 27.6 present papers on H-bonding on pure oxides (with the exception of pure silica and alumina), mixed oxides, zeolites, clays and salts, respectively. To prevent an unlimited listing of publications, the presentation of the relevant literature has been restricted to publications which appeared after 1969, with a few exceptions only. The time between January 1970 and June 1973 is covered fairly completely to the authors knowledge. The reader is requested to find older publications from the references in the papers

TABLE 27.6

Review of H-bonded systems on oxides (except SiO_2 and Al_2O_3), zeolites, clays and some salts

Adsorbate	Exp. method or property studied	References
I. Pure Oxides		
1. BeO		
Water	IR spectroscopy	Stuart and Whateley [1965]
2. MgO	IR spectroscopy	Anderson et al. [1965]
Water	IR spectroscopy	Ramsay [1971]
	IR diffuse reflectance	Takezawa et al. [1971]
	Raman-spectroscopy	Careri et al. [1972]
	NMR (wide line)	Webster et al. [1965]
	Dielectric loss	Martens et al. [1973]
Ammonia	IR spectroscopy	Tench and Giles [1972]/ Tench [1972]
Methanol	IR spectroscopy	Kagel and Greenler [1968]
	IR spectroscopy	Tench et al. [1971]
Ethanol	IR spectroscopy	Kagel and Greenler [1968]
3. CaO		
Water	IR spectroscopy	Low et al. [1971]
Isopropanol	IR spectroscopy	Iizuka et al. [1971]
Benzaldehyde	IR spectroscopy	Iizuka et al. [1971]
4. GeO		
Alcohols	IR spectroscopy	McManus et al. [1969]
Benzene	IR spectroscopy	Cusumano and Low [1972]
	Thermodynamics	Cusumano and Low [1972]
Fluorobenzene	IR spectroscopy	Sahay and Low [1974]
	Thermodynamics	Sahay and Low [1974]
5. TiO_2 (rutile)		
Water	IR spectroscopy	Yates [1961]
	IR spectroscopy	Lewis and Parfitt [1966]
	IR spectroscopy	Munuera and Stone [1971]
	IR spectroscopy	Jackson and Parfitt [1971a]
	IR spectroscopy	Jones and Hockey [1971]
	Adsorption isotherms	Dawson [1967]
	Adsorption isotherms	Morimoto et al. [1969]
	Adsorption isotherms	Munuera and Stone [1971]
	Adsorption isotherms	Day et al. [1971]
	Heat of immersion	Iwaki and Miura [1971]
	Heat of immersion	Iwaki et al. [1972]
	Heat of immersion	Dawber et al. [1972]
	NMR (wide line)	Mays and Brady [1956]

TABLE 27.6-(*contd.*)

Adsorbate	Exp. method or property studied	References
	Slow neutron inelastic scattering	Boutin et al. [1967]
		Boutin et al. [1968]
	Temperature programmed desorption	Munuera and Stone [1971]
n-alcanols		
C_1–C_{10}	Electrophoresis	Jackson and Parfitt [1971b]
C_2, C_8, C_{10}	Adsorption isotherms	Day and Parfitt [1968]
C_2, C_6, C_6–1.6-diol	Adsorption isotherms	Day et al. [1971]
C_2, C_4, C_6	IR spectroscopy	Jackson and Parfitt [1972]
Isopropanol, Acetone	Adsorption isotherm; Temperature programmed desorption	Munuera and Stone [1971]
Ammonia	IR spectroscopy	Parfitt et al. [1971]
	IR spectroscopy	Primet et al. [1971]
Pyridine	IR spectroscopy	Primet et al. [1971]
Trimethylamine	IR spectroscopy	Primet et al. [1971]
Benzene	NMR (high resolution)	Enriquez et al. [1970]
6. TiO_2(anatase)		
Water	IR spectroscopy	Yates [1961]
	IR spectroscopy	Lewis and Parfitt [1966]
	IR spectroscopy	Knözinger [1970]
	IR spectroscopy	Munuera et al. [1972]
	Adsorption isotherms	Munuera et al. [1972]
	Temperature programmed desorption	Munuera et al. [1972]
	Thermogravimetry	Munuera et al. [1972]
n-alcohols C_2	IR spectroscopy	Knözinger [1970]
C_1–C_4	IR spectroscopy	Munuera and Carrizosa [1973]
	Adsorption isotherms	Munuera and Carrizosa [1973]
Ammonia	IR spectroscopy	Primet et al. [1971]
Pyridine	IR spectroscopy	Primet et al. [1971]
Trimethylamine	IR spectroscopy	Primet et al. [1971]
7. ZrO_2		
Water	IR spectroscopy	Tretyakov et al. [1970]
Benzene	IR spectroscopy	Tretyakov et al. [1970]
Benzonitrile	IR spectroscopy	Tretyakov et al. [1970]
Pyridine	IR spectroscopy	Tretyakov et al. [1970]
Formic acid	NMR (high resolution)	Gradsztajn et al. [1971]
8. V_2O_5		
Formic acid	NMR (high resolution)	Fraissard et al. [1968]

TABLE 27.6–(*contd.*)

Adsorbate	Exp. method or property studied	References
9. γ MnO_2		
Water	Temperature programmed desorption	Lee et al. [1973]
10. α Cr_2O_3		
Pyridine	IR spectroscopy	Zecchina et al. [1972]
11. α Fe_2O_3		
Water	IR spectroscopy	Blyholder and Richardson [1964]
	Adsorption isotherms	Morimoto et al. [1969]
	Heat of immersion	Zettlemoyer and McCafferty [1969]
	Thermodynamics	McCafferty and Zettlemoyer [1970a]
	Thermodynamics	McCafferty and Zettlemoyer [1971]
	Dielectric properties	McCafferty et al. [1970]
	Dielectric properties	McCafferty and Zettlemoyer [1970b]
	Dielectric properties	McCafferty and Zettlemoyer [1971]
Ammonia	IR spectroscopy	Blyholder and Richardson [1964]
Hydrogen sulfide	IR spectroscopy	Blyholder and Richardson [1964]
n-butanol	Flow-microcalorimetry	Husbands et al. [1971]
stearic acid	Flow-microcalorimetry	Husbands et al. [1971]
12. ZnO		
Water	Adsorption isotherms	Egorov et al. [1965]
	Adsorption isotherms	Morimoto and Nagao [1970]
	Adsorption isotherms	Nagao [1971]
	Heat of immersion	Nagao and Morimoto [1969]
	Thermodynamics	Nagao [1971]
	Thermogravimetry	Mattmann et al. [1972]
	IR spectroscopy	Taylor and Amberg [1961]
	IR spectroscopy	Atherton et al. [1971]
	Raman-spectroscopy	Careri et al. [1972]
	Adsorption capacity	Huettenrauch et al. [1971]
13. ThO_2		
Water	Adsorption isotherms	Holmes et al. [1968a, b]
	Adsorption isotherms	Gammage et al. [1970]
	Heat of adsorption	Holmes et al. [1966]
	Heat of adsorption	Ron and Schindler [1972]
	Heat of immersion	Fuller et al. [1968]
	IR spectroscopy	Fuller et al. [1970]
	surface diffusion	Knözinger and Jeziorowski [1970]
	NMR (wide line)	Brey and Lawson [1964]
n-butylamine	NMR (wide line)	Brey and Lawson [1964]
Methanol	NMR (wide line)	Brey and Lawson [1964]

TABLE 27.6–(*contd.*)

Adsorbate	Exp. method or property studied	References
Ethanol	NMR (wide line)	Brey and Lawson [1964]
Ethanol	Surface diffusion	Knözinger and Jeziorowski [1970]
Hexenes	Thermodynamics	Jeziorowski [1972]
Hexenes	Thermodynamics	Jeziorowski and Knözinger [1974]
Hexenes	Diffusion	Jeziorowski and Knözinger [1975]
Benzene	Thermodynamics	Jeziorowski [1972]
14. PuO_2		
Water	Adsorption isotherm	Stakebake and Steward [1973]
	Temperature programmed desorption	Stakebake [1973]
II. Mixed Oxides		
1. SiO_2–Al_2O_3		
Water	Adsorption isotherms	Morimoto et al. [1971]
	IR spectroscopy	Peri [1966b]
	Raman spectroscopy	Careri et al. [1972]
	NMR (wide line)	Kemarec et al. [1968]
	NMR (wide line)	Kodratoff et al. [1966]
Ammonia	IR spectroscopy	Cant and Little [1966]
	IR spectroscopy	Basila and Kantner [1967]
	IR spectroscopy	Peri [1966b]
	IR spectroscopy	Pichat et al. [1969]
	Thermodesorption	Pap et al. [1972]
Pyridine	IR spectroscopy	Parry [1963]
	IR spectroscopy	Basila et al. [1964]
Trimethylamine	IR spectroscopy	Basila et al. [1964]
Trimethylamine	Calorimetrical heats of adsorption	Cook and Ross [1972]
Formic acid	NMR (high resolution)	Fraissard et al. [1968]
2. SiO_2–Cr_2O_3		
Propio-, benzo-, acryl- } nitrile	IR spectroscopy	Zecchina et al. [1969]
III. Zeolites		
1. Zeolite A		
Water	Heat of adsorption	Meerson et al. [1972]
	Heat of adsorption	Jury and Horng [1973]
	Thermodynamics	Murdmaa and Serpinskii [1972]
	Heat capacity	Haly [1972]

TABLE 27.6–(*contd.*)

Adsorbate	Exp. method or property studied	References
	Heat capacity	Basler and Lechert [1972b]
	IR spectroscopy	Kunath and Moeller [1971]
	Dielectric properties	Lohse et al. [1971a, b]
	Dielectric properties	Matron et al. [1971]
	Thermal analysis	Vučelič et al. [1973]
	NMR (wide line)	Matyash et al. [1969]
	NMR (wide line)	Brekhunets et al. [1971]
	NMR (high resolution)	Šušić et al. [1969]
	NMR (pulsed)	Deininger et al. [1970]
	Length changes	Dubinin et al. [1972]
Ammonia	Heat of adsorption	Lohse et al. [1971c]
	Dielectric properties	Lohse et al. [1971c]
	NMR (wide line)	Eremenko et al. [1966]
	Slow neutron inelastic scattering	Downes et al. [1966]
Acetonitrile	Slow neutron inelastic scattering	Downes et al. [1966]
Methanol	Slow neutron inelastic scattering	Downes et al. [1966]
	NMR (high resolution)	Šušić et al. [1969]
2. Zeolite X		
Water	Heat of adsorption	Jansen and Schoonheydt [1970]
	Diff. heat of adsorption	Dubinin et al. [1973]
	Electrical conductivity	Dubinin et al. [1973]
	Heat capacity	Basler and Lechert [1972a]
	Heat capacity	Berezin et al. [1973]
	IR spectroscopy	Bosacek and Tvaruzkova [1972]
	IR spectroscopy	Kiselev et al. [1973]
	Dielectric properties	Matron et al. [1971]
	Dielectric properties	Benzar and Rimkevich [1973]
	NMR	Resing [1972]
	NMR (wide line)	Matyash et al. [1969]
	NMR (wide line)	Matyash et al. [1971]
	NMR	Resing and Thompson [1971]
	NMR	Resing and Murday [1973]
	NMR (pulsed)	Deininger et al. [1970]
	NMR (pulsed)	Gutsze et al. [1972]
	Length changes	Dubinin et al. [1972]
Ammonia	NMR (wide line)	Eremenko et al. [1966]
3. Zeolite Y		
Water	Heat of adsorption	Jansen and Schoonheydt [1970]

TABLE 27.6–(*contd.*)

Adsorbate	Exp. method or property studied	References
	Electrical conductivity	Jansen and Schoonheydt [1970]
	NMR	Resing and Murday [1973]
	NMR (pulsed)	Riedel et al. [1973]
	Length changes	Dubinin et al. [1972]
Acetonitrile	IR spectroscopy	Angell and Howell [1969]
Ammonia	NMR (double resonance)	Vedrine et al. [1972]
	NMR	Deininger and Reimann [1972]
Pyridine	NMR	Deininger and Reimann [1972]
	NMR (double resonance)	Vedrine et al. [1972]
5-cyano-1-pentene	IR spectroscopy	Butler and Poles [1973]
IV. Clays		
1. Montmorillonites		
Water	Adsorption isotherms	Calvet [1972, 1973]
	Adsorption isotherms	Tarasevich et al. [1971]
	Heat of immersion	Tarasevich et al. [1971]
	Thermodynamics	Hecht and Geissler [1972]
	Thermodynamics	Fripiat et al. [1965]
	Thermodynamics	De Busetti and Tschapek [1972]
	Electrical conductivity	De Busetti and Tschapek [1972]
	Electrical conductivity	Metsik et al. [1973]
	Dielectric behavior	Metsik et al. [1973]
	Dielectric behavior	Hoekstra and Doyle [1971]
	Viscosity	Linares and Huertas [1970]
	IR spectroscopy	Leonard [1971]
	IR spectroscopy	Tarasevich [1971]
	IR spectroscopy	Fripiat et al. [1969]
	IR spectroscopy	Swoboda and Kunze [1968]
	NMR	Mortland et al. [1963]
	NMR	Toullaux et al. [1967]
	NMR	Hecht and Geissler [1972]
	X-ray	Low et al. [1971b]
Methanol	Adsorption capacity	Sharkina and Tarasevich [1973]
	X-ray	Sharkina and Tarasevich [1973]
Isopropanol	IR spectroscopy	Radul and Ovcharenko [1971]
	X-ray	Radul and Ovcharenko [1971]
n-alcohols C_1–C_{10}	X-ray	German and Harding [1971]
n-alcohols C_6–C_{20}	X-ray	Brindley and Ray [1964]

TABLE 27.6-(*contd.*)

Adsorbate	Exp. method or property studied	References
n-alcohols C_{10}–C_{18}	X-ray	Lagaly and Weiss [1971]
Acetamide	IR spectroscopy	Tahoun [1971]
Amides (primary, secondary, tertiary)	IR spectroscopy	Tahoun and Mortland [1966]
Acetonitrile	Adsorption capacity	Sharkina and Tarasevich [1973]
	X-ray	Sharkina and Tarasevich [1973]
Ethylenediamine	Adsorption capacity	Cloos and Laura [1972]
Amines (primary C_6–C_{20})	X-ray	Brindley [1965]
Fatty acids (C_2–C_{20})	X-ray	Brindley and Moll [1965]
2. Kaolinites		
Water	Thermodynamics	Jakobs [1970]
	Thermodynamics	Sklyankin and Kireev [1972]
	Dielectric behavior	Nelson et al. [1969]
	Dielectric behavior	Baron [1970]
	Electrical conductivity	Baron [1970]
	IR spectroscopy	Gribina and Tarasevich [1971]
Ethanol	Dielectric behavior	Nelson et al. [1969]
Acetone	Dielectric behavior	Nelson et al. [1969]
Amides:		
Formamide, *N*-methylformamide, Dimethylformamide	IR spectroscopy	Olejnik et al. [1971a]
Acetamide, *N*-methylacetamide, Dimethylacetamide	IR spectroscopy	Olejnik et al. [1971b]
3. Vermiculites		
Water	Adsorption isotherms	Tarasevich and Rudenko [1971]
	Adsorption isotherms	Hougardy et al. [1970]
	Heats of immersion	Hougardy et al. [1970]
	X-ray	Hougardy et al. [1970]
	NMR (pulsed)	Hougardy et al. [1970]
	IR spectroscopy	Hougardy et al. [1970]
n-alcohols C_{10}–C_{18}	X-ray	Lagaly and Weiss [1971]
Amines (primary C_6–C_{20})	X-ray	Brindley [1965]

TABLE 27.6-(*contd.*)

Adsorbate	Exp. method or property studied	References
4. Miscellaneous Clays		
Water	Viscosity	Linares and Huertas [1970]
	Thermogravimetry	Escoubes et al. [1972a]
	Calorimetry	Escoubes et al. [1972a]
	Calorimetry	Escoubes et al. [1972b]
	Adsorption capacity	Escoubes et al. [1972b]
	Dielectric properties	Escoubes et al. [1972b]
	Dielectric properties	Fripiat [1971b]
	Dielectric properties	Forslind and Jakobsen [1971]
	NMR	Forslind and Jakobsen [1971]
V. Salts		
1. NaCl		
Water	Adsorption isotherms	Kaiho et al. [1973]
HCl, HBr, HJ	IR spectroscopy	Smart and Sheppard [1971]
2. NaBr, NaJ		
HCl, HBr, HJ	IR spectroscopy	Smart and Sheppard [1971]
3. Li- and Cs-halides:		
HCl, HBr, HJ	IR spectroscopy	Smart and Sheppard [1971]
4. KCl		
Aliphatic amines	Adsorption isotherms	Aleksandrovich et al. [1971]
5. KBr, KJ		
Water	Adsorption isotherms	Kaiho et al. [1973]
6. AgJ		
Water	Isosteric heats	Fukuta and Paik [1973]
	IR spectroscopy	Morachevskii et al. [1973

cited in table 27.6. The adsorbate, the experimental method applied or the physical property studied and the corresponding reference is given in columns 1 to 3 of table 27.6 for the various adsorbents.

The relevant literature regarding particularly the application of dielectric methods to the study of adsorbed phases is reviewed in ch. 20.

Notes added in proof

[1] This value was redetermined recently by Stählin and Knözinger [1974] and found to be 7300 cm^{-1}. The corresponding OD stretching overtone is at 5450 cm^{-1}. Davydov et al. [1973] reported a value of 5430 cm^{-1} for this overtone.

[2] Griffiths et al. [1974], when studying the adsorption of acetone from carbontetrachloride solution, also reached the conclusion, that H-bonded neighboring silanol groups are important adsorption sites. These authors suggest two modes of adsorption of acetone. The weaker adsorption involves a single H-bond interaction between isolated silanol groups and acetone, and leads to spectral shifts $\Delta\bar{\nu}_{OH} = 306\ cm^{-1}$ and $\Delta\bar{\nu}_{CO} = 15\ cm^{-1}$. The stronger adsorption which most favorably occurs on fully hydroxylated surfaces, leads to wave number shifts $\Delta\bar{\nu}_{OH} = 246\ cm^{-1}$ and $\Delta\bar{\nu}_{CO} = 30\ cm^{-1}$ and is assigned as a species in which acetone is engaged in H-bonding with two silanol groups. These are assumed to have been mutually H-bonded before interaction with acetone. Analogous species have been described by Bertoluzza et al. [1966] and by Hertl and Hair [1969] for the interaction of acetone with silica from the gas phase.

[3] Additional recent calorimetric and IR spectroscopic data may be found in a paper by Curthoys et al. [1974].

[4] Griffiths et al. [1974] observed a decrease of the wave number shift $\Delta\bar{\nu}_{OH}$ with increasing coverage on adsorption of pyridines from carbon tetrachloride solution. They explained this observation by a solvent exclusion effect, which leads to a sterically restricted approach of solvent around the interacting silanol groups and acceptor molecules.

[5] Griffiths et al. [1974] recently reported a fairly linear correlation between the wavenumber shifts $\Delta\bar{\nu}_{OH}$ and the pK_a values of 8 substituted pyridines in CCl_4-solution.

[6] Kondo and Mūroya [1969] assigned the low temperature dispersion to a proton diffusion process in chains of H-bonded surface silanol groups.

[7] Stählin and Knözinger [1974] have very recently shown, that the above interpretation of the IR spectra of adsorbed alcohols on silica surfaces on the basis of Sandorfy's discussion of solution spectra has to be disregarded. If the assignment given in table 27.4 of the band at 3485 cm^{-1} as the OH stretching fundamental of the anchoring silanols and of the band at 3330 cm^{-1} as the corresponding vibration in H-bonded alcohol clusters was correct, the adsorption of a sterically hindered alcohol such as 2,2,4-

trimethyl-2-pentanol should only produce the band around 3500 cm^{-1}. On the contrary, a broad asymmetric band centered at 3330 cm^{-1} is observed on adsorption, while a much narrower band is observed at 3517 cm^{-1} in CCl_4-solution. Nevertheless, all experimental evidence is still in accord with the cluster model of the adsorbed phase of sterically less hindered alcohols. At the present status of our work we must assume that the different IR bands in the OH stretching region have to be related to the possible interaction of the alcohol molecules with either two vicinal OH groups or with geminal and isolated groups (see section 27.2.2). The isosteric heats of adsorption of *tert.* butanol on silica surfaces outgassed at 200 and 500°C are comparable to the heat of liquefaction in the temperature range between 50 and 150°C. This suggests the formation of clusters and species held by two H-bonds such as structures XVI and XVII. At least part of the adsorbed molecules must be mobile as shown by entropy considerations. On a silica surface outgassed at 700°C, the isosteric heat of adsorption drops to about half the heat of liquefaction at temperatures below 70°C (see also Sewell and Morgan [1969]). A 1 : 1 interaction is thus highly probable and the molecules in this adsorption state are again at least partly mobile. At temperatures higher than 70°C, the isosteric heat of adsorption is again comparable to the heat of liquefaction and the alcohol molecules in this state are less mobile. As a possible interpretation, one may assume the existence of isolated OH groups and of a certain number of vicinal OH groups (structure II) which interact strongly with each other. Alcohol molecules can then successfully attack these groups only at sufficiently high temperatures to form species XVII. Spectroscopic studies in the IR fundamental and overtone range, dielectric and NMR studies on these systems are in progress.

[8)] Structure XIX for the adsorption of alcohols on silica and the discussed dynamic behavior of these systems found some support by recent pulse deuterium magnetic resonance studies by Seymour et al. [1973]. These authors also consider the possibility of a preferred orientation even on the surface of an amorphous porous material. This preferred orientation is suggested to be a result of some regular arrangements of surface silanol groups and of the H-bond formation between these hydroxyls and clusters of partially ordered adsorbed species. Furthermore, Seymour et al. [1973] conclude that the observation of a decreased deuterium quadrupole coupling constant (90 kHz as compared to 200 kHz for the neutral CH_3OD) is good evidence for the existence of the charged species $CH_3OD_2^+$, which corresponds to the $CH_3OH_2^+$ species in structure XXI.

References

Abramov, V. N., A. V. Kiselev and V. I. Lygin, 1963, Zh. Fiz. Khim. **37**, 2783.
Abramov, V. N., A. V. Kiselev and V. I. Lygin, 1964a, Zh. Fiz. Khim. **38**, 1045.
Abramov, V. N., A. V. Kiselev and V. I. Lygin, 1964b, Zh. Fiz. Khim. **38**, 1867.
Aleksandrovich, Kh. M., S. S. Gusev and E. F. Korshuk, 1971, Zh. Prikl. Khim. **44**, 1188; **CA75**, 91519.
Anderson, Jr., J. H., 1965, Surface Sci. **3**, 290.
Anderson, D. M., 1967, J. Colloid Interface Sci. **25**, 174.
Anderson, J. H. and G. A. Parks, 1968, J. Phys. Chem. **72**, 3662.
Anderson, Jr., J. H. and K. A. Wickersheim, 1964, Surface Sci. **2**, 252.
Anderson, P. J., R. F. Horlock and J. F. Oliver, 1965, Trans. Faraday Soc. **61**, 2754.
Angell, C. L. and M. V. Howell, 1969, J. Phys. Chem. **73**, 2551.
Antipina, T. V., G. D. Chukin and O. F. Kirina, 1972, Russ. J. Phys. Chem. **46**, 1663.
Antoniou, A. A., 1964, J. Phys. Chem. **68**, 2754.
Arai, H., Y. Saito and Y. Yoneda, 1967, Bull. Chem. Soc. Japan **40**, 731.
Armistead, C. G. and J. A. Hockey, 1967, Trans. Faraday Soc. **61**, 2549.
Armistead, C. G., A. J. Tyler, F. H. Hambleton, S. A. Mitchell and J. A. Hockey, 1969, J. Phys. Chem. **73**, 3947.
Atherton, K., G. Newbold and J. A. Hockey, 1971, Disc. Faraday Soc. **52**, 33.
Avgul, N. N., A. V. Kiselev and I. A. Lyzina, 1961, Kolloid Zh. **23**, 513.
Baldwin, M. C. and J. C. Morrow, 1962, J. Chem. Phys. **36**, 1591.
Baron, A. 1970, Acta Univ. Carolinae Geol. 45.
Barto, J., J. L. Durham, K. F. Baston and W. H. Wade, 1966, J. Colloid Interface Sci. **22**, 491.
Basila, M. R., 1961, J. Chem. Phys. **35**, 1151.
Basila, M. R., 1962, J. Phys. Chem. **66**, 2223.
Basila, M. R. and T. R. Kantner, 1967, J. Phys. Chem. **71**, 467.
Basila, M. R., T. R. Kantner and K. H. Rhee, 1964, J. Phys. Chem. **68**, 3197.
Basler, W. D. and H. Lechert, 1972a, Z. Physik. Chem. (Frankfurt) **78**, 199.
Basler, W. D. and H. Lechert, 1972b, Ber. Bunsenges. Physik. Chem. **76**, 1234.
Basset, D. R., E. A. Boucher and A. C. Zettlemoyer, 1970, J. Colloid Interface Sci. **34**, 436.
Bauer, E. and M. Magat, 1938, J. Phys. Radium, Paris, **9**, 319.
Bavarez, M. and J. Bastick, 1969, J. Chim. Phys. **66**, 935.
Beckert, D., 1971, Z. Physik. Chem. (Leipzig), **246**, 113.
Beliakova, L. D. and A. V. Kiselev, 1950, Dokl. Akad. Nauk. SSSR **70**, 441.
Bellamy, L. J., and R. J. Pace, 1969, Spectrochim. Acta **25A**, 319.
Bellamy, L. J., H. E. Hallam and R. L. Williams, 1958, Trans. Faraday Soc. **54**, 1120.
Bellamy, L. J., A. R. Osborn, E. R. Lippincott and A. R. Bandy, 1969, Chem. Ind. 686.
Benesi, H. A. and A. C. Jones, 1959, J. Phys. Chem. **63**, 179.
Benzar, K. K. and I. M. Rimkevich, 1973, Russ. J. Phys. Chem. **47**, 435.
Berezin, G. I., A. V. Kiselev and V. A. Sinitsin, 1963, Zh. Fiz. Khim. **37**, 325.
Berezin, G. I., A. V. Kiselev, A. A. Kozlov and L. V. Kuznetsova, 1970, Russ. J. Phys. Chem. **44**, 879.
Berezin, G. I., A. V. Kiselev and V. A. Sinitsin, 1973, JCS Faraday I, **69**, 614.
Bermudez, V. M., 1970, J. Phys. Chem. **74**, 4160.
Bermudez, V. M., 1971, J. Phys. Chem. **75**, 3249.

Bertoluzza, A., G. B. Bonino, G. Fabbri and V. Lorenzelli, 1966, J. Chim. Phys. **63**, 395.
Blake, T. D. and W. H. Wade, 1971, J. Phys. Chem. **75**, 1887.
Blake, T. D. and W. H. Wade, 1972, J. Phys. Chem. **76**, 675.
Blake, T. D., J. L. Cayias, W. H. Wade and J. A. Zerdecki, 1971, J. Colloid Interface Sci. **37**, 678.
Bliznakov, G. and J. Polikarova, 1966, J. Catal, **5**, 18.
Blyholder, G. and E. A. Richardson, 1964, J. Phys. Chem. **68**, 3882.
Boddenberg, B., R. Haul and G. Oppermann, 1969, Naturwisschaften **56**, 635.
Boddenberg, B., R. Haul and G. Oppermann, 1972, J. Colloid Interface Sci. **38**, 210.
Boehm, H. P., 1966a, Adv. Catalysis **16**, 179.
Boehm, H. P., 1966b, Angew. Chemie **78**, 617.
Boehm, H. P., 1968, Kolloid Z. u. Z. Polymere **227**, 17.
Boehm, H. P., M. Schneider and F. Arendt, 1963, Z. Anorg. Allg. Chem. **320**, 43.
Bonardet, J. L. and J. Fraissard, 1973, Ind. Chim. Belge **38**, 370.
Borello, E., A. Zecchina and C. Morterra, 1963–64, Atti Accad. Sci. Torino A **98**, 1139.
Borello, E., A. Zecchina and C. Morterra, 1967a, J. Phys. Chem. **71**, 2938.
Borello, E., A. Zecchina, C. Morterra and G. Ghiotti, 1967b, J. Phys. Chem. **71**, 2945.
Bosacek, V. and Z. Tvaruzkova, 1972, Collect. Czech. Chem. Commun. **37**, 513.
Bourderon C. and C. S. Sandorfy, 1973, J. Chem. Phys. **59**, 2527.
Boutin, H. and H. Prask, 1964, Surface Sci. **2**, 261.
Boutin, H., H. Prask and R. D. Iyengar, 1967, J. Catal. **9**, 309.
Boutin, H., H. Prask and R. D. Iyengar, 1968, Adv. Colloid Interface Sci. **2**, 1.
Boyle, F. W., W. J. Gaw and R. A. Ross, 1965, J. Chem. Soc. 240.
Brekhunets, A. G., V. V. Mauk, F. D. Ovcharenko, M. A. Piontkovskaya and G. I. Denisenko, 1971, Teor. Eksp. Khim. **7**, 125; **CA75**, 53615j.
Brey, W. S. and K. D. Lawson, 1964, J. Phys. Chem. **68**, 1474.
Brindley, G. W., 1965, Clay Miner. **6**, 61.
Brindley, G. W. and W. F. Moll, Jr., 1965, Am. Mineralogist **30**, 1355.
Brindley, G. W. and S. Ray, 1964, Amer. Mineralogist **49**, 106.
Brodskii, I. A. and A. E. Stanevich, 1973, Kinetika i Kataliz **14**, 463.
Brodskii, I. A., A. E. Stanevich and N. G. Yaroslavskii, 1970, Russ. J. Phys. Chem. **44**, 998.
Butler, J. D. and T. C. Poles, 1973, JCS, Perkin Trans. **2**, 48.
Calvet, R., 1972, Bull. Soc. Chim. France 3097.
Calvet, R., 1973, Ann. Agron. **24**, 77; **CA79**: 10174e.
Camara, B., H. Dunken and P. Fink, 1968, Z. Chem. **8**, 155.
Cant, N. W. and L. H. Little, 1964, Can. J. Chem. **42**, 802.
Cant, N. W. and L. H. Little, 1965, Can. J. Chem. **43**, 1252.
Cant, N. W. and L. H. Little, 1966, Nature **211**, 69.
Cant, N. W., and L. H. Little, 1967, Can. J. Chem. **45**, 3055.
Cant, N. W., and L. H. Little, 1968, Can. J. Chem. **46**, 1373.
Careri, G., A. Mazzacurati, M. Sampoli and G. Signorelli, 1972, J. Catal. **26**, 494.
Carter, J. L., P. J. Lucchesi, P. Corneil, D. J. C. Yates and J. H. Sinfelt, 1965, J. Phys. Chem. **69**, 3070.
Chapman, I. D. and M. L. Hair, 1963, J. Catal. **2**, 145.
Chapman, I. D. and M. L. Hair, 1965, Trans. Faraday Soc. **61**, 1507.
Chikazawa, M., M. Kaiho and T. Kanazawa, 1972, Nippon Kagaku Kaishi **8**, 1339.

Chuang, T. T. and J. G. Dalla Lana, 1972, JCS Faraday I **68**, 773.
Clement, G., H. Knözinger and W. Stählin, 1973, unpublished data.
Cloos, P. and R. D. Laura, 1972, Clays Clay Miner. **20**, 259.
Conway, B. E., J. O. M. Bockris and H. J. Linton, 1956, J. Chem. Phys. **24**, 832.
Cook, W. G. and R. A. Ross, 1972, Canad. J. Chem. **50**, 2451.
Cook, M. A., R. O. Daniels and J. H. Hamilton, 1954, J. phys. Chem. **58**, 358.
Cornelius, E. B., T. H. Milliken, G. A. Mills and A. G. Oblad, 1955, J. Phys. Chem. **59**, 809.
Cruz, M. I., W. E. E. Stone and J. J. Fripiat, 1972a, J. Phys. Chem. **76**, 3078.
Cruz, M. I., L. Van Cangh and J. J. Fripiat, 1972b, Bull. Acad. Roy. Belg. (Classe Sci.) **58**, 439.
Cruz, M. I., K. Verdinne, J. Andrè and J. J. Fripiat, 1973, An. Quim (Madrid) **69**, 895.
Curthoys, G., V. Ya. Davydov, A. V. Kiselev, S. A. Kiselev and B. V. Kuznetsov, 1974, J. Colloid Interface Sci. **48**, 58.
Cusumano, J. A. and M. J. D. Low, 1970a, J. Phys. Chem. **74**, 1950.
Cusumano, J. A. and M. J. D. Low, 1970b, J. Phys. Chem. **74**, 792.
Cusumano, J. A. and M. J. D. Low, 1971, J. Catal. **23**, 214.
Cusumano, J. A. and M. J. D. Low, 1972, J. Colloid Interf. Sci. **38**, 245.
Davis, R. E., 1972a, in: Electron Spectroscopy, Ed. D. A. Shirley (North Holland Publ. Comp., Amsterdam), p. 903.
Davis, R. E., 1972b, in: Electron Spectroscopy, Ed. D. A. Shirley (North Holland Publ. Comp., Amsterdam), p. 909.
Davydov, V. Ya., A. V. Kiselev and V. I. Lygin, 1962, Dokl. Akad. Nauk SSSR **147**, 131.
Davydov, V. Ya., A. V. Kiselev and V. I. Lygin, 1963a, Kolloid Zh. **25**, 152.
Davydov, V. Ya., A. V. Kiselev and V. I. Lygin, 1963b, Russ. J. Phys. Chem. **37**, 243.
Davydov, V. Ya., L. T. Zhuravlev and A. V. Kiselev, 1964a, Zhur. Fiz. Khim. **38**, 2047.
Davydov, V. Ya., A. V. Kiselev and L. T. Zhuravlev, 1964b, Trans. Faraday Soc. **60**, 2254.
Davydov, V. Ya., A. V. Kiselev and B. V. Kuznetsov, 1965, Zh. Fiz. Khim. **39**, 2058.
Davydov, V. Ya., A. V. Kiselev and B. V. Kuznetsov, 1970, Russ. J. Phys. Chem. **44**, 1.
Davydov, V. Ya., A. V. Kiselev, V. A. Lokutsievskii and V. I. Lygin, 1973, Russ. J. Phys. Chem. **47**, 460.
Dawber, J. G., L. B. Guest and R. Lambourne, 1972, Thermochimica Acta **4**, 471.
Dawson, P. T., 1967, J. Phys. Chem. **71**, 838.
Day, R. E. and G. D. Parfitt, 1968, Trans. Faraday Soc. **64**, 815.
Day, R. E., G. D. Parfitt and J. Peacock, 1971, Disc. Faraday Soc. **52**, 215.
De Boer, J. H. and J. M. Vleeskens, 1958, Proc. Koninkl. Ned. Akad. Wetenschap. B, **61**, 2.
De Boer, J. H., J. M. H. Fortuin, B. C. Lippens and W. H. Meijs, 1963. J. Catal. **2**, 1.
De Busetti, S. G. and M. Tschapek, 1972, Geoderma **8**, 281; **CA78**, 102763d.
Deininger, D. and B. Reimann, 1972, Z. Physik. Chem. (Leipzig) **251**, 353.
Deininger, D., H. Pfeifer, F. Przyborowski, W. Schirmer and H. Stach, 1970, Z. Phys. Chem. (Leipzig) **245**, 68.
De La Hardrouyère, G., 1960, Arch. Sci. (Genève) **13**, 1.
Della Gatta, G., B. Fubini and G. Venturello, 1973, J. Chim. Phys. **70**, 64.
Deo, A. V. and J. G. Dalla Lana, 1969, J. Phys. Chem. **71**, 716.
Deo, A. V., T. T. Chuang and J. G. Dalla Lana, 1971, J. Phys. Chem. **75**, 234.
Derouane, E. G., 1969a, Bull. Soc. Chim. Belg. **78**, 111.
Derouane, E. G., 1969b, Bull. Soc. Chim. Belg. **78**, 101.

Deryagin, B. V. and N. N. Fredyakin, 1962, Dokl. Akad. Nauk SSSR **147**, 2403.

Deryagin, B. V. and N. V. Churaev, 1973, Nature (London) **244**, 430.

Deryagin, B. V., N. V. Churaev, N. N. Fredyakin, M. V. Talayev and I. G. Ershova, 1967, Izv. Akad. Nauk SSSR. Ser. Khim. **10**, 2178.

Di Paolo, Th., C. Bourderon and C. Sandorfy, 1972, Can. J. Chem. **50**, 3161.

Downes, J. S., J. W. White, P. A. Egelstaff and V. S. Rainey, 1966, Phys. Rev. Letters **17**, 533.

Dransfeld K., H. L. Frish and H. L. Wood, 1962, J. Chem. Phys. **36**, 1574.

Dubinin, M. M., A. I. Sakharov and V. F. Kononyuk, 1972, Dokl. Akad. Nauk. SSSR **205**, 901.

Dubinin, M. M., A. A. Isirikyan, G. U. Rakhmatkariev and V. V. Serpinskii, 1973, Izv. Akad. Nauk. SSSR. Ser. Khim. 934.

Dugger, D. L., J. H. Stanton, B. N. Irby, B. L. McConnell, W. W. Cummings and R. W. Maatman, 1964, J. Phys. Chem. **68**, 757.

Dunken, H. and P. Fink, 1966, Z. Chem. **6**, 194.

Dunken, H. and P. Fink, 1967, Acta Chim. Akad. Sci. Hung. **53**, 179.

Dzhigit, O. M., A. V. Kiselev and G. G. Muttik, 1961, Kolloid Zhur. **23**, 553.

Ebert, G. and G. Langhammer, 1961, Kolloid Z. u. Z. Polymere **174**, 5.

Ebert, G., W. Matron and F. H. Müller, 1971, Fortschr. Kolloid u. Polymere **55**, 74.

Egorov, M. M., N. N. Dobrovolskii, V. F. Kiselev, G. Furman and S. V. Khrustaleva, 1965, Zh. Fiz. Khim. **39**, 3070.

Egorov, M. M., V. J. Kvlividze, V. F. Kiselev and K. G. Krassilnikov, 1966, Kolloid Z. u. Z. Polymere **212**, 126.

Elkington, P. A. and G. Curthoys, 1968, J. Phys. Chem. **72**, 3475.

Enriquez, M. A., J. Fraissard and B. Imelik, 1970, C.R. Acad. Sci. Paris, Ser. **C270**, 1921.

Eremenko, A. M., M. A. Piontkovskaya, I. V. Matyash, V. V. Mank, M. G. Starkov and I. E. Neimark, 1966, Zh. Struct. Khim. **7**, 106.

Escoubes, M., J. F. Quinson, J. Gielly and M. Murat, 1972a, Bull. Soc. Chim. France 1689.

Escoubes, M., J. Gielly and J. F. Quinson, 1972b, Bull. Soc. Franc. Ceram. **94**, 3.

Felden, M., 1959, Compt. Rend. **249**, 682.

Feldman, U. and M. Folman, 1964, Trans. Faraday Soc. **60**, 440.

Fiat, D., M. Folman and U. Garbatski, 1961, Proc. Roy. Soc. (London) **A260**, 409.

Fiat, D., J. Reuben and M. Folman, 1967, J. Chem. Phys. **46**, 4453.

Filimonov, V. N., 1956, Optika Spektrosk. **1**, 490.

Filimonov, V. N. and A. N. Terenin, 1956, Dokl. Akad. Nauk SSSR **109**, 982.

Fink, P. and B. Camara, 1972, Z. Chem. **12**, 35.

Fink, P., B. Camara, E. Wolleschensky, E. Kellner and P. Welz, 1970, Z. Chem. **10**, 474.

Fink, P., B. Camara, P. Welz and Pham Dinh Ty, 1971, Z. Chem. **11**, 473.

Fink, P., W. Pohle and A. Köhler, 1972, Z. Chem. **12**, 117.

Fletcher, A. N., 1971, J. Phys. Chem. **75**, 1808.

Folman, M. and D. J. C. Yates, 1958a, Proc. Roy. Soc. (London) **A246**, 32.

Folman, M. and D. J. C. Yates, 1958b, Trans. Faraday Soc. **54**, 429.

Folman, M. and D. J. C. Yates, 1958c, Trans. Faraday Soc. **54**, 1684.

Folman, M. and D. J. C. Yates, 1959, J. Phys. Chem. **63**, 183.

Forslind, E. and A. Jakobsen, 1971, **CA75**, 10656f.

Fraissard, J., S. Bielikoff and B. Imelik, 1968, IV. Intern. Congr. Catalysis, Moscow, Paper No. 1, Symp. Mechanism and Kinetics of Complex Catalytic Reactions.

French, T. M. and G. A. Samorjai, 1970, J. Phys. Chem. **74**, 2489.
Freude D., 1969, Z. Physik. Chem. (Leipzig) **242**, 57.
Freude D., 1971, Z. Physik. Chem. (Leipzig) **247**, 209.
Freude, D., H. Pfeifer and H. Winkler, 1966, Z. Physik. Chem. (Leipzig) **232**, 276.
Freyman, M. and R. Freyman, 1954, J. Phys. Radium **15**, 165.
Fripiat, J. J., 1964, Proc. 12th Nat. Clay Conf. (Pergamon Press, London), p. 327.
Fripiat, J. J., 1971a, Bull. Acad. Roy. Belg. **57**, 1188.
Fripiat, J. J., 1971b, Bull. Groupe Franc. Argiles **23**, 1.
Fripiat, J. J., 1973, Ind. Chim. Belg. **38**, 404.
Fripiat, J. J. and J. Uytterhoeven, 1962, J. Phys. Chem. **66**, 800.
Fripiat, J. J. and M. van Tongelen, 1965, J. Catal. **5**, 158.
Fripiat, J. J., M. C. Gastuche and R. Brichard, 1962, J. Phys. Chem. **66**, 805.
Fripiat, J. J., A. Jelli, G. Poncelet and J. Andrè, 1965, J. Phys. Chem. **69**, 2185.
Fripiat, J. J., M. Pennequin, G. Poncelet and P. Cloos, 1969, Clay. Miner. **8**, 119.
Fripiat, J. J., C. van der Meersche, R. Touillaux and A. Jelli, 1970, J. Phys. Chem. **74**, 382.
Frohnsdorff, G. J. C. and G. L. Kington, 1959, Trans. Faraday Soc. **55**, 1173.
Fukuta, N. and Y. Paik, 1973, J. Appl. Phys. **44**, 1092.
Fuller, Jr., E. L., H. F. Holmes, C. H. Secoy and J. E. Stuckey, 1968, J. Phys. Chem. **72**, 573.
Fuller, Jr., E. L., H. F. Holmes and R. B. Gammage, 1970, J. Colloid Interface Sci. **33**, 623.
Galkin, G. A., A. V. Kiselev and V. I. Lygin, 1964, Trans. Faraday Soc. **60**, 431.
Gammage, R. B., W. S. Brey, Jr. and B. H. Davis, 1970, J. Colloid Interface Sci. **32**, 256.
German, W. L. and D. A. Harding, 1971, Clay. Miner. **9**, 167.
Geschke, D., 1969, Z. Physik. Chem. (Leipzig) **242**, 74.
Geschke, D. and H. Pfeifer, 1966, Z. Physik. Chem. (Leipzig) **232**, 127.
Ginsberg, H., W. Hüttig and G. Strunk-Lichtenberg, 1957, Z. Anorg. Allg. Chem. **293**, 33, 204.
Glemser, O. and E. Hartert, 1958, Z. Anorg. Allg. Chem. **297**, 175.
Gradsztajn, S., D. Vivien and J. Conrad, 1971, J. Phys. Chem. Solids, **32**, 1507.
Greenler, R. G., 1962, J. Chem. Phys. **37**, 2094.
Gribina, I. A. and Yu. I. Tarasevich, 1971, Geokhimia, 878; **CA75**, 145796x.
Griffiths, D. M., K. Marshall and C. H. Rochester, 1974, J. C. S. Faraday I **70**, 441.
Gutsze, A., D. Deininger, H. Pfeifer, G. Finger, W. Schirmer and H. Stach, 1972, Z. Phys. Chem. (Leipzig) **249**, 383.
Hair, M. L., 1967, Infrared Spectroscopy in Surface Chemistry, (M. Dekker, Inc., New York).
Hair, M. L. and W. Hertl, 1969, J. Phys. Chem. **73**, 2372.
Hair, M. L. and W. Hertl, 1970, J. Phys. Chem. **74**, 91.
Hair, M. L. and W. Hertl, 1973, J. Phys. Chem. **77**, 1965.
Hall, W. K., H. P. Leftin, F. J. Cheselske and D. E. O'Reilly, 1963, J. Catal. **2**, 506.
Hallam, H. E., 1969, J. Catal. **14**, 104.
Haly, A. R., 1972, J. Phys. Chem. Solids **33**, 129.
Hambleton, F. A. and J. A. Hockey, 1966, Trans. Faraday Soc. **62**, 1694.
Hankins, D., J. W. Moskowitz and F. H. Stillinger, 1970, J. Chem. Phys. **53**, 4544.
Haul, R. and B. Boddenberg, 1968, IV. Intern. Congr. Catal., Moscow, Paper No. 17, Symp. "The Porous Structure of Catalysts and the Role of Transport Processes in Heterogeneous Catalysts."
Haul, R. and B. Boddenberg, 1969, Z. Phys. Chem. (Frankfurt) **64**, 78.

Hecht, A. M. and E. Geissler, 1972, Bull. Soc. Franc. Mineral. Cristallogr. **95**, 291.
Hendriksen, B. A., D. R. Pearce and R. Rudham, 1972, J. Catal. **24**, 82.
Hertl, W. and M. L. Hair, 1968, J. Phys. Chem. **72**, 4676.
Hertl, W. and M. L. Hair, 1969, J. Phys. Chem. **73**, 4269.
Hill, T. L., 1946, J. Chem. Phys. **14**, 441.
Hino, M. and T. Sato, 1971, Bull. Chem. Soc. Japan **44**, 33.
Hockey, J. A., 1970, J. Phys. Chem. **74**, 2570.
Hoekstra, P. and W. T. Doyle, 1971, J. Colloid Interface Sci. **36**, 513.
Holmes, H. F., E. L. Fuller, Jr. and C. H. Secoy, 1966, J. Phys. Chem. **70**, 436.
Holmes, H. F., E. L. Fuller, Jr. and C. H. Secoy, 1968a, J. Phys. Chem. **72**, 2293.
Holmes, H. F., E. L. Fuller, Jr., R. B. Gammage and C. H. Secoy, 1968b, J. Colloid Interface Sci. **28**, 421.
Hougardy, J., J. M. Serratosa, W. Stone and H. Van Olphen, 1970, Spec. Disc. Faraday Soc. **1**, 187.
Huettenrauch, R., J. Jakob and E. Schupp, 1971, Pharmazie **26**, 438.
Huggins, G. M., 1961, J. Phys. Chem. **65**, 1881.
Husbands, D. I., W. Tallis, J. C. R. Waldsax, C. R. Woodings and M. J. Jaycock, 1971, Powder Technology **5**, 31.
Iizuka, T., H. Hattori, Y. Ohma, J. Sohma and K. Tanabe, 1971, J. Catal. **22**, 130.
Iwaki, T. and M. Miura, 1971, Bull. Chem. Soc. Japan **44**, 1754.
Iwaki, T., K. Komuro and M. Miura, 1972, Bull. Chem. Soc. Japan **45**, 2343.
Jackson, P. and G. D. Parfitt, 1971a, Trans. Faraday Soc. **67**, 2469.
Jackson, P. and G. D. Parfitt, 1971b, Kolloid Z. u. Z. Polymere **244**, 240.
Jackson, P. and G. D. Parfitt, 1972, JCS Faraday I **68**, 1443.
Jacobs, Th., 1970, Reunion Hispano-Belga Miner. Arcillas, An. 21.
Jansen, F. J. and R. A. Schoonheydt, 1970, Reunion Hispano-Belga Miner. Arcillas, An. 45.
Jeziorowski, H., 1972, Thesis, University of Munich.
Jeziorowski, H. and H. Knözinger, 1974, Z. Phys. Chem. (Frankfurt) **90**, 155.
Jeziorowski, H. and H. Knözinger, 1975, Ber. Bunsenges. Phys. Chem., in print.
Jeziorowski, H., H. Knözinger, W. Meye and H. D. Müller, 1973, JCS Faraday I **69**, 1744.
Jones, P. and J. A. Hockey, 1971, Trans. Faraday Soc. **67**, 2669.
Joris, J. and P. von Raguè-Schleyer, 1968, Tetrahedron **24**, 5991.
Jury, S. H. and J. S. Horng, 1973, AICHE J. **19**, 371.
Kagel, R. O., 1967, J. Phys. Chem. **71**, 844.
Kagel, R. O. and R. G. Greenler, 1968, J. Chem. Phys. **49**, 1638.
Kaiho, M., M. Chikazawa and T. Kanazawa, 1973, Nippon Kagaku Kaishi 914; **CA79**, 35357e.
Kamiyoshi, K., 1959, J. Phys. Radium **20**, 60.
Kämpf, G. and Kohlschütter, H. W., 1958, Z. Anorg. Allg. Chem. **249**, 10.
Karagounis, G., 1964, Nature **201**, 604.
Karagounis, G. and M. Gutbrod, 1967, Prakt. Akad. Athenon, p. 41.
Kemarec, J., J. Fraissard, J. Elston and B. Imelik, 1968, J. Chim. Phys. **65**, 920.
Kipling, J. J. and D. B. Peacall, 1957, J. Chem. Soc. 834.
Kirina, O. F., T. V. Antipina and G. D. Chukin, 1973, Russ. J. Phys. Chem. **47**, 248.
Kiselev, A. V., 1936, Kolloid Zh. **2**, 17.
Kiselev, A. V., 1964a, Zh. Fiz. Khim. **38**, 2753.

Kiselev, A. V., 1964b, Zh. Fiz. Khim. **38**, 2764.
Kiselev, A. V., 1965, Surface Sci. **3**, 292.
Kiselev, A. V., 1971, Disc. Faraday Soc. **52**, 14.
Kiselev, A. V. and E. V. Khrapova, 1957, Kolloid Zh. **19**, 572.
Kiselev, A. V. and V. I. Lygin, 1959, Kolloid Zh. **21**, 581.
Kiselev, A. V. and V. I. Lygin, 1960, Kolloid Zh. **22**, 403.
Kiselev, A. V. and V. I. Lygin, 1961a, Kolloid Zh. **23**, 157.
Kiselev, A. V. and V. I. Lygin, 1961b, Kolloid Zh. **23**, 574.
Kiselev, A. V. and V. I. Lygin, 1962, Russ. Chem. Rev. **31**, 175.
Kiselev, A. V. and V. I. Lygin, 1964, Surface Sci. **2**, 236.
Kiselev, A. V. and V. I. Lygin, 1972, Infracrasnie Spectry poverhnostnyh soedinenii, Nauka, Moscow.
Kiselev, A. V., V. I. Lygin and T. I. Titova, 1964, Zh. Fiz. Khim. **38**, 2730.
Kiselev, A. V., V. I. Lygin and R. V. Starodubtseva, 1973, JCS Faraday I **68**, 1793.
Klier, K., 1973, J. Chem. Phys. **58**, 737.
Klier, K., J. H. Shen and A. C. Zettlemoyer, 1973, J. Phys. Chem. **77**, 1458.
Knözinger, H., 1970, Z. Phys. Chem. (Frankfurt), **69**, 108.
Knözinger, H., 1973, unpublished data.
Knözinger, H., 1974, Surface Sci. **41**, 339.
Knözinger, H. and E. Ress, 1968, Z. Phys. Chem. (Frankfurt) **59**, 49.
Knözinger, H., E. Ress and H. Bühl, 1967, Naturwiss. **54**, 516.
Knözinger, H. and H. Jeziorowski, 1970, Surface Sci. **22**, 111.
Knözinger, H. and H. Stolz, 1970, Ber. Bunsenges. Phys. Chem. **74**, 1056.
Knözinger, H. and H. Stolz, 1971, Fortschr. Kolloide u. Polymere **55**, 16.
Knözinger, H., H. Bühl and E. Ress, 1968, J. Catal. **12**, 121.
Kodratoff, Y., G. Dalmai, C. Naccache and M. Prettre, 1966, C.R. Acad. Sci., Paris, Ser. **C.263**, 509.
Kondo, S. and M. Mūroya, 1969, Bull. Chem. Soc. Japan **42**, 1165.
Kondo, S., M. Mūroya, H. Fujiwara and N. Yamaguchi, 1973, Bull. Chem. Soc. Japan **46**, 1362.
Kozyrovski, Y. and M. Folman, 1964, Trans. Faraday Soc. **60**, 1532.
Kubelkovà, L. and P. Jirů, 1972, Collect. Czech. Chem. Commun. **37**, 2853.
Kubelkovà, L., P. Schürer and P. Jirů, 1969, Surface Sci. **18**, 245.
Kunath, D., 1971, Z. Chem. **11**, 471.
Kunath, D., 1972, Z. Phys. Chem. (Leipzig), **249**, 401.
Kunath, D. and K. Moeller, 1971, Exp. Tech. Phys. **19**, 295.
Kurbatov, L. N. and G. G. Neuimin, 1949, Dokl. Akad. Nauk SSSR **68**, 341.
Kurosaki, S., S. Saito and G. Sato, 1955, J. Chem. Phys. **23**, 1846.
Kvlividze, V. J., 1964, Dokl. Akad. Nauk. SSSR. **157**, 158.
Kvlividze, V. J., 1971, in: Magn. Resonance Rel. Phenomena, Proc. Congr. Ampère, 16th, 1970; Ed. I. Ursu, (Publ. House Acad. Soc. Repub. Rom., Bucharest, Rom.), p. 390.
Kvlividze, V. J., N. M. Ijevskaja, T. S. Egorova, V. F. Kiselev and N. D. Sokolov, 1962, Kinetika i Kataliz **3**, 91.
Kvlividze, V. J., R. A. Brants, V. F. Kiselev and G. M. Bliznakov, 1969, J. Catal. **13**, 255.
Lagaly, G. and A. Weiss, 1971, Kolloid Z. u. Z. Polymere **248**, 968.
Le Bot, J. and S. Le Montagner, 1955, J. Phys. Radium **16**, 79, 163.

Lee, J. A., C. A. Newnham and F. L. Tye, 1973, J. Colloid Interface Sci. **42**, 372.
Leonard R. A., 1971, Spectrosc. Lett. **4**, 267.
Leonard, A. J., F. Van Cauwelaert and J. J. Fripiat, 1967, J. Phys. Chem. **71**, 695.
Levy, S. and M. Folman, 1963, J. Phys. Chem. **67**, 1278.
Lewis, K. E. and G. D. Parfitt, 1966, Trans. Faraday Soc. **62**, 204.
Linares, J. and F. Huertas, 1970, Bol. Soc. Espan. Ceram. **9**, 257; **CA74**: 6752w.
Lippincott, E. R., R. R. Stromberg, W. H. Grant and G. L. Cessac, 1969, Science **164**, 1482.
Little, L. H., 1966, Infrared Spectra of Adsorbed Species, (Academic Press, London, New York).
Little, L. H. and M. V. Mathieu, 1960, Proc. II. Intern. Congr. Catal., Paris, Edition Technip, Paris, 1961, p. 771.
Litvan, G. G., 1972a, J. Colloid Interface Sci. **38**, 75.
Litvan, G. G., 1972b, J. Amer. Ceram. Soc. **55**, 38.
Lohse, U., H. Stach, M. Hollnagel and W. Schirmer, 1971a, Z. Phys. Chem. (Leipzig) **246**, 91.
Lohse, U., H. Stach, M. Hollnagel and W. Schirmer, 1971b, Z. Phys. Chem. (Leipzig) **247**, 65.
Lohse, U., K. H. Sichhart, A. Grossmann, H. Stach and W. Schirmer, 1971c, Monatsber. Deutsch. Akad. Wiss. Berlin **13**, 122.
Lorenzelli, V. and G. B. Bonino, 1965, Atti Accad. Naz. Lincei, Rend. Classe Sci. Fis., Mat. Nat. **38**, 312.
Low, M. J. D. and N. Ramasubramanian, 1965, Chem. Commun. 499.
Low, M. J. D. and N. Ramasubramanian, 1966, J. Phys. Chem. **70**, 2740.
Low, M. J. D. and Y. Harano, 1968, J. Res. Inst. Catalysis, Hokkaido Univ. **16**, 271.
Low, M. J. D. and V. V. Subba Rao, 1969, Can. J. Chem. **47**, 1281.
Low, M. J. D. and M. Hasegawa, 1968, J. Colloid Interface Sci. **26**, 95.
Low, M. J. D., N. Ramasubramanian and P. Ramamurthy, 1967a, J. Vacuum Sci. Technol. **4**, 111.
Low, M. J. D., N. Ramasubramanian and V. V. Subba Rao, 1967b, J. Phys. Chem. **71**, 1726.
Low, M. J. D., N. Ramasubramanian, P. Ramamurthy and A. V. Deo, 1968, J. Phys. Chem. **72**, 2371.
Low, M. J. D., N. Takezawa and A. J. Goodsel, 1971a, J. Colloid Interface Sci. **37**, 422.
Low, P. F., J. L. White and I. Ravina, 1971b, Nature **230**, 159.
Lubezky, I., U. Folman and M. Folman, 1965, Trans Faraday Soc. **61**, 940.
Luck, W. A. P. and W. Ditter, 1967–68, J. Mol. Structure **1**, 261.
Luck, W. A. P. and W. Ditter, 1969, Z. Naturforsch. **24b**, 482.
Martens, R., H. Nagel and F. Freund, 1973, Ind. Chim. Belge **38**, 514, 519.
Matron, W., G. Ebert and F. H. Müller, 1971, Kolloid Z. u. Z. Polymere **248**, 986.
Mattmann, G., H. R. Oswald and F. Schweizer, 1972, Helv. Chim. Acta **55**, 1249.
Matyash, I. V., V. V. Mank, Ya. Ya. Shcherbak, M. A. Piontkovskaya and R. S. Tyutynnik, 1969, Russ. J. Inorg. Chem. **14**, 2.
Matyash, I. V., D. A. Zhogolev and A. M. Kalinichenko, 1971, in: Magn. Resonance Rel. Phenomena, Proc. Congr. Ampère, 16th, 1970, Ed. I. Ursu (Publ. House Acad. Soc. Repub. Rom., Bucharest, Rom.), p. 393.
Mays, J. M. and G. W. Brady, 1956, J. Chem. Phys. **25**, 583.
McCafferty, E. and A. C. Zettlemoyer, 1970a, J. Colloid Interface Sci. **34**, 452.

McCafferty, E. and A. C. Zettlemoyer, 1970b, Trans. Faraday Soc. **66**, 1732.
McCafferty, E. and A. C. Zettlemoyer, 1971, Disc. Faraday Soc. **52**, 239.
McCafferty, E., V. Pravdic and A. C. Zettlemoyer, 1970, Trans. Faraday Soc. **66**, 1720.
McDonald, R. S., 1957, J. Amer. Chem. Soc. **79**, 850.
McDonald, R. S., 1958, J. Phys. Chem. **62**, 1168.
McManus, J. C., K. Matsushita and M. J. D. Low, 1969, Can. J. Chem. **47**, 1077.
Meerson, L. A., R. B. Dobrotin and M. V. Mikhailova, 1972, Zh. Fiz. Khim. **46**, 2962.
Mertens, G. and J. J. Fripiat, 1973, J. Colloid Interface Sci. **42**, 169.
Metsik, M. S., V. S. Perevertaev, V. A. Liopo, G. T. Timoshchenko and A. V. Kiselev, 1973, J. Colloid Interface Sci. **43**, 662.
Meye, W., 1972, Thesis, University of Munich.
Michel, D., 1966, Z. Naturforsch. **21a**, 366.
Michel, D., 1967, Z. Naturforsch. **22a**, 1751.
Morachevskii, V. G., N. A. Dubrovich, A. G. Popov and A. N. Potanin, 1973, Zh. Prikl. Khim. **46**, 545; **CA79**, 102049.
Morariu, V. V. and R. Mills, 1972a, Z. Physik. Chem. (Frankfurt) **78**, 298.
Morariu, V. V. and R. Mills, 1972b, J. Colloid Interface Sci. **39**, 406.
Morariu, V. V. and R. Mills, 1973, Z. physik. Chem. (Frankfurt) **83**, 41.
Morimoto, T. and M. Nagao, 1970, Bull. Chem. Soc. Japan **43**, 3746.
Morimoto, T., M. Nagao and F. Tokuda, 1969, J. Phys. Chem. **73**, 243.
Morimoto, T., M. Nagao and J. Imai, 1971, Bull. Chem. Soc. Japan **44**, 1282.
Morrow, B. A. and I. A. Cody, 1973, J. Phys. Chem. **77**, 1465.
Mortland, M. M., J. J. Fripiat, J. Chaussidon and J. Uytterhoeven, 1963, J. Phys. Chem. **67**, 248.
Mulliken, R., 1952, J. Am. Chem. Soc. **74**, 811.
Munuera, G. and F. S. Stone, 1971, Disc. Faraday Soc. **52**, 205.
Munuera, G. and I. Carrizosa, 1973, Acta Cient. Venezolana **24**, Supl. 2, 226.
Munuera, G., F. Moreno and F. Gonzalez, 1972, Proc. 7th Intern. Symp. Reactivity of Solids, Bristol, p. 681.
Murdmaa, K. and V. V. Serpinskii, 1972, Izv. Akad. Nauk SSSR, Ser. Khim. 465.
Nagao, M., 1971, J. Phys. Chem. **75**, 3822.
Nagao, M. and T. Morimoto, 1969, J. Phys. Chem. **71**, 3809.
Nair, N. K. and J. M. Thorp, 1965a, Trans. Faraday Soc. **61**, 962.
Nair, N. K. and J. M. Thorp, 1965b, Trans. Faraday Soc. **61**, 974.
Nekrasova, E. G. and I. V. Zhilenkov, 1973, Russ. J. Phys. Chem. **47**, 93.
Nelson, S. M., H. H. Huang and L. E. Sutton, 1969, Trans. Faraday Soc. **65**, 225.
Noller, H., B. Mayerböck and G. Zundel, 1972, Surface Sci. **33**, 82.
Olejnik, S., A. M. Posner and J. M. Quirk, 1971a, Clays Clay Miner. **19**, 83.
Olejnik, S., A. M. Posner and J. M. Quirk, 1971b, J. Colloid Interface Sci. **37**, 536.
O'Reilly, D. E. and E. M. Peterson, 1971, J. Chem. Phys. **55**, 2155.
O'Reilly, D. E., H. P. Leftin and W. K. Hall, 1958, J. Chem. Phys. **29**, 970.
Page, Jr., T. F., R. J. Jakobsen and E. R. Lippincott, 1970, Science **170**, 51.
Pap, A., P. Kröbl and P. Josza, 1972, J. Therm. Anal. **4**, 449.
Parfitt, G. D., J. Ramsbotham and C. H. Rochester, 1971, Trans. Faraday Soc. **67**, 841.
Parry, E. P., 1963, J. Catal. **2**, 371.
Peglar, R. J., F. H. Hambleton and J. A. Hockey, 1971, J. Catal. **20**, 309.

Peri, J. B., 1965a, J. Phys. Chem. **69**, 211.
Peri, J. B., 1965b, J. Phys. Chem. **69**, 220.
Peri, J. B., 1965c, J. Phys. Chem. **69**, 231.
Peri, J. B., 1966a, J. Phys. Chem. **70**, 2937.
Peri, J. B., 1966b, J. Phys. Chem. **70**, 3168.
Peri, J. B. and R. B. Hannan, 1960, J. Phys. Chem. **64**, 1526.
Peri, J. B. and A. L. Hensley, 1968, J. Phys. Chem. **72**, 2926.
Pfeifer, H. and B. Staudte, 1968, Z. Naturforsch. **23a**, 775.
Pichat, P., M.-V. Mathieu and B. Imelik, 1969, J. Chim. Phys. **66**, 845.
Pickett, J. H. and L. B. Rogers, 1967, Anal. Chem. **39**, 1872.
Plooster, M. N. and S. N. Gitlin, 1971, J. Phys. Chem. **75**, 3322.
Prigogine, M. and J. J. Fripiat, 1971a, Bull. Soc. Chim. France 4291.
Prigogine, M. and J. J. Fripiat, 1971b, Chem. Phys. Letters, **12**, 107.
Primet, P., P. Pichat and M.-V. Mathieu, 1971, J. Phys. Chem. **71**, 1221.
Puranik, P. G. and V. Kumar, 1963a, Proc. Indian Acad. Sci., Sect. **A58**, 29.
Punanik, P. G. and V. Kumar, 1963b, Proc. Indian Acad. Sci., Sect. **A58**, 327.
Radul, N. M. and F. D. Ovcharenko, 1971, Ukr. Khim. Zh. **37**, 775.
Ramsay, J. D. F., 1971, Disc. Faraday Soc. **52**, 49.
Reball, S. and H. Winkler, 1964, Z. Naturforsch. **19a**, 861.
Řeřicha, R. and K. Kochloefl, 1969, Collect. Czech. Chem. Commun. **34**, 206.
Resing, H. A., 1967–1968, Adv. Molecular Relaxation Processes **1**, 109.
Resing, H. A., 1972, Adv. Molecular Relaxation Processes **3**, 199.
Resing, H. A. and J. K. Thompson, 1971, Adv. Chem. Series **101**, 473.
Resing, H. A. and J. S. Murday, 1973, Adv. Chem. Series **121**, 414.
Reuben, J., D. Fiat and M. Folman, 1966, J. Chem. Phys. **45**, 311.
Riedel, E., J. Kärger and H. Winkler, 1973, Z. Physik. Chem. (Leipzig) **252**, 161.
Roev, L. M., 1960, Dokl. Akad. Nauk SSSR **133**, 561.
Roev, L. M., V. N. Filimonov and A. N. Terenin, 1958, Optika Spektrosk. **4**, 328.
Rolland, M. T. and R. Bernard, 1951, C.R. Acad. Sci. **232**, 1098.
Ron, T. and P. Schindler, 1972, Chimia **26**, 247.
Ron, A., M. Folman and O. Schnepp, 1962, J. Chem. Phys. **36**, 2449.
Ross, S. and J. P. Olivier, 1964, On Physical Adsorption, (Interscience Publ., New York, London, Sydney).
Rouxhet, 1974, personal communication.
Rouxhet, P. G. and R. E. Sempels, 1974, JCS Faraday I **70**, 2021.
Sahay, B. K. and M. J. D. Low, 1974, J. Colloid Interf. Sci. **48**, 20.
Sandorfy, C., 1972, Canad. J. Spectroscopy **17**, 24.
Schindler, P. and H. R. Kamber, 1968, Helv. Chim. Acta **51**, 1781.
Sewell, P. A. and A. M. Morgan, 1969, J. Am. Ceram. Soc. **52**, 136.
Seymour, S. J., M. I. Cruz and J. J. Fripiat, 1973, J. Phys. Chem. **77**, 2847.
Shapiro, I. and H. G. Weiss, 1953, J. Phys. Chem. **57**, 219.
Sharkina, E. V. and Yu. I. Tarasevich, 1973, Ukr. Khim. Zh. **39**, 309.
Sheppard, N. V. and D. J. C. Yates, 1956, Proc. Roy. Soc. (London) **A238**, 69.
Sheppard, N. V., M.-V. Mathieu and D. J. C. Yates, 1960, Z. Elektrochem. **64**, 734.
Sidebottom, E. W. and G. G. Litvan, 1971, Trans. Faraday Soc. **67**, 2726.
Sidorov, A. N., 1954, Dokl. Akad. Nauk SSSR **95**, 1235.

Sidorov, A. N., 1956, Zh. Fiz. Khim. **30**, 995.

Sidorov, A. N., 1960, Optika Spektrosk. **8**, 806.

Simkovich, G., 1963, J. Phys. Chem. **67**, 1001.

Sklyankin, A. A. and K. V. Kireev, 1972, Svyazannaya Voda Dispersnykh Sist. Nr. 2, 86; **CA78**, 76206c.

Smart, R. S. C. and N. Sheppard, 1971, Proc. Roy Soc. (London) **A320**, 417.

Smith, B. R. and J. M. Thorp, 1963, J. Phys. Chem. **67**, 2617.

Snyder, L. R. and J. W. Ward, 1966, J. Phys. Chem. **70**, 3941.

Soffer, A. and M. Folman, 1966, Trans. Faraday Soc. **62**, 3559.

Spannheimer, H. and H. Knözinger, 1966, Ber. Bunsenges. Physik. Chem. **70**, 570.

Stählin, W. and H. Knözinger, 1974, unpublished.

Stakebake, J. L., 1973, J. Phys. Chem. **77**, 581.

Stakebake, J. L. and L. M. Steward, 1973, J. Colloid Interface Sci. **42**, 328.

Stevenson, D. P., 1965, J. Phys. Chem. **69**, 2145.

Stuart, W. I. and T. L. Whateley, 1965, Trans. Faraday Soc. **61**, 2763.

Šušić, V., R. Vučelič, V. Pausak, B. Karaulić and V. Milaković-Vučelič, 1969, J. Phys. Chem. **73**, 1975.

Swoboda, A. R. and G. W. Kunze, 1968, Soil Sci. Soc. Amer. Proc. **32**, 806.

Tahoun, S. A., 1971, U. A. R. J. Chem. **14**, 123.

Tahoun, S. A. and M. M. Mortland, 1966, Soil Sci. **102**, 314.

Takezawa, N., K. Miyahara and I. Toyoshima, 1971, J. Res. Inst. Catalysis, Hokkaido Univ. **19**, 56.

Tarasevich, Yu. I, 1971, Kolloid Zh. **33**, 900.

Tarasevich, Yu. I. and V. M. Rudenko, 1971, Ukr. Khim. Zh. **37**, 777.

Tarasevich, Yu., I., A. O. Orazmuradov and F. O. Ovcharenko, 1971, Kolloid Zh. **33**, 272.

Taylor, J. H. and C. H. Amberg, 1961, Can. J. Chem. **39**, 535.

Tench, A. J., 1972, JCS Faraday I **68**, 197.

Tench, A. J. and D. Giles, 1972, JCS Faraday I **68**, 193.

Tench, A. J., D. Giles and J. F. J. Kibblewhite, 1971, Trans. Faraday Soc. **67**, 854.

Tempelhoff, K., H. Winde and K. Hennig, 1972, Z. Chem. **12**, 276.

Tertykh, V. A., A. A. Chuiko, V. V. Mashchenko and V. V. Pavlov, 1973, Russ. J. Phys. Chem. **47**, 85.

Thorp, J. M. and J. B. Woolf, 1967, Trans. Faraday Soc. **63**, 2068.

Thorp, J. M. and J. B. Woolf, 1968, Trans. Faraday Soc. **64**, 2190.

Toullaux, R., P. Salvador, C. Vandermeersch and J. J. Fripiat, 1967, Israel J. Chem. **6**, 337.

Treibmann, D. and A. Simon, 1966, Ber. Bunsenges. Phys. Chem. **70**, 562.

Tretyakov, N. E., D. V. Podznyakov, O. M. Oranskaya and V. N. Filimonov, 1970, Russ. J. Phys. Chem. **44**, 596.

Tyler, A. J., J. A. G. Taylor, B. A. Pethica and J. A. Hockey, 1971, Trans. Faraday Soc. **67**, 483.

Umeya, K. and T. Kanno, 1973, Bull. Chem. Soc. Japan **46**, 1660.

Van Beek, L. K. H., 1960, Physica **26**, 66.

Van Cauwelaert, F. H., J. B. Van Assche and J. B. Uytterhoeven, 1970, J. Phys. Chem. **74**, 4329.

Van Cauwelaert, F. H., F. Vermoortele and J. B. Uytterhoeven, 1971, Disc. Faraday Soc. **52**, 66.

Van Cauwelaert, F. H., P. A. Jacobs and J. B. Uytterhoeven, 1972, J. Phys. Chem. **76**, 1434.
Van Cauwelaert, F. H., P. A. Jacobs and J. B. Uytterhoeven, 1973, J. Phys. Chem. **77**, 1470.
Van Thiel, M., E. D. Becker and G. C. Pimentel, 1957, J. Chem. Phys. **27**, 486.
Vedrine, J. C., J. S. Hyde and D. S. Lennart, 1972, J. Phys. Chem. **76**, 2087.
Venable, R. L., W. H. Wade and N. Hackerman, 1965, J. Phys. Chem. **69**, 317.
Volkov, A. V., A. V. Kiselev, V. I. Lygin and V. B. Khlebnikov, 1972, Russ. J. Phys. Chem. **45**, 290.
Vučelič, V., D. Vučelič, D. Karaulić and M. Šušić, 1973, Thermochim. Acta **7**, 77.
Webster, R. K., T. L. Jones and P. J. Anderson, 1965, Proc. Brit. Ceram. Soc. **5**, 153.
Weller, S. W. and A. A. Montagna, 1971, J. Catal. **21**, 303.
West, W. and R. T. Edwards, 1937, J. Chem. Phys. **5**, 14.
White, J. L., A. N. Jelli, J. M. Andrè and J. J. Fripiat, 1967, Trans. Faraday Soc. **63**, 461.
Winkler, H., 1961, Z. Naturforsch. **16a**, 780.
Wirzing, G., 1963, Naturwissenschaften **50**, 466.
Woessner, D. E., 1963, J. Chem. Phys. **39**, 2783.
Woessner, D. E., 1966, J. Phys. Chem. **70**, 1217.
Woessner, D. E. and J. R. Zimmerman, 1963, J. Phys. Chem. **67**, 1590.
Yaroslavsky, N. G., 1950, Zh. Fiz. Khim. **24**, 68.
Yaroslavsky, N. G. and A. N. Terenin, 1949, Dokl. Akad. Nauk SSSR **66**, 885.
Yaroslavsky, N. G. and A. V. Karyakin, 1952, Dokl. Akad. Nauk SSSR. **85**, 1103.
Yates, D. J. C., 1961, J. Phys. Chem. **65**, 746.
Young, R. P. and N. Sheppard, 1967a, Trans. Faraday Soc. **63**, 2291.
Young, R. P. and N. Sheppard, 1967b, J. Catal. **7**, 223.
Zecchina, A., C. Versino, A. Appiano and G. Occhiena, 1968, J. Phys. Chem. **72**, 1471.
Zecchina, A. E. Guglielminotti, S. Coluccia and E. Borello, 1969, J. Chem. Soc. (A) 2196.
Zecchina, A., E. Guglielminotti, L. Cerutti and S. Coluccia, 1972, J. Phys. Chem. **76**, 571.
Zettlemoyer, A. C. and E. McCafferty, 1969, Z. Physik. Chem. (Frankfurt) **64**, 41.
Zettlemoyer, A. C. and K. Klier, 1971, Disc. Faraday Soc. **52**, 46.
Zhdanov, S. P., 1958, Zh. Fiz. Khim. **22**, 699.
Zhuravlev, L. T. and A. V. Kiselev, 1970, in: Surface Area Determination, Eds. D. H. Everett and R. H. Ottewill, (Butterworths, London), p. 155.
Zimmerman, J. R. and W. E. Brittin, 1957, J. Phys. Chem. **61**, 1328.
Zimmerman, J. R. and J. A. Lasater, 1958, J. Phys. Chem. **62**, 1157.
Zimmerman, J. R., B. G. Holmes and J. A. Lasater, 1956, J. Phys. Chem. **60**, 1157.
Zundel, G., 1969, Hydration and Intermolecular Interaction, (Academic Press, New York), p. 75.
Zundel, G., 1970, Allg. Prakt. Chem. (Wien) **21**, 329.

PART I

WATER AND ICE

CHAPTER 28

HYDROGEN BONDS IN LIQUID WATER

WERNER A. P. LUCK

Fachbereich Physikalische Chemie, Philipps-Universität Marburg,
D-355 Marburg (Lahn), Biegenstraße 12, West Germany

Contents

28.1. Introduction 1369
28.2. Statistical mechanics approach 1373
28.3. Spectroscopic methods 1377
28.4. Non-bonded OH groups and properties of water 1387
28.4.1. Melting heat 1387
28.4.2. Specific heat 1388
28.4.3. Vaporisation heat 1390
28.4.4. Density 1391
28.4.5. Surface energy 1391
28.4.6. Expansion of the areas of hydrogen bonded molecules 1393
28.4.7. Hydrogen bond energy 1395
28.4.8. Magnetic susceptibility 1398
28.4.9. NMR chemical shift 1398
28.4.10. Dielectric constant 1399
28.4.11. Ultrasonic absorption 1401
28.4.12. Dynamic properties 1401
28.5. **Electrolyte solutions** 1403
28.5.1. Spectroscopic results 1403
28.5.2. Solubility 1408
28.5.3. Partial molar volume 1412
28.5.4. Acids and bases 1414
28.6. Clathrate hydrates 1415
28.6.1. General review 1415
28.6.2. Structure 1416
28.6.3. Relations to solution structures 1418
28.6.4. Applications 1418
28.6.5. Applications 1419
28.6.6. Other clathrates 1420
Acknowledgement 1420
References 1420

The hydrogen bond – recent developments in theory and experiments
Eds. P. Schuster et al.

28.1. Introduction

Water is the most important substance. This is true for biochemistry and for industrial chemistry. Its properties depend mainly on the H-bond interaction. There is no other substance like water where the H-bonds play such an important role for the interaction energy. The whole intermolecular interaction energy of water is given by the heat of sublimation of ice $\Delta H_s - RT = \Delta U_s = 11.64$ kcal/mole. Briegleb [1937] has calculated the dispersion energy for distances between water molecules in ice-like arrangements as 3.6 kcal/mole. From this point of view and in comparison with methanol the H-bond energy per OH group can be estimated as 4 kcal/mole (Onsager [1974] and Luck [1974a]). Therefore, the H-bonds dominate the properties of liquid water. The dispersion energy is proportional to the number of electrons or to the molecular weight; accordingly, ice should melt at approximately -90 °C and water should have a normal boiling point of approximately -80 °C. The enthalpy H and the internal energy U of liquid and gaseous water under saturation conditions are shown in fig. 28.1. The difference $U(T_c) - U(T_F) = \Delta U_c$ of internal energies of liquid water between the critical temperature T_c and the melting point T_F is $\Delta U_c = 8.9$ kcal/mole. The corresponding value for methane is 1.9 kcal/mole. The change of enthalpy of water vapour in the ideal gas state from 0°C to T_c is about $\Delta H_{id} = 2.1$ kcal/mole. This value is due to the intramolecular degrees of freedom. Therefore, in liquid water there is an excess internal energy for the liquid state for $\Delta T = T_c - T_F$, $\Delta U_{liq} = \Delta U_c - \Delta U_{id} = 8.9 - 2.1$ kcal/mole $= 6.8$ kcal/mole. The melting heat of ice is $\Delta U_F = 1.44$ kcal/mole. If we take as an approximation that the effect of dispersion forces on ΔU_F of water is similar to $\Delta U_F(H_2S) = 0.57$ kcal/mole, the fraction of the H-bond effect on $\Delta U_F(H_2O)$ would be $(1.44 - 0.57$ kcal/mole$) = 0.87$ kcal/mole $= \Delta U_{FH}$. We would obtain the sum of excess effects on enthalpy, $\Delta H_{excess} = \Delta U_{liq} + \Delta U_{FH} = 6.8 + 0.87$ kcal/mole $= 7.7$ kcal/mole. This value is comparable to the value 8 kcal/mole of the H-bond interaction discussed above.

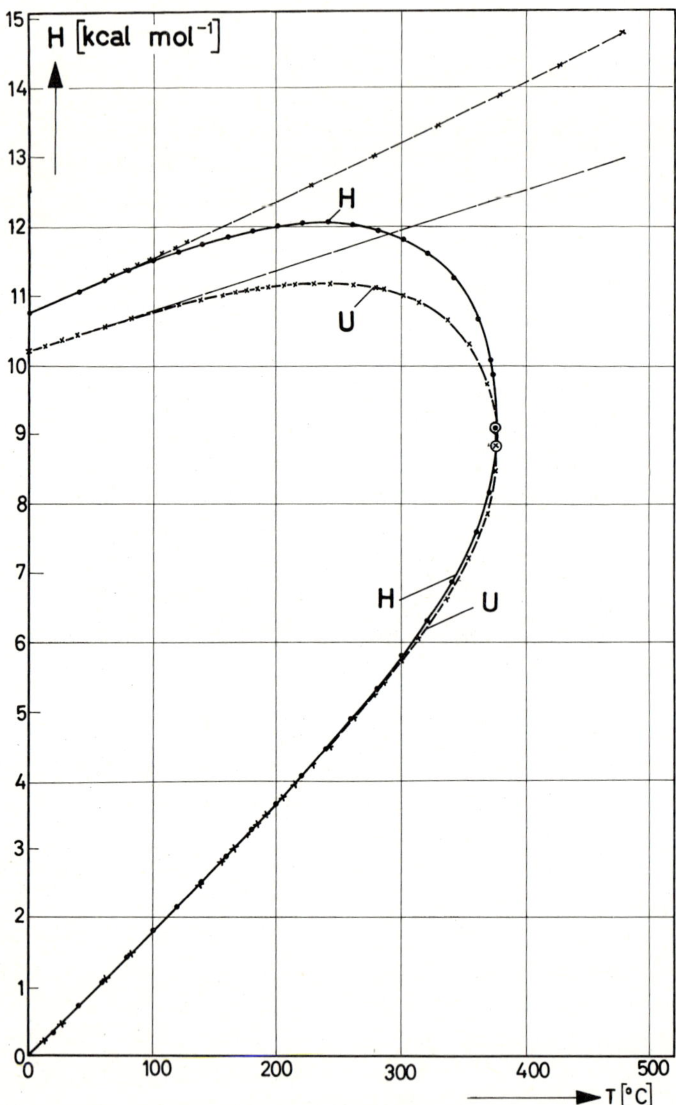

Fig. 28.1. Enthalpy H and internal energy U of water under saturation conditions. Straight lines: extrapolated properties of ideal water vapour.

In the ice state we can assume that all H-bonds are closed. This simple discussion of enthalpies indicates that the change of H-bond content during ice melting should be about $0.87/8 = 0.109$. This consequence was overlooked in some mixture theories of water. Eucken's theory [1948] would give 70% non-H-bonded–so-called free–OH groups at T_F and the Nemethy–Scheraga [1962] theory postulates 50% free OH groups. The tridymite structure of ice (fig. 28.2) can be clearly understood through the angle dependence of the H-bonds (see ch. 11). In ice the proton axes are in the direction of the axes of the orbitals of lone pair electrons. In

Fig. 28.2. Model of the tridymite structure of ice.

contradiction to non-polar molecules a structure with four first next neighbours is formed due to this angle dependence of the H-bonds. The structure of non-polar spherical molecules like Ne, Ar, Xe and CCl_4 in the liquid state near T_F can be described approximately by a simple hole-defect model (e.g. Luck and Ditter [1971]). According to this view, the coordination number in a liquid model under saturation conditions as a function of T would be expected from the saturation densities to follow the dashed line in fig. 28.3, where they are compared with the data of Eisenstein and Gingrich [1942] for argon. The crosses are values of the integrated first maximum of the correlation function up to constant distances. The circled crosses of the upper curve are values from the integral up to the T-dependent distances of the first next neighbours

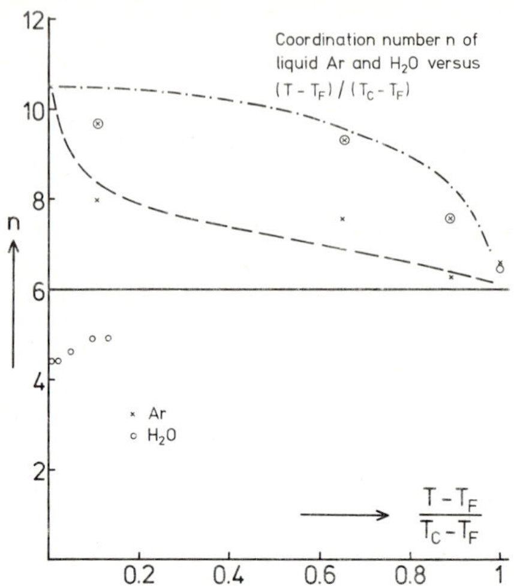

Fig. 28.3. X-ray scattering results for the coordination number n of first next neighbours of liquid Ar and H_2O as function of the quotient of the T-distance from the melting point T_F over the expansion of the liquid region $T_c - T_F$. Upper curve values of Ar calculated by an integration up to the distance of the first next neighbours using the hole theory of liquids.

calculated with the T-expansion of the hole theory. Generally, in weakly polar liquids one observes coordination numbers near the melting point of about 10 (Henshaw [1960] and Kruh [1962]) and values about 6 near T_c. In liquid water very different behaviour is observed. Near T_F $n = 4.4$ is found, and also a slow increase of n with T (Morgan and Warren [1938]; see also fig. 28.3). The limit of the area of the first next neighbours is a little doubtful (Pings [1968]). Narten [1972] gives T-independent values of $n = 4.4$ up to 200 °C for water. We can expect that the n-values of water have the asymptotic tendency to reach 6 near T_c.

In the last century there were many attempts to calculate models for the anomalous behaviour of water (see Luck [1974b]). Röntgen [1892] assumed that water is a mixture of ice-like molecules and molecules of a second type; the consideration of both species should be a function of temperature T. For five decades after Röntgen many scientists have tried to describe water as a mixture of different small aggregates (see Luck [1974b]). The physical interpretation of all these descriptions differs only in the number of adjusted constants which the different treatments need. During the last decade some papers have tried to describe water as a mixture of monomers and large aggregates. For example, Nemethy and

Scheraga [1962] gave a statistical calculation of the thermodynamic properties of water from 0 to 100°C. The assumed clusters of large aggregates of about 125 molecules at room temperature, and also some monomers.

The attempts based on the assumption of small aggregates or the calculations by Nemethy and Scheraga [1962] using statistical thermodynamics are called mixture models (Davies and Sarzynsk [1972]). One should differentiate strictly between mixture models of certain species and mixture models which discuss H-bonded and non-H-bonded OH groups. The latter do not discuss monomers. There is a distinct difference if a mixture model discusses free *molecules* or free OH *groups*. Induced by the paper of Pople [1950] there are some attempts to describe water with a continuum model, which assumed a continuum of different H-bond states with distributions of different H-bond angles and OH···O distances (Kell [1972]).

28.2. Statistical mechanics approach

The mixture models may be seen as first approximations and the attempts to find the distribution functions of the continuum models as the next step. This method has been revitalised in recent years by statistical mechanics and especially by computer experiments. For example, Rahman and Stillinger [1971, 1972, 1974] have carried out computer experiments with a model of 216 water molecules. They assumed an effective pair potential between two water molecules previously used by Ben-Naim and Stillinger [1972]. They utilise the Bjerrum four-point charge model for a water molecule with two positive charges $+\eta e$ and two negative point charges $-\eta e$ at a distance of 1 Å from the molecule centre. The directions of every two charges to the centre of the molecule give a tetrahedral angle. The pair potential is assumed as a sum of a Lennard-Jones potential V_{LJ} and an electrostatic potential V_{HB} between all eight point charges of two H_2O molecules,

$$V_{LJ} = 4\epsilon\left[\left(\frac{\sigma}{R_{12}}\right)^{12} - \left(\frac{\sigma}{R_{12}}\right)^{6}\right]. \tag{28.1}$$

The constants ϵ and σ are taken from the isoelectronic neon atom

$$\epsilon = 5.01 \times 10^{-15}\ \text{erg}, \qquad \sigma = 2.82\ \text{Å} \tag{28.2}$$

and

$$V_{HB}(1,2) = (\eta e)^2 \sum_{\alpha_1,\alpha_2=1}^{4} \frac{(-1)^{\alpha_1+\alpha_2}}{d_{\alpha_1\alpha_2}^{(1,2)}}, \tag{28.3}$$

where $d_{\alpha_1\alpha_2}$ is the distance between the point charge α_1 on particle 1 and the point charge α_2 on particle 2. The total Ben-Naim–Stillinger potential is assumed as

$$U(1,2) = V_{LJ}(R_{12}) + S(R_{12})V_{HB}(1,2). \tag{28.4}$$

$S(r_{12})$ is a switching function to consider the repulsion effects at short distances R_{12}. It is assumed equal to 0 at small distances, equal to 1 at large distances, and a function of distance in the medium region. The parameters of the pair potential $U(1,2)$ were fixed to obtain the experimental temperature-dependence of the second virial coefficient of water (e.g. $\eta = 0.19$).

Stillinger and Rahman (ST–R) [1972] calculated the properties of an ensemble of 216 molecules with the idealised potential 28.4 in a cubic cell of edge length 18.62 Å in a time $(1\text{–}2)\times 10^{12}$ s at three various T, $T_1 = -8.2$°C, $T_2 = 34.3$°C and $T_3 = 314.8$°C. The constant cubic cell corresponds to a constant density of $\rho = 1\ \text{g/cm}^3$. Spectroscopic experiments show that higher densities ρ than saturation density ρ_s induce a greater content of H-bonded molecules. Stillinger and Rahman found the coordination numbers of the first next neighbours shown in table 28.1.

TABLE 28.1

Number of first next neighbours n in the computer experiment by Stillinger and Rahman [1972]

T(°C)	$(T - T_F)/(T_c - T_F)$	n
−8.2	−0.02	5.2
34.3	0.09	5.5
314.8	0.83	8

The n-values show the increase with T, but the absolute values are higher than the experimental values. The computer experiment gives information on the energy density $p(V)$ of neighbours with which a given molecule experiences mutual interaction potential energy V (fig. 28.4). The main result of fig. 28.4 is that there are differences in the probability that one molecule has certain interaction energies. There is no continuum

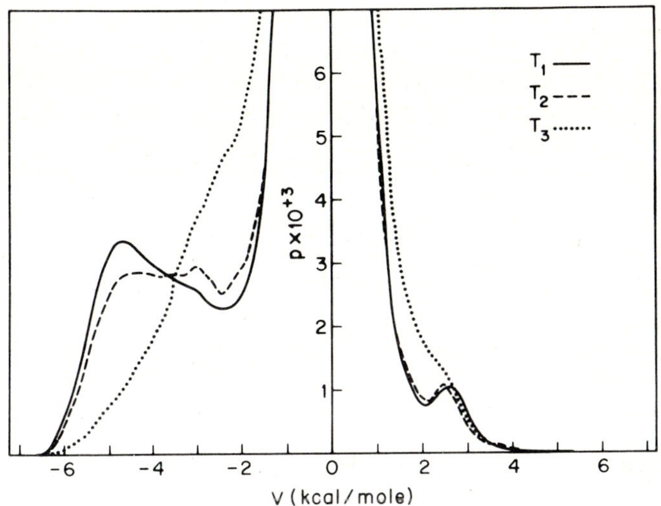

Fig. 28.4. $p(V)$: densities in mutual interaction energies V, which a given molecule experiences (Stillinger and Rahman [1972]).

of possibilities without distinct energy minima. Secondly, fig. 28.4 shows that the number of molecules with smaller interactions increases with T. In fig. 28.4 there is a medium energy region $V = 3.5$ kcal/mole $= 0.53\ V_M$ with T-independent distribution. This is in agreement with the ill-defined isosbestic points in the IR and Raman bands in a frequency region of medium H-bond interaction energy. The curves of fig. 28.4 decrease to zero above 6.5 kcal/mole because Stillinger and Rahman assumed this energy as the energy minimum V_M. Stillinger and Rahman have also calculated the distribution of the number of molecules with 0, 1, 2, 3 and 4 H-bonds. They found one T-dependent maximum of this distribution. They conclude that the existence of one maximum "conflicts with those two-state theories". But they have overlooked that a "two-state model" which discusses only bonded and non-H-bonded OH groups would not a priori lead one to expect two maxima in these plots. This means that a two-state model with bonded and non-H-bonded OH groups would not be in contradiction to the ST–R calculation, if the content of non-H-bonded OH groups is small.

Another question is if the number of 216 molecules of the computer experiment is large enough. Overtone results using a rough model give an expansion of the system of H-bonded molecules larger than 216 molecules. The n-values and the high values of the heat capacity of the ST–R model show that its structural order breaks up more rapidly than water does. The pair correlation function shows dubious shoulders at small distances. Stillinger and Rahman suppose that their point charge model gives too many configurations with two protons of one molecule interacting simultaneously with one negative point charge of a second molecule. A delocalisation of the negative charge should be attempted*. This supposition is realised better by using Gaussian functions to represent the angle dependence of H-bond interaction. This was carried out by Ben-Naim [1974]. He took a different angle dependent potential for V_{HB} in eq. (28.4)

$$V_{HB} = \epsilon_H G(R - R_H) \sum_{k,l} G(i_k \cdot \boldsymbol{n} - 1) G(j_l \cdot \boldsymbol{n} + 1). \qquad (28.5)$$

The G functions are of the Gaussian type $\exp(-x^2/2\sigma^2)$. i and j are the unit vectors in the directions of the charges in the point charge model or the directions of optimal H-bond interaction, and the $\boldsymbol{n}$ vector is the direction

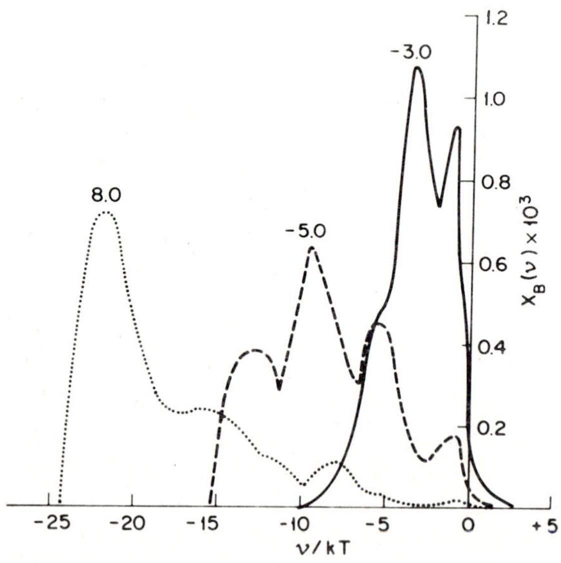

Fig. 28.5. The probability $x_B(\nu)\,d\nu$ of finding a specific molecule having binding energy between ν and $\nu + d\nu$. Parameter of the curves is ϵ_H/kT (Ben-Naim [1974]).

* Stillinger and Rahman [1974] repeated their calculation with a modified potential. They obtained similar results to those in fig. 28.4.

of the bond between two molecules. ϵ_H is an energy parameter which gives the maximum strength of the bond energy. With the potential of eqs. (28.4) and (28.5) Ben-Naim has calculated a system of water-like particles in two dimensions. He obtained a distribution function $X(\nu)$ where $X(\nu)\,d\nu$ is the fraction of molecules having binding energies between ν and $\nu + d\nu$. In the case of the Lennard-Jones potential V_{LJ} Ben-Naim obtains curves with one maximum, but with the angle-dependent potential of eq. (28.5) he obtains functions for $X(\nu)$ with some maxima (fig. 28.5). This result is in agreement with the ST–R model that there are preferences for certain H-bonded configurations. This picture receives further support from the IR spectroscopic matrix technique (see chs. 11 and 22). The matrix spectra of water show clear evidence for the presence of about four different H-bond arrangements of water under matrix conditions in solid N_2 or Ar. The Badger–Bauer rule for frequency shifts allows the interpretation that the different H-bonded states differ in the H-bond interaction energy. This is also in agreement with fig. 28.5.

28.3. Spectroscopic methods

The assumed idealised potentials of eq. (28.4) and (28.5) and the result that the mixture theories can fit the experimental data with very different models (compare Eucken [1948] and Némethy-Scheraga [1962]) indicate the necessity of experimental checks of the different theories. The best method to study H-bonds is the IR method. Latimer, Lescombe and Mecke have shown in the twenties and thirties that the IR overtone spectra are suitable for studying H-bond equilibria in solutions quantitatively. One is inclined to accept the proposal to take the concentration dependence of the extinction coefficients of the IR OH or NH bands as the definition of a H-bond. But this efficient overtone method was overtaken by the development of the automatic instruments for the mid-IR range. The quantitative study of H-bonds in the mid-IR, however, is much more difficult than in the overtone region (see ch. 11, p. 530). Infrared studies have thus been concentrated on the fundamental bands in the mid-IR with the consequent neglect of the overtone region and its considerable advantages for H-bond studies (these are discussed in ch. 11, p. 530). The disadvantage of the IR methods is that there is no solvent for water which dissolves water sufficiently without H-bond interaction. Therefore, the H-bond content of water can only be studied by examining the T-dependence of the IR spectra. Bellamy et al. [1958] have shown that the spectra of alcohols and water in a variety of solvents behave in a similar manner with regard to the H-bond interactions. Therefore, the T-dependence method for studying liquid water can reasonably be adjusted with data from alcohols. From the concentration dependence of the spectra of methanol or ethanol in CCl_4 we extrapolate that the pure alcohols should not at room temperature have more than 2% non-H-bonded, so called free, OH groups (Luck and Ditter [1968] and Luck [1974c]).

We could verify this result through measurements of the overtone spectra of pure methanol and pure ethanol from $-40\,°C$ up to $400\,°C$ under saturation conditions (Luck and Ditter [1968]). The alcohol spectra show that up to 370 °C the intensity of the free OH band is increasing and above 370 °C the content of H-bonds can be neglected. With the knowledge of the absorptivity coefficients of the free OH band we could estimate the content O_F of free OH groups as a function of T(see fig. 28.6).

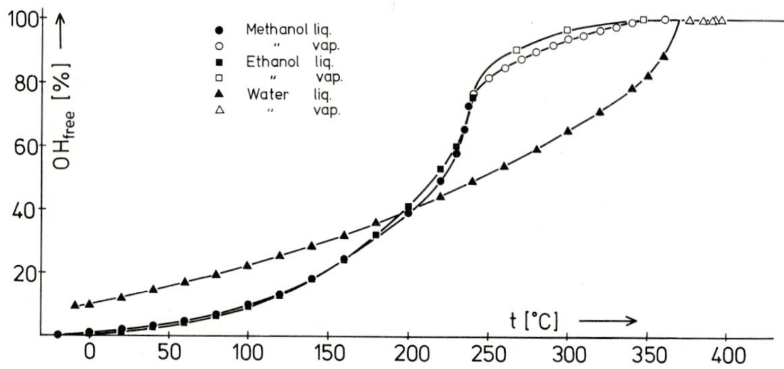

Fig. 28.6. Spectroscopically determined contents of non-H-bonded OH groups in water, methanol and ethanol under saturation conditions versus temperature. Supercritical temperature results at critical densities.

On account of the difficulties with spectra of solid alcohols and with an overlapping effect with a weak CH band the absolute values of P_F at $T < 100\,°C$ may be a little erroneous. In contradiction to similar water spectra our results in fig. 28.6 would suggest that during melting the H-bond content of alcohols would change very little. This result is in agreement with the melting heats and the comparison of ΔH_F of alcohols with ΔH_F of corresponding paraffins (table 28.2). The third line in table 28.2 gives the differences between the first and second lines. Table 28.3

TABLE 28.2

Heat of melting ΔH_F (kcal/mole)

ΔH_F	H_2O	1.43	CH_3OH	0.77	C_2H_5OH	1.11
ΔH_F	CH_4	0.22	C_2H_6	0.67	C_3H_8	0.84
$\Delta(\Delta H_F)$		1.21		0.1		0.27

TABLE 28.3

Heat of vaporisation ΔH_B (kcal/mole) at the boiling point T_B

ΔH_B	H_2O	9.7	CH_3OH	8.43	C_2H_5OH	9.25
ΔH_B	CH_4	1.9	C_2H_6	3.87	C_3H_8	4.7
$\Delta(\Delta H_B)$		7.8		4.56		4.55

gives the heat of vaporisation ΔH_B at the boiling point T_B. The third line in table 28.3 gives the differences of the values of lines one and two. These differences are in the order of magnitude of the H-bond energies. The third line in table 28.2 is in the order of magnitude of the portion of the H-bond energy dependent on ΔH_F. These values show that in water there is a remarkable influence of H-bonds on ΔH_F, but not so in alcohols.

Figure 28.7 shows the absorptivity coefficients of two water overtone bands under saturation conditions. The spectra at 396 °C are plotted at the critical density. The maximum of overcritical spectra corresponds to a vibration of non-H-bonded OH groups. This can be inferred from a comparison alcohol solution spectra and with solution spectra of water in CCl_4. The ice bands correspond to OH groups in a linear H-bond with optimal angles. The bands of free OH groups of alcohols and water have a larger half-width than in solutions. This may be an effect of the more polar environment. A comparison between the spectra of pure liquids and solutions shows that the intensity between the free OH band and the ice band is larger in liquids than in solutions. This indicates H-bonds of energetically unfavoured types because the Badger–Bauer rule states a proportionality between the frequency shift and the interaction energy. Such unfavoured H-bonds could correspond to unfavoured angles or different O···O distances. The X-ray scattering results clearly show that the distances between first next neighbours in liquid water do not change

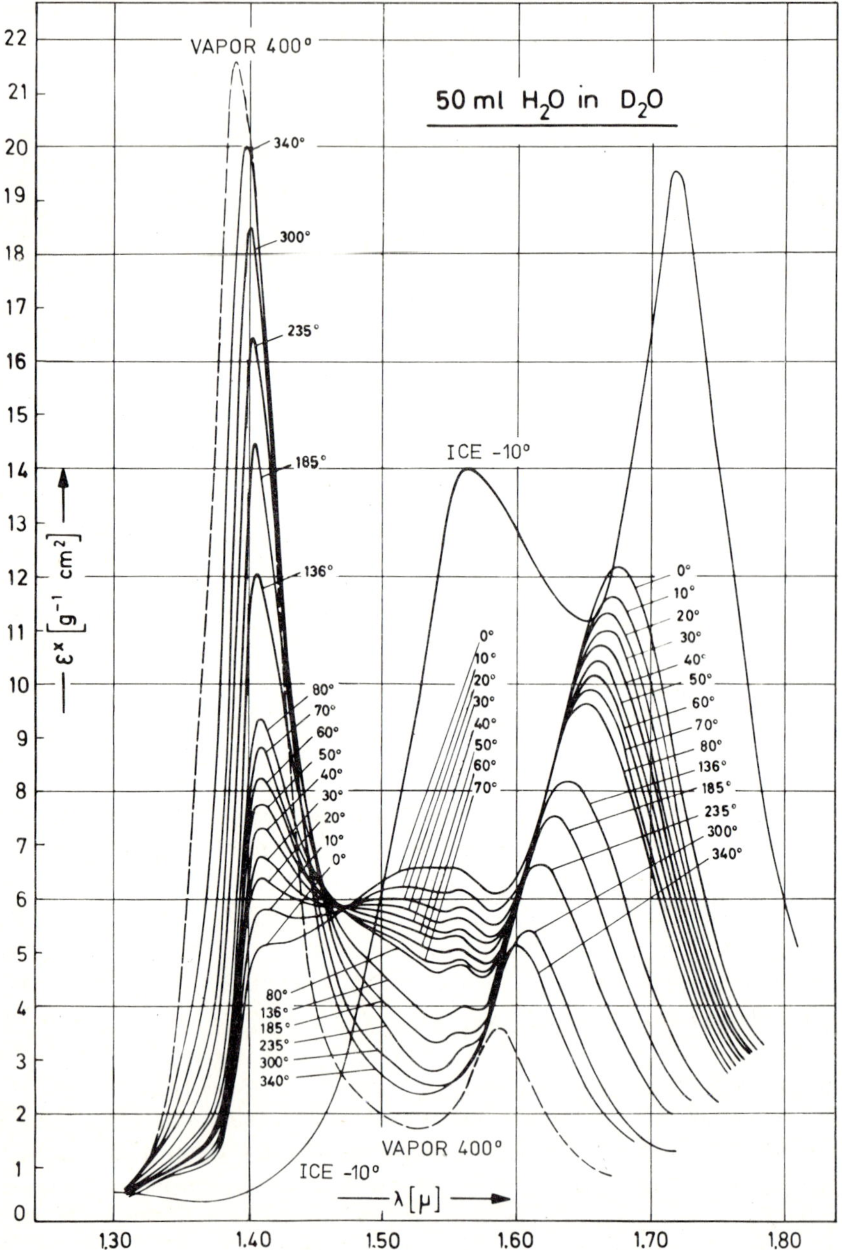

Fig. 28.7. First overtone of HOD versus T; 50 ml H_2O/l D_2O; D_2O absorption eliminated.

much with T (fig. 28.10; see also Narten [1972, 1974]). From this we favour the viewpoint that a distribution of different H-bond angles primarily affects the intensity of this medium frequency region (see ch. 11). Curve analysis can be carried out with the absorptivity coefficients of the overcritical spectra, the ice spectrum and a band with about half the wave number shift of the ice band, e.g. the first overtone of HOD: 7000 and 7150 cm^{-1} (free bands), 6450 cm^{-1} (ice band) and a medium band of 6800 cm^{-1} (energetically unfavoured H-bonds, Luck and Ditter [1969]).

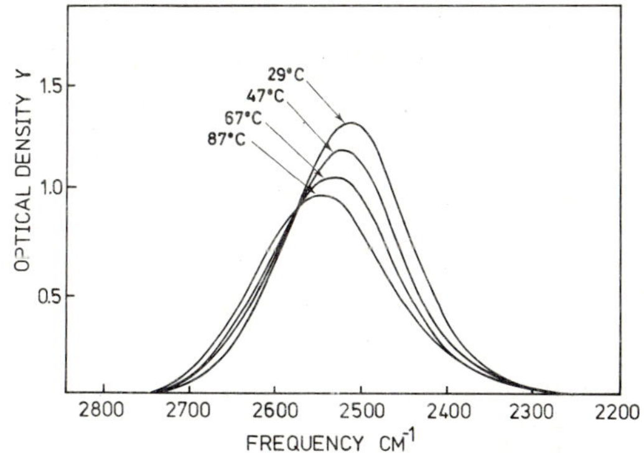

Fig. 28.8. IR fundamental stretching vibration of HOD showing an increase of intensity in the region of non-H-bonded OH with T.

The medium band of energetically unfavoured H-bonds corresponds to the dimer band in alcoholic solutions or in the matrix technique (see ch. 11). In both these cases we assume a cyclic structure. The analysis of O_F in water shows that we can neglect the content of dimers in water at lower T. Therefore, we can conclude that in water there are some cyclic H-bonds in a network of H-bonds. Near T_c the "free" OH band of water and alcohols has a wave number shift of about 10–40 cm^{-1} dependent on the order of the overtone (Luck [1974c]). This wave number shift can be understood on assuming a small disturbance of a free OH vibration if the rest of a molecule is engaged in a H-bond (see ch. 11). This indicates that far away from T_c we can observe only free OH groups of molecules in H-bonded aggregates and not free molecules. We must stress that, in general, the spectral analysis by an analog computer with a sum of curves of the Gaussian or the Lorentzian type can have infinite possibilities. But in the case of the overtone spectra we carried out this analysis with the absorptivity coefficients of the two free bands (free groups and free molecules), the ice band and a medium band of unfavoured H-bonds. This analysis does not have infinite solutions. It shows that details of the complicated spectra at all T can be quantitatively understood by an overlapping of these three bands (Luck and Ditter [1969] and Luck [1974a]). This makes it probable that we can describe liquid water or alcohols with a mixture model of these three states.

The computer analysis (Luck and Ditter [1969]) leaves out a contribution to the intensity at T_c. This probably arises from two factors: (i) the presence of unfavoured H-bonds above T_c, and (ii) deviations of the overcritical band profiles of Gaussian shape, including possibly rotational contributions. The existence of unfavoured H-bonds would show that a plot of $\%OH_{free}$ against T should become asymptotic at 100% and not vertical as shown in fig. 25.6. The maximum of the specific heat at T_c would indicate that the asymptotic approach is the more probable. The density gradient in this temperature region makes experimental realisation of this feature extremely difficult (Luck and Ditter [1966]).

We have carried out overtone experiments with four different bands of H_2O, two of HOD and one of D_2O, all of which give similar results and thus eliminate (Luck [1974a]) doubts (Falk and Ford [1966]) on this excellent method.

The fundamental bands of water have been studied with the IR technique by Hartmann [1966] and Franck and Roth [1967], and Senior and Verrall [1969a, b] and with the Raman technique by Walrafen [1968, 1972] and Lindner [1970]. The fundamental IR band (fig. 28.8) and Raman band clearly show an increase of the band of free OH groups with T. But the effects are not as distinct as in the overtone spectra. The reasons are the much higher intensity of the H-bond bands in the fundamental region (factor about 10) as well as the smaller wave number shifts due to H-bond interactions in the fundamental region (factor 1.7).* The theory of band overlapping (Luck and Ditter [1967]) shows that overtone and fundamental spectra are consistent and give similar results. Walrafen [1972] has shown that the Raman spectra can be analysed by a curve analyser similar to the overtone spectra. But he has not utilised the ice spectra and the overcritical spectra. Therefore, in our view his analysis is only sound taken in conjunction with the overtone results. Walrafen [1972] has also investigated the low frequency intermolecular vibrations using the Raman technique. He pointed out that the main intensity of these bands de-

* Compare the 1.4–1.55 μ band with the 1.6–1.75 μ band in fig. 28.7.

creases up to 100 °C. Our own measurements have shown that there is a small residual of these bands at 150 °C. Walrafen has some difficulties with the background of the Rayleigh scattering and the large half-width of the assumed 60 cm^{-1} band. Otherwise, some of these bands may correspond to larger systems of H-bonds. We have found with the IR matrix technique (H_2O in solid Ar) that the 600 cm^{-1} libration band has different concentration-dependent components (Luck et al. [1974a] and Luck [1974c]; see also fig. 28.9). In a concentration region where mainly dimers and trimers exist, we observe only a band with a maximum at approximately 530 cm^{-1}. At high concentrations where the polymer band in the fundamental region appears we found a band with a maximum at approximately 690 cm^{-1}. The presence of only two bands in this region clarified why Walrafen [1972] found two Raman bands at 720–740 cm^{-1} and 550 cm^{-1}, and in the literature only one IR band at 685 cm^{-1}. But this known IR band is asymmetric. Curnutte and Williams [1974] stressed that this band should consist of different components. The different components of

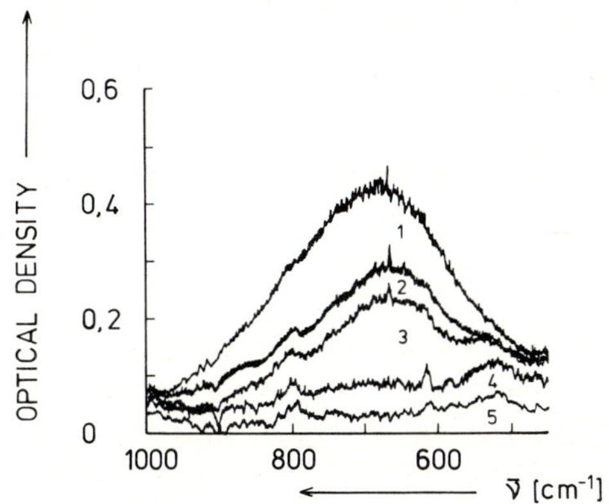

Fig. 28.9. H_2O libration band (IR matrix technique) in solid Ar at 20 K. (1) Ar/H_2O = 27; (2) Ar/H_2O = 51; (3) Ar/H_2O = 75; (4) Ar/H_2O = 595; (5) Ar/H_2O = 1448.

the IR libration bands show clearly that some bands may disappear if larger aggregates disappear.

The discussion of overtone spectra gives considerable information on the structure of H-bonded liquids. A purist may wait till an exact quantitative theory of the water structure is possible. But the knowledge of the water structure is so important that in our opinion, when quantitative interpretations are lacking, one should aim at approximate methods, like the ideal gas model, which is used knowing that corrections for intermolecular interactions are required. Such approximations can easily be effected with the overtone spectra. One can assume that the absorptivity coefficient of the free OH band is not strongly T-dependent, that the overlapping effects in the wave length region of the free OH band are not large and that the H-bond content for $T \geq 375\,°C$ at critical density is small. With these assumptions we can give the content of free OH groups O_F as a function of T quantitatively (fig. 28.6). Errors in our approximations would cause the real O_F values to be smaller than those in fig. 28.6, which means that fig. 28.6 gives the upper limits of O_F. The main result is that these experimental O_F values are smaller than all previous theories claim. This result would not be affected by our assumptions. The validity of these assumptions is reinforced by many spectroscopic measurements on water and H-bonds of organic compounds (Luck [1974a]). The quality of the O_F values of fig. 28.6 was verified by the more exact method to determine $\int \epsilon \, d\nu$ of the free OH bands. This method gave O_F values similar to the more approximate method to determine the optical density at the maximum of the free OH band. To eliminate overlapping effects may be a third species present which need not necessarily be T-with H-bond bands this integral was determined only at the one side of the band maximum toward shorter wave lengths from the maximum (Luck and Ditter [1969]). It can be shown that the $\int \epsilon \, d\nu$ of a band does not change much as a result of intermolecular interactions of non-H-bonded types (Luck [1951] and Luck and Ditter [1968]).

Statements on the different H-bonds are much more difficult than the determination of O_F. The anharmonicity may change with a change of the angle or distance distribution. This may also affect the quantitative interpretation of H-bond bands. Therefore, it cannot be decided by our work whether the mixture model or the continuum model is the better. Both models, however, should consider the latest results on non-bonded OH groups, which are of over-riding importance for the water structure. Our experience may concede that there may be distributions of different H-bond angles. But the data show that this distribution is not a continuum without any order. The IR spectra show types of isosbestic points. Walrafen [1972] has stated that Raman spectra show isosbestic points. But the quantitative intensity measurements of Raman spectra contain far more errors than the IR overtone spectra. The overtone spectra show that the isosbestic points are absent in a large T-region. But the deviations of isosbestic points are not large – with the exception of the ice spectra. Isosbestic points indicate a T-dependent equilibrium of two species; there dependent. These results are in agreement with Stillinger and Rahman's results (see fig. 28.4). It is perhaps surprising that such an approximation of two states in liquid water, non-H-bonded and H-bonded OH groups, is so efficient in describing the properties of water. We have to stress that free OH does not mean free molecules. Free molecules would be indicated by a rotational structure on the spectra. Below T_c and at or above saturation conditions we have not found any indication of such rotational spectra. This is in line with the thermodynamic properties of water. If there were large amounts of free water molecules–as many theories assume–the vapour pressure of water would be much higher. From the vapour pressure of water and the consideration of a simple model of orientation defects we do not expect monomer contents higher than 1% below 200°C. The contents O_F of non-H-bonded molecules of water and alcohols differ in one interesting way. Water reaches 100% at T_c, alcohols above T_c nearly at T_c (water) (fig. 28.6). The similar lowest T for $O_F = 100\%$ for both types of molecule is in agreement with our knowledge that the H-bond energy per bond is similar at 4 kcal/mole. This effect means that T_c of water is mainly determined by the H-bond interaction, but T_c of alcohols depends on H-bonds and on the interaction of the hydrophobic groups.

The value of $O_F(0\,°C) \approx 0.1$ gives a severe restriction for all models of liquid water, a fact that most theoreticans appear to ignore. Even if we neglect the effect of dispersion forces with water and assume that the whole heat of melting depended only on H-bonds opening, O_F could not be larger than $1.44/8 = 0.18$. This is the upper limit for $O_F(0\,°C)$. Nevertheless, many theories go beyond this limit. Especially the assumption of monomers yields too high values of $O_F(0\,°C)$. This $O_F(0\,°C)$ value considerably restricts every model of liquid water and causes difficulties for models which need monomers. The T-dependence of O_F differs between water and alcohols. This may be an effect of the different structure. X-ray scattering results in liquid alcohols and solution spectra can be understood by a chain-like model of the H-bonded systems. But in water within short domains we have to assume a network with a similar structure to ice. This assumption is based on the Raman data of liquid water which at low T show a C_{2v} symmetry of the H-bonded systems (Walrafen [1972]), and the X-ray and neutron experiments, which show the maximum of the correlation function of the molecular centres of the O-atoms at 2.84 Å at 4°C (Narten [1974]). This maximum slowly shifts with increasing T to larger values (fig. 28.10). At 200°C the maximum is at 2.94 Å. The area

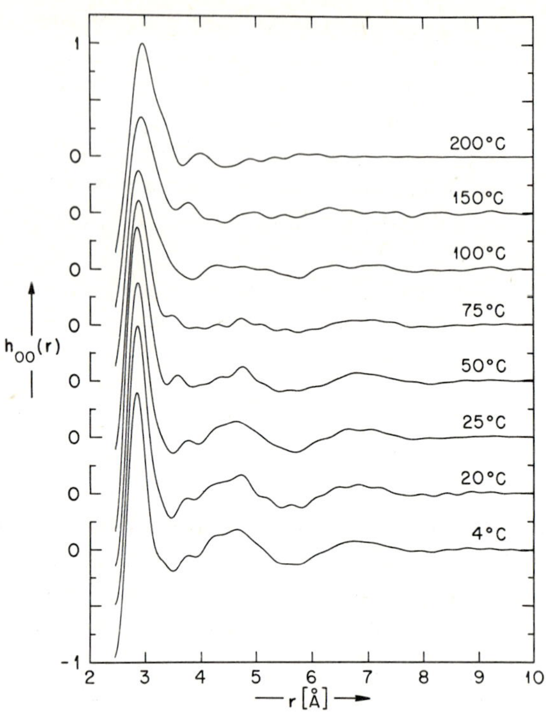

Fig. 28.10. X-ray scattering results on water show a small change of the distance of the first next neighbours (Narten [1974]).

under the first maximum corresponds to an average coordination number slightly larger than 4. The distances for the maxima of the first and second neighbour in the correlation function indicate that at lower T the average deviation from ideal tetrahedral coordination in liquid water must be quite small. At room T the correlation function shows a certain order up to 8 Å. In the three dimensions this corresponds to an order of some hundred molecules. The electron scattering experiments are in good agreement with the X-ray experiments (Kalman et al. [1974]). Fisher and Adamivich [1963] have concluded correlations of fluctuations in regions with diameters of about 15 Å.

Experiments on alcohols have shown that the anharmonicity of alcohols is reduced by H-bonding (Luck and Ditter [1968]). In supercooled water the O_F values decrease with T, one can also extrapolate the T-function, observed above the melting point (fig. 28.6), to lower T.

28.4. Non-bonded OH groups and properties of water

A two-parameter model with non-H-bonded and H-bonded groups is surprisingly successful in describing some properties of water*. The experimental function of $O_F = f(T)$ provides such a parameter and one needs no further adjusted constants. This will be shown in this section for water (Luck [1967, 1974a]). Alcohols can also be treated similarly (Luck [1967]).

We describe the gas state in a first approximation with the ideal gas law, and the ideal solid state with the ideal lattice model. Similarly, one can describe an ideal liquid assuming that in an ideal liquid the density decreases with T only by the T-dependent vibration volume (Luck and Ditter [1971]). In the second approximation with real non-polar liquid one has to assume hole defects. The ideal gas law is more valid at high T, whereas the ideal liquid is more valid at low T. For liquids with dominant H-bonds, one can additionally assume orientation defects of the H-bonds (Luck and Ditter [1972]). The spectroscopically determined O_F values are needed to understand these orientation defects. Thus a simple model for the first approximation is obtained. Deviations between the properties of this idealised model and the experiments are not larger than the deviations of the properties of gases from the ideal gas law. The O_F values offer a possibility to consider the properties of water along general lines.

28.4.1. Melting heat

Figure 28.6 yields the approximate value of $\Delta O_F(T_F) \approx 0.1$. From the H-bond energy of H_2O mole of $\Delta H_H = = 8$ kcal/mole we obtain the contribution of H-bond energy to the melting heat of $0.1 \times 8 = 0.8$ kcal/mole. This is in fair agreement with the experimental value of 1.44 kcal/mole and the effect of dispersion energy on the melting process. This was discussed in the first section. If we assume that the influence of dispersion forces on $H_F(H_2O)$ is approximately the same as on $H_F(H_2S) = 0.57$ kcal/mole, the change of O_F of water at T_F should be

$$\frac{H_F(H_2O) - H_F(H_2S)}{H_H} = \frac{1.43 - 0.57}{8} = 0.107. \qquad (28.6)$$

This is in very good agreement with the spectroscopic value.

* For an excellent review on the properties of water and aqueous solutions see Gmelin [1963a].

During melting the H-bond content changes by about 10%. If these orientation defects were statistically distributed over the whole ice lattice, the breakdown of the ice lattice and the sharp melting point could not be understood. But the theory of cooperative mechanism can describe both (Bresler [1939]). Especially in the case of liquids in which the H-bonds determine the main properties, the cooperative mechanism can easily be demonstrated by means of the angle dependence of the H-bond energy (see ch. 11).

28.4.2. Specific heat

In terms of the intermolecular forces the change in the heat content of liquid water from 0 to 100°C can be expressed by

$$\Delta U = \int_0^{100} [C_v(\text{liq}) - C_v(\text{vapour})]\,dT = 1.1\ \text{kcal/mole}.$$

If this enthalpy change can be considered to be induced mainly by the opening of the H-bonds, it follows that $1.1/8 = 0.137$ of the H-bonds rupture at this T-interval. Considering the intermolecular degrees of freedom in the specific heat corresponding to eq. (28.7), the estimate of the rupture of H-bonds with the change of heat content would give a value of $\Delta O_F(0\text{–}100\,°\text{C}) \approx 0.1$. The spectroscopic result of fig. 28.6 gives $\Delta O_F = (100\text{–}0\,°\text{C}) = 0.12$. The anomalous T-increase of the specific heat C of liquid water up to T_c can be described quantitatively within (fig. 28.11) the limits of error of the spectroscopic method by the formula

$$C = C_{vap} + (\partial O_F/\partial T)\Delta H_H + 2(1 - O_F)R, \qquad T_F < T < T_c, \qquad (28.7)$$

where C_{vap} is the specific heat of H_2O vapour in the ideal gas state as a measure of the intramolecular degrees of freedom, and R denotes the gas constant.

The last term gives an approximation of the general interaction factor of the liquid state. One can show that this fraction of C, due to non-polar interactions in the liquid state is given approximately by the number of first next neighbours times $R/2$ (Luck and Ditter [1971]). This fraction is small for water, especially at higher T. In the lower T-region the coordination number of the main domains of water is about 4. Thus we set this general term $(1 - O_F)\, R\, 4/2$. This approximation is rough but errors are second-order effects. A further similar term for the free OH term is neglected because the correct coordination number of these groups is unknown and could be T-dependent.

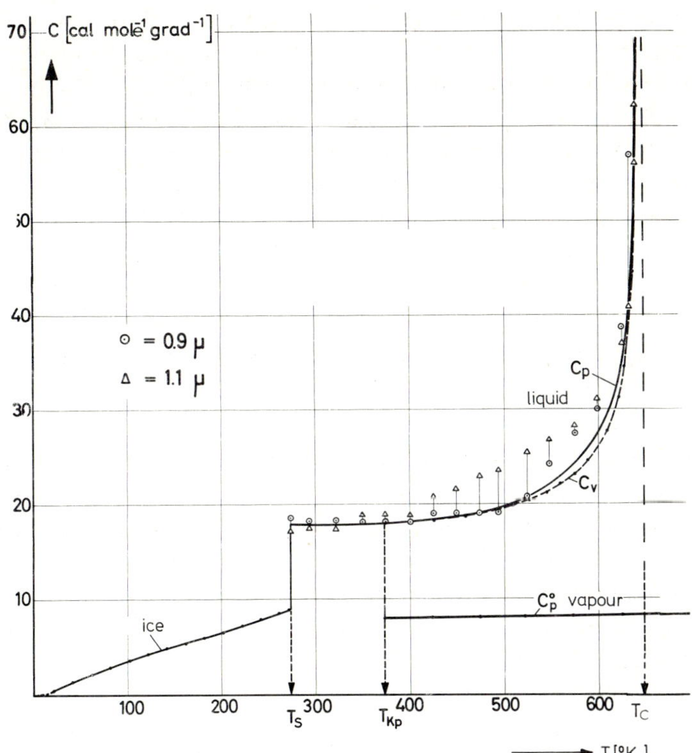

Fig. 28.11. Specific heats of liquid water, ice and water vapour at low p. Circles and triangles: theoretical values calculated by eq. (28.7) and the spectroscopically determined contents of non-H-bonded OH groups.

The simple formula (28.7) needs no adjusted parameters. It also explains the minimum of the specific heat at 38 °C. The increase of O_F with T is nearly linear up to 38 °C, therefore the differential quotient in eq. (28.7) is constant in this region. Above this T, O_F increases a little faster than linearly. The last term $1 - O_F$ decreases with T, thus both factors result in a minimum at the right T (Luck and Ditter [1972]) where it is observed.

The specific heat at T_c and ρ_c as function of T has a pressure-dependent maximum (Gmelin [1963b]). The steep slope of C_p shortly before T_c can be understood by the slope of O_F. The specific heat as a function of T has a nearly symmetrical maximum at T_c and ρ_c. It fades away at about 400 °C. Exact measurements of the properties in the supercritical region are difficult because large density gradients are observed experimentally up to $T/T_c \approx 1.1$ (Luck and Ditter [1966]). The slope of C_p for $T > T_c$ may indicate that the proportion of energetically unfavoured H-bonds decreases. With $p > p_c$ the C_p maximum moves to higher T and flattens (Gmelin [1963b]). Both effects are correlated with similar behaviour of the O_F function determined with overtone spectra at higher p.

Walrafen [1972] has given a more complicated formula to describe the specific heat in the range $0°C < T < 100°C$ assuming three types of water molecules with two, three and four H-bonds. He has to adjust some constants by a computer method with reference to his Raman measurements of water. But one has to stress that his Raman data give no absolute values for the H-bond content. The scattering intensity coefficients of the different H-bond Raman bands are still unknown. Walrafen's result agrees with our experimental O_F values (e.g. $O_F(0°C) = 0.075$). We emphasise that the spectra only give information on bonded and non-bonded OH groups and not information on molecules with one, two, three or four H-bonds. The interpretation of specific heat values with the assumption of different species seems to us unnecessary and is difficult to prove experimentally.

28.4.3. Vaporisation heat

The vaporisation heat L_v under constant volume and saturation conditions can be represented as a function of T by

$$L_v = W_{real} + (1 - O_F)\Delta U_s + O_F W - 2(1 - O_F)RT, \tag{28.8}$$

where W_{real} denotes the intermolecular interaction energy in the saturated vapour state, ΔU_s is the sublimation energy as a measure of the total intermolecular interaction of H-bonded OH groups in the liquid state, and W is the intermolecular interaction energy of non-H-bonded OH groups in the liquid state. W_{real} is obtained by the difference of the enthalpy of H_2O vapour in the ideal gas state (given by the uppermost straight line in fig. 28.1) and the enthalpy H of the saturated H_2O vapour (given by the uppermost curve in fig. 28.1). W is given by $\Delta U_s - \Delta H_H = 3.6$ kcal/mole.

Equation (28.8) represents the differences between the H-values of the saturated H_2O vapour and liquid water (see fig. 28.1). The second term of eq. (28.8) represents the energy which should be supplied to transfer the H-bonded OH groups from the liquid to the ideal gas state. The third term of eq. (28.8) is the equivalent term for the non-H-bonded OH groups. The last term is the heat content of the intermolecular degrees of freedom in a liquid (compare this term with the analogous term of specific heat in eq. (28.7)).

Applying eq. (28.8) the correct values and the correct dependence of T on the heat of vaporisation L_v are obtained within the limits of error of the spectroscopically determined O_F values (Luck [1967, 1974c]).

28.4.4. Density

The density ρ of liquid water including the anomalous maximum at 4°C and the anomalous density increase at the melting point can be given by (Luck [1967])

$$1/\rho = (1 - O_F)V_B + O_F V_F, \tag{28.9}$$

where V_B denotes the partial molar volume of H-bonded OH groups, taken as ice-like volume, V_F denotes the partial molar volume of non-H-bonded OH groups, and V_F is an adjusted parameter, which cannot be obtained from the spectra. V_B and V_F are assumed as linear functions of T. The coefficient of thermal expansion of V_B was chosen as that of ice and the coefficient of thermal expansion of V_F was adjusted. Equation 28.9 agrees well with experimental values at saturation conditions up to 350°C (Luck [1967]). Near T_c some deviations occur; this is to be expected because near T_c the constancy of β is invalid for all liquids.

We have to emphasise that the density, eq. (28.9), needs adjusted parameters in contradiction to the representation of C_p, L_v and H_σ.

28.4.5. Surface energy

The surface tension σ of water is anomalously large. This is often discussed as a result of the anomalously high intermolecular interaction; but this is only part of the picture. σ is the free energy which one needs to enlarge the surface of a liquid by 1 cm^2. For the discussion of structural problems in physical chemistry one must consider equal numbers of molecules, generally one mole molecules. Therefore we should only use the molar surface tension σ_M (Eucken [1948] and Luck and Ditter [1971]).

$$\sigma_M = \sigma N_A^{1/3} V^{2/3}, \tag{28.10}$$

where N_A is Avogadro's number, V is the molar volume, and σ_M is the energy which we need to bring one mole molecules from the inside of a liquid to its surface. σ of water is one of the largest values known. But σ_M

is of medium size, σ_M of organic molecules, like soaps, is larger than σ_M of water. That means the high value of σ depends mainly on the small size of the water molecules. Water contains in 1 cm^2 surface more molecules than most other liquids. σ_M is nearly proportional to the molecular weight of non-polar material. But σ_M of water is higher than σ_M of methanol or ethanol. This is a result of the higher intermolecular forces caused by H-bonds. We have to take into account that σ_M is a free energy, determined mechanically under isothermal conditions with heat exchange with the environment. Therefore, the surface enthalpy H_σ is given by the Gibbs–Helmholtz equation

$$H_\sigma = \sigma_M - T\left(\frac{\partial \sigma_M}{\partial T}\right)_p. \qquad (28.11)$$

For all non-polar liquids σ_M is an almost linear T-function for $T < 0.9\,T_c$. Therefore, H_σ for non-polar liquids is T-independent for $T < 0.9\,T_c$ (Luck and Ditter [1971]). (Einstein [1901] assumed that this is the consequence of the fact that the specific heat in the surface layer should be zero. But it is only necessary that C_p is equal inside the liquid and in the surface layer.) The decrease of H_σ for $T > 0.9\,T_c$ to zero at T_c depends on the high saturation pressure near T_c. The interaction energy of the surface layer with the vapour state cannot be neglected near T_c.

A specific property of liquids with a high portion of H-bonds in the intermolecular forces is the T-dependence of H_σ. Generally, H_σ has a maximum as a function of T (fig. 28.12).

The increase of H_σ of water with T and the absolute value at room T can be represented by eq. (28.12) (Luck [1967])

$$H_\sigma = [(1 - O_F)\Delta H_s + O_F W]_{in} - [(1 - O_F)\Delta H_s + O_F W]_{surf}. \qquad (28.12)$$

This equation is obtained from eq. (28.8). The suffixes *in* and *surf* signify the values in the bulk of the liquid and at the surface. H_σ in eq. (28.12) depends on a difference of two similar values. Therefore, the error is larger than in eq. (28.8). But the absolute value of H_σ at room T, $H_\sigma = 1.6$ kcal/mole, is obtained by eq. (28.12) (fig. 28.12).

A similar formula for C_p, L_v and H_σ can be deduced for alcohols from the spectroscopically determined O_F values (Luck [1967]) also in good agreement with experimental values. In this case we need a structural model for the alcohols as chain-like aggregates.

For a model of chain-like aggregates H_σ of alcohols should be (Luck [1967])

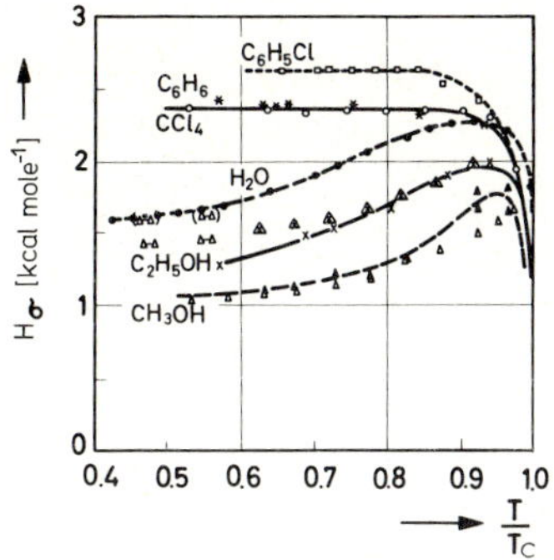

Fig. 28.12. Surface enthalpy H_σ versus reduced temperature. Triangles determined by eq. (28.12) and the spectroscopic contents of non-H-bonded OH groups.

$$H_\sigma = W_D + \tfrac{1}{3}O_F W, \tag{28.12}$$

where W_D is the dispersion forces of an alcohol molecule in a chain, and W is the dispersion energy of an alcohol molecule in the end group of a chain. According to this formula, H_σ of alcohols should become constant at low T because O_F is very small.

No measurements of H_σ of alcohols at low T are available. Therefore, eq. (28.12) was an interesting example of whether the spectroscopic method could predict values. We have made measurements of the surface tension at low T with a capillary method. Figure 28.13 shows that the expected T-independency of H_σ is obtained. At low T alcohols behave like non-polar materials with reference to the surface enthalpy. H_σ in this T-region is proportional to T_c.

28.4.6. Expansion of the areas of hydrogen bonded molecules

Starting with the Némethy and Scheraga model [1962] the large aggregates of liquid water were described by clusters. There is no spectroscopic evidence of clusters, but the low degree of non-H-bonded molecules indicates that a large number of H_2O molecules are cross-linked by H-bonds. The quantitative knowledge of O_F allows us to estimate the average size z of the group of water molecules which are H-bonded. To do this one needs an idealised model. This should take into account the ice-like tetrahedral arrangement which is claimed by the X-ray and neutron scattering experiments (Narten [1972, 1974]) and the C_{2v} symmetry claimed by the Raman data (Walrafen [1972, 1974]). The

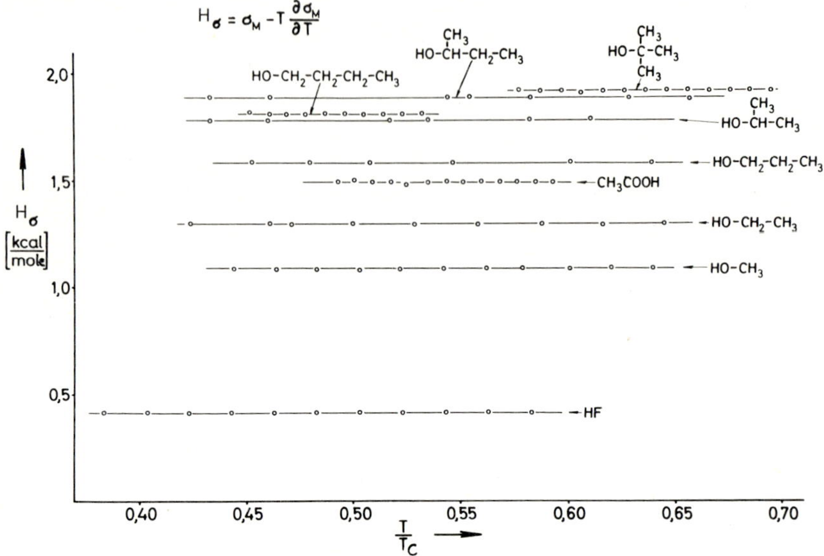

Fig. 28.13. Surface enthalpy H_σ of alcohols at low T. In this T-region alcohols have a T-independent H_σ similar to non-polar liquids.

simplest way to do this is the tridymite arrangement. In addition, one should consider the cooperative mechanism of H-bonds due to the angle dependence of the H-bond energy (see ch. 11). From this viewpoint we can assume that the orientation defects of non-H-bonded OH groups are concentrated in flickering fissure plains between ordered areas. The lifetime of these arrangements we would expect in the order of magnitude of the known relaxation time of 10^{-12} s. Figure 28.14 gives the ideal estimation of the dimensions of the area of H-bonded molecules, calculated by means of the spectroscopic O_F values. The non-H-bonded OH groups are orientation defects at the surfaces of these ordered areas. The small content of O_F values at lower T suggests that the different areas of ordered zones are in contact. The boundary areas of the ordered zones are the non-H-bonded OH groups and give the area of higher density. In this model the assumption of monomers is precluded. At high $T > 200\,°C$ these assumptions seem to be too crude since there is some spectroscopic evidence that in this T-region small aggregates may exist. The frequency of the band maxima of the free OH groups changes slightly between 200 °C and T_c (Luck [1974c]). This wave number change is of the same order as the observed perturbation of a free vibration in a molecule whose

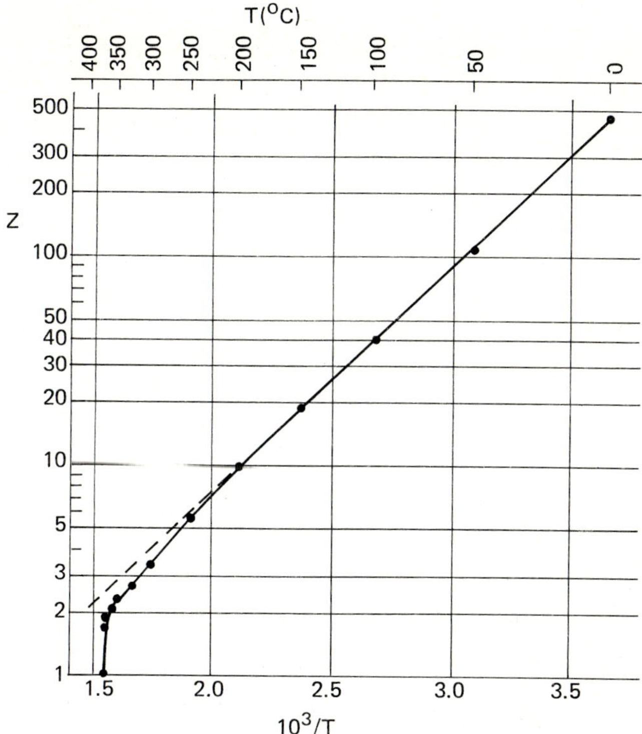

Fig. 28.14. Estimate of the size of H-bonded aggregates in liquid water at saturation conditions. Z: number of molecules per aggregate.

second OH is H-bonded (see ch. 11). In this T-region free molecules without this wave number shift become more and more dominant. We should therefore assume in this T-region an equilibrium between small aggregates (molecule number smaller than 6) and monomers.

28.4.7. Hydrogen bond energy

In the first section we assumed a H-bond energy of water per bond of about 4 kcal/mole. There are some spectroscopic results with smaller energies of 2.5 kcal/mole (Worley and Klotz [1966], Lindner [1970] and Walrafen [1972]). These authors assume an equilibrium

$$OH_{free} \rightleftharpoons OH_{bond} \qquad (28.13)$$

and

$$K = \frac{C_{\mathrm{OH,bond}}}{C_{\mathrm{OH,free}}} = \frac{\text{intensity bonded OH band}}{\text{intensity unbonded OH band}}. \qquad (28.14)$$

The overtone method with its more separated bands and with the larger T-interval from ice to overcritical T should give better values. Using this method we obtained energies from the HDO band of about 2.1 kcal/mole. This contradiction with the value of 4 kcal/mole can easily be resolved. Wc believe if one discusses the network of H-bonds in water as an equilibrium it is much better to write

$$\mathrm{OH_{free}} + \theta_{\mathrm{free}} \rightleftharpoons \mathrm{OH_{bond}}, \qquad (28.15)$$

where θ_{free} denotes the lone electron pairs not engaged in H-bonds, and

$$K = C_{\mathrm{OH,bond}}/C_{\mathrm{OH,free}}C_{\theta,\mathrm{free}}. \qquad (28.16)$$

In water the following is valid

$$C_{\theta,\mathrm{free}} = C_{\mathrm{OH,free}}. \qquad (28.17)$$

Therefore, we can write

$$K = C_{\mathrm{OH,bond}}/C^2_{\mathrm{OH,free}}. \qquad (28.18)$$

These log K-values plotted against $1/T$ are given in fig. 28.15a. Again we get a straight line up to 200 °C but with an energy of 3.8 kcal/mole. This value is consistent with the calorimetric data of water. For $T > 200$ °C we do not get a straight line, which is in agreement with fig. 28.14. In this T-region we expect equilibria of small aggregates and monomers. In this case eq. (28.15) is not valid. The reason why Walrafen and other authors could not recognise the neglected θ_{free} is that the logarithm of intensities of every band is in a certain T-region a linear function of $1/T$. In such a case every combination of quotients or quotients of squares of intensities, etc. give straight lines when plotted against $1/T$.

For alcohols eq. (28.17) is not valid. In this case we have to take into account

$$C_{\theta,\mathrm{free}} = [(1 - O_{\mathrm{F}}) + 2O_{\mathrm{F}}]C_0. \qquad (28.19)$$

With this correlation our alcohol data also give linear log K versus $1/T$ plots up to 300 °C with a slope of 5.7 kcal/mole (fig. 28.15b). We can conclude from this section that in water and alcohols under saturation conditions in the liquid state below 200 °C larger H-bonded aggregates exist, and coupled equilibria of smaller aggregates and monomers exist above 200 °C.

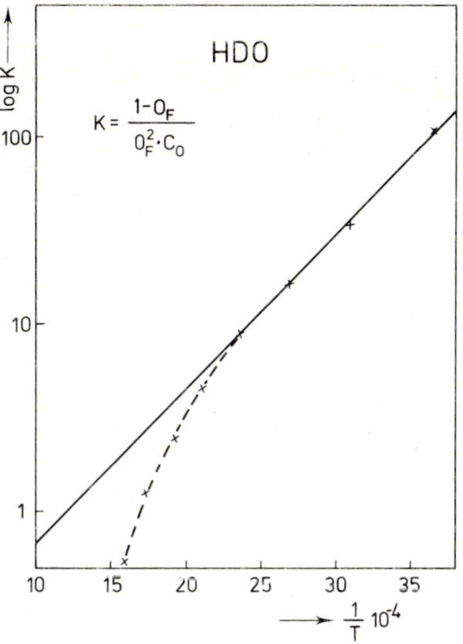

Fig. 28.15a. Equilibrium constant K of water aggregation versus $1/T$.

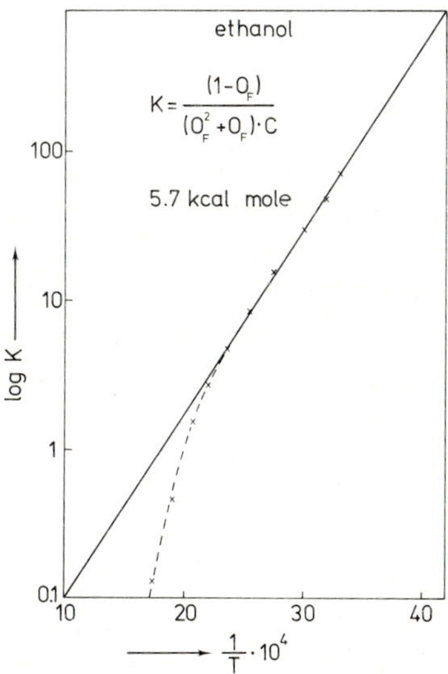

Fig. 28.15b. Equilibrium constant K of ethanol aggregation versus $1/T$.

28.4.8. Magnetic susceptibility

The magnetic susceptibility increases almost linearly from 0 to 80 °C (Cini and Torrini [1968]). These authors discuss this in terms of the statistical theory of Vand and Senior [1965] as a linear function of the different H-bonded species. This linear increase can also be related to the O_F function.

28.4.9. NMR chemical shift

The chief features of some other properties can also be understood through the O_F function. For example, there are some parallels between O_F and the chemical shift. The O_{17} chemical shift of measurements of Luz

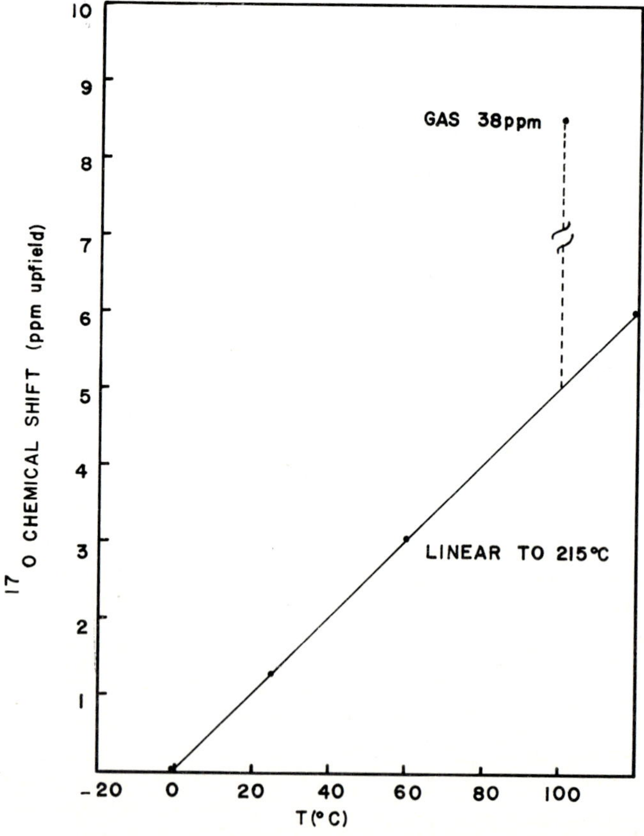

Fig. 28.16. NMR ^{17}O chemical shift of liquid water versus T (Glasel [1974]).

and Yagil [1966] and Fister and Hertz [1967] in a diagram given by Glasel [1974] is shown in fig. 28.16. We cannot compare these chemical shifts completely with the O_F function since the change of chemical shift when ice melts is missing. But if we transfer the change of O_F during melting from the overtone data or from the calorimetric data as 0.1, then we can assume that the change of oxygen chemical shift of 43 ppm between liquid water at 0 °C and the vapour state corresponds to a change of the H-bond state of 90%. The chemical shift of liquid water changes from 0 to 100 °C by 5 ppm which corresponds to 10.3% of the total. The chemical shift, too, is a linear function of T, as is approximately the case with the O_F data determined from overtone measurements. The proton chemical shift gives almost parallel results (see Glasel [1974], data by Hindman [1966]). The H-shift data of the vapour state of Hindman [1966] may be influenced by a calculation based on some assumptions on the liquid structure. Glasel [1972] estimates a decrease of 15% bonded protons from 0 to 100 °C.

28.4.10. Dielectric constant

Haggis et al. [1952] gave the first useful theory of the static dielectric constant ϵ to T_c under saturation conditions. Their theory is based on some assumptions on O_F which they obtained from the heat of vaporisation. This indirectly determined function agrees well with the spectroscopically determined values. From this viewpoint the dielectric constant can easily be understood on the basis of the first approximation. The static dielectric constant in a large p–T region calculated with data of Quist and Marshall [1965] is given in fig. 28.17. The dotted lines are values under saturation conditions. For this, the simple theory of Haggis et al. [1952] explains the main features well. The diagram shows clearly that an increase of density higher than saturation conditions at constant T increases ϵ like a decrease of T at saturation conditions. A comparison of some O_F measurements with overtone spectra at higher p shows a parallelism between O_F behaviour and the change of ϵ. Both experiments are consistent with the assumption that an increase of p at constant T increases the concentration and thus changes the equilibrium between the H-bonded and the free OH content.

Rahman and Stillinger [1971] have evaluated $\epsilon'' = f(\epsilon')$ with their computer-dynamical calculation but their result is only roughly in agreement with experiment. An attempt by Pople [1951] to calculate the static dielectric constant with a model of angle bending, specified by Del Bene

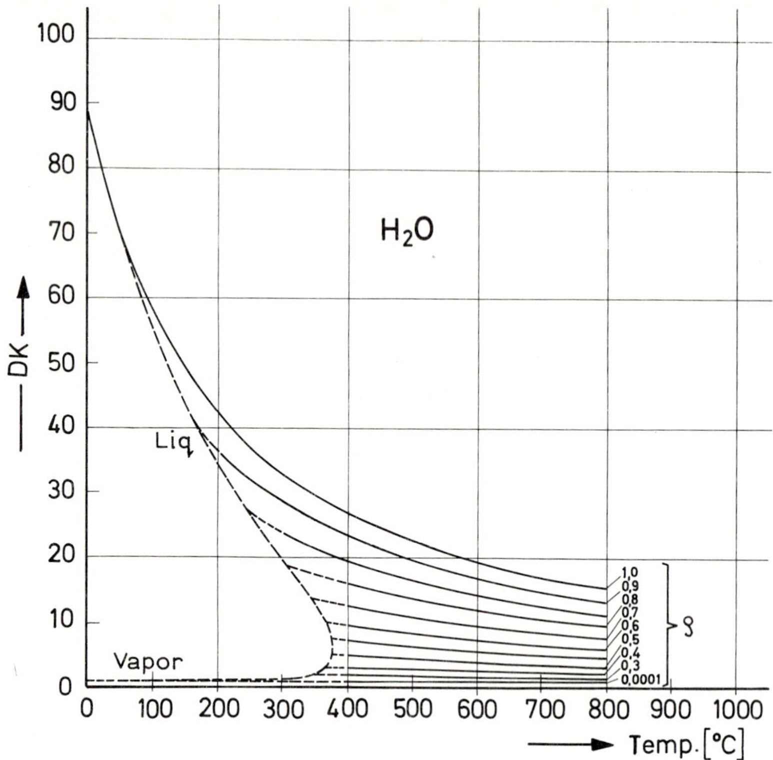

Fig. 28.17. Static dielectric constant of water as a function of T and density.

and Pople [1970], gave agreement with experiments only a little worse than the model of bond breaking by Haggis et al. [1952] (see Hasted [1974]). But Pople extended his theory only to 90°C.

28.4.11. Ultrasonic absorption

A new attempt to describe the ultrasonic properties by a mixture model is given by Breitschwerdt [1974]. He has shown that he needs three different states of liquid water molecules. This is in agreement with the spectroscopic results which show that the isosbestic points are not good ones and that the curve analysis needs three bands. But Breitschwerdt needs about 20 adjusted constants for every T to fit the water properties. Figure 28.18 gives an example of his calculated data.

We have to stress that this discussion of the water properties with the spectroscopically determined O_F function can only account for the main

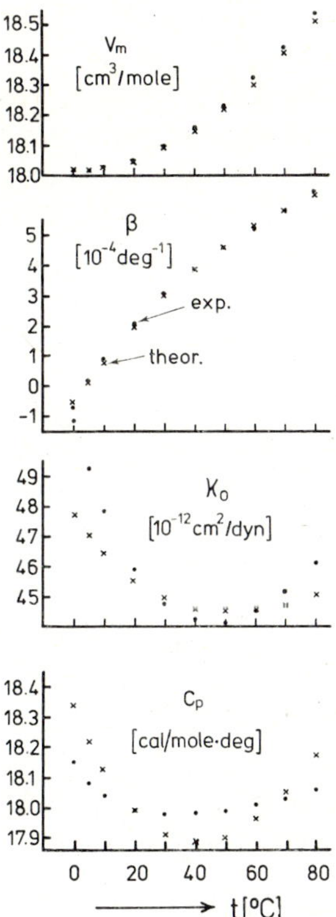

Fig. 28.18. Properties of liquid water versus T. V_m: molar volume; β: thermal expansion coefficient; κ_0: total compressibility; C_p: specific heat; circles: experimental values; crosses: calculated by sound absorption (Breitschwerdt [1974]).

features. Exact interpretation of the water properties would need more than this one parameter O_F. The modern view of water research inclines to the knowledge of distribution functions of different molecule states. These functions may also be evaluated from spectra. Such theories are still in their early development (see Gordon [1967]). However, it is remarkable how efficient the one parameter function O_F is to explain water properties. This is surprising because one needs no adjusted constants, only the experimentally determined O_F values.

28.4.12. Dynamic properties

Stillinger and Rahman's calculation [1972] is not in good agreement with static properties. But the calculated relaxation time τ (34.4°C) =

5.6×10^{-12} s agrees much better with the experimental value $\tau = 6.7 \times 10^{-12}$ s.

So far only one relaxation has been discussed, but in view of the complicated water structure one would expect more than one τ, in fact, a minimum of one τ of OH groups in ordered areas, and one τ of non-bonded OH groups. But from the viewpoint of the last section at $T < 200°C$ the bonded OH groups may dominate and thus cause the observation of only one τ. In ice τ is of the order of 10^{-6} s and yields the relation $\tau = 5.3 \times 10^{-16} e^{B/RT}$ s with $B = 13.25$ kcal/mole (Auty and Cole [1952]; see also fig. 28.19). B is in the region of the total intermolecular interaction energy. In liquid water, τ is known to 100°C and follows the relation $e^{A/RT}$ with $A = 4.5$ kcal/mole (fig. 28.19; see also Hasted [1974]). Hasted [1974] discusses this energy in relation to the energy of one H-bond. This energy agrees well with the activation energy A of the self-diffusion of $H_2{}^{16}O$ or $H_2{}^{18}O$ or HOD in D_2O in $0°C < T < 55°C$, $A = 4.6$ kcal/mole (Wang [1951, 1953] and Samoilov [1961]). The activation energy of the viscosity at room T is about 5.1 kcal/mole (Eyring et al. [1941] and Kampmeyer [1952]). Hasted [1974] stresses some curvature in the semi-logarithmic plot of the dielectric relaxation time.

The dielectric constant ϵ_∞ at infinite frequency in ice is $\epsilon_\infty(0°C) = 3.2$, and in liquid water $\epsilon_\infty(0°C) = 4$ (Hasted [1974]). Hasted discusses the T-dependence of ϵ_∞ in relation to a second relaxation time $\tau(25°C) = 0.53\ 10^{-13}$ s. The activation energy of this process is about $1 \pm$

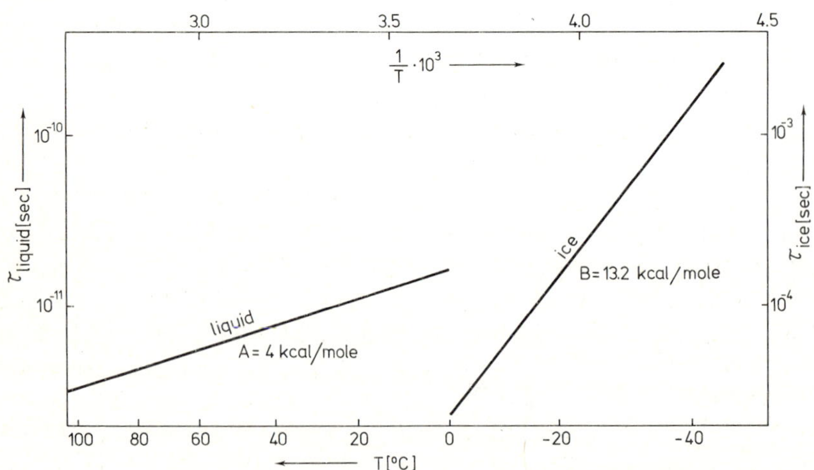

Fig. 28.19. Relaxation time τ in s of liquid water and ice versus T.

0.5 kcal/mole (Hasted [1974]). Hasted discusses this process as rotation of a molecule without breakage of H-bonds by those molecules in water which possess less than two H-bonds. The portions of these molecules are estimated to be 0.045 at 0 °C, 0.05 at 25 °C and 0.075 at 60 °C. Therefore, two distinct relaxation processes are expected at higher T. These discussions of the dielectric relaxation times are consistent with the results from overtone spectra, especially with low O_F content in the region of low T.

The static dielectric constant of ice has an interesting behaviour. It has two deep minima in ice doped with 10^{-4} mole/l HF or 10^{-2} mole/l HF (Jaccard [1974]). Jaccard discusses this as cancelling of the two different relaxation mechanisms: first the rotation mechanism of pure ice, and secondly, a proton jump mechanism due to the acid protons.

The spin–lattice relaxation time T_1 of water determined by NMR measurements increases with T from 0° to room T nearly linearly and then to 100 °C slightly more rapidly (Glasel [1972, 1974]). A molecular relaxation time $\tau_c(25\,°C) = 2.5 \times 10^{-12}$ s is given by NMR data (Hertz [1974]).

28.5. Electrolyte solutions

28.5.1. Spectroscopic results

The properties of aqueous electrolyte solutions can also be understood approximately in terms of non-H-bonded OH groups. Bernal and Fowler [1933] made the qualitative statement that some properties of water change with the addition of electrolytes, to a first approximation, like a change of T in the case of pure water. They thus introduced the term structure temperature of an electrolyte solution T_{str} as the temperature at which pure water would have similar properties. This assumption is confirmed by the overtone spectra (Suhrmann and Breyer [1933]).

Overtone spectra give a method for defining T_{str} quantitatively by the T at which pure water has the same content of non-bonded OH groups (Luck [1965, 1973]). The results of these investigations are:

1. The spectra are relatively insensitive to electrolyte additions. One needs concentrations larger than 0.5 molar to identify pronounced changes in the spectra. This is in contradiction to the salt-out or salt-in effects on the solubility of solutes in water. The small effects of electrolytes on water IR spectra are in agreement with the small effects on NMR chemical shift (Shoorley and Alder [1955]).

2. T_{str} of salts determined by O_F give an ion series similar to that of the lyotropic or Hofmeister ion series of colloid chemistry (Hofmeister [1890] and Hardy [1900]). From these experiments the cause of the lyotropic ion series is attributed to changes of the water structure.

3. The main influence on the overtone spectra are the anions. The series of anions depends roughly on the charge e and radius r of the ion through e^2/r. Similar effects on the fundamental IR spectra are observed by Falk [1970, 1971a] and Hecki [1973].

Both the dominance of the anions and the importance of e^2/r for their quantitative effect on the spectra has some parallelism with colloid effects. Figure 28.20 gives as an example the turbidity points T of aqueous solutions of p-iso-octylphenol with a linear chain of nine polyethylenoxide groups versus e^2/r_{An}, where r_{An} is the radius of the anion. The turbidity point T_K depends only very weakly on the concentration of the ethylenoxide derivative; it is influenced by the H-bonds to water and therefore depends on the water structure.

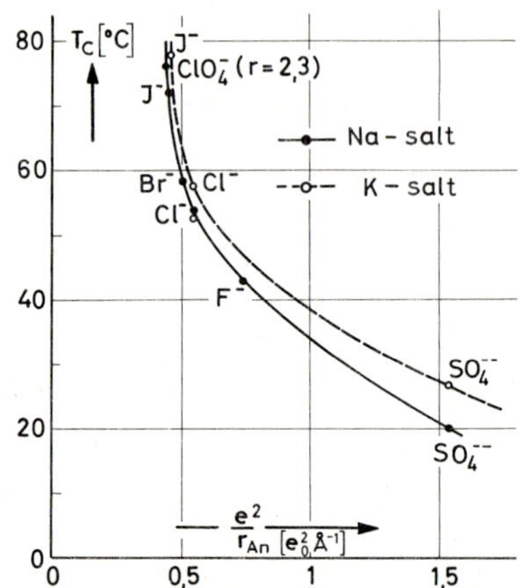

Fig. 28.20. Influence of ions on the turbidity temperature T_K of aqueous solutions of p-iso-$H_{18}C_8$ — — $(OCH_2CH_2)_9OH$ versus the quotient of charge e^2 over the radius of the anions. Salt concentrations: 1/2 mole anions per litre.

Figure 28.21 and 28.22 give examples for the differences between T_{str} and the solutions T versus solution T determined by the overtone method. The quantitative determinations of T_{str} are near the limits of the

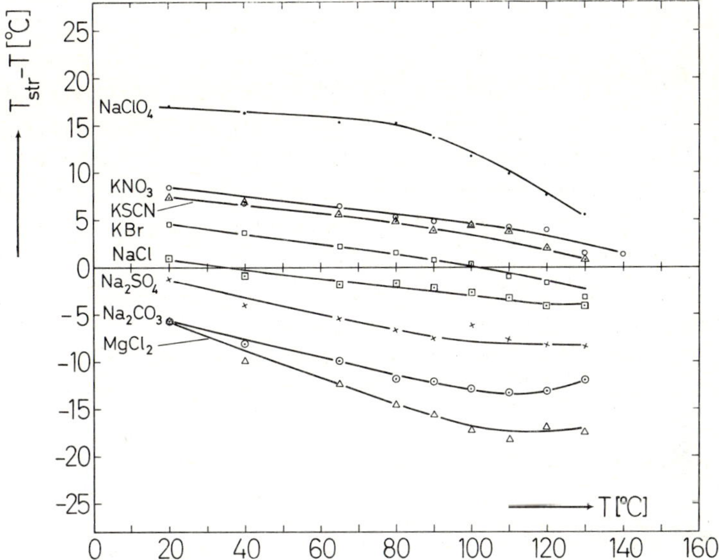

Fig. 28.21. Difference between structure temperature and solution temperature ($T_{str} - T$) for aqueous solutions of 1 M salt versus solution temperature T determined from the second overtone of water.

spectroscopic method. Therefore, the absolute values of T_{str} in figs. 28.21 and 28.22 are imprecise; the relative values of T_{str} however, are reliable. The discrepancy between the small effects of ions on the spectra and the large effects on solubility data can be resolved by considering the models discussed in section 28.4. If we assume that the non-H-bonded OH groups are not distributed statistically, due to cooperative mechanism, but are concentrated in fissure plains between the large systems of H-bonded water molecules, at room T the expansion of aggregates of H-bonded molecules changes greatly with small changes of O_F (Luck [1964, 1974c]). Figure 28.23 gives the content O_F of non-bonded OH groups versus the expansion of H-bonded aggregates at room T. O_F is of the order of 0.1. A change of the content of H-bonded molecules by ions would not change O_F and the spectra significantly at room T. If these views are correct then the ion effect on the spectra should be larger at higher T, because with higher T the content of non-H-bonded OH groups is more sensitive to the expansion of H-bonded aggregates. This prediction was made prior to the

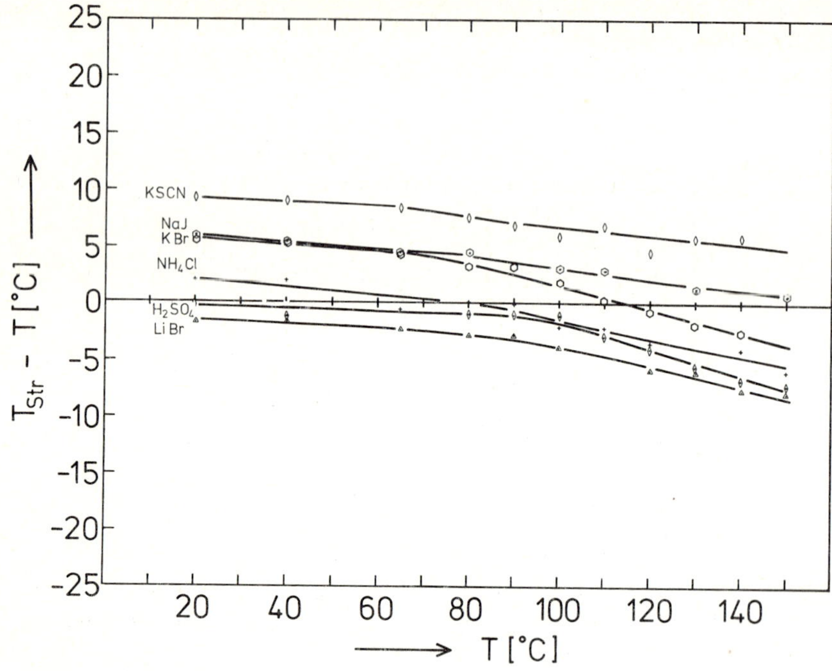

Fig. 28.22. Difference between structure temperature and solution temperature ($T_{str} - T$) for aqueous solutions of 1 M salt versus solution temperature T determined from the second overtone of water.

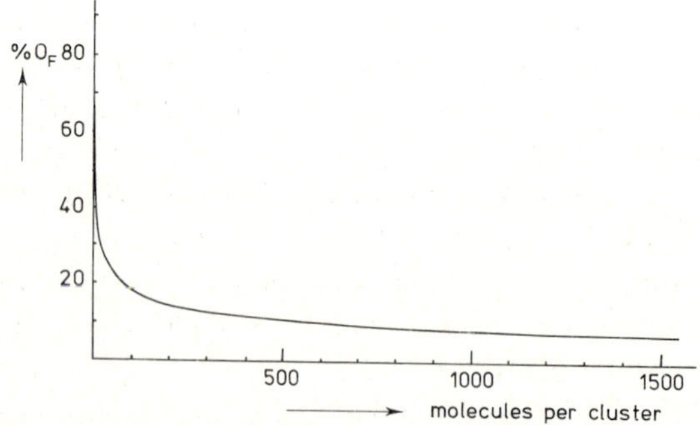

Fig. 28.23. The percent content of non-H-bonded OH groups as a function of the number of molecules per aggregate in a tridymite-like arrangement of water molecules.

salt measurements. Figures 28.21 and 28.22 show that our simple model can describe properties of water and it can give predictions.

From figs. 28.21 and 28.22 we can calculate the change of the expansion of H-bonded aggregates by ions. These values are given versus the solution temperature T in fig. 28.24. This figure shows the difference between ions which have a $T_{str} < T$ and ions with $T_{str} > T$. The first group

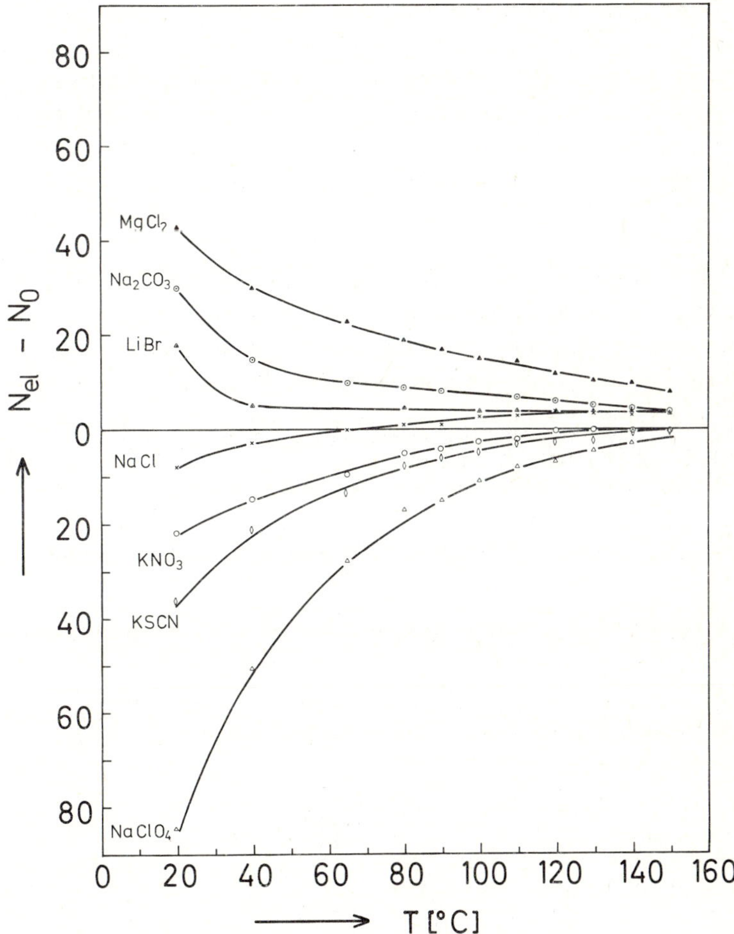

Fig. 28.24. The mean number of water molecules N_{el} per aggregate in electrolyte solutions minus the number N of molecules per aggregate in pure water versus solution temperature T.

of ions has a higher content of H-bonds. This may be the effect of a fixation of OH groups by the Coulomb field of ions. Sack [1927] and Debye [1929] have discussed the change of dielectric constant by ions. From these results they concluded that around every ion with the charge 1 e all water dipoles are fixed by the Coulomb field in an area of about 11 Å. In all three dimensions this gives an aggregate of about 180 water molecules per ion or about 380 water molecules per ion pair of a 1–1 electrolyte. This first group of ions of fig. 28.24 is called structure makers. They increase the H-bond content in water and induce salt-out effects of solutes in water. The second group of ions in fig. 28.24 has a higher T_{str} than the solution T; they reduce the content of H-bonded OH groups. Water becomes more hydrophilic by virtue of these ions, thus some solutes become more soluble in such electrolyte solutions (salt-in effect), these ions are called structure breakers.

28.5.2. Solubility*

From the viewpoint of fig. 28.24 we could predict that the salt-out effect should decrease with higher T. This was verified by measurements of the partition coefficient k of p-cresol between cyclohexane and water (Luck and Gosh [1974]). $k = c(\text{cyclohexane})/c(\text{water})$ is increased, by adding structure-making ions, to $k_s = c(\text{cyclohexane})/c(\text{electrolyte-solution})$. We can estimate the overall effect of the ions by assuming that the primary and secondary hydration spheres of ions are so structured that no solutes can invade into these hydration spheres (Luck [1964]).

We take the following definitions for the number of moles per litre of the different species:

- m' solute in electrolyte solution;
- m solute in pure water;
- m'_{H_2O} water in electrolyte solution which acts as solvent;
- m_{HN} hydrate water which does not act as solvent;
- m_{H_2O} pure water;
- m_c solute in cyclohexane in equilibrium with pure water;
- m'_c solute in cyclohexane in equilibrium with electrolyte solution.

In our simplified model we assume

$$m'_c = m_c \quad \text{and} \quad m/m_{H_2O} = m'/m'_{H_2O}. \tag{28.20}$$

* For a detailed report on solvent properties of water see Franks [1973].

Then the following relations are valid

$$k/k_s = \frac{m_c/m/m_{H_2O}}{m'_c/m'/(m'_{H_2O} + m_{HN})}, \tag{28.21}$$

$$\frac{m'/(m'_{H_2O} + m_{HN})}{m/m_{H_2O}} = \frac{m'}{m} = \frac{m_{H_2O}}{(m'_{H_2O} + m_{HN})} \tag{28.22}$$

and

$$k_s/k = 1 + m m_{HN}/m' m_{H_2O} \tag{28.23}$$

because

$$m_{H_2O} = m'_{H_2O} + m_{HN}, \tag{28.24}$$

$$k/k_s = m'/m, \tag{28.25}$$

$$k_s/k = 1 + k_s m_{HN}/k m_{H_2O} \tag{28.26}$$

and

$$m_{HN}/m_{H_2O} = (1 - k/k_s). \tag{28.27}$$

With m_s the molar concentration of salt, and HN the apparent number of hydrate water molecules per ion pair

$$\mathrm{HN} = m_{HN}/m_s = (1 - k/k_s) m_{H_2O}/m_s. \tag{28.28}$$

Figure 28.25 shows the estimates of HN from partition experiments between cyclohexane and water or electrolyte solutions at 25°C. The series of ions in fig. 28.25 is related to the lyotropic ion series. The concentration dependence of the HN values is different with different ions. Therefore, the lyotropic ion series determined with this method, differs in detail depending on the concentrations of the solute. The result of fig. 28.25 is similar to determinations of HN values with the equilibrium constants of micelle building of *p*-isooctylphenol with a linear chain of 40 polyethylenoxide molecules (Luck [1964]). Here the change of the equilibrium constants due to adding salts was assumed to arise from the reduced number of water molecules effective for dissolving the polyethylenoxide derivate. In this case, too, the concentration dependence of HN values differed with different salts, especially with Na_2SO_4. The HN values determined by partition of *p*-cresol at different concentrations decreased with increasing T in line with the predictions of the spectroscopic model (fig. 28.26). Some determined partition-HN numbers depend strongly on T; for example, with added Na_2SO_4, some HN changes only weakly with T compared with added NaCl (fig. 28.27). This T-dependence follows that

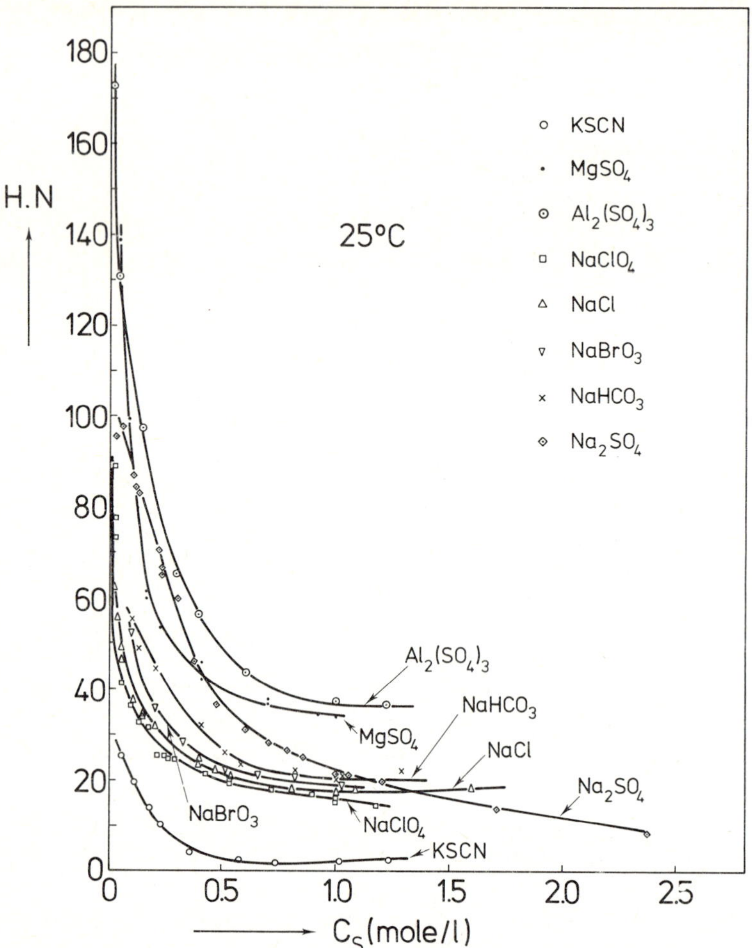

Fig. 28.25. The apparent hydration numbers per ion pair versus salt concentration determined by partition coefficients of *p*-cresol between cyclohexane and electrolyte solutions at 25 °C (Luck and Gosh [1974]).

of the spectroscopically determined T_{str}. T_{str} of Na_2SO_4 solutions change far more with T than do T_{str} of NaCl solutions (see fig. 28.21).

The HN values depend mainly on the type of ions and to a lesser extent on the type of solute. The second effect means that the influence of the fixation of water molecules around the ions depends on the interaction energy between water and the solute.

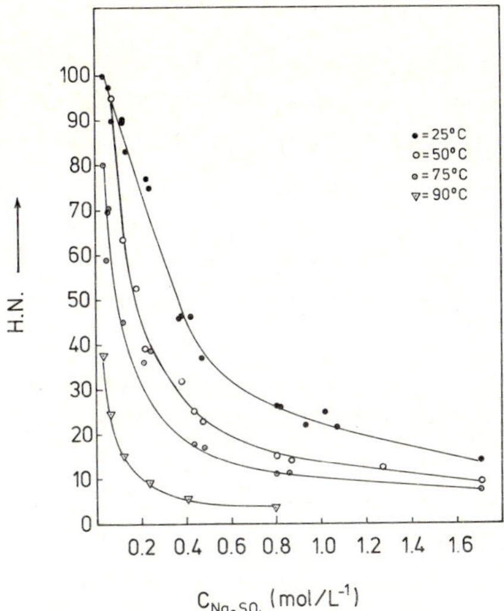

Fig. 28.26. Apparent hydration numbers per molecule of Na_2SO_4, determined by partition coefficients of *p*-cresol between cyclohexane and Na_2SO_4 solutions at different *T*. The hydration number decreases with *T* (Luck and Gosh [1974]).

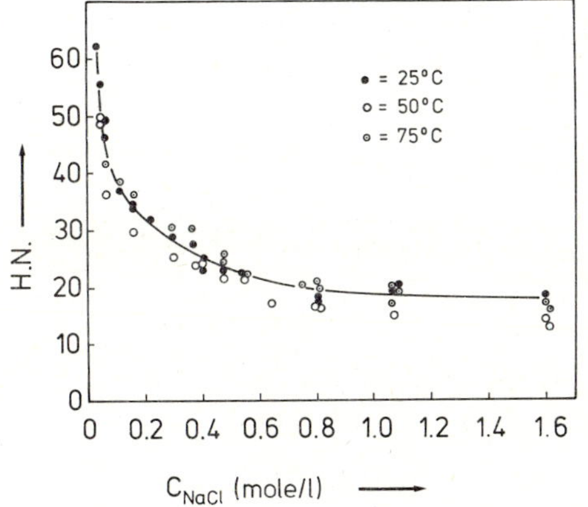

Fig. 28.27. The apparent hydration numbers of NaCl from partition coefficients of *p*-cresol between cyclohexane NaCl solutions does not change much with *T*. This is similar to the observed dependence of T_{str} on *T* (Luck and Gosh [1974]).

28.5.3. Partial molar volume

The simple model of HN or of the effect of non-H-bonded OH groups is only a rough one. It gives a sum of several effects. Around every ion an electrostriction of the water volume takes place. This can be shown by the partial molar volume V_1 of water in electrolyte solutions. A collection of partial molar volumes is given by Millero [1972]. We have determined such values by simulating the concentration dependence of the density of electrolyte solutions by a power series with a computer and determination of the derivates. The ion series, found by the influence on the partial molar volume of water, is similar to the ion found from the spectroscopic determination of T_{str}. Figure 28.28 gives the difference of the partial molar volume V_1 minus the molar volume of pure water V_{01}. The influence on V_1 at equal concentration of Cl^- ions is nearly independent of the type of electrolyte. In fig. 28.29 $V_1 - V_{01}$ is given versus salt concentration. The circles give the values of $MgCl_2$ versus molar concentration of Cl^-.

To combine both results, the partial molar volume and the spectroscopic, or the estimate of the HN, one should assume that we have to

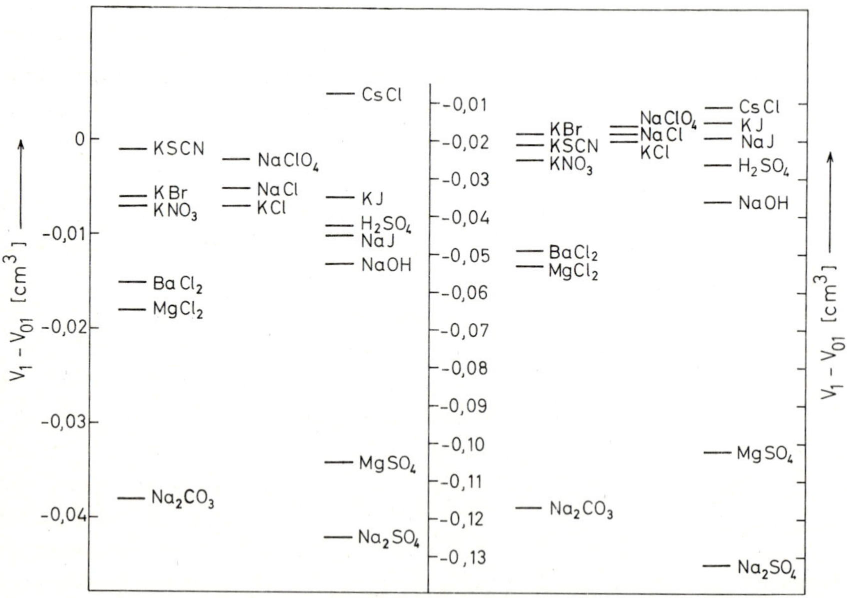

Fig. 28.28. The difference partial molar volume of water V_1 minus the mole volume of pure water V_{01} at 20°C in electrolyte solutions (Luck and Dahm [1974]). Left: salt concentration 0.5 molar; right: 1 molar.

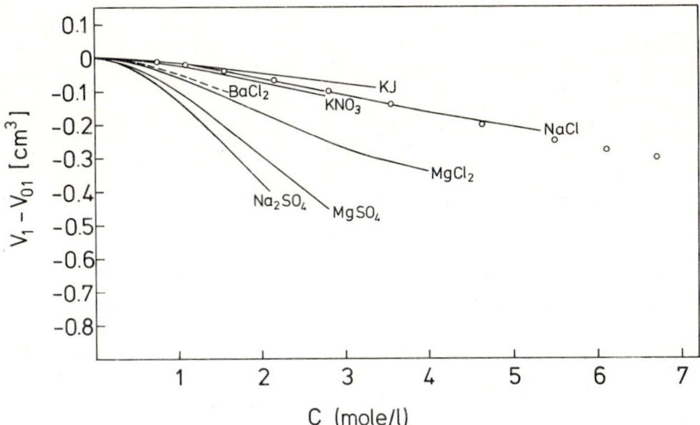

Fig. 28.29. The partial molar volume of water $V_1 - V_{01}$ as a function of salt concentration at 20 °C. Open circles: values of $MgCl_2$ versus Cl^- concentration (Luck and Dahm [1974]).

differentiate between the first hydration shell and the secondary hydration sphere. The first is determined by the orientation of the water dipoles by the ions and gives the main effect on the decrease of volume. The second hydration sphere is determined by the transition stage between the orientation effects due to the ion and the normal water structure. This secondary hydration sphere depends on the size of ions. Large ions give a distinct disturbance of the normal water structure. The spectra indicate the sum effect of the primary and the secondary hydration effects.

There is spectroscopic evidence, too, for details of ion influence on the water structure. The T_{str} values of this section were defined by the content of non-H-bonded OH groups and determined by the optical density in the wave length region of the band maxima of free OH groups. Attempts to determine T_{str} in the wave length region of the H-bond bands give slightly different values of T_{str} (Luck and Zukovskij [1974]). This may be an indication on different water structures around the ions. But there are no indications that the absorption bands of water molecules interacting with ions are much different from the ice absorption band. If this presumption is true it would indicate that the interacting fields of the lone pair electrons of water are not much weaker than the ion fields.

The curves of the partial molar volume of salts V_2 minus the molar volume of salts in solid state V_{02} also have a relation to the lyotropic ion series (fig. 28.30). The curves of Na_2SO_4 cross the other curves, which is in agreement with the HN curves.

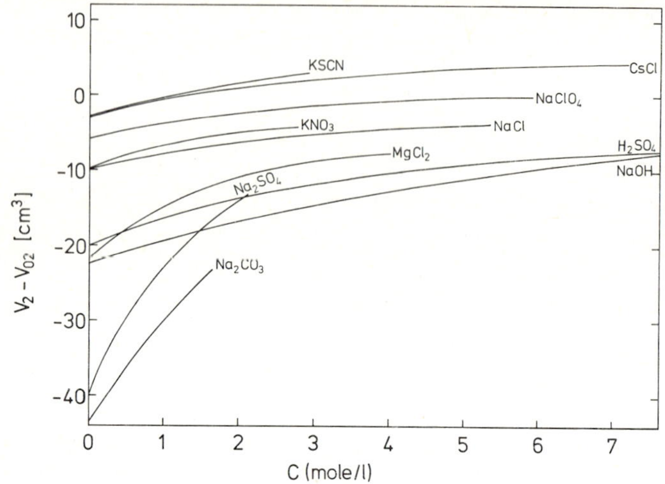

Fig. 28.30. Partial molar volume V_2 of salts in aqueous solutions minus the mole volume of solid at 20 °C (Luck and Dahm [1974]).

Relating to structure effects of water, one can determine three concentration regions of electrolyte solutions. (1) The region of very dilute concentration (Debye–Hückel theory). (2) The region of medium concentration (up to about 1–2 mole/l). (3) The region of high concentrations (larger than 2 mole/l).

In the medium concentration region we can describe the properties of solutions by means of T_{str} or by the content of non-H-bonded OH groups of water. In the high concentration region nearly all water molecules belong to hydrate structures.

The content of this section is related mainly to the region of medium concentration. An older review on the properties of salt solutions in this concentration region is given by Luck [1964].

28.5.4. Acids and bases

A special discussion is necessary for aqueous solutions of acids or bases. Jaccard [1959, 1974] has shown that in ice the relaxation mechanism changes from a rotation mechanism to a proton transfer mechanism in ice doped with HF. The spectra of aqueous solutions of acids or bases are characterised by an effect of large half-width of the water bands (Luck [1964] and Zundel [1969]). This effect may be an indication that the lifetime of the H-bonds are of the order of the timescale of measurements. Zundel has discussed this effect related to the polarisability of special H-bonds (for details see ch. 15).

In aqueous solutions of $HClO_4$ the overtone spectra show clearly the difference between the structure breaking effect of the ClO_4^- ions, observed by an increase of the intensity in the region of non-H-bonded OH groups and the broadening effect of the water spectrum induced by the proton. The properties of electrolyte solutions depend on three factors: (i) the number of non-H-bonded OH groups, (ii) special water structures near the ions, and (iii) the lifetime of H-bonds.

Therefore, it is difficult to give an exact picture of the structure of electrolyte solutions in the medium or very high concentration regions. Until we can give a quantitative model of these solutions the simplified model given in this section may be helpful, but we have to stress that this model is only a very crude one. For modern reviews on solute–solute interactions and excess properties of aqueous solutions, and on thermodynamics of mixtures of electrolytes, Raman, dielectric or NMR studies see Luck et al. [1974a]. A more detailed review on the properties of electrolyte solutions is given by Franks et al. [1973].

28.6. Clathrate hydrates

28.6.1. General review

Many gases or liquids form solid gas hydrates with water, the structure of which are of general interest. They give indications of special arrangements between water and hydrophobic groups, for example, Pauling [1959] suggested liquid water could have a structure of gas-hydrate-like arrangements. This proposal, however, has now been superceded, since firstly, X-ray data are inconsistent with this assumption (Danford and Levy [1962]) and secondly, there are no gas hydrates known with strong H-bonding substances. Gas hydrates are formed by small inorganic molecules like Ar, Kr, O_2, CO_2, Cl_2, COS and SF_6, and by small organic molecules such as C_2H_4, $(CH_3)_2O$, tetrahydrofuran, acetone, furan, cyclopentane, propylene oxide, C_2H_5Br, etc. (see the review of Davidson [1973]). The first gas hydrate was found by Faraday [1823] who suggested a solid Cl_2–$10H_2O$. The structures of the gas hydrates were determined by X-ray scattering methods (see Stackelberg and Jahns [1954] and the review given by Schlenk [1951]).

28.6.2. Structure

The simplest element of the gas hydrate structure is a pentagondodecahedron formed by 20 water molecules (fig. 11.28). A pentagondodecahedron consists of 12 two-dimensional five-membered rings of water molecules. In the centre of a pentagondodecahedron is a hole. The guest molecules are arranged inside these holes (fig. 11.28). The H-bond angles of these five-membered rings have small deviations from the angle zero. Therefore, the H-bond energy in these arrangements differs little from the optimum value. The dispersion forces between the guest molecules and the water molecules have to compensate the difference between the optimum of the H-bond interaction energy of ice with H-bond angles zero and the H-bond energy of the five-membered rings. No proton and no lone pair electron is arranged in the direction of the hole in the centre of the pentagondodecahedron. This means that the water molecules surround the hole with their most hydrophobic side.

In addition to the pentagondodecahedron there is another arrangement of 24 water molecules of 12 five-membered rings and two six-membered rings (structure I) or of 28 water molecules of five-membered rings and four six-membered rings (structure II). The holes of these arrangements are larger than those of the pentagondodecahedral. The connection between the pentagondodecahedron together with the second clathrate element forms the whole structure. All water molecules have four neighbouring water molecules. Structure I is formed by guest molecules with diameters smaller than 5.3 Å and structure II with molecules larger than 5.3 Å. Other clathrates are the Br_2 hydrate with a pentakaidecahedral cage of water molecules or the *tert*-butylamine hydrate with a heptakaidecahedron cage of water molecules.

The spectra of gas hydrates are in agreement with the indications of the X-ray scattering results that in clathrates angle-unfavoured H-bonds are formed. Figure 28.31 shows the spectrum of the first overtone of SO_2 gas hydrate compared with the ice spectrum. The maximum of the H-bond band of the SO_2 hydrate lies at shorter wave length than the ice maximum. This indicates a smaller interaction energy. The figure also shows the spectrum of liquid water at the critical T at critical density. This spectrum arises from non-H-bonded OH groups. The spectrum of 15 ml H_2O/l SO_2 has a maximum near this maximum which shows that water does not form H-bonds to SO_2. The solution spectra of CH_3OH in liquid SO_2 show a smaller H-bond interaction than in CCl_4, and SO_2 forms complexes with ketones (Winde [1967]). From both experiments we can conclude that

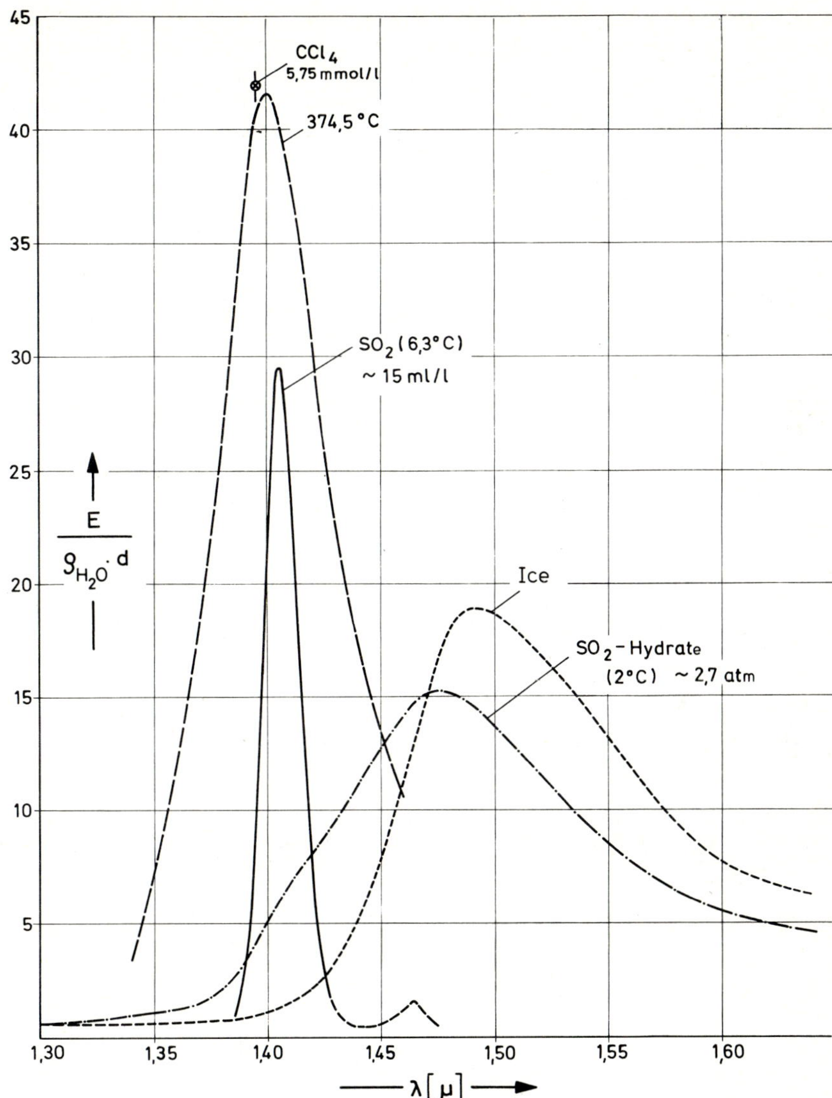

Fig. 28.31. Absorptivity coefficients of the first overtone of water. From right to left: ice; SO_2 hydrates 2°C, 2.7 atm; 5 ml H_2O/l liquid SO_2; liquid water under critical conditions; 5.75 mmole H_2O/l CCl_4.

special interactions between SO_2 and the O-atoms of OH groups are present. The fundamental vibration of SO_2 in a gas hydrate is more similar to gaseous SO_2 than to solid SO_2 (Harvey et al. [1964]). Far IR measurements of gas hydrates indicate rotational oscillations of the guest molecules (Davidson [1973]).

The structure of hexamethylenetetramine hexahydrate is an intermediate stage in aqueous solutions of H-bonded solutes. Three of the four nitrogen atoms of hexamethylenetetramine are H-bonded to water. The water molecules are arranged in six-membered rings.

There are some semi-clathrates of molecules with hydrophilic and hydrophobic groups like alkyl ammonium salts. Around an iso-amyl ammonium cation, water formed six 12-, four 14- and four 15-hedra. In all cases an optimum number of H-bonds of water is formed with four neighbouring water molecules.

28.6.3. Relations to solution structures

One could assume that in aqueous solutions of organic molecules, similar structures are formed by the water molecules around the solutes. From the anomalous entropy effect of such solutions the term iceberg structure of water around organic molecules with hydrophobic groups was proposed. However, experiments have yet to verify that the number of H-bonded OH groups of water is increased. There is no reason that the small interactions of water with hydrophobic organic molecules should increase the H-bond content of water but if we assume that in aqueous solutions around hydrophobic groups water forms clathrate-like structures, we should expect five-membered water rings. The five-membered rings of water molecules are two-dimensional as opposed to the six-membered rings in the tridymite structure of ice. In the tridymite structure the rings of water associates are not planar. The entropy effect of aqueous organic solutions may be based on forming planar rings of water molecules from non-planar rings in pure water. If this assumption is correct the nomenclature "iceberg-forming" is confusing.

A similar criticism can be made of the nomenclature, *hydrophobic forces.* The term hydrophobic groups in aqueous systems is applied if hydrophobic groups form associates in water. The main cause of the association is the tendency of forming a maximum of H-bonded OH groups. A maximum of H-bonded OH groups of water exists if, in a solution of hydrophobic molecules, a minimum of hydrophobic molecules are adjacent to water molecules*. The simplest example of hydrophobic bonds is the formation of soap micelles in water. The strong H-bond interaction between water molecules is the main cause of the "hydrophobic bonds" and not special interactions between the hydrophobic groups. The water molecules adjacent to hydrophobic groups may be arranged in gas-hydrate-like structures. The entropy changes connected with the "hydrophobic bond" formation may be explained in terms of a transition from five- to six-ring structures of water molecules. But until now there has been no direct experimental verification of these assumptions.

In contrast to the solubility of molecules with hydrophobic groups the solubility of hydrophilic molecules can be described in the cluster model of water as H-bonding at the cluster surfaces. For example, the spectra of NH_3 in H_2O indicate non-bonded NH groups and a reduction of the content of non-H-bonded OH groups (Luck and Ditter [1970]).

28.6.4. Properties of gas hydrates

Most gas hydrates have two invariant points. The first is near the melting point of ice and the second at higher temperature and higher pressure. For example: $CHCl_3$ = 1.7°C, 0.09 atm; SO_2 = 12.1°C, 2.33 atm; CO_2 = 9.9°C, 44.4 atm; C_2H_6 = 14.7°C, 33.5 atm; and H_2Se = 30°C, 11 atm. Some gas hydrates of the structure II type receive a stabilising effect caused by a *second* encageable component. Air can be such an assisting gas. Some mixed hydrates with H_2S or H_2Se as one component have a constant vapour pressure until all hydrate has disappeared. Stackelberg [1956] called these systems double hydrates. Other mixed hydrates have a gradually decreasing vapour pressure under intermittent pumping. These systems Stackelberg called mixed hydrates. A review on the properties of gas hydrates is given by Davidson [1973].

28.6.5. Applications

The production of some gas hydrates gives rise to certain problems. Some guest molecules have a higher solubility in the solid gas hydrate than in liquid water. Gas hydrates can induce blockage in natural gas pipelines (Hammerschmidt [1934]). There are some hydrate inhibitors, e.g. methanol. In permafrost regions, natural gases may exist as gas hydrates.

* With hydrated polyelectrolytes such an effect has been observed when alkylammonium cations were present (Zundel [1969] p. 92ff).

N_2 or O_2 hydrates become stable at 1 atm near −70°C. Miller [1969] proposed that the presence of air in ice cores at depths greater than 1200 m may be induced by gas hydrates. On other planets like Uranus, Neptune or Saturn Co_2, CH_4 or C_2H_6 hydrates may exist (Delsemme and Swings [1952], Miller [1961] and Delsemme and Miller [1970]).

Applications of gas hydrates have been tried for desalination processes of water or for fractionation of gases by selective enclathration (Davidson [1973]).

28.6.6. Other clathrates

Many other clathrates are known of molecules with H-bond interaction, other than water. For example, urea forms channel-adducts with hydrocarbons (see the review by Schlenk [1951]), β-quinol forms clathrates with some similarities to the water clathrates. 6 quinol molecules form a six-membered ring with a hole in the centre. In this hole HCl, H_2S, SO_2 or CO_2; Ar, Xe, CH_4, CH_3CN, CH_3OH, etc. can be trapped. McKean [1973] has given a recent review on such clathrates. In the quinol cages the guest molecules have barriers to rotation in the range 25–100 cm^{-1} (see McKean [1973]). This barrier can be determined by far IR spectroscopy or by specific heat data. The near IR spectra of the guest molecules are similar to liquid solution spectra. They indicate rotational structure.

Acknowledgement

I should like to express my debt of gratitude to Harry Hallam, University College of Swansea, for improvements in the English presentation.

References

Auty, R. P. and R. H. Cole, 1952, J. Chem. Phys. **20**, 1309.

Bellamy, L. J., H. E. Hallam and R. L. Williams, 1958, Trans. Faraday Soc. **54**, 1120.

Ben-Naim, A., 1974, in: Structure of Water and Aqueous Solutions, Ed. W. A. P. Luck (Verlag Chemie, Weinheim), ch. II, 2.

Ben-Naim, A. and F. H. Stillinger, 1972, Aspects of the Statistical-Mechanical Theory of Water, in: Structure and Transport Process in Water and Aqueous Solutions, Ed. R. A. Horne (Wiley–Interscience, New York).

Bernal, S. D. and R. H. Fowler, 1933, J. Chem. Phys. **1**, 516.

Breitschwerdt, 1974, in: Structure of Water and Aqueous Solutions, Ed. W. A. P. Luck (Verlag Chemie, Weinheim), ch. VIII, 1.

Bresler, S. E., 1939, Acta Physicochim. USSR, **10**, 491.

Briegleb, G., 1937, Zwischenmolekulare Kräfte und Molekülstruktur (Enke Verlag, Stuttgart).

Cini, R. and M. Torrini, 1968, J. Chem. Phys. **49**, 2826.
Curnutte, B. and D. Williams, 1974, in: Structure of Water and Aqueous Solutions, Ed. W. A. P. Luck (Verlag Chemie, Weinheim), ch. III, 1.
Danford, M. D. and H. A. Levy, 1962, J. Am. Chem. Soc. **84**, 3965.
Davidson, D. W., 1973, Clathrate Hydrates, in: Water, A Comprehensive Treatise, Vol. **2**, Ed. F. Franks (Plenum Press, New York).
Davies, C. M. and J. Sarzynsk, 1972, Mixture Models of Water, in: Water and Aqueous Solutions, Ed. R. A. Horne (Wiley–Interscience, New York), p. 377.
Debye, O., 1929, Polare Molekeln, Kap. VI (Hirzel Verlag, Leipzig).
Del Bene, J. and J. A. Pople, 1970, J. Chem. Phys. **52**, 4858.
Delsemme, A. H. and D. C. Miller, 1970, Planetary Space Sci. **18**, 717.
Delsemme, A. H. and P. Swings, 1952, Ann. Astrophys. **15**, 1.
Einstein, A., 1901, Ann. Phys. **4**, 513.
Eisenstein, A. and N. S. Gingrich, 1942, Phys. Rev. **62**, 261.
Eucken, A., 1948, Z. Elektrochem. **52**, 264.
Eyring, H., I. F. Kincaid and A. E. Stearn, 1941, Chem. Rev. **28**, 301.
Falk, M., 1970, Can. J. Chem. **48**, 607, 2098.
Falk, M., 1971a, Can. J. Chem. **49**, 1137, 1413.
Falk, M., 1971b, Spectrochim. Acta **27A**, 1811.
Falk, M. and T. A. Ford, 1966, Can. J. Chem. **44**, 1699.
Faraday, M., 1823, Quart. J. Sci. **15**, 71.
Fisher, I. Z. and V. Adamivich, 1963, J. Struct. Chem. USSR, **4**, 759.
Fister, F. and H. G. Hertz, 1967, Ber. Bunsenges. **71**, 1032.
Franck, E. U. and K. Roth, 1967, Discussions Faraday Soc. **43**, 108.
Franks, F., 1973, The Solvent Properties of Water, in: Water, A Comprehensive Treatise, Vol. **2**, Ed. F. Franks (Plenum Press, New York).
Franks, F. et al., 1973, Aqueous Solutions of Simple Electrolytes, in: Water, A Comprehensive Treatise, Vol. **3** (Plenum Press, New York).
Ganz, E., 1937, Z. Physik. Chem. **35**, 1.
Glasel, J. A., 1972, in: Water, A Comprehensive Treatise, Ed. F. Franks (Plenum Press, New York), p. 215.
Glasel, J. A., 1974, in: Structure of Water and Aqueous Solutions, Ed. W. A. P. Luck (Verlag Chemie, Weinheim).
Gmelin's Handbuch der anorganischen Chemie, 1963a, Nr. 5, 3, Ed. E. Pietsch, Sauerstoff (Verlag Chemie, Weinheim).
Gmelin's Handbuch der anorganischen Chemie, 1963b, Nr. 5,3, Ed. E. Pietsch, Sauerstoff (Verlag Chemie, Weinheim), p. 1389.
Gordon, R. C., 1967, Advan. Magn. Resonance **3**, 1.
Haggis, G. H., J. B. Hasted and J. Buchanan, 1952, J. Chem. Phys. **20**, 1452.
Hammerschmidt, E. G., 1934, Ind. Eng. Chem. **26**, 851.
Hardy, W. B., 1900, Z. Phys. Chem. **33**, 391.
Hartmann, K. A., 1966, J. Phys. Chem. **70**, 270.
Harvey, K. B., F. R. McCourt and H. F. Shurvell, 1964, Can. J. Chem. **42**, 960.
Hasted, J. B., 1974, in: Structure of Water and Aqueous Solutions, Ed. W. A. P. Luck (Verlag Chemie, Weinheim), ch. VII, 1.
Henshaw, D. G., 1960, Phys. Rev. **119**, 22.

Hertz, H. G., 1974, in: Structure of Water and Aqueous Solutions, Ed. W. A. P. Luck (Verlag Chemie, Weinheim), ch. VII, 2.

Hindman, J. C., 1966, J. Chem. Phys. **44**, 4582.

Hofmeister, F., 1890, Arch. exp. Path. Pharm. **25**, 295.

Jaccard, C., 1959, Helv. Chim. Acta **32**, 89.

Jaccard, C., 1974, in: Structure of Water and Aqueous Solutions, Ed. W. A. P. Luck (Verlag Chemie, Weinheim), ch. VI, 2.

Kalman, E., S. Lengyel, G. Palinkas, L. Haklik and A. Eke, 1974, in: Structure of Water and Aqueous Solutions, Ed. W. A. P. Luck (Verlag Chemie, Weinheim), ch. V, 2.

Kampmeyer, P. M., 1952, J. Appl. Phys. **23**, 99.

Keçki, Z., 1973, Advan. Mol. Relax. Proc. **5**, 137.

Kell, G. S., 1972, Continuum Theories of Liquid Water, in: Water and Aqueous Solutions, Ed. R. A. Horne (Wiley–Interscience, New York), p. 331.

Kruh, R. F., 1962, Chem. Rev. **62**, 327.

Lindner, H. A., 1970, Thesis (Karlsruhe).

Luck, W. A. P., 1951, Z. Naturforsch. **6A**, 191.

Luck, W. A. P., 1964, Fortschr. Chem. Forsch. **4**, 653.

Luck, W. A. P., 1965, Zur Stereochemie der Wasserstoffbrückenbindung, Naturwissenschaften **52**, 25, 49.

Luck, W. A. P., 1967, Discussions Faraday Soc. **43**, 115.

Luck, W. A. P., 1973, Infrared Studies of Hydrogen Bonding in Pure Liquids and Solutions, in: Water, A Comprehensive Treatise (Plenum Press, New York) **2**, 235.

Luck, W. A. P., 1974a, in: Structure of Water and Aqueous Solutions, Ed. W. A. P. Luck (Verlag Chemie, Weinheim), ch. III, 3.

Luck, W. A. P., 1974b, in: Structure of Water and Aqueous Solutions, Foreword, Ed. W. A. P. Luck (Verlag Chemie, Weinheim).

Luck, W. A. P., 1974c, in: Structure of Water and Aqueous Solutions, Ed. W. A. P. Luck (Verlag Chemie, Weinheim), ch. III, 2.

Luck, W. A. P. and W. Ditter, 1966, Ber. Bunsenges. **70**, 1113.

Luck, W. A. P. and W. Ditter, 1967, J. Mol. Struct. **1**, 339.

Luck, W. A. P. and W. Ditter, 1968, Ber. Bunsenges, **72**, 365.

Luck, W. A. P. and W. Ditter, 1969, Z. Naturforsch. **24B**, 482.

Luck, W. A. P. and W. Ditter, 1970, J. Phys. Chem. **74**, 3687.

Luck, W. A. P. and W. Ditter, 1971, Tetrahedron **27**, 201.

Luck, W. A. P. and W. Ditter, 1972, Advan. Mol. Relax. Proc. **3**, 321.

Luck, W. A. P. and R. Gosh, 1974, in press.

Luck, W. A. P. and W. Strate, 1974, in press.

Luck, W. A. P. and G. Dahm, 1974, unpublished data.

Luck, W. A. P. and A. Zukovskij, 1974, in: Molecular Physics and Biophysics of Water Systems, Vol. 2, Leningrad University Press, S. 131.

Luck, W. A. P., B. Mann and Th. Neikes, 1974a, Ber. Bunsenges.

Luck, W. A. P. et al., 1974b, in: Structure of Water and Aqueous Solutions, Ed. W. A. P. Luck (Verlag Chemie, Weinheim).

Luz, Z. and G. Yagil, 1966, J. Phys. Chem. **70**, 554.

McKean, 1973, Infrared and Raman Spectra of Clathrates, in: Vibrational Spectroscopy of Trapped Species, Ed. H. E. Hallam (John Wiley & Sons, New York), ch. 8, p. 355.

Miller, S. L., 1961, Proc. Natl. Acad. Sci. U.S. **47**, 1798.
Miller, S. L., 1969, Science **165**, 489.
Millero, F. J., 1972, Partial Molar Volumes of Electrolytes in Aqueous Solutions, in: Water and Aqueous Solutions, Ed. R. A. Norne (Wiley–Interscience, New York), p. 519.
Morgan, J. and B. E. Warren, 1938, J. Chem. Phys. **6**, 666.
Narten, A. H., 1972, in: Water, A comprehensive Treatise, Ed. F. Franks (Plenum Press, New York), p. 311.
Narten, A. H., 1974, in: Structure of Water and Aqueous Solutions, Ed. W. A. P. Luck (Verlag Chemie, Weinheim), ch. V, 1.
Némethy, G. and H. A. Scheraga, 1962, J. Chem. Phys. **36**, 3382, 3401.
Onsager, L., 1974, in: Structure of Water and Aqueous Solutions, Ed. W. A. P. Luck (Verlag Chemie, Weinheim), ch. I, 1.
Pauling, L., 1959, in: Hydrogen Bonding, Eds. D. Hadži and H. W. Thompson (Pergamon, New York), p. 1.
Pings, C. S., 1968, in: Physics of Simple Liquids (North-Holland Publ. Co., Amsterdam), p. 390.
Pople, J. A., 1950, Proc. Roy. Soc. **A202**, 323.
Pople, J. A., 1951, Proc. Roy. Soc. **A205**, 163.
Pople, J. A., 1954, Proc. Roy. Soc. **A221**, 498.
Quist, A. S. and W. L. Marshall, 1965, J. Phys. Chem. **69**, 3165.
Rahman, A. and F. H. Stillinger, 1971, J. Chem. Phys. **55**, 3336.
Röntgen, W. C., 1892, Ann. Phys. **45**, 91.
Sack, H., 1927, Z. Physik. **28**, 199.
Samoilov, O. J., 1961, Die Struktur von wässrigen Elektrolytlösungen (Teubner Verlagsges. Leipzig).
Schlenk, W., 1951, Fortschr. Chem. Forsch. **2**, 127.
Senior, G. E. and R. E. Verrall, 1969a, J. Phys. Chem. **73**, 4242.
Senior, G. E. and R. E. Verrall, 1969b, J. Chem. Phys. **50**, 2746.
Shoorley, J. N. and B. J. Alder, 1955, J. Chem. Phys. **23**, 805.
Stackelberg, M. von, 1956, Rec. Trav. Chim. Pays-Bas **75**, 902.
Stackelberg, M. von and W. Jahns, 1954, Z. Elektrochem. **58**, 162.
Stillinger, F. H. and A. Rahman, 1972, J. Chem. Phys. **57**, 1281.
Stillinger, F. H. and A. Rahman, 1974, J. Chem. Phys. **60**, 1545.
Stuart, H. A., 1967, Molekülstruktur, 3. Aufl. (Springer-Verlag, Heidelberg), p. 474.
Suhrmann, R. and F. Breyer, 1933, Z. Physik. Chem. **20**, 17; **23**, 193.
Vand, V. and W. A. Senior, 1965, J. Chem. Phys. **43**, 1896.
Walrafen, G., 1968, J. Chem. Phys. **48**, 244.
Walrafen, G., 1972, in: Water, A Comprehensive Treatise, Vol. **1**, Ed. F. Franks (Plenum Press, New York), p. 151.
Wang, J. H., 1951, J. Am. Chem. Soc. **73**, 510, 4181.
Wang, J. H., 1953, J. Am. Chem. Soc. **75**, 1769.
Winde, H., 1967, Z. Physik. Chem. (Leipzig) **244**, 225.
Worley, J. D. and I. M. Klotz, 1966, J. Chem. Phys. **45**, 2868.
Zundel, G., 1969, Hydration and Intermolecular Interaction (Academic Press, New York).

CHAPTER 29

THE HYDROGEN BOND IN ICE*

E. WHALLEY

Division of Chemistry, National Research Council, Ottawa, K1A OR9, Canada

* N.R.C. No. 14715.

Contents

29.1. Introduction 1427
29.2. Geometry 1430
29.2.1. Proton–proton distances from the second moment of the NMR spectrum 1430
29.2.2. Ice Ih and Ic 1431
29.2.3. Ice II 1437
29.2.4. Ice III and IX 1441
29.2.5. Ice V 1445
29.2.6. Ice VI 1449
29.2.7. Ice VII and VIII 1450
29.2.8. Origin of the increased density of the high-pressure phases 1452
29.3. Dissociation energy 1453
29.4. Vibrations 1458
29.4.1. O–H stretching vibrations 1458
29.4.2. Bending and rotational vibrations 1461
29.4.3. Translational vibrations 1461
29.5. Electric-field gradient of the nuclei 1466
References 1466

The hydrogen bond – recent developments in theory and experiments
Eds. P. Schuster et al.

29.1. Introduction

Ice is a uniquely valuable substance for the study of H-bonding because it is the only substance known whose nearest-neighbor interactions are all H-bonds. Consequently, measurements of the properties of the H-bonds of ice tend to be less affected than measurements in other substances by other intermolecular interactions. In addition, there are many different crystalline forms of ice, and so the H-bonds can be studied in many different aspects. Some of the forms can be made only under pressure. Others are not, strictly speaking, phases of ice, but can be considered as phases of ice that are stabilized by the presence of "impurity" molecules. These are the clathrate hydrates, in which a fully H-bonded relatively open network of water molecules is prevented from collapsing to a denser form by the presence of foreign molecules in its larger cavities. A detailed review of these interesting substances has been given by Davidson [1973].

Much work has been done on the structures, molecular vibrations, and molecular migrations of the phases of ice, and general reviews have been published recently by Eisenberg and Kauzmann [1969], Fletcher [1970], and Hobbs [1974]. Only those properties that tell the most about the effects of H-bonding will be discussed in this chapter.

The phase diagram of water is shown in fig. 29.1, in which solid and long-dashed lines represent directly measured stable and metastable lines respectively, and short-dashed and dotted lines represent boundaries that are extrapolated or estimated as well as possible from the available data. The only naturally occurring form is ice Ih, where the h indicates its hexagonal structure and distinguishes it from the closely related form ice Ic having a cubic structure. This cubic phase appears to be always metastable relative to ice Ih and can be made only from an even less stable form, such as by warming vitreous ice or any of the high-pressure phases recovered at low temperature and pressure (Bertie et al. [1963, 1964]). Ice II, III, V, VI, VII, and VIII can exist as stable phases at suitable temperatures and pressures, whereas ice IV and IX cannot.

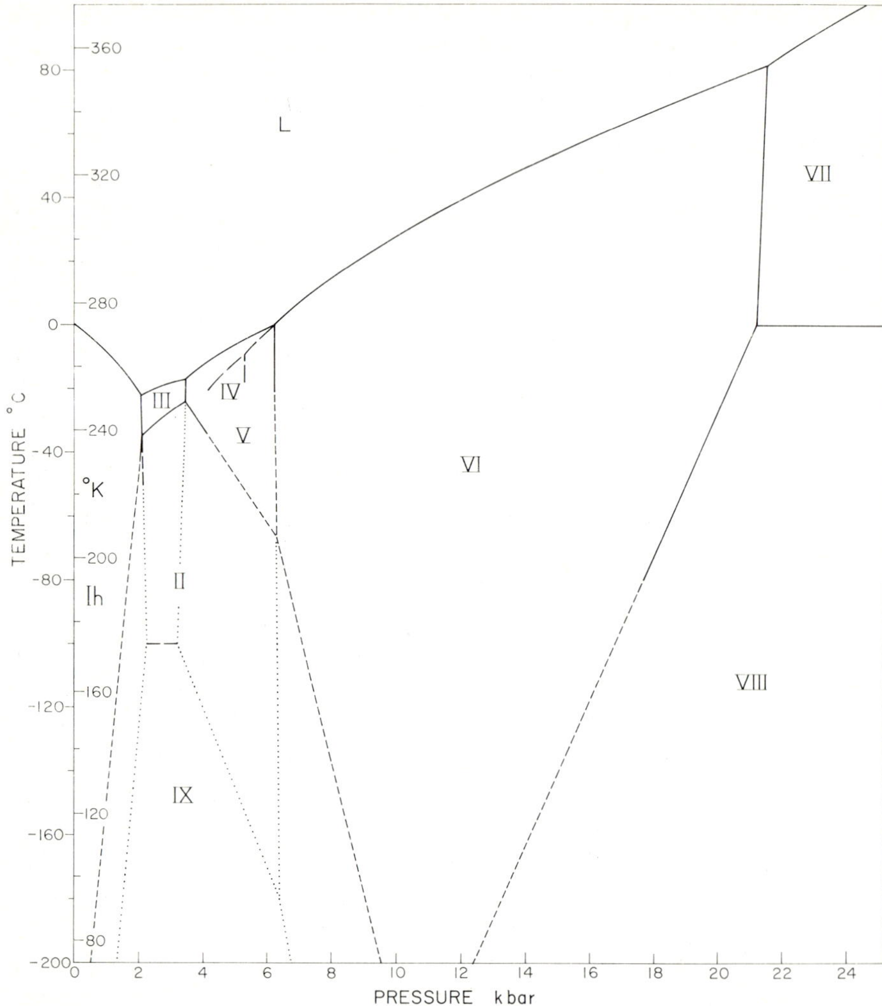

Fig. 29.1. The phase diagram of ice. The solid and long-dashed lines represent directly measured stable and metastable lines respectively, and short-dashed and dotted lines represent boundaries that are extrapolated or estimated as well as possible from the available data.

These phases can be investigated either by making them at the appropriate temperature and pressure and making measurements under pressure, or by quenching them to low temperatures under pressure and removing them from the pressure vessel. Measurements can then be made at zero pressure so long as the ice is not warmed too far (Kamb and Datta [1960] and Bertie et al. [1963, 1964]). Ice III and VII cannot be quenched in this way as they transform rapidly on cooling. Little is known about ice IV although it has been made and recovered (Engelhardt and Whalley [1972]), and so can be examined.

In all these phases the water molecules are not destroyed, but exist as clearly recognizable entities, although they are appreciably distorted from the vapor structure. The structures of all the phases are variants on two simple themes, four-coordination of water molecules and the possibility of order or disorder in their orientations. Both themes have their origin in a property that is unique to water molecules. This is their capacity to form four H-bonds in a tetrahedral arrangement. The H-bonds of a particular molecule are illustrated in fig. 29.2. The molecule donates its two hydrogen atoms to form two H-bonds to two of its neighbors, and two other neighbors each donate one hydrogen atom to an H-bond to the central molecule. It is clear that each molecule can form its H-bonds in six different ways, which is the number of arrangements of the two near and two far hydrogen atoms associated with a given oxygen atom. Consequently, orientational disorder is possible. Each water molecule in the crystal cannot, of course, occupy all orientations because each H-bond can contain only one hydrogen atom. If disorder is present, it has profound effects on some of the properties of the phase. All the phases of ice except II are disordered at high temperatures, and only ice VI and VII are not fully disordered. All except ice Ih and Ic become either partly or completely ordered at low temperature.

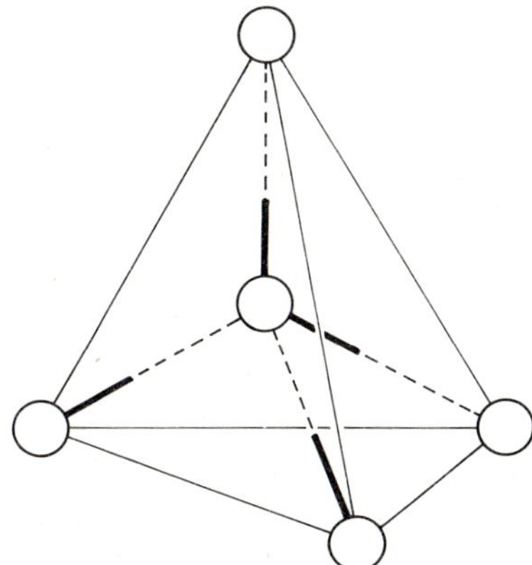

Fig. 29.2. The H-bonding of a water molecule in ice.

It is, of course, not necessary that the H-bonding be perfect tetrahedral, and indeed, the symmetry of the arrangement shown in fig. 29.2 suggests that the maximum symmetry that might be expected is C_{2v}, which is the symmetry of an isolated molecule. However, since the forces required to change the O···O bond lengths and the O···O···O bond angles are not strong, the symmetry may be appreciably distorted from the maximum.

29.2. Geometry

Eventually, the crystal structures of the phases of ice will provide much detailed information about the forces between water molecules. Any good intermolecular potential function should be able to predict the geometry and the energies of the phases, but few attempts have been made to understand the structures of even ice Ih and Ic in these terms, and none has been reported on the high-pressure phases, although elaborate calculations have been done on the liquid using empirical potential functions (Rahman and Stillinger [1971, 1973], Stillinger and Rahman [1972], and Barker and Watts [1973]). An excellent test of a potential function would be to show that it predicts that structures similar to those of the phases of ice are stable to small deformations, and have about the correct energy. Consequently, the structures of the various phases are reviewed in some detail.

29.2.1. Proton–proton distances from the second moment of the NMR spectrum

Most of the precise structural data on the phases of ice has been obtained by X-ray and neutron diffraction. Some information, however, can be obtained from the second moment M_2 of the proton magnetic resonance spectrum, in particular the intramolecular proton–proton distance. The measurements are summarized in table 29.1. The measured moment is the sum of contributions from intramolecular M_2(intra) and intermolecular M_2(inter) effects. Values of M_2(inter) for several phases of ice were calculated by Rabideau et al. [1968], and so the values of M_2(intra) listed in table 29.1 were obtained. Since the second moment is determined by the square of the mean reciprocal cube of the inter-proton distance, and the mean distance is required, the observed values of M_2(intra) must be adjusted to the value appropriate to the rigid lattice. This has been done following the discussion of Pedersen [1964] and Barnaal and Lowe [1967]. The resulting inter-proton distances are listed in table 29.1; they are, of

TABLE 29.1

Second moments of the proton magnetic resonance spectrum of the phases of ice, and the H--H distances

Phase	M_2 (G^2)	M_2(inter) (G^2)	M_2(intra) (G^2)	M_2(intra, rigid) (G^2)	r_{H--H} (Å)
Ih	32.4 ± 1.1[a), b), c)]	12.83 ± 0.11[a)]	19.6	22.9 ± 1.4	1.581 ± 0.016
Ic	32.2 ± 1.1[a)]	12.06 ± 0.11[a)]	19.1	22.3 ± 1.5	1.590 ± 0.016
II	34.1 ± 1.3[a)]	15.12 ± 0.13[a)]	19.0	22.2 ± 1.6	1.590 ± 0.017
IX	35.6 ± 1.3[a)]	15.31 ± 0.14[a)]	20.3	23.7 ± 1.6	1.571 ± 0.017
				Average	1.584 ± 0.017

a) Rabideau et al. [1968].

b) Barnaal and Lowe [1967].

c) Kume's [1960] value of 36.7 ± 1.2 G^2 seems too high.

course, means over all the molecules in the crystal. They do not differ significantly from one another, and are 1.584 ± 0.017 Å. The vapor value at the average proton position is 1.539 Å, calculated from the equilibrium O–H distance and H–O–H angle of 0.9584 Å and 104.45° and the effect of zero-point motion (Kern and Karplus [1972]), suggesting at most a small distortion of the O–H bond from the vapor value.

The values of the H–O–H angles and the O–H bond lengths that are compatible with the second moment are plotted as a band in fig. 29.3, which is modified from an earlier and less accurate version of Kume's [1960].

29.2.2. Ice Ih and Ic

Ice Ih is the only form of ice that occurs naturally on earth, and although much work has been done on it, less is known about the detailed geometry of the molecules in it than in some of the high-pressure phases. The reason for this is that the molecules in ice Ih are orientationally disordered, and consequently, X-ray and neutron diffraction, at least as presently measured, determine only an average geometry.

The structures of ice Ih and Ic are illustrated in fig. 29.4. Ice Ic is cubic with the diamond structure (space group Fd3m), and the molecules are on sites of point symmetry T_d. Ice Ih has the corresponding hexagonal structure (space group $P6_3/mmc$), and the molecules are on sites of symmetry C_{3v}. Since the water molecules have symmetry C_{2v}, their orientations must be disordered to give these average symmetries, a

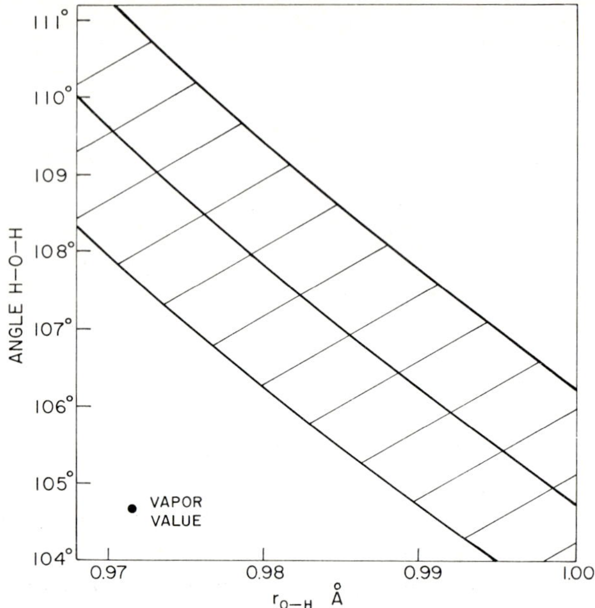

Fig. 29.3. The band of simultaneous values of the O–H distance and the H–O–H angle that are consistent with the observed second moment of the NMR spectrum.

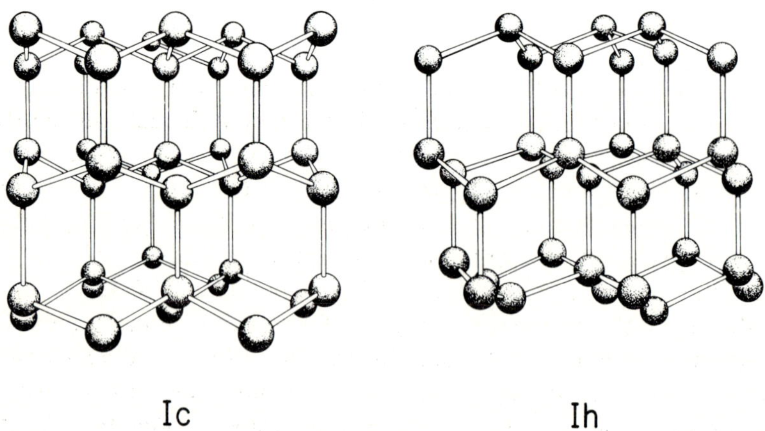

Fig. 29.4. Positions of the oxygen atoms in ice Ih and Ic.

conclusion that is well verified by many methods. The site symmetry T_d in ice Ic implies that the molecules have, on the average, perfect tetrahedral coordination, while in ice Ih the symmetry is degraded to C_{3v}. If the c/a ratio, where a and c are the lengths of the unit cell edges, of ice Ih were exactly $(8/3)^{1/2} = 1.6330\ldots$, the molecules would have, on the average, perfect tetrahedral local coordination, and the departure from T_d symmetry would occur only in the second and further neighbors. The ratio however is 1.6282, which departs significantly from the ideal, and is independent of temperature in the range 13 to 263 K (LaPlaca and Post [1960] and Brill and Tippe [1967]) and of pressure, according to its elastic constants (see data reviewed by Stephens [1958], Brockamp and Rüter [1969] and Zarembovitch and Kahane [1964]), although an increase at the rate of 0.2 or 0.3 Mbar^{-1} is not excluded.

It follows that either the bonds parallel to the c axis are, on the average, shorter than those inclined to it, or the bond angles that are in planes parallel to the c axis are smaller than the others. Unless the changes in bond lengths and angles are in the opposite senses, the maximum differences that could occur are about 0.015 Å in the length if the angles are the same or about 0.4° in the angles if the lengths are the same. The oxygen atom positions have not yet been located accurately enough to tell which occurs. The unit cell edge of ice Ic is 6.353 ± 0.001 Å (Arnold et al. [1968]) at 80 K, which corresponds to an average $O \cdots O$ distance of 2.751 Å, and the average distance in ice Ih is the same.

The average $O \cdots O$ bond length in ice Ih changes with temperature according to the linear thermal expansivity, the best values of which for ice Ih are plotted in fig. 29.5, and changes with pressure at the rate 3.8 Mbar^{-1} (Stephens [1958], Brockamp and Rüter [1969] and Zarembovitch and Kahane [1964]). The thermal expansion and compression of ice Ic have not been measured, but it seems safe to assume that they are the same as for ice Ih.

There is no direct measurement of the distribution of $O \cdots O$ bond distances and $O \cdots O \cdots O$ angles, but the infrared spectrum provides an indirect and partial measure of the distribution of $O \cdots O$ distances. The half-width of the O–H stretching vibration of a molecule of HDO in dilute solution in either ice Ih or Ic is about 50 cm^{-1} at 100 K (Bertie and Whalley [1964a]). In phases of ice with no orientational disorder the half-width of this band is only a small fraction of this value (Bertie and Whalley [1964b]), and so it seems that the orientational disorder causes the major part of the half-width in ice Ih. The wave numbers of the O–D stretching

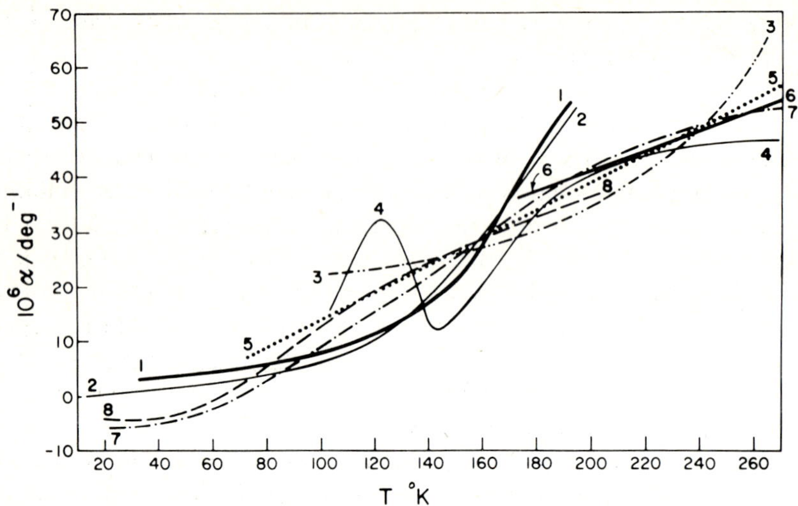

Fig. 29.5. The linear thermal expansivity of ice Ih according to various authors (Whalley [1973]): 1, Brill and Tippe [1967], *c* axis, X-ray; 2, Brill and Tippe [1967], ⊥ *c*, X-ray; 3, LaPlaca and Post [1968], *c*, X-ray; 4, LaPlaca and Post [1960], ⊥ *c*, X-ray; 5, Powell [1958], *c* and ⊥ *c*, direct; 6, Yamaji [1957], *c* and ⊥ *c*, direct; 7, Jacob and Erk [1928]; 8, Dantl [1962]. Reproduced with permission of the Royal Society of Canada.

vibration of HDO in dilute solution in H_2O phases of ice can be correlated approximately with O···O distances as shown in fig. 29.17 with a slope of $\sim 770\ \mathrm{cm^{-1}\ Å^{-1}}$, and there is no clear evidence that bending the O–H···O bond has a comparable effect. The corresponding slope for the O–H vibrations is $\sim 1040\ \mathrm{cm^{-1}\ Å^{-1}}$. If it is assumed that the width of the DO–H vibration is caused largely by a distribution of O···O distances and that this slope holds, then the distribution of O···O distances has a half-width of about 0.05 Å.

The O–H or O–D distances are not well known in either ice Ih or Ic, in ice Ih because there are reasons to doubt the measurements, and in ice Ic because they have never been done. The only detailed measurements reported are those of Peterson and Levy's [1957] on D_2O by neutron diffraction. The directly measured value is 0.997 Å at 150 K, and this was corrected for the relative motion of the deuterium and oxygen atoms perpendicular to the O···O line to 1.01 Å, compared with the vapor value in the ground vibrational state of 0.9686 Å (Kern and Karplus [1972]). Chamberlain et al. [1973] have reported the even larger values of 1.023 ± 0.005 and 1.018 ± 0.009 Å.

Peterson and Levy's value, which corresponds to an increase in bond length of 0.04 Å on H-bonding, has been widely quoted, but doubts have been expressed about it recently. Whalley [1969] pointed out that if it were true, the stored elastic energy per water molecule due to the stretching of the O–H bond would be about 20% of the heat of sublimation, and commented that it would be surprising if such a large amount of elastic energy were stored. At the same time, Hamilton et al. [1969, 1971] and LaPlaca et al. [1973], who have determined O–D distances in the ordered phases II and IX of ice to be little distorted from the vapor values, pointed out that the O–D stretching wave number of HDO changed from 2727 cm^{-1} in the vapor to about 2460 cm^{-1} in ice II and IX, whilst the average O–D distance changed from 0.970 Å to about 0.985 Å or at the average rate of about 20000 cm^{-1} $Å^{-1}$. The corresponding wave number in ice Ih and Ic is 2421 cm^{-1} (Bertie and Whalley [1964a] and Ford and Falk [1968]), and it seems unlikely that the O–D distances in ice Ih and Ic can depart greatly from those in ice II and IX. If the above rate of change of wave number holds, the difference is only of the magnitude 0.002 Å. Other evidence confirming this conclusion has recently been summarized (Whalley [1974]).

Although the O–H distances are uncertain, their distribution can be obtained approximately from the infrared spectrum. If the O–D wave number changes with O–D distance at the rate quoted above of about 20000 cm^{-1} $Å^{-1}$, a value that can be calculated from the anharmonic potential function of a gas phase molecule when the bond is lengthened by an external force that has no other effect on the properties of the molecule (LaPlaca et al. [1973]), the half-width of the band in ice Ih corresponds to a half-width of the distribution of O–H distances of the magnitude of 0.002 Å.

The H–O–H angle is not known by direct measurement for either ice Ih or Ic because the orientational disorder ensures that the spatial average of the hydrogen positions is on the O···O line, and the present neutron diffraction measurements are not able to distinguish between amplitudes of the deuteron due to this effect of the disorder and due to other effects (Chidambaram [1961]). Although this is probably obvious even without Chidambaram's detailed calculations, the angle is frequently taken to be tetrahedral and reasons for its opening are often sought.

There are two kinds of theories of the H-bond, whether in ice or in other substances. The most accurate is undoubtedly the ab initio molecular orbital theory, but it suffers from the disadvantages that the calculations are difficult to perform and the physical interpretation of the results is not always transparent. Consequently, more approximate models in which the molecules are treated as interacting non-overlapping charge distributions are often used. These models are very approximate and sometimes give quite wrong answers, but the calculations are relatively readily done and lead to simple-minded interpretations that are sometimes useful. The elongation of the H-bond can be understood on the basis of this simple model. Coulson and Eisenberg [1966] have calculated the average electric field at a molecular site in ice Ih due to the neighboring molecules, which were treated as classically polarizable non-overlapping charge distributions that could be represented by electric dipole, quadrupole, and octopole moments. The dipole moment in the absence of molecular interaction was reduced from the vapor value of 1.85 to 1.78 D to allow for a supposed opening of the H–O–H angle to tetrahedral. There is no evidence that the bond is opened so much, and as is shown later in this section, there is much evidence that it is not, so Coulson and Eisenberg fields have been adjusted to correspond to an unperturbed dipole moment of 1.85 D. The electric field acting on a molecule, averaged over all configurations of the neighbors, is 560 kesV cm^{-1} parallel to the molecular axis. If the potential energy V of the O–H bond is written as the sum of terms due to the stretching of the O–H bond and bending of the O–H···O bond and a contribution from the electric field, we can write

$$V = \tfrac{1}{2}k_r\,\delta r^2 + \tfrac{1}{2}r_e^2k_\theta\,\delta\theta^2 + e_r^*\,\delta r\,E\cos\theta + e_\theta^* r_e\,\delta\theta E\sin\theta,$$

where k_r and k_θ are the force constants, δr and $\delta\theta$ the departure of the O–H bond length and O–H···O angle from the equilibrium values in the absence of the electric field, r_e the equilibrium O–H length, e_r^* and e_θ^* the effective charges for stretching and bending, and θ the angle between the field E and the O–H line. The equilibrium distance and angle in the presence of the field are obtained from the conditions that

$$\partial V/\partial r = \partial V/\partial\theta = 0,$$

as

$$\delta r = (e_r^* E\cos\theta)/k_r, \qquad \delta\theta = (e_\theta^* E\sin\theta)/r_e k_\theta.$$

If e_r^* and k_r are taken as having the values for the crystal (see section 29.3), namely $\pm 4.0\times 10^{-10}$ esu and 6.32 mdyn Å^{-1}, then

$$\delta r = \pm 0.021\ \text{Å}.$$

The positive displacement is 0.007 Å larger than the experimental value, and if the model has any correspondence with reality e_r^* is positive. This is the sign expected on the grounds that transfer of the hydrogen atom across the bond leads to $O^- \cdots H{-}O^+$. The value of e_θ^* is not known, but is small because of the small infrared intensity of the ν_2 band of ice. If k_θ is taken as 0.75 mdyn Å^{-1}, then

$$\delta\theta / e_r^* = 4.3°(10^{-10}\ \text{esu})^{-1}.$$

In ice II and IX, the D–O–D angles depart little from the vapor value (see sections 29.2.3 and 29.2.4), and this is at least qualitatively in agreement with this calculation that the electric field would be expected to have little effect.

It is often considered that while ice Ih or Ic have a long-range order in their molecular positions, they have no long-range order in their molecular orientations. This is not quite true and, as Stillinger and Cotter [1973] have pointed out recently, there is a long-range order in the orientations. Each molecule in one of the puckered hexagonal sheets that ice Ih and Ic can be considered to be made from forms three H-bonds to neighboring molecules in the sheet and one to a molecule in a neighboring sheet. Alternate molecules bond on alternate sides of the sheet. A consequence of the ice rules is that if a sheet accepts n H-bonds on one of its sides, it must donate n H-bonds on the other side. If there are instantaneously N molecules in the sheet, the dipole moment perpendicular to the sheet is then $(\frac{1}{2}N - n)\mu_{\text{O-H}}$, where $\mu_{\text{O-H}}$ is the dipole moment of a water molecule parallel to the O–H direction, and this dipole moment propagates throughout the crystal. Stillinger and Cotter [1973] suggest that this correlation has an important effect on the dielectric constant. The occurrence of imperfections of all kinds will of course restrict the range of the correlation.

29.2.3. Ice II

The structure of ice II as determined by single crystal X-ray diffraction on H_2O (Kamb [1964]) and neutron diffraction on D_2O (Hamilton et al. [1971]) is illustrated in fig. 29.6. Its unit cell is rhombohedral, space group $R\bar{3}$, with $a = 7.78 \pm 0.01$ Å, $\alpha = 113.1 \pm 0.2°$ and contains 12 water molecules. There are six of each of two kinds of oxygen atom in the unit cell, labelled I and II in fig. 29.6, and four kinds of hydrogen atom, labelled 1, 2, 3, and 4. All the molecules are in almost flat hexagonal rings, stacked coaxially with one another along the corresponding hexagonal c axis, which is the axis that subtends the three rhombohedral axes equally.

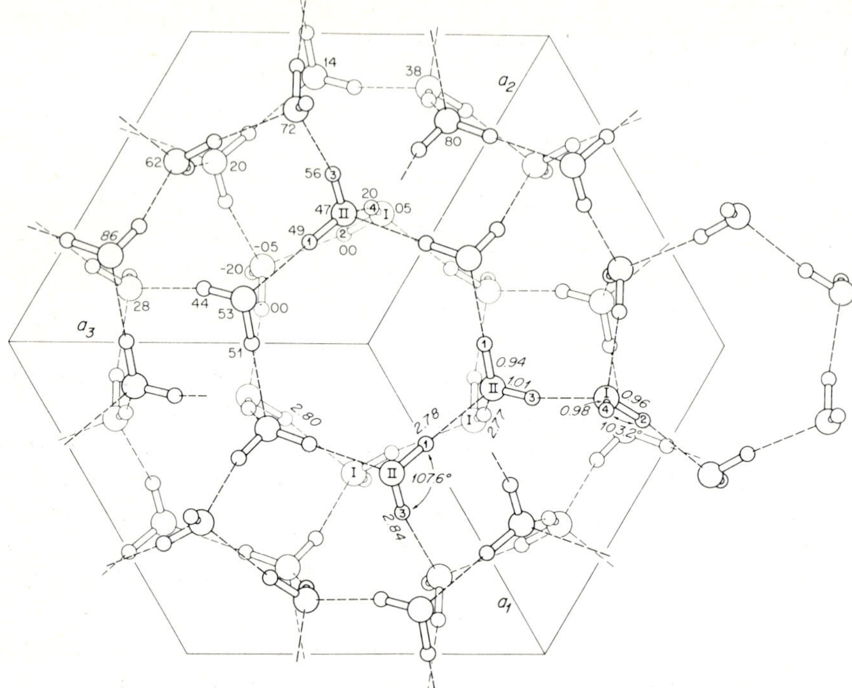

Fig. 29.6. Structure of ice II in projection along the hexagonal c axis. Reproduced with permission of the Journal of Chemical Physics.

Stereoscopic drawings of the structure are given by Hamilton et al. [1971]. Each ring forms three H-bonds to the ring below and three to the ring above, and each water molecule is H-bonded to one of the six neighboring rings.

The detailed arrangement of the bonding about each oxygen atom is illustrated in fig. 29.7 in which angles are expressed in degrees and distances in Angstroms. There are four kinds of O···O bonds of equal multiplicity, having O···O distances of 2.768, 2.781, 2.803 and 2.844 Å and O–D distances (uncorrected for the effect of deuteron motion perpendicular to the O–D bond) of 0.975, 0.937, 0.956 and 1.014 Å. The quoted experimental error is 0.02 Å, and it seems likely that the O–D distances do not differ significantly. In particular, the contraction from the length in the vapor implied by the value 0.937 Å and the large expansion implied by the value 1.014 Å are perhaps doubtful. The average value is 0.971 Å and this is the best estimate for the length of each bond (Hamilton

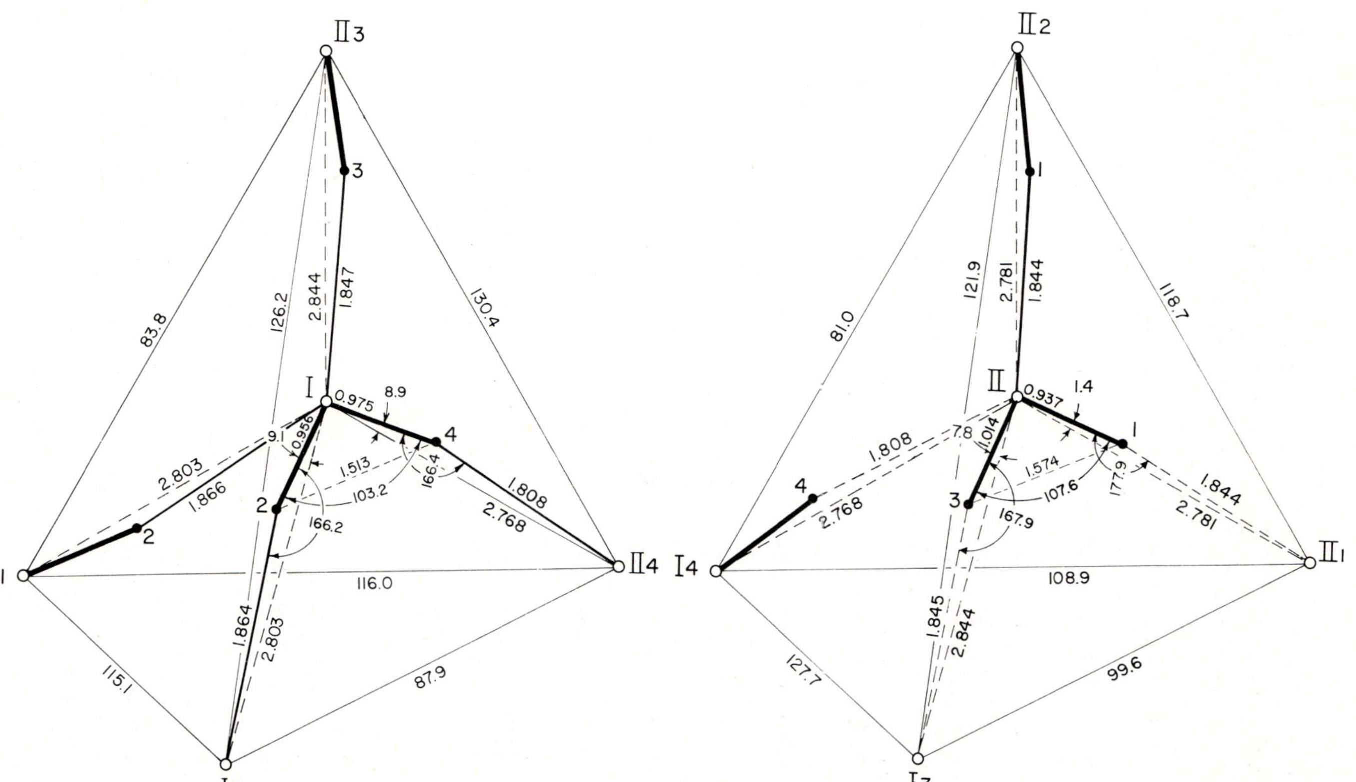

Fig. 29.7. Detailed geometry of the two independent water molecules and their four neighbors in ice II. After Hamilton et al. [1971].

et al. [1971]). When corrected for the effects of thermal motion perpendicular to the O–D bond, assuming it is similar to the effect in ice I and IX, the length becomes 0.982 Å (Hamilton et al. [1971]). The D--D distances in molecules I and II are 1.513 and 1.574 Å respectively, whose average is 1.54 ± 0.02 Å, which is not in serious disagreement with the value 1.584 ± 0.017 Å derived from the proton magnetic resonance spectrum (see section 29.2.1). The D–O–D angles are 103.2 and 107.6° with an estimated standard deviation of 2°; there is no strong evidence that they depart from the vapor value of 104.5°. The O–D···O bonds are significantly non-linear, and the O–D···O angles are 166.2, 166.4, 177.9, and 167.9°. The tetrahedral coordination of the oxygen atoms is significantly imperfect, the O···O···O angles being in the range 81.0 to 130.4°.

Neither water molecule places itself symmetrically in the O···D–O–D···O plane, as fig. 29.8 shows.

The oxygen atom positions can nearly be described by the space group $R\bar{3}c$ in which the oxygen atoms are on the general positions 12f. The hydrogen atoms can fit into this space group only if they are disordered so that the occupancy of a proton on each side of the O···O bond is $\frac{1}{2}$. The ordering of the hydrogen atoms lowers the symmetry to $R\bar{3}$.

The reasons for the adoption of this particular ordered structure, given the particular approximate arrangement of the oxygen atoms, is not

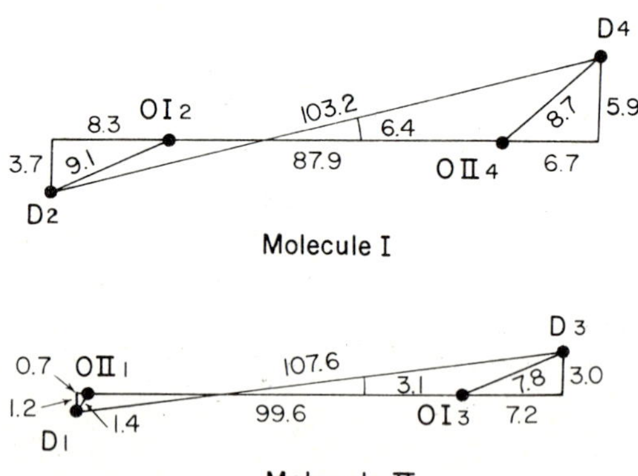

Fig. 29.8. The displacement of each D_2O molecule in ice II from the plane of the oxygen atoms of the group O···D–O–D···O, represented by the vectors from the central oxygen atoms to the deuterium and peripheral oxygen atoms. The number attached to each line gives the angle in degrees subtended by the ends of the line with the central oxygen, or the angle of rotation of the water molecule from the O···O···O plane. After Hamilton et al. [1971].

entirely clear. A major factor is the tendency of the O–H···O bonds to be approximately linear, and the orientation of the acceptor water molecule, presumably so that its "lone pair" is oriented correctly, seems to play a part (Kamb [1964]).

29.2.4. Ice III and IX

Ice III and IX are orientationally disordered and ordered versions of the same structure, and the transition between them was found by dielectric methods (Whalley et al. [1968]). There is no detectable volume change at the transition (Whalley et al. [1968]), and ice III under its equilibrium conditions and ice IX recovered at 100 K and zero pressure have essentially the same X-ray powder diffraction pattern, and so essentially the same oxygen atom positions (Bertie et al. [1963] and Calvert and Whalley [1965]).

The structure of ice III has not otherwise been investigated, but that of ice IX is well known (LaPlaca et al. [1973]). The unit cell is tetragonal, space group $P4_12_12$ with dimensions $a = 6.73 \pm 0.01$, $c = 6.83 \pm 0.01$ Å and contains 12 molecules. A c axis projection is shown in fig. 29.9. There are two kinds of oxygen atom, type I, of which there are eight on sites of no symmetry and which form two four-fold spirals, and type II, of which there are four on sites of symmetry C_2 and which join four of the spirals by H-bonds. The angles and distances are shown in fig. 29.10. There are three nonequivalent O···O distances at 2.750, 2.763, and 2.793 Å, and O···O···O angles that range from 90.4 to 143.7°. The hydrogen atoms can fit into this space group in either an ordered or a disordered arrangement. Ice III appears from the low-frequency dielectric constant (Whalley et al. [1968]) to be essentially fully disordered within the conditions imposed by the ice rules, and presumably each near-neighbor O···O bond has, on the average, half a hydrogen atom near each oxygen atom.

Ice IX is nearly completely ordered. The (so far hypothetical) fully ordered structure will be described first. There are three kinds of hydrogen atom, labelled 1, 2, and 3 in fig. 29.9, and there are eight of each in the unit cell. The local coordination, as determined by single-crystal neutron diffraction on D_2O (LaPlaca et al. [1973]), is shown in fig. 29.10.

The O–D distances, the values of which are corrected for thermal motion, differ insignificantly from one another, and their average, 0.984 Å is greater by 0.014 Å than the average distance in the vapor of 0.9686 Å (Kern and Karplus [1972]). It is close to the average distance in ice II of

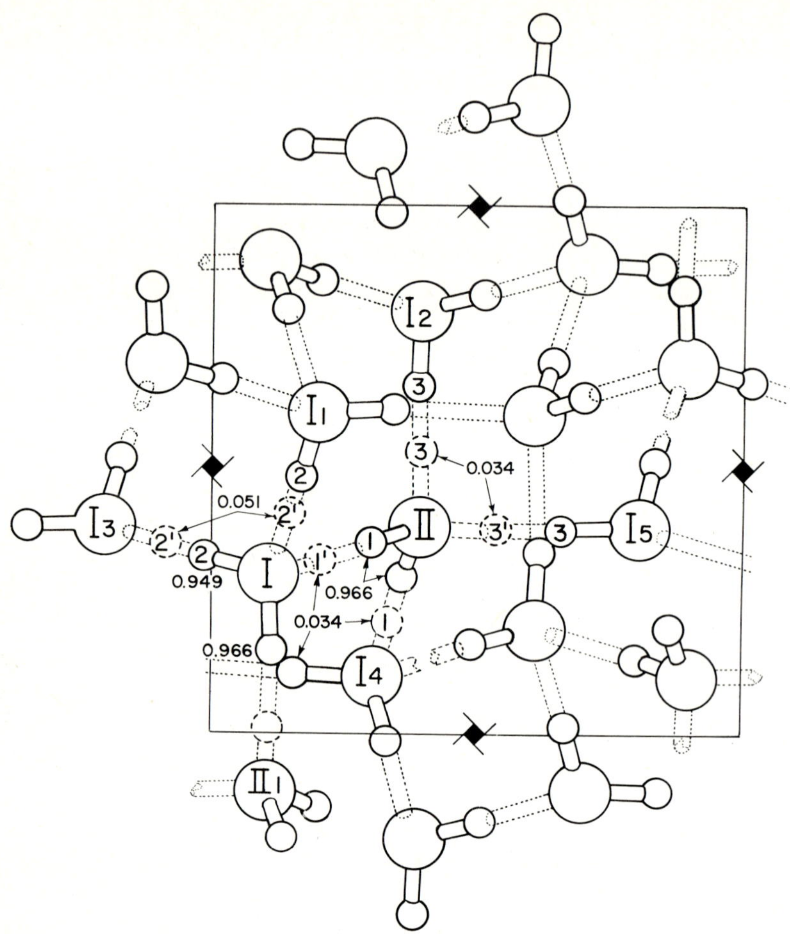

Fig. 29.9. The structure of ice IX as viewed in a projection along the *c* axis. After LaPlaca et al. [1973].

0.982 Å (see section 29.2.3). The D–OII–D angle does not differ significantly from the vapor value, but the D–OI–D angle seems to be increased slightly to 106.0°. The D--D distances of 1.565 and 1.569 Å agree within the combined experimental errors with the mean value of 1.584 ± 0.017 Å deduced from the second moment of the nuclear magnetic resonance spectrum (see section 29.2.1). It appears that in ice II and IX, in which the geometries of the water molecules are best known, the water molecules are distorted from the vapor geometry by a lengthening of the

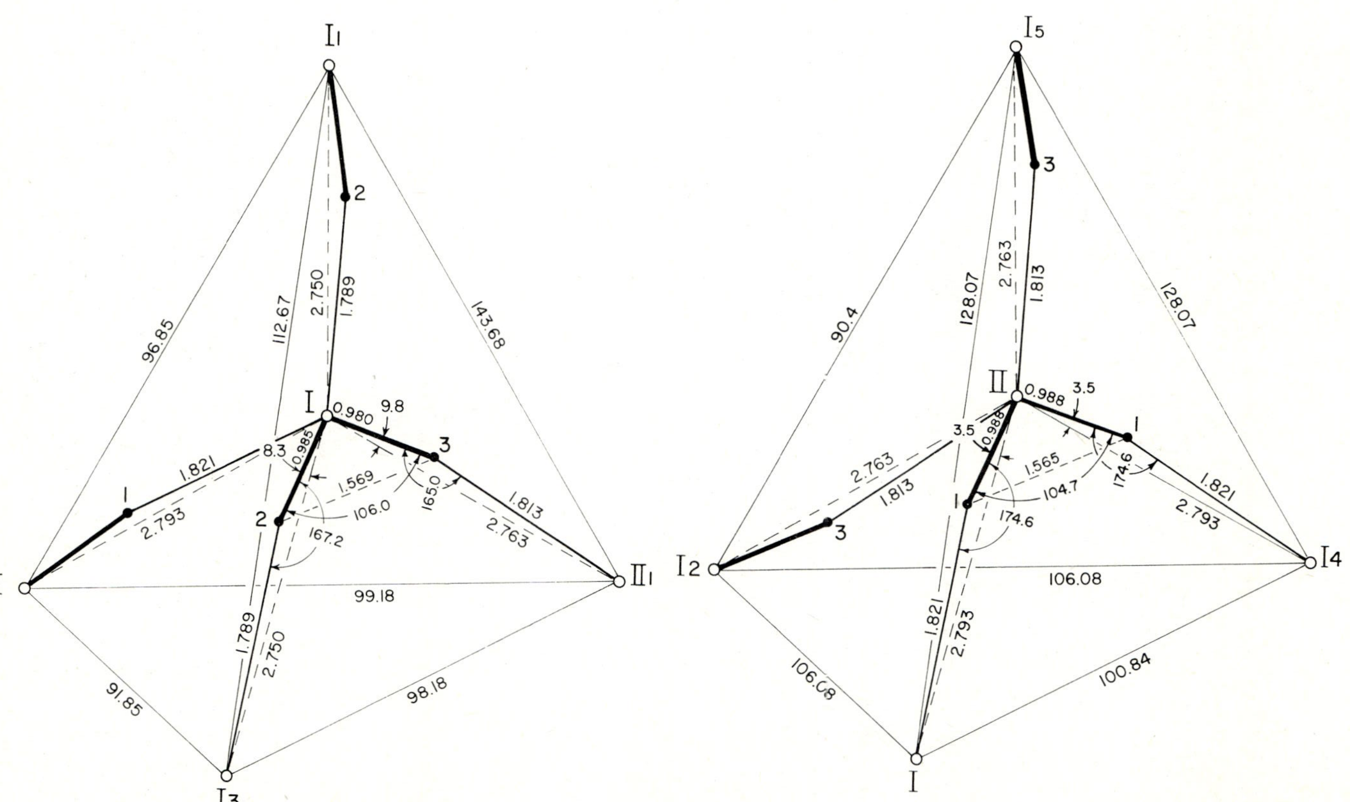

Fig. 29.10. Detailed geometry of the two independent water molecules and their four neighbors in ice IX. After LaPlaca et al. [1973].

O–D distance by about 0.014 Å and an opening of the D–O–D angle by about 1°.

The O···O···O angle of the group O···D–O–D···O for both oxygen atoms is less than the D–O–D angle, but the D–O–D angle of OI is increased by 1.5°. This cannot be due to a force that tends to make the O–D···O bond linear. It is possible that the H-bonding promotes more tetrahedral sp^3 bonding, or that the electric field of neighboring molecules distorts the D–O–D angle by interacting with the transition moment for the symmetric bending vibration (see section 29.2.2).

The fit of the D–O–D molecule into the O···O···O angle is illustrated in fig. 29.11. The only distortion of molecule II from the O···O···O plane that is allowed by the C_{2v} symmetry is a twist about the axis, as shown. Molecule I departs from a symmetrical position in the O···O···O plane by a significant rock about the axis perpendicular to the plane of the molecule and a wag about the axis in the plane that is perpendicular to the C_2 axis. There is also a small twist. The causes of these distortions are not understood in detail.

The correct approximate orientations of the water molecules in ice IX were predicted by Whalley et al. [1968] on the basis of Kamb and Datta's

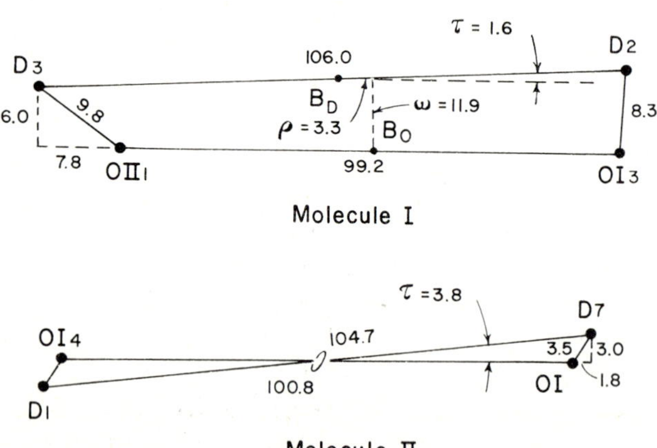

Fig. 29.11. The displacement of each D_2O molecule in ice IX from the plane of the oxygen atoms of the group O···D–O–D···O, represented by the vectors from the central oxygen to the deuterium and peripheral oxygens. B_D and B_O represent the bisectors of the D–O–D and O···O···O lines. The numbers attached to the lines give the angles in degrees subtended by ends of the line with the central oxygen. The angles ω, ρ and τ represent the wagging, rocking and twisting of the water molecule from the symmetrical position in the O···O···O plane. After LaPlaca et al. [1973].

[1960] oxygen positions, by assuming that the departure of the O···O···O angles from the D–O–D angles is minimized. It appears that a single consideration determines the orientation in ice IX, in contrast with the several that are important in ice II (Kamb [1964]).

The structure of ice IX as actually made so far contains a small amount of orientational disorder (LaPlaca et al. [1973]). This is most easily described by saying that positions 1 and 3 are occupied by a deuterium atom with a probability of 0.966, and the corresponding positions at the other end of the bond, 1′ and 3′, are occupied with a probability 0.034, and the corresponding probabilities for positions 2 and 2′ are 0.949 and 0.051. These probabilities are indicated in fig. 29.9. Calorimetric measurements of the entropy of the transition III–IX confirm this order (Nishibata and Whalley [1974]). The experimental entropy of the transformation III–IX is -0.32 cal K^{-1} mol^{-1}, compared with the expected -0.81 cal K^{-1} mol if ice III were fully disordered and ice IX fully ordered, and the calculated value for the occupancies reported by Hamilton et al. [1973] of -0.31 to -0.45 cal K^{-1} mol^{-1}.

29.2.5. Ice V

Ice V crystallizes, according to X-ray diffraction on a single crystal recovered at ~100 K and zero pressure (Kamb et al. [1967]), in a monoclinic unit cell, in space group A2/a with $a = 9.22$, $b = 7.54$, $c = 10.38$ Å, $\beta = 109.2°$, and containing 28 molecules. It is shown in projection in fig. 29.12. Single-crystal neutron diffraction under the same conditions (Hamilton et al. [1969]) shows that the hydrogen atoms are partly ordered in the same space group.

There are 28 oxygen atoms in the face-centered cell, four on equivalent sites of symmetry C_2 and three sets of eight on general positions. The oxygen atoms labelled I are on the two-fold axes, and those labelled II, III, and IV are on general positions. The structure can be visualized (Kamb et al. [1967]) as being made from zig-zag chains of water molecules parallel to *a* and linked by H-bonding. It is perhaps easier however to consider it as made of the seven-molecule units illustrated in fig. 29.13, which are situated on the C_2 axes, a point also noted by von Hippel and Farrell [1973]. They contain one OI and two each of OII, OIII, and OIV atoms. They are closely related to the units that occur in ice VI (see section 29.2.6) and in fact the ice VI unit can be obtained from them by combining the two OIV atoms to give a single atom equivalent to OI and changing the bonding so that OII and OIII become equivalent.

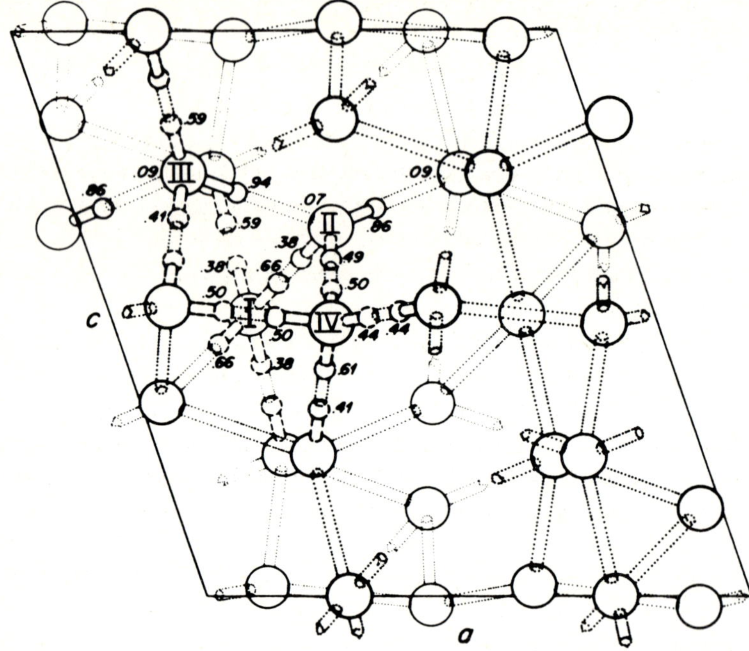

Fig. 29.12. The structure of ice V shown in projection along the *b* axis. Reproduced with permission of Plenum Press.

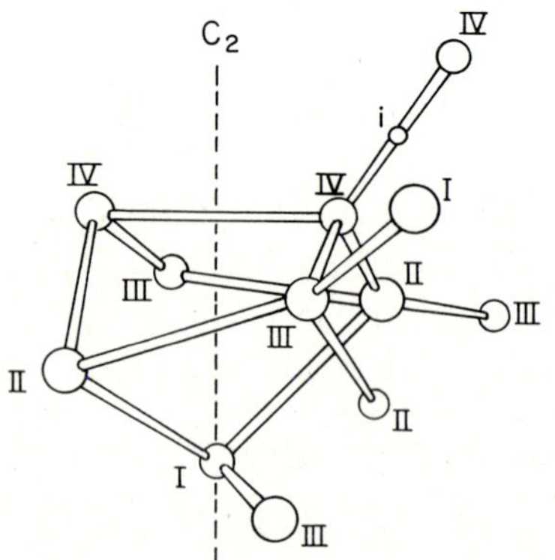

Fig. 29.13. The structural unit of ice V showing the C_2 axis and the inversion center.

Each unit has two four-membered and two five-membered rings, and forms ten H-bonds to other rings. There are two units in the primitive cell formed by joining two antiparallel units by centrosymmetric bonds between OIV atoms. The ten external connections are made to eight neighboring units, two by OIV–OIV links as above which pass through the centers of eight-membered rings, four by OIII–OI and OI–OIII links, which pass through eight-membered rings to which the OI atom is directly linked, and two by pairs of connections OIII–OII and OII–OIII between neighboring rings, which form the only six-membered rings in the structure.

The inter- and intra-group links of the OIV atoms form chains and the eight-membered rings through which the chains pass are illustrated in fig. 29.14, where the horizontal OIV–OIV bonds are intragroup and the inclined bonds are intergroup. Neighboring eight-membered rings share three oxygen atoms, so that they form chains that resemble the Christmas decoration made of paper rings stacked axially and glued to their neighbors alternately at one side and the opposing side, and then stretched. This interpenetration of the bonding, which is manifested at the lowest pressure in ice V, becomes a dominant structural feature in the higher-pressure phases.

The hydrogen atoms in ice V cannot be completely ordered in space group A2/a if the ice rules hold because the C_2 axis and the inversion center require the protons to be evenly disordered on the intragroup and intergroup OIV–OIV bonds. A partial order is however possible, that

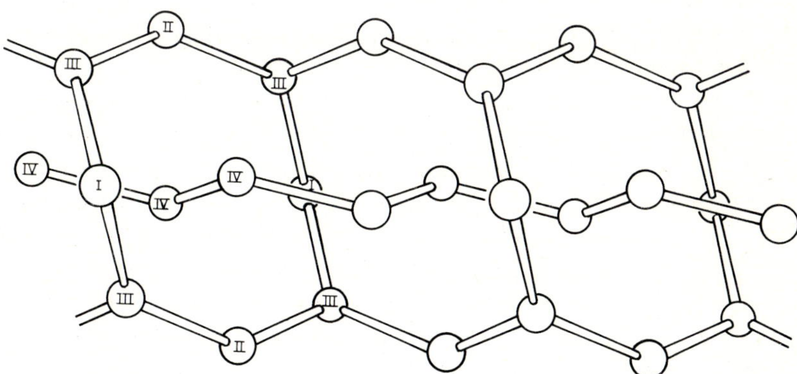

Fig. 29.14. The arrangement of the OIV–OIV chains in ice V and the rings they pass through.

would require three independent probabilities of occupancy of proton positions, which could be taken as the probabilities around either OII or OIII. This seems to occur at low temperatures, and the probabilities obtained by Hamilton et al. [1969] without forcing the ice rules are summarized in fig. 29.12. The probabilities fall into three groups, near 0, $\frac{1}{2}$, and 1, but the deviations from these ideal values are thought (Hamilton et al. [1969]) to be real. Since the space group A2/a contains a symmetry center, the partly ordered form is non-polar.

The maximum amount of order occurs when the only probabilities of occupancy different from 0 or 1 are those along the chains of OIV–OIV bonds. Each chain can have only two orientations, and the space group A2/a is obtained when both directions have equal probabilities and there is no long-range order. The configurational entropy of such an arrangement is effectively zero.

The configurational entropy of the partly ordered phase, as calculated assuming the occupancy probabilities are those given by the neutron diffraction although they do not obey the ice rules exactly, and neglecting closed rings (Nagle [1973]), is 0.57 cal K^{-1} mol^{-1}, compared with the value for complete disorder, within the limits prescribed by the ice rules, of 0.81 cal K^{-1} mol^{-1}. The entropy changes at the transitions III–V and V–VI are about -0.063 and -0.013 cal K^{-1} mol^{-1} respectively (Bridgman [1912]). If ice III and VI were fully disordered and ice V were ordered to the extent determined by neutron diffraction at ~ 100 K, the change of configurational entropy would be -0.24 and $+0.24$ cal K^{-1} mol^{-1} respectively. Ice III is almost certainly fully disordered above ~ -65°C (Whalley et al. [1968]) and the entropy change at the III–V transformation strongly suggests that ice V is fully disordered at high temperature, and this is confirmed by the dielectric properties (Whalley et al. [1968]). Ice VI is never fully disordered according to the dielectric measurements and unfortunately (see section 29.2.6), the effect of partial order on the permittivity is not known quantitatively, and so the partial order in ice VI cannot be expressed as a configurational entropy. A relation between the dielectric correlation parameter g, which is obtained from the permittivity, and the configurational entropy for two-dimension ice models is implicit in the exact calculations described by Lieb and Wu [1973], but has not yet been extracted. It seems very likely that the configurational entropy of ice VI is small. An attempt to detect the ordering transition in ice V at low temperatures by dielectric methods (Johari and Whalley [1973]) was unsuccessful because it readily transformed to the more stable ice II.

29.2.6. Ice VI

Ice VI, like ice V, becomes partly ordered at low temperatures, and indeed, the static permittivity (Johari and Whalley [1973]) suggests that it is never fully disordered even at the melting point.

The hypothetical fully disordered structure (Kamb [1965]) is illustrated in fig. 29.15. It is tetragonal, in space group $P4_2/nmc$ with $a = 6.27$ Å, $c = 5.97$ Å, and there are ten water molecules in the unit cell. Two of the oxygen atoms, labelled I in fig. 29.15, are on points of symmetry $\bar{4}$ and eight, labelled II, are on points of symmetry m. The structural unit which is shown on the left side of fig. 29.15, is an OI atom bonded to four neighboring OII atoms in a structure of symmetry $\bar{4}$. These units are stacked along the c axis, and neighboring chains are joined by H-bonds parallel to the a and b axes. The resulting structure is an open one, and is stabilized by the insertion of an exactly similar structure in its interstices. Other examples of this structural feature, the interpenetration of two substructures without cross-connections, have been known for many years, such as the parallel-ordered form of the cesium bromide structure of ammonium bromide (Levy and Peterson [1952] and earlier work).

Symmetry requires that ice VI be fully disordered because the four-fold screw axis and the mirror planes parallel to the four-fold axis require that the proton occupancy probabilities be all $\frac{1}{2}$ ($\alpha = \frac{1}{2}$ in fig. 29.15). As the

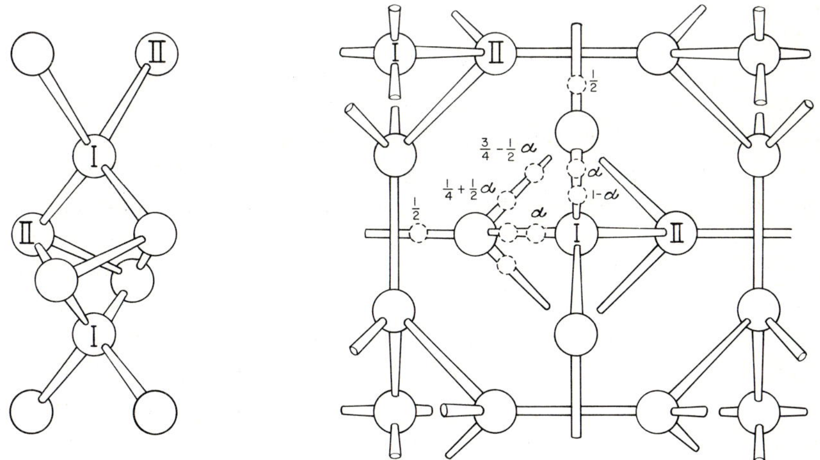

Fig. 29.15. The structural unit of ice VI (left-hand diagram) and the projection of the unit cell of ice VI along the c axis. The proton occupancies are for space group Pmmn. If $\alpha = \frac{1}{2}$ the space group is $P4_2/nmc$. After Kamb [1965].

dielectric evidence (Johari and Whalley [1973]) suggests that ice VI is never fully disordered, ice VI with the space group $P4_2/nmc$ never occurs. No neutron diffraction test of this conclusion has been reported, and the only X-ray diffraction studies of ice VI at high temperature (Kamb and Davis [1964] and Weir et al. [1965]) would probably not detect a partial ordering of the orientations.

The dielectric measurements (Johari and Whalley [1973]) indicate that ice VI becomes more ordered in the direction of a polar crystal: the permittivity increases with decreasing temperature much faster than the inverse of the temperature down to at least 130 K. A neutron diffraction study at ~100 K on samples recovered at low temperature and zero pressure (Kamb [1973]), however, indicates that it is partially ordered in the direction of a non-polar crystal, and these two observations have not yet been reconciled. The space group must degrade from $P4_2/nmc$ if there is any orientational order, and experimentally the symmetry is reduced to Pmmn, the *c* glide plane and the 4_2 axis disappearing, with the unit cell having the same contents as the fully disordered $P4_2/nmc$ cell. The deuteron occupancy probabilities required by the space group Pmmn and the ice rules are shown in fig. 29.15 in terms of the probability α. Maximum order occurs with $\alpha = 0$ or 1. Actual probabilities were determined from neutron diffraction without requiring the ice rules. They correspond approximately to $\alpha = 0.84$, which in turn corresponds to a configurational entropy of 0.64 cal K^{-1} mol^{-1}, compared with the value 0.24 cal K^{-1} mol^{-1} for the maximum order in this space group.

29.2.7. Ice VII and VIII

Ice VII is disordered, but both the permittivity and the entropy of transformation to the ordered phase ice VIII (Johari and Whalley [1973]) indicate that it is not disordered as much as the ice rules allow. The only X-ray diffraction study (Weir et al. [1965]) does not detect any order and so only the fully disordered structure can be described.

It is body-centered cubic, and although the hydrogen positions have not been directly observed, the structure must be two ice Ic structures that interpenetrate one another as shown in fig. 29.16 (Kamb and Davis [1964]). As far as is known, each ice Ic structure is H-bonded independently of the other, although there may be connections at defects. The structure closely resembles that of the cesium–chloride phase of ammonium chloride (Levy and Peterson [1952]) which can be described in the same way.

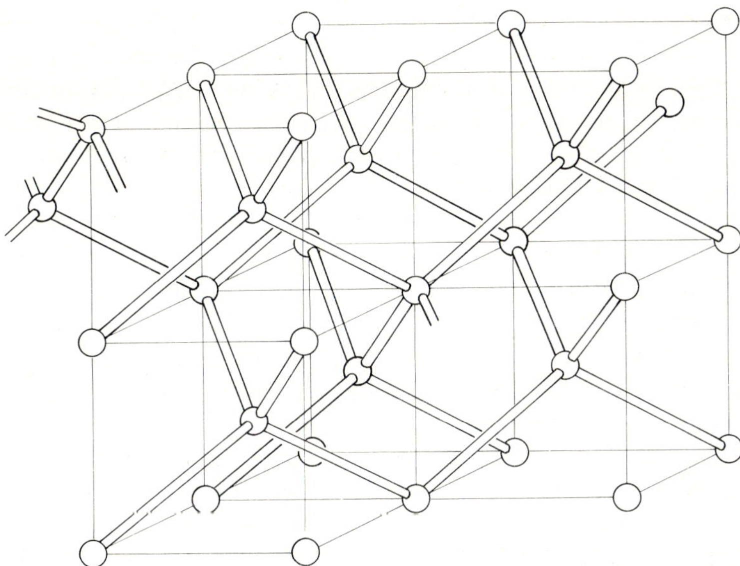

Fig. 29.16. Positions of the oxygen atoms in ice VII. After Kamb and Davis [1964]. The positions in ice VIII are obtained by moving one ice Ic framework up and the other down an amount z. The hydrogen atoms in the upward-translated framework are below their oxygen atoms, and those of the downward-translated framework are above.

Ice VIII is an ordered form of ice VII which was discovered by dielectric methods (Whalley et al. [1966]) following a prediction (Whalley and Davidson [1965]) based on Bridgman's [1937] measurements of the ice VI–VII boundary line. Single-crystal X-ray and neutron diffraction (Kamb and Prakash [1970]) show that it belongs to space group $I4_1/amd$ with eight water molecules in a cell of edges $a = 4.80$ Å, $c = 6.988$ Å. The oxygen atoms are on sites of symmetry C_{2v} (mm) at locations 0, 0, z etc. relative to an origin at the point $\bar{4}m2$. If z were zero, the oxygen atoms would be on body-centered cubic positions.

The structure can be considered as that of ice VII in which all the molecules in one ice Ic structure have their axis in the [100] direction and all those in the other have them in the [$\bar{1}$00] direction, so that the two Ic structures are completely polarized in opposite directions. The primitive unit cell contains four molecules, two having their axes in the [100] direction and two in the [$\bar{1}$00] direction, and of each pair, one has its plane in the (100) plane and the other in the (010) plane.

Each ice Ic structure is not exactly centered in its partner, but is displaced in the c direction by the amount $z = 0.022$. The O–H···O bonds are all 2.965 Å long. If the two structures were centered in one another, this would also be the nearest-neighbor non-H-bonded O···O distance, but the displacement reduces two of these distances for each oxygen atom to 2.795 Å and increases two to 3.15 Å.

29.2.8. Origin of the increased density of the high-pressure phases

There are several ways in which the increasing density of the high-pressure phases with increasing pressure can be viewed. The number of nearest neighbors remains the same in all phases except VII and VIII, but their distance in general increases with increasing density. The increasing density must therefore be obtained by the decreasing distance of second and higher neighbors. The second-neighbor distance in ice I is 4.5 Å, whereas the nearest second-neighbor distance in ice II is about 3.3 Å (Kamb [1964]) and the average value in ice III is about 3.6 Å (Kamb and Prakash [1968]). This tendency is carried to extreme in ice VII (Kamb and Prakash [1970]) in which the average second-neighbor distance is the same as the first, and in ice VIII, in which the "second" neighbor (non-H-bonded) is closer than the first.

Another way to view the increasing density is through the sizes of the rings of water molecules. In ice Ih and Ic, each molecule takes part in six six-membered rings which contain much empty space within them. This empty space can be removed in several ways, for example, by folding or contracting the rings or passing chains through their centers. The rings can also be expanded and either folded or have chains passing through their centers. All these processes are adopted in one or more of the high-pressure phases of ice.

The sizes of the smallest rings that form a closure for the six O···O···O angles about each water molecule in the various phases of ice are summarized in table 29.2. In ice II, the smallest rings are still six-membered, but they are either highly puckered in the chair form or are flattened across a diagonal so that they are more nearly rectangular. The eight-membered ring is also highly puckered. Five-membered rings appear in ice III and IX, and the seven- and eight-membered rings in these phases are highly puckered. There are two kinds of eight-membered rings in ice V, both of which have H-bonds passing through them, one without touching the ring and one with one of its molecules directly joined to the ring. In ice VI there are only four- and eight-membered rings, and the

TABLE 29.2

Sizes of the smallest rings round each kind of oxygen atom in the phases of ice

Phase	Type of O atom	No. in cell	Size of ring 4	5	6	7	8
Ih	I	4	–	–	6	–	–
Ic	I	2					
II	I	6	–	–	5	–	1[a)]
	II	6	–	–	5	–	1[a)]
III, IX	I	8	–	3	–	3	–
	II	4	–	3	–	1	2[a)]
V	I	4	–	1	2	2	1[b)]
	II	2	1	1	2	1	1[b)]
	III	2	1	–	1	1	3[c)]
	IV	2	2	1	–	–	3[c)]
VI	I	4	2	–	–	–	4[b)]
	II	8	3	–	–	–	3[b)]
VII	I	1	–	–	6[b)]	–	–
VIII	I	4					

a) Highly distorted rings.
b) Rings with a bond clear through them.
c) Two rings with bonds clear through them and one with a bond through, one molecule of which is joined to the ring.

eight-membered rings have H-bonds passing through them. Finally, in ice VII and VIII, there are only six-membered rings, each of which has a H-bond passing through it.

29.3. Dissociation energy

Molecular interaction can be divided approximately into components, of which H-bonding is one, but since the division is not exact, the interaction that can be associated with H-bonding but not with any other kind of interaction cannot be exactly described in principle and so cannot be exactly measured in practice.

Molecular interactions can be measured in many ways, and one of the commonest is to measure the energy required to distort the intermolecular distances and angles. The most extreme distortion is to break the bonds completely by separating the molecules to infinity, and the corresponding energy is the energy of vaporization to the ideal vapor at the same temperature. At all temperatures except 0 K, the crystal and the vapor have thermal energy due to excitation of translational, rotational, and perhaps internal motions. These do not occur at 0 K, and the energy of vaporization of both H_2O and D_2O ice Ih at 0 K can be obtained with high accuracy from various calorimetric data. The results (Whalley [1957]) are for H_2O:

$$E(\mathrm{H_2O\ Ih, disordered};\ T=0, p=0) - E(\mathrm{H_2O\ g};T=0, p=0) = 11.315 \pm 0.004\ \mathrm{kcal\ mol^{-1}},$$

and for D_2O:

$$E(\mathrm{D_2O\ Ih, disordered};\ T=0, p=0) - E(\mathrm{D_2O\ g};\ T=0, p=0) = 11.924 \pm 0.040\ \mathrm{kcal\ mol^{-1}}.$$

These energies of vaporization are determined in part by the difference of the zero-point energies of the intramolecular and intermolecular vibrations. If these were known accurately, the energy of vaporization from the hypothetical crystal to the hypothetical vapor, in both of which the nuclei are at rest in their equilibrium positions, could be calculated. The hydrogen and deuterium nuclei differ only in mass and in the various electric and magnetic moments. Apart from small effects due to the interaction of the moments with the fields due to the electrons and other nuclei, and to the failure of the Born-Oppenheimer approximation, which are quite negligible on the scale of energies being considered here, the energies of vaporization with the nuclei at their equilibrium positions are isotope independent. Thus the observed difference of 608 cal mol^{-1} in the energies of vaporization of H_2O and D_2O is due entirely to the difference of zero-point energy.

Insufficient is known about the vibrational properties of the crystal to allow the energies of vaporization with the nuclei in their equilibrium positions to be calculated, but some allowance must be made for the effects of zero-point energy if the observed energy of vaporization is to be compared with the current theories of the energy of ice, which always treat the nuclei as at rest. It appears that the best that can be done at present is to assume the vibrations are harmonic about equilibrium positions that differ from the true equilibrium positions due to anharmonicity, and that the experimental measurements of the vibrational spectra by optical, mechanical, thermal, and other methods can be interpreted on this assumption.

An early attempt to calculate the energy of sublimation from this non-vibrating crystal to the non-vibrating gas (Whalley [1957]) has recently been revised (Whalley [1973]) in the light of recent data. The frequencies at 0 K ought to be used, but the spectroscopic frequencies, on which all the frequencies except the translational are based, have been measured only to 100 K, and consequently, the 100 K spectroscopic frequencies have been used; this should add little to the uncertainty. The mean frequency of the translational band is taken from the heat capacity (Leadbetter [1965]). The rotational band spans the spectroscopic range about 1040–525 cm^{-1} in H_2O and about 775–390 cm^{-1} in D_2O (Bertie et al. [1969]), while the heat capacity appears to give a mean wave number of 650 ± 16 and 520 ± 12 cm^{-1} respectively, with a half-width of less than ~ 200 and ~ 160 cm^{-1}. The best mean wave number at present appears to be 750 and 557 cm^{-1} respectively, with unknown error, but probably not more than 50 and 35 cm^{-1} respectively. The means of the O–H and O–D stretching vibration wave numbers are taken to be the uncoupled O–H and O–D stretching vibration wave numbers of HDO in dilute solution in D_2O and H_2O respectively (Ford and Falk [1968]); this is a significant improvement over the earlier estimate (Whalley [1957]) that is made possible by better data and understanding, and is the main cause of the difference between this and the earlier results. The change of the wave number of the H_2O bending vibration is uncertain for experimental reasons, but the values quoted should not be too far wrong.

The zero-point energies are summarized in table 29.3. The intermolecular zero-point energy of the crystal is 35% of the experimental energy of sublimation. Hence, its thermal properties are influenced in a major way by the quantum mechanical restrictions on the motion of the whole molecules. The intramolecular vibrational energy changes by about 10.5% of the energy of sublimation.

If we consider the crystal to be vaporized by increasing all intermolecular distances uniformly until they become infinite, the decrease of the rotational kinetic energy and the increase of the O–H stretching kinetic energy constitute, in a sense, a repulsive and an attractive potential respectively. The decrease in the O–H stretching kinetic energy is a real contribution to the intermolecular potential energy (Whalley [1958]), and is one which, in principle, can be evaluated exactly if the appropriate harmonic and anharmonic contributions to the vibrational energy levels are known.

The difference of the energies of vaporization of H_2O and D_2O (H_2O–D_2O) at 0 K is 613 cal mol^{-1}. When the intermolecular zero-point energy is allowed for, the energy becomes 240 cal mol^{-1}, and when the

TABLE 29.3

Changes of vibrational zero-point energy of H_2O and D_2O ice Ih on vaporization at 0 K

Motion	No. of vibrations per molecule	H_2O		D_2O	
		$\Delta\bar{\nu}(cm^{-1})$	$E_z(kcal\ mol^{-1})$	$\Delta\bar{\nu}(cm^{-1})$	$E_z(kcal\ mol^{-1})$
Intermolecular					
Translation $\langle\nu_T\rangle$	3	171	0.73	165	0.71
Rotational $\langle\nu_R\rangle$	3	750	3.22	557	2.39
Total intermolecular zero-point energy			3.95		3.10
Intramolecular					
O–H stretching	2	−436	−1.24	311	−0.89
H_2O bending	1	50	0.07	37	0.05
Total intramolecular zero-point energy			−1.17		−0.84
Energy of vaporization at 0 K			11.315		11.924
Energy of vaporization +intermolecular zero-point energy			15.26		15.02
Energy of vaporization from non-vibrating crystal to non-vibrating gas			14.09		14.18

intramolecular energy is allowed for it becomes -90 cal mol^{-1}. There is little doubt that these quantities are real. For example, if the heat of sublimation, when all the nuclei are at rest, were to be the same for H_2O and D_2O, the H_2O wave numbers would have to be wrong by 238 cm^{-1}, which seems impossible.

It can be concluded from this that the difference in the energy required to break an O–H···O and an O–D···O bond in ice when the rotational and translational energy is zero is 120 cal mol^{-1}, which corresponds to an attractive potential between D_2O molecules that is shallower by this amount than that for H_2O. The intramolecular zero-point energy causes this relative difference, and when it is allowed for, the D_2O potential well is deeper than the H_2O well by 45 cal mol^{-1}. This corresponds, at least approximately, to the difference in intermolecular potential energy when the nuclei are at their average positions, and is a direct consequence of the anharmonicity.

Theoretical calculations of the energy of dissociation of ice are usually made with the nuclei in their equilibrium positions, perhaps even the same position in the vapor and crystal. These calculations must be compared with the values in the last row of table 29.3, i.e. when the vibrational energy is allowed for as best we can.

The energies required to break the H-bonds in the high-pressure phases of ice have not been directly measured, but approximate values of the heats of the irreversible transformation of the metastable high-pressure phases at atmospheric pressure to ice Ic have been obtained (Bertie et al. [1963, 1964]), and they are summarized in table 29.4.

TABLE 29.4

Approximate energies of transformation from the high-pressure phases of ice to ice Ic when heated from 77 K and 1 bar at about 1–1.5 K min^{-1} (Bertie et al. [1963, 1964])

Phase	Heating rate (K min^{-1})	Transformation temperature (K)	E(phase)–E(Ic) (cal mol^{-1})
II	–	174	2–10
IX	1.1	150	$70 \pm \sim 10$
V	1.5	151–155	$200 \pm \sim 60$
VI	1.5	150	$290 \pm \sim 40$
VIII	1.5	128	$560 \pm \sim 80$

29.4. Vibrations

The lattice dynamics of ice has been reviewed recently (Whalley [1973]) and only a brief account will be given here.

29.4.1. O–H STRETCHING VIBRATIONS

The most useful information on O–H and O–D stretching force constants has been obtained from the spectrum of small amounts of HDO in either H_2O or D_2O (Hornig et al. [1958]). The early work on ice I (Hornig et al. [1958] and Haas and Hornig [1960]) was partly marred by the undetected presence of vitreous ice (Zimmermann and Pimentel [1962] and Bertie and Whalley [1964a]) but later work (Bertie and Whalley [1964b], Ford and Falk [1968] and Hardin and Harvey [1973]) has cleared up the discrepancies.

The O–H stretching vibration wave nunber at 100 K is 3270 cm^{-1}, and Haas and Hornig [1960] reported the first overtone at 6300 cm^{-1} when the fundamental was at 3275 cm^{-1}. The harmonic frequency ω_e and anharmonicity constant $\omega_e x_e$ from these values, assuming no coupling to other vibrations, are 3525 and -125 cm^{-1} compared with the vapor values of 3890 and -83 cm^{-1} (Kern and Karplus [1972]). The corresponding harmonic force constant is 6.94 mdyn Å. The ratio of the O–H and O–D stretching frequencies of HDO is 1.352, compared with the expected value of 1.374 for harmonic oscillators and assuming no coupling. If the anharmonicity is allowed for, the predicted ratio is 1.348, in reasonable agreement with experiment.

Values of the uncoupled O–H and O–D frequencies are known for all the high-pressure phases of ice that can be obtained at 100 K (Bertie and Whalley [1964b] and Bertie et al. [1968]). Ice II and IX have several bands because they are ordered and contain several crystallographically distinct O–H bonds. The O–D stretching vibration wave number is plotted (Hamilton et al. [1971]) against the O–D···O bond length for various phases in fig. 29.17, assuming that within one phase the order of wave number is the order of bond length. The wave numbers deviate from a smooth curve, but perhaps by little more than corresponds to the uncertainties of $\pm\sim 5$ cm^{-1} in the wave numbers and $\pm\sim 0.01$ Å or more in the distances. It is possible that features of the geometry of the bond other than the O–D···O distance help to determine the wave numbers. One of the more obvious is the deviation of the hydrogen atom from the O···O line, and the deviations in Angstrom units are attached to the points in fig. 29.17. For comparison, the root-mean-square displacements

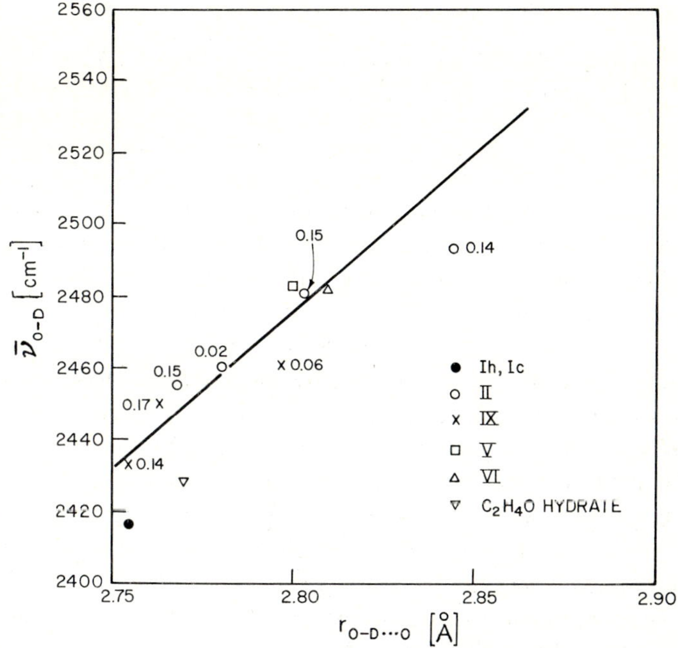

Fig. 29.17. Plot of O–D wave number against O–D bond length for the phases of ice and ethylene oxide hydrate. Modified from Hamilton et al. [1971]. The point for ethylene oxide hydrate is from Bertie et al. [1973].

of the hydrogen atoms due to the symmetric bending and the rotational vibrations are about 0.07 and 0.11 Å respectively. There is no apparent correlation between the deviations from linearity and the deviations of the points from a smooth curve. The slope of the curve is about 770 cm^{-1} Å^{-1}, which corresponds to a slope for the O–H stretching vibration of 1040 cm^{-1} Å^{-1}. If the change of O···O distance with change of phase has the same effect as the change of O···O distance within one phase by the change of pressure, the Grüneisen parameter for the O–H stretching vibration,

$$\gamma = \mathrm{d}\ln \nu / \mathrm{d}\ln \rho,$$

where ρ is the density, is -0.30. No measurements under pressure have been made to test this equivalence.

The O–H stretching vibration wave number is well known as a function of temperature (Ford and Falk [1968]). Above about 100 K it is linear within experimental error with a slope of 0.20 cm^{-1} K^{-1}. The average linear thermal expansivity of ice in this range is about 35 MK^{-1}, and so the variation of O–H wave number with O···O distance, when the distance is varied by varying the temperature at constant pressure, is about 2000 cm^{-1} $Å^{-1}$. It seems that varying the O···O distance by varying the temperature has about twice the effect on the O–H wave number that varying the distance by varying the phase has.

LaPlaca et al. [1973] have pointed out that the change of O–D wave number with average O–D distance from the vapor to ice IX (ice II gives a similar value) corresponds to a value of

$$\mathrm{d}\bar{\nu}_{OD}/\mathrm{d}r_{OD} = \sim 20000\ \mathrm{cm^{-1}\ Å^{-1}},$$

and have shown that this value can be obtained from the anharmonic potential function of the vapor molecule by applying a constant force so as to stretch the bond. It is shown in section 29.2.2 that the electric field acting on a molecule due to its neighbors can provide such a force. While this simple picture may contain some truth, it is clearly considerably simplified.

The coupling constant between neighboring O–H···O–H groups can be obtained from the separation of the in-phase and out-of-phase vibrations of the groups in dilute solution in D_2O (Haas and Hornig [1960]) and is -0.123 mdyn $Å^{-1}$.

Little detailed information about the H-bond in ice has been obtained from the vibrational analysis of O–H stretching bands in H_2O ice because the band is too little understood in detail (Whalley [1973]). The main difficulty is that all $2N$ normal vibrations, where N is the number of water molecules, are infrared and Raman active. The integrated intensity of the band, however, can be shown on the harmonic approximation to be independent of the coupling of the molecular vibrations, and equal to the sum of the integrated intensities of the uncoupled O–H symmetric and antisymmetric stretching vibrations. Both the coupled and uncoupled bands of H_2O, D_2O, and HDO have been measured by Ikawa and Maeda [1968] and the coupled band of H_2O by Bertie et al. [1969], all essentially in agreement. The integrated intensity of the band in H_2O is 140 cm^2 μmol^{-1}, which corresponds to a dipole moment derivative for the stretching of the O–H bond of

$$\mathrm{d}\mu/\mathrm{d}r_{O\text{-}H} = \pm 4.0\ \mathrm{D\ Å^{-1}},$$

or, equivalently, to an effective charge of $\pm 0.83e$ where e is the electronic charge. This large value is sufficient to explain the coupling between neighboring O–H groups (O–H···O–H) as due to the transition moments (Haas and Hornig [1960]).

The overtones of the O–H stretching vibrations have an unusually low intensity. This is due to the mutual cancellation of the effects of the cubic anharmonicity and the quadratic variation of the dipole moment with the normal coordinate (Whalley [1973]). The cancellation is in a sense due to the high transition moment of the fundamentals (DiPaolo et al. [1972]).

29.4.2. Bending and rotational vibrations

Little is known about the bending vibrations, as they are weak and overlapped by relatively strong overtones and combinations. The maximum intensity appears to be about 1650 cm^{-1} compared with the vapor wave number of 1595 cm^{-1}. It is not known whether or not the peak can be assigned to the center wave number of the coupled band.

The rotational vibrations occupy the range 1040–525 cm^{-1} (Bertie et al. [1969]). They are strong in the infrared, and the integrated intensity yields (Ikawa and Maeda [1968]) an effective moment of 1.45 D, which is appreciably smaller than the moment 1.85 D of the isolated molecule. No detailed explanation of this effect has been proposed. The Raman intensity (Wong and Whalley [1974]) is very weak, no doubt because the anisotropy of the polarizability is small.

29.4.3. Translational vibrations

Because of the disorder all the translational vibrations of ice I are infrared and Raman active (Bertie and Whalley [1967], Whalley and Bertie [1967a] and Wong et al. [1973]), and so the spectra are closely related to the density of vibrational states. Even the sound waves are infrared active as fundamentals (Whalley and Labbé [1969]). The spectra below 350 cm^{-1} (Bertie et al. [1969] and Wong et al. [1973]) at 100 K are shown in fig. 29.18. Both have strong maxima near 229 cm^{-1} and there is absorption and scattering up to about 318 cm^{-1}. The exact origin of the high-wave-number part, in particular, of the spectrum is not yet settled (see Whalley [1973] for a review), the main point at issue being whether long-range electrostatic forces consequent on the high transition moment of 0.4e for the O···O stretching motion (see below) are required (Whalley and Bertie [1967b] and Wong et al. [1973]) or not (Prask et al. [1972] and Bosi et al. [1973a, b]). Consequently, the O···O (or equivalently H···O) stretching force constant is uncertain. The compressibility of ice is about 11×10^{-6} bar^{-1} (Leadbetter [1965]), which implies that the force constant

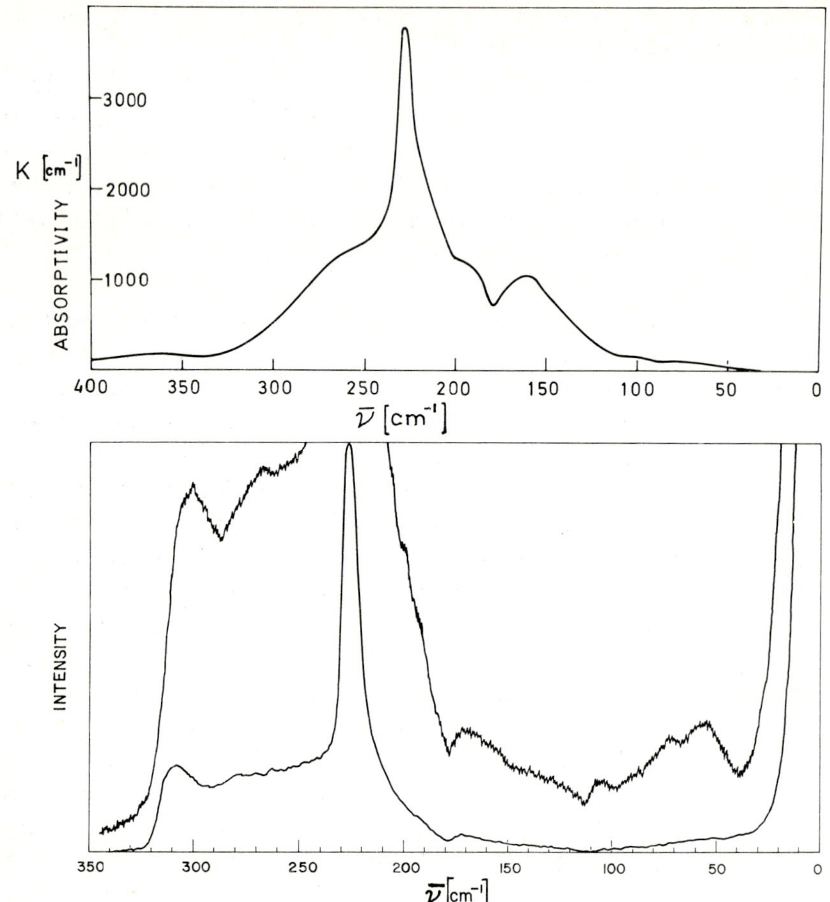

Fig. 29.18. Infrared (Bertie et al. [1969]) (upper frame) and Raman (Wong and Whalley [1974]) (lower frame) spectra of H_2O ice below 350 cm^{-1} at ~ 100 K. The wave number scales are not identical.

opposing a uniform compression is about 0.16 mdyn $Å^{-1}$. No detailed lattice dynamical calculations on disordered ice I that reproduce the observed features, particularly those at about 300 cm^{-1}, have been reported. However, recent calculations on ordered forms of ice Ic (Prask et al. [1972]) and ice Ih (Bosi et al. [1973a, b]) use O···O force constants of 0.265 and 0.291 mdyn $Å^{-1}$ respectively.

The transition moment for the stretching of O···O bonds is 0.4e (see below), and the coupling of the stretching motions of neighboring O···O bonds by means of the transition moments should contribute a positive term to the coupling constant for two pairs of bonds of the kind O···H–O–H···O and O–H···O···H–O and a negative term for bonds of the kind O–H···O–H···O. There is little direct evidence of the value of the coupling constant. Prask et al. [1972] included the negative but not the positive terms, and Bosi et al. [1973a, b] included only one positive and none of the negative terms. Bosi et al.'s [1973a, b] force field unfortunately predicts a compressibility that is about one-half the correct value, and so is seriously incomplete.

The dipole-moment derivatives for the deformation of the H-bonds can be obtained from the absorption intensity of the translational vibrations. Because of the orientational disorder, all $3N$ vibrations are infrared active (Whalley and Bertie [1967a]), and so all $3N$ vibrations must be considered. According to a quite general theory (Klug and Whalley [1972]) for harmonic oscillators, the integrated intensity $A(\boldsymbol{k}, j)$ of a normal vibration belonging to branch j of wave vector $\boldsymbol{k}$ of an orientationally disordered crystal is given by

$$A(\boldsymbol{k}, j) = \left(\frac{n^2+2}{3}\right)^2 \frac{1}{n} \frac{\pi}{3c^2} N \sum_{\boldsymbol{r},\kappa,\kappa'} m(\boldsymbol{r}; \kappa, \kappa') W(\boldsymbol{k}, j; \boldsymbol{r}; \kappa, \kappa'), \qquad (29.1)$$

where n is the refractive index at the frequency of the vibration, c the speed of light and N the number density of molecules. The quantity W is a purely mechanical term that depends only on the normal coordinate $Q(\boldsymbol{k}, j)$ of the vibration

$$W(\boldsymbol{k}, j; \boldsymbol{r}; \kappa, \kappa') = \left\langle \frac{\partial s(\boldsymbol{l}, \kappa)}{\partial Q(\boldsymbol{k}, j)} \frac{\partial s(\boldsymbol{l}+\boldsymbol{r}, \kappa')}{\partial Q(\boldsymbol{k}, j)} \right\rangle_l, \qquad (29.2)$$

where $s(\boldsymbol{l}, \kappa)$ is the κth internal coordinate in the $\boldsymbol{l}$th unit cell and the angular brackets $\langle\ \rangle_l$ indicate that the average is to be taken over all unit cells $\boldsymbol{l}$. The quantity m is a purely electrical term and does not depend on the normal coordinate,

$$m(\boldsymbol{r}; \kappa, \kappa') = \langle \boldsymbol{M}''(\boldsymbol{l}, \kappa) \cdot \boldsymbol{M}''(\boldsymbol{l}+\boldsymbol{r}; \kappa') \rangle_l, \qquad (29.3)$$

where

$$\boldsymbol{M}''(\boldsymbol{l}, \kappa) = \frac{\partial \boldsymbol{\mu}}{\partial s(\boldsymbol{l}, \kappa)} - \left\langle \frac{\partial \boldsymbol{\mu}}{\partial s(\boldsymbol{l}, \kappa)} \right\rangle_l. \qquad (29.4)$$

For ice I, symmetry requires that

$$\left\langle \frac{\partial \boldsymbol{\mu}}{\partial s(\boldsymbol{l}, \kappa)} \right\rangle_l = 0. \qquad (29.5)$$

These equations have been written assuming that the vibrations are mechanically regular or nearly regular. If they are not, the equations are still valid if the indices $\boldsymbol{k}, j$ are replaced by a label that does not imply a wave vector or a branch.

The first approximation used (Bertie et al. [1969]) assumed that

$$m(\boldsymbol{r};\kappa,\kappa')/\phi(\boldsymbol{r};\kappa,\kappa') = \text{constant}, \tag{29.6}$$

where $\phi(\boldsymbol{r};\kappa,\kappa')$ is the force constant for the internal coordinates $s(\boldsymbol{l},\kappa)$ and $s(\boldsymbol{l}+\boldsymbol{r},\kappa')$, and neglected the effects of the local correlation of molecular orientations. A purely macroscopic relation between the dipole-moment derivative, the limiting dielectric constants of the band and the compressibility was derived and it yielded

$$\partial\mu/\partial r_{\mathrm{O\cdots O}} = 1.4\ \mathrm{D\ \AA^{-1}} = 0.3e. \tag{29.7}$$

The second approximation (Klug and Whalley [1973]) removed the assumption (29.6) and took account of the effects of the orientational correlation. Then the lattice dynamics had to be solved in detail. The force field used did not reproduce the observed density of states exactly and, in particular, predicted no vibrations above about $255\ \mathrm{cm^{-1}}$. Nevertheless, the dipole-moment derivatives should be moderately accurate and are, for the central M_r and non-central $M_\perp$ displacements of the O$\cdots$O bonds,

$$M_r = 1.93\ \mathrm{D\ \AA^{-1}} = 0.40e, \qquad M_\perp = 0.19\ \mathrm{D\ \AA^{-1}} = 0.04e.$$

By taking account of the effect of the correlations, the derived effective charge for stretching the bonds is increased by $\frac{1}{3}$.

The effective charge cannot be understood on simple classical grounds. A model in which the water molecules are represented as classically polarizable non-overlapping charge distributions predicts an effective charge of only about $0.1e$, so that the calculated integrated intensity is about $\frac{1}{16}$ of the observed value (Whalley [1972]). It appears that specific valence-type interactions must be invoked.

A little is known about the pressure and temperature dependence of the vibrational properties of the H-bonds (Whalley [1973]). The frequencies in the transverse acoustic branches decrease with increasing pressure (Brockamp and Rüter [1969]). This suggests that the thermal expansivity should become negative at low temperatures, which agrees with Jacob and Erk's [1928] and Dantl's [1962] measurements, but not with Brill and Tippe's [1967] (see fig. 29.5).

TABLE 29.5

Electric-field gradient eq and its anisotropy η at the nuclei of a water molecule in the vapor and in the phase of ice at zero pressure

		D			^{17}O		
Phase	T (K)	e^2qQ/h (kHz)	η	Ref.	e^2qQ/h (MHz)	η	Ref.
Ih sc[a]	263	$213.2 \pm 0.8(D_1)$[b]	0.100 ± 0.002	**c**	6.66 ± 0.10	0.935 ± 0.01	**k, l**
	263	$216.4 \pm 1.0(D_2)$		**c**	6.525 ± 0.015	0.925 ± 0.02	**n**
Ih p	130	192 ± 2.7		**d**			
Ic p	75	185 ± 3.6		**e**			
II p	75	179 ± 5.4		**e**			
V p	75	183 ± 7.7		**e**			
IX p	75	178 ± 9.3		**e**			
Liquid		~ 230		**g, h, j**	~ 7.7		**g**
vapor		318.6 ± 2.4	0.06 ± 0.16	**f**	9.82 ± 0.2	0.407 ± 0.1	**m**
					10.17 ± 0.07	0.75 ± 0.01	**p**

a) s.c. = single crystal; p = powder.

b) D, D = deuterons in O–D bonds parallel and inclined respectively to the c axis.

References: **c** = Waldstein et al. [1964]; **d** = Jackson and Rabideau [1964]; **e** = Rabideau et al. [1968]; **f** = Thaddeus et al. [1964]; **g** = Garrett et al. [1967]; **h** = Woessner [1964]; **j** = Powles and Rhodes [1966]; **k** = Spiess et al. [1969]; **l** = Waldstein and Rabideau [1967]; **m** = Stevenson and Townes [1957]; **n** = Edmonds and Zussman [1972]; **p** = Verhoeven et al. [1969].

29.5. Electric-field gradient of the nuclei

The energy level of a nucleus having an electric quadrupole moment due to spin splits when the nucleus is placed in an electric field gradient, and the value of the gradient can be determined from the splitting of the lines of the nuclear magnetic resonance spectrum. Several measurements have been reported of the gradient at the deuterium and oxygen (^{17}O) nuclei and they are summarized in table 29.5, and compared with the values for the vapor and liquid. There is a considerable change from the vapor to ice I, but little difference between the phases of ice. A theoretical discussion (Gornostansky and Kern [1971]) based on a molecular orbital treatment of water polymers up to the pentamer concluded that the shift of the D coupling constant is more sensitive to geometry than to charge-transfer and overlap exchange forces, although the O–D distances (1.00 and 1.01 Å) were probably taken too long (see section 29.2), and this may have affected the conclusions. The ^{17}O shifts were more sensitive to exchange and charge transfer. No doubt calculations on larger fragments would be well worthwhile.

If the field gradient at the deuterium nuclei is linear in the O–D distance in the range of distances between the vapor and ice II and IX, it varies at the rate 10 MHz Å^{-1}. The small differences between the gradient in ice I, II, and IX suggests that the O–D distances do not differ by more than about 0.001 Å.

References

Arnold, G. P., E. D. Finch, S. W. Rabideau and R. G. Wenzel, 1968, Neutron-diffraction Study of Ice Polymorphs. III. Ice Ic, J. Chem. Phys. **49**, 4365.

Barker, J. A. and R. O. Watts, 1973, Monte Carlo Studies of the Dielectric Properties of Water-like Models, Mol. Phys. **26**, 789.

Barnaal, D. E. and I. J. Lowe, 1967, Experimental Free-induction-decay Shapes and Theoretical Second Moments for Hydrogen in Hexagonal Ice, J. Chem. Phys. **46**, 4800.

Bertie, J. E. and E. Whalley, 1964a, Infrared Spectra of Ices Ih and Ic in the Range 4000–350 cm^{-1}, J. Chem. Phys. **40**, 1637.

Bertie, J. E. and E. Whalley, 1964b, Infrared Spectra of Ices II, III, and V in the Range 4000–350 cm^{-1}, J. Chem. Phys. **40**, 1646.

Bertie, J. E. and E. Whalley, 1967, Optical Spectra of Orientationally Disordered Crystals, II: Infrared Spectrum of Ice Ih and Ic from 360 to 50 cm^{-1}, J. Chem. Phys. **46**, 1271.

Bertie, J. E., L. D. Calvert and E. Whalley, 1963, Transformation of Ice II, III, and V at Atmospheric Pressure, J. Chem. Phys. **38**, 840.

Bertie, J. E., L. D. Calvert and E. Whalley, 1964, Transformation of Ice VI and VII at Atmospheric Pressure, Can. J. Chem. **42**, 1373.

Bertie, J. E., H. J. Labbé and E. Whalley, 1968, Infrared Spectrum of Ice VI in the Range 4000–50 cm^{-1}, J. Chem. Phys. **49**, 2839.

Bertie, J. E., H. J. Labbé and E. Whalley, 1969, Absorptivity of Ice I in the Range 4000–30 cm^{-1}, J. Chem. Phys. **50**, 4501.

Bertie, J. E., D. A. Othen and M. Solinas, 1973, The Infrared Spectra of Ethylene Oxide Hydrate and Hexamethylenetetramine Hydrate at 100 K, in: Physics and Chemistry of Ice, Eds. E. Whalley, S. J. Jones and L. W. Gold (Royal Society of Canada, Ottawa), p. 61.

Bosi, P., R. Tubino and G. Zerbi, 1973a, Lattice Dynamics of Hydrogen Bonded Crystals: Ice Ih, in: Physics and Chemistry of Ice, Eds. E. Whalley, S. J. Jones and L. W. Gold (Royal Society of Canada, Ottawa), p. 98.

Bosi, P., R. Tubino and G. Zerbi, 1973b, On the Problem of the Vibrational Spectrum and Structure of Ice Ih: Lattice Dynamical Calculations, J. Chem. Phys. **59**, 4578.

Bridgman, P. W., 1912, Water, in the Liquid and Five Solids Forms, under Pressure, Proc. Am. Acad. Arts Sci. **47**, 441.

Bridgman, P. W., 1937, The Phase Diagram of Water to 45000 kg/cm^2, J. Chem. Phys. **5**, 964.

Brill, R. and A. Tippe, 1967, Gitterparameter von Eis I bei tiefen Temperaturen, Acta Cryst. **23**, 343.

Brockamp, B. and H. Rüter, 1969, Die Abhängigkeit der elastischen Parameter des Eises vom hydrostatischen Druck bis zu 400 bar, Z. Geophysik **35**, 277.

Calvert, L. D. and E. Whalley, 1965, unpublished work.

Chamberlain, J. S., F. H. Moore and N. H. Fletcher, 1973, Neutron-diffraction Study of H_2O Ice at 77 K, in: Physics and Chemistry of Ice, Eds. E. Whalley, S. J. Jones and L. W. Gold (Royal Society of Canada, Ottawa), p. 283.

Chidambaram, R., 1961, A Bent Hydrogen Bond Model for the Structure of Ice. I. Acta Cryst. **14**, 467.

Coulson, C. A. and D. Eisenberg, 1966, Interactions of Water Molecules in Ice. I. The Dipole Moment of H_2O Molecule in Ice, Proc. Roy. Soc. **A291**, 445.

Dantl, G., 1962, Wärmausdehung von H_2O- und D_2O-Einkristallen, Z. Phys. **166**, 115.

Davidson, D. W., 1973, Clathrate Hydrates, in: Water A Comprehensive Treatise, Vol. **2**, Ed. F. Franks (Plenum Press, New York), p. 115.

DiPaolo, T., C. Bouderon and C. Sandorfy, 1972, Model Calculations on the Influence of Mechanical and Electrical Anharmonicity on Infrared Intensities: Relation to Hydrogen Bonding, Can. J. Chem. **50**, 3161.

Edmonds, D. T. and A. Zussman, 1972, Pure Quadrupole Resonance of ^{17}O in Ice, Phys. Letters **41A**, 167.

Eisenberg, D. and W. Kauzmann, 1969, The Structure and Properties of Water (Plenum Press, Oxford).

Engelhardt, H. and E. Whalley, 1972, Ice IV, J. Chem. Phys. **56**, 2678.

Fletcher, N. H., 1970, The Chemical Physics of Ice (Cambridge University Press).

Ford, T. A. and M. Falk, 1968, Hydrogen Bonding in Water and Ice, Can. J. Chem. **46**, 3579.

Garrett, B. B., A. B. Denison and S. W. Rabideau, 1967, Oxygen-17 Relaxation in Water, J. Phys. Chem. **71**, 2606.

Gornostansky, S. D. and C. W. Kern, 1971, Analysis of the D and ^{17}O Quadrupole Coupling Constants in Ice Ih, J. Chem. Phys. **55**, 3253.

Haas, C. and D. F. Hornig, 1960, Inter- and Intramolecular Potentials and the Spectrum of Ice. J. Chem. Phys. **32**, 1763.

Hamilton, W. C., B. Kamb, S. J. LaPlaca and A. Prakash, 1969, Deuteron Arrangements in High-pressure Forms of Ice, in: Physics of Ice, Eds. N. Riehl, B. Bullemer and H. Engelhardt (Plenum Press, New York), p. 44.

Hamilton, W. C., B. Kamb, S. J. LaPlaca and A. Prakash, 1971, Ordered Proton Configuration in Ice II from Single-crystal Neutron Diffraction, J. Chem. Phys. **55**, 1934.

Hardin, A. H. and K. B. Harvey, 1973, Temperature Dependences of the Ice I Hydrogen Bond Spectral Shifts. I. The Vitreous to Cubic Ice I Phase Transformation, Spectrochim. Acta **29A**, 1139.

von Hippel, A. and E. F. Farrell, 1973, Molecular Phenomena in H_2O Systems, I: A Molecular Interpretation of the Phase Diagram of Ice, Mater. Res. Bull. **8**, 127.

Hobbs, P. V., 1975, Ice Physics (Oxford University Press).

Hornig, D. F., H. F. White and F. P. Reding, 1958, The Infrared Spectra of Crystalline H_2O, D_2O and HDO, Spectrochim. Acta **12**, 378.

Ikawa, S. I. and S. Maeda, 1968, Infrared Intensities of the Stretching and Librational Bands of H_2O, D_2O and HDO in Solids, Spectrochim. Acta **24A**, 655.

Jackson, J. A. and S. W. Rabideau, 1964, Deuteron Magnetic Resonance in Polycrystalline Heavy Ice (D_2O), J. Chem. Phys. **41**, 4008.

Jacob, M. and S. Erk, 1928, Wärmdehung des Eises zwischen O und −253°, Z. Ges. Kalteind. **35**, 125.

Johari, G. P. and E. Whalley, 1973, Orientational Order in Ice I, V, VI and VII, in: Physics and Chemistry of Ice, Eds. E. Whalley, S. J. Jones, and L. W. Gold (Royal Society of Canada, Ottawa), p. 278.

Kamb, B., 1964, Ice II: A Proton Ordered Form of Ice, Acta Cryst. **17**, 1437.

Kamb, B., 1965, Structure of Ice VI, Science **150**, 205.

Kamb, B., 1973, Crystallography of Ice, in: Physics and Chemistry of Ice, Eds. E. Whalley, S. J. Jones, and L. W. Gold (Royal Society of Canada, Ottawa), p. 28.

Kamb, W. B. and S. K. Datta, 1960, Crystal Structures of the High-pressure Forms of Ice: Ice III, Nature **187**, 140.

Kamb, B. and B. L. Davis, 1964, Ice VII: The Densest Form of Ice, Proc. Natl. Acad. Sci. U.S. **52**, 1433.

Kamb, B. and A. Prakash, 1968, Structure of Ice III, Acta Cryst. **B24**, 1317.

Kamb, B. and A. Prakash, 1970, Structure of Ice VIII, unpublished manuscript, communicated by B. Kamb, June 1970.

Kamb, B., A. Prakash and C. Knobler, 1967, Structure of Ice V, Acta Cryst. **22**, 706.

Kern, C. W. and M. Karplus, 1972, The Water Molecule, in: Water A Comprehensive Treatise, Vol. **1**, Ed. F. Franks (Plenum Press, New York), ch. 2.

Klug, D. D. and E. Whalley, 1972, Optical Spectra of Orientationally Disordered Crystals. IV. Effects of Short-range Correlation of Orientations, J. Chem. Phys. **56**, 553.

Klug, D. D. and E. Whalley, 1973, The Effect of Orientational Correlation on the Far-infrared Spectrum of Ice, in: Physics and Chemistry of Ice, Eds. E. Whalley, S. J. Jones and L. W. Gold (Royal Society of Canada, Ottawa), p. 93.

Kume, K., 1960, Proton Magnetic Resonance in Pure and Doped Ice, J. Phys. Soc. Japan **15**, 1493.

LaPlaca, S. J. and B. Post, 1960, Thermal Expansion of Ice, Acta Cryst. **13**, 503.

LaPlaca, S. J., W. C. Hamilton, B. Kamb and A. Prakash, 1973, A Nearly Proton-ordered Structure for Ice IX, J. Chem. Phys. **58**, 567.

Leadbetter, A. J., 1965, The Thermodynamic and Vibrational Properties of H_2O and D_2O Ice, Proc. Roy. Soc. **A287**, 403.

Levy, H. A. and S. W. Peterson, 1952, Neutron Diffraction Determination of the Crystal Structure of Ammonium Bromide in Four Phases, J. Am. Chem. Soc. **86**, 1536.

Lieb, E. H. and F. Y. Wu, 1973, Two-dimensional Ferroelectric Models, in: Phase Transitions and Critical Phenomena, Eds. C. Domb and M. Green (Academic Press, New York), p. 331.

Nagle, J. F., 1973, in: Physics and Chemistry of Ice, Eds. E. Whalley, S. J. Jones and L. W. Gold (Royal Society of Canada, Ottawa), p. 70.

Nishibata, K. and E. Whalley, 1974, Thermal Effects of the Transformation Ice III–IX, J. Chem. Phys. **60**, 3189.

Pedersen, B., 1964, NMR in Hydrate Crystals: Correction for Vibrational Motions, J. Chem. Phys. **41**, 92.

Peterson, S. W. and A. L. Levy, 1957, A Single-crystal Neutron Diffraction Study of Heavy Ice, Acta Cryst. **10**, 70.

Powell, R. W., 1958, Preliminary Measurements of the Thermal Conductivity and Expansion of Ice, Proc. Roy. Soc. **A247**, 464.

Powles, J. G. and M. Rhodes, 1966, Deuteron Spin Relaxation in Benzene, Bromobenzene, Water and Ammonia, Mol. Phys. **11**, 515.

Prask, H. J., S. F. Trevino, J. D. Gault and K. W. Logan, 1972, Ice I: Lattice Dynamics and Incoherent Neutron Scattering, J. Chem. Phys. **56**, 3217.

Rabideau, S. W., E. D. Finch and A. B. Denison, 1968, Proton and Deuteron NMR of Ice Polymorphs, J. Chem. Phys. **49**, 4660.

Rahman, A. and F. H. Stillinger, 1971, Molecular Dynamics of Liquid Water, J. Chem. Phys. **55**, 3336.

Rahman, A. and F. H. Stillinger, 1973, Hydrogen Bond Patterns in Liquid Water, J. Am. Chem. Soc. **95**, 7943.

Spiess, H. W., B. B. Garrett, R. K. Sheline and S. W. Rabideau, 1969, Oxygen-17 Quadrupole Coupling Parameters for Water in Its Various Phases, J. Chem. Phys. **51**, 1201.

Stephens, R. W. B., 1958, The Mechanical Properties of Ice. II. The Elastic Constants and Mechanical Relaxation of Single-crystal Ice, Advan. Phys. **7**, 266.

Stevenson, M. J. and C. H. Townes, 1957, Quadrupole Moment of ^{17}O, Phys. Rev. **107**, 635.

Stillinger, F. H. and A. Rahman, 1972, Molecular Dynamics Study of Temperature Effects on Water Structure and Kinetics, J. Chem. Phys. **57**, 1281.

Stillinger, F. H. and M. A. Cotter, 1973, Local Orientational Order in Ice, J. Chem. Phys. **58**, 2532.

Thaddeus, P., L. C. Krisher and T. H. N. Loubser, 1964, Hyperfine Structure in the Microwave Spectrum of HDO, HDS, CH_2O and CHDO: Beam-maser Spectroscopy of Asymmetric-top Molecules, J. Chem. Phys. **40**, 257.

Verhoeven, J., A. Dymanus and H. Bluyssen, 1969, Hyperfine Structure of $HD^{17}O$ by Beam-maser Spectroscopy, J. Chem. Phys. **50**, 3330.

Waldstein, P. and S. W. Rabideau, 1967, ^{17}O Nuclear Magnetic Resonance of Single Crystals of D_2O Ice, J. Chem. Phys. **47**, 5338.

Waldstein, P., S. W. Rabideau and J. A. Jackson, 1964, Nuclear Magnetic Resonance of Single Crystals of D_2O Ice, J. Chem. Phys. **41**, 3407.

Weir, C., S. Block and G. Piermarini, 1965, Single-crystal X-ray Diffraction at High Pressure, J. Res. Natl. Bur. Std. C **69**C, 275.

Whalley, E., 1957, The Difference in the Intermolecular Forces of H_2O and D_2O, Trans. Faraday Soc. **53**, 1578.

Whalley, E., 1958, Zero-point Energy: A Contribution to Intermolecular Forces, Trans. Faraday Soc. **54**, 1613.

Whalley, E., 1969, Structure Problems of Ice, in: Physics of Ice, Eds. N. Riehl, B. Bullemer and H. Engelhardt (Plenum Press, New York), p. 19.

Whalley, E., 1972, Dipole-moment Derivative of the Hydrogen Bond in Ice, Can. J. Chem. **50**, 310.

Whalley, E., 1973, Lattice Dynamics of Ice, in: Physics and Chemistry of Ice, Eds. E. Whalley, S. J. Jones and L. W. Gold (Royal Society of Canada, Ottawa).

Whalley, E., 1974, The O–H Distance in Ice, Mol. Phys. **28**, 1105.

Whalley, E. and J. E. Bertie, 1967a, Optical Spectra of Orientationally Disordered Crystals. I. Theory for Translational Lattice Vibrations, J. Chem. Phys. **46**, 1264.

Whalley, E. and J. E. Bertie, 1967b, Far-infrared Spectrum and Long-range Forces in Ice, J. Colloid Interface Sci. **25**, 161.

Whalley, E. and D. W. Davidson, 1965, Entropy Changes at the Phase Transitions in Ice, J. Chem. Phys. **43**, 2148.

Whalley, E. and H. J. Labbé, 1969, Optical Spectra of Orientationally Disordered Crystals, III. Infrared Spectra of the Sound Waves, J. Chem. Phys. **51**, 3120.

Whalley, E., D. W. Davidson and J. B. R. Heath, 1966, Dielectric Properties of Ice VII. Ice VIII: A New Phase of Ice, J. Chem. Phys. **45**, 3976.

Whalley, E., J. B. R. Heath and D. W. Davidson, 1968, Ice IX: An Antiferroelectric Phase Related to Ice III, J. Chem. Phys. **48**, 2362.

Woessner, D. E., 1964, Molecular Reorientation in Liquids: Deuteron Quadrupole Relaxation in Liquid Deuterium Oxide and Perdeuterobenzene, J. Chem. Phys. **40**, 2341.

Wong, P. T. T. and E. Whalley, 1974, Optical Spectra of Orientationally Disordered Crystals. V. Raman Spectrum of Ice in the Range 4000–350 cm^{-1}, J. Chem. Phys. **62**, 2418.

Wong, P. T. T., D. D. Klug and E. Whalley, 1973, The Raman Spectrum of the Translational Lattice Vibrations of Ice Ih, in: Physics and Chemistry of Ice, Eds. E. Whalley, S. J. Jones and L. W. Gold (Royal Society of Canada, Ottawa), p. 97.

Yamaji, K., 1957, Thermal Expansion of Ice in the Temperature Range 0–100 °C. Low Temp. Sci. Ser. A**16**, 73.

Zarembovitch, A. and A. Kahane, 1964, Détermination des Vitesses de Propagation d'Ondes Ultrasonores Longitudinales dans la Glace: Etude de Leur Variation avec la Temperature, Compt. Rend. **258**, 2529.

Zimmermann, R. and G. C. Pimentel, 1962, The Infrared Spectrum of Ice: Temperature Dependence of the Hydrogen Bond Potential Function, Advan. Mol. Spect., Vol. **2**, Ed. A. Mangini (Pergamon Press, New York), p. 726.

AUTHOR INDEX TO VOLUMES I–III

Italic numbers refer to the pages on which references to the respective chapters have been listed

Abe, R., 1142, 1158, *1160*, *1165*

Abragam, A., 329, *353*, 1029, 1031, 1034, 1037, *1058*

Abraham, D.B., 1137, *1160*

Abramov, V.N., 1277, 1278, 1290, 1292, 1296, *1353*

Ackermann, T., 687, 688, 693, 696, 700, 753, *762*, *763*

Ackermann, T. see Wicke, E., 475, *525*, 677, *682*, 688, 753, *766*

Acquista, N. see Mann, D.E., 1101, *1105*

Adamivich, V. see Fisher, I.Z., 1386, *1421*

Adelman, S.A., 378, *387*

Adriaenssens, G.J., 1130, 1140, *1160*

Ady, E., 117, 151, *155*, 219, *242*

Afanas'ev, M.L. see Gavrilova-Podol'skaya, G.V., 1136, 1139, *1162*

Affsprung, H.E., 809, *827*

Affsprung, H.E. see Christian, S.D., 1203, *1223*

Agrawal, A.K., 932, *933*

Ahlrichs, R., 36, *155*

Ahlrichs, R. see Jungen, M., 36, *158*

Ahmad, D. see Tellgren, R., 1139, *1164*

Aikins, J. see Adriaenssens, G.J., 1130, *1160*

Aizu, K., 1143, 1146, 1158, 1159, *1160*, *1165*

Aksnes, G., 1201, *1223*

Alagona, G., 82, 85, 86, 121, *155*, 190, *214*

Alagona, G. see Pullman, A., 82, 85, *161*

Albert, N., 601, *607*

Albertsson, J., 445, *453*

Albrecht, G., 409, *453*

Alder, B.J. see Shoorley, J.N., 1403, *1423*

Aldred, B.K., 927, 932, *933*

Aldridge, W.N., 762, *763*

Aldridge, W.N. see Rose, M.S., 762, *765*

Alefeld, B., 896, 933, *933*, 1178, *1195*

Alei, M., 814, *827*

Alei, M. see Litchman, W.M., 822, *829*

Aleksandrov, K.S., 1146, *1160*

Aleksandrova, I.P., 1146, *1160*

Aleksandrova, I.P. see Aleksandrov, K.S., 1146, *1160*

Aleksandrovich, Kh.M., 1350, *1353*

Alexeeva, T.L. see Bogachev, Y.S., 797, *827*

Alger, T.D., 1045, *1058*

Algie, J.E., 1014, *1017*

Allavena, M. see Fournier, M., 661, *681*

Allen, G., 533, *560*

Allen, G.R., 1160, *1165*

Allen, G.R. see Nagle, J.F., 1137, 1138, *1163*

Allen, L.C. see Kollman, P.A., *20*, 27, 48, 51, 52, 57, 60, 61, 70, 72, 75, 80, 111, 113, 121, 150, *159*, 186, 190, 206, 214, *215*, 320, *355*, 517, 519, *524*, 723, *764*, 793, *828*, 946, *1021*

Allerhand, A., 573, *607*, 951, *1017*, 1207, *1223*

Allies, M. see Kálmán, E., 744, *764*

Almenningen, A., 323, 324, 333, *353*, 466, *468*, 771, *788*

Almlöf, J., 119, 121, 123, *155*, 183, 185, 186, 208, *214*, 297, 300, 352, *353*, 399, 417, 425, 438, 439, *453*, 482, 490, 493, 508, 511, 512, 518, 519, *523*, 676, *681*

Altenburg, H. see Mootz, D., 482, 495, *525*

Ambard, L. see Trautmann, S., 744, *765*

Amberg, C.H. see Taylor, J.H., 1345, *1363*

Amos, A.T. see Musher, J.I., 39, *160*

Andersen, E., 483, 488, 493, 500, 501, *523*

Andersen, H.C., 368, 386, *387*, *389*

Andersen, H.C. see Chandler, D., 386, *389*

Andersen, K.K., 711, *763*

Anderson, A., 909, *933*

Anderson, A. see Savoie, R., 1073, 1074, 1077, *1105*

Anderson, A. see Torrie, B.H., 1145, *1165*

Anderson, D.K. see Hardt, A.P., 1048, *1059*

Anderson, D.M., 1004, *1017*, 1316, *1353*

Anderson, G.R., 123, *155*, 203, *214*, *353*

Anderson, G.R. see Jiang, G.J., 123, *158*, 185, *215*

Anderson, J.E., 940, *1017*, 1042, 1043, 1049, *1058*

Anderson Jr. J.H., 1267, 1285, 1306, 1307, 1308, 1313, 1314, *1353*

Anderson, M.R., 1152, *1160*

Anderson, P.J., 1343, *1353*

Anderson, P.J. see Webster, R.K., 1343, *1364*

Andrè, J. see Cruz, M.I., 1320, 1322, 1324, 1325, 1326, *1355*

Andrè, J. see Fripiat, J.J., 1305, 1313, 1314, 1315, 1317, 1348, *1357*

André, J.M., 142, *155*

André, J.M. see Clementi, E., 116, *156*
André, J.M. see White, J.L., 1285, *1364*
André, M.C. see Clementi, E., 116, *156*
Andrew, E.R., 477, *523*, 871, *885*, 1029, 1034, 1037, *1058*
Andrews, L., 460, *468*
Andrews, L.J. see Keefer, R.M., 1205, *1224*
Angadiyavar, C.S. see R. Srinivasan, 18, *22*
Angell, C.L., 1348, *1353*
Anistratov, A.T. see Aleksandrov, K.S., 1146, *1160*
Antipina, T.V., 1334, *1353*
Antipina, T.V. see Kirina, O.F., 1334, *1358*
Antonini, M., 932, *933*
Antoniou, A.A., 1316, *1353*
Appiano, A. see Zecchina, A., 1276, 1293, *1364*
Arai, H., 1337, *1353*
Arend, H., 1151, *1161*
Arendt, F. see Boehm, H.P., 1272, *1354*
Arita, M. see Satake, I., 827, *830*
Armbruster, A.M. see Pullman, A., *161*
Armistead, C.G., 1267, 1272, *1353*
Armstrong, G.T., 1246, *1257*
Arnett, E.M., 708, 711, *763*, 1201, 1203, 1205, *1223*
Arnold, D. see Camp, P.R., 973, *1017*
Arnold, G.M., 909, *933*
Arnold, G.P., 1433, *1466*
Arnold, J., 224, *242*, 330, *353*, 634, 635, 640, *652*
Arora, S.K., 483, 495, *523*
Arrighini, G.P., 35, *155*
Arshadi, M., 121, 139, 140, *155*
Arshadi, M. see Kebarle, P., 121, 139, 140, *159*, 475, 481, 519, *524*, 980, *1021*
Arsić-Eskinja, M., 921, 922, 924, 925, 928, 932, *933*, 1187, 1189, 1191, *1195*
Ashida, T., 855, *885*
Ashley, M. see Giaque, W.F., 363, *387*
Ashmore, J.P. see Kerr, K.A., 460, *469*
Asselin, M., 336, *353*, 570, 571, 573, *607*, 622, 625, 633, 637, 638, 642, 643, 645, 646, *652*
Assink, R.A., 1039, *1058*
Atherton, K., 1345, *1353*
Atoji, M., 80, *155*
Atwood, M.R., 1083, *1104*
Aubry, S. see Villain, J., 1181, 1193, *1195*
Augdahl, E. see Mollendal, H., 1203, *1224*
Ault, B.S., 603, *607*, 665, *681*
Aung, S. see Dunning, Th.H., 35, *157*
Auty, R.P., 967, *1017*, 1402, *1420*
Avgul, N.N., 1338, *1353*
Ažman, A., 573, *607*
Ažman, A. see Koller, J., 601, *609*
Ažman, A. see Zaučer, M., 841, 842, *887*
Azumi, see Ito, M., 15, *21*

Baba, H. see Kitamura, M., 15, *21*
Babb, A.L. see Hardt, A.P., 1048, *1059*
Backéus, M. see Kvick, Å., 398, *454*
Bacon, G.E., 336, *353*, 529, 555, *560*, 584, 597, *607*, 920, 923, 924, *933*, 1118, 1121, 1124, *1161*, 1179, 1180, 1190, *1195*
Bacon, J., 141, *155*
Baddur, A. see Strukov, B.A., 1154, 1157, *1166*
Bade, W.L. see Jehle, H., 254, *292*, 729, *764*
Badger, R.M., 302, 305, *353*, 1201, *1223*
Badger, R.M. see Albert, N., 601, *607*
Badger, R.M. see Newman, R., 583, *609*
Bagster, L.S., 473, 474, *523*
Baier, G. see Schuster, P., 219, *243*
Bailey, A.D. see Narcisi, R.S., 475, *525*
Bajorek, A., 902, 906, 929, 930, 931, 932, *933*
Bajorek, A. see Janik, J.M., 902, 930, 931, *934*
Bak, B., 858, *885*
Baker, A.N., 172, *214*
Balasubramanian, R. see Padmanabhan, V.M., 1151, *1163*
Baldwin, M.C., 1313, 1336, *1353*
Ballester, L. see Schaarschmidt, K., 955, *1023*
Bambagiotti, M. see Marzocchi, M.P., 582, 609
Ban, N.T. see Cox, G.W., 449, *453*
Bando, S. see Ashida, T., 855, *885*
Bandy, A.R. see Bellamy, L.J., 1317, *1353*
Barakat, T.M., 1201, *1223*
Baranov, A.I. see Shirokov, A.M., 1136, 1140, *1164*
Barber, M.N., 384, 385, *387*
Bardet, L., 771, *788*
Barker, A.S., 347, *353*, *354*
Barker, J.A., 368, 379, 380, *387*, 826, 827, *829*, 1228, *1257*, 1430, *1466*
Barker, P.G., 1080, *1104*
Barkla, H.M., 1135, *1161*
Barnaal, D.E., 1430, 1431, *1466*
Barnes, A.J., 214, *214*, 545, 546, 551, *560*, 580, 606, 607, *611*, 1068, 1071, 1073, 1074, 1075, 1076, 1077, 1079, 1080, 1082, 1083, 1087, 1088, 1089, 1092, 1094, 1095, 1096, 1097, 1098, 1101, 1102, *1104*
Baron, A., 1004, *1017*, 1349, *1353*

Baron, D., 797, 826, *827*, *829*
Barrow, G.M., 949, *1017*
Barrow, G.M. see Datta, P., 1253, 1254, *1257*
Barrow, G.M. see Yenger, E.A., 304, *357*
Bartell, L.S. see Janzen, J., 80, 135, *158*
Bartell, L.S. see Kuchitsu, K., 400, 436, 437, *454*
Barto, J., 1338, *1353*
Baryshanskaya, F.S. see Landsberg, G.S., 584, 588, *609*
Basch, H. see Snyder, L.C., 101, *163*
Basila, M.R., 1275, 1276, 1277, 1278, 1280, 1282, 1285, 1301, 1346, *1353*
Basile, L., 663, 664, 665, 666, 667, 670, 675, 676, 679, *681*
Baskin, C.P. see Yarkony, D.R., 61, 72, 129, 132, *163*
Basler, W.D., 1347, *1353*
Bassani, F., 142, *155*
Basset, D.R., 1316, *1353*
Bastiansen, O. see Almenningen, A., 323, 324, 333, *353*, 466, *468*, 771, *788*
Bastick, J. see Bavarez, M., 1318, *1353*
Baston, K.F. see Barto, J., 1338, *1353*
Bateman, L.R. see Bateman, R.J., 484, *523*
Bateman, R.J., 484, *523*
Bates, J.B., 671, 672, 675, *681*
Batuev, M.I., 305, *354*
Baxter, R.J., 363, 385, *387*
Baxter, R.J. see Barber, M.N., 384, 385, *387*
Bauer, E., 1284, *1353*
Bauer, O. see Wolff, H., 1228, 1229, 1230, 1235, 1236, 1238, 1243, *1260*
Bauer, S.H. see Badger, R.M., 302, 305, *353*, 1201, *1223*
Baur, W.H., 399, 402, *453*
Bavarez, M., 1318, *1353*
Bearman, R.J. see Freasier, B., 386, *389*
Bearman, R.J. see Isbister, D., 386, *389*
Beck, W., 601, *607*
Becker, E.D., 540, *560*, 576, *607*, 796, *827*, 1083, *1104*, 1201, *1223*
Becker, E.D. see Farrar, T.C., 1029, 1031, 1034, 1036, *1058*
Becker, E.D. see Gramstad, T., 821, 822, *828*, 1209, *1224*
Becker, E.D. see Liddell, U., 326, *356*, 1233, *1258*
Becker, E.D. see Paolillo, L., 822, 823, *829*
Becker, E.D. see Tucker, E.E., 797, 798, 799, 801, 803, 804, *829*, *830*, 1233, *1259*
Becker, E.D. see Van Thiel, M., 79, *163*, 545, 546, 548, 551, *562*, 580, 595, *610*, 637, *654*, 1070, 1086, 1088, 1090, 1093, *1105*, 1335, *1364*
Becker, F., 826, *829*
Beckert, D., 1042, 1043, *1058*, 1291, *1353*
Beers, Y. see Treacy, E.B., 1051, *1060*
Beersman, J., 1246, *1257*
Beezhold, W., 862, 863, *885*
Behrson, R. see Andrew, E.R., 871, *885*
Beier, G. see Schuster, P., *20*, 126, 149, 150, 151, *162*
Belanger, G. see Asselin, M., 336, *353*, 637, *652*
Bélanger, G. see Bernard-Houplain, M.-C., 625, 648, *652*
Beliakova, L.D., 1318, *1353*
Belin, C., 601, *607*
Bell, G.M., 368, 373, 374, 375, 378, *387*
Bell, R.P., 222, 240, *242*, 475, *523*
Bellamy, L.J., 302, *354*, 541, 552, *560*, 634, *652*, 661, *681*, 1095, *1104*, 1233, *1257*, 1288, 1317, *1353*, 1377, *1420*
Bellemans, A. see Van Loon, R., 541, *562*
Bellocq, A.M., 302, *354*
Bellocq, A.M. see Perchard, Cl., 302, *356*
Belozerskaya, L.P., 635, *652*
Bender, C.F., 37, *155*
Bender, C.F. see Kollman, P.A., 146, *159*, 500, 518, *524*
Bender, C.F. see Yarkony, D.R., 61, 72, 129, 132, *163*
Bender, H.J., 1038, *1058*
Bender, M.L., 761, *763*
Benedetti, E., 462, *468*
Benedict, W.S., 436, 437, 439, *453*, 1254, *1257*
Benesi, H.A., 1205, *1223*, 1267, *1353*
Ben-Naim, A., 377, 378, 381, *387*, 529, *560*, 1373, 1376, *1420*
Benson, G.C. see Dacre, B., 827, *829*
Bentley, F.F. see Carlson, G.L., 780, 781, *789*
Bentrude, W.G. see Arnett, E.M., 1203, *1223*
Benzar, K.K., 1347, *1353*
Beretta, E., 56, *155*
Berezin, G.I., 1306, 1316, 1347, *1353*
Berglund, B., 437, *453*
Bergman, A., 16, 18, *20*
Bergmann, E.D., 124, *155*
Berkebile, C.A. see Sequeira, A., 445, *455*
Bermudez, V.M., 1270, 1271, 1272, 1273, *1353*
Bernal, I. see Jones, D.D., 460, *468*
Bernal, J.D., 368, *387*

Bernal, S.D., 1403, *1420*
Bernard, R. see Rolland, M.T., 1315, *1362*
Bernard-Houplain, M.-C., 621, 625, 627, 628, 629, 630, 631, 632, 641, 646, 648, *652*
Berney, C.L., 1099, *1104*
Bernstein, H.J. see Clague, A.D.H., 116, 117, *156*, 819, *828*
Bernstein, H.J. see Govil, G., 818, *828*
Bernstein, H.J. see Murphy, W.F., 626, *653*
Bernstein, H.J. see Pople, J.A., 1029, 1034, 1037, *1060*
Berry, R.S., 237, *242*
Bertagnolli, H., *155*
Berthier, G., 82, *155*
Berthier, G. see Meunier, A., 63, 64, *160*
Berthod, H., 111, 116, *155*
Berthod, H. see Marius, W., 74, *160*
Berthod, H. see Pullman, A., 117, *161*
Berthomieu, C., 629, *652*
Bertie, J.E., 177, 181, 203, *214*, 224, *242*, 324, 330, *354*, 581, *607*, 635, 637, *652*, 972, *1017*, 1427, 1429, 1433, 1435, 1441, 1455, 1457, 1458, 1459, 1460, 1461, 1462, 1464, *1466*, *1467*
Bertie, J.E. see Whalley, E., 972, *1025*, 1461, 1463, *1470*
Bertin, D.M., 941, *1017*
Bertoluzza, A., 1308, 1351, *1354*
Bertoncini, P.J., 56, *155*
Bertrand, G.L. see Duer, W.C., 1205, 1206, *1223*
Bethell, D.E., 477, *523*, 660, 663, *681*
Bettelheim, F.A. see Lubezky, I., 1016, *1021*
Beveridge, D.L. see Pople, J.A., 37, 40, 125, *161*, 842, *887*
Beyer, A., 154, 155, *163*
Bezrukov, O.F., 1049, *1058*
Bhat, S.N. see Murthy, A.S.N., 125, 127, *160*
Biasotti, J.B. see Andersen, K.K., 711, *763*
Bicca de Alencastro, R., 625, 632, 648, *652*
Biczó, G. see Del Re, G., 142, *157*
Biczó, G. see Ladik, J., 142, *159*
Bielikoff, S. see Fraissard, J., 1344, 1346, *1356*
Biermann, W.J., 696, *763*
Bigeleisen, J., 1244, 1246, *1257*
Bigotto, A., 594, 605, *607*
Bilgram, J.H., 983, *1017*
Binbrek, O.S. see Torrie, B.H., 1145, *1165*
Biran, A. see Montano, P.A., 1147, *1163*
Bird, R.B. see Hirschfelder, J.O., 29, 83, *158*
Birr, M. see Alefeld, B., 896, 933, *933*, 1178, *1195*
Bishop, D.M., 518, *523*
Bishop, P.G., 979, *1017*
Bist, H.D. see Jain, Y.S., 1159, *1166*
Bittrich, H.-J. see Krug, Kl., 1228, 1229, *1258*
Bixon, M., 85, *155*
Bjerrum, N., 341, *354*, 362, 376, 377, 379, *387*, 966, *1017*
Bjorkstam, J.L., 843, 862, 866, *885*, 1115, 1119, 1121, 1127, *1161*
Bjorkstam, J.L. see Adriaenssens, G.J., 1130, *1160*
Bjorkstam, J.L. see Blinc, R., 866, *885*, 1129, 1134, 1154, *1161*
Blake, T.D., 1338, *1354*
Blanckenhagen, P., 898, 899, 900, 901, 912, 913, 914, 929, *933*
Blinc, R., 247, 265, *292*, 340, 347, 348, *354*, 574, 596, 605, *607*, 839, 840, 842, 843, 844, 846, 848, 851, 852, 853, 855, 857, 859, 860, 861, 862, 863, 866, 867, 868, 871, 885, *885*, *886*, 1115, 1121, 1122, 1127, 1128, 1129, 1130, 1131, 1132, 1134, 1138, 1140, 1141, 1144, 1145, 1150, 1154, 1155, 1157, *1161*, *1165*, 1181, 1183, 1185, 1193, *1195*
Blinc, R. see Lavrencie, B., *356*
Blinc, R. see Miller, S.R., 908, *935*, 991, *1022*
Blinc, R. see Stepišnik, J., 1158, *1166*
Blinc, R. see Žekš, B., 1143, *1165*
Bliznakov, G., 1298, *1354*
Bliznakov, G.M. see Kvlividze, V.J., 1298, *1359*
Block, S. see Weir, C., 1450, *1470*
Bloembergen, N., 875, *886*, 1053, *1058*
Bloux, P. see Gaillard, J., 1155, *1165*
Blum, L., 378, *387*
Blumen, A. see Godzik, K.D., 353, *357*, 1157, *1165*, 1193, *1195*
Bluyssen, H. see Verhoeven, J., 851, *887*, 1465, *1469*
Blyholder, G., 1345, *1354*
Bock, E. see Tomchuk, E., 1057, *1061*
Bockris, J.O'M. see Conway, B.E., 220,222, 240, *242*, 289, *292*, 475, *523*, 1314, *1355*
Boddenberg, B., 1294, 1295, *1354*
Boddenberg, B. see Haul, R., 1294, 1295, *1357*
Boehm, H.P., 1266, 1270, 1272, 1318, 1330, *1354*
Boesch, H., 1151, *1161*
Bogachev, Y.S., 797, *827*
Boika, A.A. see Strukov, B.A., 1145, *1164*
Bokii, N.G. see Mikhailov, I.D., 333, *356*, 581, *609*, 622, *653*

Boldrini, P. see Rocaries, C., 1150, *1163*
Bolles, T.F., 1205, *1223*
Bonaccorsi, R., 48, 60, 62, 82, 84, 85, *155*
Bonaccorsi, R. see Berthier, G., 82, *155*
Bonamy, J., 99, *155*
Bonardet, J.L., 1297, 1299, *1354*
Bond, R.A. see Saxton, J.A., 1047, *1060*
Bondybey, V., 602, *607*
Bonera, G., 1153, 1160, *1165*
Bonino, G.B. see Bertoluzza, A., 1308, 1351, *1354*
Bonino, G.B. see Lorenzelli, V., 1308, *1360*
Boos, H. see Lippert, E., 9, *21*
Borah, B., 606, *611*
Bordewijk, P., 1233, *1257*
Borello, E., 1269, 1318, 1319, 1320, 1321, 1322, 1324, 1325, 1327, *1354*
Borello, E. see Zecchina, A., 1346, *1364*
Born, M., 342, *354*
Bornarel, J., 1120, *1161*
Borsa, F. see Bonera, G., 1153, *1165*
Borštnik, B. see Ažman, A., 573, *607*
Bosacek, V., 1347, *1354*
Bosi, P., 1461, 1462, 1463, *1467*
Botschwina, P., 211, *214*
Böttcher, C.J.F., *292*
Botter, R. see Merlivat, L., 1252, *1258*
Boucher, E.A. see Basset, D.R., 1316, *1353*
Bourdéron, C., 580, *607*, 622, 623, 624, 625, 626, 627, 647, 648, *652*, 1324, 1325, *1354*
Bourdéron, C. see Bernard-Houplain, M.-C., 621, *652*
Bourderon, C. see Di Paolo, T., 182, *215*, 648, 650, *653*, 1290, 1326, *1356*, 1461, *1467*
Bourdéron, C. see Péron, J.J., 622, *654*
Bournay, J., 214, *214*, 305, 332, 336, *354*, 353, *357*, 576, 606, *607*, *611*
Bourre-Maladière, P., 478, 482, *523*
Boutin, M., 909, 930, 931, 932, *934*, 1306, 1333, 1344, *1354*
Boutin, H. see Brajović, V., 931, 932, *934*
Boutin, H. see Prask, H.J., 902, 929, 930, *935*
Bowers, M.J.T., 71, *156*, 1071, 1079, 1100, *1104*
Bowman, J.D. see Chipman, D.M., 56, *156*
Boyle, F.W., 1299, *1354*
Brady, G.W. see Mays, J.M., 1343, *1360*
Brajović, V., 931, 932, *934*
Brand, J.C.D., 87, *156*
Brants, R.A. see Kvlividze, V.J., 1298, *1359*
Brasch, J.W. see Jacobsen, R.J., 581, *609*, 622, 633, 638, *653*, 770, *789*
Bratož, S., *20*, 27, *156*, 221, *242*, 307, 317, 325, *354*, 583, 589, 590, *607*, 639, 642, 643, *652*, 1222, *1223*,
Braunholtz, J.T., 584, *607*
Brausse, G., 1013, *1017*
Bray, P.J. see Schempp, E., 837, 857, 858, 859, 860, *887*
Breitschwerdt, 1400, 1401, *1420*
Brekhunets, A.G., 1347, *1354*
Brenman, M. see Miller, S.R., 908, *935*, 991, *1022*
Bresler, S.E., 553, *560*, 1388, *1420*
Brey, W.S., 1345, 1346, *1354*
Brey Jr.,W.S. see Gammage, R.B., 1345, *1357*
Breyer, F. see Suhrmann, R., 687, 696, 700, *765*, 1403, *1423*
Brichard, R. see Fripiat, J.J., 1272, 1304, *1357*
Brickenkamp, C.S. see Negran, T.J., 1159, *1166*
Brickmann, J., 173, 175, *214*, *215*, 220, 221, 222, 223, 225, 226, 228, 229, 230, 231, 232, 236, 237, 239, 240, 241, *242*, 247, 289, *292*
Brickmann, J. see Ady, E., 117, 151, *155*, 219, *242*
Brickwedde, F.G. see Armstrong, G.T., 1246, *1257*
Brickwedde, F.G. see Wooley, H.W., 1246, *1260*
Bridgman, P.W., 1448, 1451, *1467*
Briegleb, G., 1369, *1420*
Brill, R., 1433, 1434, 1464, *1467*
Brill, R. see Zaromb, S., 982, *1025*
Brindley, G.W., 1348, 1349, *1354*
Brittin, W.E. see Zimmerman, J.R., 1042, 1043, *1060*, 1311, *1364*
Broberg, T.W., 1124, *1161*
Broberg, T.W. see She, C.Y., 348, *357*, 1124, 1131, *1164*
Brockamp, B., 1433, 1464, *1467*
Brockhouse, B., 1174, *1195*
Brockhouse, B.N. see Sakamoto, M., 929, *936*
Brockhouse, B.N. see Teh, C., 906, 907, *936*
Brockhouse, B.N. see Woods, A.D.B., 929, 930, *936*
Brockway, L.O. see Karle, J., 533, *561*
Brodskii, I.A., 784, 785, *789*, 1267, 1308, *1354*
Brody, E.M., 1156, *1165*, 1193, *1195*
Brosowski, G., 1160, *1165*
Brot, C., 1012, *1017*
Brown, D.G. see Slejko, F.L., 818, *829*, 1209, *1224*

Brown, G.M., 442, 446, *453*
Brown, G.M. see El Saffar, Z.M., 440, *453*
Brown, G.M. see Williams, J.M., 449, *456*
Brown, H.G., 1202, *1223*
Brown, I.D. see Anderson, M.R., 1152, *1160*
Brown, K.G., 787, *789*
Brown, R.F. see Golomb, D., 1103, *1105*
Brown, R.J.S., 1050, *1058*
Brugger, R.M., 929, *934*
Bruinink, J., 964, *1017*
Brunk, S.D. see Purcell, K.F., 1204, 1207, 1211, 1212, 1214, 1215, 1216, *1224*
Brunton, G.D., 671, 672, *681*
Brunton, G.D. see Johnson, C.K., 483, 492, 500, 502, 505, *524*
Brunton, G.D. see Steinfink, H., 483, *525*
Brzezinski, B., 605, *607*
Buchachenko, A.L. see Kabankin, A.S., 820, *828*
Buchanan, J. see Haggis, G., 1252, *1258*, 1399, 1400, *1421*
Buchheit, W. see Brosowski, G., 1160, *1165*
Buckingham, A.D., 616, 617, 647, *652*, 725, *763*
Buenker, R.J., 838, *886*
Buenker, R.J. see Merlet, P., 121, 150, *160*, 247, *292*
Bugayong, R.R., 459, *468*
Bühl, H. see Knözinger, H., 1337, 1339, *1359*
Buijs, K., 370, *387*
Bulanin, M.O. see Pimentel, G.C., 1090, *1105*
Bullemer, B. see Engelhardt, H., 758, *763*
Bullemer, B. see Riehl, N., 341, 342, *356*, 939, *1023*
Bullemer, B. see Schneider, K.E., 973, *1024*
Bulmer, J.T., 576, *607*, 806, *827*
Bunzl, K., 1011, *1017*
Bunzl, K. see Dickel, G., 1010, 1011, *1018*
Burfoot, J.C., 1115, *1161*
Burgar, M. see Blinc, R., 1130, *1161*
Burgman, J.O., 902, 929, *934*
Burker, J.J. see Arnett, E.M., 1203, *1223*
Burneau, A., 622, 625, 633, *652*
Busch, G.E. see Rentzepis, P.M., 15, *22*
Bushfield, W.K., *242*
Busing, W.R., 463, *468*
Busing, W.R. see Kostansek, E.C., 428, 449, *454*
Busing, W.R. see Worsham, J.E., 449, *456*
Butler, J.D., 1348, *1354*
Butler, J.N., 814, *828*
Butler, R.A. see Miller, P.J., 599, *609*
Buyers, W., 1187, *1195*
Buyers, W.L.J. see Paul, G.L., 1123, 1131, *1163*
Byers Brown, W. see Hirschfelder, J.O., 49, *158*
Bystrov, V.F., 1231, *1257*

Caballol, R., 83, *156*
Cabana, A., 582, 583, *607*, 639, *652*
Cade, P.E., 855, *886*
Caldin, E.F. see Allen, G., 533, *560*
Calleri, M., 602, *607*
Calvert, L.D., 1441, *1467*
Calvert, L.D. see Bertie, J.E., 1427, 1429, 1441, 1457, *1466*
Calvet, R., 1348, *1354*
Camara, B., 1272, 1273, *1354*
Camara, B. see Fink, P., 1273, 1303, *1356*
Camp, P.R., 973, *1017*
Campano, P., see Tapia, O., 132, *163*
Camplin, G.C., 981, 983, *1017*
Cant, N.W., 699, 717, *763*, 1277, 1295, 1296, 1297, 1301, 1302, 1346, *1354*
Carbo, R. see Caballol, R., 83, *156*
Carbonell, R.G., 221, 222, 223, 225, 226, 227, 229, 235, *242*, *354*
Careri, G., 1335, 1343, 1345, 1346, *1354*
Carlson, G.L., 780, 781, *789*
Carpenter, G.B. see Lee, F.S., 478, 482, 495, *524*, 663, *681*, 904, *935*
Carpenter, G.B. see Yoon, Y.K., 478, 482, 495, 519, *526*, 663, *682*
Carpenter, J.M. see Sampson, T.E., 932, *936*
Carr, H.Y. see Simpson, J.H., 912, *936*, 1050, *1060*
Carrizosa, I. see Munuera, G., 1344, *1361*
Carstens, E.L. see Koide, G.T., 987, *1021*
Carter, J.L., 1330, 1331, 1332, *1354*
Casabella, P.A. see Fitzgerald, M.E., 1143, *1162*
Cayias, J.L. see Blake, T.D., 1338, *1354*
Cerutti, L. see Zecchina, A., 1345, *1364*
Cessac, G.L. see Lippincott, E.R., 1317, *1360*
Cevc, P. see Blinc, R., 885, *885*, 1122, *1161*
Chamberlain, J.S., 979, *1017*, 1434, *1467*
Chan, R.K., 969, *1017*
Chan, R.K. see Wilson, G.J., 970, 971, *1025*
Chandler, D., 386, *389*
Chandler, D. see Andersen, H.C., 368, *387*
Chandler, D. see Lowden, L.W., 386, *389*
Chandler, P.C. see Sung, S., 386, *389*
Chandler, W.L., 797, *827*
Chang, C.Y. see Graham, L.L., 806, *828*

Chang, H.C., 976, *1017*
Chang, W.M. see Schug, J.C., 827, *830*
Chapman, I.D., 1297, 1299, *1354*
Chapoton, A., 1002, *1017*, *1018*
Chapoton, A. see Wacrenier, J.M., 1009, *1024*
Charles, S.W. see Pimentel, G.C., 545, *562*, 1090, 1091, *1105*
Chatterjee, N. see Tomchuk, E., 1057, *1061*
Chaussidon, J., 1005, *1018*
Chaussidon, J. see Mamy, J., 1005, *1022*
Chaussidon, J. see Mortland, M.M., 1348, *1361*
Chelkowski, A. see Piekara, A., 960, *1023*
Chemouni, E., 668, 670, *681*, 747, *763*
Chenon, B., 716, *763*
Cheselske, F.J. see Hall, W.K., 1268, 1269, *1357*
Chiba, T., 839, 843, *886*
Chiba, T. see Soda, G., 841, *887*, 1137, 1148, *1164*
Chidambaram, R., 434, 440, *453*, 1435, *1467*
Chidambaram, R. see Brown, G.M., 442, *453*
Chidambaram, R. see Bugayong, R.R., 459, *468*
Chidambaram, R. see Gupta, S.C., 459, *468*
Chidambaram, R. see Ramanadham, M., 459, *469*
Chidambaram, R. see Sequeira, A., 444, 451, *455*, 459, 460, *469*
Chidambaram, R. see Sikka, S.K., 441, *455*
Chihara, H., 856, 857, *886*, 1160, *1165*
Chihara, H. see Yamamoto, T., 1150, *1165*
Chikazawa, M. see Kaiho, M., 1350, *1358*
Chipman, D.M., 56, *156*
Choi, C.S., 447, *453*
Chojnacki, H., 994, 995, *1018*
Chojnacki, H. see Pigon, K., 993, 994, *1023*
Choppin, G.R. see Buijs, K., 370, *387*
Choquet, M. see Chapoton, A., 1002, *1018*
Christian, S.D., 805, 813, 814, 821, *828*, 1203, *1223*
Christian, S.D. see Affsprung, H.E., 809, *827*
Christian, S.D. see Tucker, E.E., 801, 826, *829*, *830*, 1233, *1259*
Christensen, A.N. see Lehmann, M.S., 443, *454*
Christoffersen, R.E., 123, *156*
Christoffersen, R.E. see Shipman, L.L., 123, 124, *162*
Chuang, T.T., 1337, 1339, *1355*
Chuang, T.T. see Deo, A.V., 1337, 1339, *1355*
Chuiko, A.A. see Tertykh, V.A., 1272, *1363*
Chukin, G.D. see Antipina, T.V., 1334, *1353*
Chukin, G.D. see Kirina, O.F., 1334, *1358*
Chulànovskii, V.M., 305, *354*
Chulanovskii, V.M. see Sokolov, N.D., *20*
Chung, M.K. see Cox, G.W., 449, *453*
Churaev, N.V. see Deryagin, B.V., 1317, 1318, *1356*
Cichanowski, S.W., 1160, *1165*
Cimiraglia, R. see Alagano, G., 121, *155*, 190, *214*
Cimiraglia, R. see Bonaccorsi, R., 84, 85, *155*
Cini, R., 1398, *1421*
Clague, A.D.H., 116, 117, *156*, 819, *828*
Clague, A.D.H. see Govil, G., 818, *828*
Clague, D., 771, 778, 779, *789*
Claverie, P., 56, *156*
Claverie, P. see Daudey, J.P., 49, 59, *156*
Claverie, P. see Diner, S., 123, *157*
Claverie, P. see Malrieu, J.P., 123, *159*
Claydon, M.F., 337, *354*, 585, *607*
Clement, G., 1333, 1336, *1355*
Clementi, E., 32, 58, 111, 117, 118, 150, 151, *156*, 214, *215*, 219, *242*
Clementi, E. see Kistenmacher, H., 84, 121, 122, *159*
Clementi, E. see Popkie, H., 59, 60, 62, 77, 80, 84, 111, 129, *161*, 369, 379, 381, *388*
Clementi, E. see Watts, R.O., 387, *389*
Clements, R., 587, 601, *607*, 706, 716, *763*
Clerbaux, T., 955, *1018*
Clerbaux, T. see Duterme, P., 955, *1018*
Clever, H.L., 475, *523*
Cleverdon, D., 942, *1018*
Clifford, J., 1056, *1058*
Cloos, P., 1349, *1355*
Cloos, P. see Fripiat, J.J., 1348, *1357*
Clusius, K., 1246, *1257*
Coats, G.T. see Saxton, J.A., 1047, *1060*
Coburn Jr., W.C. see Grunwald, E., 1048, *1059*
Coburn, W.L., 540, *560*
Cochran, W., 340, 347, 348, *354*, 467, *468*, 1115, 1124, 1128, 1131, 1134, *1161*, 1180, 1184, 1186, *1195*
Cochran, W. see Buyers, W., 1187, *1195*
Cochran, W. see Paul, G.L., 1123, 1131, *1163*
Cocking, S.J. see Egelstaff, P.A., 912, 929, *934*
Cody, I.A. see Morrow, B.A., 1267, *1361*
Coggeshall, N.D., 541, *560*, 801, 805, *828*
Cogley, D.R., 814, *828*
Coignac, J., 1187, *1195*

Cole, R.H., 1054, *1058*
Cole, R.H. see Auty, R.P., 967, *1017*, 1402, *1420*
Cole, R.H. see Davidson, D.W., 1047, *1058*
Cole, R.H. see Wörz, O., 967, 969, *1025*
Cole, R.W. see Cichanowski, S.W., 1160, *1165*
Collins, G.B. see Cleverdon, D., 942, *1018*
Collins, M.F., 336, *354*, 932, *934*
Coluccia, S. see Zecchina, A., 1345, 1346, *1364*
Condon, E.U., 101, *156*
Condon, E.U. see Gurney, R.W., 222, *242*
Connick, R.E., 1052, *1058*
Connor, T.M. see Loewenstein, A., 1050, *1060*
Conrad, J. see Gradsztajn, S., 1344, *1357*
Constant, E. see Fauquembergue, R., 945, *1019*
Conway, A., 56, *156*
Conway, B.E., 220, 222, 240, *242*, 289, *292*, *354*, 475, *523*, 1314, *1355*
Cook, M.A., 1313, *1355*
Cook, W.G., 1346, *1355*
Cooling, G. see Bagster, L.S., 474, *523*
Coombs, G.J. see Cowley, R.A., 1132, 1155, 1156, *1161*, *1165*
Coppens, P., 399, 416, 424, 437, 444, 446, *453*
Corey, R.B. see Albrecht, G., 409, *453*
Corneil, P. see Carter, J.L., 1330, 1331, 1332, *1354*
Cornelius, E.B., 1333, *1355*
Cornut, J.C. see Grenie, Y., 335, *355*, 582, 583, *608*, 1100, *1105*
Cornut, J.C. see Lassegues, J.C., 601, *609*
Cornwell, C.D. see Hindermann, D.K., 819, *828*
Corset, J., 1253, 1254, 1255, *1257*
Corset, J. see Burneau, A., 622, 625, *652*
Costa, G. see Bigotto, A., 594, 605, *607*
Cotrait, M. see Novak, A., 601, *610*
Cotter, M.A. see Stillinger, F.H., 1437, *1469*
Cottington, R.L. see Hardy, R.C., 1052, *1059*
Cotts, R.M. see Murday, J.S., 1050, 1052, *1060*
Coulson, C.A., 71, *156*, 324, 352, *354*, *357*, 368, *387*, 589, *607*, 1436, *1467*
Coupry, C. see Baron, D., 797, *827*
Couzi, M., 570, *607*, 635, *652*
Couzi, M. see Huong, P.V., 595, *609*
Cox, G.W., 449, *453*
Cow, G.W. see Sabine, T.M., 444, *455*
Covington, A.K., *20*
Cowley, E.R., 906, 907, *934*
Cowley, R., 1188, *1195*
Cowley, R. see Buyers, W., 1187, *1195*
Cowley, R.A., 1132, 1155, 1156, *1161*, *1165*
Cowley, R.A. see Dietrich, O.W., 1155, *1165*
Cowley, R.A. see Paul, G.L., 1123, 1131, *1163*
Cracco, F., 819, *828*
Cramer, R.E., 826, *829*
Craven, B.M., 408, 446, *453*
Craven, B.M. see Sabine, T.M., 444, *455*
Creitz, E.C. see Smith, F.A., 540, *562*, 580, *610*
Cremaschi, P., 143, *156*
Crestfield, A.M., 761, *763*
Cribier, D., 912, 929, *934*
Cross, L.E., 1139, 1146, *1161*
Crossley, J., 939, *1018*, 1047, *1058*
Crowe, R.W., 142, *156*
Cruz, M., 740, *763*
Cruz, M.I., 1320, 1322, 1324, 1325, 1326, 1327, 1328, 1329, *1355*
Cruz, M.I. see Seymour, S.J., 1352, *1362*
Csizmadia, I.G. see Hopkinson, A.C., 113, 146, 147, *158*
Cummings, D.L., 324, *354*, 569, *607*
Cummings, H. see Brody, E., 1193, *1195*
Cummings, W.W. see Dugger, D.L., 1269, *1356*
Cummins, H.Z. see Brody, E.M., 1156, *1165*
Cummins, H.Z. see Lagakos, N., 1156, 1157, *1166*
Cummins, H.Z. see Reese, R.L., 1125, *1163*
Cummins, H.Z. see Wilson, G.V.H., 347, *357*
Cummins, S.E. see Cross, L.E., 1139, 1146, *1161*
Cung, M.T., 124, *156*
Cunningham, A.J., 121, 139, 140, *156*, 980, *1018*
Cunningham, A.J. see Payzant, 121, *161*
Curnutte, B., 1383, *1421*
Currie, M., 445, 448, *453*
Currie, M. see McAdam, A., 445, *455*
Curry, N.A., 442, *453*
Curry, N.A. see Currie, M., 448, *453*
Curthoys, G., 1285, 1351, *1355*
Curthoys, G. see Elkington, P.A., 1276, 1277, 1278, 1279, 1280, *1356*
Curtis, N.F., 668, *681*
Curtiss, Ch.F. see Hirschfelder, J.O., 29, 83, *158*
Cusumano, J.A., 1276, 1277, 1278, 1279,

1286, 1287, 1288, 1292, 1294, 1343, *1355*
Cuthbert, J.D. see Vanderkooy, J., 989, 993, *1024*, 1151, *1165*
Czubryt, J.J. see Tomchuk, E., 1057, *1061*

Dacre, B., 827, *829*
Dahlborg, U., 915, *934*
Dahlborg, U. see Larsson, K.E., 912, 929, 932, 933, *935*
Dahlstrom, P.L. see Cramer, R.E., 826, *829*
Dahm, G. see Luck, W.A.P., 1412, 1413, 1414, *1422*
Dailey, B.P. see Townes, C.H., 835, 836, *887*
Dalal, N.S., 882, 883, *886*, 1123, *1161*
Dalbert, R., 1047, *1058*
Dale, A.J., 797, *828*
Dalla Lana, J.G. see Chuang, T.T., 1337, 1339, *1355*
Dalla Lana, J.G. see Deo, A.V., 1337, 1339, *1355*
Dalley, N.K., 441, *453*
Dalmai, G. see Kodratoff, Y., 1346, *1359*
Damen, T.C. see Kaminow, I.P., 347, *355*, 1124, *1162*, 1186, *1195*
Danford, H.D. see Narten, A.H., 368, *388*
Danford, M.D., 1415, *1421*
Danielewicz-Ferchmin, J., 962, *1018*
Daniels, R.O. see Cook, M.A., 1313, *1355*
Danielson, U. see Coulson, C.A., 71, *156*
Danner, H.R. see Boutin, H., 930, *934*
Danner, H.R. see Stiller, H.H., 912, 929, *936*
Dantl, G., 1434, 1464, *1467*
Danyluk, S.S., 1056, *1058*
Darling, B.T., 635, *652*
Dasannacharya, B.A., 932, *934*
Dasannacharya, B.A. see Kim, H.J., 918, 919, 920, 931, *935*
Dasannacharya, B.A. see Thaper, C.L., 902, 903, 930, *936*
Dashevskii, see Sarkisov, G.N., 381, *388*
Datt, I.D. see Ranner, N.B., 902, *935*
Datta, P., 1253, 1254, *1257*
Datta, S.K. see Kamb, B., 1429, 1445, *1468*
Dauchot, J.P. see Van Loon, R., 541, *562*
Daudey, J.P., 48, 49, 50, 59, 63, 64, *156*
David, R. see Bergman, A., 18, *20*
Davidovskii, P.N. see Gamayunov, N.I., 1017, *1019*
Davidson, D.W., 985, 986, *1018*, 1047, *1058*, 1415, 1418, 1419, 1420, *1421*, 1427, *1467*
Davidson, D.W. see Chan, R.K., 969, *1017*
Davidson, D.W. see Gough, S.R., 967, 985, 986, *1020*
Davidson, D.W. see Hawkins, R.E., 985, *1020*
Davidson, D.W. see Majid, Y.A., 985, *1022*
Davidson, D.W. see Morris, B., 985, *1022*
Davidson, D.W. see Venkateswaran, A., 985, *1024*
Davidson, D.W. see Whalley, E., 970, 971, 973, *1025*, 1441, 1444, 1448, 1451, *1470*
Davidson, D.W. see Wilson, G.J., 970, 971, *1025*
Davidson, E.R. see Bender, C.F., 37, *155*
Davies, C.M., 1373, *1421*
Davies, J.B., 1071, 1076, 1078, 1101, *1105*
Davies, J.B. see Barnes, A.J., 1071, 1073, 1075, 1079, 1080, 1102, *1104*
Davies, M., 583, *607*, 805, *828*, 939, 972, *1018*
Davies, M. see Davidson, D.W., 985, *1018*
Davies, M. see Hill, N.E., 939, 995, *1020*
Davies Jr., R.T., 1246, *1257*
Davis, B.H see Gammage, R.B., 1345, *1357*
Davis, B.L. see Kamb, B., 1450, 1451, *1468*
Davis, J.C., 793, 797, 800, *828*
Davis, M.L., 989, 992, *1018*
Davis, R.E., 1317, *1355*
Davydov, A.S. 333, *354*
Davydov, V.Ya., 1267, 1270, 1274, 1276, 1277, 1278, 1279, 1280, 1281, 1284, 1285, 1303, 1320, 1351, *1355*
Davydov, V.Ya. see Curthoys, G., 1285, 1351, *1355*
Davydova, T.S. see Tovbis, A.B., 1139, *1165*
Dawber, J.G., 1343, *1355*
Dawson, P.T., 1343, *1355*
Day, D.H., 929, *934*
Day, R.E., 1343, 1344, *1355*
De Abeledo, M.J. see De Benyacar, M.A.R., 1149, *1161*
De Alti, G. see Bigotto, A., 594, 605, *607*
Dean, R.C., 606, *611*
Dean, R.L. see Clements, R., 587, 601, *607*, 706, 716, *763*
Deb, K.K. see Davis, J.C., 793, 800, *828*
Debecker, G., 948, 957, 958, *1018*
De Benyacar, M.A.R., 1149, *1161*
De Boer, J.H., 1270, 1334, *1355*
De Busetti, S.G., 1348, *1355*
Debye, O., 1408, *1421*
De Dussel, H.L. see De Benyacar, M.A.R., 1149, *1161*
Defay, R. see Prigogine, I., 1227, 1231, 1232, 1233, 1234, *1258*
Dega-Szafran, Z., 602, *607*
Dega-Szafran, Z. see Szafran, M., 703, 741, *765*

De Gennes, P.G., 347, *354*, 1127, 1133, 1157, *1162*, 1180, 1183, *1195*
Deininger, D., 1347, 1348, *1355*
Deininger, D. see Gutsze, A., 1347, *1357*
De Jeu, W.H., *156*
De La Hardrouyère, G., 1319, 1339, *1355*
Delaplane, R.G., 186, *215*, 336, *354*, 422, 424, *453*, 482, 483, 492, 495, 500, *523*, 602, *607*, 663, *681*, 877, *886*
Delaplane, R.G. see Taesler, I., 482, 493, 508, *525*, 676, *682*
De la Vega, J.R. see Fang, Y., 150, *157*
Del Bene, J.E., 61, 62, 72, 92, 93, 96, 104, 111, 131, 134, 135, 137, 138, 144, 145, *156*, *157*, 369, 370, 372, 379, *387*, 1400, *1421*
Delhalle, J. see André, J.M., 142, *155*
Della Gatta, G., 1333, 1334, 1335, *1355*
Del Re, G., 142, *157*
Delsemme, A.H., 1420, *1421*
De Maeyer, L., 995, *1018*
De Maeyer, L. see Eigen, M.L., 342, *354*, 966, 976, *1018*, *1019*
Dementyeva, L.A., 584, *607*
Dempster, A.B., 581, 587, *607*, 778, *789*
Denisenko, G.I. see Brekhunets, A.G., 1347, *1354*
Denison, A.B. see Garret, B.B., 1052, *1059*, 1465, *1467*
Denison, A.B. see Rabideau, S.W., 1430, 1431, 1465, *1469*
Denisov, G.S. see Gusakova, G.V., 737, *763*, 949, *1020*
Dennison, D.M., 222, *242*
Dennison, D.M. see Darling, B.T., 635, *652*
Deo, A.V., 1337, 1339, *1355*
Deo, A.V. see Low, M.J.D., 1278, *1360*
De Paz, M., 121, *156*, 475, *523*, 980, *1018*
Deranleau, D.A., 815, *828*
Derbyshire, W., 879, *886*
Derouane, E.G., 1316, 1339, *1355*
Derry, J.E., 483, *523*
Deryagin, B.V., 1005, *1018*, 1317, 1318, *1356*
De Tar de Los, F., 703, 741, *763*
Detoni, S., 569, 570, 571, 572, 576, 577, 584, 585, 604, *607*, *608*, 941, *1018*
Deutch, J.N. see Adelman, S.A., 378, *387*
De Villepin, J., 599, 600, 601, *610*, *611*
Devonshire, A.F., 1114, *1162*
De Vreese, J.T., *354*
Dexter, D.D., 483, *523*
Dickel, G., 1010, 1011, 1012, *1018*
Dickinson, R.M. see Saxton, J.A., 1047, *1060*
Dienes, A., 16, *20*
Diercksen, G. see Janoschek, R., 183, *215*
Diercksen, G.H.F., 51, 61, 62, 65, 72, 73, 75, 80, 111, 121, 146, 154, 155, *157*, *163*, 203, 207, *215*, 415, 416, *453*, 1056, *1058*
Diercksen, G.H.F. see Kraemer, W.P., 121, 135, 140, *159*, 190, *215*, 519, 520, *524*, 724, 764
Dietrich, O.W., 1155, *1165*
DiGennaro, T.M. see Jonas, J., 1039, *1059*
Dill, L. see Wolff, H., 1230, 1236, 1238, 1243, *1259*
Diner, S., 123, *157*
Diner, S. see Malrieu, J.P., 123, *159*
Dinius, R.H. see Chandler, W.L., 797, *827*
Di Paolo, T., 182, *215*, 648, 650, *653*, 1290, 1326, *1356*, 1461, *1467*
Ditchfield, R., 61, *157*
Ditchfield, R. see Hehre, W.J., *158*
Ditter, W. see Luck, W.A.P., 540, 554, *561*, 621, 624, 625, 648, *653*, 1253, *1258*, 1308, 1326, *1360*, 1371, 1378, 1381, 1382, 1384, 1386, 1387, 1388, 1389, 1390, 1391, 1392, 1419, *1422*
Dixon, H.P., 121, *157*
Dixon, W.B., 797, 821, *828*
Doak, G.O. see Jaffé, H.H., 783, *789*
Dobosh, P.A. see Pople, J.A., 37, *161*, 842, *887*
Dobrotin, R.B. see Meerson, L.A., 1346, *1361*
Dobrovolskii, N.N. see Egorov, M.M., 1345, *1356*
Dogonadze, R.R. see Vorotyntsev, M.A., 266, *292*
Dollimore, D., 668, *681*
Domb, C., 362, 383, *387*
Donckt, E.V. see Nasielski, J., 949, *1022*
Donohue, J., 409, 417, *453*
Donohue, J. see Marsh, R.E., 459, 462, 463, *469*
Dopierala, Z. see Malecki, J., 960, *1022*
Dorda, G., 1008, *1018*
Douglass, D.C. see McCall, D.W., 1055, *1060*
Downes, J.S., 1347, *1356*
Dows, D.A. see Whittle, E., 1067, *1105*
Doyle, W.T. see Hoekstra, P., 1004, *1020*, 1336, 1348, *1358*
Drago, R.S., 1202, 1204, 1206, 1209, 1211, 1212, 1213, 1215, 1216, 1217, 1220, 1222, *1223*
Drago, R.S. see Bolles, T.F., 1205, *1223*

Drago, R.S. see Epley, T.D., 1202, 1209, 1211, *1223*
Drago, R.S. see Eyman, D.P., 1208, 1215, *1223*
Drago, R.S. see Guidry, R.M., 1211, *1224*
Drago, R.S. see Joesten, M.D., 1201, 1209, *1224*
Drago, R.S. see Lim, Y.Y., 819, 820, *828*
Drago, R.S. see Nozari, M.S., 1203, 1206, 1207, 1211, 1212, *1224*
Drago, R.S. see Purcell, K.F., 571, *610*, 1214, *1224*
Drago, R.S. see Sacks, L.J., 1202, *1224*
Drago, R.S. see Slejko, F.L., 818, *829*, 1208, 1209, 1217, 1220, *1224*
Drago, R.S. see Vogel, G.C., 1204, 1215, 1220, *1224*
Dransfeld, K., 1313, 1336, *1356*
Dreyfus, M., 57, 85, 89, 90, 111, 116, 117, *157*, *354*
Dreyfus, R.W. see Srinivasan, R., 18, *22*
Drumheller, J.E. see Boesch, H., 1151, *1161*
Drumheller, J.E. see Schmidt, V.H., 989, 993, *1023*, *1024*, 1152, *1164*
Dubinin, M.M., 1002, *1018*, 1347, 1348, *1356*
Dubinin, M.M. see Glazun, B.A., 1001, *1019*
Dubinin, M.M. see Kopylova, V.M., 1001, *1021*
Dubrovich, N.A. see Morachevskii, V.G., 1350, *1361*
Duecker, H.C. see Haller, W., 964, *1020*
Duer, W.C., 1205, 1206, *1223*
Dugger, D.L., 1269, *1356*
Duggleby, P.McC. see Arnett, E.M., 1203, *1223*
Dunford, H.B. see Robertson, E.B., 475, *525*
Dunken, H., 1331, 1332, 1342, *1356*
Dunken, H. see Camara, B., 1272, 1273, *1354*
Dunken, H. see Wolf, K.L., 533, *562*
Dunning, F.B. see Stokes, E.D., 18, *22*
Dunning, Th.H.Jr., 35, *157*
Dupuis, M. see Onsager, L., 340, 341, *356*, 365, 372, *388*, 966, *1023*
Durden, D.A. see Good, A., 475, *524*
Durham, J.L. see Barto, J., 1338, *1353*
Durocher, G., 302, *354*, 589, *608*, 617, 621, 622, *653*
Dushchenko, V.P., 1016, *1018*
Duterme, P., 955, *1018*
Duterme, P. see Clerbaux, T., 955, *1018*
Dvorák, V., 1115, 1139, *1162*
Dyczmons, V. see Lischka, H., 146, *159*
Dyke, T.R., 61, 72, 79, 80, 115, *157*
Dymanus, A. see Verhoeven, J., 851, *887*, 1465, *1469*
Dzhigit, O.M., 1274, *1356*
Dziembowska, T. see Brzezinski, B., 605, *607*
Dzyaloshinskii, I.Y., 254, *292*

Eames, T.B. see Hoffman, B.M., 820, *828*
Easterfield, J. see Venkateswaran, A., 985, *1024*
Eberhard, J.W., 1154, *1165*
Ebert, G., 1006, *1018*, 1315, 1316, 1335, *1356*
Ebert, G. see Matron, W., 1002, *1022*, 1347, *1360*
Eckener, U., 980, 997, *1018*
Eda, B., 953, *1018*
Edelmann, K. see Murto, J., 1096, *1105*
Eden, R.G. see Aldred, B.K., 927, 932, *933*
Edmonds, D.J., 851, *886*
Edmonds, D.T., 1052, *1058*, 1465, *1467*
Edmonds, W.H. see Andersen, K.K., 711, *763*
Edwards, D.F. see Broberg, T.W., 1124, *1161*
Edwards, D.F. see She, C.Y., 348, *357*, 1124, 1131, *1164*
Edwards, R.T. see West, W., 1284, *1364*
Egelstaff, P.A., 894, 898, 912, 929, *934*
Egelstaff, P.A. see Downes, J.S., 1347, *1356*
Egelstaff, P.A. see Franks, F., 929, *934*
Egorov, M.M., 1268, 1297, 1345, *1356*
Egorova, T.S. see Kvlividze, V.J., 1311, *1359*
Ehrenberg, L. see Blinc, R., 859, 861, *885*, 1144, *1161*
Ehrenson, S. see Newton, M.D., 121, 139, 140, *161*, 190, *215*, 519, 520, 521, *525*, 688, 753, *764*
Ehrenson, S. see Weston, R.E., 711, *766*
Eigen, M., 219, *242*, 342, *354*, 475, *524*, 692, 757, *763*, 966, 976, 1018, *1019*
Eigen, M. see Wicke, E., 475, *525*, 677, *682*, 688, 753, *766*
Einstein, A., 1392, *1421*
Eiriksson, V.R. see Nelmes, R.J., 1118, 1134, 1137, *1163*
Eisenberg, D., 79, *157*, 371, *387*, 939, *1019*, 1427, *1467*
Eisenberg, D. see Coulson, C.A., 1436, *1467*
Eisenschitz, H., 39, *157*
Eisenstein, A., 1371, *1421*
Eke, A. see Kálmán, E., 1386, *1422*
Eley, D.D., 988, 995, *1019*
Elkington, P.A., 1276, 1277, 1278, 1279, 1280, *1356*

Elliott, R.J., 1133, *1162*
Elliott, R.J. see Young, A.P., 1156, 1157, *1167*
Ellison, R.D., 430, 431, 445, *453*, 604, *608*
El Saffar, Z.M., 440, *453*, 1143, *1162*
El Saffar, Z.M. see O'Reilly, D.E., 1150, *1163*
Elston, J. see Kemarec, J., 1346, *1358*
Emerson, K. see Boesch, H., 1151, *1161*
Emerson, M.T. see Howard, B.B., 1208, *1224*
Endo, K. see Morishima, I., 820, *829*
Endom, L., 1055, *1058*
Endtinger, F. see Clusius, K., 1246, *1257*
Engel, G., 1055, 1056, *1058*
Engelbrecht, A., see Rode, B.M., 116, 117, *162*
Engelhardt, H., 341, 342, *354*, 758, *763*, 966, 973, 979, 997, *1019*, 1429, *1467*
Engelhardt, H. see Eckener, U., 980, 997, *1018*
Engelhardt, H. see Nedetzka, T., 978, 979, 996, *1022*
Engelhardt, H. see Pick, M., 979, *1023*
Engelhardt, H. see Riehl, N., 341, 342, *356*, 939, *1023*
Engelhardt, H. see Schneider, K.E., 973, *1024*
Engelsmann, K., 1038, *1058*
Ennis, M.P. see Bushfield, W.K., *242*
Enemoto, Y. see Kobayashi, J., 1145, 1154, *1162*, *1166*
Enriquez, M.A., 1344, *1356*
Epley, T.D., 1202, 1209, 1211, *1223*
Epley, T.D. see Drago, R.S., 1204, 1212, 1220, *1223*
Epstein, S.T. see Hirschfelder, J.O., 49, *158*
Erdey-Gruz, T., 758, *763*
Eremenko, A.M., 1347, *1356*
Erfurth, S.C. see Brown, K.G., 787, *789*
Eric, B., 940, 941, *1019*
Ericksson, V. see Nehnes, R., 1190, *1195*
Eriksson, A., 747, *763*
Erk, S. see Jacob, M., 1434, 1464, *1468*
Ermer, O., 85, 87, *157*
Ermler, W.C., 35, *157*
Ernst, H., 877, *886*
Ershova, I.G. see Deryagin, B.V., 1317, *1356*
Escoubes, M., 1350, *1356*
Eucken, A., 899, *934*, 1253, *1257*, 1371, 1377, 1391, *1421*
Evans, D.F., 601, *608*, 823, *828*
Evans, J.C., 211, *215*, 585, 602, *608*, 637, *653*
Evans, J.C. see Davies, M., 583, *607*
Evans, J.R.N. see Thomas, J.M., 708, 758, 760, *765*, 995, *1024*
Evans, M.W. see Frank, H.S., 1057, *1059*
Evans, W.G. see McDaniel, D.H., 121, *160*
Excoffon, P., 214, *215*, 305, 306, 332, *354*
Eyman, D.P., 1208, 1215, *1223*
Eyman, D.P. see Sacks, L.J., 1202, *1224*
Eyring, H., 371, *387*, 962, *1019*, 1255, *1257*, 1402, *1421*
Eyring, H. see Glasstone, S., 917, *934*
Eyring, H. see Hobbs, M.E., 962, *1020*
Eyring, H. see Jhon, M.S., 368, 372, *388*
Eyring, H. see Marchi, R.P., 1255, *1258*
Ezumi, K. see Mataga, N., 11, *21*

Fabbri, G. see Bertoluzza, A., 1308, 1351, *1354*
Fabrègue, E. see Bardet, L., 771, *788*
Fairall, C.W., 1117, 1120, 1129, 1154, 1157, *1162*, *1165*
Faithful, B.D., 484, *524*
Fajans, K., 474, *524*
Falk, H., 125, 127, *157*
Falk, M., 336, *354*, 433, 434, 438, *453*, 1382, 1404, *1421*
Falk, M. see Cogley, D.R., 814, *828*
Falk, M. see Ford, T.A., 336, *355*, 1435, 1455, 1458, 1459, *1467*
Falk, M.V. see Bertie, J.E., 581, *607*, 635, *652*
Falzone, A.J. see Silvidi, A.A., 1146, *1164*
Fanconi, B., 787, *789*
Fang, Y., 150, *157*
Faraday, M., 963, *1019*, 1415, *1421*
Farkas, R. see Andrews, L., 460, *468*
Farnham, S.B. see Tucker, E.E., 801, *829*, *830*, 1233, *1259*
Farrar, T.C., 1029, 1031, 1034, 1036, *1058*
Farrell, E.F. see Von Hippel, A., 1445, *1468*
Fateley, W.G. see Carlson, G.L., 780, 781, *789*
Fatuzzo, E., 1054, *1058*, 1115, *1162*
Fauquembergue, R., 945, *1019*
Faure, P., *355*
Fayos, J. see Mootz, D., 483, 492, 508, 509, *525*
Federighi, F.D. see Goldman, D.T., 929, *934*
Fedorov, V.M., 1001, *1019*
Fedorov, V.M. see Dubinin, M.M., 1002, *1018*
Fedorov, V.M. see Glazun, B.A., 1001, *1019*
Fedorov, V.M. see Kopylova, V.M., 1001, *1021*
Feeney, J., 1231, *1257*, *1258*

Felden, M., 1299, *1356*
Feldman, U., 1298, 1299, *1356*
Felsche, J. see Arend, H., 1151, *1161*
Fenzke, D. see Ernst, H., 877, *886*
Ferraris, G., 433, 434, 436, 441, 442, 443, *453*
Ferraro, J.R. see Basile, L., 663, 664, 665, 666, 667, 670, 675, 676, 679, *681*
Ferraro, J.R. see Delaplane, R.G., 336, *354*, 602, *607*
Ferriso, C.C., 477, *524*, 660, 661, 663, *681*
Ferro, D.R. see Pelligrini, A., 778, *789*
Few, A.V., 940, *1019*
Fiat, D., 1298, 1327, 1328, *1356*
Fiat, D. see Reuben, J., 1298, *1362*
Field, F.H. see Lampe, F.W., 475, *524*
Figgins, R., 1039, *1058*
Fild, M., 819, *828*
Filimonov, V.N., 1290, 1322, *1356*
Filimonov, V.N. see Roev, L.M., 1278, 1301, *1362*
Filimonov, V.N. see Tretyakov, N.E., 1344, *1363*
Finch, E.D. see Arnold, G.P., 1433, *1466*
Finch, E.D. see Rabideau, S.W., 1430, 1431, 1465, *1469*
Finch, J.N. see Lippincott, E.R., *356*, 574, *609*
Finch, N.D. see Andrew, E.R., 477, *523*
Finger, G. see Gutsze, A., 1347, *1357*
Finholt, J.E., 663, 664, *681*
Fink, D.W., 17, *21*
Fink, P., 1273, 1276, 1277, 1303, 1337, *1356*
Fink, P. see Camara, B., 1272, 1273, *1354*
Fink, P. see Dunken, H., 1331, 1332, 1342, *1356*
Finlayson, D.M. see Barkla, H.M., 1135, *1161*
Fischer, A. see Clusius, K., 1246, *1257*
Fischer, C.O., 926, 927, 932, 933, *934*
Fischer, S.F., 219, 220, *242*, 247, *292*, 297, 301, 305, 309, 311, 313, 318, 320, 324, 331, 337, 342, 345, 346, 348, *354*, *355*, 590, *608*, 644, *653*, 981, *1019*, 1194, *1195*
Fisher, I.Z., 1386, *1421*
Fister, F., 1052, 1055, 1056, *1058*, 1399, *1421*
Fitzgerald, M.E., 1143, *1162*
Fletcher, A.N., 626, *653*, 821, *828*, 1233, *1258*, 1320, *1356*
Fletcher, N.H., 374, *387*, 939, 966, 974, *1019*, 1427, *1467*
Fletcher, N.H. see Chamberlain, J.S., 979, *1017*, 1434, *1467*
Flick, C. see El Saffar, Z.M., 1143, *1162*
Florin, A.E. see Alei, M., 814, *827*
Florin, A.E. see Litchman, W.M., 822, *829*
Flygare, W.H. see Bowers, M.T., 1071, 1079, 1100, *1104*
Foglizzo, R., 582, 583, 606, *608*, *611*, 633, 638, *653*
Foldes, A., 336, *355*, 570, *608*, 617, 619, 637, *653*
Follner, H., 484, *524*
Folman, M., 1277, 1279, 1280, 1296, 1297, 1304, *1356*
Folman, M. see Feldman, U., 1298, 1299, *1356*
Folman, M. see Fiat, D., 1298, 1327, 1328, *1356*
Folman, M. see Kozyrovski, Y., 1278, *1359*
Folman, M. see Levy, S., 1328, *1360*
Folman, M. see Lubezky, I., 1016, *1021*, 1298, 1299, *1360*
Folman, M. see Reuben, J., 1298, *1362*
Folman, M. see Ron, A., 1292, 1293, 1294, *1362*
Folman, M. see Soffer, A., 1313, *1363*
Fong, Dodd-Wing., 749, *763*
Fontaine, J., 1010, *1019*
Fontaine, J. see Wacrenier, J.M., 1009, *1024*
Ford, T.A., 336, *355*, 1435, 1455, 1458, 1459, *1467*
Ford, T.A., see Falk, M., 336, *354*, 1382, *1421*
Forel, M.-T. see Rey-Lafon, M., 771, *789*
Forrester, J.D., 867, *886*
Forslind, E., 902, *934*, 1350, *1356*
Förster, Th., 13, *21*
Fortuin, J.M.H. see De Boer, J.H., 1334, *1355*
Fournier, M., 660, 661, 663, *681*
Fournier, M. see Chemouni, E., 668, 670, *681*, 747, *763*
Fousková, A. see Cross, L.E., 1139, 1146, *1161*
Fousková, A. see Dvorák, V., 1115, *1162*
Fowler, R.A. see Bernal, J.D., 368, *387*
Fowler, R.H. see Bernal, S.D., 1403, *1420*
Fraissard, J., 1344, 1346, *1356*
Fraissard, J. see Bonardet, J.L., 1297, 1299, *1354*
Fraissard, J. see Enriquez, M.A., 1344, *1356*
Fraissard, J. see Kemarec, J., 1346, *1358*
Franchini-Angela, M. see Ferraris, G., 433, 434, 436, *453*
Franck, E.U., 61, 80, 135, *157*, 1382, *1421*
Frank, H.S., 362, 368, 370, 372, *387*, 553, *560*, 1057, *1059*

Franklin, J.L. see Lampe, F.W., 475, *524*
Franks, F., *20*, 929, *934*, 939, *1019*, 1057, *1059*, 1408, 1415, *1421*
Frazer, B.C., 843, *886*, 1179, *1195*
Frazer, B.C. see Skalyo Jr. J., 1124, *1164*, 1187, 1189, 1190, 1192, *1195*
Freasier, B., 386, *389*
Fredyakin, N.N. see Deryagin, B.V., 1317, *1356*
Freeman, H.C., 451, *453*
French, T.M., 1332, *1357*
Frenkel, J., 553, *560*
Freude, D., 1291, 1294, 1295, *1357*
Freund, F. see Gieseke, W., 758, *763*, 995, *1019*
Freund, F. see Martens, R., 1343, *1360*
Frey, M.N., 451, 452, *453*, *454*, 460, 461, *468*
Frey, M.N. see Andrews, L., 460, *468*
Frey, M.N. see Hamilton, W.C., 459, *468*
Frey, M.N. see Jones, D.D., 460, *468*
Frey, M.N. see Koetzle, T.F., 460, *469*
Freyman, M., 1313, 1315, *1357*
Freyman, R. see Freyman, M., 1313, 1315, *1357*
Friedman, L. see De Paz, M., 121, *156*, 475, *523*, 980, *1018*
Friedrich, H.B., 945, *1019*
Fripiat, J.J., 758, *763*, 1272, 1274, 1295, 1299, 1300, 1301, 1302, 1303, 1304, 1305, 1313, 1314, 1315, 1317, 1328, 1348, 1350, *1357*
Fripiat, J.J. see Cruz, M., 740, *763*
Fripiat, J.J. see Cruz, M.I., 1320, 1322, 1324, 1325, 1326, 1327, 1328, 1329, *1355*
Fripiat, J.J. see Leonard, A.J., 1330, *1360*
Fripiat, J.J. see Mertens, G., 1318, *1361*
Fripiat, J.J. see Mortland, M.M., 1348, *1361*
Fripiat, J.J. see Prigogine, M., 1317, *1362*
Fripiat, J.J. see Seymour, S.J., 1352, *1362*
Fripiat, J.J. see Toullaux, R., 1348, *1363*
Fripiat, J.J. see White, J.L., 1285, *1364*
Frish, H.L. see Dransfeld, K., 1313, 1336, *1356*
Fritsche, H., 760, *763*
Fritz, I.J., 1125, *1162*
Fritz, I.J. see Reese, R.L., 1125, *1163*
Fritzsche, F., 580, *608*
Fröhlich, H., 343, *355*
Frohnsdorff, G.J.C., 1281, *1357*
Frolov, V.I., 1012, *1019*
Fromm, J. see Watts, R.O., 387, *389*
Fruwert, S. see Geiseler, G., 533, 534, *560*
Fryer, C.W. see Marzocchi, M.P., 582, *609*
Fryer, F.A. see Anderson, J.E., 1042, 1043, *1058*
Fubini, B. see Della Gatta, G., 1333, 1334, 1335, *1355*
Fujimoto, H. see Yamabe, S., 150, *163*
Fujishiro, R. see Kimura, K., 1203, *1224*
Fujiwara, F.Y., 820, 821, 826, *828*, *829*
Fujiwara, F.Y. see Martin, J.S., 821, 826, *829*, *830*
Fujiwara, H. see Kondo, S., 1336, *1359*
Fukuda, A., 968, *1019*
Fukui, K. see Yamabe, S., 150, *163*
Fukushima, K., 325, *355*, 568, *608*
Fukuta, N., 1350, *1357*
Fuliński, A. see Szkatula, A., 900, 901, 929, *936*
Fuller Jr., E.L., 1345, *1357*
Fuller Jr., E.L. see Holmes, H.F., 1345, *1358*
Fuller, R.G., 989, 992, *1019*
Funck, T. see Schuster, P., 116, 117, *162*
Fung, B.M. see Olympia, P.L., 840, 841, *887*
Furman, G. see Egorov, M.M., 1345, *1356*
Fuson, N., 583, *608*
Fuson, N. see Josien, M.-L., 604, *609*

Gailar, N. see Benedict, W.S., 436, 437, 439, *453*
Gaillard, J., 1155, *1165*
Galasso, V. see Bigotto, A., 594, 605, *607*
Galkin, G.A., 1274, 1275, 1276, 1277, 1282, 1283, 1284, 1285, 1290, 1292, 1293, *1357*
Gallagher, K.J., 600, *608*
Gallagher, K.J. see Ubbelohde, A.R., 323, 324, *357*
Gallifa, R. see Caballol, R., 83, *156*
Gamayunov, N.I., 1017, *1019*
Gamba, A. see Cremaschi, P., 143, *156*
Gamer, G. see Wolff, H., 630, *654*, 1237, 1251, *1259*
Gammage, R.B., 1345, *1357*
Gammage, R.B. see Fuller Jr., E.L., 1345, *1357*
Gammage, R.B. see Holmes, H.F., 1345, *1358*
Gamow, G., 222, 228, *242*
Garbatski, U. see Fiat, D., 1298, *1356*
Garcia, A. see Chidambaram, R., 440, *453*
Garg, S.K. see Majid, Y.A., 985, *1022*
Garing, J.S., 1254, *1258*
Garland, C.W., 1120, *1162*
Garland, C.W. see Litov, E., 1130, *1163*
Garnovskij, A.D. see Rustanova, S.N., 953, *1023*

Garret, B.B., 1052, *1059*, 1465, *1467*
Garrett, B.B. see Spiess, H.W., 1465, *1469*
Gärtner, K., 1011, *1019*
Gash, B.W. see Christian, S.D., 813, 814, *828*
Gastuche, N.C. see Fripiat, J.J., 1272, 1304, *1357*
Gault, J.D. see Prask, H.J., 929, *935*, 1461, 1462, 1463, *1469*
Gavrielides, A.T. see Yi, P.-N., 1148, *1165*
Gavrilova-Podol'skaya, G.V., 1136, 1139, *1162*
Gaw, W.J. see Boyle, F.W., 1299, *1354*
Gayles, J.N. see Clementi, E., 111, *156*
Gebbie, H. see Anderson, A., 909, *933*
Geiseler, G., 533, 534, *560*, *561*
Geissler, E. see Hecht, A.M., 1348, *1358*
Genin, D.J., 1122, 1146, *1162*
Gerdil, E. see Heilbronner, E., 173, *215*, 221, *242*
Germain, G., 461, *468*
German, W.L., 1348, *1357*
Gerritz, w.H. see Anderson, J.E., 1049, *1058*
Gerson, F., 173, 178, *215*
Gerson, F. see Heilbronner, E., 173, *215*, 221, *242*
Gerstein, M. see Brown, H.C., 1202, *1223*
Geschke, D., 1319, 1339, *1357*
Gesemann, H.J. see Geiseler, G., 533, *561*
Gesemann, R. see Geiseler, G., 533, *561*
Gesi, K., 1151, 1159, 1160, *1162*, *1165*
Ghersetti, S., 782, *789*, 1201, *1223*
Ghiotti, G. see Borello, E., 1319, 1320, 1321, 1324, 1325, *1354*
Ghosh, R.E., 909, 932, *934*
Giaque, W.F., 363, *387*
Gibby, M. see Pines, A., 882, *887*
Gibby, M.G. see Waugh, J.S., 880, 881, *887*
Gielly, J. see Escoubes, M., 1350, *1356*
Gieseke, W., 758, *763*, 995, *1019*
Giessner-Prettre, C., 83, *157*
Giguère, P.A., 477, *524*, 661, *681*
Giguère, P.A. see Pavia, A.C., 203, *215*, 668, 669, 670, 671, *682*, 747, *765*
Giguère, P.A. see Savoie, R., 663, *682*
Gilbert, A.S., 601, *608*, 669, 670, 673, 676, 677, 679, *681*, 747, *763*
Gilbert, M. see Diner, S., 123, *157*
Giles, D. see Tench, A.J., 1342, 1343, *1363*
Gillard, R.D., 665, 668, 670, 676, 679, *681*, 747, *763*
Gillard, R.D. see Dollimore, D., 668, *681*
Gillen, K.T., 1038, 1039, *1059*
Gillespie, R.J., 7, *21*
Gilmour, J.B. see Biermann, W.J., 696, *763*
Gingrich, N.S. see Eisenstein, A., 1371, *1421*
Ginn, S.G.W., 305, *355*, 568, *608*
Ginsberg, H., 1330, *1357*
Giorgianni, S. see Ghersetti, S., 782, *789*
Girardet, C., 1080, 1082, 1083, *1105*
Gitlin, S.N. see Plooster, M.N., 1311, *1362*
Gladkii, V.V. see Sidnenko, E.V., 1154, 1157, *1166*
Gladkii, V.V. see Zheludev, I.S., 1130, *1165*
Glasel, J.A., 1052, *1059*, 1398, 1399, 1403, *1421*
Glass, A.M. see Negran, T.J., 1159, *1166*
Glasser, L., 964, *1019*
Glasstone, S., 917, *934*
Glazun, B.A., 1001, 1002, *1019*
Glazun, B.A. see Fedorov, V.M., 1001, *1019*
Glemser, O., 1335, *1357*
Glen, J.W. see Bishop, P.G., 979, *1017*
Glen, J.W. see Camplin, G.C., 981, 983, *1017*
Glen, J.W. see Jones, S.J., 983, *1021*
Glockner, G. see Wall, F.T., 172, *216*, 222, *244*
Glogar, P. see Dvorák, V., 1115, *1162*
Gobush, W., 969, *1019*
Godzik, K.D., 341, 345, 353, *355*, *357*, 1157, *1165*, 1193, *1195*
Godycki, L.E., 605, *608*
Gold, L.W. see Whalley, E., 939, *1025*
Goldblith, S.A. see Roebuck, B.D., 1017, *1023*
Golden, S., 228, *242*
Goldman, D.T., 929, *934*
Goldring, H. see Verstegen, J.M.P.J., 1071, 1083, *1105*
Goldschmidt, H., 473, *524*
Goldstein, I.P., 948, 955, *1019*
Goldstein, I.P. see Guryanova, E.N., 955, 956, *1020*
Golebiewski, A. see Chojnacki, H., 994, *1018*
Golič, L., 599, 602, *608*
Golič, L. see Hamilton, W.C., 459, *468*
Golič, L. see Koetzle, T.F., 460, *469*
Golikov, V.V., 912, 929, *934*
Göller, R., 1047, *1059*
Gölles, F., 1228, *1258*
Golomb, D., 1103, *1105*
Gonzalez, F. see Munuera, G., 1344, *1361*
Gonzalo, J.A. see Mercado, A., 1144, *1163*
Good, R.E. see Golomb, D., 1103, *1105*
Goode, E.V., 940, *1019*
Goode, E.V. see Eric, B., 940, 941, *1019*
Goodenow, J.M., 1202, *1223*

Goodsel, A.J. see Low, M.J.D., 1343, *1360*
Gorbatyi, L.V., 1138, *1162*
Gordeyeva, N.V. see Shuvalov, L.A., 1135, 1136, 1139, *1164*
Gordon, M.S., 125, 127, *157*
Gordon, R.G., 317, *355*, 773, 783, *789*, 1401, *1421*
Gordon, R.G. see Agrawal, A.K., 932, *933*
Gordy, W. see Jones, G., 854, 855, *886*
Gore, E.S. see Danyluk, S.S., 1056, *1058*
Gornostansky, S.D., 851, *886*, 1466, *1467*
Gorrie, T.M. see Arnett, E.M., 1201, 1205, *1223*
Gorvin, T.C. see Derbyshire, W., 879, *886*
Gosar, P., 324, 329, 341, 345, 348, *355*, 981, *1020*
Gosh, R. see Luck, W.A.P., 1408, 1410, 1411, *1422*
Götz, R. see Wolff, H., 1228, 1229, 1230, 1235, 1236, 1238, 1243, 1246, *1259*, *1260*
Gough, S.R., 955, 967, 985, 986, *1020*
Govil, G., 818, *828*
Govil, G. see Clague, A.D.H., 819, *828*
Goyal, P.S. see Basannacharya, B.A., 932, *934*
Goyal, P.S. see Kim, R.J., 918, 919, 920, 931, *935*
Gradsztajn, S., 1344, *1357*
Graham, L.L., 806, *828*
Grahn, R., 517, *524*
Gramstad, T., 818, 821, 822, *828*, 1201, 1209, *1224*
Gramstad, T. see Aksnes, G., 1201, *1223*
Gramstad, T. see Dale, A.J., 797, *828*
Granger, R. see Bardet, L., 771, *788*
Gränicher, H., 341, 342, *355*, 966, *1020*
Gränicher, H. see Kind, R., 862, *886*
Gränicher, H. see Taubenberger, R., 969, *1024*
Grannel, P.K., 882, *886*
Grant, D.M. see Alger, T.D., 1045, *1058*
Grant, D.M. see Lyerla Jr., J.R., 1044, 1045, *1060*
Grant, W.H. see Lippincott, E.R., 1317, *1360*
Gräslund, C. see Dahlborg, U., 915, *934*
Gray, C.G. see Gubbins, K.E., 378, *387*
Grech, E. see Dega-Szafran, Z., 602, *607*
Grech, E. see Szafran, M., 741, *765*
Greenler, R.G., 1337, 1339, *1357*
Greenler, R.G. see Kagel, R.O., 1342, 1343, *1358*
Gregory, T.M. see Josien, M.-L., 604, *609*
Greinacher, E.G., 326, 336, *355*
Grenie, Y., 335, *355*, 582, 583, *608*, 1100, *1105*
Grenie, Y. see Lassegues, J.C., 601, *609*
Grenthe, I. see Albertsson, J., 445, *453*
Gribina, I.A., 1349, *1357*
Griengl, H., 131, *157*
Griffiths, D.M., 1351, *1357*
Grigas, J. see Petzelt, J., 1159, *1166*
Grimaldi, F., 37, *157*
Grimm, H., 224, *242*, 921, 932, *934*, 1179, 1181, 1190, *1195*
Grimm, H. see Arsić-Eskinja, M., 921, 922, 924, 925, 928, 932, *933*, 1187, 1189, 1191, *1195*
Grimsrud, E.P., 140, *157*
Grindlay, J., 348, *355*, 1115, *1162*
Gronau, B., 15, *21*
Grosescu, R., 1057, *1061*
Grosescu, R. see Trepadus, V., 1057, *1061*
Grosh, J. see Jhon, M.S., 368, 372, *388*
Gross, G.W., 983, *1020*
Gross, P.M. see Weith, A.J., 941, *1025*
Grossmann, A. see Lohse, U., 1347, *1360*
Groth, W., 1246, 1254, *1258*
Groves, M.R., 17, *21*
Gruen, D.M. see Strommen, D.P., 1086, 1088, *1105*
Grundnes, J. see Christian, S.D., 1203, *1223*
Grundnes, J. see Mollendal, H., 1203, *1224*
Grüner, M., 927, *934*, 1041, 1044, 1048, 1049, *1059*
Grunwald, E., 758, *763*, 1048, *1059*
Grunwald, E. see Coburn, W.L., 540, *560*
Grunwald, E. see Cogley, D.R., 814, *828*
Grunwald, E. see Fong, Dodd-Wing., 749, *763*
Gubbins, K.E., 378, *387*
Güdel, H.U., 123, *157*, 569, 601, *608*
Guest, L.B. see Dawber, J.G., 1343, *1355*
Guggenheim, E.A., 826, *829*
Guglielminotti, E. see Zecchina, A., 1345, 1346, *1364*
Guha, S., 1160, *1165*
Guidoni-Giardini, A. see De Paz, M., 121, *156*
Guidotti, C., 137, 138, 139, *157*
Guidotti, C. see Arrighini, G.P., 35, *155*
Guidry, R.M., 1211, *1224*
Guissani, Y. see Bratoz, S., 317, *354*
Günthard, H.H., 18, *22*
Günthard, H.H. see Heilbronner, E., 173, *215*, 221, *242*

Gupta, S.C., 459, *468*
Gurikov, Yu.V., 368, 374, *387*
Gurka, D., 816, 817, 822, *828*
Gurka, D. see Taft, R.W., 818, *829*
Gurney, R.W., 222, *242*
Gurskaya, G.V., 459, *468*
Guryanova, E.N., 956, *1020*
Guryanova, E.N. see Goldstein, I.P., 948, 955, *1019*
Gusakova, G.V., 737, *763*, 949, *1020*
Gusev, S.S. see Aleksandrovich, Kh.M., 1350, *1353*
Gutbrod, M. see Karagounis, G., 1319, *1358*
Gutowsky, H.S., 794, *828*
Gutsze, A., 1347, *1357*

Ha, T. see O'Konsky, C.T., 859, *886*
Haas, C., 173, *215*, 221, *242*, 1458, 1460, 1461, *1467*
Haberland, D. see Krug, Kl., 1228, 1229, *1258*
Habuda, S.P., 1149, *1162*
Hackerman, N. see Venable, R.L., 1333, *1364*
Hadži, D., *20*, 119, *157*, 298, 304, 337, 338, *355*, 568, 569, 571, 574, 584, 587, 594, 595, 596, 599, 600, 601, 602, 603, 604, 605, *608*, 741, *763*, 957, *1020*, 1201, *1224*
Hadži, D. see Ažman, A., 573, *607*
Hadži, D. see Blinc, R., *354*, 574, 605, *607*, 839, 840, 842, *885*
Hadži, D. see Bratož, S., 221, *242*, 302, 325, *354*, 583, 589, *607*, 639, 642, 643, *652*
Hadži, D. see Detoni, S., 569, 570, 571, 572, 576, 577, 584, 585, 604, *607*, *608*, 941, *1018*
Hadži, D. see Golič, L., 602, *608*
Hadži, D. see Koller, J., 601, *609*
Hadži, D. see Lavrencie, B., *356*
Hadži, D. see Macdonald, A.L., 422, 445, *455*, 600, *609*
Hadži, D. see Obradović, M., 569, *610*
Hadži, D. see Orel, B., 584, 586, 599, *610*
Hadži, D. see Stepišnik, J., 600, *610*
Hadži, D. see Zaučer, M., 841, 842, *887*
Haeberlen, U., 873, 877, 878, 879, 880, *886*
Haeberlen, U. see Waugh, J.S., 873, 874, *887*
Hafner, K., 604, *608*
Hagemark, K.I., 826, *829*
Haggis, G., 1252, *1258*, 1399, 1400, *1421*
Hagler, A.T., 86, 87, 88, 89, 90, *157*, *158*
Hahn, D. see Clementi, E., 116, *156*
Hahn, E.L. see Hartmann, S.R., 849, *886*
Hair, M.L., 1266, 1269, 1272, 1273, 1288, *1357*
Hair, M.L. see Chapman, I.D., 1297, 1299, *1354*
Hair, M.L. see Hertl, W., 1272, 1276, 1277, 1278, 1279, 1280, 1286, 1351, *1358*
Haklik, L. see Kálmán, E., 1386, *1422*
Hall, A., 302, 303, *355*, 583, *608*, 782, 783, *789*
Hall, G.E. see Braunholtz, J.T., 584, *607*
Hall, G.G., 84, *158*
Hall, G.G. see Tait, A.D., 84, *163*
Hall, L., 914, *934*
Hall, P.G., 1007, *1020*
Hall, W.K., 1268, 1269, *1357*
Hall, W.K. see O'Reilly, D.E., 1268, 1327, *1361*
Hallam, H.E., 545, *561*, 569, *608*, 1068, 1069, 1086, 1090, 1104, *1105*, 1201, *1224*, 1331, *1357*
Hallam, H.E. see Barnes, A.J., 214, *214*, 545, 546, 551, *560*, 580, 606, *607*, *611*, 1068, 1071, 1073, 1074, 1075, 1076, 1077, 1079, 1080, 1082, 1083, 1092, 1094, 1095, 1096, 1097, 1098, 1101, 1102, *1104*
Hallam, H.E. see Bellamy, L.J., 1288, *1353*, 1377, *1420*
Hallam, H.E. see Davies, J.B., 1071, 1076, 1078, 1101, *1105*
Haller, W., 964, *1020*
Haly, A.R., 1346, *1357*
Hambleton, F.A., 1272, *1357*
Hambleton, F.H. see Armistead, C.G., 1267, 1272, *1353*
Hambleton, F.H. see Peglar, R.J., 1272, *1361*
Hamilton, J.H. see Cook, M.A., 1313, *1355*
Hamilton, W.C., *20*, 323, 336, 340, *355*, 397, 402, 410, 421, 425, 431, 433, *454*, 459, 461, *468*, 658, 659, *681*, 833, *886*, 894, *934*, 970, 991, *1020*, 1145, *1162*, 1435, 1437, 1438, 1439, 1440, 1445, 1448, 1458, 1459, *1468*
Hamilton, W.C. see Frey, M.N., 451, 452, *453*, *454*, 460, 461, *468*
Hamilton, W.C. see Golič, L., 599, *608*
Hamilton, W.C. see Jönsson, P.-G., 445, 448, *454*
Hamilton, W.C. see Koetzle, T.F., 451, 452, *454*, 460, *469*
Hamilton, W.C. see La Placa, S.J., 437, 440, *454*, 1435, 1441, 1442, 1443, 1444, 1445, 1460, *1468*
Hamilton, W.C. see Lehmann, M.S., 450, 452, *454*, 460, *469*

Hamilton, W.C. see Park, Y.J., 446, *455*
Hamilton, W.C. see Schlemper, E.O., 336, *357*, 447, 449, *455*, 465, *469*, 931, *936*
Hamilton, W.C. see Sequeira, A., 445, *455*
Hamilton, W.C. see Verbist, J.J., 450, *456*, 460, 461, *469*
Hammaker, R.M. see Patterson, L.K., 804, *829*
Hammerschmidt, E.G., 1419, *1421*
Hamor, T.A. see Derry, J.E., 483, *523*
Hampton, M., 882, 884, *886*
Hankins, D., 62, 73, 75, 133, 134, *158*, 207, *215*, 369, 379, *388*, 1320, *1357*
Hanna, M.W., 815, *828*, 946, *1020*
Hanna, M.W. see Lippert, J.L., 946, *1021*
Hanna, R.B. see Peri, J.B., 1330, 1332, *1362*
Hantzsch, A., 5, *21*, 473, *524*
Harada, I., 786, *789*
Harano, Y. see Low, M.J.D., 1318, 1320, *1360*
Hardin, A.H., 632, *653*, 1458, *1468*
Harding, D.A. see German, W.L., 1348, *1357*
Hardt, A.P., 1048, *1059*
Hardy, R.C., 1052, *1059*
Hardy, W.B., 1404, *1421*
Harget, A.J. see Murrell, J.N., 37, 40, *160*
Harling, O.K., 929, *934*
Harrington, D., 18, *21*
Harris, F.E. see Rein, R., 220, *243*
Harrison, J.F., 837, 838, *886*
Harrison, M.C. see Moskowitz, J.W., 517, *525*
Hartert, E. see Glemser, O., 1335, *1357*
Hartmann, H., 182, 183, 186, *215*
Hartmann, J.W. see Dickel, G., 1012, *1018*
Hartmann, K.A., 1382, *1421*
Hartmann, S.R., 849, *886*
Harvey, K.B., 1418, *1421*
Harvey, K.B. see Hardin, A.H., 1458, *1468*
Harvey, S.C., 1014, *1020*
Hasegawa, M. see Low, M.J.D., 1278, *1360*
Haser, R., 484, *524*
Haskell, R.W., 1233, 1235, *1258*
Haslett, J.C. see Schmidt, V.H., 989, *1023*, *1024*
Hassinen, E. see Murto, J., 1096, *1105*
Hasted, J.B., 972, *1020*, 1400, 1402, 1403, *1421*
Hasted, J.B. see Haggis, G., 1252, *1258*, 1399, 1400, *1421*
Hattori, H. see Iizuka, T., 1343, *1358*
Haubenreisser, U., 877, *886*
Haul, R., 1294, 1295, *1357*
Haul, R. see Boddenberg, B., 1294, 1295, *1354*
Haurie, M., 302, 303, 334, *355*, 583, *608*, 639, *653*, 1100, *1105*
Hauser, M. see Klein, U., 15, *21*
Hausser, R., 1049, 1053, *1059*
Hautecler, S. see Brockhouse, B., 1174, *1195*
Havens, W.W. see Rush, J.J., 925, 926, 931, *935*
Havlin, S., 1157, *1165*
Hawkins, R.E., 985, *1020*
Hawranek, J.P., 724, *763*, 942, 948, 957, 1012, *1020*
Hawranek, J.P. see Detoni, S., 569, 570, 572, 576, 584, 585, *608*, 941, *1018*
Hayd, A., 257, *292*, 733, *763*
Haydon, S.C. see Groves, M.R., 17, *21*
Haywood, B.C. see Collins, M.F., 336, *354*, 932, *934*
Haywood, B.C. see Egelstaff, P.A., 929, *934*
Heastie, R. see Arnold, G.M., 909, *933*
Heath, J.B.R. see Whalley, E., 970, 971, 973, *1025*, 1441, 1444, 1448, 1451, *1470*
Hecht, A.M., 1348, *1358*
Hechter, O., 558, *561*
Hedestrand, G., 943, *1020*
Hehre, W.J., 61, *158*
Hehre, W.J. see Ditchfield, R., 61, *157*
Heidemann, H. see Alefeld, B., 896, 933, *933*, 1178, *1195*
Heilbronner, E., 173, *215*, 221, *242*
Heink, W. see Ernst, H., 877, *886*
Heinzinger, K. see Weston, R.E., 711, *766*
Heller, C.A. see Fletcher, A.N., 626, *653*, 1233, *1258*
Helmreich, D., 979, *1020*
Helmreich, D. see Eckener, U., 980, 997, *1018*
Henderson, D. see Barker, J.A., 368, 379, *387*
Hendriksen, B.A., 1333, 1334, *1358*
Hennel, J.W., 660, *681*, 872, *886*
Hennig, K. see Tempelhoff, K., 1305, 1311, *1363*
Henshaw, D.G., 1372, *1421*
Hensley, A.L. see Peri, J.B., 1272, *1362*
Herman, R.C., 649, *653*
Herman, R.M., 99, *158*
Hernandez, G. see Huyskens, P., 948, 950, *957*
Herring, F.G. see Hampton, M., 882, 884, *886*

Herrington, T.M., 989, 991, *1020*
Hertl, W., 1272, 1276, 1277, 1278, 1279, 1280, 1286, 1351, *1358*
Hertl, W. see Hair, M.L., 1266, 1269, 1272, 1273, 1288, *1357*
Hertz, H.G., 122, *158*, 724, *763*, 914, *934*, 1042, 1043, 1050, 1053, 1054, 1055, 1056, 1057, 1058, *1059*, *1061*, 1403, *1422*
Hertz, H.G. see Endom, L., 1055, *1058*
Hertz, H.G. see Engel, G., 1055, 1056, *1058*
Hertz, H.G. see Engelsmann, K., 1038, *1058*
Hertz, H.G. see Fister, F., 1052, 1055, 1056, *1058*, 1399, *1421*
Hertz, H.G. see Göller, R., 1047, *1059*
Hertz, H.G. see Grüner, M., 1041, 1044, 1048, 1049, *1059*
Hertz, H.G. see Langer, H., 1058, *1061*
Hertz, H.G. see Van Goldammer, E., 1039, 1044, 1056, *1059*
Herzberg, G., 222, *242*, 304, 310, *355*, 1254, *1258*
Heya, H. see Cramer, R.E., 826, *829*
Heydweiler, A. see Kohlrausch, F., 963, *1021*
Hiebert, G.G. see Hornig, D.F., 581, *609*
Higashi, A., 968, 979, *1020*
Higashi, A. see Fukuda, A., 968, *1019*
Hilbers, C.W. see Mackor, E.L., 819, *829*
Hildebrand, J.H. see Benesi, H.A., 1205, *1223*
Hill, N.E., 939, 995, *1020*
Hill, R.M., 586, *608*, 849, *886*, 1119, 1129, *1162*
Hill, T.L., 1293, *1358*
Hindermann, D.K., 819, *828*
Hindman, J.C., 1052, 1053, 1058, *1059*, *1061*, 1399, *1422*
Hineno, M., 781, *789*
Hino, M., 1267, *1358*
Hirai, N., see Eyring, H., 371, *387*
Hirano, E., 819, *828*
Hirotsu, K. see Takusagawa, K., 433, *455*
Hirschfelder, J.O., 29, 30, 49, 83, *158*
Hirschfelder, J.O. see Chipman, D.M., 56, *156*
Hitterman, R.L. see Taylor, J.C., 441, *455*
Hobbs, M.E., 962, *1020*
Hobbs, M.E. see Weith, A.J., 941, *1025*
Hobbs, P.V., 939, *1020*, 1427, *1468*
Hockey, J.A., 1267, *1358*
Hockey, J.A. see Armistead, C.G., 1267, 1272, *1353*
Hockey, J.A. see Atherton, K., 1345, *1353*
Hockey, J.A. see Hambleton, F.A., 1272, *1357*
Hockey, J.A. see Jones, P., 1343, *1358*
Hockey, J.A. see Peglar, R.J., 1272, *1361*
Hockey, J.A. see Tyler, A.J., 1309, *1363*
Hoekstra, P., 1004, *1020*, 1336, 1348, *1358*
Hoekstra, P. see Harvey, S.C., 1014, *1020*
Hoeve, C.A. see Gobush, W., 969, *1019*
Hofacker, G.L., 219, 220, 221, 222, 224, 225, 228, *242*, *355*
Hofacker, G.L. see Fischer, S.F., 219, 220, *242*, 247, *292*, 297, 301, 305, 309, 313, 318, 324, 331, 337, 342, 345, *355*, 590, *608*, 644, *653*, 981, *1019*, 1194, *1195*
Hofacker, G.L. see Godzik, K., 341, 345, *355*
Hofacker, U.A.H. see Hofacker, G.L., *355*
Hofer, O. see Falk, H., 125, 127, *157*
Hoffman, B.M., 820, *828*
Hoffmann, E.G., 541, *561*
Hofmann, K.P. see Schiöberg, D., 698, 701, 708, *765*
Hofmann, R. see Arend, H., 1151, *1161*
Hofmeister, F., 1404, *1422*
Högfeldt, E. see Ödberg, L., 814, *829*
Holakovský, J., 1158, *1166*
Holbrook, N.K. see Hopkinson, A.C., 113, 146, 147, *158*
Hollinger, H.B. see Haskell, R.W., 1233, 1235, *1258*
Hollinger, H.B. see Van Ness, H.C., 1233, *1259*
Hollnagel, M. see Lohse, U., 1002, 1003, *1021*, 1347, *1360*
Holmes, B.G. see Zimmerman, J.R., 1311, *1364*
Holmes, H.F., 1345, *1358*
Holmes, H.F. see Fuller Jr., E.L., 1345, *1357*
Holmes, R.R. see Fild, M., 819, *828*
Holmryd, S. see Larsson, K.E., 912, 929, *935*
Holz, M. see Hertz, H.G., 122, *158*
Holzapfel, W.B., 980, *1020*
Hopfield, J.J. see Barker, A.S. 347, *353*
Höpfner, A. see Wolff, H., 1234, 1236, 1238, 1240, 1243, 1246, 1248, 1250, 1254, *1259*, *1260*
Höpfner, H.-M. see Wolff, H., 1236, *1260*
Hopkinson, A.C., 113, 146, 147, *158*
Höppel, H.-E. see Wolff, H., 1231, 1233, 1243, 1249, *1259*
Horák, M., 573, *609*
Horlock, R.F. see Anderson, P.J., 1343, *1353*
Horn, D. see Wolff, H., 1237, *1259*
Horn, P.M. see Eberhard, J.W., 1154, *1165*
Horne, R.A., 939, *1020*

Horng, J.S. see Jury, S.H., 1346, *1358*
Hornig, D.F., 581, *609*, 1458, *1468*
Hornig, D.F. see Ferriso, C.C., 477, *524*, 660, 661, 663, *681*
Hornig, D.F. see Haas, C., 173, *215*, 221, *242*, 1458, 1460, 1461, *1467*
Hornig, D.F. see Mullhaupt, J.T., 477, *525*, 660, *681*
Hornig, D.F. see Plumb, R.C., 639, *653*
Hornig, D.F. see Ron, A., 622, *654*
Hornig, D.F. see Somorjai, R.L., 173, *216*, 221, *243*, 247, *292*
Hornig, D.F. see Schiffer, J., 580, *610*, 638, *654*
Hornig, D.F. see Wall, T.T., 580, *611*
Hoshino, S., 843, 860, *886*
Hoshino, S. see Niimura, N., 1147, *1163*
Hoshino, S. see Pepinsky, R., 1151, *1163*
Hougardy, J., 1349, *1358*
Howard, B.B., 1208, *1224*
Howard, B.J. see Dyke, T.R., 61, 72, 80, 115, *157*
Howell, F.L. see Schmidt, V.H., 989, 993, *1023*, *1024*, 1152, *1164*
Howell, M.V. see Angell, C.L., 1348, *1353*
Howells, J.D.R. see Barnes, A.J., 214, *214*, 1071, 1073, 1075, 1079, 1080, 1087, 1088, 1089, 1097, 1098, 1102, *1104*
Høye, J.S. see Rushbrooke, G.S., 386, *389*
Hoyland, J.R., 137, 138, *158*, 946, *1020*
Huang, H.H. see Nelson, S.M., 1003, *1023*, 1349, *1361*
Huang, K. see Born, M., 342, *354*
Huber, J.R., 16, *21*
Huber, L. see Waugh, J.S., 873, 874, *887*
Hubmann, M. see Taubenberger, R., 969, *1024*
Hückel, E., 474, *524*
Hudson, R.A. see Vinogradov, S.N., 737, *766*
Huertas, F. see Linares, J., 1348, 1350, *1360*
Huettenrauch, R., 1345, *1358*
Huggins, C.M., 595, *609*, 795, *828*, 1266, *1358*
Huggins, M.L., 5, *20*, *21*, 474, *524*
Hughes, D.J., 912, 929, *934*
Huler, E. see Hagler, A.T., 86, 87, 88, 89, 90, *157*, *158*
Humprey, F.B. see Waugh, J.S., 867, *887*
Hund, F., 289, *292*, 517, *524*, 722, *763*
Hunt, M.J., 841, 862, *886*
Huntress, Jr. W.T., 1038, 1049, *1059*
Huong, P.V., 353, *357*, 573, 595, 606, *609*, *611*
Huong, P.V. see Corset, J., 1253, *1257*
Huong, P.V. see Couzi, M., 570, *607*, 635, *652*
Huong, P.V. see Lascombe, J., 581, *609*
Huong, P.V. see Lassegues, J.C., 601, *609*, 635, *653*
Husar, J., 701, *763*
Husbands, D.I., 1345, *1358*
Hussein, M.A., 606, *611*
Hussein, M.A. see Millen, D.J., 19, *22*
Hustrulid, A. see Mann, M.M., 475, *525*
Hüttig, W. see Ginsberg, H., 1330, *1357*
Huyskens, P., 947, 948, 950, 955, 957, *1020*
Huyskens, P. see Clerbaux, T., 955, *1018*
Huyskens, P. see Cracco, F., 819, *828*
Huyskens, P. see Debecker, G., 948, 957, 958, *1018*
Huyskens, P. see Duterme, P., 955, *1018*
Huyskens, P. see Nouwen, R., 948, *1023*
Huzinaga, S., 106, *158*
Hyde, J.S. see Vedrine, J.C., 1348, *1364*
Hyne, J.B. see Saunders, M., 796, 799, *829*

Ibbitson, D.A., 541, *561*, 940, 941, *1020*, *1021*, 1233, *1258*
Ibbitson, D.A. see Eric, B., 940, 941, *1019*
Ibbitson, D.A. see Goode, E.V., 940, *1019*
Ibers, J.A., 173, 185, 186, *215*, 225, *243*, 324, 340, *355*, 425, *454*, 571, 597, *609*, 621, *653*
Ibers, J.A. see Delaplane, R.G., 186, *215*, 336, *354*, 422, 424, *453*, 602, *607*, 877, *886*
Ibers, J.A. see Hamilton, W.C., *20*, 323, 336, 340, *355*, 397, 410, 421, 425, 431, 433, *454*, 658, 659, *681*, 833, *886*, 894, *934*, 991, *1020*
Ibers, J.A. see Schroeder, L.W., 483, 484, 495, *525*
Ibers, J.A. see Snyder, R.G., 320, *357*
Ichikawa, M., 856, *886*
Ichiki, S.K. see Hill, R.M., 586, *608*, 849, *886*, 1119, 1129, *1162*
Ida, M., 1009, *1021*
Ihle, H. see Groth, W., 1246, 1254, *1258*
Ihon, M.S. see Eyring, H., 962, *1019*
Ihon, M.S. see Hobbs, M.E., 962, *1020*
Iizuka, T., 1343, *1358*
Ijevskaja, N.M. see Kvlividze, V.J., 1311, *1359*
Ikawa, S.I., 1460, 1461, *1468*
Illy, H. see Kiss, I., 1252, *1258*
Imai, J. see Morimoto, T., 1304, 1335, 1346, *1361*
Imelik, B. see Enriquez, M.A., 1344, *1356*
Imelik, B. see Fraissard, J., 1344, 1346, *1356*

Imelik, B. see Kemarec, J., 1346, *1358*
Imelik, B. see Pichat, P., 1346, *1362*
Imry, J. see Pelah, I., 908, 912, 929, 932, *935*
Inaba, A. see Chihara, H., 856, 857, *886*, 1160, *1165*
Ingraham, L.L., 151, *158*
Inskeep, R.G., 1092, 1094, *1105*
Iogansen, A.V., 584, 596, *609*
Iogansen, A.V. see Dementyeva, L.A., 584, *607*
Iogansen, A.V. see Odinkov, S.E., 585, *610*
Iogansen, A.V. see Rosenberg, M.Sh., 596, *610*
Ionov, S.P., 761, *764*
Ionova, G.V. see Ionov, S.P., 761, *764*
Irby, B.B. see Dugger, D.L., 1269, *1356*
Ireland, P., see Gillespie, R.J., 7, *21*
Isbister, D., 386, *389*
Ishibashi, Y., 1148, 1155, 1158, *1162*, *1166*
Ishibashi, Y. see Kameyama, H., 1160, *1166*
Ishibashi, Y. see Sato, K., 1158, *1166*
Ishibashi, Y. see Sawada, A., 1144, *1164*
Ishibashi, Y., see Takagi, Y., 1143, 1158, *1164*, *1166*
Ishmukhametpva, N.N. see Sedykh, N.V., 1016, *1024*
Isirikyan, A.A. see Dubinin, M.M., 1347, *1356*
Itagaki, K., 969, 979, *1021*
Ito, K. see Eda, B., 953, *1018*
Ito, M., 15, *21*
Itoh, K., 771, 772, 773, 778, 787, *789*
Itoh, U. see Takakusa, M., 18, *22*
Ivanov, N.R. see Shirokov, A.M., 1135, *1164*
Ivanov, N.R. see Shuvalov, L.A., 1135, 1136, 1137, 1139, 1140, *1164*
Ivanov, N.R. see Stepišnik, J., 1158, *1166*
Iwaki, T., 1343, *1358*
Iwasaki, H. see Saito, Y., 484, *525*
Iwata, S., 144, 145, *158*
Iwata, S., see Morokuma, K., *20*, 27, 123, 127, 144, 145, *160*
Iyengar, P.K. see Thaper, C.L., 902, 903, 930, *936*
Iyengar, P.K. see Venkataraman, G., 930, 931, *936*
Iyengar, R.D. see Boutin, H., 1344, *1354*
Izrailenko, A.N. see Strukov, B.A., 1145, *1164*

Jaccard, C., 341, 342, *355*, 966, 984, *1021*, 1403, 1414, *1422*
Jackson, J.A., 1465, *1468*
Jackson, J.A. see Waldstein, P., 1051, *1060*, 1465, *1469*
Jackson, P., 1342, 1343, 1344, *1358*
Jacob, M., 1434, 1464, *1468*
Jacobs, P.A. see Van Clauwelaert, F.H., 1267, *1364*
Jacobs, Th., 1349, *1358*
Jacobsen, R.J., 581, *609*, 622, 633, 638, *653*, 770, *789*
Jacobsen, R.J. see Page, Jr., T.F., 1317, *1361*
Jacrot, B. see Cribier, D., 912, 929, *934*
Jadzyn, J., 754, *764*, 950, 951, 952, 953, *1021*
Jaffe, G. see Chang, H.C., 976, *1017*
Jaffé, H.H., 782, 783, *789*
Jagendorf, A.T., 761, *764*
Jahns, W. see Von Stackelberg, M., 1415, *1423*
Jain, K. see Dienes, A., 16, *20*
Jain, Y.S., 1159, *1166*
Jakly, H. see Kiss, I., 1252, *1258*
Jakob, J. see Huettenrauch, R., 1345, *1358*
Jakobsen, A. see Forslind, E., 1350, *1356*
Jakubetz, W., 131, *158*
Jakubetz, W. see Rode, B.M., 116, 117, *162*
Jakubetz, W. see Schuster, P., *20*, 74, 126, 139, 149, 150, 151, *162*, 219, *243*
Jamsek-Vilfan, M. see Blinc, R., 843, 844, 846, 848, *885*, 1121, 1130, 1150, *1161*
Jancso, G., 1244, *1258*
Janik, J.A., 902, 905, 930, 931, *934*
Janik, J.A. see Bajorek, A., 902, 929, 930, *933*
Janik, J.A. see Golikov, V.V., 912, 929, *934*
Janik, J.A. see Janik, J.M., 904, 905, 915, 917, 930, 931, *934*
Janik, J.A. see Mikuli, E., 902, 930, *935*
Janik, J.M., 904, 905, 915, 917, 930, 931, *934*
Janik, J.M. see Bajorek, A., 902, 929, 930, *933*
Janik, J.M. see Janik, J.A., 905, 930, 931, *934*
Janik, J.M. see Mikuli, E., 902, 930, *935*
Janoschek, R., 80, 116, 122, *158*, 175, 176, 177, 178, 179, 180, 181, 182, 183, 191, 193, 203, 207, 209, *215*, 221, 224, 225, 231, *243*, 247, 251, 253, 255, 260, *292*, 297, 300, 304, 324, 329, *355*, 619, 644, *653*, 688, 722, 723, 724, 725, 728, 731, 733, 746, *764*
Janoschek, R. see Pecul, K., 111, *161*, 212, *215*, 1098, *1105*
Janoschek, R. see Pernoll, I., 689, 696, 733, *765*
Janoschek, R. see Swanstrom, P., 182, *216*
Janoschek, R., see Weidemann, E.G., 177, 204, 205, *216*

Jansen, F.J., 1347, 1348, *1358*
Jansoone, V.M., 381, 385, *389*
Janzen, J., 80, 135, *158*
Jarboe, C.H. see Kito, T., 1201, *1224*
Jaszunski, M., 80, 81, *158*
Jaycock, M.J. see Husbands, D.I., 1345, *1358*
Jean-Louis, M. see Atwood, M.R., 1083, *1104*
Jeffrey, G.A., 433, *454*
Jeffrey, G.A. see McMullan, R.K., 985, *1022*
Jeffrey, G.A. see Park, Y.J., 446, *455*
Jehle, H., 254, *292*, 729, *764*
Jelli, A. see Fripiat, J.J., 758, *763*, 1295, 1299, 1300, 1302, 1305, 1313, 1314, 1315, 1317, 1348, *1357*
Jelli, A.N. see White, J.L., 1285, *1364*
Jellinek, H.H.G., 974, *1021*
Jeneveau, A. see Sixou, P., 978, *1024*
Jenkins, H.D.B. see Dixon, H.P., 121, *157*
Jensen, C.D. see Nozari, M.S., 1206, 1207, *1224*
Jentschura, U., 6, *21*, 807, 808, 809, 812, 813, 822, *828*
Jentschura, U. see Ziessow, D., 811, *829*
Jeziorowski, H., 1267, 1304, 1320, 1322, 1323, 1324, 1325, 1327, 1338, 1339, 1340, 1346, *1358*
Jeziorowski, H. see Knözinger, H., 1345, 1346, *1359*
Jeziorski, B., 39, *158*
Jhon, M.S., 368, 372, *388*
Jiang, G.J., 123, *158*, 185, *215*
Jiang, G.J., see Anderson, G.E., 123, *155*, 203, *214*
Jirů, P. see Kubelkovà, L., 1267, 1303, 1318, 1320, *1359*
Joesten, M.D., *20*, 154, *163*, 1201, 1209, *1224*
Johansson, A., 63, 111, 113, 135, *158*
Johansson, A., see Kollman, P.A., 111, *159*
Johari, G.P., 970, *1021*, 1448, 1449, 1450, *1468*
Johnson, C.K., 400, *454*, 483, 485, 492, 500, 502, 505, *524*, 665, 671, *681*
Johnson, C.K. see Brunton, G.D., 671, 672, *681*
Johnson, C.K. see Ellison, R.D., 445, *453*
Johnson, E.W., 533, *561*
Johnson, J.R. see Christian, S.D., 1203, *1223*
Johnson, R.G. see Sakamoto, M., 929, *936*
Jolly, D. see Freasier, B., 386, *389*
Jona, F., 340, *355*, 1114, 1117, 1150, *1162*
Jonas, J., 1039, *1059*
Jonas, J. see Assink, R.A., 1039, *1058*
Jonas, J. see Lee, Y., 1058, *1061*
Jones, A.C. see Benesi, H.A., 1267, *1353*
Jones, D. see Barnes, A.J., 606, *611*, 1092, 1096, *1104*
Jones, D. see Evans, D.F., 601, *608*
Jones, D.A.K., 1201, *1224*
Jones, D.D., 460, *468*
Jones, D.W. see Curry, N.A., 442, *453*
Jones, D.W. see Ferraris, G., 441, 442, 443, *453*
Jones, G., 854, 855, *886*
Jones, P., 1343, *1358*
Jones, P., see Covington, A.K., *20*
Jones, R.V. see Sandy, F., 1142, *1164*
Jones, S.J., 983, *1021*
Jones, S.J. see Nakamura, T., 968, 983, *1022*
Jones, S.J. see Whalley, E., 939, *1025*
Jones, T.L. see Webster, R.K., 1343, *1364*
Jones, W.J., 635, *653*
Jones, W.M., 1252, 1253, *1258*
Jönsson, P.-G., 399, 409, 418, 444, 445, 448, 450, *454*, 460, *468*, 497, *524*
Jönsson, P.-G. see Kvick, Å., 449, *454*
Joos, G. see Fajans, K., 474, *524*
Jordan, F. see Diner, S., 123, *157*
Joris, J., 1284, *1358*
Joris, L., 818, *828*
Joris, L. see Arnett, E.M., 1201, 1205, *1223*
Joris, L. see Taft, R.W., 818, *829*
Jortner, J. see Bergman, A., 16, 18, *20*
Joshi, B.D., 518, *524*
Josien, M.-L., 569, 604, *609*
Josien, M.-L., see Bellocq, A.M., 302, *354*
Josien, M.-L. see Fuson, N., 583, *608*
Josza, P. see Pap, A., 1346, *1361*
Joussot-Dubien, J. see Novak, A., 601, *610*
Jumper, C.F. see Grunwald, E., 758, *763*
Jumper, C.F. see Howard, B.B., 1208, *1224*
Juneau, R.J., see Yakatan, G.J., 16, *22*
Jungen, M., 36, *158*
Jungers, J.C. see Beersman, J., 1246, *1257*
Juranji, M. see Detoni, S., 570, 571, 604, *608*
Jurga, J., 1058, *1061*
Jurga, K. see Jurga, J., 1058, *1061*
Jurga, S. see Jurga, J., 1058, *1061*
Jurga, S. see Szczesniak, E., 1058, *1061*
Jury, S.H., 1346, *1358*

Kaatze, U. see Pottel, R., 1032, *1060*
Kabankin, A.S., 820, *828*
Kadaba, P.K., 1148, *1162*
Kadanoff, L.P., 384, *388*

Kaderka, M. see Dorda, G., 1008, *1018*
Kagarise, R.E. see Whetsel, K.B., 825, *829*
Kagel, R.O., 1337, 1339, 1342, 1343, *1358*
Kahane, A. see Faure, P., *355*
Kahane, A. see Zarembovitch, A., 1433, *1470*
Kaifu, Y. see Mataga, N., 10, *21*
Kaiho, M., 1350, *1358*
Kakudo, M. see Ashida, T., 855, *885*
Kakiuchi, Y., 476, *524*, 871, *886*
Kalinichenko, A.M. see Matyash, I.V., 1347, *1360*
Kálmán, E., 744, 759, *764*, 1386, *1422*
Kamb, B., 724, *764*, 970, 971, 973, *1021*, 1429, 1437, 1441, 1445, 1449, 1450, 1451, 1452, *1468*
Kamb, B. see Hamilton, W.C., 970, *1020*, 1435, 1437, 1438, 1439, 1440, 1445, 1448, 1458, 1459, *1468*
Kamb, B. see La Placa, S.J., 437, 440, *454*, 1435, 1441, 1442, 1443, 1444, 1445, 1460, *1468*
Kamber, H.R. see Schindler, P., 1288, *1362*
Kameyama, H., 1160, *1166*
Kaminow, I.P., 347, *355*, 1119, 1124, 1132, *1162*, 1186, *1195*
Kamiyoshi, K., 1269, *1358*
Kämpf, G., 1315, *1358*
Kampmeyer, P.M., 1402, *1422*
Kampschulte-Scheuing, I., 695, 696, 697, *764*
Kan, L.S., 823, *828*
Kanaskova, Iu.D., 596, *609*
Kanazawa, T. see Kaiho, M., 1350, *1358*
Kan'no, K. see Okada, K., 1119, 1120, 1154, *1163*
Kanno, T. see Umeya, K., 1336, *1363*
Kanters, J.A. see Kroon, J., 403, 404, 405, 419, *454*
Kantner, T.R. see Basila, M.R., 1346, *1353*
Känzig, W., 1114, 1117, 1136, 1141, *1162*
Kao, S. see Kan, L.S., 823, *828*
Kaplan, N.O. see Raszka, M., 824, *829*
Kaplan, S. see Waugh, J.S., 880, 881, *887*
Kaplan, S.F., 1137, *1162*
Kaplansky, M., 855, *886*
Kaplow, R. see Lee, S.C., 724, *764*
Kapsomenos, S. see André, J.M., 142, *155*
Karagounis, G., 1293, 1319, *1358*
Karaulić, B. see Šušić, V., 1347, *1363*
Karaulić, D. see Vučelič, V., 1347, *1364*
Kärger, J. see Riedel, E., 1348, *1362*
Karle, J., 533, *561*
Karpfen, A., 135, 136, 137, 138, 142, *158*, *159*
Karplus, M., 81, *159*
Karplus, M. see Kern, C.W., 1431, 1434, 1441, 1458, *1468*
Karplus, M. see Kolker, H.J., 81, *159*
Karplus, S., 85, *159*
Karyakin, A.V. see Yaroslavsky, N.G., 1290, *1364*
Kasahara, M., 1159, *1166*
Kasha, M., 5, *21*
Katiyar, R., see Scott, J.F., 347, *357*
Katiyar, R.S., 1124, 1131, 1156, *1162*
Katiyar, R.S. see Cowley, R.A., 1132, *1161*
Kato, S. see Yamabe, S., 150, *163*
Katritzky, A.R., 710, *764*
Katsman, L.A., 749, *764*
Katsman, L.A. see Vargaftik, M.N., 749, *766*
Katz, B., 1074, *1105*
Katz, B. see Verstegen, J.M.P.J., 1071, 1083, *1105*
Kauffman, D.L. see Walsh, K.A., 761, *766*
Kauzmann, W., 30, *135*
Kauzmann, W. see Eisenberg, D., 79, *157*, 371, *387*, 939, *1019*, 1427, *1467*
Kawada, A., 758, *764*, 989, 994, *1021*
Kawada, S., 983, *1021*
Kawamori, A. see Kuroda, N., 1140, *1163*
Kawano, T., 1155, *1166*
Kay, M.I., 1145, 1148, *1162*
Kay, M.I. see Kaplan, S.F., 1137, *1162*
Kebarle, P., 121, 139, 140, *159*, 475, 481, 519, *524*, 980, *1021*
Kebarle, P. see Arshadi, M., 121, 139, 140, *155*
Kebarle, P. see Cunningham, A.J., 121, 139, 140, *156*, 980, *1018*
Kebarle, P. see Good, A., 475, *524*
Kebarle, P. see Grimsrud, E.P., 140, *157*
Kebarle, P. see Payzant, J.D., 121, *161*
Keçki, Z., 1404, *1422*
Keefer, R.M., 1205, *1224*
Kehiaian, H., 1231, *1258*
Kell, G.S., 1373, *1422*
Keller, B., 773, *789*
Keller, G. see Hertz, H.G., 1055, 1056, *1059*
Kelliher, J.M. see Inskeep, R.G., 1092, 1094, *1105*
Kellner, E. see Fink, P., 1273, *1356*
Kemarec, J., 1346, *1358*
Kempf, J. see Haeberlen, U., 877, 878, 879, 880, *886*
Kempter, H., 541, *561*

Kent, M., 1015, *1021*
Kerley, G.I. see Bowers, M.T., 1071, *1104*
Kern, C.W., 1431, 1434, 1441, 1458, *1468*
Kern, C.W. see Ermler, W.C., 35, *157*
Kern, C.W. see Gornostansky, S.D., 851, *886*, 1466, *1467*
Kerr, K.A., 460, *469*
Kestner, N.R., 29, 97, 98, *160*
Ketelaar, J.A.A., 183, 184, 186, *215*
Keyser, L.F., 1071, 1073, *1105*
Kezdy, F.J. see Bender, M.L., 761, *763*
Kheiker, D.M. see Gorbatyi, L.V., 1138, *1162*
Khlebnikov, V.B. see Volkov, A.V., 1319, *1364*
Khrapova, E.V. see Kiselev, A.V., 1280, *1359*
Khrustaleva, S.V. see Egorov, M.M., 1345, *1356*
Kibblewhite, J.F.J. see Tench, A.J., 1342, 1343, *1363*
Kiefer, M. see Becker, F., 826, *829*
Kielich, S., 959, *1021*
Kielich, S. see Piekara, A., 960, *1023*
Kier, L.B. see Hoyland, J.R., 137, 138, *158*, 946, *1020*
Kigoshi, K. see Kakiuchi, Y., 476, *524*
Kilpatrick, P.J. see Christian, S.D., 1203, *1223*
Kim, H.J., 918, 919, 920, 931, *935*
Kim, D.Y., 981, *1021*
Kimel, S. see Verstegen, J.M.P.J., 1071, 1083, *1105*
Kimizuka, H. see Satake, I., 827, *830*
Kimtys, I., 827, *829*
Kimura, K., 1203, *1224*
Kincaid, I.F. see Eyring, H., 1402, *1421*
Kind, R., 862, *886*, 1136, 1141, *1162*
Kind, R. see Roos, J., 1141, *1163*
Kindt, T., 16, *21*
King, C.M., 1074, 1084, 1085, 1095, *1105*
King, J.A. see Power, L.F., 441, *455*
King, S.T., 1070, *1105*
Kington, G.L. see Frohnsdorff, G.J.C., 1281, *1357*
Kintzinger, J.P., 1039, 1047, *1059*, *1060*
Kipling, J.J., 1330, *1358*
Kireev, K.V. see Sklyankin, A.A., 1349, *1363*
Kirina, O.F., 1334, *1358*
Kirina, O.F. see Antipina, T.V., 1334, *1353*
Kirpichnikova, L.F. see Shuvalov, L.A., 1135, 1136, 1139, *1164*
Kirshenbaum, I., 1246, 1252, 1254, *1258*
Kiselev, A.V., 1266, 1271, 1273, 1276, 1277, 1278, 1279, 1280, 1281, 1284, 1285, 1290, 1291, 1296, 1303, 1304, 1305, 1306, 1307, 1308, 1310, 1347, *1358*, *1359*
Kiselev, A.V. see Abramov, V.N., 1277, 1278, 1290, 1292, 1296, *1353*
Kiselev, A.V. see Avgul, N.N., 1338, *1353*
Kiselev, A.V. see Beliakova, L.D., 1318, *1353*
Kiselev, A.V. see Berezin, G.I., 1306, 1316, 1347, *1353*
Kiselev, A.V. see Curthoys, G., 1285, 1351, *1355*
Kiselev, A.V. see Davydov, V.Ya., 1267, 1270, 1274, 1276, 1277, 1278, 1279, 1280, 1281, 1284, 1285, 1303, 1320, 1351, *1355*
Kiselev, A.V. see Dzhigit, O.M., 1274, *1356*
Kiselev, A.V. see Galkin, G.A., 1274, 1275, 1276, 1277, 1282, 1283, 1284, 1285, 1290, 1292, 1293, *1357*
Kiselev, A.V. see Metsik, M.S., 1348, *1361*
Kiselev, A.V. see Volkov, A.V., 1319, *1364*
Kiselev, A.V. see Zhuravlev, L.T., 1270, *1364*
Kiselev, S.A. see Curthoys, G., 1285, 1351, *1355*
Kiselev, V.F. see Egorov, M.M., 1268, 1297, 1345, *1356*
Kiselev, V.F. see Kvlividze, V.J., 1298, 1311, *1359*
Kishida, S., 568, *609*
Kiss, I., 1246, 1252, 1254, *1258*
Kistenmacher, H., 84, 121, 122, *159*, 204, 214, *215*, 369, 379, 381, *388*
Kistenmacher, H. see Popkie, H., 59, 60, 62, 77, 80, 84, 111, 129, *161*, 369, 379, 381, *388*
Kister, A. see Redlich, O., 1228, 1233, *1259*
Kiszenick, W. see Camp, P.R., 973, *1017*
Kitamura, M., 15, *21*
Kito, T., 1201, *1224*
Kivinen, A. see Murto, J., 1096, *1105*
Kjällman, T., 422, *454*, 483, 492, 500, *524*
Klein, U., 15, *21*
Kleinberg, R., 408, *454*
Kleinberg, R. see Kay, M.I., 1145, 1148, *1162*
Klemperer, W. see Dyke, T.R., 61, 72, 80, 115, *157*
Kley, W. see Hughes, D.J., 912, 929, *934*
Kley, W. see Schenk, C.H., 932, *936*
Klier, K., 1267, 1304, 1307, 1308, 1309, 1310, 1311, 1313, *1359*
Klier, K. see Zettlemoyer, A.C., 1307, 1308, 1310, *1364*

Klint, D. see Clementi, E., 116, *156*
Klotz, I.M. see Worley, J.D., 1253, *1260*, 1395, *1423*
Klug, D.D., 972, *1021*, 1463, 1464, *1468*
Klug, D.D. see Wong, P.T.T., 1461, *1470*
Klute, R. see Hertz, H.G., 122, *158*, 1055, *1059*
Kneubühl, F., 777, *789*
Kneubühl, F. see Keller, B., 773, *789*
Knewstubb, P.F., 475, *524*
Knispel, R.R., 1138, *1162*
Knispel, R.R. see Torrie, B.H., 1157, *1167*
Knobler, C. see Kamb, B., 1445, *1468*
Knoll, D.B. see Von Hippel, A., 967, *1024*
Knop, O. see Falk, M., 433, 434, 438, *453*
Knözinger, H., 699, 718, *764*, 1277, 1278, 1279, 1286, 1287, 1289, 1301, 1302, 1303, 1332, 1334, 1335, 1337, 1339, 1342, 1344, 1345, 1346, *1359*
Knözinger, H. see Clement, G., 1333, 1336, *1355*
Knözinger, H. see Jeziorowski, H., 1267, 1304, 1320, 1322, 1323, 1324, 1325, 1327, 1338, 1339, 1340, 1346, *1358*
Knözinger, H. see Spannheimer, H., 1333, *1363*
Knözinger, H. see Stählin, W., 1351, *1363*
Kobayashi, J., 1145, 1154, *1162*, *1166*
Kobayashi, K.K., 347, 348, *355*, 1128, 1129, 1131, 1155, 1156, 1157, *1162*, 1180, 1184, 1186, *1195*
Kobilarov, N. see Hadži, D., 584, 585, 596, 602, *608*, 741, *763*
Kochloefl, K. see Rericha, R., 1339, *1362*
Kodratoff, Y., 1346, *1359*
Koehler, W.R. see Fink, D.W., 17, *21*
Koetzle, T.F., 451, 452, *454*, 460, *469*
Koetzle, T.F. see Andrews, L., 460, *468*
Koetzle, T.F. see Frey, M.N., 451, 452, *453*, *454*, 460, 461, *468*
Koetzle, T.F. see Hamilton, W.C., 459, *468*
Koetzle, T.F. see Jones, D.D., 460, *468*
Koetzle, T.F. see Kerr, K.A., 460, *469*
Koetzle, T.F. see Kvick, Å., 427, 429, *454*
Koetzle, T.F. see Lehmann, M.S., 450, 452, *454*, 460, *469*
Koetzle, T.F. see Verbist, J.J., 450, *456*, 460, 461, *469*
Köhler, A. see Fink, P., 1273, 1276, 1277, *1356*
Kohler, F., 386, *389*
Kohlrausch, F., 963, *1021*
Kohlschuter, D. see Haeberlen, U., 877, 878, 879, 880, *886*
Kohlschütter, H.W. see Kämpf, G., 1315, *1358*
Kohn, R.L. see Dienes, A., 16, *20*
Koide, G.T., 987, *1021*
Kölkenbeck, K. see Zundel, G., 697, 760, *766*
Kolker, H.J., 81, *159*
Kolker, H.J. see Karplus, M., 81, *159*
Koll, A. see Hawranek, J., 1012, *1020*
Koll, A., 702, 719, 739, *764*, 941, 953, *1021*
Koller, J., 601, *609*
Koller, J. see Zaučer, M., 841, 842, *887*
Kollmann, P.A., *20*, 27, 48, 51, 52, 57, 60, 61, 62, 70, 71, 72, 75, 80, 111, 113, 121, 129, 146, 150, *159*, 186, 190, 206, 214, *215*, 320, *355*, 500, 517, 518, 519, *524*, 723, *764*, 793, *828*, 946, *1021*
Kollmann, P.A. see Johansson, A., 63, 111, 113, 135, *158*
Kolodyashnij, Yu.V. see Rustanova, S.N., 953, *1023*
Kolodziej, H.H. see Hawranek, J., 1012, *1020*
Kolos, W., 101, *159*
Komarov, V.E. see Bajorek, A., 902, 929, 930, *933*
Komatsu, H. see Kakiuchi, Y., 476, *524*, 871, *886*
Komiyama, Y. see Ooi, S., 478, 484, *525*
Komuro, K. see Iwaki, T., 1343, *1358*
Kondo, S., 1315, 1336, 1351, *1359*
Kondo, S. see Muroya, M., 1009, *1022*
Kononyuk, V.F. see Dubinin, M.M., 1347, 1348, *1356*
Konsin, P.I., 1128, *1162*
Koob, R.D. see Gordon, M.S., 125, 127, *157*
Kopp, M., 973, *1021*
Koptsik, V.A. see Krasnikova, A.Ya., 1149, *1163*
Koptsik, V.A. see Strukov, B.A., 1130, 1145, 1154, 1157, *1164*, *1166*
Kopylova, V.M., 1001, *1021*
Korshuk, E.F. see Aleksandrovich, Kh.M., 1350, *1353*
Kortzeborn, R.N. see Noble, P.N., 121, *161*, 185, *215*
Korzhuev, M.A. see Strukov, B.A., 1130, *1164*
Kosaly, G., 931, *935*
Kostansek, E.C., 428, 449, *454*
Kostin, M.D. see Carbonell, R.G., 221, 222, 223, 225, 226, 227, 229, 235, *242*, *354*

Kotcheshkov, K.A. see Goldstein, I.P., 948, 955, *1019*
Kottwitz, D.A., 912, 929, *935*
Kouvarellis, G.K. see Hall, P.G., 1007, *1020*
Kozima, K. see Hirano, E., 819, *828*
Kozlov, A.A. see Berezin, G.I., 1316, *1353*
Kozyrovski, Y., 1278, *1359*
Krainik, N.N. see Smolenski, G.A., 1153, *1166*
Kramer, H.E.A. see Hafner, K., 604, *608*
Kraemer, W.P., 121, 135, 140, *159*, 190, *215*, 519, 520, *524*, 724, *764*
Krämer, W.P. see Diercksen, G.H.F., 61, 65, 72, 80, 111, 121, 146, 154, 155, *157*, *163*, 203, *215*, 1056, *1058*
Krasnikova, A.Ya., 1149, *1163*
Krasnobokii, Y.N. see Miglyachenko, A.F., 1016, *1022*
Krassilnikov, K.G. see Egorov, M.M., 1268, 1297, *1356*
Kreevoy, M.M. see Husar, J., 701, *763*
Kreevoy, M.M. see Williams, J.M., 695, 697, *766*
Kremer, F., 254, *292*
Kriegsmann, H. see Storek, W., 797, 799, *829*, 1233, *1259*
Krisher, L.C. see Thaddeus, P., 1465, *1469*
Kriss, E.E. see Sheka, I.A., 955, *1024*
Kristofel', N.N. see Konsin, P.I., 1128, *1162*
Kröbl, P. see Pap, A., 1346, *1361*
Kroh, A. see Mikke, K., 930, 931, *935*
Kromhout, R.A., 778, *789*
Kromhout, R.A. see Moulton, W.G., 595, *609*
Kroon, J., 403, 404, 405, 419, *454*
Krug, Kl., 1228, 1229, *1258*
Kruglik, A.I., 1159, *1166*
Kruh, R.F., 1255, *1258*, 1372, *1422*
Krylov, N.A. see Deryagin, B.V., 1005, *1018*
Krynicki, K., 1050, *1060*
Kubarev, S.I., 297, 342, 345, *355*
Kubelkovà, L., 1267, 1303, 1318, 1320, *1359*
Kubo, R., 306, 317, *355*
Kubo, R. see Suzuki, M., 1131, *1164*
Kubota, T. see Mataga, N., 11, 13, *21*
Kuchitsu, K., 400, 436, 437, *454*
Kugler, E. see Erdey-Gruz, T., 758, *763*
Kuhn, L.P., 540, 542, *561*, 1091, *1105*
Kumar, V. see Puranik, P.G., 1222, *1224*, 1286, *1362*
Kume, K., 1431, *1468*
Kunath, D., 1303, 1318, 1347, *1359*
Kunath, D. see Möller, K.D., 785, *789*
Kuniyoshi, H. see Matsuo, S., 1252, *1258*
Kunst, M. see Bordewijk, P., 1233, *1257*
Kunze, G.W. see Swoboda, A.R., 1348, *1363*
Kuper, C.G., 305, 310, *356*
Kurbatov, L.N., 1290, *1359*
Kuriakose, A.K., 982, *1021*
Kurkchi, G.A. see Dementyeva, L.A., 584, *607*
Kuroda, N., 1140, *1163*
Kurosaki, S., 1313, *1359*
Kuroya, H. see Nakahara, A., 478, 484, *525*, 667, *681*
Kuroya, H. see Ooi, S., 478, 484, *525*
Kurz, J.L., 266, *292*
Kurz, L.C. see Kurz, J.L., 266, *292*
Kutzelnigg, W., 33, 36, *159*
Kutzelnigg, W. see Ahlrichs, R., 36, *155*
Kuzmany, P. see Neckel, A., 121, *160*
Kuznetsov, A.M. see Vorotyntsev, M.A., 266, *292*
Kuznetsov, N.V. see Curthoys, G., 1285, 1351, *1355*
Kuznetsov, B.V. see Davydov, V.Ya., 1274, 1276, 1277, 1278, 1279, 1280, 1281, 1284, 1285, *1355*
Kuznetsova, L.V. see Berezin, G.I., 1316, *1353*
Kvick, Å., 398, 427, 429, 449, *454*
Kvick, Å. see Almlöf, J., 399, 417, *453*
Kvick, Å. see Jönsson, P.-G., 399, 409, 450, *454*, 460, *468*
Kvlividze, V.J., 1269, 1298, 1311, 1316, *1359*
Kvlividze, V.J. see Egorov, M.M., 1268, 1297, *1356*
Kydon, D.W., 1145, *1163*

Labbé, H.J. see Bertie, J.E., 972, *1017*, 1455, 1458, 1460, 1461, 1462, 1464, *1467*
Labbé, H.J. see Whalley, E., 972, *1025*, 1461, *1470*
Labes, M.M. see Kawada, A., 758, *764*, 989, 994, *1021*
LaBonville, P., 665, *681*
LaBonville, P. see Basile, L., 663, 664, 665, 666, 667, 670, 675, 676, 679, *681*
Ladell, J., 117, *159*
Ladik, A.J. see Karpfen, A., 135, 136, 137, 138, 142, *159*
Ladik, J., 142, *159*
Ladik, J. see Del Re, G., 142, *157*
Ladik, J. see Rein, R., 220, *243*
Ladik, J. see Suhai, S., 142, *163*
Lady, J.H. see Whetsel, K.B., 797, 824, *829*

Lagakos, N., 1156, 1157, *1166*
Lagaly, G., 1349, *1359*
Lahajnar, G. see Blinc, R., 1131, 1150, *1161*
Laidler, K.J. see Glasstone, S., 917, *934*
Lajzerowicz, J. see Bornarel, J., 1120, *1161*
Lamanna, U., see Alagona, G., 121, *155*, 190, *214*
Lamanna, U. see Guidotti, C., 137, 138, 139, *157*
Lambourne, R. see Dawber, J.G., 1343, *1355*
Lamotte, B. see Gaillard, J., 1155, *1165*
Lampe, F.W., 475, *524*
Landa, B. see Gillespie, R.J., 7, *21*
Landau, L., 554, *561*
Landau, L.D., 229, 237, 240, 241, *243*
Landeck, H. see Wolff, H., 1228, 1229, 1230, 1235, 1236, 1238, 1243, *1260*
Landis, D.H., 267, *292*
Landsberg, G.S., 584, 588, *609*
Lane, E.H. see Christian, S.D., 821, *828*
Langer, H., 1058, *1061*
Langhammer, G. see Ebert, G., 1335, *1356*
Lapinskii, F.L. see Dubinin, M.M., 1002, *1018*
La Placa, S.J., 437, 440, *454*, 1433, 1434, 1435, 1441, 1442, 1443, 1444, 1445, 1460, *1468*
La Placa, S.J. see Hamilton, W.C., 970, *1020*, 1435, 1437, 1438, 1439, 1440, 1445, 1448, 1458, 1459, *1468*
La Placa, S.J. see Schlemper, E.O., 336, *357*, 449, *455*, 465, *469*
LaPlanche, L.A., 805, 806, 808, *828*
Larsen, F.K., 443, *454*
Larsen, F.K. see Lehmann, M.S., 422, 443, *454*, 1138, *1163*
Larsson, K.E., 911, 912, 919, 929, 932, 933, *935*
Larsson, K.E. see Dahlborg, U., 915, *934*
Larsson, K.E. see Sköld, K., 912, 929, *936*
Lasater, J.A. see Zimmerman, J.R., 1311, *1364*
Lascombe, J., 581, *609*, 647, *653*
Lascombe, J. see Corset, J., 1253, 1254, 1255, *1257*
Lascombe, J. see Couzi, M., 635, *652*
Lascombe, J. see Huong, P.V., 595, *609*
Lascombe, J. see Novak, A., 601, *610*
Lassegues, J.C., 334, 336, *356*, 601, *609*, 635, *653*
Lassegues, J.C. see Grenie, Y., 335, *355*, 582, 583, *608*, 1100, *1105*
Lassegues, J.C. see Huong, P.V., 573, *609*
Lassegues, J.C. see Lascombe, J., 581, *609*
Lassier, B. see Brot, C., 1012, *1017*
Lathan, W.A. see Morokuma, K., *20*, 27, 123, 127, 144, 145, *160*
Latimer, W.M., 5, *21*
Laura, R.D. see Cloos, P., 1349, *1355*
Lautie, A., 336,*356*, 859, *886*
Lautie, E.R., 576, *609*
Lavis, D.A. see Bel, G.M., 368, 373, 374, 378, *387*
Lavrenčič, B., *356*
Lavrenčič, B. see Blinc, R., 1155, *1165*
Lawson, K.D. see Brey, W.S., 1345, 1346, *1354*
Lazarini, F. see Golič, L., 602, *608*
Leadbetter, A.J., 1455, 1461, *1469*
Lebas, J.-M. see Josien, M.-L., 604, *609*
Le Bot, J., 1313, *1359*
Lebowitz, J.L., 378, *388*
Lebrun, A. see Chapoton, A., 1002, *1017*, *1018*
Lebrun, A. see Wacrenier, J.M., 1009, *1024*
Le Calve, J. see Couzi, M., 635, *652*
Lechert, H. see Basler, W.D., 1347, *1353*
Lecourt, A. see Grimaldi, F., 37, *157*
Lee, C.S., 92, *159*
Lee, F.S., 478, 482, 495, *524*, 663, *681*, 904, *935*
Lee, J.A., 1345, *1360*
Lee, M.K.T. see West, R., 1201, *1224*
Lee, S.C., 724, *764*
Lee, T.D., 370, *388*
Lee, Y., 1058, *1061*
Leftin, H.P. see Hall, W.K., 1268, 1269, *1357*
Leftin, H.P. see O'Reilly, D.E., 1268, 1327, *1361*
Lehmann, M.S., 422, 443, 450, 452, *454*, 460, *469*, 1138, *1163*
Lehmann, M.S. see Andrews, L., 460, *468*
Lehmann, M.S. see Frey, M.N., 451, 452, *453*, *454*, 460, 461, *468*
Lehmann, M.S. see Hamilton, W.C., 459, *468*
Lehmann, M.S. see Koetzle, T.F., 451, 452, *454*, 460, *469*
Lehmann, M.S. see Larsen, F.K., 443, *454*
Lehmann, M.S. see Lindqvist, O., 443, *454*
Lehmann, M.S. see Verbist, J.J., 450, *456*, 460, 461, *469*
Lehn, J.M. see Kintzinger, J.P., 1039, *1059*, *1060*
Lehner, H. see Falk, H., 125, 127, *157*
Lehrer, S.S., 858, *886*

Le Montagner, S. see Le Bot, J., 1313, *1359*
Lengyel, S. see Kálmán, E., 1386, *1422*
Lennart, D.S. see Vedrine, J.C., 1348, *1364*
Lentz, B.R., 133, 134, *159*
Leonard, A.J., 1330, *1360*
Leonard, B.R. see Kottwitz, D.A., 912, 929, *935*
Leonard, R.A., 1348, *1360*
Leonidova, G.G., 1158, *1166*
Leroy, G. see André, J.M., 142, *155*
Leung, P.S., 931, 932, *935*
Leung, P.S. see Rush, J.J., 929, 930, 931, *935*
Leung, P.S. see Safford, G.J., 912, 929, *935*
Leung, R.C. see Lowndes, R.P., 1155, *1166*
Leventhal, J.J., see De Paz, M., 121, *156*, 475, *523*, 980, *1018*
Leviel, J.L., 179, 214, *215*, 305, 308, 331, 332, *356*, 644, *653*
Levine, S., 371, 372, *388*
Levine, S. see Perram, J.W., 371, 372, *388*
Levitt, M. see Warshel, A., 85, *163*
Levstek, I. see Blinc, R., 1155, *1165*
Levstek, I. see Lavrenčič, B., *356*
Levstik, A. see Blinc, R., 1130, *1161*
Levy, A.L. see Peterson, S.W., 1434, *1469*
Levy, B. see Meunier, A., 63, 64, *160*
Levy, G., 824, *828*
Levy, G.C., 1049, *1060*
Levy, H.A., 1449, 1450, *1469*
Levy, H.A. see Brown, G.M., 446, *453*
Levy, H.A. see Busing, W.R., 463, *468*
Levy, H.A. see Danford, M.D., 1415, *1421*
Levy, H.A. see Ellison, R.D., 430, 431, 445, *453*, 604, *608*
Levy, H.A. see Narten, A.H., 368, *388*, 554, *561*
Levy, H.A. see Smith, H.G., 478, 482, 495, *525*
Levy, H.A. see Williams, J.M., 427, *456*, 483, 508, 509, *526*, 675, *682*
Levy, S., 1328, *1360*
Lewis, K.E., 1343, 1344, *1360*
Lewis, T.J. see Thomas, J.M., 708, 758, 760, *765*, 995, *1024*
Lezina, V.P. see Bystrov, V.F., 1231, *1257*
Li, N.C. see Kan, L.S., 823, *828*
Li, N.C. see Shaw, Y.L., 814, *829*
Licha, A., see Brickmann, J., 175, *215*, 225, *242*
Lichter, R.L., 823, *828*
Lichtfus, G., 782, 783, *789*
Liddel, U., 326, *356*, 1233, *1258*
Liddel, U. see Becker, E.D., 540, *560*
Lieb, E.H., 362, 364, 366, 367, 383, *388*, 1448, *1469*
Lieb, E.H. see Abraham, D.B., 1137, *1160*
Lifshitz, E.M. see Landau, L.D., 229, 237, 240, 241, *243*
Lifshitz, Y.M., 254, *292*
Lifshitz, Y.M. see Dzyaloshinskii, I.Y., 254, *292*
Lifson, S., 85, *159*
Lifson, S. see Bixon, M., 85, *155*
Lifson, S. see Ermer, O., 85, 87, *157*
Lifson, S. see Hagler, A.T., 86, 87, 88, 89, 90, *157*, *158*
Lifson, S. see Karplus, S., 85, *159*
Lifson, S. see Warshel, A., 85, *163*
Lim, Y.Y., 819, 820, *828*
Limbach, H.H., 825, *830*
Liminga, R., 414, *454*
Liminga, R. see Jönsson, P.-G., 448, *454*
Liminga, R. see Kvick, Å., 449, *454*
Liminga, R. see Nilsson, Å., 449, *455*
Liminga, R. see Tellgren, R., 437, 438, 443, 444, *455*, 1138, 1139, *1164*
Lin, C. see Dienes, A., 16, *20*
Lin, C.C. see Torrie, B.H., 1145, *1165*
Lin, K.C. see Berney, C.L., 1099, *1104*
Lin, K.C. see Redington, R.L., 335, *356*, 582, *610*, 1099, *1105*
Lin, W.C. see Hampton, M., 882, 884, *886*
Linares, J., 1348, 1350, *1360*
Lindemann, R., 703, 705, 709, 737, 741, *764*, 949, *1021*
Lindgren, J., 438, *454*
Lindgren, J. see Almlöf, J., 119, 123, *155*, 208, *214*, 438, 439, *453*
Lindgren, J. see Berglund, B., 437, *453*
Lindgren, J. see Eriksson, A., 747, *763*
Lindmann, B. see Hertz, H.G., 1056, *1059*
Lindner, H.A., 1382, 1395, *1422*
Lindqvist, O., 443, *454*
Linggoatmodjo, K. see Chidambaram, R., 440, *453*
Linnell, R.H., 641, *653*
Linnell, R.H., see Vinogradov, S.N., *20*, *163*, *357*
Linton, H. see Conway, B.E., 220, 222, 240, *242*, 289, *292*, 475, *523*, 1314, *1355*
Liopo, V.A. see Metsik, M.S., 1348, *1361*
Lippens, B.C. see De Boer, J.H., 1334, *1355*
Lippert, E., 8, 9, 11, 13, 18, *21*, 798, 806, *828*
Lippert, E. see Gronau, B., 15, *21*
Lippert, E. see Jentschura, U., 6, *21*, 807, 808, 809, 812, 813, 822, *828*

Lippert, E. see Kindt, T., 16, *21*
Lippert, E. see Prigge, H., 9, *22*
Lippert, E. see Schroer, W., 825, 827, *830*
Lippert, E. see Ziessow, D., 811, 822, 823, *829*
Lippert, J.L., 946, *1021*
Lippincott, E.R., 172, *215*, 221, *243*, 305, 335, *356*, 571, 573, 574, *609*, 621, *653*, 841, *886*, 1085, *1105*, 1202, 1214, 1218, *1224*, 1317, *1360*
Lippincott, E.R. see Anderson, G.R., *353*
Lippincott, E.R. see Bellamy, L.J., 1317, *1353*
Lippincott, E.R. see Miller, P.J., 599, *609*
Lippincott, E.R. see Page Jr., T.F., 1317, *1361*
Lipscomb, W.N., see Atoji, M., 80
Lipscomb, W.N. see Nordman, C.E., 425, *455*
Lischka, H., 35, 57, 61, 63, 64, 65, 66, 67, 72, 77, 93, 96, 97, 98, 99, 100, 101, 102, 103, 111, 146, 147, 148, 154, *159*
Lischka, H. see Beyer, A., 154, 155, *163*
Lischka, H. see Russegger, P., 137, *162*
Litchman, W.M., 822, *829*
Litov, E., 1120, 1127, 1130, 1131, *1163*
Litov, E. see Havlin, S., 1157, *1165*
Little, L.H., 1266, 1285, 1295, *1360*
Little, L.H. see Cant, N.W., 699, 717, *763*, 1277, 1295, 1296, 1297, 1301, 1302, 1346, *1354*
Litvan, G.G., 1317, *1360*
Litvan, G.G. see Sidebottom, E.W., 1317, *1362*
Lo, G.Y.S. see Evans, J.C., 602, *608*, 637, *653*
Loewenstein, A., 1050, *1060*
Logan, K.W. see Prask, H.J., 929, *935*, 1461, 1462, 1463, *1469*
Lohse, U., 1002, 1003, *1021*, 1347, *1360*
Lokutsievskii, V.A. see Davydov, V.Ya., 1351, *1355*
London, F. see Eisenschitz, H., 39, *157*
Long, F.T., 1202, *1224*
Longini, R.L. see Seidensticker, R.G., 983, *1024*
Lord, R.C., 531, *561*, 661, *681*
Lord, R.C. see Harada, I., 786, *789*
Lorenzelli, V., 1308, *1360*
Lorenzelli, V. see Bertoluzza, A., 1308, 1351, *1354*
Losonczy, M., 93, 96, 100, 102, *159*
Loubser, T.H.N. see Thaddeus, P., 1465, *1469*
Lovesey, S. see Marshall, W., 1173, 1174, 1177, *1195*
Low, M.J.D., 699, 718, *764*, 1273, 1277, 1278, 1297, 1299, 1301, 1302, 1318, 1320, 1343, *1360*
Low, M.J.D. see Cusumano, J.A., 1276, 1277, 1278, 1279, 1286, 1287, 1288, 1292, 1294, 1343, *1355*
Low, M.J.D. see McManus, J.C., 1342, 1343, *1361*
Low, M.J.D. see Sahay, B.K., 1343, *1362*
Low, P.F., 1343, 1348, *1360*
Lowden, L.W., 386, *389*
Löwdin, P.-O., 83, *159*, 219, 220, 221, 222, 240, 241, *243*
Lowe, I.J. see Barnaal, D.E., 1430, 1431, *1466*
Lown, D.A., 758, *764*
Lowndes, R.P., 1155, *1166*
Lubezky, I., 1016, *1021*, 1298, 1299, *1360*
Lubos, W.D. see Zundel, G., 697, 760, *766*
Lucazeau, G., 573, 582, 583, *609*
Lucchesi, P.J. see Carter, J.L., 1330, 1331, 1332, *1354*
Luck, W.A.P., *20*, 130, *159*, 530, 531, 533, 534, 536, 540, 542, 543, 544, 546, 549, 550, 551, 554, 555, 556, 557, 558, *561*, 621, 624, 625, 648, *653*, 696, 700, *764*, 1086, *1105*, 1253, *1258*, 1308, 1326, *1360*, 1369, 1371, 1372, 1378, 1381, 1382, 1383, 1384, 1386, 1387, 1388, 1389, 1390, 1391, 1392, 1394, 1403, 1405, 1408, 1409, 1410, 1411, 1412, 1413, 1414, 1415, 1419, *1422*
Luck, W.A.P. see Mann, B., 1088, *1105*
Lucken, E.A.C., 1035, *1060*
Luckin, A.E., 576, *609*
Lüder, W. see Lippert E., 9, *21*
Ludupov, Ts.-Zh., 1119, 1127, *1163*
Lumbroso, H., 941, 957, *1022*
Lumbroso, H. see Bertin, D.M., 941, *1017*
Lumbroso, H. see Pigenet, C., 941, *1023*
Lumbroso-Bader, N. see Baron, D., 797, 826, *827*, *829*
Lundgren, J.O., 446, *454*, 473, 482, 483, 485, 486, 487, 488, 489, 492, 493, 495, 496, 499, 500, 501, 504, 505, 508, 509, 512, *524*, *525*, 663, 664, 675, 676, 677, 679, *681*
Lundgren, J.O. see Almlöf, J., 482, 493, 511, 512, *523*
Lundgren, J.O. see Delaplane, R.G., 483, 492, 500, *523*
Lundgren, J.O. see Spencer, J.B., 483, 495, *525*

Lundin, A.G. see Gavrilov-Podol'skaya, G.V., 1136, 1139, *1162*
Lundin, A.G. see Habuda, S.P., 1149, *1162*
Lundin, P. see Lundgren, J.O., 483, 492, 508, *524*
Lurie, F.M., 849, *886*
Lusa, A. see Ghersetti, S., 1201, *1223*
Lüttke, W. see Greinacher, E.G., 326, 336, *355*
Luz, Z., 1052, *1060*, 1399, *1422*
Luzzati, V., 477, 478, 482, *525*, 663, 676, *681*
Lych, A.M. see Gamayunov, N.I., 1017, *1019*
Lyerla Jr., J.R., 1044, 1045, *1060*
Lyerla Jr., J.R. see Alger, T.D., 1045, *1058*
Lygin, V.I. see Abramov, V.N., 1277, 1278, 1290, 1292, 1296, *1353*
Lygin, V.I. see Davydov, V.Ya., 1274, 1276, 1277, 1279, 1285, 1320, 1351, *1355*
Lygin, V.I. see Galkin, G.A., 1274, 1275, 1276, 1277, 1282, 1283, 1284, 1285, 1290, 1292, 1293, *1357*
Lygin, V.I. see Kiselev, A.V., 1266, 1276, 1277, 1278, 1279, 1290, 1291, 1296, 1303, 1304, 1306, 1347, *1359*
Lygin, V.I. see Volkov, A.V., 1319, *1364*
Lyzina, I.A. see Avgul, N.N., 1338, *1353*

Maatman, R.W. see Dugger, D.L., 1269, *1356*
Macdonald, A.C., 440, *455*
Macdonald, A.L., 422, 445, *455*, 600, *609*
MacDonald, J.R., 976, *1022*
Macewen, I.Y. see Bushfield, W.K., *242*
Machekhina, T.A. see Bajorek, A., 906, 930, 931, *933*
Maciel, G.E., 823, *829*
Maciel, G.E. see Nakashima, T.T., 825, *830*
Mackay, A.L. see Hunt, M.J., 841, 862, *886*
Mackor, E.L., 819, *829*
MacLean, C. see Mackor, E.L., 819, *829*
Maeda, S. see Ikawa, S.I., 1460, 1461, *1468*
Maeno, N., 983, *1022*
Maestro, M. see Arrighini, G.P., 35, *155*
Maestro, M. see Guidotti, C., 137, 138, 139, *157*
Magat, M., 939, *1022*
Magat, M. see Bauer, E., 1284, *1353*
Magat, M. see Dalbert, R., 1047, *1058*
Magataev, V.K. see Zheludev, I.S., 1130, *1165*
Maguire, M.M., 541, *561*
Mahnke, H. see Lippert, E., 8, *21*
Mahr, B.W. see Hardt, A.P., 1048, *1059*
Maidique, M.A., 973, 980, *1022*
Maier, U. see Pernoll, I., 689, 696, 733, *765*
Maigret, B. see Dreyfus, M., 116, 117, *157*
Maigret, B. see Pullman, A., 124, *161*
Maillols, J. see Bardet, L., 771, *788*
Main, P. see Germain, G., 461, *468*
Majid, Y.A., 985, *1022*
Majoube, M., 1253, *1258*
Mak, T.C.W., 985, *1022*
Makita, Y., 1136, 1137, 1139, 1147, 1158, *1163*
Makita, Y. see Gesi, K., 1159, *1165*
Malarski, Z., 941, *1022*
Malecki, J., 959, 960, 961, 962, 963, *1022*
Malecki, J. see Jadzyn, J., 754, *764*, 950, 951, 952, 953, *1021*
Mali, M. see Blinc, R., 855, 857, 859, 860, 861, 862, 863, 866, 885, *885*, *886*, 1122, 1132, 1144, 1154, *1161*, *1165*
Malmstadt, H.V. see Harrington, D., 18, *21*
Malrieu, J.P., 123, *159*
Malrieu, J.P. see Daudey, J.P., 49, 59, *156*
Malrieu, J.P. see Diner, S., 123, *157*
Mamy, J., 1005, *1022*
Mangini, A. see Ghersetti, S., 782, *789*
Mank, V.V. see Eremenko, A.M., 1347, *1356*
Mank, V.V. see Matyash, I.V., 1347, *1360*
Mann, B., 1088, *1105*
Mann, B. see Luck, W.A.P., 549, 550, *561*, 1383, 1415, *1422*
Mann, D.E., 1101, *1105*
Mann, F.G. see Braunholtz, J.T., 584, *607*
Mann, M.M., 475, *525*
Manning, M., 172, *215*, 222, *243*
Manocha, A.S. see Carlson, G.L., 780, 781, *789*
Mansfield, P. see Grannel, P.K., 882, *886*
Mapes, J.E. see Choi, C.S., 447, *453*
Marchese, F.T. see Del Bene, J.E., 111, *157*
Marchi, R.P., 1255, *1258*
Marchi, R.P. see Eyring, H., 1255, *1257*
Maréchal, Y., 173, 179, 214, *215*, 221, *243*, 297, 303, 305, 307, 323, 324, 332, 334, 341, *356*, 590, 591, *609*, 643, *653*, 1099, *1105*
Maréchal, Y. see Bournay, J., 214, *214*, 305, 332, 336, 353, *354*, *357*, 576, 606, 607, *611*
Maréchal, Y. see Excoffon, P., 214, *215*, 305, 306, 332, *354*
Maréchal, Y. see Hofacker, G.L., 219, 220, 221, 224, 225, *242*
Maréchal, Y. see Leviel, J.L., 179, 214, *215*, 305, 308, 331, 332, *356*, 644, *654*

Maréchal, Y. see Zellsman, H.R., 606, *611*
Margenau, H., 29, 97, 98, *160*
Margoshes, M. see Nakamoto, K., 410, *455*, 659, 661, *681*
Maričić, S., 989, 993, *1022*
Marino, R.A., 859, *886*
Marius, W., 63, 74, 100, 101, *160*
Marius, W. see Schuster, P., 74, 139, *162*
Marraud, M. see Cung, M.T., 124, *156*
Marsh, R.E., 459, 462, 463, *469*
Marshall, K. see Griffiths, D.M., 1351, *1357*
Marshall, W., 1173, 1174, 1177, *1195*
Marshall, W.L. see Quist, A.S., 1399, *1423*
Martens, R., 1343, *1360*
Martin, J.S., 821, 826, *829*, *830*
Martin, J.S. see Fujiwara, F.Y., 820, 821, 826, *828*, *829*
Martin, M. see Caballol, R., 83, *156*
Marzocchi, M.P., 582, *609*
Mascarenhas, S., 979, *1022*
Mascherpa, G. see Fournier, M., 660, 661, 663, *681*
Mashchenko, V.V. see Tertykh, V.A., 1272, *1363*
Maslenkova, G.L. see Valkov, V.I., 636, *654*
Mason, P.R. see Fatuzzo, E., 1054, *1058*
Masri, F.N., 601, *609*
Masri, F.N. see Dean, R.C., 606, *611*
Massara, C.I. see Bonera, G., 1160, *1165*
Mataga, N., 10, 11, 13, *21*
Mathias, D. see Wolff, H., 1237, *1259*
Mathieu, M.-V. see Little, L.H., 1295, *1360*
Mathieu, M.-V. see Pichat, P., 1346, *1362*
Mathieu, M.-V. see Primet, P., 1344, *1362*
Mathieu, M.-V. see Sheppard, N.V., 1290, *1362*
Matron, W., 1002, *1022*, 1347, *1360*
Matron, W. see Ebert, G., 1006, *1018*, 1315, 1316, *1356*
Matsubara, T. see Tokunaga, M., 247, 266, *292*, *357*, 1128, 1133, *1164*, 1180, 1181, *1195*
Matsubara, T. see Yoshimitsu, K., 1130, *1165*
Matsuda, M. see Abe, R., 1158, *1165*
Matsuo, S., 1252, *1258*
Matsushita, K. see McManus, J.C., 1342, 1343, *1361*
Matsuzaki, A., 14, *21*
Matthies, M., 703, 761, *764*
Mattmann, G., 1345, *1360*
Matuara, R. see Satake, I., 827, *830*
Matus, L. see Kiss, I., 1246, 1254, *1258*
Matyash, I.V., 1347, *1360*
Matyash, I.V. see Eremenko, A.M., 1347, *1356*
Mauk, V.V. see Brekhunets, A.G., 1347, *1354*
Maurin, M. see Philippot, E., 442, *455*
Mavrin, B.N., 1155, *1166*
May, L.F. see Reese, W., 1120, 1129, *1163*
Mayer, A. see Brausse, G., 1013, *1017*
Mayer, A. see Reichle, M., 1013, *1023*
Mayer, J. see Janik, J.A., 905, 931, *934*
Mayer, W. see Schuster, P., 219, *243*
Mayerböck, B. see Noller, H., 699, 710, 718, 738, *765*, 1285, 1301, *1361*
Mayerová, I. see Petzelt, J., 1159, *1166*
Mays, J.M., 1343, *1360*
Mazurkiewicz, A. see Mikuli, E., 902, 930, *935*
Mazzacurati, A. see Careri, G., 1335, 1343, 1345, 1346, *1354*
McAdam, A., 445, *455*
McBeth, R.L. see Strommen, D.P., 1086, 1088, *1105*
McCafferty, E., 1008, *1022*, 1336, 1345, *1360*, *1361*
McCafferty, E. see Zettlemoyer, A.C., 1345, *1364*
McCall, D.W., 1055, *1060*
McClellan, A.L. see Pimentel, G.C., 7, *20*, 112, 115, 116, 117, *161*, 299, 305, 319, 323, 324, *356*, 401, 404, *455*, 552, *562*, 567, 570, 576, 602, *610*, 658, 659, 661, *682*, 1201, 1220, *1224*
McConnell, B.L. see Dugger, D.L., 1269, *1356*
McCourt, F.R. see Harvey, K.B., 1418, *1421*
McDaniel, D.H., 121, *160*
McDonald, I.R., 386, *389*
McDonald, R.S., 1267, 1269, 1276, 1279, 1280, 1282, 1284, 1318, *1361*
McDowell, C.A. see Dalal, N.S., 882, 883, *886*, 1123, *1161*
McDowell, C.A. see Hampton, M., 882, 884, *886*
McGhie, A.R. see Kawada, A., 758, *764*, 989, 994, *1021*
McGuire, R.F. see Momany, F.A., 124, *160*
McKean, 1420, *1422*
McKeever, L.D. see Taft, R.W., 817, *829*
McKenzie, E.D. see Dollimore, D., 668, *681*
McLaughlin, D.R., 56, *160*
McLean, A.D., 101, 112, 118, *160*, 185, *215*
McLean, A.D. see Clementi, E., 118, *156*

McMahon, P.E. see Inskeep, R.G., 1092, 1094, *1105*
McManus, J.C., 1342, 1343, *1361*
McMullan, R.K., 985, *1022*
McMullan, R.K. see Mak, T.C.W., 985, *1022*
McWeeny, R., 51, *160*
McWeeny, R. see Diercksen, G.H.F., 51, *157*
Meath, W.J. see Hirschfelder, J.O., 29, 30, *158*
Mecke, R., 541, *561*, 1233, *1258*
Mecke, R. see Greinacher, E.G., 326, 336, *355*
Mecke, R. see Kempter, H., 541, *561*
Meerson, L.A., 1346, *1361*
Megaw, H.D., 1114, *1163*
Mehl, J. see Clementi, E., 116, 117, 150, 151, *156*, 214, *215*, 219, *242*
Meiboom, S., 1052, *1060*
Meiboom, S. see Grunwald, E., 758, *763*
Meiboom, S. see Luz, Z., 1052, *1060*
Meijs, W.H. see De Boer, J.H., 1334, *1355*
Mellor, J. see Janik, J.A., 905, 930, 931, *934*
Melnick, A.M. see Affsprung, H.E., 809, *827*
Melveger, A.J. see Rush, J.J., 587, *610*, 932, *935*
Mercado, A., 1144, *1163*
Merckel, K. see Wolf, K.L., 533, *562*
Merlet, P., 121, 150, *160*, 247, *292*
Merlivat, L., 1252, *1258*
Merrifield, R.E. see Lord, R.C., 661, *681*
Mertens, G., 1318, *1361*
Merz, W.J. see Fatuzzo, E., 1115, *1162*
Mesnard, G., 1012, *1022*
Messiah, A., 30, *160*, 223, 239, 240, *243*, 316, *356*
Metsik, M.S., 1348, *1361*
Metzger, G. see Wolf, K.L., 533, *562*
Metzger, H. see Zundel, G., 687, 689, 697, 744, *766*
Meunier, A., 63, 64, *160*
Meye, W., 1332, *1361*
Meye, W., see Jeziorowski, H., 1267, 1304, 1320, 1322, 1323, 1324, 1325, 1327, 1338, 1339, 1340, *1358*
Meyer, B., 1067, 1104, *1105*
Meyer, F. see Franck, E.U., 61, 80, 135, *157*
Meyer, W., 36, 122, 148, 149, 151, *160*
Meyer, W. see Schuster, P., *20*, 126, 149, 150, 151, *162*
Mezei, F., 1178, *1195*
Michel, D., 1305, 1306, 1311, 1312, 1313, 1337, *1361*
Michel, H. see Lippert, E., 8, *21*
Middlemiss, K.M., 141, *160*
Miglyachenko, A.F., 1016, *1022*
Mikawa, Y. see Jacobsen, R.J., 581, *609*, 622, 633, 638, *653*, 770, *789*
Mikhailov, I.D., 333, *356*, 581, *609*, 622, *653*
Mikhailov, V.K. see Strukov, B.A., 1154, *1166*
Mikhailova, M.V. see Meerson, L.A., 1346, *1361*
Miki, H. see Makita, Y., 1136, 1137, 1158, *1163*
Mikke, K., 929, 930, 931, *935*
Mikolaj, P.G., 368, *388*
Mikuli, E., 902, 930, *935*
Mikulskis, P. see Kimtys, I., 827, *829*
Milaković-Vučelič, V. see Šušić, V., 1347, *1363*
Millen, D.J., 19, *22*, 224, *243*, 330, *356*, 477, *525*, 636, *653*, 660, *681*
Millen, D.J. see Arnold, J., 224, *242*, 330, *353*, 634, 635, 640, *652*
Millen, D.J. see Bertie, J.E., 177, 181, 203, *214*, 224, *242*, 324, 330, *354*, 581, *607*, 635, *652*
Millen, D.J. see Hussein, M.A., 606, *611*
Miller, D.C. see Delsemme, A.H., 1420, *1421*
Miller, P.J., 599, *609*
Miller, S.L., 1420, *1423*
Miller, S.R., 908, *935*, 991, *1022*
Millero, F.J., 1412, *1423*
Millikan, R.C., 1099, *1105*
Milliken, T.H. see Cornelius, E.B., 1333, *1355*
Mills, G.A. see Cornelius, E.B., 1333, *1355*
Mills, R., 912, *935*, 1032, *1060*
Mills, R. see Morariu, V.V., 1270, 1305, 1313, 1316, *1361*
Mines, G.W. see Millen, D.J., 19, *22*
Mishchenko, A.V. see Mavrin, B.N., 1155, *1166*
Mishchenko, A.V. see Strukov, B.A., 1157, *1166*
Mitchell, E. see Arnett, E.M., 1201, 1205, *1223*
Mitchell, P., 761, *764*
Mitchell, S.A. see Armistead, C.G., 1267, 1272, *1353*
Mitsky, J. see Joris, L., 818, *828*
Mitsui, T., 340, *356*, 1142, 1153, 1154, 1158, *1163*, *1166*

Miura, M. see Iwaki, T., 1343, *1358*
Miyahara, K. see Takezawa, N., 1343, *1363*
Miyake, Y. see Matsuo, S., 1252, *1258*
Miyazawa, T., 1088, 1099, *1105*
Moccia, M. see Arrighini, G.P., 35, *155*
Moeller, K. see Kunath, D., 1347, *1359*
Mognaschi, E.R. see Bonera, G., 1160, *1165*
Moll, F. see Lippert, E., 9, *21*
Moll Jr., W.F. see Brindley, G.W., 1349, *1354*
Mollendal, H., 1203, *1224*
Möller, K.D., 770, 785, 786, *789*
Momany, F.A., 124, *160*
Momin, S.N. see Sikka, S.K., 441, *455*
Montagna, A.A. see Weller, S.W., 1332, *1364*
Montano, P.A., 1147, *1163*
Mookerji, A., 749, *764*
Moor, S. see Crestfield, A.M., 761, *763*
Moore, F.H., 444, *455*
Moore, F.H. see Chamberlain, J.S., 1431, *1467*
Moore, F.H. see Power, L.F., 441, *455*
Moore, L.F. see Ibbitson, D.A., 541, *561*, 940, *1020*, *1021*, 1233, *1258*
Moore, L.S., 92, *160*
Moore, T.S., 1201, *1224*
Mootz, D., 482, 483, 492, 495, 508, 509, *525*
Morachevskii, V.G., 1350, *1361*
Morariu, V.V., 1270, 1305, 1313, 1316, *1361*
Moravec, J. see Horák, M., 573, *609*
Morcom, K.W., 1204, *1224*
Moreno, F. see Munuera, G., 1344, *1361*
Morgan, A.M. see Sewell, P.A., 1319, 1352, *1362*
Morgan, J., 368, *388*, 1372, *1423*
Morgan, L.O. see Bloembergen, N., 1053, *1058*
Morimoto, T., 1304, 1335, 1343, 1345, 1346, *1361*
Morimoto, T. see Nagao, M., 1345, *1361*
Morishima, I., 820, *829*
Morita, H., 127, *160*, *356*, 570, *609*
Morokuma, K., *20*, 27, 48, 49, 57, 60, 62, 73, 75, 111, 123, 127, 131, 135, 144, 145, 154, *160*, 206, *215*
Morokuma, K. see Iwata, S., 144, 145, *158*
Morosin, B., 1117, 1125, *1163*
Morosin, B. see Kaplan, S.F., 1137, *1162*
Morris, B., 985, 999, 1000, *1022*
Morrow, B.A., 1267, *1361*
Morrow, J., 533, *561*
Morrow, J.C. see Baldwin, M.C., 1313, 1336, *1353*
Morse, P.M., 222, *243*
Morse, P.M. see Rosen, N., 222, *243*
Morterra, C. see Borello, E., 1269, 1318, 1319, 1320, 1321, 1322, 1324, 1325, 1327, *1354*
Mortland, M.M., 1348, *1361*
Mortland, M.M. see Tahoun, S.A., 1349, *1363*
Moser, C. see Grimaldi, F., 37, *157*
Moskowitz, J.W., 517, *525*
Moskowitz, J.W. see Hankins, D., 62, 73, 75, 133, 134, *158*, 207, *215*, 369, 379, *388*, 1320, *1357*
Moskowitz, J.W. see Losonczy, M., 93, 96, 100, 102, *159*
Moskowitz, J.W. see Neumann, D., 35, *161*
Moskvichev, B.V. see Frolov, V.I., 1012, *1019*
Motegi, H. see Niimura, N., 1147, *1163*
Motzfeld, T. see Almenningen, A., 323, 324, 333, *353*, 771, *788*
Moulton, W.G., 595, *609*
Moulton, W.G. see Kromhout, R.A., 778, *789*
Mounier, S., 983, *1022*
Mueller, M.H. see Dalley, N.K., 441, *453*
Mueller, M.H. see Taylor, J.C., 441, *455*
Muenter, J.S. see Dyke, T.R., 79, 115, *157*
Mühlinghaus, J. see Zundel, G., 695, 697, 698, 708, 709, 761, *766*
Müller, D. see Brosowski, G., 1160, *1165*
Müller, F.H. see Ebert, G., 1006, *1018*, 1315, 1316, *1356*
Müller, F.H. see Matron, W., 1002, *1022*, 1347, *1360*
Müller, H. see Wolff, H., 19, *22*, 214, *216*, 606, *611*
Müller, H.D. see Jeziorowski, H., 1267, 1304, 1320, 1322, 1323, 1324, 1325, 1327, 1338, 1339, 1340, *1358*
Muller, N., 822, *829*
Mullhaupt, J.T., 477, *525*, 660, *681*
Mulliken, R.S., 73, 85, 89, *160*, 1286, *1361*
Mundheim, O. see Gramstad, T., 818, *828*
Munuera, G., 1343, 1344, *1361*
Murat, M. see Escoubes, M., 1350, *1356*
Murday, J.S., 1050, 1052, *1060*
Murday, J.S. see Resing, H.A., 1347, 1348, *1362*
Murdmaa, K., 1346, *1361*
Muroya, M., 1009, *1022*
Muroya, M. see Kondo, S., 1315, 1336, 1351, *1359*

Murphy, E.J., 989, 991, *1022*
Murphy, W.F., 626, *653*
Murr, A. see Zundel, G., 748, 749, *766*
Murrell, J.N., 37, 39, 40, 49, 56, 112, *160*, 946, *1022*
Murrell, J.N. see Condon, E.U., 101, *156*
Murrell, J.N., see Van Duijneveldt, F.B., 39, 45, 49, 60, *163*
Murrenhoff, A. see Groth, W., 1246, 1254, *1258*
Murto, J., 1096, *1105*
Murto, J. see Barnes, A.J., 1096, *1104*
Murthy, A.S.N., *20*, 27, 70, 113, 125, 127, *160*, *356*, 567, 603, *609*, 793, 804, *829*
Murty, T.S.S.R. see Arnett, E.M., 1201, 1205, *1223*
Müser, H.E., 1115, *1163*
Musher, J.I., 39, 49, *160*
Musso, H. see Hafner, K., 604, *608*
Muszik, J.A., *21*
Muttik, G.G. see Dzhigit, O.M., 1274, *1356*
Mykolajewycz, R. see Von Hippel, A., 967, 968, 983, *1024*
Mylov, V.P. see Leonidova, G.G., 1158, *1166*
Myrat, C.D., see Lee, C.S., 92, *159*

Naccache, C. see Kodratoff, Y., 1346, *1359*
Naegerl, H. see Gieseke, W., 758, *763*
Nagakura, S. see Matsuzaki, A., 14, *21*
Nagakura, S. see Morita, H., 127, *160*, *356*, 570, *609*
Nagamiya, T., 863, 864, *886*, 1154, 1155, *1166*
Nagao, M., 1345, *1361*
Nagao, M. see Morimoto, T., 1304, 1335, 1343, 1345, 1346, *1361*
Nagata, S. see Saitô, H., 823, *829*
Nagel, H. see Martens, R., 1343, *1360*
Nägele, W. see Lippert, E., 9, *21*
Nägerl, H. see Gieseke, W., 995, *1019*
Nagle, J.F., 362, 365, 366, 368, 383, *388*, 969, *1022*, 1137, 1138, 1148, *1163*, 1448, *1469*
Nahringbauer, I., 417, *455*
Nair, N.K., 1304, 1315, *1361*
Nakahara, A., 478, 484, *525*, 667, *681*
Nakahara, Y., 929, *935*
Nakamoto, K., 410, *455*, 659, 661, *681*
Nakamoto, K. see Kishida, S., 568, *609*
Nakamura, E. see Mitsui, T., 1153, 1154, *1166*
Nakamura, E. see Nakano, J., 1118, *1163*
Nakamura, N. see Chihara, H., 856, 857, *886*, 1160, *1165*
Nakamura, N. see Yamamoto, T., 1150, *1165*
Nakamura, T., 968, 983, *1022*
Nakamura, T. see Sugawara, F., 1187, *1195*
Nakano, J., 1118, *1163*
Nakashima, M. see Huber, J.R., 16, *21*
Nakashima, T.T., 825, *830*
Narajanamurti, V., 251, *292*, 723, *764*
Narang, H. see Stell, G., 378, 381, *388*
Narasimhan, P.T., 1208, *1224*
Narcisi, R.S., 475, *525*
Narten, A.H., 368, 381, *388*, 554, *561*, 724, *764*, 1372, 1381, 1385, 1386, 1393, *1423*
Narten, A.H. see Triolo, R., 724, *765*
Nash, L.K. see Johnson, E.W., 533, *561*
Nasielski, J., 949, *1022*
Naskret-Barciszewska, M.Z. see Dega-Szafran, Z., 602, *607*
Nations, R. see Pope, N.K., 912, 929, *935*
Natkaniec, I. see Bajorek, A., 902, 929, 930, *933*
Natkaniec, I. see Janik, J.A., 902, 930, *934*
Natkaniec, I. see Janik, J.M., 915, 917, 930, *934*
Natterstad, J.J. see Maciel, G.E., 823, *829*
Naumann, A.W. see Safford, G.J., 912, 929, *935*
Navarro, Q.O. see Chidambaram, R., 440, *453*
Nawrocik, W. see Janik, J.M., 915, 917, 930, *934*
Neckel, A., 121, *160*
Nedetzka, T., 978, 979, 996, *1022*
Nedetzka, T. see Brausse, G., 1013, *1017*
Nedetzka, T. see Reichle, M., 1013, *1023*
Nee, T.W., 1054, *1060*
Needham, T.E. see Drago, R.S., 1215, 1216, 1222, *1223*
Neel, J. see Cung, M.T., 124, *156*
Neff, B.L. see Van Hecke, P., 877, *887*
Negran, T.J., 1159, *1166*
Nehnes, R., 1190, *1195*
Neikes, T. see Luck, W.A.P., 549, 550, *561*, 1383, 1415, *1422*
Neikes, T. see Mann, B., 1088, *1105*
Neimark, I.E. see Eremenko, A.M., 1347, *1356*
Nekrasova, E.G., 1006, *1022*, 1315, 1316, *1361*
Nelmes, R.J., 1118, 1134, 1137, 1146, *1163*
Nelson, M.J. see Barakat, T.M., 1201, *1223*

Nelson, S.M., 1003, *1023*, 1349, *1361*
Nelson, S.M. see Barakat, T.M., 1201, *1223*
Némethy, G., 368, 370, 371, 372, 374, *388*, 1371, 1373, 1377, 1393, *1423*
Nesbet, R.K., 36, *160*
Nettleton, R.E., 1115, *1163*, 1183, *1195*
Neugebauer, F.A. see Otting, W., 604, *610*
Neuimin, G.G. see Kurbatov, L.N., 1290, *1359*
Neumann, D., 35, *161*
Neurath, H. see Walsh, K.A., 761, *766*
Newbold, G. see Atherton, K., 1345, *1353*
Newman, R., 583, *609*
Newnham, C.A. see Lee, J.A., 1345, *1360*
Newton, M.D., 121, 139, 140, *161*, 190, *215*, 519, 520, 521, *525*, 688, 753, *764*
Ng, S. see Pang, T.S., 825, *830*
Nibler, J.W., 186, 187, 190, *215*, 595, 602, *609*
Nief, G. see Merlivat, L., 1252, *1258*
Nielsen, H.H., 619, *653*
Nielsen, H.H. see Garing, J.S., 1254, *1258*
Niimura, N., 1147, *1163*
Nilsson, Å., 449, *455*
Nishibata, K., 1445, *1469*
Nishimura, T. see Ida, M., 1009, *1021*
Nixon, E.R. see King, C.M., 1074, 1084, 1085, 1095, *1105*
Nixon, E.R. see Tursi, A.J., 549, *562*, 1086, 1087, 1088, *1105*
Noack, F. see Hausser, R., 1053, *1059*
Noble, P.N., 121, *161*, 185, *215*, 571, 587, 602, *609*
Nogales, A., see Tapia, O., 132, *163*
Noggle, J.H. see Gillen, K.T., 1038, 1039, *1059*
Noll, G., 968, 982, *1023*
Noller, H., 699, 710, 718, 738, *765*, 1285, 1301, *1361*
Noller, H. see Zundel, G., 687, 697, *766*, 1011, *1025*
Nordman, C.E., 425, *455*, 478, 482, 495, *525*, 660, 661, 663, *681*, 904, *935*
Norman, I., 1067, *1105*
Nouwen, R., 948, *1023*
Novak, A., 352, *357*, 571, 574, 575, 587, 595, 599, 601, *610*, 859, *886*
Novak, A. see Bellocq, A.M., 302, *354*
Novak, A. see Blinc, R., 574, *607*
Novak, A. see Clague, D., 771, 778, 779, *789*
Novak, A. see De Villepin, J., 599, 600, 601, *610*, *611*
Novak, A. see Foglizzo, R., 582, 583, 606, *608*, *611*, 633, 638, *653*
Novak, A. see Hadži, D., 569, 587, 594, 599, 601, 605, *608*
Novak, A. see Haurie, M., 302, 303, 334, *355*, 583, *608*, 639, *653*, 1100, *1105*
Novak, A. see Lautie, A., 336, *356*, 859, *886*
Novak, A. see Lautie, E.R., 576, *609*
Novak, A. see Lucazeau, G., 573, 582, 583, *609*
Novak, A. see Perchard, Cl., 302, *356*
Novak, R.W. see De Tar de Los, F., 703, 741, *763*
Novik, V.F. see Deryagin, B.V., 1005, *1018*
Novotny, D.B. see Garland, C.W., 1120, *1162*
Nozari, M.S., 1203, 1206, 1207, 1211, 1212, *1224*
Nozari, M.S. see Drago, R.S., 1206, *1223*
Nukada, K. see Saitô, H., 822, 823, *829*
Nygaard, L. see Bak, B., 858, *885*

Oakes, J., 1057, *1060*
Oblad, A.G. see Cornelius, E.B., 1333, *1355*
Obradović, M., 569, 594, *610*
Obradovic, M. see Hadži, D., 568, *608*
O'Bryan, N. see Drago, R.S., 1211, 1212, 1213, *1223*
Occhiena, G. see Zecchina, A., 1276, 1293, *1364*
O'Connell, J.P., 92, *161*
O'Connell, J.P. see Lee, C.S., 92, *159*
O'Connell, J.P. see Moore, L.S., 92, *160*
O'Connell, J.P. see Rigby, M., 92, *162*
Ödberg, L., 814, *829*
Odinkov, S.E., 585, *610*
Odishaw, H. see Condon, E.U., 101, *156*
Oettel, R.E. see Bjorkstam, J.L., 1119, 1127, *1161*
Ohma, Y. see Iizuka, T., 1343, *1358*
Ohya, S. see Ishibashi, Y., 1148, 1155, *1162*, *1166*
Oja, T., 1147, *1163*
Oja, T. see Marino, R.A., 859, *886*
Okada, K., 1119, 1120, 1148, 1154, *1163*
Okaya, Y. see Hoshino, S., 843, 860, *886*
Okaya, Y. see Pepinsky, R., 1151, *1163*
O'Keeffe, 989, 990, *1023*
O'Konsky, C.T., 859, *886*
O'Konsky, C.T. see Lehrer, S.S., 858, *886*
Olejnik, S., 1349, *1361*
Olenko, Y.V. see Dushchenko, V.P., 1016, *1018*
Oliver, J.F. see Anderson, P.J., 1343, *1353*
Olivier, J.P. see Ross, S., 1329, *1362*

Olovsson, I., 414, 432, *455*, 482, 487, 492, 500, *525*, 668, *681*, 1255, *1258*
Olovsson, I. see Almlöf, J., 482, 493, 511, 512, *523*
Olovsson, I. see Delaplane, R.G., 482, 483, 492, 495, 500, *523*, 663, *681*
Olovsson, I. see Kjällman, T., 422, *454*, 483, 492, 500, *524*
Olovsson, I. see Liminga, R., 414, *454*
Olovsson, I. see Lundgren, J.O., 482, 483, 486, 488, 489, 492, 493, 495, 496, 499, 500, 508, 509, 512, *524*, *525*, 676, 677, 679, *681*
Olovsson, I. see Nilsson, Å., 449, *455*
Olovsson, I. see Taesler, I., 482, 483, 486, 493, 495, 508, 519, *525*, 663, 676, *682*
Olovsson, I. see Tellgren, R., 422, 423, 439, *455*
Olovsson, I. see Thomas, J.O., 422, 426, *455*, *456*
Olympia, P.L., 840, 841, *887*
O'Neill, S.V. see Yarkony, D.R., 61, 72, 129, 132, *163*
Onsager, L., 340, 341, 342, *356*, 365, 370, 372, 383, *388*, 966, 979, *1023*, 1369, *1423*
Ooi, S., 478, 484, *525*
Oosterhoff, L.J., 9, *21*
Opauszki, I. see Kiss, I., 1246, 1254, *1258*
Oppermann, G. see Boddenberg, B., 1294, 1295, *1354*
Oranskaya, O.M. see Tretyaķov, N.E., 1344, *1363*
Orazmuradov, A.G. see Tarasevich, Yu.I., 1348, *1363*
O'Reilly, D.E., 497, 506, *525*, 660, 673, *681*, *682*, 854, 872, *887*, 905, *935*, 1038, 1039, 1046, *1060*, 1145, 1147, 1150, *1163*, 1268, 1327, *1361*
O'Reilly, D.E. see Blinc, R., 1132, *1161*
O'Reilly, D.E. see El Saffar, Z.M., 1143, *1162*
O'Reilly, D.E. see Genin, D.J., 1122, 1146, *1162*
Orel, B. see Hadži, D., 568, 569, 571, 574, 587, 594, 599, 600, 603, 605, *608*
O'Reilly, D.E. see Hall, W.K., 1268, 1269, *1357*
O'Reilly, D.E. see Kadaba, P.K., 1148, *1162*
Orel, B., 584, 586, 599, *610*
Orentlicher, N., 372, *388*
Orville-Thomas, W.J. see Ratajczak, H., 76, *162*, *356*, 943, *1023*
Osaka, T. see Gesi, K., 1159, *1165*
Osawa, E., 573, *610*
Osawa, E. see Yoshida, Z., 1201, *1224*
Osborn, A.R. see Bellamy, L.J., 1317, *1353*
Osborn, K. see Yip, S., 929, *936*
Osipov, O.A. see Rustanova, S.N., 953, *1023*
Osredkar, R. see Blinc, R., 851, 852, 853, 855, 859, 860, 861, 862, 863, *885*, 1144, 1154, *1161*, *1165*
Oster, G., 1233, *1258*
Osterheld, R.K. see Negran, T.J., 1159, *1166*
Oswald, H.R. see Mattmann, G., 1345, *1360*
Oszust, J. see Hawranek, J.P., 724, *763*, 948, 957, *1020*
Othen, D.A. see Bertie, J.E., 1459, *1467*
Otnes, K. see Larsson, K.E., 912, 929, *935*
Otnes, K. see Palevsky, H., 930, 932, *935*
Otting, W., 604, *610*
Ovcharenko, F.D. see Brekhunets, A.G., 1347, *1354*
Ovcharenko, F.D. see Radul, N.M., 1348, *1362*
Ovcharenko, F.O. see Tarasevich, Yu.I., 1348, *1363*
Overend, J., 643, *653*
Overend, J. see Parizeau, M.A., 619, *653*
Owen, A.J. see Bellamy, L.J., 661, *681*
Ozawa, K. see Gesi, K., 1159, 1160, *1165*
Ozerov, R.P. see Bajorek, A., 902, 929, 930, *933*
Ozerov, R.P. see Ranner, N.B., 902, *935*

Pace, R.J. see Bellamy, L.J., 541, 552, *560*, 1095, *1104*, 1233, 1257, 1288, *1353*
Padmanabhan, V.M., 1151, *1163*
Padmanabhan, V.M. see Cox, G.W., 449, *453*
Page, P.I. see Franks, F., 929, *934*
Page Jr., T.F., 1317, *1361*
Paik, Y. see Fukuta, N., 1350, *1357*
Pajak, Z. see Jurga, J., 1058, *1061*
Pajak, Z. see Szczesniak, E., 1058, *1061*
Pak, K.N., 1157, *1166*
Palevsky, H., 930, 932, *935*
Palevsky, H. see Brajović, V., 931, 932, *934*
Palevsky, H. see Hughes, D.J., 912, 929, *934*
Palevsky, H. see Janik, J.A., 905, 930, 931, *934*
Palinkas, G. see Kálmán, E., 1386, *1422*
Pang, T.S., 825, *830*
Paolillo, L., 822, 823, *829*
Pap, A., 1346, *1361*
Papakostidis, G., 698, 703, 720, 760, *765*
Papon, P. see Theveneau, H., 1160, *1167*
Parasol, M. see Rundle, R.E., 605, *610*

Paren, J.G., 983, *1023*
Parfitt, G.D., 1344, *1361*
Parfitt, G.D. see Day, R.E., 1343, 1344, *1355*
Parfitt, G.D. see Jackson, P., 1342, 1343, 1344, *1358*
Parfitt, G.D. see Lewis, K.E., 1343, 1344, *1360*
Parizeau, M.A., 619, *653*
Park, Y.J., 446, *455*
Parke, W. see Jehle, H., 254, *292*, 729, *764*
Parker, R. see Blinc, R., 862, 863, *885*, 1154, *1165*
Parker, R.S. see Schmidt, V.H., 1152, *1164*
Parks, G.A. see Anderson Jr., J.H., 1313, 1314, *1353*
Parliński, K., 906, *935*
Parliński, K. see Bajorek, A., 902, 906, 929, 930, 931, 932, *933*
Parliński, K. see Janik, J.A., 902, 930, *934*
Parliński, K. see Janik, J.M., 930, 931, *934*
Parmigiani, A., 813, *829*
Parry, E.P., 1346, *1361*
Parthasarathy, R. see Koetzle, T.F., 451, *454*
Patey, G.N., 381, 386, *388*, *389*
Patten, F.W. see Fuller, R.G., 989, 992, *1019*
Patterson, L.K., 804, *829*
Paul, G.L., 1123, 1131, *1163*
Paul, G. see Buyers, W., 1187, *1195*
Paul, G.L. see Freeman, H.C., 451, *453*
Pauling, L., 362, 363, *388*, 401, *455*, 559, *561*, 1178, *1195*, 1255, *1258*, 1415, *1423*
Pauncz, R. see Srebrenik, S., 83, *163*
Pausak, V. see Šušić, V., 1347, *1363*
Pavia, A.C., 203, *215*, 668, 669, 670, 671, *682*, 747, *765*
Pavlov, V.V. see Tertykh, V.A., 1272, *1363*
Pawelka, Z., 952, *1023*
Pawelka, Z. see Sobczyk, L., 948, 949, *1024*
Pawlak, Z., 601, *610*, 695, 698, 701, 703, 706, 707, 720, 735, *765*
Pawley, G.S., 400, *455*
Payzant, J.D., 121, *161*
Payzant, J.D., see Cunningham, A.J., 121, 139, 140, *156*, 980, *1018*
Peacall, D.B. see Kipling, J.J., 1330, *1358*
Peacock, J. see Day, R.E., 1343, 1344, *1355*
Pearce, D.R. see Hendriksen, B.A., 1333, 1334, *1358*
Pearson, R.G., 1214, 1216, *1224*
Pease, R.S., see Bacon, G.E., 336, *353*, 584, 597, *607*, 920, 923, 924, *933*, 1118, 1121, 1124, *1161*, 1179, 1180, 1190, *1195*
Pecul, K., 111, *161*, 212, *215*, 1098, *1105*
Pedersen, B., 404, 405, 406, 434, 435, *455*, 1430, *1469*
Pedersen, L., see Morokuma, K., 62, *160*, 206, *215*
Pedone, C. see Benedetti, E., 462, *468*
Peercy, P.S., 1125, 1156, 1157, *1163*, *1166*
Peglar, R.J., 1272, *1361*
Peierls, R., 554, *562*
Pelah, I., 908, 912, 929, 932, *935*
Pelligrĩni, A., 778, *789*
Penel, C. see Haser, R., 484, 524
Penfold, B.R. see Cochran, W., 467, *468*
Penneman, R.A., 484, 492, 500, *525*
Penneman, R.A. see Ryan, R.R., 483, *525*
Pennequin, M. see Fripiat, J.J., 1348, *1357*
Pentin, Iu.A. see Kanaskova, Iu.D., 596, *609*
Pepinsky, R., 340, *356*, 1146, 1151, *1163*
Pepinsky, R. see Frazer, B.C., 843, *886*, 1179, *1195*
Pepinsky, R. see Hoshino, S., 843, 860, *886*
Perchard, C., 302, *356*, 606, *611*, 622, *654*
Perchard, C., see Bellocq, A.M., 302, *354*
Perchard, J.P. see Perchard, C., 606, *611*, 622, *654*
Percus, J.K. see Lebowitz, J.L., 378, *388*
Perepelkova, T.I. see Goldstein, I.P., 948, 955, *1019*
Perevertaev, V.S. see Metsik, M.S., 1348, *1361*
Peri, J.B., 1267, 1272, 1297, 1330, 1331, 1332, 1333, 1334, 1342, 1346, *1362*
Pernoll, I., 689, 696, 733, *765*
Péron, J.-J., 622, *654*
Péron, J.-J. see Bernard-Houplain, M.-C., 621, *652*
Péron, J.-J. see Bourdéron, C., 622, 625, 626, 627, 647, *652*
Perotti, A. see Parmigiani, A., 813, *829*
Perram, J.W., 368, 371, 372, 373, 375, 378, 386, *388*
Perram, J.W., see Kohler, F., 386, *389*
Perram, J.W., see Levine, S., 371, 372, *388*
Perram, J.W. see Smith, E.R., 385, 386, *389*
Perrin, D.D., 710, *765*
Perrino, C.T. see O'Keeffe, 989, 990, *1023*
Persoons, A. see De Maeyer, L., 995, *1018*
Person, W.B., 815, *829*
Person, W.B. see Friedrich, H.B., 945, *1019*
Perst, H., 18, *22*
Perutz, R.N., 1096, *1105*
Peshikov, E.V., 1158, *1166*
Petch, H.E. see Kydon, D.W., 1145, *1163*
Petch, H.E. see Vanderkooy, J., 989, 993, *1024*, 1151, *1165*

Petch, H.E. see Watton, A., 1149, 1160, *1165*
Peterson, E.M. see Blinc, R., 1132, *1161*
Peterson, E.M. see El Saffar, Z.M., 1143, *1162*
Peterson, E.M. see O'Reilly, D.E., 497, 506, *525*, 660, 673, *681*, *682*, 872, *887*, 905, *935*, 1038, 1039, 1046, *1060*, 1150, *1163*, 1268, 1327, *1361*
Peterson, E.R., 1151, *1163*
Peterson, S.W., 1434, *1469*
Peterson, S.W. see Levy, H.A., 1449, 1450, *1469*
Peterson, S.W. see Williams, J.M., 427, 449, *456*, 483, 484, 493, 500, 501, 502, 508, 509, *526*, 657, 673, 674, 675, *682*, 692, 747, *766*, 872, *887*
Petersson, J. see Brosowski, G., 1160, *1165*
Petersson, J. see Müser, H.E., 1115, *1163*
Pethica, B.A. see Clifford, J., 1056, *1058*
Pethica, B.A. see Tyler, A.J., 1309, *1363*
Peticolas, W.L. see Brown, K.G., 787, *789*
Peticolas, W.L. see Fanconi, B., 787, *789*
Petrongolo, C., 82, *161*
Petrongolo, C., see Bonaccorsi, R., 48, 60, 62, 82, 84, 85, *155*
Petrov, I. see Hadži, D., 594, 595, *608*
Petz, J.I. see Kruh, R.F., 1255, *1258*
Petzelt, J., 1159, *1166*
Peyerimhoff, S.D. see Buenker, R.J., 838, *886*
Peyerimhoff, S.D., see Merlet, P., 121, 150, *160*, 247, *292*
Pfeifer, H., 1053, *1060*, 1295, *1362*
Pfeifer, H. see Beckert, D., 1042, 1043, *1058*
Pfeifer, H. see Deininger, D., 1347, *1355*
Pfeifer, H. see Freude, D., 1295, *1357*
Pfeifer, H. see Geschke, D., 1319, 1339, *1357*
Pfeifer, H. see Gutsze, A., 1347, *1357*
Pfeiffer, H., 277, *292*, 754, *765*
Pfeiffer, H., see Janoschek, R., 122, *158*, 175, 178, 179, 180, 182, 191, *215*, 221, 225, *243*, 247, 251, 253, 255, 260, *292*, *355*, 688, 722, 723, 724, 728, 731, *764*
Pfeiffer, P., 5, *22*
Pham Dinh Ty, see Fink, P., 1273, *1356*
Philippot, E., 442, *455*
Phillips, J., 621, *653*
Pichat, P., 1346, *1362*
Pichat, P. see Primet, P., 1344, *1362*
Pichler, E. see Sköld, K., 912, 929, *936*
Pick, M., 979, *1023*
Pickett, H.M., 604, *610*
Pickett, J.H., 1293, *1362*
Piekara, A., 959, 960, 962, *1023*
Piela, L., 39, 111, 119, 121, *161*
Piela, L., see Jeziorski, B., 39, *158*
Piermarini, G. see Weir, C., 1450, *1470*
Pierrot, M. see Haser, R., 484, *524*
Piesberger, U. see Clusius, K., 1246, *1257*
Pietronero, L. see Bassani, F., 142, *155*
Pigenet, C., 941, *1023*
Pigon, K., 993, 994, *1023*
Pimentel, G.C., 7, *20*, 112, 115, 116, 117, *161*, 299, 305, 319, 323, 324, *356*, 401, 404, *455*, 545, 552, *562*, 567, 570, 576, 602, *610*, 658, 659, 661, *682*, 1090, 1091, *1105*, 1201, 1220, *1224*
Pimentel, G.C. see Ault, B.S., 603, *607*, 665, *681*
Pimentel, G.C. see Becker, E.D., 1083, *1104*
Pimentel, G.C. see Bondybey, V., 602, *607*
Pimentel, G.C. see Huggins, C.M., 595, *609*, 795, *828*
Pimentel, G.C., see Nibler, J.W., 186, 187, 190, *215*, 595, 602, *609*
Pimentel, G.C. see Noble, P.N., 571, 587, 602, *609*
Pimentel, G.C. see Van Thiel, M., 79, *163*, 545, 546, 548, 551, *562*, 580, 595, *610*, 637, *654*, 1070, 1086, 1088, 1090, 1093, *1105*, 1335, *1364*
Pimentel, G.C. see Whittle, E., 1067, *1105*
Pimentel, G.C. see Zimmermann, 1458, *1470*
Pines, A., 882, *887*
Pines, A. see Waugh, J.S., 880, 881, *887*
Pings, C.S., 1372, *1423*
Pings, C.J., see Mikolaj, P.G., 368, *388*
Pintar, M. see Blinc, R., 862, 885, *885*
Pintar, M. see Gosar, P., 981, *1020*
Pintar, M. see Kydon, D.W., 1145, *1163*
Pintar, M.M. see Watton, A., 1149, 1160, *1165*
Piontkovskaya, M.A. see Brekhunets, A.G., 1347, *1354*
Piontkovskaya, M.A. see Eremenko, A.M., 1347, *1356*
Piontkovskaya, M.A. see Matyash, I.V., 1347, *1360*
Pirš, J. see Blinc, R., 866, *886*, 1121, 1141, 1154, *1161*, *1165*
Pirš, J. see Žumer, S., 1160, *1167*
Pitayevskii, L.P., see Dzyaloshinskii, I.Y., 254, *292*
Pitner, T.P., 823, *829*
Pitzer, K.S. see Davis, J.C., 797, *828*
Pitzer, K.S. see Millikan, R.C., 1099, *1105*

Pitzer, K.S. see Miyazawa, T., 1099, *1105*
Pitzer, R.M., see Bowers, M.T., 71, *156*
Pitzer, R.M., see Dunning, Th. H., 35, *157*
Planckaert, A.A., 646, *653*
Plesser, T., 135, *161*, 894, 921, 932, *935*
Plesser, T., see Grimm, H., 224, *242*, 921, 932, *934*, 1179, 1181, 1190, *1195*
Plint, C.A., 1255, *1258*
Pliva, J. see Horák, M., 573, *609*
Plooster, M.N., 1311, *1362*
Ploss, G. see Hafner, K., 604, *608*
Plumb, R.C., 639, *653*
Plyler, E.K. see Benedict, W.S., 436, 437, 439, *453*, 1254, *1257*
Poberrzhets, I.-I. see Dushchenko, V.P., 1016, *1018*
Podznyakov, D.V. see Tretyakov, N.E., 1344, *1363*
Pohl, R.O., see Narajanamurti, V., 251, *292*, 723, *764*
Pohle, W. see Fink, P., 1273, 1276, 1277, *1356*
Pokotilovsky, Y.N. see Bajorek, A., 902, 929, 930, *933*
Poland, D., 117, *161*
Polandov, I.N. see Krasnikova, A.Ya., 1149, *1163*
Poles, T.C. see Butler, J.D., 1348, *1354*
Polikarova, J. see Bliznakov, G., 1298, *1354*
Pollak-Stachura, M., 871, *887*
Pollak-Stachura, M. see Hennel, J.W., 660, *681*, 872, *886*
Pollock, J.M., 988, 989, 991, 993, *1023*
Poncelet, G. see Fripiat, J.J., 1305, 1313, 1314, 1315, 1317, 1348, *1357*
Pongs, O. see Rüterjans, H., 761, *765*
Ponomarev, V.I. see Gorbatyi, L.V., 1138, *1162*
Pope, N.K., 912, 929, *935*
Pope, N.K. see Sakamoto, M., 929, *936*
Popkie, H., 59, 60, 62, 77, 80, 84, 111, 129, *161*, 369, 379, 381, *388*
Popkie, H., see Clementi, E., 32, 58, *156*
Popkie, H., see Kistenmacher, H., 84, 121, 122, *159*, 369, 379, 381, *388*
Pople, J.A., 37, 40, 84, 125, *161*, 362, 368, 369, *388*, 842, *887*, 1029, 1034, *1060*, 1373, 1399, *1423*
Pople, J.A., see Del Bene, J.E., 61, 62, 71, 72, 92, 111, 134, 135, 137, 138, *157*, 369, 370, 372, 379, *387*, 1400, *1421*
Pople, J.A. see Ditchfield, R., 61, *157*
Pople, J.A., see Hehre, W.J., 61, *158*
Pople, P.N. see Bondybey, V., 602, *607*
Popov, A.G. see Morachevskii, V.G., 1350, *1361*
Porro, T.J. see Lord, R.C., 531, *561*
Port, G.N.J., 82, 85, *161*
Port, G.N.J., see Pullman, A., 85, *161*
Porter, G. see Norman, I., 1067, *1105*
Posener, D.W., 1051, *1060*
Posner, A.M. see Olejnik, S., 1349, *1361*
Post, B. see Ladell, J., 117, *159*
Post, B. see La Placa, S.J., 1433, 1434, *1468*
Potanin, A.N. see Morachevskii, V.G., 1350, *1361*
Potier, J. see Chemouni, E., 668, 670, *681*, 747, *763*
Potier, J. see Fournier, M., 660, 661, 663, *681*
Potier, J. see Rosière, J., 601, *610*, 669, 670, 671, 676, *682*
Pottel, R., 1032, *1060*
Poulet, H. see Coignac, J., 1187, *1195*
Poulsen, F.R. see Lehmann, M.S., 443, *454*
Powell, D.L. see West, R., 1201, *1224*
Powell, R.W., 1434, *1469*
Power, L.F., 441, *455*
Power, L.F. see Moore, F.H., 444, *455*
Powles, J.G., 1039, 1049, *1060*, 1465, *1469*
Powles, J.G. see Smith, D.W.G., 1049, *1060*
Praefke, K., 18, *22*
Prakash, A. see Hamilton, W.C., 970, *1020*, 1435, 1437, 1438, 1439, 1440, 1445, 1448, 1458, 1459, *1468*
Prakash, A. see Kamb, B., 970, *1021*, *1445*, 1451, 1452, *1468*
Prakash, A. see La Placa, S.J., 437, 440, *454*, 1435, 1441, 1442, 1443, 1444, 1445, 1460, *1468*
Prask, H.J., 902, 912, 929, 930, *935*, 1461, 1462, 1463, *1469*
Prask, H. see Boutin, H., 931, *934*, 1306, 1333, 1344, *1354*
Prask, H. see Trevino, S., 909, 932, *936*
Prausnitz, J.M., 1227, 1228, 1233, *1258*
Prausnitz, J.M., see Lee, C.S., 92, *159*
Prausnitz, J.M., see O'Connell, J.P., 92, *161*
Prausnitz, J.M. see Renon, H., 1233, *1259*
Prausnitz, J.M., see Rigby, M., 92, *162*
Pravdić, V. see Maričić, S., 989, 993, *1022*
Pravdić, V. see McCafferty, E., 1008, *1022*, 1345, *1361*
Prelesnik, A. see Blinc, R., 851, 852, 853, 859, 860, 861, *885*, 1144, *1161*
Prettre, M. see Kodratoff, Y. 1346, *1359*

Preuss, H., see Janoschek, R., 183, *215*
Preuss, H. see Swanstrom, P., 182, *216*
Price, A.H. see Gough, S.R., 955, *1020*
Price, A.H. see Hill, N.E., 939, 995, *1020*
Prigge, H., 9, *22*
Prigge, H. see Lippert, E., 9, *21*
Prigogine, I., 1227, 1231, 1232, 1233, 1234, *1258*
Prigogine, M., 1317, *1362*
Primet, P., 1344, *1362*
Prince, E. see Choi, C.S., 447, *453*
Proctor, W., 873, *887*
Pryor, A.W., 449, *455*
Przyborowski, F. see Deininger, D., 1347, *1355*
Psenichnov, E.A., *356*
Pullin, A.D.E. see Barakat, T.M., 1201, *1223*
Puranik, P.G., 1222, *1224*, 1286, *1362*
Purcell, K.F., 571, *610*, 1204, 1207, 1211, 1212, 1214, 1215, 1216, *1224*
Purcell, K.F. see Sherry, A.D., 1203, 1204, 1206, 1211, 1212, 1214, 1215, 1216, 1217, 1220, 1221, *1224*
Purnell, J.H. see Barker, P.G., 1080, *1104*
Pullman, A., see Alagano, G., 82, 85, 86, *155*
Pullman, A. see Berthod, H., 111, 116, *155*
Pullman, A. see Bonaccorsi, R., 82, *155*
Pullman, A. see Dreyfus, M., 57, 89, 90, 111, 116, 117, *157*, *354*
Pullman, A. see Giessner-Prettre, C., 83, *157*
Pullman, A., see Marius, W., 74, *160*
Pullman, A. see Pullman, B., 219, *243*
Pullman, B., 82, 85, 117, 124, *161*, 219, *243*
Pullman, B. see Bergmann, E.D., 124, *155*
Pytasz, G. see Janik, J.M., 915, 917, 930, *934*
Pytasz, G. see Mikuli, E., 902, 930, *935*

Quinson, J.F. see Escoubes, M., 1350, *1356*
Quirk, J.M. see Olejnik, S., 1349, *1361*
Quist, A.S., 1399, *1423*
Quitsch, K. see Geiseler, G., 533, *561*

Rabideau, S.W., 1430, 1431, 1465, *1469*
Rabideau, S.W. see Arnold, G.P., 1433, *1466*
Rabideau, S.W. see Garret, B.B., 1052, *1059*, 1465, *1467*
Rabideau, S.W. see Jackson, J.A., 1465, *1468*
Rabideau, S.W. see Spiess, H.W., 1465, *1469*
Rabideau, S.W. see Waldstein, P., 1051, *1060*, 1465, *1469*
Rabinovich, I.B., 1243, 1246, 1255, 1256, *1258*
Rachwalska, M. see Janik, J.M., 904, 905, 915, 917, 930, *934*
Raczy, L. see Fauquembergue, R., 945, *1019*
Rädle, C., see Hertz, H.G., 122, *158*, 724, *763*, 1050, 1056, 1057, *1059*
Radul, N.M., 1348, *1362*
Rae, A.I.M., 112, *161*
Rahman, A., 84, *161*, 372, 380, *388*, 901, 915, *935*, 970, *1023*, 1373, 1399, *1423*, 1430, *1469*
Rahman, A. see Singwi, K.S., 929, *936*
Rahman, A., see Stillinger, F.H., 84, *163*, 386, *389*, 1374, 1375, 1376, 1401, *1423*, 1430, *1469*
Rainey, V.S. see Downes, J.S., 1347, *1356*
Rainey, V.S. see Saunderson, D.H., 932, 933, *936*
Rajagopal, H. see Sequeira, A., 451, *455*, 459, 460, *469*
Rajagopal, H. see Sikka, S.K., 441, *455*
Rajnvajn, J. see Hadži, D., 584, *608*
Rakhmatkariev, G.U. see Dubinin, M.M., 1347, *1356*
Rakityanskaya, M.F. see Glazun, B.A., 1001, *1019*
Rakshys, J.W. see Taft, R.W., 818, *829*
Ramachandran, G.N., 124, *161*
Ramamurthy, P. see Low, M.J.D., 1273, 1278, *1360*
Ramanadham, M., 459, *469*
Ramanujam, P.S. see Tellgren, R., 437, 438, 444, *455*
Ramasubramanian, N. see Low, M.J.D., 1273, 1277, 1278, 1297, 1299, *1360*
Ramsay, J.D.F., 1343, *1362*
Ramsbotham, J. see Parfitt, G.D., 1344, *1361*
Ramsey, N.F., 838, *887*
Randić, M., see Murrell, J.N., 39, 49, 56, 112, *160*
Ranner, N.B., 902, *935*
Rao, C.N.A. see Murthy, A.S.V., 567, 603, *609*
Rao, C.N.R., 27, *161*
Rao, C.N.R. see Davis, J.C., 797, *828*
Rao, C.N.R., see Murthy, A.S.N., *20*, 27, 70, 113, 125, 127, *160*, *356*, 793, 804, *829*
Rao, K.N. see Garing, J.S., 1254, *1258*
Rapeanu, S. see Trepadus, V., 1057, *1061*
Rapp, W., see Gronau, B., 15, *21*
Rapp, W., see Kindt, T., 16, *21*
Rashkovich, L.N. see Mavrin, B.N., 1155, *1166*
Rasiah, J. see Stell, G., 378, 381, *388*

Rasmussen, S.E. see Lehmann, M.S., 443, *454*
Raszka, M., 824, *829*
Ratajczak, H., 76, *161*, *162*, *356*, 586, *610*, 724, 737, 740, *765*, 943, 947, 949, 950, 955, 956, *1023*
Ratajczak, H. see Hadži, D., 957, *1020*
Ratajczak, H. see Koll, A., 941, 953, *1021*
Rathburn, R. see Hardt, A.P., 1048, *1059*
Ratner, M.A., 305, 306, 320, 324, 326, 330, 350, *356*
Ratner, M.A., see Fischer, S.F., 219, 220, *242*, 247, *292*, 301, 305, 309, 311, 318, 331, 337, 345, *355*, 590, *608*, 644, *653*, 1194, *1195*
Ratner, M.A., see Hofacker, G.L., 219, 220, 221, 224, 225, *242*
Ravalitera, G. see Chapoton, A., 1002, *1017*, *1018*
Ravenhill, J. see Franks, F., 929, *934*
Ravina, I. see Low, P.F., 1343, 1348, *1360*
Ray, S. see Brindley, G.W., 1348, *1354*
Reball, S., 1336, 1337, *1362*
Rebič, M. see Trontelj, Z., 1146, *1165*
Reding, F.P. see Hornig, D.F., 1458, *1468*
Redington, R.L., 335, *356*, 582, *610*, 1099, *1105*
Redington, R.L. see Berney, C.L., 1099, *1104*
Redlich, O., 1228, 1233, *1259*
Ree, T., see Eyring, H., 371, *387*
Ree, T.K., see Jhon, M.S., 368, 372, *388*
Reese, R.L., 1125, *1163*
Reese, W., 1117, 1120, 1129, *1163*
Reese, W. see Fairall, C.W., 1117, 1120, 1129, 1154, 1157, *1162*, *1165*
Reeves, L.W., 533, *562*
Rehm, H.J., 1014, *1023*
Reichardt, C., 951, *1023*
Reichle, M., 1013, *1023*
Reid, C., 172, *216*, 225, *243*, 305, 335, *356*, 621, *653*, 1134, *1163*
Reimann, B. see Deininger, D., 1348, *1355*
Rein, R., 220, *243*
Reiter, R.C. see Muller, N., 822, *829*
Renker, B., 902, *935*
Renon, H., 1233, *1259*
Rentzepis, P.M., 15, *22*
Rericha, R., 1339, *1362*
Resing, H.A., 1294, 1313, 1347, 1348, *1362*
Ress, E. see Knözinger, H., 1335, 1337, 1339, *1359*
Resta, R., see Bassani, F., 142, *155*
Reuben, J., 1298, *1362*
Reuben, J. see Fiat, D., 1327, 1328, *1356*
Reuter, A. see Mecke, R., 541, *561*
Rey-Lafon, M., 771, *789*
Reynolds, S.I. see Brot, C., 1012, *1017*
Rhee, K.H. see Basila, M.R., 1346, *1353*
Rhenius, P. see Becker, F., 826, *829*
Rhodes, M. see Figgins, R., 1039, *1058*
Rhodes, M. see Powles, J.G., 1039, *1060*, 1465, *1469*
Rhys, F., 362, 383, *388*
Ribarič, H., see Blinc, R., 347, *354*
Ribarič, M. see Blinc, R., 247, 265, *292*, 1181, *1195*
Ribnikar, S.V. see Bigeleisen, J., 1246, *1257*
Rice, S.A., 221, *243*, *356*, 573, *610*, 647, *654*
Rice, S.A. see Weres, O., 368, 375, *388*
Richard, P. see Philippot, E., 442, *455*
Richards, R.E., 476, 477, 478, *525*, 659, *682*, 871, *887*
Richards, R.E. see Smith, J.A.S., 476, *525*
Richardson, E.A. see Blyholder, G., 1345, *1354*
Richtol, H.H. see Van Ness, H.C., 1233, *1259*
Riedel, E., 1348, *1362*
Riehl, N., *20*, 219, *243*, 341, 342, *356*, 939, *1023*
Riehl, N. see Engelhardt, H., 758, *763*, 979, 997, *1019*
Rigamonti, A. see Bonera, G., 1153, 1160, *1165*
Riganti, D. see Parmigiani, A., 813, *829*
Rigby, M., 92, *162*
Rimkevich, I.M., see Benzar, K.K., 1347, *1353*
Rios, J., see Bratoz, S., 317, *354*
Rip, A. see Bordewijk, P., 1233, *1257*
Robert, D., see Bonamy, J., 99, *155*
Robert, D. see Girardet, C., 1080, 1082, 1083, *1105*
Roberts, J.D. see Lichter, R.L., 823, *828*
Robertson, E.B., 475, *525*
Robertson, G.N., see Coulson, C.A., 352, *357*, 589, *607*
Robertson, J.H., 417, *455*
Robertson, J.M., 426, *455*
Robinson, E.A., 781, *789*
Robinson, G.W. see Keyser, L.F., 1071, 1073, *1105*
Rocaries, C., 1150, *1163*
Rochester, C.H. see Griffiths, D.M., 1351, *1357*
Rochester, C.H. see Parfitt, G.D., 1344, *1361*

Rode, B.M., 116, 117, *162*
Rode, B.M., see Schuster, P., *20*, 126, 149, 150, 151, *162*, 219, *243*
Rodebush, W.H. see Latimer, W.M., 5, *21*
Rodgers, M.T. see Narasimhan, P.T., 1208, *1224*
Roebuck, B.D., 1017, *1023*
Roev, L.M., 699, 717, *765*, 1278, 1301, *1362*
Rogasch, P.E. see Bellamy, L.J., 302, *354*
Rogers, E., 989, *1023*
Rogers, L.B. see Pickett, J.H., 1293, *1362*
Rogers, M.T. see LaPlanche, L.A., 805, 806, 808, *828*
Rojas, O. see Daudey, J.P., 49, *156*
Rolland, M.T., 1315, *1362*
Rollar, H.-G. see Wolff, H., 1253, *1260*
Romanet, R., 622, *654*
Romanovskii, I.A. see Dushchenko, V.P., 1016, *1018*
Ron, A., 622, *654*, 1292, 1293, 1294, *1362*
Ron, A. see Katz, B., 1074, *1105*
Ron, T., 1345, *1362*
Röntgen, W.C., 1372, *1423*
Roos, B.O., see Diercksen, G.H.F., 154, 155, *163*
Roos, J., 1141, *1163*
Roothaan, C.C.J., 31, *162*
Rösch, M., 556, *562*
Rościszewski, K., 911, *935*
Rösch, N., 220, 222, 225, 226, *243*, 301, 326, 335, *356*
Rösch, N., see Ratner, M.A., 306, 324, 326, 330, 336, 350, *356*
Rose, D.G. see Hanna, M.W., 815, *828*
Rose, M.S., 762, *765*
Rose, M.S. see Aldridge, W.N., 762, *763*
Rosen, N., 222, *243*
Rosenberg, M.Sh., 596, *610*
Rosenberg, M.Sh. see Iogansen, A.V., 596, *609*
Rosengren, Kj. see Pimentel, G.C., 1090, 1091, *1105*
Rosenstein, R.A., 326, *356*
Rosenstein, R.D. see Negran, T.J., 1159, *1166*
Rospenk, M. see Koll, A., 702, 719, 739, *764*
Ross, R.A. see Boyle, F.W., 1299, *1354*
Ross, R.A. see Cook, W.G., 1346, *1355*
Ross, S., 1329, *1362*
Roth, E. see Bigeleisen, J., 1246, *1257*
Roth, K. see Franck, E.U., 1382, *1421*
Rothenberg, S. see Johansson, A., 63, 111, 113, 135, *158*
Rothenberg, S. see Kollman, P.A., 111, *159*
Rothschild, W.G., 773, 777, 779, 780, 788, *789*
Rothschild, W.G. see Möller, K.D., 770, 786, *789*
Roudault, R. see Philippot, E., 442, *455*
Roupe, J.L. see Silvidi, A.A., 1146, *1164*
Rouse, K.D. see Nelmes, R.J., 1118, 1134, 1137, *1163*, 1190, *1195*
Rousselet, D. see Fournier, M., 660, 661, 663, *681*
Rouxhet, P.G., 1286, 1288, *1362*
Rowland, T.J. see Bloembergen, N., 875, *886*
Rowlinson, J.S., 362, 376, 377, 379, 381, *388*
Roy, A.P. see Venkataraman, G., 930, 931, *936*
Royston, R. see Egelstaff, P.A., 912, 929, *934*
Rozière, J. see Chemouni, E., 668, 670, *681*, 747, *763*
Rozière, J. see Fournier, M., 660, 661, 663, *681*
Rozière, J., 601, *610*, 668, 669, 670, 671, 676, *682*
Rudenko, V.M. see Tarasevich, Yu.I., 1349, *1363*
Rudham, R. see Hendriksen, B.A., 1333, 1334, *1358*
Rudolph, J., 677, 678, 679, 680, *682*
Rudolph, J. see Zimmermann, H., 221, *244*
Ruepp, R., 967, 973, 979, *1023*
Runck, A.H. see Von Hippel, A., 967, 968, 975, 976, 978, 983, *1024*
Rundel, R.D. see Stokes, E.D., 18, *22*
Rundle, R.E., 424, *455*, 503, 507, *525*, 605, *610*
Rundle, R.E. see Godycki, L.E., 605, *608*
Rundle, R.E. see Nakamoto, K., 410, *455*, 659, 661, *681*
Rush, J.J., 587, *610*, 908, 925, 926, 929, 930, 931, 932, *935*
Rush, J.J. see Delaplane, R.G., 336, *354*, 602 *607*
Rush, J.J. see Leung, P.S., 931, 932, *935*
Rush, J.J. see Schlemper, E.O., 931, *936*
Rushbrooke, G.S., 386, *389*
Russegger, P., 74, 125, 126, 137, *162*
Russegger, P. see Karpfen, A., 135, 136, 137, 138, 142, *159*
Rustanova, S.N., 953, *1023*
Rüter, H. see Brockamp, B., 1433, 1464, *1467*
Rüterjans, H., 761, *765*
Rutishauser, H. see Heilbronner, E., 173, *215*, 221, *242*

Ryan, J.F. see Cowley, R.A., 1132, *1161*
Ryan, J.F. see Katiyar, R.S., 1124, 1131, 1156, *1162*
Ryan, R.R., 483, *525*
Ryan, R.R. see Penneman, R.A., 484, 492, 500, *525*

Sabin, J.R., 111, 121, 122, *162*, *357*
Sabin, J.R. see Fischer, S.F., 247, *292*, 297, 305, 313, 324, 342, *355*, 981, *1019*
Sabin, J.R. see Ratner, M.A., 320, *356*
Sabine, T.M., 444, *455*
Sabine, T.M. see Coppens, P., 424, 437, 444, *453*
Sabine, T.M. see Cox, G.W., 449, *453*
Sabine, T.M. see Craven, B.M., 446, *453*
Sabine, T.M. see Freeman, H.C., 451, *453*
Sabine, T.M. see Taylor, J.C., 447, *455*
Sack, H., 1408, *1423*
Sacks, L.J., 1202, *1224*
Sadlej, A.J., see Jaszunski, M., 80, 81, *158*
Safford, G.J., 912, 929, *935*
Safford, G.J. see Boutin, H., 909, 930, 932, *934*
Safford, G.J. see Brajovič, V., 931, 932, *934*
Sagnowski, S. see Pollak-Stachura, M., 871, *887*
Sahay, B.K., 1343, *1362*
Saier, E.L. see Coggeshall, N.D., 541, *560*, 801, 805, *828*
Saika, A. see Gutowsky, H.S., 794, *828*
Sailor, E., 1158, *1166*
Saitô, H., 822, 823, *829*
Saito, S. see Kurosaki, S., 1313, *1359*
Saito, Y., 484, *525*
Saito, Y. see Arai, H., 1337, *1353*
Saito, Y. see Nakahara, A., 478, 484, *525*, 667, *681*
Saito, Y. see Ooi, S., 478, 484, *525*
Sakabe, Y. see Ida, M., 1009, *1021*
Sakai, Y. see Huzinaga, S., 106, *158*
Sakamoto, M., 929, *936*
Sakamoto, M., 929, *936*
Sakamoto, M. see Woods, A.D.B., 929, 930, *936*
Sakharov, A.I. see Dubinin, M.M., 1347, 1348, *1356*
Salem, L., 837, *887*
Salem, L., see Musher, J.I., 39, *160*
Salimov, M.A. see Rustanova, S.N., 953, *1023*
Salomon, M., see Conway, B.E., *354*
Salomon, R.E. see Young, I.G., 983, *1025*
Salthouse, J.A., 602, *610*, 736, *765*
Salvador, P. see Toullaux, R., 1348, *1363*
Salvetti, O., see Arrighini, G.P., 35, *155*
Salyers, A. see Jehle, H., 254, *292*, 729, *764*
Samara, G.A., 1117, 1118, 1125, 1128, 1136, 1138, 1140, 1154, *1163*, *1164*
Samara, G.A. see Morosin, B., 1117, 1125, *1163*
Samara, G.A. see Peercy, P.S., 1156, *1166*
Samoilov, O.J., 1402, *1423*
Samorjai, G.A. see French, T.M., 1332, *1357*
Samorjai, R.L., 221, *243*
Sampathkumar, K.S.V. see Walsh, K.A., 761, *766*
Sampoli, M. see Careri, G., 1335, 1343, 1345, 1346, *1354*
Sampson, T.E., 932, *936*
Samsonov, G.V. see Frolov, V.I., 1012, *1019*
Samsonov, O.A. see Millen, D.J., 224, *243*, 330, *356*, 636, *653*
Sandall, J.P.B. see Ibbitson, D.A., 940, 941, *1020*, *1021*
Sandorfy, C., 176, 178, *216*, 325, *357*, 583, *610*, 622, *654*, 1320, 1321, 1325, *1362*
Sandorfy, C. see Asselin, M., 336, *353*, 570, 571, 573, *607*, 622, 625, 633, 637, 638, 642, 643, 645, 646, *652*
Sandorfy, C. see Bernard-Houplain, M.-C., 621, 625, 627, 628, 629, 630, 631, 632, 641, 646, 648, *652*
Sandorfy, C. see Berthomieu, C., 629, *652*
Sandorfy, C. see Bicca de Alencastro, R., 625, 632, 648, *652*
Sandorfy, C. see Bourdéron, C., 580, *607*, 622, 623, 624, 625, 626, 627, 647, 648, *652*, 1324, 1325, *1354*
Sandorfy, C. see Cabana, A., 582, 583, *607*, 639, *652*
Sandorfy, C. see Chenon, B., 716, *763*
Sandorfy, C. see Di Paolo, T., 182, *215*, 648, 650, *653*, 1290, 1326, *1356*, 1461, *1467*
Sandorfy, C. see Durocher, G., 302, *354*, 589, *608*, 617, 621, *653*
Sandorfy, C. see Foldes, A., 336, *355*, 570, *608*, 617, 619, 637, *653*
Sandorfy, C. see Péron, J.J., 622, *654*
Sandorfy, C. see Planckaert, A.A., 646, *653*
Sandler, I. see Steele, W.A., 386, *389*
Sandstrom, J. see Gramstad, T., 1201, *1224*
Sandy, F., 1142, *1164*
Sanfeliz, M. see Schaarschmidt, K., 955, *1023*
Sang-Il Choi see Gosar, P., 329, *355*

Sanger, P.L. see Pryor, A.W., 449, *455*
Sansoni, B. see Bunzl, K., 1011, *1017*
Santry, D.P., 141, *162*
Santry, D.P. see Bacon, J., 141, *155*
Santry, D.P. see Crowe, R.W., 142, *156*
Santry, D.P. see Middlemiss, K.M., 141, *160*
Santry, D.P. see Pople, J.A., 37, *161*
Sarkisov, G.N., 381, *388*
Sarzynsk, J. see Davies, C.M., 1373, *1421*
Sasisekharan, V. see Ramachandran, G.N., 124, *161*
Satake, I., 827, *830*
Sato, G. see Kurosaki, S., 1313, *1359*
Sato, K., 1158, *1166*
Sato, S., 1139, *1164*
Sato, T. see Hino, M., 1267, *1358*
Sato, Y., 1134, *1164*
Sato, Y. see Kobayashi, J., 1145, *1162*
Saunders, M., 796, 799, *829*
Saunderson, D.H., 932, 933, *936*
Savelev, V.A. see Mikhailov, I.D., 333, *356*, 581, *609*, 622, *653*
Savini, C.G., 1237, *1259*
Savoie, R., 663, *682*, 1073, 1074, 1077, *1105*
Sawada, A., 1143, 1144, *1164*
Sawada, A. see Sato, K., 1158, *1166*
Sawada, A. see Takagi, Y., 1158, *1166*
Saxton, J.A., 1047, *1060*
Scarborough, J. see Kebarle, P., 121, 139, 140, *159*, 475, 481, 519, *524*, 980, *1021*
Schaad, L.J. see Joesten, M.D., *20*, 154, *163*
Schaarschmidt, K., 955, *1023*
Schaefer III, H.F., 31, *162*
Schaefer, H.F. see McLaughlin, 56, *160*
Schaefer III, H.F. see Yarkony, D.R., 61, 72, 129, 132, *163*
Schäfer, E.P., 15, 22
Schaffer, P.C. see Safford, G.J., 912, 929, *935*
Schagina, N.M. see Shuvalov, L.A., 1135, 1136, *1164*
Schagina, N.M. see Stepišnik, J., 1158, *1166*
Schara, M. see Blinc, R., 885, *885*, 1122, *1161*
Scheie, C.E. see O'Reilly, D.E., 506, *525*, 660, 673, *681*, *682*, 872, *887*
Schempp, E., 837, 857, 858, 859, 860, *887*
Schenk, C.H., 932, *936*
Schenk, D. see Schneider, K.E., 973, *1024*
Scheraga, H.A., 124, *162*
Scheraga, H.A. see Lentz, B.R., 133, 134, *159*
Scheraga, H.A. see Momany, F.A., 124, *160*
Scheraga, H.A. see Némethy, G., 368, 370, 371, 372, 374, *388*, 1371, 1373, 1377, 1393, *1423*
Scheraga, H.A. see Poland, D., 117, *161*
Schiessler, R.W. see Davies Jr., R.T., 1246, *1257*
Schiffer, J., 580, *610*, 638, *654*
Schiller, O. see Wolff, H., 1228, 1229, 1230, 1235, 1236, 1238, 1243, *1259*, *1260*
Schimpf, L. see Wolff, H., 1228, 1229, 1230, 1235, 1236, 1238, 1243, *1260*
Schindler, P., 1288, *1362*
Schindler, P. see Ron, T., 1345, *1362*
Schiöberg, D., 688, 695, 696, 698, 700, 701, 703, 708, 709, 711, 716, 718, 719, 740, 743, 744, 745, 751, 753, 754, 759, *765*
Schirmer, W. see Deininger, D., 1347, *1355*
Schirmer, W. see Gutsze, A., 1347, *1357*
Schirmer, W. see Lohse, U., 1002, 1003, *1021*, 1347, *1360*
Schlaak, M. see Huong, P.V., 353, *357*, 606, *609*, *611*
Schlecht, P. see Brausse, G., 1013, *1017*
Schleich, K. see Clusius, K., 1246, *1257*
Schlemper, E.O., 336, *357*, 465, *469*, 447, 449, *455*, 931, *936*
Schlenk, W., 1415, 1420, *1423*
Schleyer, P.V.R. see Allerhand, A., 573, *607*, 1207, *1223*
Schleyer, P.V.R. see Arnett, E.M., 1201, 1205, *1223*
Schleyer, P.V.R. see West, R., 1201, *1224*
Schmidt, E. see Mann, B., 1088, *1105*
Schmocker, U. see Boesch, H., 1151, *1161*
Schmidt, U. see Wolff, H., 1237, *1259*
Schmidt, V.H., 989, 990, 991, 993, 997, *1023*, *1024*, 1121, 1123, 1127, 1130, 1132, 1134, 1152, 1154, *1164*
Schmidt, V.H. see Kim, D.Y., 981, *1021*
Schmidt, V.H. see Silsbee, H.B., 990, *1024*, 1126, 1129, 1130, 1132, 1154, 1156, 1157, *1164*
Schnabel, B. see Haubenreisser, U., 877, *886*
Schneemeyer, L.F. see Williams, J.M., 118, *163*, 431, *456*, 748, *766*
Schneider, K.E., 973, *1024*
Schneider, M. see Boehm, H.P., 1272, *1354*
Schneider, W.G., 129, *162*, 535, *562*
Schneider, W.G. see Pople, J.A., 1029, 1034, 1037, *1060*
Schneider, W.G. see Reeves, L.W., 533, *562*
Schnepp, O. see Katz, B., 1074, *1105*
Schnepp, O. see Ron, A., 1292, 1293, 1294, *1362*
Schofield, P., 895, *936*
Schomaker, V., 463, *469*

Schön, G., 1005, *1024*
Schoonheydt, R.A. see Jansen, F.J., 1347, 1348, *1358*
Schreiber, H.D. see Robinson, E.A., 781, *789*
Schreiber, V.A. see Gusakova, G.V., 737, *763*
Schrobilgen, G.J. see Gillespie, R.J., 7, *21*
Schroeder, D.C., 19, *22*
Schroeder, J.P. see Schroeder, D.C., 19, *22*
Schroeder, L.W., 483, 484, 495, *525*
Schroeder, L.W. see Rush, J.J., 587, *610*, 932, *935*
Schroeder, R. see Lippincott, E.R., 172, *215*, 221, *243*, 305, 335, *356*, 571, 573, 574, *609*, 621, *653*, 841, *886*, 1085, *1105*, 1202, 1214, 1218, *1224*
Schröer, W., 825, 827, *830*
Schröer, W. see Lippert E., 8, *21*
Schug, J.C., 827, *830*
Schulman, S.C. see Yakatan, G.J., 16, *22*
Schulz, G. see Hafner, K., 604, *608*
Schupp, E. see Huettenrauch, R., 1345, *1358*
Schupp, R.L., 541, *562*
Schupp, R.L. see Mecke, R., 541, *561*
Schürer, P. see Kubelkovà, L., 1267, 1318, 1320, *1359*
Schuster, H.D. see Zigan, F., 443, *456*
Schuster, P., *20*, 27, 72, 73, 74, 75, 76, 113, 114, 116, 117, 122, 124, 125, 126, 127, 128, 129, 130, 131, 132, 135, 139, 146, 149, 150, 151, *162*, 219, 220, 224, *243*, 1222, *1224*
Schuster, P. see Beyer, A., 154, 155, *163*
Schuster, P. see Griengl, H., 131, *157*
Schuster, P. see Jakubetz, W., 131, *158*
Schuster, P. see Karpfen, A., 135, 136, 137, 138, 142, *159*
Schuster, P. see Marius, W., 74, *160*
Schuster, P. see Russegger, P., 74, 125, 126, 137, *162*
Schwab, G.-M. see Zundel, G., 687, 693, 697, *766*, 1011, *1025*
Schwabe, K., 744, 759, *765*
Schwartz, M. see Gillen, K.T., 1038, *1059*
Schwarz, G., 940, *1024*
Schwarzenbach, G. see Landis, D.H., 267, *292*
Schwarzmann, E., 433, *455*
Schweighardt, F.K. see Kan, L.S., 823, *828*
Schweizer, F., see Mattmann, G., 1345, *1360*
Ściesińska, E. see Mikuli, E., 902, 930, *935*
Ściesiński, J. see Burgman, J.O., 902, 929, *934*
Ściesiński, J. see Mikuli, E., 902, 930, *935*
Scott, J.F., 347, *357*, 1132, *1164*
Scott, J.F. see Cowley, R.A., 1132, *1161*
Scott, J.F. see Katiyar, R.S., 1124, 1131, 1156, *1162*
Scott, R.B. see Armstrong, G.T., 1246, *1257*
Scott, R.B. see Wooley, H.W., 1246, *1260*
Scott, R.L., 1204, 1205, *1224*
Scott, R.M. see Vinogradov, S.N., 737, *766*
Scrimshaw, G.F. see Barnes, A.J., 1068, 1071, 1073, 1074, 1075, 1076, 1077, 1080, 1082, 1083, 1101, *1104*
Scrocco, E., 82, *162*, 885, *887*
Scrocco, E. see Alagano, G., 82, 85, 86, *155*
Scrocco, E. see Berthier, G., 82, *155*
Scrocco, E. see Bonaccorsi, R., 48, 60, 62, 82, 84, 85, *155*
Searles Jr., S. see Tamres, M., 1202, *1224*
Searles, S.K. see Kebarle, P., 121, 139, 140, *159*, 475, 481, 519, *524*, 980, *1021*
Secker, P. see Thomas, J.M., 708, 758, 760, *765*, 995, *1024*
Secoy, C.H. see Fuller Jr., E.L., 1345, *1357*
Secoy, C.H. see Holmes, H.F., 1345, *1358*
Sedykh, N.V., 1016, *1024*
Seel, R.M., 584, 586, *610*
Seel, R.M. see Jones, W.J., 635, *653*
Segal, G.A., see Pople, J.A., 37, *161*
Seibold-Blankenstein, I. see Lippert E., 9, *21*
Seidensticker, R.G., 983, *1024*
Seiffert, W. see Limbach, H.H., 825, *830*
Seliger, J. see Blinc, R., 851, 852, 853, 855, 862, 863, *885*, 1154, *1165*
Sempels, R.E. see Rouxhet, P.G., 1288, *1362*
Senior, G.E., 1382, *1423*
Senior, W.A. see Vand, V., 371, 372, *388*, 1398, *1423*
Senko, M.E., 990, *1024*, 1126, *1164*
Senter, M.E. see Forrester, J.D., 867, *886*
Sequeira, A., 444, 445, 451, *455*, 459, 460, *469*
Sequeira, A. see Bugayong, R.R., 459, *468*
Sequeira, A. see Chidambaram, R., 343, 440, *453*
Sequeira, A. see Gupta, S.C., 459, *468*
Sequeira, A. see Thaper, C.L., 902, 903, 930, *936*
Serpinskii, V.V. see Dubinin, M.M., 1347, *1356*
Serpinskii, V.V. see Murdmaa, K., 1346, *1361*
Serratosa, J.M. see Hougardy, J., 1349, *1358*
Sessler, W., *357*, 698, 701, 702, 708, 709, 710, 711, 719, 731, 738, 740, *765*

Sewell, P.A., 1319, 1352, *1362*
Seymour, S.J., 1352, *1362*
Shakhparonov, M.I., *357*
Shank, C.V. see Dienes, A., 16, *20*
Shapet'ko, N.N. see Bogachev, Y.S., 797, *827*
Shapet'ko, N.N. see Kimtys, I., 827, *829*
Shapiro, F.L. see Golikov, V.V., 912, 929, *934*
Shapiro, I., 1272, *1362*
Shapiro, S.M. see Dietrich, O.W., 1155, *1165*
Sharan, M. see Pollock, J.M., 989, 991, 993, *1023*
Sharbaugh, A.H. see Brot, C., 1012, *1017*
Sharkina, E.V., 1348, 1349, *1362*
Shavitt, I., 36, *162*
Shaw, E.L. see Tsang, T., 1148, *1165*
Shaw, G. see Murrell, J.N., 39, 40, *160*
Shaw, Y.L., 814, *829*
Shchepkin, D.N. see Belozerskaya, L.P., 635, *652*
Shcherbak, Ya.Ya. see Matyash, I.V., 1347, *1360*
Shcherbakov, V.N. see Aleksandrova, I.P., 1146, *1160*
She, C.Y., 348, *357*, 1124, 1131, *1164*
She, C.Y. see Broberg, T.W., 1124, *1161*
Shechter, H. see Montano, P.A., 1147, *1163*
Shedrin, M.I. see Kubarev, S.I., 297, 342, 345, *355*
Sheka, I.A., 955, *1024*
Sheline, R.K. see Spiess, H.W., 1465, *1469*
Shen, J.H. see Klier, K., 1267, 1304, 1307, 1308, 1309, 1310, 1311, *1359*
Sheppard, N., 302, 325, *357*, 602, *610*, 640, 642, *654*, 1290, *1362*
Sheppard, N. see Bethell, D.E., 477, *523*, 660, 663, *681*
Sheppard, N. see Bratož, S., 302, 325, *354*, 583, *607*, 639, *652*
Sheppard, N. see Braunholtz, J.T., 584, *607*
Sheppard, N. see Claydon, M.F., 337, *354*, 585, *607*
Sheppard, N. see Gilbert, A.S., 601, *608*, 669, 670, 673, 676, 677, 679, *681*, 747, *763*
Sheppard, N. see Hadži, D., 604, *608*
Sheppard, N. see Jones, W.J., 635, *653*
Sheppard, N. see Seel, R.M., 584, 586, *610*
Sheppard, N. see Smart, R.S.C., 1350, *1363*
Sheppard, N. see Young, R.P., 1279, *1364*
Sherry, A.D., 1203, 1204, 1206, 1211, 1212, 1214, 1215, 1216, 1217, 1220, 1221, *1224*
Shi-Chien, L. see Chidambaram, R., 440, *453*
Shimada, A. see Takusagawa, F., 433, *455*
Shimanouchi, T. see Itoh, K., 771, 772, 773, 778, 787, *789*
Shimaoka, K. see Niimaru, N., 1147, *1163*
Shinitzky, M., 708, 761, *765*
Shiozaki, Y. see Nakano, J., 1118, *1163*
Shipman, L.L., 123, 124, *162*
Shirane, G. see Jona, F., 340, *355*, 1114, 1117, 1150, *1162*
Shirane, G. see Skalyo Jr., J., 1124, *1164*, 1187, 1189, 1190, 1192, *1195*
Shirokov, A.M., 1135, 1136, 1140, *1164*
Shirokov, A.M. see Shuvalov, L.A., 1135, 1136, *1164*
Shono, H. see Kakiuchi, Y., 476, *524*
Shoolery, J.N., 1403, *1423*
Shoolery, J.N. see Becker, E.D., 540, *560*
Shoolery, J.N. see Huggins, C.M., 795, *828*
Shukla, G.C. see Žekš, B., 1143, *1165*
Shuler, K.E. see Herman, R.C., 649, *653*
Shulman, R.G., 760, *765*
Shurvell, H.F. see Bulmer, J.T., 576, *607*, 806, *827*
Shurvell, H.F. see Harvey, K.B., 1418, *1421*
Shuvalov, L.A., 1135, 1136, 1137, 1139, 1140, *1164*
Shuvalov, L.A. see Shirokov, A.M., 1135, 1136, 1140, *1164*
Shuvalov, L.A. see Stepišnik, J., 1158, *1166*
Sichhart, K.H. see Lohse, U., 1347, *1360*
Sidebottom, E.W., 1317, *1362*
Sidnenko, E.V., 1154, 1157, *1166*
Sidnenko, E.V. see Zheludev, I.S., 1130, *1165*
Sidorov, A.N., 1267, 1276, 1277, 1279, 1296, 1297, 1303, 1318, *1362*, *1363*
Siepe, V. see Hertz, H.G., 1056, *1059*
Signorelli, G. see Carceri, G., 1335, 1343, 1345, 1346, *1354*
Sikka, S.K., 441, *455*
Sikka, S.K. see Chidambaram, R., 434, *453*
Sikka, S.K. see Macdonald, A.C., 440, *455*
Sikka, S.K. see Ramanadham, M., 459, *469*
Silsbee, H.B., 990, *1024*, 1126, 1129, 1130, 1132, 1154, 1156, 1157, *1164*
Silvidi, A.A., 1146, *1164*
Simkovich, G., 1313, *1363*
Simon, A., 1147, *1164*
Simon, A. see Treibmann, D., 1337, *1363*
Simon, H. see Praefke, K., 18, *22*
Simonetta, M. see Cremaschi, P., 143, *156*
Simonov, V.I. see Kruglik, A.I., 1159, *1166*
Simonov, V.I. see Tovbis, A.B., 1139, *1165*
Simonsen, S.H. see Dalley, N.K., 441, *453*

Simova, P.D. see Chulànovskii, V.M., 305, *354*
Simpson, J.H., 912, *936*, 1050, *1060*
Sinanoğlu, O., 36, *162*
Sinclair, R.N. see Day, D.H., 929, *934*
Sinclair, R.N. see Woods, A.D.B., 929, 930, *936*
Sinfelt, J.H. see Carter, J.L., 1330, 1331, 1332, *1354*
Singh, T.R., 174, 175, 178, 181, *216*, 221, 225, *244*, *357*, 424, 425, *455*, 507, *525*, 621, *654*
Singh, T.R. see Clements, R., 587, 601, *607*, 706, 716, *763*
Singurel, L., 568, *610*
Singwi, K.S., 910, 929, *936*
Sinitsin, V.A. see Berezin, G.I., 1306, 1347, *1353*
Sirigo, A. see Benedetti, E., 462, *468*
Sixou, P., 978, *1024*
Sixou, P. see Mounier, S., 983, *1022*
Sjölander, A. see Singwi, K.S., 910, 929, *936*
Skalyo Jr., J., 1124, *1164*, 1187, 1189, 1190, 1192, *1195*
Sklyankin, A.A., 1349, *1363*
Sköld, K., 911, 912, 919, 929, *936*
Sköld, K. see Burgman, J.O., 902, 929, *934*
Skomorokhova, T.L. see Strukov, B.A., 1145, *1164*
Skulski, L., 940, *1024*
Slak, J. see Blinc, R., 1132, 1145, *1161*
Slak, J. see Stepišnik, J., 1158, *1166*
Slater, J.C., 347, *357*, 362, 382, *388*, 863, 864, *887*, 990, *1024*, 1126, 1133, *1164*, 1178, *1195*
Slejko, F.L., 818, *829*, 1208, 1209, 1217, 1220, *1224*
Slichter, C.P., 874, *887*, 1029, 1031, 1034, *1060*
Slichter, C.P. see Lurie, F.M., 849, *886*
Small, E.W. see Brown, K.G., 787, *789*
Small, E.W. see Fanconi, B., 787, *789*
Small, R.M.B. see Plint, C.A., 1255, *1258*
Smart, R.S.C., 1350, *1363*
Smedal, H.S. see Beck, W., 601, *607*
Smerkolj, R. see Detoni, S., 569, 570, 572, 576, 577, 584, 585, *608*, 941, *1018*
Smith, B.R., 1338, *1363*
Smith, D.F., 61, 80, 115, 135, *162*, *163*
Smith, D.W.G., 1049, *1060*
Smith, D.W.G. see Powles, J.G., 1049, *1060*
Smith, E.R., 385, 386, *389*
Smith, F. see Barker, J.A., 827, *829*
Smith, F.A., 540, *562*, 580, *610*
Smith, H.G., 478, 482, 495, *525*
Smith, J.A.S., 476, *525*
Smith, J.A.S. see Richards, R.E., 476, 477, 478, *525*, 659, *682*, 871, *887*
Smith, J.W., 940, *1024*
Smith, J.W. see Cleverdon, D., 942, *1018*
Smith, J.W. see Few, A.V., 940, *1019*
Smith, R.B. see Kottwitz, D.A., 912, 929, *935*
Smith, W.R. 386, *389*
Smolej, V. see Blinc, R., 1155, *1165*
Smolenski, G.A., 1153, *1166*
Smoljanski, A.L. see Gusakova, G.V., 737, *763*, 949, *1020*
Snowden Jr., B.S. see Woessner, D.E., 1038, *1060*
Snyder, L.C., 101, *163*
Snyder, L.R., 1271, 1272, *1363*
Snyder, R.G., 320, *357*
Sobczyk, L., 741, *765*, 939, 946, 948, 949, *1024*
Sobczyk, L. see Detoni, S., 941, *1018*
Sobczyk, L. see Hadži, D., 957, *1020*
Sobczyk, L. see Hawranek, J.P., 724, *763*, 942, 948, 957, 1012, *1020*
Sobczyk, L. see Koll, A., 702, 719, 739, *764*, 941, 953, *1021*
Sobczyk, L. see Malarski, Z., 941, *1022*
Sobczyk, L. see Pawelka, Z., 952, *1023*
Sobczyk, L. see Pawlak, Z., 695, 698, 701, 703, 706, 707, 720, 735, *765*
Sobczyk, L. see Ratajczak, H., 724, 737, 740, *765*, 947, *1023*
Sobetzko, W. see Gärtner, K., 1011, *1019*
Soda, G., 841, *887*, 1137, 1148, *1164*
Soffer, A., 1299, 1313, *1363*
Sohma, J. see Iuzuka, T., 1343, *1358*
Sokolov, N.D., *20*, *357*, 595, *610*
Sokolov, N.D. see Kvlividze, V.J., 1311, *1359*
Sokolov, N.D. see Mikhailov, I.D., 333, *356*, 581, *609*, 622, *653*
Sokolov, N.D. see Psenichnov, E.A., *356*
Solinas, M. see Bertie, J.E., 1459, *1467*
Šolmajer, T. see Hadži, D., 568, *608*
Šolmajer, T. see Koller, J., 601, *609*
Šolmajer, T. see Obradović, M., 569, *610*
Solovev, S.P. see Bajorek, A., 902, 929, 930, *933*
Solt, G. see Kosaly, G., 931, *935*
Somers, B.G. see Inskeep, R.G., 1092, 1094, *1105*

Somorjai, R.L., 173, *216*, 247, *292*, *357*, 601, *610*
Sompolinsky, H. see Havlin, S., 1157, *1165*
Sosnkowska-Kehiaian, K. see Kehiaian, H., 1231, *1258*
Sosnowska, I. see Antonini, M., 932, *933*
Søtofte, I. see Larsen, F.K., 443, *454*
Sousa, J.A. see Huber, J.R., 16, *21*
Spangenberg, H.-J. see Möller, K.D., 785, *789*
Spannheimer, H., 1333, *1363*
Spatz, H.Ch. see Eigen, M.L., 342, *354*, 976, *1019*
Speakman, J.C., 429, *455*, 522, *525*, 587, 597, *610*
Speakman, J.C. see Brandt, J.C.D., 87, *156*
Speakman, J.C. see Calleri, M., 602, *607*
Speakman, J.C. see Currie, M., 445, 448, *453*
Speakman, J.C. see Macdonald, A.L., 422, 445, *455*, 600, *609*
Speakman, J.C. see McAdam, A., 445, *455*
Spencer, J.B., 483, 495, *525*
Spences, J.N. see Robinson, E.A., 781, *789*
Spiess, H.W., 1465, *1469*
Spiess, H.W. see Haeberlen, U., 877, 878, 879, 880, *886*
Spivey, D.I. see Eley, D.D., 988, 995, *1019*
Spunta, G. see Ghersetti, S., 782, *789*
Srebrenik, S., 83, *163*
Srikanta, S. see Chidambaram, R., 440, *453*
Srikanta, S. see Sequeira, A., 444, *455*
Srinivasan, R., 18, *22*
Srinivasan, R. see Dalal, N.S., 882, 883, *886*
Stach, H. see Deininger, D., 1347, *1355*
Stach, H. see Gutsze, A., 1347, *1357*
Stach, H. see Lohse, U., 1002, 1003, *1021*, 1347, *1360*
Stählin, W., 1351, *1363*
Stählin, W. see Clement, G., 1333, 1336, *1355*
Stakebake, J.L., 1346, *1363*
Stalidis, G. see Hertz, 122, *158*
Stamenković, S., 1153, *1166*
Stanevich, A.E. see Brodskii, I.A., 784, 785, *789*, 1267, 1308, *1354*
Stanton, J.H. see Dugger, D.L., 1269, *1356*
Starkov, M.G. see Eremenko, A.M., 1347, *1356*
Starodubtseva, R.V. see Kiselev, A.V., 1347, *1359*
Staschewski, D. see Wolff, H., 1237, *1259*
Stasyuk, I.V., 341, *357*
Statz, G. see Lippert, E., 806, *828*
Staudte, B. see Pfeifer, H., 1295, *1362*
Staveley, L.A.K. see Herrington, T.M., 989, 991, *1020*
Stearn, A.E. see Eyring, H., 1402, *1421*
Stebbings, R.F. see Stokes, E.D., 18, *22*
Steele, B.D. see Bagster, L.S., 473, *523*
Steele, W.A., 386, *389*
Stein, W.H. see Crestfield, A.M., 761, *763*
Steinemann, A., 984, *1024*
Steinfink, H., 483, *525*
Steinman, D.K., 932, *936*
Stell, G., 378, 381, *388*
Stell, G. see Rushbrooke, G.S., 386, *389*
Stenhouse, I.A. see Von Dreele, P.H., 826, *830*
Stepanov, B.I., 221, 225, *244*, *357*, 589, *610*, 641, 643, *654*
Stephens, R.W.B., 1433, *1469*
Stepišnik, J., 600, *610*, 1158, *1166*
Stepišnik, J. see Blink, R., 843, 844, 846, 848, *885*, 1121, 1130, 1132, 1138, 1140, 1145, *1161*
Sterin, Kh.E. see Mavrin, B.N., 1155, *1166*
Stern, M.J., 1246, *1259*
Stevenson, D.P., 1335, *1363*
Stevenson, M.J., 1052, *1060*, 1465, *1469*
Steward, L.M. see Stakebake, J.L., 1346, *1363*
Stewart, R.F. see Hehre, W.J., 61, *158*
Stikeleather, J.A. see Purcell, K.F., 1204, 1207, 1211, 1212, 1214, 1215, 1216, *1224*
Stiller, H., 219, 224, *244*, 336, 347, *357*, 912, 929, *936*
Stiller, H. see Arsis-Eskinja, M., 921, 922, 924, 925, 928, 932, *933*, 1187, 1189, 1191, *1195*
Stiller, H. see Brockhouse, B., 1174, *1195*
Stiller, H. see Grimm, H., 224, *242*, 921, 932, *934*, 1179, 1181, 1190, *1195*
Stiller, H. see Plesser, T., 894, 921, 932, *935*
Stillinger, F.H., 84, *163*, 386, *389*, 1374, 1375, 1376, 1401, *1423*, 1430, 1437, *1469*
Stillinger, F.H. see Ben-Naim, A., 377, 381, *387*, 529, *560*, 1373, *1420*
Stillinger, F.H. see Hankins, D., 62, 73, 75, 133, 134, *158*, 207, *215*, 369, 379, *388*, 1320, *1357*
Stillinger, F.H. see Losonczy, M., 93, 96, 100, 102, *159*
Stillinger, F.H. see Rahman, A., 84, *161*, 372, 380, *388*, 901, 915, *935*, 970, *1023*, 1373, 1399, *1423*, 1430, *1469*
Stirling, G.C. see Collins, M.F., 932, *934*

Stockmayer, W.H., 92, *163*
Stokes, E.D., 18, *22*
Stolz, H. see Knözinger, H., 1332, 1334, 1337, *1359*
Stone, F.S. see Munuera, G., 1343, 1344, *1361*
Stone, W. see Hougardy, J., 1349, *1358*
Stone, W.E.E. see Cruz, M.I., 1327, 1328, *1355*
Storek, W., 797, 799, *829*, 1233, *1259*
Storms, R.D. see Wang, C.H., 775, 776, *789*
Stout, J.W. see Giaque, W.F., 363, *387*
Stover, E.D. see Connick, R.E., 1052, *1058*
Strecker, R.A. see Andersen, K.K., 711, *763*
Street, B.W. see Aldridge, W.N., 762, *763*
Strom, E.T. see Woessner, D.E., 1038, *1060*
Stromberg, R.R. see Lippincott, E.R., 1317, *1360*
Strommen, D.P., 1086, 1088, *1105*
Strong, R.L. see Long, F.T., 1202, *1224*
Strukov, B.A., 1130, 1145, 1154, 1157, *1164*, *1166*
Strunk-Lichtenberg, G. see Ginsberg, H., 1330, *1357*
Stuart, W.I., 1343, *1363*
Stuchly, S.S., 1010, *1024*
Stückelberg, E.C.G. see Morse, P.M., 222, *243*
Stuckey, J.E. see Fuller Jr., E.L., 1345, *1357*
Stymne, B., 576, *610*
Stymne, H. see Stymne, B., 576, *610*
Subbarao, E.C., 1153, *1166*
Subba Rao, V.V. see Low, M.J.D., 699, 718, *764*, 1277, 1278, 1297, 1299, 1301, 1302, *1360*
Sudnik-Hrynkiewicz, M. see Bajorek, A., 902, 929, 930, 932, *933*
Sudnik-Hrynkiewicz, M. see Janik, J.A., 902, 930, *934*
Sugawara, F., 1187, *1195*
Sugié, H. see Okada, K., 1119, 1120, 1154, *1163*
Suh, I.-H. see Chidambaram, R., 440, *453*
Suhai, S., 142, *163*
Suhai, S. see Karpfen, A., 135, 136, 137, 138, 142, *159*
Suhorukhov, I.B. see Kanaskova, Iu.D., 596, *609*
Suhrmann, R., 687, 696, 700, 744, *765*, 1403, *1423*
Sumita, M. see Makita, Y., 1147, *1163*
Summerfield, G.C. see Steinman, D.K., 932, *936*
Sundaralingam, M. see Arora, S.K., 483, 495, *523*
Sung, S., 386, *389*
Surdut, A. see Dalbert, R., 1047, *1058*
Surjadi, A.J. see Cox, G.W., 449, *453*
Šušić, M. see Vučelič, V., 1347, *1364*
Šušić, V., 1347, *1363*
Susott, R. see Negran, T.J., 1159, *1166*
Sussmann, J.A., 341, 348, *357*
Sustman, R., 51, *163*
Susuki, I. see Parizeau, M.A., 619, *653*
Sutcliffe, L.H. see Feeney, J., 1231, *1257*, *1258*
Sutton, L.E. see Nelson, S.M., 1003, *1023*, 1349, *1361*
Suzuki, M., 1131, *1164*
Svetina, S. see Blinc, R., 1127, 1128, *1161*
Svirmickas, A. see Hindman, J.C., 1052, 1053, 1058, *1059*, *1061*
Swanstrom, P. 182, *216*
Swings, P. see Delsemme, A.H., 1420, *1421*
Swiniarski, M.F. see Fild, M., 819, *828*
Swoboda, A.R., 1348, *1363*
Syrkin, Ja.K. see Katsman, L.A., 749, *764*
Syrkin, J.K. see Sobczyk, L., 946, 949, *1024*
Syrkin, Ja.K. see Vargaftik, M.N., 749, *766*
Syrkin, Ya.K. see Vassiliev, V.G., 940, *1024*
Szafran, M., 601, *610*, 703, 741, *765*
Szafran, M. see Brzezinski, B., 605, *607*
Szafran, M. see Dega-Szafran, Z., 602, *607*
Szczesniak, E., 1058, *1061*
Szkatula, A., 900, 901, 929, *936*
Szkatula, A. see Golikov, V.V., 912, 929, *934*

Tabata, Y. see Kuroda, N., 1140, *1163*
Taesler, I., 482, 483, 486, 493, 495, 508, 519, *525*, 663, 676, *682*
Taesler, I. see Delaplane, R.G., 482, 495, *523*, 663, *681*
Taft, R.W., 817, 818, *829*
Taft, R.W. see Gurka, D., 816, 817, 822, *828*
Taft, R.W. see Joris, L., 818, *828*
Taha, A.A. see Christian, S.D., 813, 814, *828*
Tahoun, S.A., 1349, *1363*
Tait, A.D., 84, *163*
Takagi, Y., 383, *388*, 990, *1024*, 1126, 1128, 1143, 1158, *1164*, *1166*, 1179, *1195*
Takagi, Y. see Ishibashi, Y., 1148, 1155, 1158, *1162*, *1166*
Takagi, Y. see Kameyama, H., 1160, *1166*
Takagi, Y. see Sato, K., 1158, *1166*
Takagi, Y. see Sawada, A., 1143, 1144, *1164*
Takakusa, M., 18, *22*

Takei, W.J. see Craven, B.M., 408, *453*
Takezawa, N., 1343, *1363*
Takezawa, N. see Low, M.J.D., 1343, *1360*
Takusagawa, F., 433, *455*
Takusagawa, F. see Kvick, Å., 427, 429, *454*
Talayev, M.V. see Deryagin, B.V., 1317, *1356*
Tallis, W. see Husbands, D.I., 1345, *1358*
Tamres, M., 1212, *1224*
Tamres, M. see Goodenow, J.M., 1202, *1223*
Tanabe, K. see Iizuka, T., 1343, *1358*
Tanaka, Y. see Saitô, H., 822, 823, *829*
Tandon, S.P. see Mookerji, A., 749, *764*
Tanguy, M. see Baron, D., 797, *827*
Tapia, O., 132, *163*
Tarasevich, Yu.I., 1348, 1349, *1363*
Tarasevich, Yu.I. see Gribina, I.A., 1349, *1357*
Tarasevich, Yu. I. see Sharkina, E.V., 1348, 1349, *1362*
Tarte, P., 594, *610*
Tate, J.T. see Mann, M.M., 475, *525*
Tatsuzaki, I. see Kasahara, M., 1159, *1166*
Tatsuzaki, I. see Mitsui, T., 1153, *1166*
Tatsuzaki, I. see Yagi, T., 1136, *1165*
Taubenberger, R., 969, 973, *1024*
Taylor, J.A.G. see Tyler, A.J., 1309, *1363*
Taylor, J.C., 441, 447, *455*
Taylor, J.H., 1345, *1363*
Taylor, M.J., 636, *654*
Taylor, R.C., 477, *525*, 660, 661, 663, *682*
Taylor, T.I. see Leung, P.S., 931, 932, *935*
Taylor, T.I. see Rush, J.J., 908, 925, 926, 929, 930, 931, *935*
Te Beek, J.B. see Wegdam, G.H., 774, *789*
Tegenfeldt, J. see Almlöf, J., 119, 123, *155*, 208, *214*, 438, 439, *453*
Tegenfeldt, J. see Berglund, B., 437, *453*
Tegenfeldt, J. see Lindgren, J., 438, *454*
Teh, C., 906, 907, *936*
Tellgren, R., 422, 423, 437, 438, 439, 443, 444, *455*, 1138, 1139, *1164*
Tellgren, R. see Lundgren, J.-O., 446, *454*, 483, 493, 495, 496, 500, 501, 504, 505, *525*
Tellgren, R. see Thomas, J.O., 422, 426, *455*, *456*
Temme, F.P. see Ghosh, R.E., 909, 932, *934*
Tempelhoff, K., 1305, 1311, *1363*
Templeton, D.H. see Forrester, J.D., 867, *886*
Templeton, D.H. see Olovsson, I., 414, *455*, 1255, *1258*
Tench, A.J., 1342, 1343, *1363*
Teranoto, Y. see Higashi, A., 968, 979, *1020*
Terenin, A.N. see Filimonov, V.N., 1290, 1322, *1356*
Terenin, A.N. see Roev, L.M., 699, 717, *765*, 1278, *1362*
Terenin, A.N. see Yaroslavsky, N.G., 1290, *1364*
Ter Haar, D. see Grindlay, J., 348, *355*
Tertykh, V.A., 1272, *1363*
Texeira-Dias, J.J.C. see Murrell, J.N., 40, *160*
Thaddeus, P., 1465, *1469*
Thaper, C.L., 902, 903, 930, *936*
Thaper, C.L. see Dasannachary, B.A., 932, *934*
Thaper, C.L. see Kim, H.J., 918, 919, 920, 931, *935*
Theveneau, H., 1160, *1167*
Thirsk, H.R. see Lown, D.A., 758, *764*
Thomas, D.K. see Davies, M., 805, *828*
Thomas, J. see Ratajczak, H., *356*
Thomas, J.M., 708, 758, 760, *765*, 995, *1024*
Thomas, J.O., 422, 423, 426, 439, *455*, *456*
Thomas, J.O. see Almlöf, J., 399, 417, *453*
Thomas, J.O. see Tellgren, R., 439, *455*
Thomas, R. see Kvick, Å., 427, 429, *454*
Thomas, R.K., 334, *357*, 582, *610*, 635, 641, *654*
Thompson, H.W. see Hadži, D., *20*, *355*
Thompson, H. see Thomas, R.K., 334, *357*, 582, *610*, 635, *654*
Thompson, H.B. see LaPlanche, L.A., 805, 806, 808, *828*
Thompson, J.K. see Resing, H.A., 1347, *1362*
Thomson, W., 963, *1024*
Thorp, J.M., 1292, *1363*
Thorp, J.M. see Nair, N.K., 1304, 1315, *1361*
Thorp, J.M. see Smith, B.R., 1338, *1363*
Thorson, I. see Egelstaff, P.A., 912, 929, *934*
Thül, B. see Endom, L., 1055, *1058*
Tichy, M., 603, *610*
Tickner, A.W. see Knewstubb, P.F., 475, *524*
Timoshchenko, G.T. see Metsik, M.S., 1348, *1361*
Tinkham, M. see Barker, A.S., 347, *354*
Tippe, A. see Brill, R., 1433, 1434, 1464, *1467*
Tokuda, F. see Morimoto, T., 1343, 1345, *1361*
Tokumaru, Y. see Abe, R., 1142, *1160*
Tokunaga, M., 247, 266, *292*, *357*, 1128, 1133, *1164*, 1180, 1181, *1195*
Tokunaga, M. see Mitsui, T., 1154, *1166*

Tomasi, J. see Alagano, G., 82, 85, 86, *155*
Tomasi, J. see Berthier, G., 82, *155*
Tomasi, J. see Bonaccorsi, R., 48, 60, 62, 82, 84, 85, *155*
Tomasi, J. see Petrongolo, C., 82, *161*
Tomasi, J. see Pullman, A., 82, 85, *161*
Tomasi, J. see Scrocco, E., 82, *162*, 855, *887*
Tomchuk, E., 1057, *1061*
Tornberg, N.E. see Lowndes, R.P., 1155, *1166*
Torrey, H.C., 1050, *1060*
Torrie, B.H., 1145, 1157, *1165*, *1167*
Torrini, M. see Cini, R., 1398, *1421*
Torruella see Blum, L., 378, *387*
Toth, L.M. see Bates, J.B., 671, 672, 675, *681*
Touillaux, R., 1348, *1363*
Touillaux, R. see Fripiat, J.J., 758, *763*, 1295, 1299, 1300, 1302, *1357*
Tovbis, A.B., 1139, *1165*
Tovbis, A.B. see Ranner, N.B., 902, *935*
Townes, C.H., 835, 836, *887*
Townes, C.H. see Stevenson, M.J., 1052, *1060*, 1465, *1469*
Toyoda, K., 1115, *1165*
Toyoshima, I. see Takezawa, N., 1343, *1363*
Traficante, D.D. see Nakashima, T.T., 825, *830*
Trautmann, S., 744, *765*
Travers, D.N. see Morcom, K.W., 1204, *1224*
Treacy, E.B., 1051, *1060*
Treibmann, D., 1337, *1363*
Trepadus, V., 1057, *1061*
Treszczanowicz, A. see Kehiaian, H., 1231, *1258*
Tretyakov, N.E., 1344, *1363*
Trevino, S., 909, 932, *936*
Trevino, S. see Boutin, H., 931, *934*
Trevino, S.F. see Prask, H.J., 929, *935*, 1461, 1462, 1462, 1463, *1469*
Triolo, R., 724, *765*
Trontelj, Z., 1146, *1165*
Trontelj, Z. see Blinc, R., 857, 867, 868, 871, 885, *885*
Trotter, P.J. see Lippert, J.L., 946, *1021*
Truby, F.K., 968, *1024*
Trueblood, K.N. see Schomaker, V., 463, *469*
Tsallis, C., 1129, *1165*
Tsang, T., 1148, *1165*
Tsang, T. see Genin, D.J., 1122, *1162*
Tsang, T. see O'Reilly, D.E., 1145, *1163*
Tschapek, M. see De Busetti, S.G., 1348, *1355*
Tsubomura, M., 71, *163*
Tsuno, S. see Mataga, N., 10, *21*
Tubino, R., *357*, 932, *936*
Tubino, R. see Bosi, P., 1461, 1462, 1463, *1467*
Tucker, E.E., 797, 798, 799, 801, 803, 804, 819, 826, *829*, *830*, 1233, *1259*
Tucker, E.E. see Christian, S.D., 805, 821, *828*
Tunkelo, E. see Hughes, D.J., 912, 929, *934*
Turchin, W.F., 915, *936*
Turner, J.J. see Perutz, R.N., 1096, *1105*
Turrell, G. see Huong, P.V., 595, *609*
Tursi, A.J., 549, *562*, 1086, 1087, 1088, *1105*
Tutsch, R., 1045, *1060*
Tutsch, R. see Göller, R., 1047, *1059*
Tutsch, R. see Hertz, H.G., 1045, 1055, *1059*
Tvaruzkova, Z. see Bosacek, V., 1347, *1354*
Tye, F.L. see Lee, J.A., 1345, *1360*
Tyler, A.J., 1309, *1363*
Tyler, A.J. see Armistead, C.G., 1267, 1272, *1353*
Tyutynnik, R.S. see Matyash, I.V., 1347, *1360*

Ubbelohde, A.R., 323, 324, *357*
Ubbelohde, A.R. see Pollock, J.M., 988, *1023*
Ubbelohde, A.R. see Rogers, E., 989, *1023*
Ubbelohde, A.R. see Robertson, J.M., 426, *455*
Udby, O. see Goldschmidt, H., 473, *524*
Uedaira, H., 1045, *1060*
Uehling, E.A. see Beezhold, W., 862, 863, *885*
Uehling, E.A. see Bjorkstam, J.L., 843, *885*, 1121, *1161*
Uehling, E.A. see Havlin, S., 1157, *1165*
Uehling, E.A. see Litov, E., 1120, 1127, 1130, 1131, *1163*
Uehling, E.A. see Schmidt, V.H., 989, 990, *1023*, *1024*, 1121, 1123, 1127, *1164*
Uehling, E.A. see Silsbee, H.B., 990, *1024*, 1126, 1129, 1130, 1132, 1154, 1156, 1157, *1164*
Uesu, Y. see Kobayashi, J., 1154, *1166*
Uhlenbeck, G.E., 222, *242*
Umeya, K., 1336, *1363*
Unruh, H.G. see Sailor, E., 1158, *1166*
Urban, Z., 1011, *1024*
Urey, H.C. see Kirshenbaum, I., 1246, 1254, *1258*
Uribe, E. see Jagendorf, A.T., 761, *764*

Urry, D.W. see Pitner, T.P., 823, *829*
Usha, K. see Venkataraman, G., 930, 931, *936*
Usha-Deniz, K. see Venkataraman, G., 931, *936*
Uytterhoeven, J. see Fripiat, J.J., 1272, *1357*
Uytterhoeven, J. see Mortland, M.M., 1348, *1361*
Uytterhoeven, J.B. see Van Clauwelaert, F.H., 1267, 1272, 1274, 1277, 1278, 1280, 1286, 1287, 1289, 1304, *1363*, *1364*

Vaal, E.G. see Millen, D.J., 477, *525*, 660, *681*
Vacacchino, M. see Antonini, M., 932, *933*
Vahrenholt, F. see Sustman, R., 51, *163*
Vaks, V.G., 1157, 1158, *1167*
Vaks, V.G. see Shuvalov, L.A., 1137, 1140, *1164*
Valkov, V.I., 636, *654*
Valleau, J.D. see Patey, N., 381, *388*
Valleau, J.P. see Patey, G.N., 386, *389*
Van Assche, J.B. see Van Cauwelaert, F.H., 1272, 1280, 1287, *1363*
Van Beek, L.K.H., 1313, *1363*
Van Caugh, L. see Cruz, M., 740, *763*, 1328, 1329, *1355*
Van Cauwelaert, F.H., 1267, 1272, 1274, 1277, 1278, 1280, 1286, 1287, 1289, 1304, *1363*, *1364*
Van Cauwelaert, F. see Leonard, A.J., 1330, *1360*
Vand, V., 371, 372, *388*, 1398, *1423*
Vander Donckt, E., 11, 13, *22*
Van der Elsken, J. see Wegdam, G.H., 774, *789*
Vanderkooy, J., 989, 993, *1024*, 1151, *1165*
Van der Linden, H. see Wegdam, G.H., 774, *789*
Van der Meersche, C. see Fripiat, J.J., 758, *763*, 1295, 1299, 1300, 1302, *1357*
Vandermeersche, C. see Touillaux, R., 1348, *1363*
Vandorpe, B. see Chapoton, A., 1002, *1018*
Vandorpe, B. see Fontaine, J., 1010, *1019*
Van Duijneveldt, F.B., 39, 45, 49, 60, *163*
Van Duijneveldt, F.B. see Kroon, J., 403, 404, 405, 419, *454*
Van Duijneveldt, F.B. see Van Duijneveldt-Van de Rijdt, J.G.C.M., 39, 45, 46, 47, 48, 129, *163*, 403, *456*
Van Duijneveldt-Van de Rijdt, J.G.C.M., 39, 45, 46, 47, 48, 129, *163*, 403, *456*
Van Duijneveldt-Van de Rijdt, J.G.C.M. see Kroon, J. 403, 404, 405, 419, *454*
Van Goldammer, E., 1039, 1044, 1045, 1050, 1051, 1056, *1059*
Van Hecke, P., 877, *887*
Van Hook, W.A., 1246, 1253, *1259*
Van Hook, W.A. see Bigeleisen, J., 1246, *1257*
Van Hook, W.A. see Jancso, G., 1244, *1258*
Van Hook, W.A. see Stern, M.J., 1246, *1259*
Van Hove, L., 1171, *1195*
Van Loon, R., 541, *562*
Van Ness, H.C., 1233, *1259*
Van Ness, H.C. see Haskell, R.W., 1233, 1235, *1258*
Van Ness, H.C. see Savini, C.G., 1237, *1259*
Van Olphen, H. see Hougardy, J., 1349, *1358*
Van Thiel, M., 79, *163*, 545, 546, 548, 551, *562*, 580, 595, *610*, 637, *654*, 1070, 1086, 1088, 1090, 1093, *1105*, 1335, *1364*
Van Thiel, M. see Pimentel, G.C., 1090, *1105*
Van Tongelen, M. see Fripiat, J.J., 1272, *1357*
Van Winkle, J. see Van Ness, H.C., 1233, *1259*
Vargaftik, M.N., 749, *766*
Vargaftik, M.N. see Katsman, L.A., 749, *764*
Varikash, V.M. see Leonidova, G.G., 1158, *1166*
Vasianian, L.K. see Bogachev, Y.S., 797, *827*
Vasilescu, D. see Mesnard, G., 1012, *1022*
Vassiliev, V.G., 940, *1024*
Vaughan, W.E. see Hill, N.E., 939, 995, *1020*
Vecchi, M. see Clusius, K., 1246, *1257*
Vedam, K. see Pepinsky, 1146, 1151, *1163*
Vedder, W. see Ketelaar, J.A.A., 183, 186, *215*
Vedrine, J.G., 1348, *1364*
Veksli, Z. see Maričić, S., 989, 993, *1022*
Venable, R.L., 1333, *1364*
Venkataraman, G., 930, 931, *936*
Venkataraman, G. see Kim, H.J., 918, 919, 920, 931, *935*
Venkateswaran, A., 985, *1024*
Venturello, G. see Della Gatta, G., 1333, 1334, 1335, *1355*
Verbist, J.J., 450, *456*, 460, 461, *469*
Verbist, J.J. see Hamilton, W.C., 459, *468*
Verbist, J.J. see Koetzle, T.F., 451, *454*, 460, *469*
Verbist, J.J. see Lehmann, M.S., 452, *454*, 460, *469*
Verdinne, K. see Cruz, M.I., 1320, 1322, 1324, 1325, 1326, *1355*

Verhoeven, J., 851, *887*, 1465, *1469*
Verlet, L., 386, *389*
Vermoortele, F. see Van Clauwelaert, F.H., 1274, 1277, 1278, 1286, 1287, 1289, 1304, *1363*
Verrall, R.E. see Senior, G.E., 1382, *1423*
Versino, C. see Zecchina, A., 1276, 1293, *1364*
Versmold, H., 1044, *1060*
Versmold, H. see Hertz, H.G., 122, *158*, 1055, 1056, *1059*
Verstegen, J.M.P.J., 1071, 1083, *1105*
Verwey, E.J.W., 84, *163*
Vetrano, F. see Beretta, E., 56, *155*
Vidale, G.L. see Taylor, R.C., 477, *525*, 660, 661, 663, *682*
Vijayaraghavan, P.R. see Venkataraman, G., 930, 931, *936*
Vikane, O. see Gramstad, T., 818, *828*
Villain, J., 1181, 1193, *1195*
Vinek, G. see Neckel, A., 121, *160*
Vineyard, G.H., 910, *936*
Vinogradov, S.N., *20*, *163*, *357*, 737, *766*
Vivien, D. see Gradsztajn, S., 1344, *1357*
Vleeskens, J.M. see De Boer, J.H., 1270, *1355*
Vliegenhart, J.A. see Kroon, J., 403, 404, 405, 419, *454*
Vogel, G.C., 1204, 1215, 1220, *1224*
Vogel, G.C. see Drago, R.S., 1206, 1211, 1212, 1213, 1215, 1216, 1222, *1223*
Vogel, H. see Brausse, G., 1013, *1017*
Vogel, H. see Reichle, M., 1013, *1023*
Vogelhut, P.O. see Orentlicher, N., 372, *388*
Volavšek, B. see Blinc, R., 867, 868, 871, *885*
Volkov, A.V., 1319, *1364*
Volmer, M., 474, *525*
Von Dreele, P.H., 826, *830*
Von Gutfeld, R.J. see Srinivasan, R., 18, *22*
Von Hippel, A., 966, 967, 968, 975, 976, 978, 983, *1024*, 1445, *1468*
Von Hippel, A. see Maidique, M.A., 973, 980, *1022*
Von Niessen, W., 67, 68, 69, 93, 96, 100, *163*
Von Niessen, W. see Clementi, E., 116, 117, 150, 151, *156*, 214, *215*, 219, *242*
Von Niessen, W. see Diercksen, G.H.F., 111, 121, 146, *157*
Von Raguè-Schleyer, P. see Joris, J., 1284, *1358*
Von Schleyer, P. see Allerhand, A., 951, *1017*
Von Schleyer, P. see Taft, R.W., 818, *829*
Von Stackelberg, M., 1415, 1419, *1423*
Vorotyntsev, M.A., 266, *292*
Vos, A. see Coppens, P., 446, *453*
Vu, H. see Atwood, M.R., 1083, *1104*
Vučelič, D. see Vučelič, V., 1347, *1364*
Vučelič, R. see Šušić, V., 1347, *1363*
Vučelič, V., 1347, *1364*
Vuks, M.F. see Bezrukov, O.F., 1049, *1058*

Waclawek, W. see Skulski, L., 940, *1024*
Wacrenier, J.M., 1009, *1024*
Waddington, T.C. see Dixon, H.P., 121, *157*
Waddington, T.C. see Ghosh, R.E., 909, 932, *934*
Wade, W.H. see Barto, J., 1338, *1353*
Wade, W.H. see Blake, T.D., 1338, *1354*
Wade, W.H. see Venable, R.L., 1333, *1364*
Waddington, T.C. see Salthouse, J.A., 602, *610*, 736, *765*
Wahl, A.C. see Bertoncini, P.J., 56, *155*
Wahlgren, U. see Almlöf, 119, 123, *155*, 518, 519, *523*
Wakuta, Y. see Palevsky, H., 930, 932, *935*
Waldner, F. see Boesch, H., 1151, *1161*
Waldron, R.D., 639, *654*
Waldsax, J.C.R. see Husbands, D.I., 1345, *1358*
Waldstein, P., 1051, *1060*, 1465, *1469*
Walker, J.C.F. see Paren, J.G., 983, *1023*
Wall, F.T., 172, *216*, 222, *244*
Wall, L.S. see Broberg, T.W., 1124, *1161*
Wall, L.S. see She, C.Y., 348, *357*, 1124, 1131, *1164*
Wall, T. see Trevino, S., 909, 932, *936*
Wall, T.T., 580, *611*
Wallace, R.A. see Urban, Z., 1011, *1024*
Wallwork, S.C. see Faithful, B.D., 484, *524*
Walmsby, S. see Anderson, A., 909, *933*
Walrafen, G.E., 208, *216*, 554, *562*, 972, *1025*, 1253, *1259*, 1382, 1383, 1384, 1385, 1390, 1393, 1395, *1423*
Walsh, K.A., 761, *766*
Walshaw, S.M. see Smith, J.W., 940, *1024*
Walters, G.K. see Stokes, E.D., 18, *22*
Wang, C.H., 775, 776, 777, *789*
Wang, J.H., 1402, *1423*
Wang, S.M. see Shaw, Y.L., 814, *829*
Ward, J.W. see Snyder, L.R., 1271, 1272, *1363*
Warner, D.T., 558, *562*
Warren, B.E. see Morgan, J., 368, *388*, 1372, *1423*
Warshel, A., 85, *163*
Warshel, A. see Lifson, S., 85, *159*

Watermark, H. see Stymne, B., 576, *610*
Watkinson, J.G. see Jones, D.A.K., 1201, *1224*
Watton, A., 1149, 1160, *1165*
Watts, R.O., 379, 381, 386, 387, *388*, *389*
Watts, R.O. see Barker, J.A., 380, *387*, 1430, *1466*
Waugh, J.S., 867, 873, 874, 880, 881, *887*
Waugh, J.S. see Haeberlen, U., 873, *886*
Waugh, J.S. see Miller, S.R., 908, *935*, 991, *1022*
Waugh, J.S. see Pines, A., 882, *887*
Waugh, J.S. see Van Hecke, P., 877, *887*
Wayland, B.B. see Drago, R.S., 1202, 1216, *1223*
Weast, R.C., 710, *766*
Weaver, J.C. see Van Hecke, P., 877, *887*
Webster, R.K., 1343, *1364*
Weckermann, B. see Schenk, C.H., 932, *936*
Weeks, J.D. see Andersen, H.C., 368, *387*
Wegdam, G.H., 774, *789*
Weidemann, E.G., 177, 181, 182, 204, 205, *216*, 231, 232, 236, *244*, 247, 248, 251, 254, 260, 290, *293*, 640, *654*, 687, 688, 689, 722, 723, 729, 730, 757, *766*
Weidemann, E.G. see Hayd, A., 257, *292*, 733, *763*
Weidemann, E.G. see Janoschek, R., 122, *158*, 175, 176, 178, 179, 180, 181, 182, 183, 191, 193, *215*, 221, 225, *243*, 247, 251, 253, 255, 260, *292*, 297, 300, 304, 324, 329, *355*, 619, 644, *653*, 688, 722, 723, 724, 728, 731, 733, 746, *764*
Weidemann, E.G. see Kremer, F., 254, *292*
Weidemann, E.G. see Pfeiffer, H., 277, *292*, 754, *765*
Weidemann, E.G. see Zundel, G., 248, *293*, 699, 730, 748, 761, *766*
Weidner, J.U. see Bertagnolli, H., *155*
Weinstein, H. see Srebrenik, S., 83, *163*
Weir, C., 1450, *1470*
Weis, J.-J. see Verlet, L., 386, *389*
Weiss, A. see Lagaly, G., 1349, *1359*
Weiss, A. see Schön, G., 1005, *1024*
Weiss, H.G. see Shapiro, I., 1272, *1362*
Weisskopf, V., 321, *357*
Weissman, M., 839, 841, *887*
Weith, A.J., 941, *1025*
Weller, A., 13, *22*
Weller, E.F., 1115, *1165*
Weller, S.W., 1332, *1364*
Welsh, H.L. see Plint, C.A., 1255, *1258*
Welz, P. see Fink, P., 1273, *1356*
Wen, W.Y. see Frank, H.S., 362, 368, 370, 372, *387*, 553, *560*, 1057, *1059*
Wen, W.Y. see Hertz, H.G., 1057, 1058, *1059*, *1061*
Wenzel, R.G. see Arnold, G.P., 1433, *1466*
Wenzl, H. see Pick, M., 979, *1023*
Weres, O., 368, 375, *388*
Werner, A., 5, *22*
Werner, D. see Derbyshire, W., 879, *886*
Wertheim, M.S., 378, 381, *388*
West, R., 1201, *1224*
West, W., 1284, *1364*
Weston, R.E., 711, *766*
Westphal, W.B. see Maidique, M.A., 973, 980, *1022*
Westphal, W.B. see Roebuck, B.D., 1017, *1023*
Westphal, W.B. see Von Hippel, A., 967, 968, 975, 976, 978, 983, *1024*
Westrum, E.F. see Davis, M.L., 989, 992, *1018*
Whalley, E., 651, *654*, 939, 970, 971, 972, 973, *1025*, 1434, 1435, 1441, 1444, 1448, 1451, 1454, 1455, 1458, 1460, 1461, 1463, 1464, *1470*
Whalley, E. see Bertie, J.E., 637, *652*, 972, *1017*, 1427, 1429, 1433, 1435, 1441, 1455, 1457, 1458, 1460, 1461, 1462, 1464, *1466*, *1467*
Whalley, E. see Calvert, L.D., 1441, *1467*
Whalley, E. see Chan, R.K., 969, *1017*
Whalley, E. see Engelhardt, H., 1429, *1467*
Whalley, E. see Gough, S.R., 985, *1020*
Whalley, E. see Johari, G.P., 970, *1021*, 1448, 1449, 1450, *1468*
Whalley, E. see Klug, D.D., 972, *1021*, 1463, 1464, *1468*
Whalley, E. see Kuriakose, A.K., 982, *1021*
Whalley, E. see Nishibata, K., 1445, *1469*
Whalley, E. see Taylor, M.J., 636, *654*
Whalley, E. see Wilson, G.J., 970, 971, *1025*
Whalley, E. see Wong, P.T.T., 778, *789*, 1461, 1462, *1470*
Whateley, T.L. see Stuart, W.I., 1343, *1363*
Whatley, L.S. see West, R., 1201, *1224*
Whetsel, K.B., 797, 824, 825, *829*
Whitaker, M.A.B. see Grannel, P.K., 882, *886*
White, D. see Mann, D.E., 1101, *1105*
White, D.N. see Brot, C., 1012, *1017*
White, H.F. see Hornig, D.F., 1458, *1468*
White, J.L., 1285, *1364*
White, J.L. see Low, P.F., 1343, 1348, *1360*

White, J.W. see Aldred, B.K., 927, 932, *933*
White, J.W. see Downes, J.S., 1347, *1356*
White, L.R. see Perram, J.W., 379, 386, *388*
Whitefield, G.D. see Kuper, C.G., 305, 310, *356*
Whitehead, M.A. see Kaplansky, M., 855, *886*
Whitten, J.L. see Buenker, R.J., 838, *886*
Whittle, E., 1067, *1105*
Wicke, E., 475, *525*, 677, *682*, 688, 753, *766*
Wickersheim, K.A. see Anderson Jr., J.H., 1267, 1306, 1307, 1308, *1353*
Widom, B., 384, *388*
Wiedersich, I. see Suhrmann, R., 744, *765*
Wiener, E. see Pelah, I., 908, 932, *935*
Wiener, E. see Schenk, C.H., 932, *936*
Wigner, E. see Weisskopf, V., 321, *357*
Wilkinson, G. see Evans, D.F., 601, *608*
Wilkinson, G. see Gillard, R.D., 665, 668, 670, 676, 679, *681*, 747, *763*
Will, G., 247, *293*
Willett, R.D. see Peterson, E.R., 1151, *1163*
Williams, D. see Curnutte, B., 1383, *1421*
Williams, D.E. see Hanna, M.W., 946, *1020*
Williams, D.R. see Murrell, J.N., 39, 49, 56, 112, *160*
Williams, G., 940, *1025*
Williams, J.M., 118, *163*, 427, 431, 449, *456*, 483, 484, 492, 493, 500, 501, 502, 506, 508, 509, *526*, 657, 668, 673, 674, 675, *682*, 692, 695, 697, 747, 748, *766*, 872, *887*
Williams, J.M. see Basile, L., 663, 664, 665, 666, 667, 670, 675, 676, 679, *681*
Williams, J.M. see Finholt, J.E., 663, 664, *681*
Williams, J.M. see LaBonville, P., 665, *681*
Williams, J.M. see Lundgren, J.-O., 446, *454*, 483, 495, 496, *525*, 663, 664, 675, *681*
Williams, J.M. see O'Reilly, D.E., 497, 506, *525*, 660, 673, *681*, *682*, 872, *887*, 905, *935*
Williams, J.M. see Rozière, J., 668, *682*
Williams, K. see Davidson, D.W., 985, *1018*
Williams, O.M. see Groves, M.R., 17, *21*
Williams, R.C., 573, *611*
Williams, R.L. see Bellamy, L.J., 1288, *1353*, 1377, *1420*
Willis, B.T.M. see Pawley, G.S., 400, *455*
Wilson, G.J., 970, 971, *1025*
Wilson, G.M., 1228, *1259*
Wilson, G.V.H., 347, *357*
Winde, H., 1416, *1423*
Winde, H. see Tempelhoff, K., 1305, 1311, *1363*
Winick, J.R. see Morokuma, K., 49, 60, 62, 73, 75, 111, *160*, 206, *215*
Winkler, H., 1336, 1337, *1364*
Winkler, H. see Freude, D., 1295, *1357*
Winkler, H. see Reball, S., 1336, 1337, *1362*
Winkler, H. see Riedel, E., 1348, *1362*
Winmill, T.F. see Moore, T.S., 1201, *1224*
Winterhalter, D.R. see Savini, C.G., 1237, *1259*
Wirzing, G., 1306, *1364*
Witkop, B., 716, *766*
Witkowski, A., 297, 309, 322, 325, 331, 346, *357*, 590, 606, *611*, 643, *654*
Witkowski, A. see Maréchal, Y., 173, 179, 214, *215*, 221, *243*, 297, 305, 307, 324, 332, 341, *356*, 590, 591, *609*, 643, *653*, 1099, *1105*
Witt, H.T., 761, *766*
Witzel, H. see Rüterjans, H., 761, *765*
Woessner, D.E., 1038, 1051, *1060*, 1269, 1294, 1311, *1364*, 1465, *1470*
Wojcik, M., 606, *611*
Wojcik, M. see Witkowski, A., 306, 309, 322, 325, 346, *357*, 590, 606, *611*
Wolf, K.L., 533, *562*
Wolfbauer, O. see Gölles, F., 1228, *1258*
Wolff, E. see Wolff, H., 302, *357*, 1231, 1252, 1253, *1259*, *1260*
Wolff, H., 19, *22*, 214, *216*, 302, *357*, 606, *611*, 630, 631, *654*, 1228, 1229, 1230, 1231, 1233, 1234, 1235, 1236, 1237, 1238, 1240, 1243, 1246, 1248, 1249, 1250, 1251, 1252, 1253, 1254, *1259*, *1260*
Wolfsberg, M. see Stern, M.J., 1246, *1254*
Wolleschensky, E. see Fink, P., 1273, *1356*
Wolniewicz, L. see Kolos, W., 101, *159*
Wong, P.T.T., 778, *789*, 1461, 1462, *1470*
Wood, H.L. see Dransfeld, K., 1313, 1336, *1356*
Wood, J.L. see Dean, R.L., 606, *611*
Wood, J.L., 19, *22*, 587, 605, 606, *611*, 716, *766*
Wood, J.L. see Borah, B., 606, *611*
Wood, J.L. see Clements, R., 587, 601, *607*, 706, 716, *763*
Wood, J.L. see Cummings, D.L., 324, *354*, 569, *607*
Wood, J.L. see Ginn, S.G.W., 305, *355*, 568, *608*
Wood, J.L. see Hall, A., 302, 303, *355*, 583, *608*, 782, 783, *789*
Wood, J.L. see Masri, F.N., 601, *609*
Wood, J.L. see Rice, S.A., 221, *243*, *356*, 573, *610*, 647, *654*

Wood, J.L. see Singh, T.R., 174, 175, 178, 181, *216*, 221, 225, *244*, *357*, 424, 425, *455*, 507, *525*, 621, *654*
Wood, M. see Hindman, J.C., 1052, 1053, 1058, *1059*, *1061*
Woodings, C.R. see Husbands, D.I., 1345, *1358*
Woods, A.D.B., 929, 930, *936*
Wooley, H.W., 1246, *1260*
Woolf, J.B. see Thorp, J.M., 1292, *1363*
Woolfson, M.M. see Germain, G., 461, *468*
Worley, J.D., 1253, *1260*, 1395, *1423*
Worlock, J.H. see Scott, J.F., 1132, *1164*
Worsham, J.E., 449, *456*
Wörz, O., 967, 969, *1025*
Wright, R.B. see Wang, C.H., 777, *789*
Wu, F.Y., 385, *388*
Wu, F.Y. see Baxter, R.J., 385, *387*
Wu, F.Y. see Lieb, E.H., 1448, *1469*
Wunderlich, H. see Mootz, D., 483, 495, *525*
Würtz, R. see Wolff, H., 631, *654*, 1230, 1236, 1237, 1238, 1240, 1243, 1246, 1249, 1250, *1259*
Wüthrich, K. see Conninck, R.E., 1052, *1058*

Yagi, T., 1136, *1165*
Yagil, G. see Luz, Z., 1399, *1422*
Yakatan, G.J., 16, *22*
Yamabe, S., 150, *163*
Yamaguchi, N. see Kondo, S., 1336, *1359*
Yamaji, K., 1434, *1470*
Yamamoto, T., 1150, *1165*
Yamamoto, T. see Chihara, H., 856, 857, *886*, 1160, *1165*
Yamdagni, R. see Arshadi, M., 121, *155*
Yan, J.F. see Momany, F.A., 124, *160*
Yang, C.N. see Lee, T.D., 370, *388*
Yarkony, D.R., 61, 72, 129, 132, *163*
Yaroslavsky, N.G., 1267, 1290, *1364*
Yaroslavskii, N.G. see Brodskii, I.A., 784, 785, *789*, 1267, 1308, *1354*
Yasaitis, E.L. see O'Reilly, D.E., 1038, *1060*
Yates, D.J.C., 1335, 1343, 1344, *1364*
Yates, D.J.C. see Carter, J.L., 1330, 1331, 1332, *1354*
Yates, D.J.C. see Folman, M., 1277, 1279, 1280, 1296, 1297, 1304, *1356*
Yates, D.J.C. see Sheppard, N.V., 1290, *1362*
Yates, K. see Hopkinson, A.C., 113, 146, 147, *158*
Yenger, E.A., 304, *357*
Yerkess, J. see Ferraris, G., 441, 442, 443, *453*
Yi, F.-N., 1148, *1165*
Yip, S., 929, *936*
Yip, S. see Agrawal, A.K., 932, *933*
Yip, S. see Prask, H.J., 929, *935*
Yip, S. see Trevino, S., 909, 932, *936*
Yoneda, Y. see Arai, H., 1337, *1353*
Yonezawa, T. see Morishima, I., 820, *829*
Yoon, C., 1055, *1060*
Yoon, C. see Hertz, H.G., 1055, *1059*
Yoon, C. see Versmold, H., 1044, *1060*
Yoon, Y.K., 478, 482, 495, 519, *526*, 663, *682*
Yos, J.M. see Jehle, H., 254, *292*, 729, *764*
Yoshida, Z., 1201, *1224*
Yoshida, Z.I. see Osawa, E., 573, *610*
Yoshihara, K., see Matsuzaki, A., 14, *21*
Yoshimine, M. see McLean, A.D., 185, *215*
Yoshimine, V. see McLean, A.D., 101, 112, 118, *160*
Yoshinaga, H. see Hineno, M., 781, *789*
Yost, P.M. see Waugh, J.S., 867, *887*
Young, A.P., 1156, 1157, *1167*
Young, A.P. see Elliott, R.J., 1133, *1162*
Young, I.G., 983, *1025*
Young, R.P., 1279, *1364*
Yudin, A.L. see Gavrilova-Podol'skaya, G.V., 1136, 1139, *1162*
Yuzvak, V.I. see Aleksandrova, I.P., 1146, *1160*
Yuzvak, V.I. see Kruglik, A.I., 1159, *1166*

Zabicky, J. see Millen, D.J., 224, *243*, 330, *356*
Zachariasen, W.H., 461, *469*
Zaitseva, M.P. see Aleksandrov, K.S., 1146, *1160*
Zakrajšek, E. see Žaucer, M., 841, 842, *887*
Zalkin, A. see Forrester, J.D., 867, *886*
Zarembovitch, A., 1433, *1470*
Zaromb, S., 982, *1025*
Zaucer, M., 841, 842, *887*
Zbinden, R., 786, 787, *789*
Zecchina, A., 1276, 1293, 1345, 1346, *1364*
Zecchina, A. see Borello, E., 1269, 1318, 1319, 1320, 1321, 1322, 1324, 1325, 1327, *1354*
Zeegers-Huyskens, Th. see Clerbaux, T., 955, *1018*
Zeegers-Huyskens, Th. see Duterme, P., 955, *1018*
Zeegers-Huyskens, Th. see Huyskens, P., 947, *1020*

Zeegers-Huyskens, Th. see Lichtfus, G., 782, 783, *789*
Zeer, É.P. see Habuda, S.P., 1149, *1162*
Zeer, É.P. see Krasnikova, A.Ya., 1149, *1163*
Zeidler, M.D., 1038, 1039, 1057, *1060*
Zeidler, M.D. see Bender, H.J., 1038, *1058*
Zeidler, M.D. see Endom, L., 1055, *1058*
Zeidler, M.D. see Engelsmann, K., 1038, *1058*
Zeidler, M.D. see Hertz, H.G., 1056, *1059*
Zeidler, M.D. see Kintzinger, J.P., 1047, *1059*, *1060*
Zeidler, M.D. see Langer, H., 1058, *1061*
Zeidler, M.D. see Van Goldammer, E., 1044, 1045, 1050, 1051, 1056, *1059*
Zein, N.E. see Shuvalov, L.A., 1137, 1140, *1164*
Zein, N.E. see Vaks, V.G., 1158, *1167*
Žekš, H., 1143, *1165*
Žekš, B. see Blinc, R., 340, 348, *354*, 1115, 1127, 1134, 1155, *1161*, *1165*, 1183, 1185, 1193, *1195*
Žekš, B. see Lavrencie, B., *356*
Zellsman, H.R., 606, *611*
Zerbi, G. see Bosi, P., 1461, 1462, 1463, *1467*
Zerbi, G. see Dempster, A.B., 581, 587, *607*, 778, *789*
Zerbi, G. see Pelligrini, A., 778, *789*
Zerbi, G. see Tubino, R., *357*, 932, *936*
Zerdecki, J.A. see Blake, T.D., 1338, *1354*
Zettlemoyer, A.C., 1307, 1308, 1310, 1345, *1364*
Zettlemoyer, A.C. see Basset, D.R., 1316, *1353*
Zettlemoyer, A.C. see Klier, K., 1267, 1304, 1307, 1308, 1309, 1310, 1311, *1359*
Zettlemoyer, A.C. see McCafferty, E., 1008, *1022*, 1336, 1345, *1360*, *1361*
Zhdanov, S.P., 1304, *1364*
Zheludev, I.S., 1130, 1153, *1165*, *1167*
Zherebtsova, L.I. see Aleksandrov, K.S., 1146, *1160*
Zhidomirov, G.M. see Kabankin, A.S., 820, *828*
Zhilenkov, I.V. see Dubinin, M.M., 1002, *1018*
Zhilenkov, I.V. see Fedorov, V.M., 1001, *1019*
Zhilenkov, I.V. see Glazun, B.A., 1001, 1002, *1019*
Zhilenkov, I.V. see Kopylova, V.M., 1001, *1021*
Zhilenkov, I.V. see Nekrasova, E.G., 1006, *1022*, 1315, 1316, *1361*
Zhogolev, D.A. see Matyash, I.V., 1347, *1360*
Zhuravlev, L.T., 1270, *1364*
Zhuravlev, L.T. see Davydov, V.Ya., 1267, 1270, 1303, *1355*
Zielen, A.J. see Hindman, J.C., 1052, 1053, *1059*
Ziessow, D., 811, 822, 823, *829*
Zigan, F., 443, *456*
Zimermann, H. see Haeberlen, U., 877, 878, 879, 880, *886*
Zimmerman, J.R., 1042, 1043, *1060*, 1311, *1364*
Zimmerman, J.R. see Woessner, D.E., 1311, *1364*
Zimmermann, H., *20*, 219, 220, 221, *244*
Zimmermann, H. see Brickmann, J., 173, 175, *214*, *215*, 220, 221, 222, 223, 225, 228, 229, 230, 236, 237, 239, 241, *242*, 247, 289, *292*
Zimmermann, H. see Rudolph, J., 677, 678, 679, 680, *682*
Zimmermann, H.W. see Bertagnolli, H., *155*
Zimmermann, R., 1458, *1470*
Zinenko, V.I. see Vaks, V.G., 1157, *1167*
Zinenko, V.N. see Strukov, B.A., 1154, 1157, *1166*
Žitko, M. see Hadži, D., 594, 595, *608*
Zolla, A. see Kebarle, P., 121, 139, 140, *159*, 475, 481, 519, *524*, 980, *1021*
Zukovskij, A. see Luck, W.A.P., 1413, *1422*
Zukowska, I. see Golikov, V.V., 912, 929, *934*
Žumer, S., 1160, *1167*
Žumer, S. see Blinc, R., 843, 844, 846, 848, 862, 863, 866, *885*, *886*, 1121, 1130, 1131, 1132, 1141, 1154, *1161*, *1165*
Zundel, G., *20*, 247, 248, *293*, 673, *682*, 687, 689, 692, 693, 695, 696, 697, 698, 699, 700, 708, 709, 717, 729, 730, 744, 748, 749, 753, 754, 757, 760, 761, *766*, 1011, *1025*, 1302, 1320, 1333, *1364*, 1414, 1419, *1423*
Zundel, G. see Ackermann, T., 693, 700, *762*, *763*
Zundel, G. see Hayd, A., 257, *292*, 733, *763*
Zundel, G. see Janoschek, R., 122, *158*, 175, 176, 178, 179, 180, 181, 182, 191, 193, *215*, 221, 225, *243*, 247, 251, 253, 255, 260, *292*, 297, 300, 304, 324, 329, *355*, 619, 644, *653*, 688, 722, 723, 724, 728, 731, 733, 746, *764*
Zundel, G. see Kampschulte-Scheuing, I., 695, 696, 697, *764*
Zundel, G. see Kremer, F., 254, *292*
Zundel, G. see Lindemann, R., 703, 705, 709, 737, 741, *764*, 949, *1021*

Zundel, G. see Matthies, M., 703, 761, *764*
Zundel, G. see Noller, H., 699, 710, 718, 738, *765*, 1285, 1301, *1361*
Zundel, G. see Papakostidis, G., 698, 703, 720, 760, *765*
Zundel, G. see Pernoll, I., 689, 696, 733, *765*
Zundel, G. see Pfeiffer, H., 277, *292*, 754, *765*
Zundel, G. see Schiöberg, D., 688, 695, 696, 698, 700, 701, 703, 708, 709, 711, 716, 718, 719, 740, 743, 744, 745, 751, 753, 754, 759, *765*
Zundel, G. see Sessler, W., *357*, 698, 701, 702, 708, 709, 710, 711, 719, 731, 738, 740, *765*
Zundel, G. see Weidemann, E.G., 177, 181, 182, 204, 205, *216*, 231, 232, 236, *244*, 247, 248, 251, 254, 260, 290, *293*, 687, 688, 689, 722, 723, 729, 730, 757, *766*
Zupančič, I. see Blinc, R., 859, 860, 861, 862, 885, *885*, 1138, 1140, 1144, *1161*
Zussman, A. see Edmonds, D.T., 851, *886*, 1052, *1058*, 1465, *1467*
Zwanzig, R. see Nee, T.W., 1054, *1060*
Zwernemann, K. see Ackermann, T., 693, 700, *762*, *763*
Zwolinski, B.Z. see Fukushima, K., 325, *355*, 568, *608*

SUBJECT INDEX TO VOLUMES I – III

Ab initio calculations 34 ff, 61f, 65f, 73, 75f, 94 ff, 101, 108ff, 117, 120, 147, 183, 186, 188, 191, 199, 210, 220, 379, 415, 425, 438, 517, 723 ff
  method 30 ff, 167, 175
Acceptor atom 401, 736
Acetic acid 932, 1099, 1100
  cyclic dimer 92f, 735
  trichloroacetic acid, mixed cyclic dimer 93
  trifluoroacetic acid, mixed cyclic dimer 93
Acetoaceticacidester 5
Acetone 942
  adsorbed on silica 1274, 1275, 1279
  adsorption on oxides 1274, 1275, 1279, 1344, 1349
  complexes of 119
Acetonitrile 8
  complex with hydrogenfluoride 642
Acetylacetone, enol 125, 127
Acetylene, complexes with polar molecules 94ff, 104ff
Acids 687, 689ff, 742ff, 750ff, 1414
  conductivity of solids 979ff
  conductivity of solutions 757ff
  dissociation process 753
  salt effects on polarizable H-bonds 744f
  solvate structures 689ff, 753ff
Acid salts 422, 427, 429
  of carboxylic acids 587, 597ff
Acoustic phonon branches 902
Acrolein, complexes of 145
Acridine 10
Activation energy 693, 966, 1113, 1121, 1123, 1126, 1146, 1149, 1155
  of the viscosity of H_2O 1402
Activation enthalpy 917, 972
Activation entropy 917
Activity coefficients 1227ff
  of H^+ and neutral salts 744f
  of variously deuterated methylamines and methanols in solutions with *n*-hexane 1239
Additivity, pairwise in intermolecular forces 132ff, 136, 140, 142
Adiabatic approximation 174, 180
Adiabatic photoreaction 14ff
Adsorbed molecules 1263ff
  polarizable H-bonds between 717
  tetrahydrofuran 784
  water, on zeolite 785, 786
Adsorbed water 998ff, 1303ff, 1333ff
Aggregates of molecules 1040
Aging effects 966
Alcohol spectra 1378
Alcohols 540, 942, 1070, 1092, 1093, 1096
  adsorbed on alumina 1337ff
    adsorption isotherms 1338
    autophobicity 1338, 1339
    coordinately bonded 1339, 1341
    heats of adsorption 1338, 1342
    IR fundamentals 1339ff
    NMR chemical shifts 1339
  adsorbed on oxides 1318ff, 1337ff, 1343ff
  adsorbed on silica 1318ff
    adsorption isotherms 1322
    cluster formation 1320, 1321ff
    deuterium NMR 1328
    deuterium quadrupole coupling constant 1328
    dielectric behavior 1322, 1328
    dynamic behavior 1327ff
    electrical conductivity 1328
    entropy of adsorption 1327
    equation of state 1329
    heat of adsorption 1319, 1321, 1322, 1327, 1328
    IR fundamental 1320ff
    IR overtone 1322ff
    NMR chemical shifts 1319
    NMR relaxation 1327
    proton transfer in 1328
    self association 1320
    surface diffusion 1327, 1328, 1329
Alcohols, H-bonding in 622ff

with acid IR continuum 694, 704
with base IR continuum 705
Aldehydes, adsorption on oxides 1279, 1343
Aliphatic dialcohols 542
Alkene–HI complexes 1102, 1103
Alumina
adsorption on 1329ff
hydroxyl group on 1330ff
characterization of surface 1329ff
LEED pf-surface 1332
surface structure 1329ff, 1332
Alums
methylammonium aluminum sulfate dodecahydrate, ferric ammonium sulfate dodecahydrate 1146f, 1159
Amides 85, 89, 109, 111, 632
adsorption on oxides 1349
Amines 627ff
adsorbed on oxides 1274, 1278, 1281, 1286, 1289, 1301, 1344ff, 1349
adsorbed on silica 1274, 1278, 1281, 1286, 1289
Amino acids 459ff
Ammonia 414, 932
adsorbed on alumina 1342
adsorbed on silica 1274, 1295ff
continuous absorption 717ff, 1301
dielectric behavior 1298
electrical conductivity 1299
heat of adsorption 1298
IR spectra of 1295
length changes of adsorbent 1296ff
NMR chemical shifts 1299ff
NMR relaxation time 1298ff
NMR wide line 1297
proton transfer 1299
adsorption on oxides 1274, 1295ff, 1341ff
water complex 64, 68, 89, 109, 131
dimer 92, 108
complexes of 108ff, 131, 150
Ammonium
azide 931
berylliumtetrafluoride 931
bisulfate 1146, 1159
bromide 931
cadmium sulfate 1145f
carbonate 931
chromate 932
chloride 16, 17, 40, 906
compounds 930
dichromate 931
dihydrogen arsenate 1117, 1123, 1155
dihydrogen phosphate 362, 382f, 1119, 1122, 1124f
fluoberyllate (AFE) 1145
fluoride 931
hydrogensulfate 931
hydroxide 5
iodide 901, 931
ion, complex with ammonia 120, 150, 718, 1295
nitrate 931
perchlorate 931
phosphorhexafluoride 931
Rochelle salt (ARS) 1142, 1158
silicohexafluoride 931
sulfate 908, 912, 918, 920, 926, 931, 1145, 1159
thiocyanate 931
trihydrogen periodate 1141, 1136
vanadate 932
Anatase, adsorption on 1335, 1342, 1344
Anesthetic effect 559
Angle dependence of H-bond 403ff, 528ff, 536, 545, 1081, 1087, 1095
energy 533, 536
Anharmonic coupling 213, 305, 353, 347, 315, 571, 593
Angle dependent potential 1377
Anharmonicity 570, 571, 589, 592, 614–652
constants 616, 624, 625, 633
electrical 177
mechanical 177, 204
of alcohols 622ff, 1386
oscillator
diatomic 615
triatomic 618
Aniline, adsorbed on silica 1281, 1301
Anions 1404
Anionic H-bonds 7
'Anomalous' water 1317
Anisole, adsorbed on silica 1277, 1279, 1281
Antiferroelectrics 1114ff, 1136, 1139, 1141, 1142, 1148–1155
Apparent hydration numbers per ion pair 1410

Association constants
from vapor pressure measurements 1234
of variously deuterated alcohols, amines and methanethiols in solutions with *n*-hexane 1236, 1238
Association energies from vapor pressure measurements 1235
Association equilibria 1231, 1234ff
Association models 1231, 1233, 1235, 1253
Asymmetry parameter, definition 835
Atomic
charges 73
displacements 1176, 1180, 1184, 1190
units 30
ATP hydrolysis 760f
Auto-correlation function 901
Average degree of association 1235
^{75}As quadrupole coupling and electrical field gradient (EFG) tensor
temperature dependence 865
in antiferroelectric $NH_4H_2AsO_4$ 863
in ferroelectric KH_2AsO_4 863
effect of proton ordering 864
effect of formation of clusters of polarized H-bonds 866

Badger–Bauer relationship 575
Badger–Bauer rule 543, 563, 1379
Band broadening 577ff, 637ff, 750, 1099
Band shape 577, 588
Band width 577ff, 588ff, 637ff
Band splitting pattern 577ff, 1080
Bare proton model 178
Barium chlorate hydrate 930
Barium chloride dihydrate 930
Barker method 1228
Barrier height 904, 905, 908, 918, 926
Bases 687, 693ff, 742ff, 1414
conductivity base solutions 689, 757ff
salt effects on polarizable H-bonds 744f
solvate structures 693ff, 753ff
Basis sets (in MO calculations) 31ff, 46, 56, 59ff, 63ff, 70, 106, 112f, 139, 176
Bending vibration (*see also* H-bond bending mode) 175, 594ff
Ben-Naim–Stillinger Potential 1374
Bent H-bonds 362, 369, 401ff, 466, 529ff, 1079
Benzene, adsorption on oxides 1273, 1274, 1276ff, 1286, 1291ff, 1342, 1343, 1346
Benzene, adsorption on silica 1273, 1274, 1276ff, 1286, 1291ff
dielectric behavior 1292
electronic spectra 1292
entropy of adsorption 1293ff
intramolecular vibrations 1292
NMR relaxation times 1294ff
set diffusion coefficients 1294ff
Benzene HI complexes 1102
Benzoic acids 533
Beryllium oxide, adsorption on 1343
Beryllium sulfate tetrahydrate 930
$BeHBe^+$ 182
BET monolayer 1008, 1015
Bichloride ion ($ClHCl^-$) 186, 697
Bifluoride ion, *see* Hydrogen fluoride ion
Bifurcated H-bonds 408
Bigeleisen equation 1244
Binding energy, H-bond 182
Biological membranes 558, 760
Biopolymers 759ff, 1012
Bjerrum faults 966
Boltzmann distribution 230
Bond angles in hydrates 530
Bond energy analysis (BEA) 58
Bond length 401ff, 574, 1113, 1118, 1121, 1125, 1132f, 1137, 1140
Bond orbital analysis 71
Bond polarity 77
Born approximation 1173
Born–Oppenheimer approximation 30, 168
Born–von Karman periodic boundary condition 142
Boundary structures of proton 692, 693, 706 and structure diffusion 757f
Brillouin scattering 1125, 1156, 1158
Brillouin zone 893, 909
Buckingham's theory 616
1,4-Butanediol 543
β-Naphthol 10
β-Naphthylamine 10
β-Quinol 1420

Calcium oxide, adsorption on 1343
Calcium sulfate dihydrate 930
Carbocations 131

Carbonyl groups as H-bond acceptor 417
Carboxyl group 467, 1011
Carboxylic acids 303, 307, 330, 923, 1070, 1089, 1099, 1100
  adsorption on oxides 1344, 1349
Carboxylic acid dimers 5ff, 577ff, 735, 779f
Cartesian coordinates 170
Centered H-bonds 426, 428, 503
Cesium dihydrogen arsenate 1117, 1124, 1131f, 1154–56
Cesium dihydrogen phosphate 1117, 1121
Cesium trihydrogen selenite 1121, 1136, 1139
$CsHCl_2$ 932
Change of dielectric constant by ions 1408
Channel adducts 1420
Charge transfer 73ff, 88, 92, 129
  complex 942ff
    dipole moments of 942ff
  effects 6, 9ff
  energy 40, 44, 59, 89f, 97
  model 1286ff
Chemical potentials of binary mixtures of a selfassociating compound and an inert solvent 1234
Chemical shifts 809ff, 1208ff
  ^{13}C 822, 825
  ^{19}F 809, 826
  ^{14}N and ^{15}N 823
Chloride anion, complex with water 119
Chromia, adsorption on 1345
Critical and tricritical points 1119f, 1130, 1153, 1154
Critical exponents 363, 383–385
Critical fluctuations 1115, 1119, 1121f, 1123f, 1129f, 1135, 1140, 1141, 1144, 1149, 1158, 1160
Cross-over of the vapor pressure ratios 1243
Cross section 896
Cross section for neutron scattering 894, 909, 926
Cryogenic temperatures 1067
Crystallites 1080
Crystal orbital method 142f
Clays, adsorption on 1348ff
Clathrates 964, 985ff
Cluster approximation 1128, 1131, 1140, 1157, 1158
Clusters 1393f
  in water 1053, 1393f
Coercive field 1119, 1152
Colemanite 1149, 1160
Collision complexes 573
Collisional interactions 581
Colloid effects 1404
Combination tones 583, 619, 633, 635
Combination transitions 181, 199
Complete neglect of differential overlap (CNDO) 37–40, 51ff, 70, 83, 113, 116ff, 123ff, 129ff, 135ff, 141ff, 143ff, 146f
Complexes
  free radical 819, 826
  bihalide 820
  vapor 818
  higher order 824, 826, 942, 944, 949, 952, 955
Conductivity, anomalous proton 757ff, 973ff
  salt effects on 758
Configuration interaction 33, 49f, 70, 143, 148, 182
Conformation of biopolymers 786f
Conjugate–chelate H-bonds 603ff
Consistent force field 85ff
Constant of ionization 12ff
Content
  of free OH groups 1378
  of non-H-bonded OH groups in water 1378
  of non-bonded OH groups of H_2O 1405, 1406
Continuous association 1231
Continuous IR absorption (*see* IR continuum)
Continuous rotational diffusion 914
Continuum model 1373
Continuum (*see* IR continuum)
Coordinate systems, for molecular interactions 128f
Coordination numbers 1372, 1386
Coordination number in a liquid model 1371
Coordination numbers of the first next neighbours 1374
Cooperative bonds 362–372
Cooperative effects 1335
Cooperative mechanism 553f, 1388, 1405
Copper (II) chloride dihydrate 930
Copper formate tetrahydrate 1148, 1160
Copper (II) sulfate pentahydrate 930

Correlation
  energy 33, 49, 97ff
  function 591, 895, 1171, 1173, 1313
Correlations 407ff, 557ff, 1209ff
  between N–H-bond length and N···O distance 412
  between N–H-bond length and H···O distance 412
  between O–H-bond length and H···O distance 410, 411
  between O–H-bond length and O···O distance 410, 411
  between N–H···O angle and H···O distance 407
  between O–H···O angle and H···O distance 406
  bond length and wave number shifts 574
  IR intensity and bond strength 576
  of protons in H-bonds 248, 253, 259ff
  spectroscopic shifts and enthalpy 1209ff
  wave number shift and bond energy 574f
Correlation time 591
  rotational 1032ff
Coulomb energy 42, 47, 52, 59, 63
Coulomb integral 37, 52f
Counterpoise method (correction of basis set error) 63
Coupled electron pair approximation (CEPA) 35, 36, 66f, 70, 148
Coupled equilibrium of cyclic dimers and trimers 534
Coupled equilibrium of different aggregates 541
Coupled mode theories: Kobayashi, etc. 1128f, 1131, 1133f, 1155f
Coupling constants 329, 345, 346
Coupling
  electrical 201
  mechanical 201
  of vibrations and IR continuum 728, 731
Covalent energy 88
Covalent H-bond contributions 1216–1219
4-Cyanopyridine, adsorbed on silica 1282
Cycle structures 537ff, 1074ff
Cyclic aggregates of methanol, water 547
Cyclic dimers 531, 535, 550
Cyclic structures 537ff, 1074ff
Cyclohexanone oxime 535
Cyclopentanone oxime 538, 539
Curie point, T_c (ferroelectric transition temperature) 1114, 1116f, 1125, 1135–1151, 1153, 1159
Curie–Weiss law 1117, 1119, 1120, 1132, 1136, 1142–51, 1153, 1159
Coumarin dyes 16
^{13}C chemical shift tensors
  for solid acetic and thioacetic acid 881
^{35}Cl quadrupole coupling
  NMR frequency 854
  temperature dependence 856
  and the strength of the H-bond 855
  theory of 855
  in $NH_4H(ClCH_2COO)_2$ 856, 857
C_{2v} symmetry of liquid water 554

Damped proton motion in H-bonds 233
Debye–Hückel theory 1414
Debye–Waller factor 898
Defects in H-bonded networks 341ff, 345
Deformation modes 594ff
Delocalization energy 55, 91
Density of vibrational states 897
Desalination processes of water 1420
DNA 759ff, 1012
Deuteration 181
Deuteration shift of γ_{AH} 572, 600
Deuterated compounds 186, 205, 421ff, 690f, 696, 844, 847, 848, 850, 926, 927, 967, 1009, 1071, 1077, 1082, 1084–87, 1090, 1094, 1095, 1099, 1102, 1230, 1235–40, 1242–1244, 1251, 1254–57, 1380, 1381
Deutero-acetic acid 1099
Deutero-bifluoride ion 186
Deutero-methanol 1094, 1095
Deuteron dynamics
  and the form of the potential surface 842–846
  and ferroelectric phase transition 845–846
  and spin lattice relaxation 847–849
Deuteron quadrupole coupling and electric field gradient tensor
  theory of 838, 841, 842
  proportionality of deuteron quadrupole coupling to square of OH stretching frequency 889
  deuteron dynamics 843–849

in KD_2PO_4, KD_2AsO_4 and CsD_2AsO_4 844
and strength of H-bond 838–842
and the strength of the O–H···O bond 838, 840, 841
and the strength of the N–D···O bond 841
and the strength of the O–D···Cl bond 841
and the strength of the B–D···Cl bond 841
Deuteron quadrupole coupling in symmetric H-bonds 838, 842
Deuterium bonding 1227ff
Deuterium chloride 696, 1077, 1082, 1102
Deuterium cyanide 1084, 1085
D-defects 965, 966, 969
Deuterium halides 1071, 1077
Deuterium isotope shift 572, 600, 1085, 1095
Deuterium oxide 967, 1009, 1086, 1087
Deuterium sulphide 1090
Dialcohols 544
Diborane 7
Diaquahydrogen ion (*see* $H_5O_2^+$ *and* Hydronium hydrates)
Diaquaoxonium ion
structure 481, 507, 674ff
theoretical studies 520
Dielectric absorption in adsorbates 998ff, 1292, 1298, 1313, 1314, 1315ff, 1322, 1328, 1335, 1343, 1345, 1347ff
Dielectric constant in ice 1402
Dielectric constant of H_2O 1399
Dielectric dispersion 967ff
Dielectric polarization 940, 963
Dielectric relaxation 1113, 1119, 1127f, 1129f, 1134
Dielectric susceptibility 1118, 1126, 1129f, 1132, 1137, 1139, 1142–43, 1145, 1147–48, 1150–52, 1156–59
Dielectric titration 955, 956
Diethylamine $(C_2H_5)_2NH$ 1230, 1236, 1237, 1243
Diethylamine $(C_2H_5)_2ND$ 1230, 1238, 1239, 1243
Difference map 398
Diffusion 910, 912, 913
Dihydrated proton ($H_5O_2^+$) (*see* Hydronium hydrates *and* $H_5O_2^+$)
Dihydrated deuteron 203, 696
Dihydrate of polyethylene oxide derivatives 557
Diisopropyl phenol 647
Dimensions of the area of H-bonded molecules 1394f
Dimers 531ff, 537, 539, 1068, 1069, 1073–75, 1078, 1083–90, 1092, 1096, 1098, 1104
open chain 1074, 1076, 1084, 1087, 1088, 1090–92, 1095, 1096, 1098, 1099, 1104
cyclic 107, 116ff, 531ff, 1074, 1075, 1079, 1086, 1088, 1091, 1094, 1096, 1104
pseudodimers 1081
linear 107ff, 129
bifurcated 107
Dimethylamine $(CH_3)_2NH$ 627, 1230, 1236, 1237, 1243ff
Dimethylamine $(CH_3)_2ND$ 1230, 1238, 1239, 1243ff
Dimethylamine $(CD_3)_2NH$ 1230, 1236, 1237, 1243ff
Dimethylamine $(CD_3)_2ND$ 1230, 1237, 1238, 1243ff
DMSO with acid, IR continuum 694, 704, 735
Dioxane 942
Dioxane effect 940
Dipole–dipole interactions 552, 1081
Dipole energy 950
Dipole–quadrupole interactions 1081
Dipole moment 32, 33, 36, 49, 64, 65, 77, 85, 89, 98ff, 107, 139, 169, 177, 940, 942, 950, 954, 960, 962, 969
of $H_5O_2^+$ 175, 724
of hydrogen bonds 946ff
of alcohols 541
Dispersion energy 40, 44, 45, 56, 97ff
Displacive transition 113f, 1128, 1144–1145, 1150, 1160
Dissociation of acids
true degree of 689
processes with 753ff
Distance criterion for H-bonding 401f
Domain walls 1119, 1129, 1135, 1153, 1160
Dopant molecules (in matrices) 1101
Double hydrates 1419
Double minimum potential well 3, 172, 175, 196, 220f, 240, 247, 249, 261, 263, 266,

279, 288f, 291, 692ff, 917, 918, 920, 921, 923, 925, 928
and polarizability of H-bonds 706f, 722ff
Double perturbation theory 49, 56
Double scale enthalpy equation 1212, 1216
Dynamic reorientation (*see also* Molecular dynamics studies) 4
Dynamic structure factor 1176, 1189, 1190
DL-serine 462

Effective potential depth in water 1054, 1055
Effects of H-bonding on vibrational modes 568ff, 620ff, 1209ff
Eight-vertex problem 363, 383–385
Elastic incoherent neutron scattering 908, 910f, 921, 923
Electrical anharmonicity 299, 303, 315, 353, 620, 648
Electronic spectra of adsorbates 1290, 1292ff
Electrical (protonic) conductivity
proton interbond jumps 757ff, 761ff, 973ff
of adsorbates 1299ff, 1313ff, 1327, 1330, 1336, 1347ff
Electric field gradient tensor
definition 835
relation to electronic structure 835
relation to point symmetry at nuclear site 839
Electric quadrupole moment of the nucleus 834
Electrolyte solutions 1403
Electron affinity 10
Electron correlation 33, 47, 49, 70, 130, 148, 151
intramolecular 49, 64
intermolecular 64
Electron density 73, 74, 81, 82, 149, 187
Electron density difference function 73, 74
Electron exchange 41, 47
Electron excitation spectra 143ff
Electron nuclear double resonance (ENDOR) 1155, 1223
Electron hole potential theory 145
Electron pair theory 36
Electron paramagnetic resonance 1122f
Electron excited states 8ff
Electrostatic fields 179, 187
effects on H-bonds 724ff
Electrostatic H-bond contributions 1216–1219
Electrostatic energy (*see also* Coulomb energy) 42, 49, 51, 56, 59, 107, 119, 129
Electrostatic potential, molecular 81ff
Enclathration 1420
Energetically unfavoured H-bonds 1381
Energy levels 199, 249, 253ff
continuous distribution of 720ff
shifts due to electrical fields 202, 724ff
shifts due to interactions 728ff
Energy partitioning 46, 50, 52, 57ff
Energy in intramolecular H-bonds 124ff
Energy resolution 896, 898, 904, 910, 911, 916, 918, 921
Energy surface (*see also* Potential energy function) 25ff, 168, 183, 188, 199, 210, 220, 301, 379, 693, 736ff, 842ff
Entropy change at the transition 920
Entropy of adsorption 1293ff, 1305, 1315, 1327, 1343, 1345, 1348
Equilibrium constant 814, 826, 951, 1397
of ethanol aggregation 1397
of water aggregation 1397
Equilibrium geometry 28, 34, 35, 39, 40, 56, 68, 77, 82, 88, 94ff, 107, 108ff, 112, 114, 117, 120f, 125, 129ff, 140, 144, 403ff, 530ff
Ethane thiol 1097, 1098
Ethanol C_2H_5OH 926, 927, 932, 1069, 1230, 1236, 1243ff, 1378
Ethanol CH_3CD_2OH 926
Ethanol C_2H_5OD 926, 1230, 1238, 1243ff
Ethanol C_2D_5OD 1230, 1238, 1243ff
Ethanol C_2D_5OH 1230, 1236, 1243f
Ethanol CD_3CH_2OH 926
Ethers, adsorbed on silica 1277, 1279, 1281, 1291
Ether hydrogen chloride 635
Ether hydrogen fluoride 634
Ethyl alcohol (*see* Ethanol)
Ethylamine $C_2H_5NH_2$ 1230, 1236
Ethylamine $C_2H_5ND_2$ 1230, 1238
Ethylene, complexes with polar molecules 95ff, 104ff
Ethylenoxide derivatives 1404
Ethyl methylamine $CH_3C_2H_5NH$ 1230, 1236
Ethyl methylamine $CH_3C_2H_5ND$ 1230, 1238
Evans effect 592, 599, 602

Exchange energy 41, 42, 52, 59, 97
Exchange, slow and fast 1041f, 1048, 1054
Exciplexes 10
Excited states, H-bonds in 143ff
Expansion of the areas of H-bonded molecules of H_2O 1393
Expansion of H-bonded aggregates of electrolyte solutions 1407
Eyring formula 917
E and *C* parameters 1212, 1217

Fermi interaction 19
Fermi resonance 181, 201, 256, 261f, 302, 303, 325, 346, 571, 586, 590, 592, 632, 639, 643, 732
Ferroelectricity and ferroelectric substances 326, 340, 346, 382–385, 908, 920, 924, 965, 1113ff, 1178ff
  and polarizability of H-bonds 748
First hydration shell 748ff, 1413
Fissure plains 1394
Five-membered rings of water 558f
Flickering clusters 370–372, 1053, 1393f
Fluctuating potential 225, 233
Fluctuation of proton 693
  and $H_7O_4^-$ 753ff
  and $H_9O_4^+$ 753ff
  charge fluctuation forces 729
Fluorescence quenching 10
Fluoride anion
  complex with water 119ff, 203
  complex with monodeuterated water 205
Fluorine bridging 7
Fluorine hydroxide, complex of 108ff
Fluoro alcohols 1096
Fluoro carbon solvents 621, 628
Förster cycle 13
Force constants 33ff, 39, 66, 77, 170, 174, 178, 902, 909, 910, 943, 1080
Formaldehyde 9
  complex of 108ff, 125, 127, 131, 135, 143ff
Formamide
  cyclic dimer 209, 116f
  hydration 85, 86, 109
  linear dimer 111
  two-dimensional networks 142
  translatory and rotatory modes 772–774
  far infrared dipole correlation function and phonon lifetimes 775
Formic acid 932, 1099
  dimer (vapor)
    structure 771
    infrared relaxation 778–780
    proton tunneling in H-bonds 779–780
  cyclic dimer 116f, 150f
Fourier map 398
Franck–Condon state 13
Franck–Condon transitions 316, 338
Freon solvents 621, 628
Frequency distribution 897, 898
Frequency spectrum 893, 901, 902, 906
Frost damage 1317
Fundamental spectrum of water in solid argon 548
Fundamental stretching vibration of HOD 1381
Furan, adsorbed on silica 1277, 1281

Gas hydrates 559, 1415, 1419
Gas phase 19
  H-bonding in 634ff
Gaussian approximation 915
Gaussian distribution 227
Geometrical requirements in H-bond formation 413
Geometry of water in hydrates 436
Germanium oxide, adsorption on 1342, 1343
Gibbs energy of binary mixtures from a self associating component and an inert solvent 1232
Gibbs–Helmholtz equation 1392
Glycerol 932
Glycole 548
Grotthus conductivity (*see also* Electrical (protonic) conductivity) 757ff, 979ff
  salt effects on 759
Guanine–cytosine base pair 116, 118, 759f

Hall effect 997
Harmonic approximation 169, 170, 182, 205
Hartree equations 182
Hartree–Fock approximation 31, 379
Hartree–Fock limit 32, 33, 37
Hartree–Fock orbitals 33
Heat bath of harmonic oscillators 225

Heat capacity of adsorbates 1306, 1309ff, 1342, 1345
Heat content of the intermolecular degrees of freedom in a liquid 1391
Heats of adsorption 1273ff, 1285ff, 1291, 1298, 1319, 1322, 1327, 1328, 1333, 1335, 1338, 1342, 1343, 1345ff
Heats of immersion 1309, 1343, 1345, 1349
Heat of melting of H_2O, CH_3OH, C_2H_5OH 1379
Heat of sublimation of ice 1369
Heat of vaporization of H_2O, CH_3OH, C_2H_5OH 1379
Helium dimer 54
Helium, complexes with polar molecules 93ff
Hemoglobin 1015
Heptakaidecahedron cage of water molecules 1416
Hetero-associated species 1100–1102
Hetero excimer 10
Hexafluoroacetyl acetone, enol 125, 127
Hexafluoro propanol 1096
High resolution NMR in solids
  Hamiltonian 842
  experimental technique 874, 881–882
  in oxalic acid dihydrate 877
  in malonic acid 879
  proton chemical shift tensors 878
  carbon-13 chemical shift tensors 881
  chemical shift tensor and the strength of the H-bond 880
Hindered rotation 1122, 1146f, 1149, 1159
H_3O^+ (*see* Hydronium ion)
$H_3O_2^-$ (*see also* Hydroxyde hydrate) 120, 688, 693ff, 753ff
  concentration and IR continuum 742ff
$H_5O_2^+$ (*see also* Hydronium hydrates) 474f, 498ff, 667ff, 687, 689ff, 753ff
  concentration and IR continuum 742
  dipole moment 195, 250, 724
  Grotthus conductivity 757f
  polarizability 722
  potential energy surface in 194, 726
  salt effects on 744f
  SCF calculations 191ff, 723ff
  structure and spectra, solid state 474f, 498ff, 667ff
  structure diffusion 692
$H_7O_4^-$ 688, 753ff
$H_9O_4^+$ (*see also* Hydroniumtrihydrate) 510ff, 677ff, 688, 753ff
  fluctuating structure 753ff
  structure and spectra, solid state 510ff, 677ff
Hofmeister ion series 1404
Hole defects 1387
Hole-defect model 1371
Hot bands 640, 641
Hybridization of lone pairs 129
Hydrated proton (*see* Hydronium ion)
Hydrates 433, 902, 964
Hydrate structures 593, 1414
  H^+ and OH^- 491ff, 659ff, 753ff
Hydration 998ff, 1000
  and polarizability of H-bonds 748
Hydrazoic acid 1090, 1091
Hydrocarbons 85
Hydrocyanic acid (*see* Hydrogen cyanide)
H-bond
  angle 128ff, 549
  angle distribution 404
  bending mode 175, 594ff, 1074, 1076, 1084, 1086–1088, 1091, 1094, 1095, 1100, 1102
  content during ice melting 1371
  definition 3, 401
  disorder 667, 673, 674
  energies 539, 1092, 1395
  energy per OH group 1369
  geometry 401ff, 529ff
  in ice 1425ff
  infrared force constants 658
  linearity 403, 434, 529ff
  proton transfer in (*see* Proton transfer)
  polarizable (*see* Polarizable H-bonds, Polarizability of H-bonds)
  potential function (*see also* Potential energy functions) 220, 693, 736ff, 842ff, 1085
  rupture and molecular mobility 25ff
    of NH_3^+ in L-alanine 776
    of glycerol 777
  shift and anharmonicity 644
  temperature and shift 646
  and weakness of overtones 648
H-bonded crystals 395ff, 1425ff
H-bonded cyclic dimers 303, 307, 330, 531ff, 735, 1074ff

H-bonding conferences 6
Hydrogen
  cyanide 1083–1085
    complex of 108ff, 112, 135, 141
    complex with hydrogen fluoride 642
  chloride 909, 932, 1071, 1072, 1075, 1080, 1083, 1101, 1102, 1103
    complexes 107, 108, 110, 112
  difluoride anion 431, 118, 120, 183
  fluoride 932, 1071
    crystal 141ff
    dimer 27, 55, 57, 60, 61, 64–67
    molecule 35, 89
    oligomers 135f
  fluoronium cation, complex with hydrogen fluoride 121
  halides, HCl, HBr, HI 1147f, 1160, 1071, 1072, 1073, 1100, 1103
  iodide 1071, 1073, 1075, 1080, 1101–1103
  maleate anion 125, 127
  molecule, complexes with polar molecules 93ff
  peroxide, complexes of 108ff
  position from X-ray and neutron diffraction 398, 399, 400
  sulphide 1087–1090, 1103
    anion, complex with hydrogen sulfide 120
    dimer 107, 108
Hydrolysis of H_2O due to ions 749
Hydronium
  compounds 930
  hydrates (*see also* $H_5O_2{}^+$, $H_9O_4{}^+$ *and* Diaquaoxonium ion) 118ff, 150f, 190, 474f, 498ff, 659ff, 667f, 687, 689ff, 753ff
    structure and spectra (liquid state) 687, 689ff, 753ff
    structure and spectra (solid state) 474f, 498ff, 667ff
  ion (H_3O^+) 472ff, 480, 659ff, 691f
    dissociation of 148ff
    fundamental vibrational modes 665
    NMR 476, 497f
    structure and spectra (solid state) 477f, 491ff, 661ff
    theoretical studies 517f
    X-ray 485ff
  dihydrate (solid state) 507ff, 674ff
  nitrate 930
  perchlorate 903ff, 911, 915, 930
  polyhydrates 139ff, 753ff
  trihydrate (*see also* $H_9O_4{}^+$)
    fluctuating structure 753ff
    structure and spectra (solid state) 677ff
Hydrophobic
  forces 1418
  groups 1418
  hydration 1056
  side 1416
Hydroxide hydrates (*see also* $H_3O_2{}^-$ *and* $H_7O_2{}^-$) 120, 688, 693ff, 753ff
Hydroxide, polyhydrates 139ff, 753ff
Hydroxonium hydrate (*see* Hydronium hydrate)
Hydroxonium ion (*see* Hydronium ion)
Hydroxylamine, complexes of 108ff
Hydroxyl groups, on alumina 1328ff
Hydroxyl groups, on silica (*see* Silanol groups)
3-Hydroxy propionic aldehyde 124ff
Hyperpolarizability 193
  of $H_5O_2{}^+$ 725, 727
  of H-bonds 725, 727
Hysteresis 1151–1152, 1158

Ice 138, 141f, 634, 898ff, 902, 929, 964, 966ff, 1425ff
  bending vibrations 1461ff
  doped with HF 1403
  dissociation energy 1453ff
  geometries 1430ff
  nucleation of, on surfaces 1311, 1316, 1317, 1336
  OH stretching vibrations 1458ff
  residual entropy of 363, 368
  rotational vibrations 1461
  iceberg forming 1418
Iceberg structure 1418
Iceberg-like structure of water 558f
Ideal associated solutions 1229
Ideal liquid 1387
Imidazole solutions 708f
  Grotthus conductivity of semi-protonated 761ff
Imperfections 967

Improper ferroelectrics 1139–1140, 1143, 1145, 1146, 1158
Impurities 983
Incoherent scattering 893ff, 1177, 1179, 1181, 1190
Independent electron pair approximation 34–36, 66ff, 70, 98, 103, 147
Induced bands 1101, 1102
Induced dipole interactions 688, 729ff
  and IR continuum 731ff
Induced dipole moment 179, 197
Inductive effects 945
Induction effects in H-bonds 248, 252, 270, 291, 688, 729ff
  and IR continuum 731ff
Inelastic neutron scattering 18, 893ff, 1177ff
IR absorption intensities 77, 78, 80, 177, 178, 200, 256ff, 576, 648
Infrared bandshapes 314, 319, 337, 577, 583, 637ff
  of adsorbed molecules 1290
IR continuous absorption (*see* IR continuum)
IR continuum 247, 277, 281, 687ff
  a more general effect 694
  and polarizability of H-bonds 728ff
  dependence on concentration of polarizable H-bonds 742f
  $H_3O_2^-$ 693ff, 753ff
  $H_5O_2^+$ 689ff, 753ff
  $H_9O_4^+$ 753ff
  influence of temperature 693, 745ff
  non-aqueous systems 694ff
  reasons for 731f
  salt effect on 744f
  table for systems with 646–703
  wave number dependence 734ff
  with adsorbed molecules 717, 1301ff
  with asymmetrical H-bonds 718ff
  with dissolved molecules 707ff
Infrared shift techniques 1207
IR spectra 537ff, 566ff, 614ff, 684ff, 689ff, 768ff, 945, 972, 979, 1134, 1145, 1148, 1159, 1377ff
IR transitions, intensity of (*see* IR absorption intensities)
Initial proton states 227
Initial wave packet 223
Intensity of γ_{AH} (*see also* IR absorption intensities) 576
Interaction energy of water 1369
Interaction potential 1080, 1081
Interaction between H-bond protons 253ff, 272, 276f, 281ff, 290, 687f, 729f
Interaction of H-bonds with environment and IR continuum 728ff
  induced dipole 688, 729
  proton dispersion forces (*see also* Interaction between H-bond protons) 687f, 729f
Intermediate neglect of differential overlap (INDO) 37–40, 71, 83, 123ff
Intermediate scattering function 895
Intermolecular relaxation rate 1032, 1046, 1048
Internal coordinates 169, 170
Internal energies of liquid water 1369
Intramolecular degrees of freedom 224
Intramolecular H-bonds 3, 123ff, 542, 544, 603, 720, 953
  ortho-halophenols, torsional vibrations and electronegativities of halogen substituents 780f
  polarizable 720
  saccharides 781
Intramolecular relaxation rate 1031, 1047, 1050
Inverse vapor pressure isotope effect 1240ff
Iodoalkane–HI complexes 1102f
Ion exchange resins 1010
  polarizable H-bonds in 689ff
Ionization potential 10
Iron oxide, adsorption on 1336, 1345
Ising model 1131, 1148, 1152
Isomerization 220
Isomerization rates 219
Isosbestic points 140
Isotope effect 324, 333, 338, 348, 350, 420, 507, 600, 1113, 1116f, 1124, 1127, 1129, 1132, 1133, 1135–40, 1141, 1143–51, 1154, 1155, 1157, 1159
  IR continuum 690
  on γ_{AH} 572, 600
Isotopic isolation 581, 624, 635, 638

Jacobian coordinates 171
Jump diffusion 914f

Jump length 919
Jump rotation 911, 912, 916–19, 923

Kaolinite, adsorption on 1349

Lactams 531
Lactim–lactam equilibrium 12
Lagrange equations 169, 170
Laser pulses
  intensity effects 15
  ultrashort, and time resolved spectra 15
Latent heat, entropy of transition 1120, 1129, 1134f, 1137, 1141, 1145, 1147
Lattice modes of
  solid formamide 772
  solid methanol 778
  biopolymer model substances 786
$La_2Mg_3(NO_3)_{12} \cdot 24H_2O$ 930
Lead monetite 1159
Lecontite 1146, 1159
Lennard-Jones potential 86ff
Librational levels 1082
Librational mode 1076, 1085, 1088
Libration band (*see also* Torsional vibrations) 1383
Lifetimes of H-bonds 4, 1035, 1041f, 1046f, 1053, 1415
Lifetime of proton in XH group 1030, 1050, 1052, 1055
Linear combination 173
Linear enthalpy, spectroscopic shift relations 1209, 1210
Linear H-bond 403ff, 530ff, 1079, 1379
Line profile 893
Lingering times in potential wells 247, 266, 279, 285, 288f
Lippincott–Schröder model 1214
Lithium
  ammonium tartrate 1143, 1158
  chloride hydrate 930
  hydrazinium sulfate 1151
  perchlorate trihydrate 930
  sulphate hydrate 930
  thallium tartrate 1143
  trihydrogen selenite 1121, 1138, 1140, 1136
Localized molecular orbitals (LMOs) 68, 141
Lone pair electrons 5, 413
Lorentzian–Gaussian convolution 910, 911
Low temperature effects 582, 587
Lyotropic ion series 1409
L-defects 965f, 969

Manganese oxide, adsorption on 1345
Magnesium oxide, adsorption on 1342, 1343
Magnetic dipole–dipole interaction 1030
Magnetic ferroelectrics 1148, 1151
Magnetic susceptibility of H_2O 1398
Malonic dialdehyde, enol 124ff, 150, 151
Malonic acid difluoride, enol 125ff
Mass tensor 898
Matrices
  noble gases 1067, 1069
  nitrogen 79, 1067
  carbon monoxide 1067
Matrix isolation technique 545ff, 580, 582f, 603, 1067ff
Maxwell – Wagner dispersion 998, 1006, 1009, 1013
Mean association number 543
Mean field theories (Slater, Tagaki et al.) 1126f, 1129f, 1132f, 1134, 1137, 1143, 1144, 1148, 1153f, 1157
Mechanical anharmonicity 177, 204, 298, 303, 321, 614-42
Melting heat of water 1387
Membranes 558
  proton conductivity with 761ff
Membrane mechanism 558
Mesomorphism 19
Metastable proton states in H-bonds 279f, 755
Methanethiol CH_3SH 1097, 1098, 1103, 1230, 1236, 1243
Methanethiol CH_3SD 1230, 1238, 1243
Methanethiol dimer 108
Methanol (CH_3OH) 545, 546, 926ff, 927, 932, 1093, 1094-96, 1103, 1230, 1236, 1243ff, 1378, 1379
  complexes of 108ff, 143ff, 1230, 1234f, 1238
  protonated and deprotonated 694f, 704f
  trideuterated 926, 927, 1230, 1236, 1243ff
2-Methoxypyridine, adsorbed on silica 1278, 1279, 1282
Methyl alcohol (*see* Methanol)
Methylamine 1230, 1238ff, 1280, 1236, 1239

Methylchloroform 8
4-Methyl umbelliferone 16ff
Methylmercaptane (*see* Methanethiol)
Micelle building 1409
Microviscosity 1039
Microwave spectroscopy 115
Minerals 999, 1003
Minimum energy pathway 151
Minimum of dipole moment 541
Minimum of the specific heat of water 1389
Mixed association 3
Mixed hydrates 1419
Mixture models 1373
  of certain species 1373
  of water 1401
Molar surface tension 1391
Molecular clusters 1103
Molecular dynamics calculation 380, 902
Molecular dynamics studies
  of collective proton motion 1178ff
  with dielectric methods 938ff
  with inelastic neutron scattering 901, 915
  with nuclear magnetic resonance relaxation 1028ff
Molecular liquids 911
Molecules-in-molecules (MIM) method 67ff
Modulation method 18
Monomers of H_2O 1394
Monte-Carlo calculations 380f
Montmorillonite, adsorption on 1348ff
Morse curve 172
Morse potential 88ff, 615
Multimers 1068-70, 1075, 1080, 1084, 1089, 1090, 1092-95, 1098f
Multipole site trapping 1084, 1086, 1095
Multipole expansion 83

N–H distances 456, 466
N–H · · · O bonds 464–467
N–Methylacetamide 632
N–Methylaniline 631
N–Methylpivalamide 632
^{14}N quadrupole coupling
  NQR frequency 858
  and the strength of the NH· · · N bond 859
  and the N–H bond stretching frequency 859
  temperature dependence 860-862
  effect of deuteration 862
  theory of 859
  in TGS 860f
Neutron elastic incoherent scattering 908, 910f, 921, 923
Neutron inelastic scattering 892ff, 1169ff
  coherent 895, 902, 906, 908, 921, 928, 1169
  incoherent 893ff, 1177, 1179, 1181, 1190f
  of adsorbates 1305, 1333, 1347
Neutron diffraction 395, 459, 494, 502, 510, 1113, 1118, 1121, 1134
Neutron momentum transfer 893
Neutron scattering (*see also* Neutron elastic and neutron inelastic scattering) 597, 1113, 1123f, 1153, 1155
Nickel sulphate heptahydrate 930
Nickel sulphate hexahydrate 930
Nitriles, adsorption on oxides 1278, 1344, 1346ff
Nitrogen, adsorbed on silica 1276, 1280
Noble gases 1067
Noble-gas atoms, complexes 54, 93ff
Non-additivity of intermolecular energies 132ff, 136, 140, 142
Non-H-bonded liquids, correlation time for 1036
Non-H-bonded OH groups 1373, 1415
Non-linear effects 959
Non-stationary proton states 227
Normal coordinates 169, 171
Normal coordinate analysis 568
Normal modes of vibration, computer drawn 664, 666
Normal vapor pressure isotope effect 1240ff
Nuclear double resonance
  in the laboratory frame 853
  in the rotating frame 861
Nuclear magnetic resonance NMR
  adsorption line shape
    in case of an axially symmetric chemical shift tensor 876
    in case of a non-axially symmetric chemical shift tensor 875, 876
    dipolar line shape of an A_3B_2 ion 871
    of KH_2F_3 868
  chemical shifts 77, 81, 139

of H_2O 1398
of adsorbates 1299, 1319, 1339, 1244, 1346, 1347
deuterium pulsed 1328
in solids
wide line 867
high resolution 872
proton resonance shifts 1208
quadrupole interactions, spin–lattice relaxation 1113, 1120f, 1127, 1129f, 1132, 1137f, 1140, 1141, 1143, 1145-50, 1153-54, 1157-60
relaxation times
of adsorbed molecules 1305, 1306, 1311, 1316, 1327ff, 1336, 1347ff, 1391, 1394ff, 1398ff
wide line of adsorbates 1297, 1343, 1345ff
Number of non-H-bonded OH groups 1415
Nuclear quadrupole resonance (NQR) 834-848
Nuclear quadrupole interaction 1034
Nylon 558

Octet theory 5
o-hydroxyphenanthrene aldehydeanil 11
Olefins, adsorption on oxides 1346, 1348
Oleic acid 932
One-minimum potential 3
One-phonon approximation 897
Optical properties (*see also* Brillouin and Raman scattering, *and* Infrared spectra) 1113, 1139, 1151
^{17}O quadrupole coupling
dependence of the transition frequencies on the asymmetry parameter 852
and the strength of the O–H···O bond 850
effect of proton ordering 851
in ice, HDO and KH_2PO_4 850
Orbital basis set (*see also* Basis set) 32, 59ff
Order–disorder transition 1113f, 1115, 1119, 1122, 1126, 1128f, 1131, 1137, 1138f, 1141, 1144, 1150f, 1152-53, 1158-60
Organic hydrates 558
Orientation defects 1387, 1394
Oscillator function 173, 176
Overlap integral 37, 42, 47, 52ff, 56
Overtones 185, 205, 207, 616, 621, 625, 648
IR continuum 687
of HOD 1380
Oxalic acid dihydrate 555
Oxygen, adsorbed on silica 1276, 1280
Oxonium ion (*see* Hydronium ion)

π-bonded complexes 1102, 1103
π-electron systems, H-bonds to 104ff
$\pi^* \rightarrow n$ states 5, 9
Pair interacting potential 906
Pair natural orbitals 36
Pair potential of water 1373
Parabolic potential 921
Paraelectric phase 908, 920, 922, 924
Partial molar volume
of electrolyte solutions 1412
of salts 1413
of water 1413
Partial pressure ratio (*see also* Vapor pressure ratio) 1240ff
Partition coefficient 1408, 1411
Pentagon–dodecahedron 558f, 1416
Pentakaidecahedral cage of water molecules 1416
Peptides 91, 124
Perovskites, two–dimensional 1151
Perturbation theory
intermolecular 39, 99, 129, 137
second order 167, 177, 180, 204
Phase transitions 382-85
Phenol 942
Phenol + 3,6-aminoacridine 10
Phonon dispersion curves 906, 908
Phonon dispersion relations 902
Phosphate ion as H-bond acceptor 419, 718
Phosphatidyl serine 760
Phosphoric acid, polarizable H-bonds 718, 751
Point charge model 84
p-iso-octylphenol 1404
Poisson equation 83
pK_a
and acid water bonds 750
and formation of polarizable H-bonds 708
and potential well 736ff, 750
Plastic crystals 911

Plutonium oxide, adsorption on 1346
Polar solvents, influence on proton motion in H-bonds 248, 265f, 270, 272, 277ff, 281f, 284, 291f, 754ff
Polarity 940, 947, 951
Polarizability 37, 63, 98ff, 102
  of H-bonds (*see also* Polarizable H-bonds) 180, 197, 247, 249ff, 253, 263ff, 285, 291, 688, 722ff
    and IR continuum 731ff
Polarizable H-bonds
  alcohols
    with acids 694, 704
    with bases 705
  and continuum, (*see* IR continuum)
  and ferroelectricity 748
  and Grotthus conductivity 757f
  asymmetrical 718ff, 738ff
  biological systems 759ff
  between adsorbed molecules 717ff
  between dissolved molecules 707ff
  DMSO with acids 694, 704, 735
  $H_3O_2^-$ 639
  $H_5O_2^+$ 689ff, 723ff, 750ff, 753ff
  imidazole semi-protonated 705ff
  interaction with environment 728ff, 753ff
  salt effects on 744f
  table of systems with 696-703
  with crystals 747
Polarization 942, 945, 946
Polarization energy 44, 50, 59, 63, 91, 97ff
Polarization factor 897
Polarization of H-bonds in the first excited proton state 249, 264, 279, 286, 289, 291, 755f
Polarization of water by cations 748ff
Polarized IR spectra 586
Polyelectrolytes 687, 689ff
  acidic 689ff
  basic 693ff
Polyethylene oxide derivatives 556
Polyhistidine 761f
  polarizable H-bonds 761f
  proton conductivity 761f
Polynomial representation 173, 175
Polynucleotides 759f
Polypeptide 558
Positions, of hydrogens within bonds 397ff, 1113, 1116f, 1124, 1126, 1137-38, 1141, 1142
Potassium
  bifluoride 932
  dihydrogen arsenate 1117, 1119f, 1122f, 1129f, 1132, 1154, 1155, 1157
  dihydrogen phosphate 362, 382f, 908, 912, 920-22, 924f, 932, 1115, 1116f, 1123f, 1132–1157
  ferrocyanide trihydrate 1149, 1160
  fluoride dihydrate 930
  hexacyanoferrate (II)-trihydrate 930
  trihydrogen selenite 1121, 1138, 1140, 1136
Potential barrier 222
  and OO distance with $H_5O_2^+$ 194, 729
Potential energy surfaces 169, 172, 196, 199, 221, 228, 249, 338, 420, 570, 590, 596, 614ff, 692, 724ff
  and environment of H-bonds 737, 754ff
  and nature of acceptors 736
  and polarizability of H-bonds 722ff
  and wave number regions of continua 736f
  calculated $H_5O_2^+$ 194, 723ff
  with acid water bonds 750ff
Potential parameters 221, 238, 614ff
Predissociation theory of band width 589
Preference for angle 0° for H-bonds 403ff, 559
Pressure effects 1113, 1117, 1125, 1128, 1135–38, 1140, 1149, 1151, 1154, 1156–59
Primary hydrates of organic molecules 558
Probability density 223, 231
Probability distribution function 234
  for a proton in a symmetrical H-bond 232
  for a proton in a weakly unsymmetrical H-bond 233
Projection operators 228
Properties of water 1387
Propionic acid 533
Proteins 91, 124, 1013, 1015
Protomeric structures (*see also* Proton boundary structures) 219f, 223, 227f, 235, 241
Proton (*see also* Hydronium ion, Hydronium hydrates, $H_5O_2^+$ and $H_9O_4^+$) 687f, 689ff
  boundary structures 692f, 706
  catalytic activities and salt effects 744f
  formation of easily polarizable H-bonds 684ff

fluctuation of 693, 753ff
tunneling (*see also* Tunneling of protons) 693
Proton acceptors 736f, 942
Proton affinities 113, 146ff
Protonation site 81
Proton chemical shift tensors
and the strength of the H-bond 879-880
in solid malonic acid 848-879
in oxalic acid dihydrate 874
Protonic conduction (*see also* Electrical (protonic) conductivity) 757ff, 761ff, 973ff
Proton dispersion forces 248, 254, 687, 729f
Proton donors 941
Proton donor–acceptor complexes
electronic effects on H-bond stretching frequencies 782-783
Proton hyperfine coupling tensors
in X-irradiated KH_2AsO_4 883
and the covalent character of the O–H···O bonds 884
and proton dynamics in O–H···O bonds 885
Proton intrabond jumps 1121f, 1123, 1127, 1130, 1132, 1134f, 1138, 1142, 1144, 1158
Proton jump mechanism of ice 979ff, 1403
Proton mobility (*see* Proton motion)
Proton motion in H-bonds (*see also* Proton intrabond jumps) 173, 217, 225, 231, 246ff, 692ff, 757ff, 761ff, 979ff
Proton states 175, 196, 221, 223, 720
Proton transfer 146ff, 219, 224, 235, 723, 946ff, 956
and dipole moment of H-bonds 946ff
and $H_5O_2^+$ dipole moment 195, 250, 723
with carboxylic acid N-base bonds 737, 949
in H-bonded adsorbates 718, 1299ff, 1301ff, 1311, 1313, 1318, 1328, 1336
mechanisms in ice 979ff, 1414
Proton wave functions 173f, 249f, 921
and polarizability of H-bonds 722f
Pseudodimers 1081
Pseudo spins 1181–1183
tunneling models 1127f, 1133
Purine bases 86, 698, 759f
Pyrazole 536, 698
Pyridine 10, 942
adsorbed on oxides 1273, 1274, 1277, 1281, 1282, 1286ff, 1289, 1301ff, 1334, 1344ff, 1348
adsorbed on silica 718, 740f, 1273, 1274, 1277, 1281, 1286ff, 1289, 1301ff
complexes of 131
bases 85
Pyroelectricity 1114, 1139, 1152
Pyrrole 631, 942
Pyrrolidone 532
Pyrromethene 124ff

Quadrupole coupling constant 1035, 1051f
definition 835
relation to electronic structure 835-37
deuteron QCC 838-42, 844
^{17}O QCC 850
^{35}Cl QCC 857
^{14}N QCC 859
^{75}As QCC 862
deuterium 1328
Quadrupole moment 101
Quadrupole–quadrupole interactions 1081
Quantum–mechanical tunneling 222, 228, 693
Quasielastic neutron scattering 893ff
Quasistationary state 228f
Quasiclassical WKB approximation 237, 240

Raman scattering 1124, 1131, 1134, 1145, 1156, 1157
continuum 689f
broadening of excitation line 732
Raman spectra, of adsorbates 1335, 1343, 1345, 1346
Raman transitions 179, 187
Rate of proton transfer 233, 236, 946ff
Reaction potential 271, 274
Rearrangement coordinate 220
Redlich–Kister equations 1228
Reduced masses 171
Reduplication of DNA 219
Relation
bond energy and wave number shift 574ff
H-bond length and wave number shift 574
spectroscopic shifts and enthalpy 1209ff
Relaxation mechanisms in ice 1414
Relaxation rate, nuclear magnetic 1030

Relaxation time T_1 of ^{13}C 824
Relaxation time of H_2O (*see also* Relaxation times) 1394, 1401
Relaxation times
  dielectric measurements 967ff, 971, 998
  inelastic neutron scattering 893
  nuclear magnetic resonance 1030
Reoccurrence time 224, 235
Reorientation time (*see also* Relaxation times) 1032, 1054
Resonance interaction 259, 287f, 291
Resonance interaction model 1073
Rigid glasses 621
Rigid ion model 906
Rochelle salt (RS) 1142f, 1158
Rocking 903
Rotational diffusion 915
Rotational freedom 905, 906, 926
Rubidium
  bisulfate 1146, 1159
  dihydrogen arsenate 1117, 1122, 1130, 1154, 1155
  dihydrogen phosphate 1116, 1117, 1121, 1130, 1155–57
  trihydrogen selenite 1139, 1140, 1136
Rutile, adsorption on 1335, 1343

Salt effects on polarizable H-bonds 745f
Salt-in effects, in water 1408
  on the solubility in water 1403
Salt-out effects, in water 1408
  on the solubility in water 1403
Salts, adsorption on 1350
Saturation effects 959
Scattering 396, 397
  function 895, 910, 915f
  law 915
Secondary hydration sphere 1413
Second moment of the NMR absorption line
  definition 867f
  in KH_2F_3 869
  of H_3O^+ ions 872
Second relaxation time of H_2O 1402
Second row elements, H-bonds to 107f, 112
Self association 3
  in alcohols 622, 626, 795
  in amines 627
  in amides 632, 805
  in thiols 632
  carboxylic acid 806
  water 813
Self-consistent field (SCF) 31, 33, 57ff, 94ff, 100
SCF perturbation theory 50, 51
  for molecular crystals 141
SCF calculations $H_5O_2^+$ 118ff, 190ff, 723ff
Self-correlation function 896, 915
Self-diffusion coefficient 1031, 1050f
  of adsorbed molecules 1291, 1294, 1327, 1328, 1329, 1345f
Semi-carbazide hydrochloride 1150
Semi-clathrates 1418
Semi-empirical MO theories 37, 40, 70, 113f, 116f, 123, 129, 141f
Shear mode, spontaneous shear 1125, 1156
Shell model 906, 908
Shift of AH stretching band 569ff
Short H-bonds 426, 432, 503, 587
Silica-alumina, adsorption on 1346
Silica-chromia, adsorption on 1346
Silanol groups 633, 1266ff
  geminal 1271ff
  IR combination bands 1267
  IR fundamental 1266ff, 1270, 1271, 1274, 1275ff, 1282ff
  IR overtone bands 1267, 1290
  isolated 1271, 1272, 1274
  NMR of 1268, 1270
  number of 1269ff
  perturbation of 1274ff, 1296ff, 1303ff, 1319
  pK_a value of 1269, 1288
  vicinal 1271, 1274
Silver trihydrogen periodate 1136, 1141
Simultaneous excitation 622
Simultaneous proton motion (*see also* Proton motion in H-bonds) 220
Single scale enthalpy equations 1220
Site splitting 1095
Six-vertex problem 364, 382
Slater groups 1126, 1129, 1133-35, 1154, 1155, 1157
Slater rules 1178, 1181
Soap micelles 1419
Sodium bifluoride 909, 932

Sodium trihydrogen selenite 1121, 1136–38, 1140, 1157
Soft modes, ferroelectric mode 1113, 1115f, 1122-1125, 1128f, 1131f, 1134, 1140, 1154, 1156, 1158f
SO_2 gas hydrate 1416
Solids 936
Solubility in H_2O 1408
Solution calorimetry 1202
Solvation energy from vapor pressure measurements 1239
Solvent effects 821
  on band shift 572
  on polarizable H-bonds 737
  on proton transfer 737, 940ff
Space charge effects 967, 997
Specific effects in biochemistry 529
Specific heat 899, 901, 1120, 1126, 1129, 1137, 1139, 1147, 1151
  of H_2O 1401, 1388
Spin–lattice relaxation of water 1403
Spin–lattice relaxation time
  of deuterons in KD_2PO_4 847-848
  and deuteron dynamics 847-848
Spontaneous polarization 1114, 1117, 1119, 1120, 1126, 1128f, 1130, 1133, 1136, 1138f, 1142–44, 1146f, 1149, 1153–1159
Stark effect 180
Statistical mechanics approach 1373
Stereochemistry 71, 82, 401ff, 466, 529ff
Sterically hindered alcohols 626
Steric restrictions 1289ff, 1303
Stochastic
  function 591
  motions 893
  tunneling 1194
  variable 226
Stockmayer potential 92
Strong coupling theory 300, 305, 306ff, 338
Strong electric fields 959
Strontium chloride hexahydrate 930
Structure breakers 1408
Structure breaking effect 1055
Structural description of hydrogen bond networks 1116f, 1137, 1138-42, 1147-48, 1151-1152
Structure diffusion 692
  and Grotthus conductivity 757f
Structure makers 1408
Structure reinforcement 1057
Structure temperature of an electrolyte solution 1403
Substituted acetates 1150, 1160
Sulfate ion as H-bond acceptor 419
Supercooled water 1386
Surface
  energy of H_2O, alcohols 1391
  enthalpy H_σ of alcohols 1394
  tension 1391
Symmetrical double well potential (*see also* Double minimum potential well) 231f
Symmetrical H-bond 3, 171, 178, 231f, 426, 428, 503ff, 587, 596ff, 604, 605
  influence of environment on 275ff, 724ff, 754ff
Synthetic troegerite 1149

Tagaki groups 1126, 1130, 1132, 1134f, 1157
Tautomeric equilibria 11
T_c of alcohols 1385
T_c of water 1385
Temperature dependence of $\Delta\gamma$ 573
Terminal OH group 1095, 1099
Tetraaquahydrogen ion (*see* $H_9O_4^+$ *and* Hydronium hydrates) 1279, 1281
Tetrahydrofuran, adsorbed on silica 1279, 1281
Tetramers 1073–1075, 1080, 1089, 1092, 1094
Thermal expansion coefficient of H_2O 1401
Thermally stimulated polarization 995
Thermal motion, effect on bond lengths 400f
Thermodynamic analysis 115, 1157, 1158
Thermodynamic initial state 230
Thiols 632, 1097
Thiourea 1150
Thorium oxide, adsorption on 1345
Tight binding approximation 142
Time–dependent Schrödinger equation 223, 225
Time–energy uncertainty relation 239
Time evolution of a quantum system 223
Time-of-flight spectra 898
Time-of-flight technique 896, 904, 908, 918
Torsional frequency 904, 906, 912, 926
Torsional mode 1076, 1091, 1099

Torsional vibrations 901-8, 911
Total compressibility of H_2O 1401
Total energy 168
Total neutron scattering 925, 927
Transfer matrix 362, 367, 383
Transition moments 177, 198
Transition polarizability (*see also* Polarizable H-bonds) 177ff
Transition temperature 1179, 1193
Translational motions 898, 902, 903, 905, 906, 915
Translatory diffusion 911
Translatory jump 914
Translatory mode 901
Translatory peaks 904-6
Transmission coefficients 239
Transport theory 225
Transversal nuclear magnetic relaxation 1030
Transverse optical (TO) phonons 1123f, 1128, 1131, 1135, 1196
Triaquahydrogen ions 481, 674
Triaquaoxonium ion 481, 753ff, 766ff
  structure 510f
Trichloroacetic acid 10
  acetic acid, mixed cyclic dimer 117
Triethylamine 10ff
Trifluoroacetic acid 1099, 932
  cyclic dimer 116f
  mixed cyclic dimer with acetic acid 117
Trifluoroethanol 1096
Trifluoropropanol 1096
Trifurcated H-bond 408
Triglycine
  fluoberyllate 1144
  selenate 1144, 1158
  sulfate 1144, 1150, 1158, 1160
Trimers 537, 539, 1086, 1073-78, 1080, 1083
  cyclic 537ff, 1075, 1077, 1078f, 1094
Tunneling
  approximation 248, 251, 259ff, 263, 289, 723
  frequency 173, 199, 247, 279, 285, 288ff, 291f, 1183, 1193
  mode 1181, 1183, 1184, 1186, 1190, 1191-1193
  of protons 222, 247ff, 304, 349, 693, 1113, 1124, 1127f, 1131, 1132, 1135, 1143, 1154, 1155, 1157, 1178ff
  time 236, 238
Twisting 903
Two-constant model of alcohol and amine association 1235
Two-level approximation 240
Two-level system 230
Two-minimum potential (*see* Double-minimum potential well)
Two-state function of the vapor pressure ratio 1250

Ultrasonic absorption of H_2O 1400
Ultrasonic attenuation and velocity measurements 1120, 1125, 1130, 1131, 1160, 1172
Univalent redox reactions 10
Unrestricted Hartree–Fock method (UHF) 143

Valence electron distribution 416
Vanadium oxide, adsorption on 1344
Van der Waals forces 964
Van der Waals radii 28, 87, 401
Vaporization energies of variously deuterated alcohols, amines and methanethiols 1235ff
Vaporization heat of H_2O 1390
Vapor phase
  chemical shifts 819
Vapor pressure isotherms 1227, 1231ff, 1247
Vapor pressure isotope effect 1240ff
  concentration dependence of 1240
  frequency dependence of 1247ff
  temperature dependence of 1242
Vapor pressure ratios
  of chloroform and deuterochloroform on dilution with acetone 1256
  of highly diluted ammonia and trideuterated ammonia 1254
  of hydrogen fluoride and deuterium fluoride 1257
  of H_2O and D_2O, ice and water 1251
  of H_2O and D_2O on dilution with pyridine 1255
  of solid and liquid ammonia and trideuteroammonia 1254
  of variously deuterated carboxylic acids 1256

of variously deuterated dimethylamines 1244
of variously deuterated ethanols 1243
of variously deuterated methanols 1243
of variously deuterated methylamines 1240, 1242
Variational method 173
Velocity of sound 556
Vermiculite, adsorption on 1349
Very weak H-bonding 7
Vibrational
coupling 174, 196, 224, 731, 1082, 1096, 1098, 1099
frequencies of water 540
mode 169, 568
normal coordinates 221
splitting 1095
Virial coefficients, second 92

Water 415, 433, 748ff, 898, 911-913ff, 928, 929, 1049, 1070, 1086, 1087, 1103, 1368ff
H-bonds in clathrate hydrates 1415
H-bonds in electrolyte solutions 748ff, 1403ff
H-bonds in ice 1426ff
H-bonds in liquid water 1369ff
adsorbed on alumina 1333
cooperative effects 1335
coordinately bonded 1333
dielectric behavior 1335
electrical conductivity 1336
heat of adsorption 1333, 1335
neutron inelastic scattering 1333
NMR relaxation 1336
proton transfer 1336
Raman scattering 1335
adsorption on oxides 1285, 1303ff, 1333ff, 1343ff
adsorbed on silica 1285, 1303
cluster formation 1306ff
correlation function 1313
dielectric behavior 1313, 1314, 1315ff
electrical conductivity 1313ff
entropy of adsorption 1305, 1314
far IR 1308
frost damage 1317
heat capacity 1306, 1309ff
heat of immersion 1309
IR fundamental bands 1303, 1306
IR overtone and combination bands 1306ff
neutron inelastic scattering 1305, 1306
NMR relaxation 1305, 1306, 1311ff, 1316
nucleation of ice 1311, 1316, 1317
proton transfer 1311, 1313ff
computer simulation 372, 379-81
complexes of 108ff, 131, 143ff
continuum theories 375-79, 1368ff
dielectric constant 369, 381
dimer $(H_2O)_2$ 27, 46-48, 50, 55, 57, 59, 60, 62, 73ff, 78, 85, 109, 115, 129, 206
film 1005
intermolecular potential 368, 371, 375-81
lattice theories 371, 373, 375
molecule 34, 84, 89, 146
tetrahedral clusters 137f
of cristallization 902, 930
of oligomers 133ff
perturbation theories 379
Wave number shifts 172, 180, 213
and anharmonicity 644f
and bond energy 574f
and bond enthalpy 1209ff
and bond length 574
with polarizable H-bonds due to electrical fields 202, 724ff
Wave packet 220ff, 227
Waving 903
Wideline NMR and H-bonding 867
determination of proton–proton or proton–fluorine distances 867-872
in KHF_2 867
in KH_2F_3 868-871
in H_3O^+ ions 872
in $H_5O_2^+$ ions 872
Wien effect 966
Wilson equations 1228
Wilson parameters, physical interpretation 1229

X–N difference maps 416f
X–N maps 503

X-ray diffraction 395, 1113, 1118, 1138f, 1146, 1147, 1154, 1159
X-ray studies, adsorbed systems 1348, 1349

Zeolites, adsorption on 1346ff
Zero differential overlap (ZDO) 37, 52, 55
Zero phonon line 339
Zinc oxide, adsorption on 1345
Zirconium oxide, adsorption on 1344
Zwitterion 461